Advances in Experimental Medicine and Biology

Volume 1480

Advances in Experimental Medicine and Biology provides a platform for scientific contributions in the main disciplines of the biomedicine and the life sciences. This series publishes thematic volumes on contemporary research in the areas of microbiology, immunology, neurosciences, biochemistry, biomedical engineering, genetics, physiology, and cancer research. Covering emerging topics and techniques in basic and clinical science, it brings together clinicians and researchers from various fields.

Advances in Experimental Medicine and Biology has been publishing exceptional works in the field for over 40 years, and is indexed in SCOPUS, Medline (PubMed), EMBASE, Reaxys, EMBiology, the Chemical Abstracts Service (CAS), and Pathway Studio.

2023 CiteScore: 5.9

Kostas Pantopoulos
Editor

Iron Metabolism in Human Health and Disease

Springer

Editor
Kostas Pantopoulos
Department of Medicine
McGill University
Montreal, QC, Canada

ISSN 0065-2598 ISSN 2214-8019 (electronic)
Advances in Experimental Medicine and Biology
ISBN 978-3-031-92032-5 ISBN 978-3-031-92033-2 (eBook)
https://doi.org/10.1007/978-3-031-92033-2

This Springer imprint is published by the registered company Springer Nature Switzerland AG
The registered company address is: Gewerbestrasse 11, 6330 Cham, Switzerland

Preface

Iron is an essential micronutrient for nearly all living organisms. From an evolutionary perspective, iron's role in biology is deeply ancient and foundational. Life is believed to have emerged on Earth approximately 3.5 billion years ago, in an oxygen-poor environment where iron was soluble and readily available. Hydrothermal vents, rich in iron-sulfur (Fe-S) minerals, likely provided the ideal conditions for early biochemical reactions and carbon dioxide fixation, fostering the synthesis of fundamental organic compounds. Iron-sulfur clusters became indispensable cofactors for primitive enzymes such as ferredoxins, ancient electron carriers central to energy metabolism. As life evolved, iron's role expanded. Contemporary cells and organisms have harnessed iron's unique chemical properties to carry out vital biological functions.

Thus, iron's ability to alternate between ferrous (Fe^{2+}) and ferric (Fe^{3+}) redox states and to participate in diverse coordination complexes with cellular macromolecules makes this metal indispensable for oxygen transport, electron transfer, and numerous enzymatic and catalytic reactions. However, these same redox capabilities render iron potentially harmful, as unregulated iron can catalyze the formation of reactive oxygen species (ROS) and contribute to oxidative stress. As such, iron metabolism must be tightly regulated to meet physiological demands while avoiding toxicity.

The biomedical relevance of iron metabolism is evident by the global burden of "classic" iron-related disorders. Iron deficiency, with or without anemia, affects nearly a quarter of the world's population. Most commonly resulting from blood loss and/or inadequate dietary iron absorption, it is associated with increased morbidity, impaired cognitive and physical performance, and reduced quality of life. Management typically involves iron replacement therapy, administered orally in mild cases and intravenously when rapid correction is required, or oral formulations are ineffective or poorly tolerated.

At the opposite end of the spectrum lies hemochromatosis, a genetic disorder characterized by inappropriately high intestinal iron absorption, leading to progressive systemic iron overload. It is the most prevalent inherited condition among individuals of Northern European ancestry. Transfusion-dependent secondary iron overload, compounded by increased dietary iron absorption, is common in thalassemias and other anemias associated with ineffective erythropoiesis. Systemic iron overload leads to iron deposition in various organs—including the liver, heart, and endocrine glands—causing tissue damage and contributing to the development of liver cirrhosis, diabetes, cardiomyopathy, and other complications. While hemochromatosis is effectively managed through therapeutic phlebotomy (venesection), iron accumulation in transfusion-dependent anemias requires iron chelation therapy.

Several other conditions are associated with deregulation of systemic or local iron metabolism. These include but are not restricted to infectious and inflammatory diseases, anemia of inflammation, chronic kidney disease, several cancers, metabolic syndrome pathologies, cardiovascular diseases, and various neurological disorders. Advances in our understanding of the molecular and regulatory pathways governing iron metabolism are not only deepening our insight into disease pathogenesis but are also paving the way for the development of novel, targeted therapeutic strategies for iron-related disorders in a broad sense.

For many years, only a few proteins were known to participate in iron metabolism: ferritin (discovered in 1937), which safely stores excess iron, and transferrin (discovered in 1947), the plasma iron transporter. The identification of the transferrin receptor in the 1970s shed light on the mechanism of cellular iron uptake from circulating transferrin. A landmark advancement came in the 1980s with the discovery of the iron-responsive element/iron regulatory protein (IRE/IRP) system, including IRP1 and IRP2—key cytosolic sensors that control intracellular iron homeostasis.

Subsequent discoveries shifted the focus from cellular to systemic regulation. The characterization of iron transporters such as DMT1 and ferroportin laid the groundwork for a more integrated understanding. The field was transformed in the early 2000s by the discovery of hepcidin, the liver-derived peptide hormone that serves as the master regulator of systemic iron homeostasis, along with its receptor, ferroportin. Hepcidin's critical role in iron regulation was quickly appreciated: insufficient hepcidin leads to iron overload (as seen in hemochromatosis), while excessive hepcidin contributes to anemia of inflammation.

Beyond being a biomarker, hepcidin has emerged as a promising pharmacological target. The elucidation of its role in iron disorders has driven the development of hepcidin agonists and antagonists, some of which are now undergoing clinical evaluation. These breakthroughs have fueled a remarkable expansion in iron biology, transforming it into one of the most dynamic fields in biomedical sciences.

We hope this book not only consolidates current knowledge but also inspires future research and innovation in the field of iron metabolism.

Montreal, QC, Canada Kostas Pantopoulos

About This Book

Iron Metabolism in Human Health and Disease brings together leading experts to provide a comprehensive overview of iron biology—from fundamental chemistry to clinical applications. The chapters span topics including molecular regulation, disease mechanisms, diagnostic tools, and standard and emerging therapies. Whether exploring iron's role in homeostasis, cell death, nutrition, anemia, pregnancy, immunity, cancer, neurodegeneration, bone pathophysiology, intestinal microbiome, and cardiovascular or metabolic diseases, this volume aims to be an essential reference for researchers, clinicians, and students alike.

Contents

1 **BioIron: Origin, Chemical Properties, and Biological Functions** 1
Kostas Pantopoulos

2 **Cellular Iron Homeostasis** 17
Caroline C. Philpott and Esther G. Meyron-Holtz

3 **Systemic Iron Metabolism** 33
Tomas Ganz

4 **The Labile Side of Iron in Health and Disease: A Narrative Review** 47
Ioav Z. Cabantchik, Ola Karmi, and Rachel Nechushtai

5 **Iron and Cell Death** 61
Giorgia Federico, Marialuisa Moccia, and Francesca Carlomagno

6 **Dietary Iron Absorption: Biochemical and Nutritional Aspects** 75
David M. Frazer, Gregory J. Anderson, and James F. Collins

7 **Hepcidin and Tissue-Specific Iron Regulatory Networks** 89
Samira Lakhal-Littleton and Carole Peyssonnaux

8 **Control of Systemic Iron Homeostasis: Insights Gained from Studying Mouse Models** 103
Sandro Altamura and Martina U. Muckenthaler

9 **Diagnosis and Treatment of *HFE*^{C282Y}-Linked Hemochromatosis** 119
Paul C. Adams

10 **Diagnosis and Management of Non-*HFE* Hemochromatosis, Ferroportin Disease, and Rare Hereditary Iron-Loading Disorders** 131
Antonello Pietrangelo

11 **Iron-Loading Anemias** 145
Maayan V. Levy and Yelena Z. Ginzburg

12 Iron Deficiency Anemia 163
Xiuqi Li and Karin E. Finberg

13 Anemia of Inflammation 179
Lukas Lanser and Günter Weiss

14 Iron, Hepcidin, and Immunity 197
Daisy Crisell, Nick Talbot, and Hal Drakesmith

15 Iron Metabolism in Cardiovascular Disease 217
Anthony Kalousdian, Mariam Ardehali, and Hossein Ardehali

16 Iron and Liver Disease 237
Heinz Zoller

17 The Impact of Iron Homeostasis in Insulin-Sensitive Tissues and Gut Microbiome on Obesity-Driven Metabolic Disorders 253
José María Moreno-Navarrete, Isa Galvao Rodrigues, and José Manuel Fernández-Real

18 Iron and Cancer 271
Suzy V. Torti, Lia Tesfay, and Frank M. Torti

19 Neurodegeneration with Brain Iron Accumulation 291
David A. Schipper and Hyman M. Schipper

20 Iron and Bone Pathophysiology 311
Martina Rauner

21 Iron and Pregnancy 327
Veena Sangkhae and Elizabeta Nemeth

22 Iron and the Intestinal Microbiome 345
Huihui Li

23 Iron Chelation Therapy 361
Maria Domenica Cappellini, Natalia Scaramellini, Simona Leoni, and Irene Motta

24 Oral and Intravenous Iron Therapy 371
Andrea U. Steinbicker and Kostas Pantopoulos

25 Diagnostics: Markers of Body Iron Status 387
Domenico Girelli, Giacomo Marchi, Fabiana Busti, Fabio Chesini, and Annalisa Castagna

26 Current Landscape of Hepcidin Therapeutics 399
Antonella Nai, Laura Silvestri, Michela Asperti, and Francesca Vinchi

1

BioIron: Origin, Chemical Properties, and Biological Functions

Kostas Pantopoulos

Abstract

Iron is a highly abundant element, essential for life on earth and integral to various biological processes. It originated from stellar nucleosynthesis, particularly in supernovae type Ia, which produced iron used in planetary formation. Life's emergence on earth, about 3.5 billion years ago, occurred in an oxygen-free atmosphere, with iron playing a pivotal role in early biochemical reactions. Hydrothermal vents, rich in iron-sulfur (Fe-S) minerals, provided the conditions for prebiotic chemistry and CO_2 fixation, potentially driving the synthesis of key organic compounds. Iron-sulfur clusters became crucial cofactors for enzymes involved in energy metabolism, such as ferredoxins, which are ancient electron carriers. Iron's unique redox properties, allowing electron transfer between ferrous (Fe^{2+}) and ferric (Fe^{3+}) states, enabled its vital role in oxygen transport, cellular respiration, DNA synthesis, replication, and repair, as well as in epigenetic, transcriptional, and translational gene regulatory mechanisms, and in several metabolic pathways. However, the prevalence of insoluble ferric ions under aerobic conditions limits iron's bioavailability. Moreover, iron's reactivity poses further challenges, as unshielded iron can undergo redox cycling, leading to oxidative stress. Thus, biological systems have evolved intricate mechanisms to regulate iron acquisition, utilization, and storage. These mechanisms ensure that iron remains bioavailable while preventing toxicity, underscoring its significance throughout the evolution of life.

Keywords

Iron-sulfur clusters · Redox reactions · Prebiotic chemistry · Iron toxicity · Iron metabolism

K. Pantopoulos (✉)
Lady Davis Institute for Medical Research, Jewish General Hospital and Department of Medicine, McGill University, Montreal, QC, Canada
e-mail: kostas.pantopoulos@mcgill.ca

1.1 Origin of BioIron

Iron (Fe) is the most prevalent element in rocky planets like Earth, making up about 80% of the Earth's core. The earth's crust contains approximately 6.3% iron, making it the fourth most abundant element (after oxygen, silicon, and aluminum), and the second most abundant metal (after aluminum) (Frey and Reed 2012). Iron and other heavy elements are generated in stars during the fusion of light elements present in the early universe, such as hydrogen (H) and helium (He). At the end of their lifespan, stars explode in supernovae. The explosion scatters heavy ele-

K. Pantopoulos (ed.), *Iron Metabolism in Human Health and Disease*, Advances in Experimental Medicine and Biology 1480, https://doi.org/10.1007/978-3-031-92033-2_1

ments throughout space, which can be eventually used as building blocks for the formation of rocky planets. The high abundance of iron on Earth has been attributed to a specific kind of supernovae, known as "supernovae type Ia", which are derived from small stars and are major iron producers (Burrows 2000; Tsebrenko and Soker 2015).

Life on Earth started in an oxygen-free atmosphere some 3.5 years ago (Martin et al. 2014). The emergence of living cells required the transfer of electrons from mineral sources to reduce organic compounds in a way that energy is produced and harnessed. Crucial prebiotic chemistry that resulted in early forms of life may have occurred in submarine alkaline hydrothermal vents. These are geological formations found on the ocean floor, where seawater interacts with the earth's crust. They release warm, alkaline fluids and other minerals into the surrounding, colder, and more acidic ocean waters.

It has been hypothesized that alkaline hydrothermal vents have functioned as natural electrochemical flow reactors during the mixing of H_2-enriched alkaline fluids with CO_2-enriched acidic ocean waters within a network of thin interconnected micropores composed of Fe(Ni)S (Martin et al. 2008; Sojo et al. 2016). Favorable conditions emerged during the process of serpentinization, where iron- and magnesium-containing minerals such as olivine ($(Mg,Fe)_2SiO_4$) are hydrolyzed and transformed to serpentine group minerals including magnetite (Fe_2O_3) and Ni-Fe alloys (Holm et al. 2015). During serpentinization, the oxidation of olivine iron causes a reduction of water with the formation of H_2, while the presence of dissolved sulfide ions results in precipitation of Fe(Ni)S minerals.

In alkaline hydrothermal vents, Fe(Ni)S may have acted as a catalyst for the reduction of CO_2 by H_2. Conceivably, the pH difference across Fe(Ni)S barriers generated natural proton gradients (Fig. 1.1), similar in magnitude and direction to the proton-motive force utilized by several present-day anaerobic methanogens and acetogens for carbon fixation. It has also been proposed that the pH difference across Fe(Ni)S barriers may drive not only early steps of CO_2 reductive fixation to CO, CH_3SH, or CH_3COSH, but also further carbonylation and hydrogenation reactions yielding longer chain carbon compounds (Camprubi et al. 2017). These include pyruvate, oxaloacetate, and α-ketoglutarate, which can also be generated during an incomplete reverse tricarboxylic acid (TCA) cycle in several archaeal and bacterial phyla (Garritano et al. 2022). The concept of iron and pH-dependent CO_2 fixation and involvement in functional non-enzymatic metabolic networks has been validated in experimental models (Akbari and Palsson 2023; Keller et al. 2016; Beyazay et al. 2023).

The reduction of succinate to α-ketoglutarate in the reverse TCA cycle is mediated by ferredoxins, ubiquitous proteins that transfer electrons via a Fe-S cluster. Ferredoxins are involved in several metabolic reactions and constitute the most potent biological reducing agents. A flavin-based electron bifurcation mechanism allows the thermodynamically unfavorable generation of reduced low-potential ferredoxin from relatively high-potential electron donors such as NADH or H_2; reduced ferredoxin then shuttles electrons to energy-yielding anaerobic biosynthetic reactions (Muller et al. 2018). Interestingly, experimental data show that only iron, but not other metals, can transfer electrons from H_2 to ferredoxin in cell-free conditions at pH 8.5 and 40 °C, in a primitive flavin-based electron bifurcation setting (Brabender et al. 2024).

Ferredoxins are relatively small proteins that have evolved by duplication events from a tetrapeptide (alanine, aspartic acid or proline, serine, glycine) that binds a Fe-S cluster (Eck and Dayhoff 1966; Meyer 1988). They share common evolutionary conserved structural motifs with other metal-binding proteins (Bromberg et al. 2022; Raanan et al. 2018) and are considered among the most ancient proteins that may have played a crucial role in the development of early forms of life (Hall et al. 1971). Conceivably, a primordial ferredoxin could have mediated the reduction of CO_2 in proto-cells evolved in alkaline hydrothermal vents (Sojo et al. 2016). Analogous early "chemistry of life" may have occurred on iron pyrite (FeS_2) surfaces (Wachtershauser 1990). This model has been at

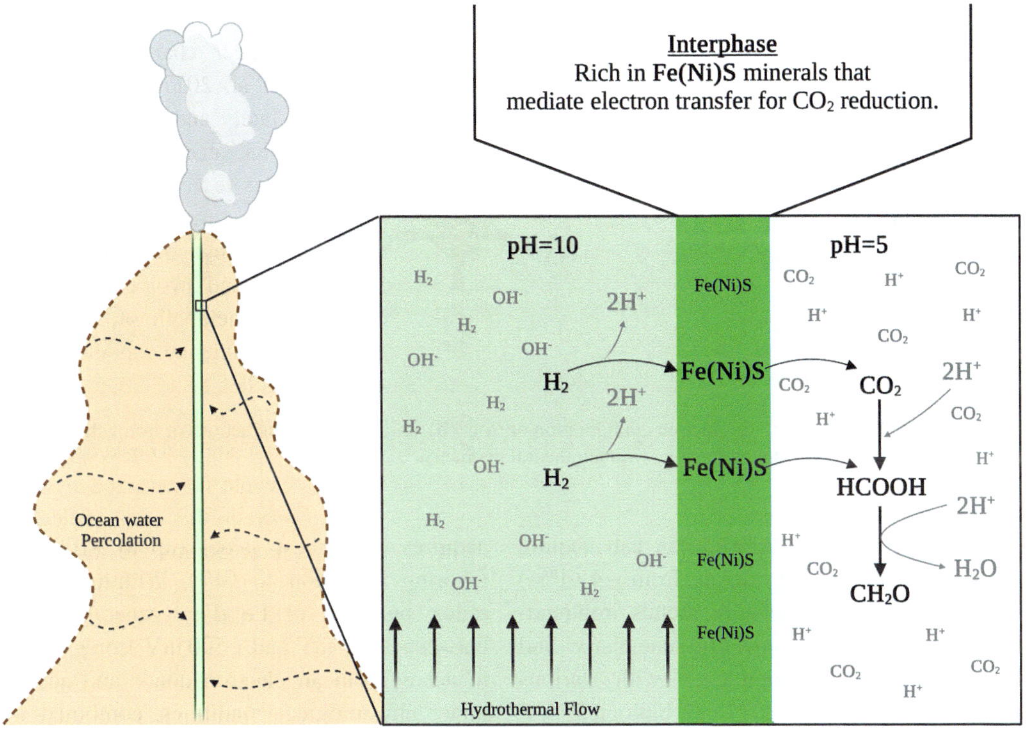

Fig. 1.1 A model of prebiotic chemistry in alkaline hydrothermal vents. Electron transfer from H_2-enriched alkaline fluids to CO_2-enriched acidic ocean waters is mediated by Fe(Ni)S present in interphase minerals

least partially supported by in vitro experiments modeling hydrothermal conditions, which showed the formation of acetate and amino acids by fixation of CO (but not of less reactive CO_2) on Fe(Ni)S (Huber and Wachtershauser 1997, 1998). More recent studies showed that iron can catalyze simple anabolic and catabolic reactions from the CO_2 fixation products pyruvate and glyoxylate, which give rise to TCA cycle intermediates under abiotic conditions (Muchowska et al. 2019).

These data highlight the important role of iron in autotrophic theories for the origin of life (Camprubi et al. 2017). Nevertheless, iron was not only crucial during early evolution, but has remained vital for almost all forms of life. Thus, iron is an essential nutrient and constituent of almost all cells and organisms. There are only minor exceptions of bacteria not requiring iron, such as some *Lactobacillus* species (Weinberg 1997), the Lyme disease pathogen *Borrelia burgdorferi* (Posey and Gherardini 2000) and, possibly, some human mycoplasmas (Peralvarez-Marin et al. 2021). These are unique examples where iron has been functionally substituted in a living cell by other metals, mainly magnesium or manganese.

The term "BioIron" was coined and has been adopted within the scientific community to emphasize the crucial roles of iron in various life processes. It refers to a focus on the study of iron within biological and medical contexts. BioIron research explores the biological functions, medical significance, and biomedical applications of iron.

1.2 Chemical Properties of BioIron

Iron is a transition metal with atomic number 26 that is positioned in the d-block of the periodic table. Its electron configuration is $[Ar]3d^6 4s^2$ (Fig. 1.2a). Since the 3d and 4s electrons are

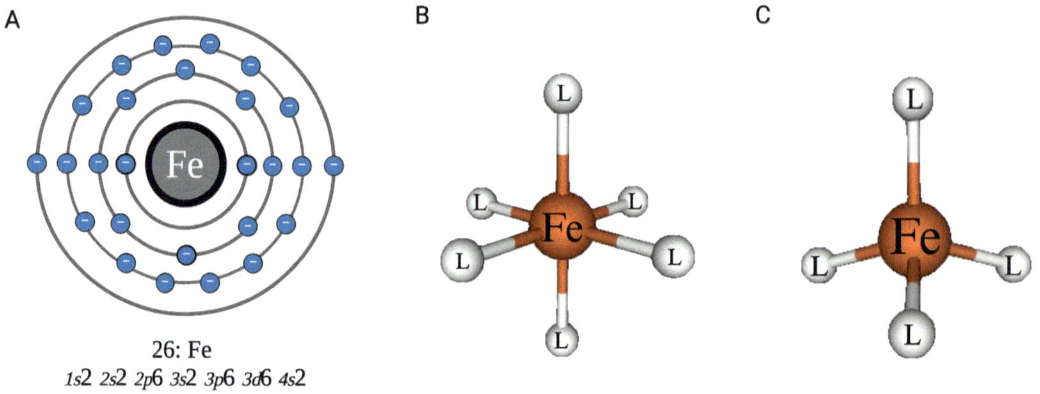

Fig. 1.2 Iron chemistry. (**a**) The electron configuration of iron; (**b**) six-coordinate interactions of iron with octahedral geometry; (**c**) four-coordinate interactions with tetrahedral geometry

energetically relatively close, iron can acquire several oxidation states ranging from −4 ($d^{10}s^2$) to +7 (d^1). Like other transition metals, iron readily engages in organometallic chemistry and forms coordination complexes with various ligands (Sanchez et al. 2017). In biological systems, iron is a key component of several metalloproteins, and its most common oxidation states are Fe^{2+} (ferrous) and Fe^{3+} (ferric). Both Fe^{2+} and Fe^{3+} can adopt various coordination environments with either fully occupied or vacant coordination sites. Within metalloproteins, Fe^{2+} or Fe^{3+} are mostly found in six-coordinate interactions with octahedral geometry (Fig. 1.2b); however, within iron-sulfur clusters, they form four-coordinate interactions with tetrahedral geometry (Fig. 1.2c).

The redox potential of the Fe^{3+}/Fe^{2+} couple is +771 mV relative to the standard hydrogen electrode and under standard conditions (pH 0, concentration 1M, temperature 25 °C). However, this is affected by the presence of ligands and, in a cellular environment, also by the pH (Sanchez et al. 2017). Thus, the redox potential of Fe^{3+}/Fe^{2+} acquires its lowest negative value (−700 mV; favoring oxidation to Fe^{3+}) when bound to the siderophore enterobactin, the molecule with the highest known affinity for Fe^{3+} (~10^{49} M^{-1}). On the other hand, in the presence of ligands that act as soft bases and stabilize Fe^{2+}, such as the N-heterocyclic compounds phenanthroline or 2.2′-bipiridyl, the redox potential of Fe^{3+}/Fe^{2+} acquires its highest values (up to +1000 mV; favoring reduction to Fe^{2+}). Within cells, the redox potential of Fe^{3+}/Fe^{2+} generally ranges between −500 mV and + 500 mV. Iron's capacity to act as both an electron donor and acceptor under physiological conditions, combined with its versatile coordination chemistry, enables it to perform essential biochemical functions. These include redox reactions and oxygen transport, explaining why iron is vital.

In the aerobic environment, soluble Fe^{2+} undergoes spontaneous oxidation to Fe^{3+}. In aqueous solution, Fe^{3+} initially gets dissolved to form $[Fe(H_2O)_6]^{3+}$. However, at physiological pH, this gets quickly hydrolyzed to insoluble ferric hydroxide $[Fe(OH)_3]$, which readily polymerizes upon dehydration to polynuclear ferric-oxo and -hydroxo complexes $[-(FeOOH)_n-]$ (Aisen et al. 2001). As a result, the maximum concentration of soluble Fe^{3+} is miniscule (~10^{-17} M) and can only increase significantly under strongly acidic conditions (pH < 2). Hence, even though iron is highly abundant in nature, it is poorly bioavailable. Iron's limited bioavailability dates to the appearance on earth of molecular oxygen (O_2), a potent oxidant, around 2.1–2.4 billion years ago (Lyons et al. 2014). The rise in oxygen levels led to the oxidation of ferrous ions, which were abundant in aquatic environments, to insoluble ferric complexes that precipitated out.

The abundance of oxygen not only reduced iron's solubility but also provided a niche for iron to engage in potentially toxic redox-cycling reactions (Koppenol and Hider 2019). Iron's toxicity primarily stems from Fenton chemistry, originally described by H.J.H. Fenton as the oxidation of tartaric acid in the presence of hydrogen peroxide (H_2O_2) and Fe^{2+}, resulting in a violet product under strongly alkaline conditions (Fenton 1894). It was later shown by F. Haber and J. Weiss that the interaction of Fe^{2+} with H_2O_2 yields the hydroxyl radical ($OH^{\bullet}$), a potent oxidant (Haber and Weiss 1934), which in turn may trigger secondary oxidation of other organic compounds. Today, the Fenton reaction is defined as the oxidation of Fe^{2+} to Fe^{3+} by H_2O_2, resulting in the generation of $OH^{\bullet}$:

$$Fe^{2+} + H_2O_2 \rightarrow Fe^{3+} + OH^{\cdot} + OH^{-}$$

The exact mechanism of the Fenton reaction remains incompletely understood, but there is evidence that it involves unstable intermediates, which decompose to yield $OH^{\cdot}$, but also reactive oxo-ferryl (FeO^{2+}) species at neutral pH (Enami et al. 2014; Koppenol 2022; Meyerstein 2022). In vivo, the Fenton reaction occurs when unshielded Fe^{2+}, possibly in a loose complex with glutathione inside cells or with citrate in plasma, interacts with H_2O_2, an abundant, diffusible byproduct of aerobic respiration and other metabolic processes (Sies et al. 2022).

Fe^{3+} can be reduced back to Fe^{2+} in a redox cycle. Thus, even miniscule amounts of unshielded iron can catalyze the formation of $OH^{\cdot}$, a short-lived but highly reactive free radical. Hydroxyl radicals can readily attack all cellular macromolecules present in proximity causing protein oxidation, lipid peroxidation, and nucleic acid mutagenesis. This is the reason why iron overload is associated with oxidative stress, cell death, tissue injury, and disease, as described in detail in Chaps. 5, 9, 10 and 11.

It is often written in textbooks that superoxide anion ($O_2^{\cdot -}$), a primary byproduct of aerobic respiration, can reduce Fe^{3+} during the Haber-Weiss catalytic cycle:

$$Fe^{3+} + O_2^{\cdot -} \rightarrow Fe^{2+} + OH^{\cdot} + OH^{-} \quad (1.1)$$

$$(\text{Fenton reaction}) \quad Fe^{2+} + H_2O_2 \rightarrow Fe^{3+} + OH^{\cdot} + OH^{-} \quad (1.2)$$

$$(\text{Haber-Weiss reaction}) \quad O_2^{\cdot -} + H_2O_2 \rightarrow OH^{-} + OH^{\cdot} + O_2 \quad (1.1) + (1.2)$$

However, there is no evidence supporting that this cycle is operational in vivo, and it is questionable whether O_2^{-} can reduce physiologically relevant Fe^{3+} complexes (Koppenol and Hider 2019). Based on kinetic and thermodynamic arguments, it has been proposed that glutathione and ascorbate may account for the reduction of cytosolic and circulating plasma Fe^{3+}, respectively (Koppenol and Hider 2019; Merkofer et al. 2006; Hider and Kong 2011; Hamed et al. 1983).

1.3 Biological Implications of BioIron Chemistry

The limited bioavailability of iron poses a challenge to living cells and organisms. Presumably, it also created a selective pressure during the appearance of atmospheric oxygen and the subsequent oxidation and precipitation of iron (Wade et al. 2021). Thus, it appears that survival under these conditions not only required the development of antioxidant defence systems, but also of mechanisms for efficient iron acquisition. Evidence for this is provided by phylogenetic data showing that the decline in iron solubility during evolution was accompanied by a modest but significant increase in bacterial iron metabolic genes (David and Alm 2011).

The virulence of contemporary pathogens frequently relies on their capacity to acquire iron, a vital element for their growth. In response, the host typically employs iron-withholding strategies to limit the availability of this essential nutrient, as detailed in Chap. 14. Consequently, iron plays a pivotal role at the intersection of infection and immune defense.

1.3.1 Iron Assimilation in Bacteria and Fungi

Bacteria, fungi, and some plants have evolved iron assimilation mechanisms based on siderophores. These are chemically diverse molecules with relatively low molecular weight that chelate Fe^{3+} with high affinity (Hider and Kong 2010). They are classified into four types: catecholates (e.g., enterobactin and vibriobactin), hydroxamates (e.g., desferrioxamine and ferrichrome), phenolates (e.g., mycobactin and pyochelin), and carboxylates (e.g. staphyloferrins and oxalochelin). Siderophores are biosynthesized and secreted from cells by efflux pumps to capture and solubilize Fe^{3+} from the surrounding environment. Subsequently, the iron-loaded siderophores are internalized via specific receptors into the cytoplasm, where Fe^{3+} is released and reduced to Fe^{2+} for metabolic utilization.

In Gram-negative bacteria, iron-loaded siderophores are initially bound to the outer membrane TonB-dependent transporters (Noinaj et al. 2010). Energy for their trans-membrane transport is provided by the TonB-ExbB-ExbD protein complex, which spans the periplasmic space and couples the proton motive force of the inner membrane to the outer membrane receptors. Inside the periplasm, iron-siderophore complexes are captured by periplasmic binding proteins, which shuttle them to inner membrane transporters. Subsequently, the complexes are transported across the inner membrane into the cytoplasm via ATP-binding cassette (ABC) transporters, in a process powered by ATP hydrolysis. In Gram-positive bacteria, which lack an outer membrane, iron-siderophore complexes bind to lipoproteins anchored in the cytoplasmic membrane for delivery to the cytoplasm via ABC transporters. Iron can be utilized in the bacterial cytoplasm, following its release from siderophores (Fig. 1.3a).

Some bacteria and fungi can acquire iron by hijacking siderophores synthesized by other bacteria or fungi. In the baker's yeast *Saccharomyces cerevisiae*, iron-loaded xenosiderophores bind to specific transporters (Arn1-4) on the plasma membrane and are internalized into the cytoplasm via receptor-mediated endocytosis (Fig. 1.3b). Endocytosis and intracellular trafficking of xenosiderophores require energy in the form of ATP, which powers various molecular motors and membrane dynamics involved in vesicle formation and transport. Intracellular iron is released from the endosome by reductive or hydrolytic mechanisms (Ramos-Alonso et al. 2020). Some pathogenic bacteria may even hijack transferrin (see below) from the host for iron acquisition via a TonB-dependent transporter (Cornelissen and Sparling 1994).

Direct reductive iron transport is an alternative iron acquisition strategy and involves the reduction of Fe^{3+} to Fe^{2+} by membrane-bound or soluble ferric reductases using NADH or NADPH as electron donors (Schroder et al. 2003). Subsequently, Fe^{2+} is directly internalized by bacteria via specific ferrous transporters or promiscuous (with broad specificity) metal transporters (Brown et al. 2021). The Feo system offers the most common bacterial ferrous transport pathway (Lau et al. 2016) (Fig. 1.3a). Its main component is FeoB, the actual ferrous transporter. The C-terminal domain of FeoB forms a transmembrane channel for direct transport of Fe^{2+} to the bacterial cytoplasm. Energy is provided via GTP hydrolysis from the cytoplasmic N-terminal GTPase domain of FeoB. FeoA and FeoC are co-factors controlling the transport mechanism.

Saccharomyces cerevisiae utilizes a specific mechanism for direct reductive iron transport involving surface ferric reductases (Fre1-4), which convert Fe^{3+} to Fe^{2+} (Ramos-Alonso et al. 2020). Counterintuitively, high-affinity iron uptake requires re-oxidation of Fe^{2+} to Fe^{3+} by the multicopper oxidase Fet3, which allows the transport of Fe^{3+} across the plasma membrane via the permease Ftr1, an interacting partner of Fet3. Iron transport relies on the electrochemical gradient maintained by ATP-dependent proton pumps (Fig. 1.3b). A low-affinity iron uptake mechanism is mediated by Fet4, a transporter of several divalent metals including Fe^{2+}.

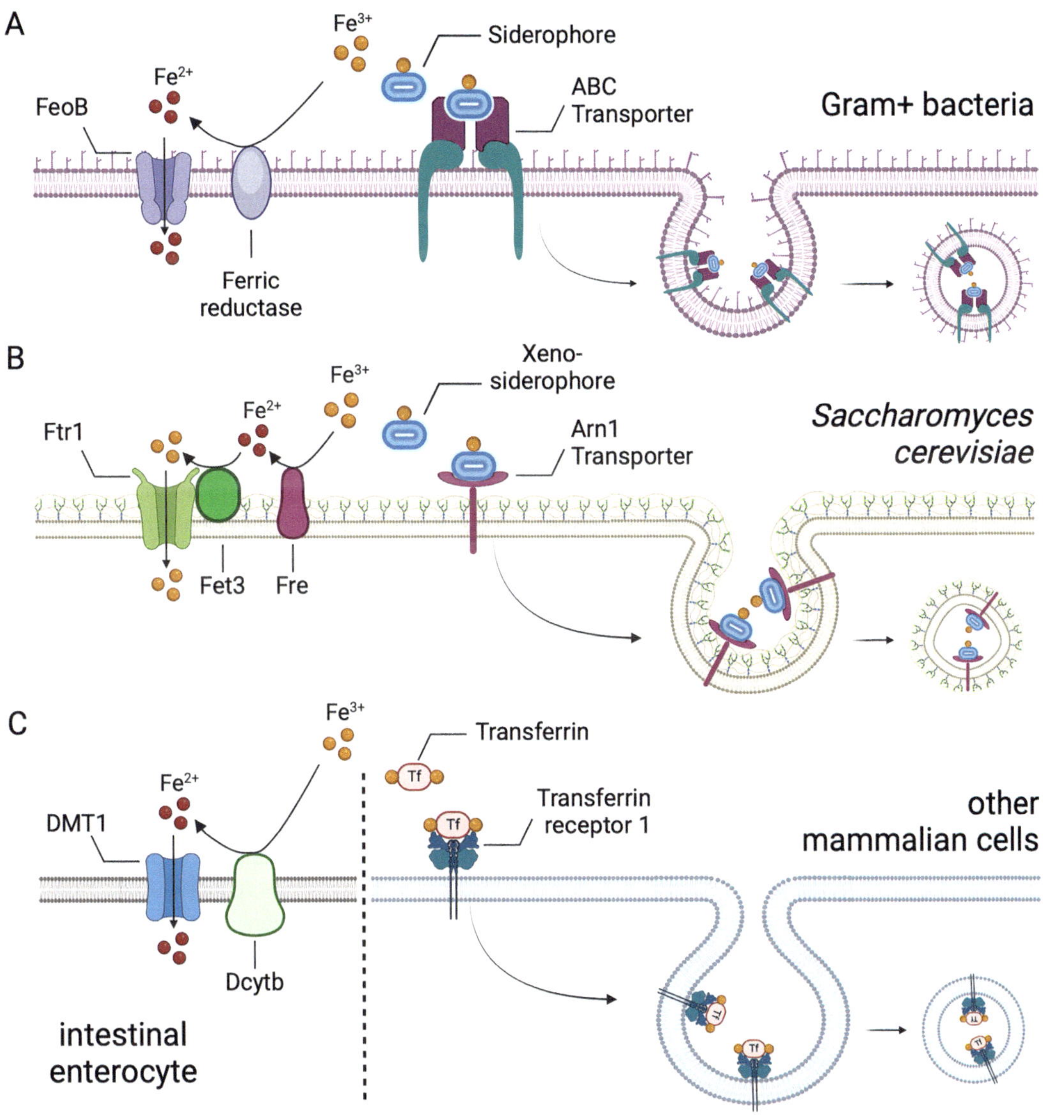

Fig. 1.3 Iron transport mechanisms in prokaryotic and eukaryotic cells. (**a**) Direct reductive iron transport via FeoB and siderophore-mediated iron uptake in Gram-positive bacteria; (**b**) direct iron transport via Fet3/Ftr1 and xenosiderophore-mediated iron uptake in the yeast *Saccharomyces cerevisiae*; (**c**) direct reductive iron transport in intestinal enterocytes via DMT1, and uptake of transferrin bound iron in other mammalian cells via receptor-mediated endocytosis

1.3.2 Iron Assimilation in Mammals

Mammals absorb dietary iron via direct reductive transport (Piskin et al. 2022); see also Chap. 6. Fe^{3+} present in the intestinal lumen undergoes reduction to Fe^{2+} via ferric reductases such as Dcytb (duodenal cytochrome b), expressed on the brush border of duodenal enterocytes. Ascorbate (vitamin C) may act as an electron donor. Reduced Fe^{2+} is then internalized via DMT1 (divalent metal transporter 1), a broad-specificity transporter for divalent metal ions (Fig. 1.3c, left). The metal transport activity of DMT1 relies on a proton gradient established mainly by intestinal Na^+/H^+ exchanger 3 (NH3) (Shawki et al. 2016). Enterocytes can utilize internalized iron for metabolic purposes, store it, or release it to the bloodstream via ferroportin, a Fe^{2+} exporter expressed on the basolateral membrane. The export of Fe^{2+} to plasma is coupled by its re-oxidation to Fe^{3+} via the multicopper ferroxidases hephaestin (membrane-associated) or ceruloplasmin (soluble), both distal homologs of yeast Fet3. Fe^{3+} is then captured by plasma transferrin for delivery to tissues.

Transferrin is a secreted glycoprotein that has evolved to bind, solubilize, detoxify, and deliver plasma Fe^{3+} to cells in distal tissues. It interacts with two Fe^{3+} ions within its C- and N-terminal lobes, respectively, and maintains them redox inert; the apparent stability constant for this interaction at pH 7.4 is ~10^{20} M^{-1} (Aisen et al. 1978). Differic transferrin binds to cell surface transferrin receptors, followed by endocytosis of the complex (Fig. 1.3c, right) via clathrin-coated vesicles (Gkouvatsos et al. 2012). Assembly of the clathrin coat and vesicle formation is facilitated by ATP-dependent adaptor proteins and dynamin, a GTPase that pinches off the vesicle from the membrane. Acidification of the endosome by vacuolar V-ATPases triggers the release of Fe^{3+} from transferrin, which undergoes reduction by the ferric reductase STEAP3. Liberated and reduced Fe^{2+} is exported across the endosomal membrane to the cytoplasm via DMT1, in a process dependent on the proton gradient maintained by V-ATPases. Finally, apo-transferrin is released from transferrin receptors on the cell surface, ready to capture Fe^{3+} and engage in another iron cycle.

Under physiological conditions, only a fraction (~30%) of circulating transferrin is occupied with iron. However, in iron overload disorders (see Chaps. 5, 9, 10 and 11) transferrin's iron saturating capacity can be exceeded, resulting in the accumulation of non-transferrin-bound iron (NTBI) in plasma. NTBI is redox-active and thus, potentially toxic. Hepatocytes and other tissue parenchymal cells can take up NTBI by a direct reductive transport mechanism via the divalent metal transporter ZIP14 (Jenkitkasemwong et al. 2015). ZIP14 operates by utilizing a proton gradient across the plasma membrane, which is maintained by proton pumps.

1.3.3 Analogies in Bacterial and Mammalian Iron Uptake Mechanisms

Indirect iron uptake via siderophores in bacteria and via transferrin in mammalian cells has many analogies (Fig. 1.3). Both siderophores and transferrin are biosynthesized and secreted to act as iron solubilization and transport vehicles. Moreover, iron bound to either siderophores or transferrin cannot undergo redox cycling. Thus, transferrin could be viewed as a more "sophisticated" siderophore. Nevertheless, siderophores are chemically diverse low molecular weight iron chelators, while transferrin is an evolutionary conserved (in metazoans) iron-binding polypeptide that evolved via duplication of an ancestral gene encoding a mono-lobal transferrin (Lambert 2012).

1.3.4 Coping with Excess Iron

Iron's redox reactivity poses another challenge to living cells and organisms. Thus, the accumulation of excess redox-active iron inside cells may lead to oxidative stress (Galaris et al. 2019) and cell death (Chap. 5). Ferroptosis is a type of iron-dependent cell death triggered by lipid peroxidation (Stockwell 2022). The low bioavailability of iron and the complexity and high energetic cost of iron assimilation by unicellular and multicellular organisms have prevented the development of specific iron excretion mechanisms as a major means for iron detoxification. Mammalian cells expressing the iron exporter ferroportin can pump excess iron out into plasma; however, most of the circulating will be eventually re-taken up in tissues. Tiny amounts of iron lost via cell desquamation are replenished through dietary iron absorption.

Excess intracellular iron can be neutralized within ferritin, the iron storage protein (Arosio et al. 2009). It consists of 24 H- and L- subunits that assemble into a nanocage structure, capable of storing up to 4000 iron atoms. Iron is deposited in the nanocage following its oxidation to Fe^{3+} by H-ferritin, which possesses functionally important ferroxidase activity. Iron sequestered in ferritin is detoxified and cannot undergo redox cycling. Nevertheless, it can be mobilized for metabolic purposes via ferritinophagy, an autophagic mechanism for ferritin degradation via the cargo receptor NCOA4 (Hoelzgen et al. 2024). Ferritin and ferritin-like molecules, such as the bacterioferritins and Dps proteins, are derived from a rubrerythrin-like ancestor and are found in metazoans and a wide range of bacteria and monocellular eukaryotic organisms (Andrews 2010). *Saccharomyces cerevisiae* is an exception and lacks any ferritin-like molecule; instead, it directs excess iron to the vacuole, an organelle that acts as a storage and recycling center (Ramos-Alonso et al. 2020).

Iron is delivered to ferritin for storage via PCBP1, a poly C-binding protein that also operates as an iron chaperone (Shi et al. 2008). PCBP1 and its homolog PCBP2 shuttle iron to further target proteins (Philpott et al. 2023), preventing oxidative stress during intracellular iron traffic. However, a fraction of cytosolic iron, known as labile iron pool (LIP), is not bound to proteins and remains redox-active and thus, potentially toxic. Maintenance of "physiological" LIP levels is critical for health, as discussed in Chap. 4.

1.3.5 Regulation of Iron Homeostasis

Cells and organisms have responded to the challenges of iron acquisition and detoxification by tightly regulating iron metabolism. The goal is to acquire sufficient iron to satisfy metabolic needs, and concomitantly, minimize the risks of iron overload. For instance, many bacteria use the iron-responsive transcription factor Fur (ferric uptake regulator). When iron is abundant, Fur binds Fe^{2+} and represses transcription of iron-regulated genes. When iron is scarce, Fur is inactivated, allowing the expression of iron uptake genes, such as those encoding siderophores and iron transporters (Andrews et al. 2003). Likewise, in *Saccharomyces cerevisiae* the transcription factors Aft1 and Aft2 activate the expression of iron uptake and transport genes under iron deficiency. Conversely, under high iron conditions, Aft1 and Aft2 remain inactive (Ramos-Alonso et al. 2020).

Iron regulatory pathways have acquired unique characteristics across the evolutionary scale. In mammals, iron homeostasis is regulated by diverse mechanisms at the cellular and systemic levels. Thus, iron regulatory proteins IRP1 and IRP2, coordinately regulate the expression of several mRNAs encoding proteins for cellular iron uptake (transferrin receptor 1), storage (ferritin), utilization (5-aminolevulinic acid synthase 2), and efflux (ferroportin). The mechanism involves the binding of IRPs to conserved "iron responsive elements" (IREs) in the untranslated regions of these mRNAs, which is induced under iron deficiency (Wilkinson and Pantopoulos 2014); see Fig. 12.2a in Chap. 12. Systemic iron balance is controlled by hepcidin, a liver-derived peptide hormone that inactivates ferroportin in target cells (Nemeth and Ganz 2023). Details on the regulation of cellular and systemic iron metabolism are provided in Chaps. 2 and 3, respectively.

1.4 BioIron at Work

1.4.1 Chemical Forms of Iron as a Protein Cofactor

The most common iron cofactors of proteins are heme and iron-sulfur clusters (ISCs). Both are biosynthesized by ancient and highly conserved pathways. Notably, heme biosynthesis is tightly connected to ISC biogenesis, considering that the activities of the heme biosynthetic enzymes ferrochelatase (Dailey et al. 1994) and ALA dehydratase (Liu et al. 2020) depend on an ISC.

Heme is a complex of the tetrapyrrole protoporphyrin IX with Fe^{2+} (Fig. 1.4a). There are several variants (a, b, c, o, d, s, l, and m) that differ on the side chains of the porphyrin ring. Each variant is tailored to a specific biological function, such as oxygen transport (heme b in hemoglobin), electron transfer (heme c in cytochrome c), or substrate activation (heme d in catalases). Heme biosynthesis starts with the generation of

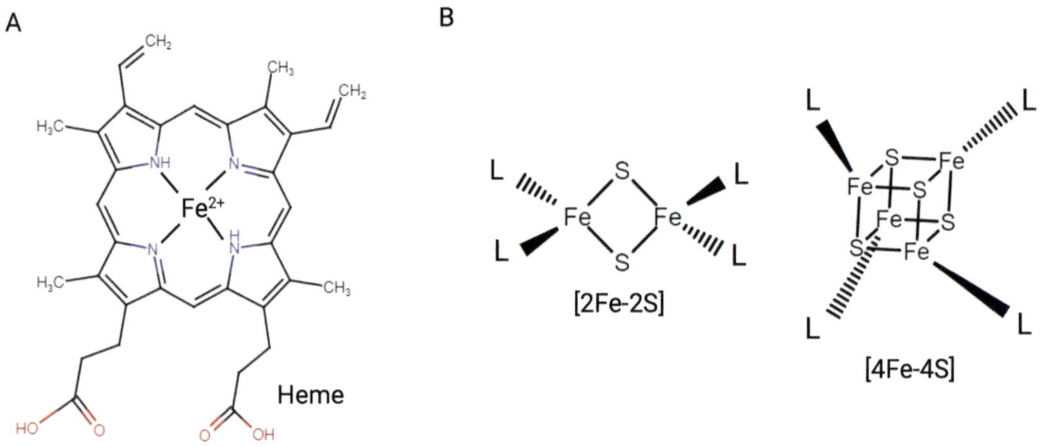

Fig. 1.4 Iron prosthetic groups. (**a**) Heme; (**b**) common iron-sulfur clusters [2Fe-2S] and [4Fe-4S]

the universal precursor 5-aminolevulinic acid (ALA) (Heinemann et al. 2008). In metazoans, fungi, and some bacteria, ALA is formed by the condensation of glycine with succinyl-CoA, catalyzed by aminolevulinic acid (ALAS). In plants and most bacteria, including archaea, ALA is derived from glutamate. In a series of subsequent enzymatic reactions, ALA is converted to porphobilinogen, uroporphyrinogen III, coproporphyrinogen III, protoporphyrinogen IX, and protoporphyrin IX; see Fig. 18.2 in Chap. 18. The last reaction, insertion of iron into protoporphyrin IX, is catalyzed by ferrochelatase. In eukaryotic cells, the initial and final steps of the pathway are compartmentalized in mitochondria and intermediate steps in the cytoplasm. Archaea utilize alternative pathways for heme biosynthesis, independent of uroporphyrinogen III formation (Bali et al. 2014).

ISCs consist of Fe^{2+} or Fe^{3+} ions bridged by S atoms and are coordinated to proteins via cysteine residues. They can have various configurations, again tailored to biological activities, with rhombic [2Fe-2S] and cubane [4Fe-4S] clusters being the most prevalent (Fig. 1.4b). For example, the Rieske proteins mediate the transport of electrons via [2Fe-2S] clusters in the respiratory chain (Trumpower and Edwards 1979). Aconitases catalyze the conversion of citrate to iso-citrate via a [4Fe-4S] cluster that provides electrons for the dehydration of citrate (Lauble et al. 1994). More complex ISCs like [8Fe-7S], [7Fe-9S], and ISCs with other metals are rare and are mostly found in bacterial enzymes such as nitrogenases or hydrogenases (Sippel and Einsle 2017; Byer et al. 2015). ISC biogenesis starts with the conversion of L-cysteine to alanine, catalyzed by the cysteine desulfurase NFS1, which releases an S atom (Lill 2020), see Fig. 18.2 in Chap. 18. This is used for the assembly of [2Fe-2S] on the scaffold protein ISCU2, with electrons donated by the ferredoxin FDX2. Iron is delivered to the scaffold via the PCBP1 chaperone, in complex with BOLA2 (Patel et al. 2019). The assembled [2Fe-2S] cluster is then released from the scaffold with the aid of HSP40-HSP70 chaperone proteins and shuttled to glutaredoxin 5 (GLRX5) for direct delivery to acceptor apo-proteins. [4Fe-4S] clusters are generated by reductive fusion of [2Fe-2S] clusters with the aid of GLRX5. Subsequently, the [4Fe-4S] clusters are targeted to recipient apo-proteins via dedicated auxiliary factors. In eukaryotes, ISC biogenesis primarily takes place in mitochondria, while a cytosolic ISC assembly system accounts for the maturation of ISCs targeted to acceptor proteins in the cytoplasm.

Other types of non-heme and non-ISC iron are also found as protein cofactors. For instance, ribonucleotide reductase contains a dinuclear Fe-O-Fe center that is used to produce a catalytically essential stable tyrosyl radical (Eklund et al. 2001). Cyclooxygenases and lipoxygenases contain a mononuclear iron center for substrate activation (Prigge et al. 1997). Furthermore, 2-oxoglutarate-dependent oxygenases utilize Fe^{2+}, in complex with 2-oxoglutarate (α-ketoglutarate), as a catalyst in a series of reactions including hydroxylations, N-demethylations, desaturations, etc. (Islam et al. 2018).

1.4.2 Iron-Dependent Biological Functions

In quantitative terms, >70% of body iron is utilized in hemoglobin of red blood cells for oxygen transport to tissues (see Chap. 3). Significant amounts of iron are present in myoglobin for oxygenation of skeletal muscles, in ferritin of hepatocytes that serve as iron stores, and in tissue macrophages, which recycle iron from senescent red blood cells. However, minuscule amounts of iron are also present in every other cell type.

It has been estimated that 2% of the human proteome consists of iron-binding proteins (Andreini et al. 2018). Among these ~400 proteins, 48% bind a heme prosthetic group, 17% an ISC, and 35% other forms of iron. In terms of function, ~37% of all human iron-binding proteins are oxidoreductases. Iron is the most common cofactor of this type of enzymes in general, considering that it is found in almost 70% of all oxidoreductases annotated in protein databases (Kim et al. 2013).

Several iron-containing enzymes and proteins are involved in physiological cellular processes. For instance, ribonucleotide reductase is essential for DNA synthesis (Eklund et al. 2001), phosphoribosyl pyrophosphate aminotransferase for purine synthesis (Itakura and Holmes 1979), the helicase DNA2 for replication (Mariotti et al. 2020), and the helicases XPD and FANCJ for DNA repair (Rudolf et al. 2006). The histone demethylase KDM3B is an epigenetic regulator of the mTORC1 pathway (Shapiro et al. 2023). The DNA-binding protein BACH1 negatively regulates gene transcription (Ogawa et al. 2001), while heme-regulated eIF2a kinase (HRI) controls protein synthesis (Chen and Zhang 2022). Mitochondrial aconitase and succinate dehydrogenase are critical components of the TCA cycle, where electrons derived from NADH and $FADH_2$ are shuttled to the respiratory chain via several iron-containing proteins (Ward and Cloonan 2019). Furthermore, iron-containing enzymes participate in the saturation of fatty acids, cholesterol metabolism, and downstream pathways (for the synthesis of steroid hormone, vitamin D, and bile acids), synthesis of dopamine, serotonin, and noradrenalin, synthesis of thyroid hormones, etc. (Teh et al. 2024).

The wide spectrum of iron-dependent functions in cells and organisms highlights the immense biological significance of iron, which is based on its chemical properties. Nevertheless, chemistry makes iron a double-edged sword, and either iron deficiency or iron overload can be detrimental. Thus, maintenance of appropriate iron-dependent functions requires tight regulation of iron metabolism.

1.5 Conclusions

Iron is an indispensable element that has shaped the evolution of life, from its role in prebiotic chemistry to its essential functions in cellular metabolism. Its redox versatility underpins crucial biological processes, including oxygen transport, DNA synthesis, and enzymatic activity. However, the delicate balance between iron's necessity and toxicity has driven the evolution of sophisticated regulatory mechanisms to ensure bioavailability while minimizing oxidative stress. Understanding the intricate chemistry and biology of iron not only provides insights into fundamental life processes but also paves the way for understanding the pathophysiology and improving the management of iron-related disorders.

Acknowledgements I wish to thank Dr. Apostolos Galaris for assistance with the figures.

References

Aisen P, Leibman A, Zweier J (1978) Stoichiometric and site characteristics of the binding of iron to human transferrin. J Biol Chem 253(6):1930–1937

Aisen P, Enns C, Wessling-Resnick M (2001) Chemistry and biology of eukaryotic iron metabolism. Int J Biochem Cell Biol 33:940–959

Akbari A, Palsson BO (2023) Metabolic homeostasis and growth in abiotic cells. Proc Natl Acad Sci USA 120(19):e2300687120. https://doi.org/10.1073/pnas.2300687120

Andreini C, Putignano V, Rosato A, Banci L (2018) The human iron-proteome. Metallomics 10(9):1223–1231. https://doi.org/10.1039/c8mt00146d

Andrews SC (2010) The ferritin-like superfamily: evolution of the biological iron storeman from a rubrerythrin-like ancestor. Biochim Biophys Acta 1800(8):691–705. https://doi.org/10.1016/j.bbagen.2010.05.010

Andrews SC, Robinson AK, Rodriguez-Quinones F (2003) Bacterial iron homeostasis. FEMS Microbiol Rev 27:215–237

Arosio P, Ingrassia R, Cavadini P (2009) Ferritins: a family of molecules for iron storage, antioxidation and more. Biochim Biophys Acta 1790(7):589–599. S0304--4165(08)00215-8 [pii]10.1016/j.bbagen.2008.09.004

Bali S, Palmer DJ, Schroeder S, Ferguson SJ, Warren MJ (2014) Recent advances in the biosynthesis of modified tetrapyrroles: the discovery of an alternative pathway for the formation of heme and heme d 1. Cell Mol Life Sci 71(15):2837–2863. https://doi.org/10.1007/s00018-014-1563-x

Beyazay T, Belthle KS, Fares C, Preiner M, Moran J, Martin WF, Tuysuz H (2023) Ambient temperature CO(2) fixation to pyruvate and subsequently to citramalate over iron and nickel nanoparticles. Nat Commun 14(1):570. https://doi.org/10.1038/s41467-023-36088-w

Brabender M, Henriques Pereira DP, Mrnjavac N, Schlikker ML, Kimura ZI, Sucharitakul J, Kleinermanns K, Tuysuz H, Buckel W, Preiner M, Martin WF (2024) Ferredoxin reduction by hydrogen with iron functions as an evolutionary precursor of flavin-based electron bifurcation. Proc Natl Acad Sci USA 121(13):e2318969121. https://doi.org/10.1073/pnas.2318969121

Bromberg Y, Aptekmann AA, Mahlich Y, Cook L, Senn S, Miller M, Nanda V, Ferreiro DU, Falkowski PG (2022) Quantifying structural relationships of metal-binding sites suggests origins of biological electron transfer. Sci Adv 8(2):eabj3984. https://doi.org/10.1126/sciadv.abj3984

Brown JB, Lee MA, Smith AT (2021) Ins and outs: recent advancements in membrane protein-mediated prokaryotic ferrous iron transport. Biochemistry 60(44):3277–3291. https://doi.org/10.1021/acs.biochem.1c00586

Burrows A (2000) Supernova explosions in the universe. Nature 403(6771):727–733. https://doi.org/10.1038/35001501

Byer AS, Shepard EM, Peters JW, Broderick JB (2015) Radical S-adenosyl-L-methionine chemistry in the synthesis of hydrogenase and nitrogenase metal cofactors. J Biol Chem 290(7):3987–3994. https://doi.org/10.1074/jbc.R114.578161

Camprubi E, Jordan SF, Vasiliadou R, Lane N (2017) Iron catalysis at the origin of life. IUBMB Life 69(6):373–381. https://doi.org/10.1002/iub.1632

Chen JJ, Zhang S (2022) Translational control by heme-regulated eIF2alpha kinase during erythropoiesis. Curr Opin Hematol 29(3):103–111. https://doi.org/10.1097/MOH.0000000000000704

Cornelissen CN, Sparling PF (1994) Iron piracy: acquisition of transferrin-bound iron by bacterial pathogens. Mol Microbiol 14(5):843–850. https://doi.org/10.1111/j.1365-2958.1994.tb01320.x

Dailey HA, Finnegan MG, Johnson MK (1994) Human ferrochelatase is an iron-sulfur protein. Biochemistry 33(2):403–407. https://doi.org/10.1021/bi00168a003

David LA, Alm EJ (2011) Rapid evolutionary innovation during an Archaean genetic expansion. Nature 469(7328):93–96. https://doi.org/10.1038/nature09649

Eck RV, Dayhoff MO (1966) Evolution of the structure of ferredoxin based on living relics of primitive amino acid sequences. Science 152(3720):363–366. https://doi.org/10.1126/science.152.3720.363

Eklund H, Uhlin U, Farnegardh M, Logan DT, Nordlund P (2001) Structure and function of the radical enzyme ribonucleotide reductase. Prog Biophys Mol Biol 77(3):177–268

Enami S, Sakamoto Y, Colussi AJ (2014) Fenton chemistry at aqueous interfaces. Proc Natl Acad Sci USA 111(2):623–628. https://doi.org/10.1073/pnas.1314885111

Fenton HJH (1894) LXXIII.—Oxidation of tartaric acid in presence of iron. J Chem Soc Trans 65:899–910

Frey PA, Reed GH (2012) The ubiquity of iron. ACS Chem Biol 7(9):1477–1481. https://doi.org/10.1021/cb300323q

Galaris D, Barbouti A, Pantopoulos K (2019) Iron homeostasis and oxidative stress: an intimate relationship. Biochim Biophys Acta, Mol Cell Res 1866(12):118535. https://doi.org/10.1016/j.bbamcr.2019.118535

Garritano AN, Song W, Thomas T (2022) Carbon fixation pathways across the bacterial and archaeal tree of life. PNAS Nexus 1(5):pgac226. https://doi.org/10.1093/pnasnexus/pgac226

Gkouvatsos K, Papanikolaou G, Pantopoulos K (2012) Regulation of iron transport and the role of transferrin. Biochim Biophys Acta 1820(3):188–202. https://doi.org/10.1016/j.bbagen.2011.10.013

Haber F, Weiss J (1934) The catalytic decomposition of hydrogen peroxide by iron salts. Proc R Soc Lond Ser A Math Phys Sci 147(861):332–351

Hall DO, Cammack R, Rao KK (1971) Role for ferredoxins in the origin of life and biological evolution. Nature 233(5315):136–138. https://doi.org/10.1038/233136a0

Hamed MY, Silver J, Wilson MT (1983) Studies of the reactions of ferric iron with glutathione and some related thiols. Inorg Chim Acta 78:1–11

Heinemann IU, Jahn M, Jahn D (2008) The biochemistry of heme biosynthesis. Arch Biochem Biophys 474(2):238–251. https://doi.org/10.1016/j.abb.2008.02.015

Hider RC, Kong X (2010) Chemistry and biology of siderophores. Nat Prod Rep 27(5):637–657. https://doi.org/10.1039/b906679a

Hider RC, Kong XL (2011) Glutathione: a key component of the cytoplasmic labile iron pool. Biometals 24(6):1179–1187. https://doi.org/10.1007/s10534-011-9476-8

Hoelzgen F, Nguyen TTP, Klukin E, Boumaiza M, Srivastava AK, Kim EY, Zalk R, Shahar A, Cohen-Schwartz S, Meyron-Holtz EG, Bou-Abdallah F, Mancias JD, Frank GA (2024) Structural basis for the intracellular regulation of ferritin degradation. Nat Commun 15(1):3802. https://doi.org/10.1038/s41467-024-48151-1

Holm NG, Oze C, Mousis O, Waite JH, Guilbert-Lepoutre A (2015) Serpentinization and the formation of H2 and CH4 on celestial bodies (planets, moons, comets). Astrobiology 15(7):587–600. https://doi.org/10.1089/ast.2014.1188

Huber C, Wachtershauser G (1997) Activated acetic acid by carbon fixation on (Fe,Ni)S under primordial conditions. Science 276(5310):245–247. https://doi.org/10.1126/science.276.5310.245

Huber C, Wachtershauser G (1998) Peptides by activation of amino acids with CO on (Ni,Fe)S surfaces: implications for the origin of life. Science 281(5377):670–672. https://doi.org/10.1126/science.281.5377.670

Islam MS, Leissing TM, Chowdhury R, Hopkinson RJ, Schofield CJ (2018) 2-Oxoglutarate-Dependent Oxygenases. Annu Rev Biochem 87:585–620. https://doi.org/10.1146/annurev-biochem-061516-044724

Itakura M, Holmes EW (1979) Human amidophosphoribosyltransferase. an oxygen-sensitive iron-sulfur protein. J Biol Chem 254(2):333–338

Jenkitkasemwong S, Wang CY, Coffey R, Zhang W, Chan A, Biel T, Kim JS, Hojyo S, Fukada T, Knutson MD (2015) SLC39A14 is required for the development of hepatocellular iron overload in murine models of hereditary hemochromatosis. Cell Metab 22(1):138–150. https://doi.org/10.1016/j.cmet.2015.05.002

Keller MA, Zylstra A, Castro C, Turchyn AV, Griffin JL, Ralser M (2016) Conditional iron and pH-dependent activity of a non-enzymatic glycolysis and pentose phosphate pathway. Sci Adv 2(1):e1501235. https://doi.org/10.1126/sciadv.1501235

Kim JD, Senn S, Harel A, Jelen BI, Falkowski PG (2013) Discovering the electronic circuit diagram of life: structural relationships among transition metal binding sites in oxidoreductases. Philos Trans R Soc Lond Ser B Biol Sci 368(1622):20120257. https://doi.org/10.1098/rstb.2012.0257

Koppenol WH (2022) Ferryl for real. the Fenton reaction near neutral pH. Dalton Trans 51(45):17496–17502. https://doi.org/10.1039/d2dt03168j

Koppenol WH, Hider RH (2019) Iron and redox cycling. Do's and don'ts. Free Radic Biol Med 133:3–10. https://doi.org/10.1016/j.freeradbiomed.2018.09.022

Lambert LA (2012) Molecular evolution of the transferrin family and associated receptors. Biochim Biophys Acta 1820(3):244–255. https://doi.org/10.1016/j.bbagen.2011.06.002

Lau CK, Krewulak KD, Vogel HJ (2016) Bacterial ferrous iron transport: the Feo system. FEMS Microbiol Rev 40(2):273–298. https://doi.org/10.1093/femsre/fuv049

Lauble H, Kennedy MC, Beinert H, Stout CD (1994) Crystal structures of aconitase with trans-aconitate and nitrocitrate bound. J Mol Biol 237:437–451

Lill R (2020) From the discovery to molecular understanding of cellular iron-sulfur protein biogenesis. Biol Chem 401(6–7):855–876. https://doi.org/10.1515/hsz-2020-0117

Liu G, Sil D, Maio N, Tong WH, Bollinger JM Jr, Krebs C, Rouault TA (2020) Heme biosynthesis depends on previously unrecognized acquisition of iron-sulfur cofactors in human amino-levulinic acid dehydratase. Nat Commun 11(1):6310. https://doi.org/10.1038/s41467-020-20145-9

Lyons TW, Reinhard CT, Planavsky NJ (2014) The rise of oxygen in Earth's early ocean and atmosphere. Nature 506(7488):307–315. https://doi.org/10.1038/nature13068

Mariotti L, Wild S, Brunoldi G, Piceni A, Ceppi I, Kummer S, Lutz RE, Cejka P, Gari K (2020) The iron-sulphur cluster in human DNA2 is required for all biochemical activities of DNA2. Commun Biol 3(1):322. https://doi.org/10.1038/s42003-020-1048-4

Martin W, Baross J, Kelley D, Russell MJ (2008) Hydrothermal vents and the origin of life. Nat Rev Microbiol 6(11):805–814. https://doi.org/10.1038/nrmicro1991

Martin WF, Sousa FL, Lane N (2014) Evolution. Energy at life's origin. Science 344(6188):1092–1093. https://doi.org/10.1126/science.1251653

Merkofer M, Kissner R, Hider RC, Brunk UT, Koppenol WH (2006) Fenton chemistry and iron chelation under physiologically relevant conditions: electrochemistry and kinetics. Chem Res Toxicol 19(10):1263–1269. https://doi.org/10.1021/tx060101w

Meyer J (1988) The evolution of ferredoxins. Trends Ecol Evol 3(9):222–226. https://doi.org/10.1016/0169-5347(88)90162-0

Meyerstein D (2022) What are the oxidizing intermediates in the Fenton and Fenton-like reactions? A

perspective. Antioxidants (Basel) 11(7). https://doi.org/10.3390/antiox11071368
Muchowska KB, Varma SJ, Moran J (2019) Synthesis and breakdown of universal metabolic precursors promoted by iron. Nature 569(7754):104–107. https://doi.org/10.1038/s41586-019-1151-1
Muller V, Chowdhury NP, Basen M (2018) Electron bifurcation: a long-hidden energy-coupling mechanism. Ann Rev Microbiol 72:331–353. https://doi.org/10.1146/annurev-micro-090816-093440
Nemeth E, Ganz T (2023) Hepcidin and iron in health and disease. Annu Rev Med 74:261–277. https://doi.org/10.1146/annurev-med-043021-032816
Noinaj N, Guillier M, Barnard TJ, Buchanan SK (2010) TonB-dependent transporters: regulation, structure, and function. Ann Rev Microbiol 64:43–60. https://doi.org/10.1146/annurev.micro.112408.134247
Ogawa K, Sun J, Taketani S, Nakajima O, Nishitani C, Sassa S, Hayashi N, Yamamoto M, Shibahara S, Fujita H, Igarashi K (2001) Heme mediates derepression of Maf recognition element through direct binding to transcription repressor Bach1. EMBO J 20(11):2835–2843. https://doi.org/10.1093/emboj/20.11.2835
Patel SJ, Frey AG, Palenchar DJ, Achar S, Bullough KZ, Vashisht A, Wohlschlegel JA, Philpott CC (2019) A PCBP1-BolA2 chaperone complex delivers iron for cytosolic [2Fe-2S] cluster assembly. Nat Chem Biol 15(9):872–881. https://doi.org/10.1038/s41589-019-0330-6
Peralvarez-Marin A, Baranowski E, Bierge P, Pich OQ, Lebrette H (2021) Metal utilization in genome-reduced bacteria: do human mycoplasmas rely on iron? Comput Struct Biotechnol J 19:5752–5761. https://doi.org/10.1016/j.csbj.2021.10.022
Philpott CC, Protchenko O, Wang Y, Novoa-Aponte L, Leon-Torres A, Grounds S, Tietgens AJ (2023) Iron-tracking strategies: chaperones capture iron in the cytosolic labile iron pool. Front Mol Biosci 10:1127690. https://doi.org/10.3389/fmolb.2023.1127690
Piskin E, Cianciosi D, Gulec S, Tomas M, Capanoglu E (2022) Iron absorption: factors, limitations, and improvement methods. ACS Omega 7(24):20441–20456. https://doi.org/10.1021/acsomega.2c01833
Posey JE, Gherardini FC (2000) Lack of a role for iron in the Lyme disease pathogen. Science 288(5471):1651–1653. https://doi.org/10.1126/science.288.5471.1651
Prigge ST, Boyington JC, Faig M, Doctor KS, Gaffney BJ, Amzel LM (1997) Structure and mechanism of lipoxygenases. Biochimie 79(11):629–636
Raanan H, Pike DH, Moore EK, Falkowski PG, Nanda V (2018) Modular origins of biological electron transfer chains. Proc Natl Acad Sci USA 115(6):1280–1285. https://doi.org/10.1073/pnas.1714225115
Ramos-Alonso L, Romero AM, Martinez-Pastor MT, Puig S (2020) Iron regulatory mechanisms in saccharomyces cerevisiae. Front Microbiol 11:582830. https://doi.org/10.3389/fmicb.2020.582830
Rudolf J, Makrantoni V, Ingledew WJ, Stark MJ, White MF (2006) The DNA repair helicases XPD and FancJ have essential iron-sulfur domains. Mol Cell 23(6):801–808. S1097-2765(06)00516-8 [pii]10.1016/j.molcel.2006.07.019
Sanchez M, Sabio L, Galvez N, Capdevila M, Dominguez-Vera JM (2017) Iron chemistry at the service of life. IUBMB Life 69(6):382–388. https://doi.org/10.1002/iub.1602
Schroder I, Johnson E, de Vries S (2003) Microbial ferric iron reductases. FEMS Microbiol Rev 27(2–3):427–447. https://doi.org/10.1016/S0168-6445(03)00043-3
Shapiro JS, Chang HC, Tatekoshi Y, Zhao Z, Waxali ZS, Hong BJ, Chen H, Geier JA, Bartom ET, De Jesus A, Nejad FK, Mahmoodzadeh A, Sato T, Ramos-Alonso L, Romero AM, Martinez-Pastor MT, Jiang SC, Sah-Teli SK, Li L, Bentrem D, Lopaschuk G, Ben-Sahra I, O'Halloran TV, Shilatifard A, Puig S, Bergelson J, Koivunen P, Ardehali H (2023) Iron drives anabolic metabolism through active histone demethylation and mTORC1. Nat Cell Biol 25(10):1478–1494. https://doi.org/10.1038/s41556-023-01225-6
Shawki A, Engevik MA, Kim RS, Knight PB, Baik RA, Anthony SR, Worrell RT, Shull GE, Mackenzie B (2016) Intestinal brush-border Na+/H+ exchanger-3 drives H+-coupled iron absorption in the mouse. Am J Physiol Gastrointest Liver Physiol 311(3):G423–G430. https://doi.org/10.1152/ajpgi.00167.2016
Shi H, Bencze KZ, Stemmler TL, Philpott CC (2008) A cytosolic iron chaperone that delivers iron to ferritin. Science 320(5880):1207–1210. 320/5880/1207 [pii]10.1126/science.1157643
Sies H, Belousov VV, Chandel NS, Davies MJ, Jones DP, Mann GE, Murphy MP, Yamamoto M, Winterbourn C (2022) Defining roles of specific reactive oxygen species (ROS) in cell biology and physiology. Nat Rev Mol Cell Biol 23(7):499–515. https://doi.org/10.1038/s41580-022-00456-z
Sippel D, Einsle O (2017) The structure of vanadium nitrogenase reveals an unusual bridging ligand. Nat Chem Biol 13(9):956–960. https://doi.org/10.1038/nchembio.2428
Sojo V, Herschy B, Whicher A, Camprubi E, Lane N (2016) The origin of life in alkaline hydrothermal vents. Astrobiology 16(2):181–197. https://doi.org/10.1089/ast.2015.1406
Stockwell BR (2022) Ferroptosis turns 10: emerging mechanisms, physiological functions, and therapeutic applications. Cell 185(14):2401–2421. https://doi.org/10.1016/j.cell.2022.06.003
Teh MR, Armitage AE, Drakesmith H (2024) Why cells need iron: a compendium of iron utilisation. Trends Endocrinol Metab 35(12):1026–1049. https://doi.org/10.1016/j.tem.2024.04.015
Trumpower BL, Edwards CA (1979) Purification of a reconstitutively active iron-sulfur protein (oxidation factor) from succinate . cytochrome c reductase complex of bovine heart mitochondria. J Biol Chem 254(17):8697–8706

Tsebrenko D, Soker N (2015) Type Ia supernova remnants: shaping by iron bullets. Mon Not R Astron Soc 453(1):166–171. https://doi.org/10.1093/mnras/stv1641

Wachtershauser G (1990) Evolution of the first metabolic cycles. Proc Natl Acad Sci USA 87(1):200–204. https://doi.org/10.1073/pnas.87.1.200

Wade J, Byrne DJ, Ballentine CJ, Drakesmith H (2021) Temporal variation of planetary iron as a driver of evolution. Proc Natl Acad Sci USA 118(51). https://doi.org/10.1073/pnas.2109865118

Ward DM, Cloonan SM (2019) Mitochondrial iron in human health and disease. Annu Rev Physiol 81:453–482. https://doi.org/10.1146/annurev-physiol-020518-114742

Weinberg ED (1997) The lactobacillus anomaly: total iron abstinence. Perspect Biol Med 40(4):578–583. https://doi.org/10.1353/pbm.1997.0072

Wilkinson N, Pantopoulos K (2014) The IRP/IRE system in vivo: insights from mouse models. Front Pharmacol 5:176. https://doi.org/10.3389/fphar.2014.00176

Cellular Iron Homeostasis

2

Caroline C. Philpott and Esther G. Meyron-Holtz

Abstract

Iron is an essential nutrient for all eukaryotic cells, yet it is potentially toxic. Cells have elegant systems coordinating the uptake, use, storage, and export of this metal while also responding to extracellular cues for regulating uptake and efflux. Here we describe the basic systems that operate in all cells with the goal of maintaining iron homeostasis in a highly dynamic system. In mammalian tissues and organs, multiple cell types are highly specialized in their iron management with the goal of maintaining cellular homeostasis subordinate to achieving a more specific function, e.g., dietary iron absorption in the intestinal epithelium, hemoglobin synthesis in the erythron, and iron recycling in splenic macrophages. The alterations in cellular iron balance that allow for these specializations will be briefly discussed.

Keywords

Cellular iron trafficking · Iron-sulfur clusters · Heme · Ferritin · Iron regulatory proteins

2.1 The Essential Functions of Iron

Iron is used by cells to form cofactors that are required for the activity of hundreds of proteins within every cell. This reliance on iron arises from the presence of unpaired electrons in the outer *d* orbital of the iron atom, as these electrons can readily be lost or reacquired in the presence of an appropriate electron acceptor/donor (Chap. 1). In an aerobic environment, oxygen is the preferred electron acceptor, and iron is readily interconverted between the ferrous (Fe^{2+}) and the ferric (Fe^{3+}) states. Thus, biochemical processes involving oxygen binding, oxygen activation, oxidation-reduction, and electron transfer often involve proteins with iron cofactors (Crichton 2001).

2.1.1 Heme Iron Cofactors

The iron cofactors of biological systems are heme, iron-sulfur (Fe-S) clusters, and mononuclear or dinuclear iron centers (mono- or di-iron centers). In the form of heme, iron can bind oxygen (hemoglobin, myoglobin), activate oxygen

C. C. Philpott (✉)
Genetics and Metabolism Section, Liver Diseases Branch, NIDDK, NIH, Bethesda, MD, USA
e-mail: carolinep@intra.niddk.nih.gov

E. G. Meyron-Holtz
Laboratory of Molecular Nutrition, Faculty of Biotechnology and Food Engineering, Technion, Haifa, Israel

K. Pantopoulos (ed.), *Iron Metabolism in Human Health and Disease*, Advances in Experimental Medicine and Biology 1480, https://doi.org/10.1007/978-3-031-92033-2_2

(catalases, peroxidases), transport electrons (cytochromes), and regulate gene expression (transcription and translation factors). Heme is synthesized in an eight-step reaction that begins in the mitochondria, where succinyl-CoA condenses with glycine to form δ-aminolevulinic acid, which is then exported to the cytosol. Subsequent steps are carried out in the cytosol until the tetrapyrrole coproporphyrinogen III is imported back into the mitochondria, with the final step being iron insertion into protoporphyrin IX by ferrochelatase; see Fig. 18.2 in Chap. 18. Because ferrochelatase requires an Fe-S cluster for activity, the process of heme synthesis is dependent on the mitochondrial Fe-S cluster assembly system as well (Reddi and Hamza 2016).

2.1.2 Iron-Sulfur Cluster Cofactors

Fe-S clusters are cofactors in which iron and sulfur atoms in a variety of oxidation states are assembled into clusters that are coordinated by amino acid side chains within folded proteins of cells. Most Fe-S clusters have either rhombic [2Fe-2S] or cubane [4Fe-4S] structures, but other types of structures are also described. Because the electrons associated with the clusters are distributed over multiple iron atoms, these clusters can readily accept or donate electrons while remaining stably coordinated by cellular proteins. Fe-S clusters serve a variety of functions, including electron transfers, catalysis, protein structural stability, cofactor transfer, and sensing of iron, oxygen, and redox stress (Johnson et al. 2005). Fe-S clusters are primarily synthesized in the mitochondrial matrix on scaffold proteins, which capture inorganic sulfur from cysteine desulfurase and facilitate the assembly of [2Fe-2S] clusters; see Fig. 18.2 in Chap. 18. These can be directly incorporated into mitochondrial proteins or undergo condensation and reduction to form [4Fe-4S] clusters, which are then transferred to recipient mitochondrial apoenzymes. A sulfur-containing Fe-S precursor from mitochondria is exported to the cytosol where, along with some de novo cluster synthesis, [2Fe-2S] and [4Fe-4S] clusters are assembled and distributed to proteins of the cytosol and nucleus (Braymer et al. 2021).

2.1.3 Non-heme Iron Centers

The simplest iron cofactors are the mononuclear and dinuclear iron centers of the non-heme iron enzymes. These consist of simple Fe^{2+} ions coordinated by oxygen and nitrogen ligands in the amino acid side chains of folded proteins. No biosynthetic or assembly factors are required for their formation, and iron is directly acquired from the cytosolic or mitochondrial labile iron pools. Iron centers of these enzymes are typically involved in binding and activating oxygen to be consumed in mono- or di-oxygenase reactions. Typical reactions are hydroxylation, desaturation, and demethylation and are catalyzed by a large number (>140) of enzymes (Herr and Hausinger 2018; Islam et al. 2018; Jasniewski and Que Jr. 2018). These iron-center proteins are found in essentially all compartments of the cell, including the cytosol, nucleus, mitochondria, and secretory pathway. Each of these cofactor species is synthesized using iron derived from the intracellular iron pools. These pools derive, in turn, from iron import and intracellular trafficking systems.

2.2 Cellular Iron Trafficking

2.2.1 Iron Import

Mammalian cells are bathed in an extracellular milieu that contains iron primarily in the form of diferric transferrin (Tf), but in some instances may also contain non-transferrin-bound iron (Knutson 2019) (Fig. 2.1). Tf-Fe is captured by cells expressing the transferrin receptor (TfR) on the plasma membrane. The main TfR, TfR1 (TFRC), is a homodimeric, type II integral membrane protein, expressed on most cells, and functions by capturing circulating iron-loaded transferrin, which triggers clathrin-mediated endocytosis of the complex. At the acidic pH of

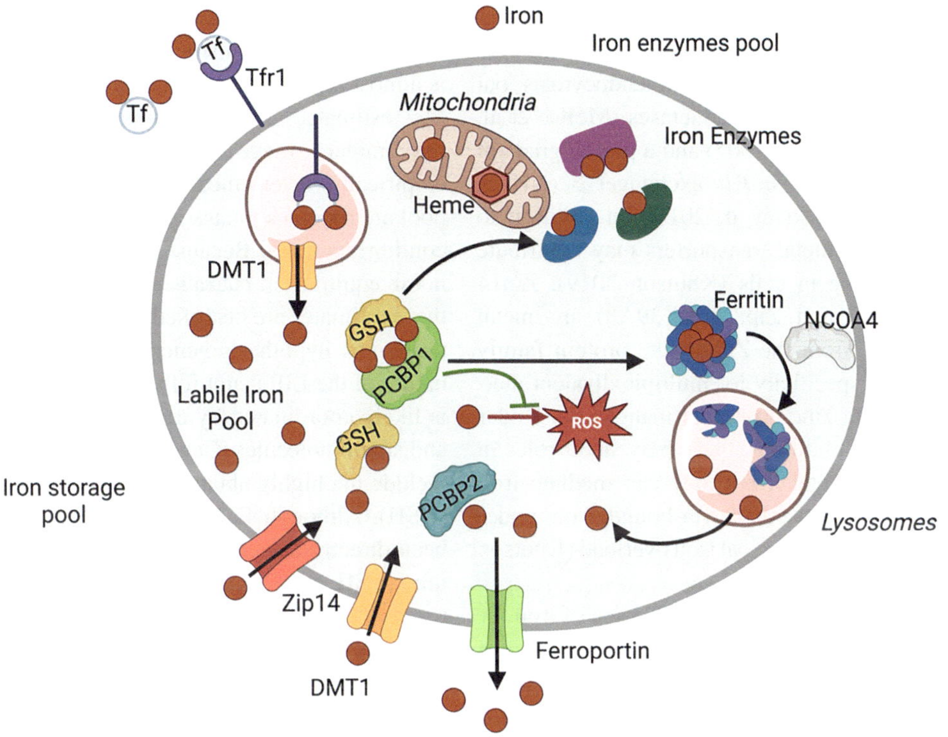

Fig. 2.1 Trafficking of iron in cells. Iron ions (red spheres) enter cells via receptor-mediated endocytosis on Tf-TfR1 complexes or directly via cell surface transporters. Intracellular iron is captured on chaperones PCBP1 or PCBP2 and trafficked to sites of utilization (mitochondria, iron enzymes), storage (ferritin), or efflux (ferroportin). (Modified from Philpott et al. 2023)

the endosome, the stability of the Tf-Fe complex is lowered and the iron is both released and reduced via the six-transmembrane epithelial antigen of prostate 3, a STEAP family reductase (Ohgami et al. 2005; Chen et al. 2023). Iron is then imported via the divalent metal transporter 1 (DMT1, NRAMP2, SLC11A2), and the Tf and TfR1 dissociate and recycle to the plasma membrane (Galy et al. 2024). Most cells rely on TfR1 for Tf-Fe uptake, and this pathway is essential in many tissues, notably the erythroid cells of the bone marrow and the crypt cells of the duodenal epithelium, which require high amounts of iron uptake to support proliferating precursor cells (Levy et al. 1999; Chen et al. 2015). Cardiomyocytes, neurons, skeletal myocytes, and lymphocytes also depend on TfR1 for iron uptake (Xu et al. 2015; Jabara et al. 2016; Matak et al. 2016; Wang et al. 2020a, b). Surprisingly, the liver can acquire iron independent of TfR1 and may rely on the highly expressed TfR2 or other iron sources for uptake (Fillebeen et al. 2019). TfR2 is 45% identical and 66% similar to TfR1 in its ectodomain. TfR2 binds Tf-Fe and can mediate endocytosis and uptake of Tf-Fe, although it binds with a much lower affinity than TfR1 (Kawabata et al. 1999). Yet, primarily TfR2 functions as a regulator of iron uptake in cells via its effects on hepatic expression of the iron regulatory hormone, hepcidin (Chap. 3).

DMT1 is the major transporter facilitating Tf-Fe uptake in most cell types (Gunshin et al. 1997). Consistent with its role in transferring iron across the endosomal membrane, it is localized to endosomes and other vesicles of the late secretory pathway. Maximal DMT1 activity occurs at acidic pH, as is present in endosomes and lysosomes, and reflects the transporter's requirement of proton symport for activity (Mackenzie et al. 2006; Liziczai et al. 2025). DMT1 also mediates dietary iron uptake in a Tf-independent manner on the apical surface of the gut epithelium. This iron uptake does not involve endocytosis but relies on surface ferric reductases (McKie et al. 2001; Gunshin et al. 2005) and a proton gradient generated by the Na^{+}/H^{+} exchanger 3 (NHE3, SLC9A3) (Shawki et al. 2016). In addition to DMT1, other metal transporters may contribute to iron uptake in cells (Knutson 2019). Zip14 (SLC39A14) and Zip8 (SLC39A8) are metal transporters from the Zrt-Irt-like protein family and exhibit specificity for multiple divalent metals, including zinc, iron, and manganese. Under physiologic conditions, these play small roles in cellular iron uptake, but they can mediate iron uptake from non-transferrin-bound iron under conditions of pathological iron overload (Knutson 2019).

Although Tf-Fe is the major source of iron for most cells, there are alternative systems for iron uptake. Macrophages of the spleen and reticuloendothelium take up whole cells and remnants of senescent erythrocytes as part of the red cell iron recycling pathway (Klei et al. 2017) and accumulate heme and hemoglobin within the phagolysosome. Heme is transported across the lysosomal membrane to the cytosol via heme-regulated gene 1 (White et al. 2013), where it is degraded by heme oxygenases and iron is released to the cytosolic iron pool (Poss and Tonegawa 1997). Additionally, in humans, secreted ferritin can be taken up by cells through binding of the ferritin H-subunit to TfR1, which triggers endocytosis and trafficking of ferritin to endosomes and lysosomes (Li et al. 2010; Montemiglio et al. 2019). In these cellular compartments, the ferritin nanocage may be degraded, its ferric iron contents reduced (Meng et al. 2022), and then transported into the cytosol (La et al. 2018).

2.2.2 Labile Iron Pool

Once iron is taken up into the cell, it enters the cytosolic labile iron pool (LIP) (see Chap. 4). The cytosolic LIP is operationally defined as the chemically reactive, kinetically exchangeable pool of iron used for the synthesis, assembly, or delivery of iron cofactors in cells. Both theoretical and empirical evidence indicate that the LIP is almost entirely reduced Fe^{2+} with a concentration estimated in the high-nanomolar to low-micromolar range at baseline. However, empirically determined concentrations of this pool are rough estimates based on the kinetics of equilibrium states. Because iron pools in cells are not in equilibrium but rather are highly dynamic, these estimates are best used for comparison purposes and hypothesis generation. The chemical nature of the LIP is not fully known, but the Fe^{2+} is likely coordinated by a mixed buffer of large and small molecules. Candidate small molecules include the highly abundant reduced glutathione (GSH). Although Fe-GSH complexes have not been directly detected in cells, depletion of cellular GSH results in changes to the chemical reactivity and intracellular trafficking of the LIP (Aron et al. 2016; Patel et al. 2021). The major large molecule coordinating cytosolic iron is the iron chaperone poly C binding protein 1 (PCBP1) (Philpott et al. 2023).

2.2.3 Iron Chaperones—Poly C Binding Proteins PCBP1 and PCBP2

Iron chaperones are proteins that specifically coordinate iron ions or iron cofactors in a kinetically labile form, allowing the metal to be transferred to a recipient apoprotein. This type of cellular protein typically does not have catalytic activity but serves in the intracellular trafficking of iron cofactors. PCBP1 was identified as an iron chaperone based on its ability to bind cytosolic iron and transfer it to ferritin, the major cytosolic iron storage protein, through a direct, metal-mediated, protein-protein interaction (Shi et al. 2008). PCBP1 was originally identified as a heterogeneous nuclear ribonucleoprotein (hnRNP) that binds cytosine-rich sequences of both RNA and single-stranded DNA (Dreyfuss et al. 1988; Grelet and Howe 2019), and its functions as an iron chaperone were discovered decades later. It is structurally comprised of three conserved RNA-binding domains called hnRNP

K-homology (KH) domains. Each of these KH domains harbors a separate iron-binding site, and KH domains 1 and 3 also contain an adjacent GSH-binding site (Patel et al. 2019). In cells, PCBP1 coordinates Fe^{2+} and requires a molecule of noncovalently bound GSH for iron binding. The residues required for GSH binding are adjacent to the iron-binding site on KH3. In vitro, KH3 binds GSH, which increases the affinity of the KH3 domain for iron. KH3-Fe complexes have a higher affinity for GSH than apo-KH3, and thus KH domains of PCBP1 have the capacity to independently bind both single-stranded nucleic acid and Fe-GSH ligands. Because each PCBP1 molecule has three KH domains, a single PCBP1 may simultaneously bind both types of ligands.

Data derived from cells and tissues indicate that PCBP1 coordinates iron as it enters the cytosolic LIP and constitutes the major protein coordinating iron in the cell. Cells lacking PCBP1 exhibit defects in iron trafficking to ferritin, to mononuclear iron enzymes [e.g., the prolyl and asparaginyl hydroxylases that regulate hypoxia-inducible factor (HIF)] (Nandal et al. 2011), to dinuclear iron enzymes (e.g., the deoxyhypusine hydroxylase that hypusinates eIF5a) (Frey et al. 2014), and to the iron-sulfur cluster chaperone complex formed by BolA2 and GLRX3 (Frey et al. 2016; Patel et al. 2019). An intermediate iron co-chaperone complex is formed between the PCBP1 KH3 domain and BolA2, which promotes the assembly of the BolA2-[2Fe-2S]-GLRX3 complex. Thus, PCBP1 also plays a role in iron trafficking into the cytosolic Fe-S cluster assembly system.

The characteristics of the iron storage protein ferritin enable the cell to sequester ferric iron, which is insoluble at physiological pH, within a soluble protein shell and allow for controlled iron release when required (Harrison and Arosio 1996). This highly conserved protein complex consists of 24 subunits of two types: L-subunits (FTL), predominantly found in the liver, and H-subunits (FTH1), more abundant in the heart. The ratio of these subunits varies across cell types. Cells specializing in iron storage tend to have higher FTL levels, while most other cell types maintain a near 1:1 ratio, and many cancer cell lines exhibit FTH1 enrichment.

Ferrous iron from the LIP enters the ferritin complex at one of the threefold symmetry pores and is oxidized to ferric iron by the ferroxidase activity of FTH1. The resulting Fe^{3+} is then stored within the ferritin cavity, where FTL contributes to the nucleation and organization of the iron core. This iron-loading process is facilitated by the iron chaperone PCBP1, which delivers Fe^{2+} to the ferritin pore via a direct protein-protein interaction, a process that is facilitated by cellular GSH (Patel et al. 2021). While this sequestration primarily occurs in the cytosol, both ferritin and PCBP1 are also found in the nucleus, suggesting that iron sequestration may also take place there.

PCBP1 is highly homologous (83% identical) to its paralog PCBP2, which also exhibits some iron chaperone activity in tissues and cells (Nandal et al. 2011; Frey et al. 2014). Both PCBP1 and PCBP2 are expressed at high levels in most cell types, and PCBP2 was found to facilitate iron import, export, and heme-iron scavenging in cells (Yanatori et al. 2014; Yanatori et al. 2016; Yanatori et al. 2017). In developing erythroid cells, PCBP1 depletion impaired iron trafficking to ferritin and subsequent heme synthesis, while PCBP2 depletion enhanced iron trafficking to ferritin and heme synthesis, suggesting that PCBP1 and PCBP2 may compete for binding the LIP (Ryu et al. 2017). PCBP1 and PCBP2 can bind to each other, raising the possibility of direct iron exchange between the two chaperones (Yanatori et al. 2016) (Fig. 2.2).

2.2.4 Iron Trafficking to Mitochondria

PCBP1 mediates iron trafficking in the cytosol to ferritin, non-heme iron enzymes, and the cytosolic Fe-S cluster assembly systems. In most cells, the majority of imported iron is directed to the mitochondria for heme and Fe-S cluster synthesis. The systems responsible for delivering iron to mitochondria are not well-defined at this point. Cells lacking PCBP1 or PCBP2 do not have obvious defects in mitochondrial iron

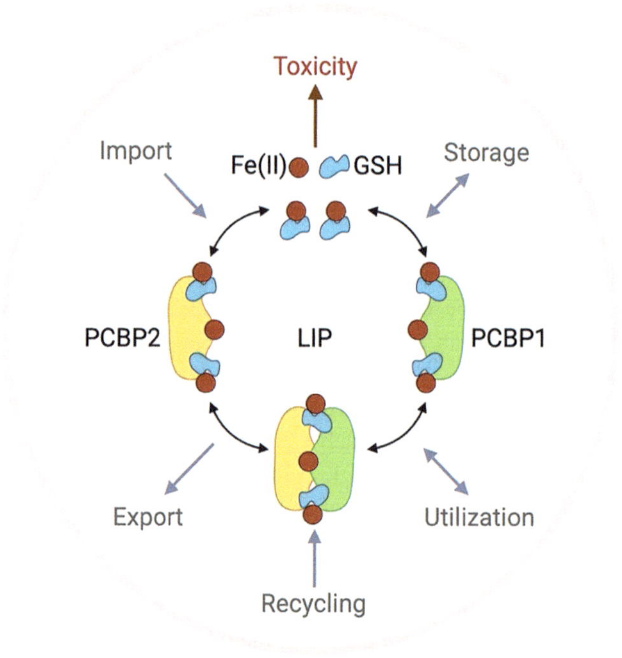

Fig. 2.2 Iron chaperones and the labile iron pool. Cytosolic labile iron is imported from the extracellular milieu or intracellular sites of recycling. The LIP may be coordinated by reduced glutathione (GSH) and captured by and/or exchanged between iron chaperones PCBP1 or PCBP2. Small (GSH) and large (PCBP) molecules of the LIP buffer and limit the availability of free iron to be transferred to sites of storage, utilization, and export. Unchaperoned iron is toxic to cells. (Modified from Philpott et al. 2020)

content (Jadhav et al. 2021). Direct transfer of iron from endo-lysosomal compartments adjacent to mitochondria has been proposed; however, evidence at the biochemical or genetic level remains lacking (Das et al. 2016; Hamdi et al. 2016). Iron entering the mitochondria must pass through both the outer and inner mitochondrial membranes. Inner membrane transporters mitoferrin 1 (MFRN1, Slc25a37) and mitoferrin-2 (MFRN2, Slc25a28) play roles in mitochondrial iron import (Kafina and Paw 2017). Together, MFRN1 and MFRN2 do not fully account for mitochondrial iron import in mouse liver (Seguin et al. 2020); however, MFRN1 alone is essential for normal erythroid development (Shaw et al. 2006).

2.2.5 Iron Cofactor Trafficking

Intracellular trafficking of heme and iron-sulfur clusters depends on protein chaperones that operate via associative transfer. Heme is a highly hydrophobic and chemically reactive molecule and exits the mitochondria after its synthesis by mechanisms that are not entirely clear (Belot et al. 2024). FLVCR1b was proposed as a mitochondrial heme exporter, but more recent work suggests that FLVCR primarily transports choline (Kenny et al. 2023; Ri et al. 2024). Heme-binding cytosolic proteins have been described (e.g., GAPDH) and may function as chaperones, but stronger genetic evidence suggests a role for the progesterone receptor membrane component 2 in heme trafficking from the mitochondria to the nucleus (Galmozzi et al. 2019) and transport and Golgi organization 2 in the transfer of intracellular heme from mitochondria to cytoplasmic apoproteins (Sun et al. 2022). The steps of mitochondrial and cytosolic Fe-S cluster assembly have been described and reviewed in detail (Lill et al. 2015; Lill and Freibert 2020; Maio and Rouault 2020). Briefly, [4Fe-4S] clusters are assembled from [2Fe-2S] clusters on scaffold proteins of the cytosolic iron-sulfur cluster assembly machinery. These clusters are then transferred to the targeting complex composed of MMS19, CIAO1, and CIAO2b, which directly interacts with apoproteins. Trafficking of Fe-S clusters to recipient apoproteins in the cytosol is thought to occur co-translationally during protein folding but may occur in the cytosol or nucleus after folding is complete (Novoa-Aponte et al. 2024).

2.2.6 Cellular Iron Export

While there are many ways for iron to get into the cell, as described above, ferrous iron is exported from the cytosol by ferroportin (also called earlier Ireg-1 or MTP-1). The export of Fe^{2+} is coupled to the ferroxidase activity, which is needed to load Fe^{3+} onto transferrin. The multicopper oxidases responsible for extracellular ferroxidation are the plasma protein ceruloplasmin; the membrane-bound hephaestin, expressed mainly in the small intestine; and the glycosyl phosphatidylinositol-anchored ceruloplasmin, expressed mainly in immune cells, hepatocytes, the brain, and the testis (Fortna et al. 1999; Frazer et al. 2001; Kono et al. 2006; Marques et al. 2012). Thus, ferroportin-mediated iron export is associated with systemic iron distribution through plasma transferrin. Ferroportin regulation on transcriptional and translational levels is mediated by the cellular iron, oxygen, and inflammatory status. Yet, most dominantly, ferroportin function and degradation are regulated by the peptide hormone hepcidin, which senses and responds to systemic iron levels, inflammation, and hypoxia (Chap. 3). The ferroportin levels thus integrate many signals, including cellular and systemic iron homeostasis. In addition, iron-containing ferritin has been shown to be secreted from cells through various direct and vesicle-mediated export mechanisms (Kimura et al. 2017; Truman-Rosentsvit et al. 2018; Brown et al. 2019; Yanatori et al. 2021). However, ferritin iron export is not sufficient to compensate for the absence of functional ferroportin (Drakesmith et al. 2015). Receptor-mediated ferritin uptake, especially through the iron-regulated TfR1, could be a way to redistribute iron between cells within a tissue.

Taken together, iron can be exported from cells through ferroportin and ferritin, but likely the two pathways have different functions and are not complementary. While iron secreted through ferroportin is mainly loaded onto transferrin and from there is distributed systemically, ferritin is most likely taken up by neighboring cells. We suggest this based on the fact that ferritin that reaches the bloodstream hardly contains iron, while ferritin secreted from cells in culture is high in its iron content. Furthermore, the subunit composition of serum ferritin does not resemble the subunit composition of cellular or secreted ferritin, suggesting that most of the ferritin secreted by cells never reaches the bloodstream (Cohen et al. 2010; Yanatori et al. 2021).

2.3 Iron Recycling

Intracellular iron recycling is an important component of cellular iron homeostasis, considering that iron-dependent enzymes may either turn over their prosthetic group or undergo degradation, releasing their iron cofactor. The return of iron from functional proteins to the LIP and PCBPs to be redistributed according to the cellular needs has not been studied yet extensively.

2.3.1 Mitochondrial Recycling—Mitophagy

Mitochondria are high consumers of iron. They require iron to metallate the complexes of the electron transport chain and other enzymes such as ferrochelatase, which depends on a Fe-S cluster to complete the final step of heme biosynthesis. Furthermore, mitochondria require iron from a mitochondrial labile iron pool for the de novo assembly of Fe-S clusters and the insertion of iron into protoporphyrin IX. Mitochondria are highly dynamic organelles that undergo fusion and fission processes to segregate damaged components, which are targeted for mitophagy, the selective autophagy of mitochondria (Terešak et al. 2022). Following mitophagy, damaged mitochondria are degraded in the lysosomes, where iron is reduced and released back into the cytosol.

2.3.2 Nuclear and Cytosolic Iron Recycling

Cytosolic quality control systems largely rely on ubiquitin-mediated proteasomal degradation, which will necessarily result in the return of metal cofactors into the LIP. Non-heme iron cofactors enter the Fe^{2+} pool, where they can be coordinated by PCBP1 and GSH. Fe-S cofactors may disassemble and be reacquired as Fe^{2+} by chaperones as well, but could also be scavenged by [2Fe-2S] chaperones such as BolA2-Glrx3 or CISD proteins. Nuclear proteins containing iron cofactors are also subject to quality control and may be either degraded in the cytosol or in the nucleus, where both proteasomes and iron chaperones are abundant (Enenkel 2014). Iron enzymes may also lose their iron cofactors without being degraded (see IRP1, below), in which case the cytosolic/nuclear iron cofactor delivery systems may replace or repair iron cofactors in folded proteins in situ. This aspect of iron trafficking has not been thoroughly defined.

2.3.3 Ferritin Iron Recycling

Ferritin enables the cell to sequester ferric iron and allow its controlled iron release when required. In addition to its ferroxidase activity, the FTH1 subunit is also responsible for ferritin interactions with the nuclear receptor coactivator (NCOA)4, human TfR1, and murine ferritin receptor TIM2. NCOA4 plays an active role in trafficking ferritin to the autophagosomal pathway and eventually to the lysosome, a pathway termed ferritinophagy (Gehrer et al. 2023). The lysosomal environment facilitates iron release from ferritin, its reduction in the lysosome, and transport back to the cytosol through one of the ferrous iron transporters such as DMT1 or TRPML1 (mucolipin 1, also known as MCOLN1) (Dong et al. 2008). Exposure of tissues and cell types to environmental stresses can affect the FTH1/FTL ratio and, with this, the dynamics of ferritin iron storage and trafficking. Transcription and translation of both subunits, as well as the NCOA4-mediated ferritinophagy, are regulated by the cellular redox and iron state, as described later in this chapter.

2.4 Regulation of Cellular Iron Trafficking

Pathways that are involved in the cellular iron import, export, and storage of iron are tightly regulated, which secures continuous iron supply and at the same time protects cells from iron overload. This regulation takes place on a systemic level through the peptide hormone hepcidin and on the cellular level by several systems including the transcription factors nuclear factor erythroid 2-related factor 2 (NRF2), and HIF, the RNA-regulation by the "iron-responsive element/iron regulatory proteins" (IRE/IRP) system and the regulation of ferritin trafficking and degradation by NCOA4. Hepcidin is a peptide hormone derived from the liver. It exerts its control by blocking and inducing the internalization and degradation of the iron exporter ferroportin, thus affecting iron recycling from macrophages and intestinal absorption in the duodenum (Aschemeyer et al. 2018; Nemeth and Ganz 2023).

2.4.1 Nuclear Factor Erythroid 2-Related Factor 2

The transcription factor NRF2 senses cellular oxidative stress (Duarte et al. 2021). Under homeostatic conditions, it is constitutively expressed and forms complexes with ubiquitin ligases that mediate its continuous degradation by the proteasome. Increased oxidative stress mediates the dissociation of NRF2 from the complex, allowing its ubiquitination and degradation and its translocation to the nucleus. NRF2 recognizes over 200 DNA-binding sites and regulates many genes involved in antioxidant defense and redox balance and some genes that are important for iron homeostasis; these include heme oxygenase 1, both subunits of ferritin, ferroportin, and the bone morphogenetic protein (BMP)6, which takes part in the induction of hepcidin in the liver. In addition, NRF2 was shown to induce HERC2, an E3 ubiquitin ligase that mediates the degradation of FBXL5 and NCOA4 (Crichton et al. 2023).

2.4.2 Hypoxia-Inducible Factor

In contrast to NRF2, HIF is a transcription factor that senses and is activated by low oxygen levels (Duarte, Talbot and Drakesmith 2021). HIF consists of a stable HIFß subunit that forms a complex with one of the labile HIF1α or HIF2α subunits in the nucleus, and the complex functions as a DNA-binding unit and transcription factor. Similar to NRF2, the HIFα subunits are constitutively produced and degraded by the proteasomal system in the cytosol when they are marked for ubiquitination by the iron-dependent prolyl hydroxylases. In several cell types, HIF2α plays an important role in regulating iron homeostasis, particularly in intestinal epithelial cells and the erythropoietin-producing interstitial fibroblasts in the kidney. HIF2α has an IRE in its 5′ untranslated region (UTR), which is regulated predominantly by IRP1, thereby integrating iron and oxygen regulation in a fine-tuned system (Sanchez et al. 2007; Camaschella et al. 2022; Zhang et al. 2025).

2.4.3 IRE/IRP System

At the translational level, the LIP is tightly regulated by the IRE/IRP system (Rouault 2006). The cytosolic RNA-binding proteins, IRP1 and IRP2, maintain iron homeostasis by interacting with IRE stem-loop structures in the 5′ or 3′UTRs of target mRNAs (see Fig. 18.2a in Chap. 12). When active in RNA binding, IRPs modulate the expression of key proteins involved in iron transport and storage, such as TfR1, DMT1, ferroportin, and ferritin subunits. Notably, IRP1 and IRP2 recognize distinct subsets of IREs, suggesting that each may regulate additional, unique mRNA targets beyond those involved in iron metabolism (Sanchez et al. 2011). Both IRPs are responsive to LIP levels, increasing RNA-binding activity under low iron conditions, but they sense the LIP through different mechanisms. IRP1 functions as a bifunctional protein: in its Fe-S cluster state, it acts as cytosolic aconitase (cAco), converting citrate to isocitrate. In the absence of the Fe-S cluster, it assumes its RNA-binding role (Haile et al. 1992). This switch depends on the Fe-S cluster's stability, which is influenced by oxygen and reactive oxygen/nitrogen species (ROS/RNS) (Hanson and Leibold 1999). Under physiological oxygen conditions, about 95% of IRP1 remains in its enzymatic cAco form due to the cluster's stability (Meyron-Holtz et al. 2004a, b). In contrast, IRP2 is regulated through ubiquitination by the iron- and oxygen-sensitive SKP1-CUL1-FBXL5 ubiquitin ligase complex and undergoes proteasomal degradation in response to high LIP, oxygen, and ROS/RNS levels (Wang et al. 2020a, b). Thus, while both IRPs respond to cellular iron and oxygen availability, their regulation differs: IRP2 is rapidly degraded under high oxygen and oxidative stress, whereas IRP1 requires extreme oxidative conditions to shift from cAco to its RNA-binding form (Phillips et al. 1996).

2.4.4 NCOA4/Ferritin Trafficking

Another key factor in cellular iron homeostasis is the dynamic turnover of ferritin. Increased ferritin production lowers the LIP as ferritin oxidizes and sequesters iron within its core. When iron levels are low, ferritin serves as an iron source, releasing stored iron to replenish the LIP (Arosio et al. 2017; Ryu et al. 2017). Ferritin expression is controlled at the transcriptional level by NRF2 (Cairo et al. 1995; Pietsch et al. 2003) and at the translational level by the IRE/IRP system. The process of ferritinophagy, mediated by NCOA4, enables the degradation of ferritin and the mobilization of stored iron, particularly under iron-deficient conditions. NCOA4, which is largely intrinsically disordered, plays a key role in the formation of biomolecular condensates with and without ferritin, supporting both macro- and micro-ferritinophagy (Ohshima et al. 2022).

2.5 Specialized Iron-Handling Cells

2.5.1 Intestinal Epithelium

All the iron required for the growth and development of mammals is absorbed from dietary sources through the intestinal epithelium, primarily in the duodenum and the proximal jejunum. Although iron import and iron efflux are reciprocally regulated in most cells, the intestinal epithelium is unique in that both iron import systems (apical DMT1) and export systems (basolateral ferroportin) are highly expressed, especially when systemic iron is low (Wilkinson and Pantopoulos 2014; Zhang et al. 2014). PCBP1 plays a crucial role in buffering the cytosolic LIP in enterocytes, as this allows the cells to retain and store iron in ferritin and limit the size of the free iron pools. This limitation of free iron allows for the regulation of efflux through ferroportin (Wang et al. 2023). HIF2α plays an important role in the regulation of intestinal iron homeostasis by activating the expression of genes required for dietary iron uptake in intestinal epithelial cells (Ramakrishnan and Shah 2016). Notably, intestinal epithelial cells express an alternative ferroportin transcript that lacks an IRE, allowing ferroportin to escape IRP-mediated repression of translation (Zhang et al. 2009).

2.5.2 Erythroid Precursors

More than 70% of total body iron is present in circulating red blood cells, which are continuously produced in the bone marrow at a rate of two million reticulocytes per second (Palis 2014; Ryu et al. 2017). During terminal differentiation, erythroid precursors exhibit gigantic increases in iron uptake, initially through the upregulation of TfR1 and ferritin, and ultimately devoting most of the cellular machinery to the coordinated synthesis of heme and globin chains. Despite the overall goal of accumulating large amounts of iron, the developing erythrocyte also expresses a significant amount of ferroportin, which protects the red cell from oxidative damage by excess iron (Zhang et al. 2018).

2.5.3 Splenic/Liver Macrophages

Macrophages in the spleen and liver play essential roles in iron recycling, storage, and regulation. Splenic red pulp macrophages are responsible for erythrophagocytosis, the breakdown of aged or damaged RBCs. Once RBCs are engulfed, hemoglobin is degraded, and iron is extracted from heme by heme oxygenase-1, releasing free iron. This iron is either stored in macrophages as ferritin or hemosiderin (a product of ferritin degradation) to prevent oxidative damage, and exported into circulation via ferroportin. Similarly, Kupffer cells, the resident macrophages of the liver, participate in iron homeostasis by clearing senescent RBCs and engulfing iron-containing proteins such as ferritin and transferrin. Like splenic macrophages, Kupffer cells store iron in ferritin or release it via ferroportin when needed. These macrophages can also secrete iron-loaded ferritin, which may be taken up locally by hepatocytes. Additionally, Kupffer cells play an important role in immune responses and inflammation, both of which can affect iron metabolism. Together, splenic and liver macrophages ensure efficient iron recycling and storage, preventing both deficiency and overload while responding dynamically to systemic iron demands and inflammatory signals (Nairz et al. 2017; Recalcati and Cairo 2021; Galy et al. 2024).

2.6 Conclusions

Additional cell types and tissues are specialized for iron regulation at the systemic level, such as hepatocytes, liver sinusoidal endothelial cells, renal peritubular fibroblasts, and erythroblasts. Much remains to be explored regarding mitochondrial iron trafficking and the reuse and recycling of iron cofactors within cells. Nuclear iron enzymes include many Fe-S cluster, heme, and non-heme iron proteins, and the coordination between cytosolic and nuclear iron pools and cofactors is an important area for further investigation, as iron plays an integral role in maintaining the integrity of the genome.

Acknowledgments This research was supported in part by the Intramural Research Program of the NIH, the National Institute of Diabetes and Digestive and Kidney Diseases (NIDDK), and an NSF-BSF grant—BSF Award 2022614 and NSF Award 2231900 to EGMH. We thank Dr. Olga Protchenko and Dr. Andrés Leon-Torres for assistance in preparing this manuscript, and we thank Dr. Noga Guttmann-Raviv, Bar Azoulay-Botzer, Liora Frech, and Shaked Shoris for critical reading and assistance.

References

Aron AT, Loehr MO, Bogena J, Chang CJ (2016) An endoperoxide reactivity-based FRET probe for ratiometric fluorescence imaging of labile iron pools in living cells. J Am Chem Soc 138(43):14338–14346

Arosio P, Elia L, Poli M (2017) Ferritin, cellular iron storage and regulation. IUBMB Life 69(6):414–422

Aschemeyer S, Qiao B, Stefanova D, Valore EV, Sek AC, Ruwe TA, Vieth KR, Jung G, Casu C, Rivella S, Jormakka M, Mackenzie B, Ganz T, Nemeth E (2018) Structure-function analysis of ferroportin defines the binding site and an alternative mechanism of action of hepcidin. Blood 131(8):899–910

Belot A, Puy H, Hamza I, Bonkovsky HL (2024) Update on heme biosynthesis, tissue-specific regulation, heme transport, relation to iron metabolism and cellular energy. Liver Int 44(9):2235–2250

Braymer JJ, Freibert SA, Rakwalska-Bange M, Lill R (2021) Mechanistic concepts of iron-sulfur protein biogenesis in biology. Biochim Biophys Acta, Mol Cell Res 1868(1):118863

Brown CW, Amante JJ, Chhoy P, Elaimy AL, Liu H, Zhu LJ, Baer CE, Dixon SJ, Mercurio AM (2019) Prominin2 drives ferroptosis resistance by stimulating iron export. Dev Cell 51(5):575–586.e574

Cairo G, Tacchini L, Pogliaghi G, Anzon E, Tomasi A, Bernelli-Zazzera A (1995) Induction of ferritin synthesis by oxidative stress. Transcriptional and post-transcriptional regulation by expansion of the "free" iron pool. J Biol Chem 270(2):700–703

Camaschella C, Pagani A, Silvestri L, Nai A (2022) The mutual crosstalk between iron and erythropoiesis. Int J Hematol 116(2):182–191

Chen AC, Donovan A, Ned-Sykes R, Andrews NC (2015) Noncanonical role of transferrin receptor 1 is essential for intestinal homeostasis. Proc Natl Acad Sci USA 112(37):11714–11719

Chen K, Wang L, Shen J, Tsai AL, Zhou M, Wu G (2023) Mechanism of stepwise electron transfer in six-transmembrane epithelial antigen of the prostate (STEAP) 1 and 2. elife 12:RP88299

Cohen LA, Gutierrez L, Weiss A, Leichtmann-Bardoogo Y, Zhang DL, Crooks DR, Sougrat R, Morgenstern A, Galy B, Hentze MW, Lazaro FJ, Rouault TA, Meyron-Holtz EG (2010) Serum ferritin is derived primarily from macrophages through a nonclassical secretory pathway. Blood 116(9):1574–1584

Crichton R (2001) Solution chemistry of iron in biological media. In: Inorganic biochemistry of iron metabolism, pp 1–15

Crichton R, Anandhan A, Dodson M, Shakya A, Chen J, Liu P, Wei Y, Tan H, Wang Q, Jiang Z, Yang K, Garcia JG, Chambers SK, Chapman E, Ooi A, Yang-Hartwich Y, Stockwell BR, Zhang DD (2023) NRF2 controls iron homeostasis and ferroptosis through HERC2 and VAMP8. Sci Adv 9(5):eade9585

Das A, Nag S, Mason AB, Barroso MM (2016) Endosome-mitochondria interactions are modulated by iron release from transferrin. J Cell Biol 214(7):831–845

Dong XP, Cheng X, Mills E, Delling M, Wang F, Kurz T, Xu H (2008) The type IV mucolipidosis-associated protein TRPML1 is an endolysosomal iron release channel. Nature 455(7215):992–996

Drakesmith H, Nemeth E, Ganz T (2015) Ironing out Ferroportin. Cell Metab 22(5):777–787

Dreyfuss G, Philipson L, Mattaj IW (1988) Ribonucleoprotein particles in cellular processes. J Cell Biol 106(5):1419–1425

Duarte TL, Talbot NP, Drakesmith H (2021) NRF2 and hypoxia-inducible factors: key players in the redox control of systemic iron homeostasis. Antioxid Redox Signal 35(6):433–452

Enenkel C (2014) Proteasome dynamics. Biochim Biophys Acta 1843(1):39–46

Fillebeen C, Charlebois E, Wagner J, Katsarou A, Mui J, Vali H, Garcia-Santos D, Ponka P, Presley J, Pantopoulos K (2019) Transferrin receptor 1 controls systemic iron homeostasis by fine-tuning hepcidin expression to hepatocellular iron load. Blood 133(4):344–355

Fortna RR, Watson HA, Nyquist SE (1999) Glycosyl phosphatidylinositol-anchored ceruloplasmin is expressed by rat Sertoli cells and is concentrated in detergent-insoluble membrane Fractions1. Biol Reprod 61(4):1042–1049

Frazer DM, Vulpe CD, McKie AT, Wilkins SJ, Trinder D, Cleghorn GJ, Anderson GJ (2001) Cloning and gastrointestinal expression of rat hephaestin: relationship to other iron transport proteins. Am J Physiol Gastrointest Liver Physiol 281(4):G931–G939

Frey AG, Nandal A, Park JH, Smith PM, Yabe T, Ryu MS, Ghosh MC, Lee J, Rouault TA, Park MH, Philpott CC (2014) Iron chaperones PCBP1 and PCBP2 mediate the metallation of the dinuclear iron enzyme deoxyhypusine hydroxylase. Proc Natl Acad Sci USA 111(22):8031–8036

Frey AG, Palenchar DJ, Wildemann JD, Philpott CC (2016) A glutaredoxin·BolA complex serves as an iron-sulfur cluster chaperone for the cytosolic cluster assembly machinery. J Biol Chem 291(43):22344–22356

Galmozzi A, Kok BP, Kim AS, Montenegro-Burke JR, Lee JY, Spreafico R, Mosure S, Albert V, Cintron-Colon R, Godio C, Webb WR, Conti B, Solt LA, Kojetin D, Parker CG, Peluso JJ, Pru JK, Siuzdak G, Cravatt BF, Saez E (2019) PGRMC2 is an intracellular haem chaperone critical for adipocyte function. Nature 576(7785):138–142

Galy B, Conrad M, Muckenthaler M (2024) Mechanisms controlling cellular and systemic iron homeostasis. Nat Rev Mol Cell Biol 25(2):133–155

Gehrer CM, Mitterstiller AM, Grubwieser P, Meyron-Holtz EG, Weiss G, Nairz M (2023) Advances in ferritin physiology and possible implications in bacterial infection. Int J Mol Sci 24(5):4659

Grelet S, Howe PH (2019) hnRNP E1 at the crossroads of translational regulation of epithelial-mesenchymal transition. J Cancer Metastasis Treat 5:16

Gunshin H, Mackenzie B, Berger UV, Gunshin Y, Romero MF, Boron WF, Nussberger S, Gollan JL, Hediger MA (1997) Cloning and characterization of a mammalian proton-coupled metal-ion transporter. Nature 388(6641):482–488

Gunshin H, Starr CN, Direnzo C, Fleming MD, Jin J, Greer EL, Sellers VM, Galica SM, Andrews NC (2005) Cybrd1 (duodenal cytochrome b) is not necessary for dietary iron absorption in mice. Blood 106(8):2879–2883

Haile DJ, Rouault TA, Tang CK, Chin J, Harford JB, Klausner RD (1992) Reciprocal control of RNA-binding and aconitase activity in the regulation of the iron-responsive element binding protein: role of the iron-sulfur cluster. Proc Natl Acad Sci USA 89(16):7536–7540

Hamdi A, Roshan TM, Kahawita TM, Mason AB, Sheftel AD, Ponka P (2016) Erythroid cell mitochondria receive endosomal iron by a "kiss-and-run" mechanism. Biochim Biophys Acta 1863(12):2859–2867

Hanson ES, Leibold EA (1999) Regulation of the iron regulatory proteins by reactive nitrogen and oxygen species. Gene Expr 7(4–6):367–376

Harrison PM, Arosio P (1996) The ferritins: molecular properties, iron storage function and cellular regulation. Biochim Biophys Acta 1275(3):161–203

Herr CQ, Hausinger RP (2018) Amazing diversity in biochemical roles of Fe(II)/2-Oxoglutarate Oxygenases. Trends Biochem Sci 43(7):517–532

Islam MS, Leissing TM, Chowdhury R, Hopkinson RJ, Schofield CJ (2018) 2-Oxoglutarate-dependent oxygenases. Annu Rev Biochem 87:585–620

Jabara HH, Boyden SE, Chou J, Ramesh N, Massaad MJ, Benson H, Bainter W, Fraulino D, Rahimov F, Sieff C, Liu ZJ, Alshemmari SH, Al-Ramadi BK, Al-Dhekri H, Arnaout R, Abu-Shukair M, Vatsayan A, Silver E, Ahuja S, Davies EG, Sola-Visner M, Ohsumi TK, Andrews NC, Notarangelo LD, Fleming MD, Al-Herz W, Kunkel LM, Geha RS (2016) A missense mutation in TFRC, encoding transferrin receptor 1, causes combined immunodeficiency. Nat Genet 48(1):74–78

Jadhav S, Protchenko O, Li F, Baratz E, Shakoury-Elizeh M, Maschek A, Cox J, Philpott CC (2021) Mitochondrial dysfunction in mouse livers depleted of iron chaperone PCBP1. Free Radic Biol Med 175:18–27

Jasniewski AJ, Que L Jr (2018) Dioxygen activation by nonheme diiron enzymes: diverse dioxygen adducts, high-valent intermediates, and related model complexes. Chem Rev 118(5):2554–2592

Johnson DC, Dean DR, Smith AD, Johnson MK (2005) Structure, function, and formation of biological iron-sulfur clusters. Annu Rev Biochem 74:247–281

Kafina MD, Paw BH (2017) Intracellular iron and heme trafficking and metabolism in developing erythroblasts. Metallomics 9(9):1193–1203

Kawabata H, Yang R, Hirama T, Vuong PT, Kawano S, Gombart AF, Koeffler HP (1999) Molecular cloning of transferrin receptor 2. A new member of the transferrin receptor-like family. J Biol Chem 274(30):20826–20832

Kenny TC, Khan A, Son Y, Yue L, Heissel S, Sharma A, Pasolli HA, Liu Y, Gamazon ER, Alwaseem H, Hite RK, Birsoy K (2023) Integrative genetic analysis identifies FLVCR1 as a plasma-membrane choline transporter in mammals. Cell Metab 35(6):1057–1071.e1012

Kimura T, Jia J, Kumar S, Choi SW, Gu Y, Mudd M, Dupont N, Jiang S, Peters R, Farzam F, Jain A, Lidke KA, Adams CM, Johansen T, Deretic V (2017) Dedicated SNAREs and specialized TRIM cargo receptors mediate secretory autophagy. EMBO J 36(1):42–60

Klei TR, Meinderts SM, van den Berg TK, van Bruggen R (2017) From the cradle to the grave: the role of macrophages in erythropoiesis and erythrophagocytosis. Front Immunol 8:73

Knutson MD (2019) Non-transferrin-bound iron transporters. Free Radic Biol Med 133:101–111

Kono S, Suzuki H, Oda T, Miyajima H, Takahashi Y, Shirakawa K, Ishikawa K, Kitagawa M (2006) Biochemical features of ceruloplasmin gene mutations linked to aceruloplasminemia. NeuroMolecular Med 8(3):361–374

La A, Nguyen T, Tran K, Sauble E, Tu D, Gonzalez A, Kidane TZ, Soriano C, Morgan J, Doan M, Tran K, Wang CY, Knutson MD, Linder MC (2018) Mobilization of iron from ferritin: new steps and details. Metallomics 10(1):154–168

Levy JE, Jin O, Fujiwara Y, Kuo F, Andrews NC (1999) Transferrin receptor is necessary for development of erythrocytes and the nervous system. Nat Genet 21(4):396–399

Li L, Fang CJ, Ryan JC, Niemi EC, Lebrón JA, Björkman PJ, Arase H, Torti FM, Torti SV, Nakamura MC, Seaman WE (2010) Binding and uptake of H-ferritin are mediated by human transferrin receptor-1. Proc Natl Acad Sci USA 107(8):3505–3510

Lill R, Freibert SA (2020) Mechanisms of mitochondrial iron-sulfur protein biogenesis. Annu Rev Biochem 89:471–499

Lill R, Dutkiewicz R, Freibert SA, Heidenreich T, Mascarenhas J, Netz DJ, Paul VD, Pierik AJ, Richter N, Stümpfig M, Srinivasan V, Stehling O, Mühlenhoff U (2015) The role of mitochondria and the CIA machinery in the maturation of cytosolic and nuclear iron-sulfur proteins. Eur J Cell Biol 94(7–9):280–291

Liziczai M, Fuchs A, Manatschal C, Dutzler R (2025) Structural basis for metal ion transport by the human SLC11 proteins DMT1 and NRAMP1. Nat Commun 16(1):761

Mackenzie B, Ujwal ML, Chang MH, Romero MF, Hediger MA (2006) Divalent metal-ion transporter DMT1 mediates both H+ -coupled Fe2+ transport and uncoupled fluxes. Pflugers Arch 451(4):544–558

Maio N, Rouault TA (2020) Outlining the complex pathway of mammalian Fe-S cluster biogenesis. Trends Biochem Sci 45(5):411–426

Marques L, Auriac A, Willemetz A, Banha J, Silva B, Canonne-Hergaux F, Costa L (2012) Immune cells and hepatocytes express glycosylphosphatidylinositol-anchored ceruloplasmin at their cell surface. Blood Cells Mol Dis 48(2):110–120

Matak P, Matak A, Moustafa S, Aryal DK, Benner EJ, Wetsel W, Andrews NC (2016) Disrupted iron homeostasis causes dopaminergic neurodegeneration in mice. Proc Natl Acad Sci USA 113(13):3428–3435

McKie AT, Barrow D, Latunde-Dada GO, Rolfs A, Sager G, Mudaly E, Mudaly M, Richardson C, Barlow D, Bomford A, Peters TJ, Raja KB, Shirali S, Hediger MA, Farzaneh F, Simpson RJ (2001) An iron-regulated ferric reductase associated with the absorption of dietary iron. Science 291(5509):1755–1759

Meng F, Fleming BA, Jia X, Rousek AA, Mulvey MA, Ward DM (2022) Lysosomal iron recycling in mouse macrophages is dependent upon both LcytB and Steap3 reductases. Blood Adv 6(6):1692–1707

Meyron-Holtz EG, Ghosh MC, Iwai K, LaVaute T, Brazzolotto X, Berger UV, Land W, Ollivierre-Wilson H, Grinberg A, Love P, Rouault TA (2004a) Genetic ablations of iron regulatory proteins 1 and 2 reveal why iron regulatory protein 2 dominates iron homeostasis. EMBO J 23(2):386–395

Meyron-Holtz EG, Ghosh MC, Rouault TA (2004b) Mammalian tissue oxygen levels modulate iron-regulatory protein activities in vivo. Science 306(5704):2087–2090

Montemiglio LC, Testi C, Ceci P, Falvo E, Pitea M, Savino C, Arcovito A, Peruzzi G, Baiocco P, Mancia F, Boffi A, des Georges A, Vallone B (2019) Cryo-EM structure of the human ferritin-transferrin receptor 1 complex. Nat Commun 10(1):1121

Nairz M, Theurl I, Swirski FK, Weiss G (2017) "Pumping iron"-how macrophages handle iron at the systemic, microenvironmental, and cellular levels. Pflugers Arch 469(3–4):397–418

Nandal A, Ruiz JC, Subramanian P, Ghimire-Rijal S, Sinnamon RA, Stemmler TL, Bruick RK, Philpott CC (2011) Activation of the HIF prolyl hydroxylase by the iron chaperones PCBP1 and PCBP2. Cell Metab 14(5):647–657

Nemeth E, Ganz T (2023) Hepcidin and iron in health and disease. Annu Rev Med 74:261–277

Novoa-Aponte L, Leon-Torres A, Philpott CC (2024) Guardians of the genome: iron–sulfur proteins in the nucleus. Inorganics 12(12):316

Ohgami RS, Campagna DR, Greer EL, Antiochos B, McDonald A, Chen J, Sharp JJ, Fujiwara Y, Barker JE, Fleming MD (2005) Identification of a ferrireductase required for efficient transferrin-dependent iron uptake in erythroid cells. Nat Genet 37(11):1264–1269

Ohshima T, Yamamoto H, Sakamaki Y, Saito C, Mizushima N (2022) NCOA4 drives ferritin phase separation to facilitate macroferritinophagy and microferritinophagy. J Cell Biol 221(10):e202203102

Palis J (2014) Primitive and definitive erythropoiesis in mammals. Front Physiol 5:3

Patel SJ, Frey AG, Palenchar DJ, Achar S, Bullough KZ, Vashisht A, Wohlschlegel JA, Philpott CC (2019) A PCBP1-BolA2 chaperone complex delivers iron for cytosolic [2Fe-2S] cluster assembly. Nat Chem Biol 15(9):872–881

Patel SJ, Protchenko O, Shakoury-Elizeh M, Baratz E, Jadhav S, Philpott CC (2021) The iron chaperone and nucleic acid-binding activities of poly(rC)-binding protein 1 are separable and independently essential. Proc Natl Acad Sci USA 118(25):e2104666118

Phillips JD, Kinikini DV, Yu Y, Guo B, Leibold EA (1996) Differential regulation of IRP1 and IRP2 by nitric oxide in rat hepatoma cells. Blood 87(7):2983–2992

Philpott CC, Patel SJ, Protchenko O (2020) Management versus miscues in the cytosolic labile iron pool: the varied functions of iron chaperones. Biochim Biophys Acta, Mol Cell Res 1867(11):118830

Philpott CC, Protchenko O, Wang Y, Novoa-Aponte L, Leon-Torres A, Grounds S, Tietgens AJ (2023) Iron-tracking strategies: chaperones capture iron in the cytosolic labile iron pool. Front Mol Biosci 10:1127690

Pietsch EC, Chan JY, Torti FM, Torti SV (2003) Nrf2 mediates the induction of ferritin H in response to

xenobiotics and cancer chemopreventive dithiolethiones. J Biol Chem 278(4):2361–2369
Poss KD, Tonegawa S (1997) Heme oxygenase 1 is required for mammalian iron reutilization. Proc Natl Acad Sci USA 94(20):10919–10924
Ramakrishnan SK, Shah YM (2016) Role of intestinal HIF-2α in health and disease. Annu Rev Physiol 78:301–325
Recalcati S, Cairo G (2021) Macrophages and iron: a special relationship. Biomedicines 9(11):1585
Reddi AR, Hamza I (2016) Heme mobilization in animals: a metallolipid's journey. Acc Chem Res 49(6):1104–1110
Ri K, Weng TH, Claveras Cabezudo A, Jösting W, Zhang Y, Bazzone A, Leong NCP, Welsch S, Doty RT, Gursu G, Lim TJY, Schmidt SL, Abkowitz JL, Hummer G, Wu D, Nguyen LN, Safarian S (2024) Molecular mechanism of choline and ethanolamine transport in humans. Nature 630(8016):501–508
Rouault TA (2006) The role of iron regulatory proteins in mammalian iron homeostasis and disease. Nat Chem Biol 2(8):406–414
Ryu MS, Zhang D, Protchenko O, Shakoury-Elizeh M, Philpott CC (2017) PCBP1 and NCOA4 regulate erythroid iron storage and heme biosynthesis. J Clin Invest 127(5):1786–1797
Sanchez M, Galy B, Muckenthaler MU, Hentze MW (2007) Iron-regulatory proteins limit hypoxia-inducible factor-2alpha expression in iron deficiency. Nat Struct Mol Biol 14(5):420–426
Sanchez M, Galy B, Schwanhaeusser B, Blake J, Bähr-Ivacevic T, Benes V, Selbach M, Muckenthaler MU, Hentze MW (2011) Iron regulatory protein-1 and -2: transcriptome-wide definition of binding mRNAs and shaping of the cellular proteome by iron regulatory proteins. Blood 118(22):e168–e179
Seguin A, Jia X, Earl AM, Li L, Wallace J, Qiu A, Bradley T, Shrestha R, Troadec MB, Hockin M, Titen S, Warner DE, Dowdle PT, Wohlfahrt ME, Hillas E, Firpo MA, Phillips JD, Kaplan J, Paw BH, Barasch J, Ward DM (2020) The mitochondrial metal transporters mitoferrin1 and mitoferrin2 are required for liver regeneration and cell proliferation in mice. J Biol Chem 295(32):11002–11020
Shaw GC, Cope JJ, Li L, Corson K, Hersey C, Ackermann GE, Gwynn B, Lambert AJ, Wingert RA, Traver D, Trede NS, Barut BA, Zhou Y, Minet E, Donovan A, Brownlie A, Balzan R, Weiss MJ, Peters LL, Kaplan J, Zon LI, Paw BH (2006) Mitoferrin is essential for erythroid iron assimilation. Nature 440(7080):96–100
Shawki A, Engevik MA, Kim RS, Knight PB, Baik RA, Anthony SR, Worrell RT, Shull GE, Mackenzie B (2016) Intestinal brush-border na+/H+ exchanger-3 drives H+-coupled iron absorption in the mouse. Am J Physiol Gastrointest Liver Physiol 311(3):G423–G430
Shi H, Bencze KZ, Stemmler TL, Philpott CC (2008) A cytosolic iron chaperone that delivers iron to ferritin. Science 320(5880):1207–1210
Sun F, Zhao Z, Willoughby MM, Shen S, Zhou Y, Shao Y, Kang J, Chen Y, Chen M, Yuan X, Hamza I, Reddi AR, Chen C (2022) HRG-9 homologues regulate haem trafficking from haem-enriched compartments. Nature 610(7933):768–774
Terešak P, Lapao A, Subic N, Boya P, Elazar Z, Simonsen A (2022) Regulation of PRKN-independent mitophagy. Autophagy 18(1):24–39
Truman-Rosentsvit M, Berenbaum D, Spektor L, Cohen LA, Belizowsky-Moshe S, Lifshitz L, Ma J, Li W, Kesselman E, Abutbul-Ionita I, Danino D, Gutierrez L, Li H, Li K, Lou H, Regoni M, Poli M, Glaser F, Rouault TA, Meyron-Holtz EG (2018) Ferritin is secreted via 2 distinct nonclassical vesicular pathways. Blood 131(3):342–352
Wang H, Shi H, Rajan M, Canarie ER, Hong S, Simoneschi D, Pagano M, Bush MF, Stoll S, Leibold EA, Zheng N (2020a) FBXL5 regulates IRP2 stability in iron homeostasis via an oxygen-responsive [2Fe2S] cluster. Mol Cell 78(1):31–41.e35
Wang S, He X, Wu Q, Jiang L, Chen L, Yu Y, Zhang P, Huang X, Wang J, Ju Z, Min J, Wang F (2020b) Transferrin receptor 1-mediated iron uptake plays an essential role in hematopoiesis. Haematologica 105(8):2071–2082
Wang Y, Protchenko O, Huber KD, Shakoury-Elizeh M, Ghosh MC, Philpott CC (2023) The iron chaperone poly(rC)-binding protein 1 regulates iron efflux through intestinal ferroportin in mice. Blood 142(19):1658–1671
White C, Yuan X, Schmidt PJ, Bresciani E, Samuel TK, Campagna D, Hall C, Bishop K, Calicchio ML, Lapierre A, Ward DM, Liu P, Fleming MD, Hamza I (2013) HRG1 is essential for heme transport from the phagolysosome of macrophages during erythrophagocytosis. Cell Metab 17(2):261–270
Wilkinson N, Pantopoulos K (2014) The IRP/IRE system in vivo: insights from mouse models. Front Pharmacol 5:176
Xu W, Barrientos T, Mao L, Rockman HA, Sauve AA, Andrews NC (2015) Lethal cardiomyopathy in mice lacking transferrin receptor in the heart. Cell Rep 13(3):533–545
Yanatori I, Yasui Y, Tabuchi M, Kishi F (2014) Chaperone protein involved in transmembrane transport of iron. Biochem J 462(1):25–37
Yanatori I, Richardson DR, Imada K, Kishi F (2016) Iron export through the transporter ferroportin 1 is modulated by the iron chaperone PCBP2. J Biol Chem 291(33):17303–17318
Yanatori I, Richardson DR, Toyokuni S, Kishi F (2017) The iron chaperone poly(rC)-binding protein 2 forms a metabolon with the heme oxygenase 1/cytochrome P450 reductase complex for heme catabolism and iron transfer. J Biol Chem 292(32):13205–13229
Yanatori I, Richardson DR, Dhekne HS, Toyokuni S, Kishi F (2021) CD63 is regulated by iron via the IRE-IRP system and is important for ferritin secretion by extracellular vesicles. Blood 138(16):1490–1503
Zhang DL, Hughes RM, Ollivierre-Wilson H, Ghosh MC, Rouault TA (2009) A ferroportin transcript that lacks an iron-responsive element enables duodenal and ery-

throid precursor cells to evade translational repression. Cell Metab 9(5):461–473

Zhang DL, Ghosh MC, Rouault TA (2014) The physiological functions of iron regulatory proteins in iron homeostasis—an update. Front Pharmacol 5:124

Zhang DL, Wu J, Shah BN, Greutélaers KC, Ghosh MC, Ollivierre H, Su XZ, Thuma PE, Bedu-Addo G, Mockenhaupt FP, Gordeuk VR, Rouault TA (2018) Erythrocytic ferroportin reduces intracellular iron accumulation, hemolysis, and malaria risk. Science 359(6383):1520–1523

Zhang DL, Ollivierre H, Qi CF, Rouault TA (2025) A bulge uridine in the HIF2α IRE allows IRP1 but not IRP2 to selectively regulate HIF2α expression and ensuing EPO levels. Blood 145(5):533–542

Systemic Iron Metabolism

3

Tomas Ganz

Abstract

Iron, a versatile electron donor and acceptor, is essential for life as a component of enzymes and oxygen transporters but also has the potential to cause toxicity when in excess. Its dietary absorption as well as plasma and tissue concentrations are subject to close regulation, systemically coordinated by the interaction of the hepatic peptide hormone hepcidin and its cellular receptor, the iron exporter ferroportin. Ferroportin mediates the delivery of iron to plasma from absorptive enterocytes, from macrophages that digest senescent erythrocytes, and from iron-storing hepatocytes. Hepcidin inhibits the transport of iron into plasma by occluding ferroportin and inducing its endocytosis and proteolysis. The production of hepcidin is feedback-regulated by the concentration of extracellular iron, as well as by the amount of iron in hepatic storage. Hepcidin production is also modulated by erythroid regulators (e.g., erythroferrone) that act to match the supply of iron to the fluctuating requirements of hemoglobin synthesis and erythropoiesis. During infections, hepcidin is induced by IL6 and restricts the growth of invading microorganisms by limiting the production of non-transferrin-bound iron. Although many of the underlying molecular and cellular mechanisms have been elucidated through the study of human diseases and their laboratory models, substantial gaps remain to be filled by further investigations.

Keywords

Systemic iron homeostasis · Hepcidin-ferroportin regulation · Iron absorption and recycling · Erythropoiesis and iron homeostasis · Iron metabolism in infection and immunity

3.1 Overview

Iron is an essential trace metal, which functions as the active moiety of proteins involved in oxygen transport, enzymes that catalyze energy metabolism, intermediary metabolism, and redox reactions, as well as nucleic acid synthesis and modification (Chap. 1). Although iron is one of the most abundant elements in the biosphere, in the current high-oxygen environment it has become oxidized, insoluble, and poorly accessible to most biological organisms. In this environment of scarcity, organismal iron conservation mechanisms have evolved so that losses of iron from the organism are very small and total body

T. Ganz (✉)
Center for Iron Disorders, Department of Medicine, David Geffen School of Medicine, University of California, Los Angeles, Los Angeles, CA, USA
e-mail: TGanz@mednet.ucla.edu

K. Pantopoulos (ed.), *Iron Metabolism in Human Health and Disease*, Advances in Experimental Medicine and Biology 1480, https://doi.org/10.1007/978-3-031-92033-2_3

iron content is almost entirely regulated by dietary iron absorption.

Although iron scarcity is the norm for most humans and animals, iron is abundant in some foods, supplements, and medications and has the potential to accumulate and result in organismal iron excess. The high chemical reactivity of excess iron within tissues could cause local tissue injury (McCord 1998; Ramm and Ruddell 2005). Mitigating this threat, the complexation of iron within proteins controls and modulates its highly reactive chemistry. Additionally, iron concentrations in tissues are subject to close physiological control, which normally prevents iron toxicity.

3.2 Hormonal Regulation of Iron by Hepcidin and Its Receptor Ferroportin

The maintenance of a stable internal body environment despite an ever-changing external environment has long been noted as an essential function of all living organisms (Bernard 1878). For iron, total body iron content (3–5 g in adults) and its extracellular concentration (10–30 μM) are both closely controlled, and chronic deviations from these normal ranges cause disease manifestations. Regulation of total body iron and extracellular or plasma iron concentrations is mediated by the interaction of the small peptide hormone hepcidin, secreted by the liver, and its cell membrane receptor ferroportin, which is also the sole known cellular exporter of iron (Nemeth et al. 2004b). Ferroportin is expressed in tissues that export iron to extracellular fluid and plasma: on basolateral membranes of duodenal enterocytes that absorb dietary iron, on splenic and hepatic macrophages that recycle iron from senescent erythrocytes, and on hepatocytes that store iron and release it when needed. Proportionally to its concentration in plasma, hepcidin occludes the iron transporting cavity of ferroportin (Billesbolle et al. 2020; Aschemeyer et al. 2018) and induces ferroportin internalization and proteolysis (Fig. 3.1). The effect of hepcidin is to inhibit the flow of iron into blood plasma transferrin from all the main sources of iron: duodenal enterocytes absorbing dietary iron, macrophages engaged in recycling iron from erythrocytes and hepatocytes that store iron. In the face of decreased iron delivery into plasma,

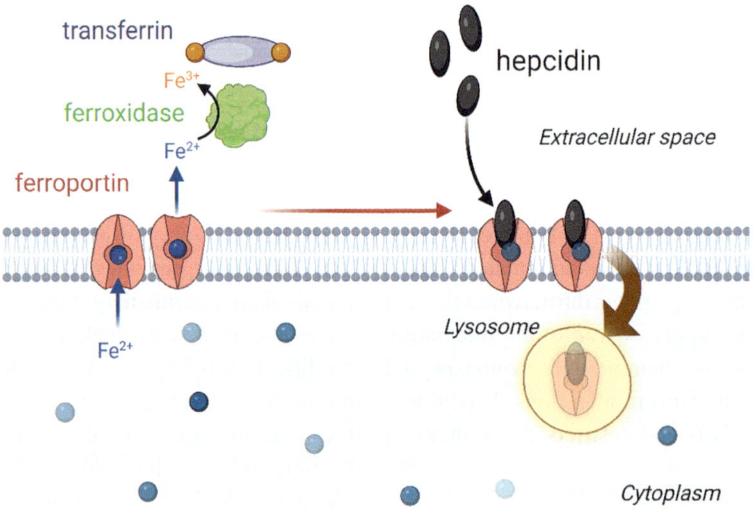

Fig. 3.1 Ferroportin exports cellular iron+ by binding Fe^{2+} in the open-in conformation and then changing conformations to open-out, releasing the iron moiety for oxidation by a ferroxidase (ceruloplasmin or hephaestin). Fe^{3+} then binds to transferrin and is transported to the rest of the organism. Hepcidin inhibits iron export by binding to the open-out conformation of ferroportin. Hepcidin binding induces the endocytosis of the hepcidin-ferroportin complex and its proteolysis in lysosomes

the continued consumption of iron from the small plasma compartment (containing only 2–3 mg of iron) rapidly depletes it, decreasing iron concentrations in much the same manner as insulin, another peptide hormone, decreases glucose concentrations.

3.3 Iron Absorption

Here we focus on the systemic regulation of iron absorption, with a more general discussion of iron absorption available in Chap. 6. Dietary iron is present in food in two main forms, mostly ferric iron complexed with other biomolecules and inefficiently absorbed, and smaller and variable amounts of the more absorbable heme iron, especially enriched in red meat (Sharp 2010). The absorption of heme iron is not well understood, but it appears that once the iron is extracted in enterocytes from absorbed heme, its systemic regulation is similar to that of nonheme iron. Nonheme iron, mostly in the form of ferric iron, is inefficiently solubilized even in the acid milieu of the stomach and the proximal duodenum. In the duodenal mucosa, ferric iron is reduced to ferrous by the ferric reductase duodenal cytochrome B (dCytB) and by other mechanisms before uptake by enterocytes, through the apical divalent metal transporter 1 (DMT1) (Fuqua et al. 2012). Mucosal acidity (high proton concentration) drives DMT1 iron absorption with the transmembrane proton gradient generated not by stomach acid but by the enterocyte sodium-dependent proton exporter NHE3 (Shawki et al. 2016). After transiting the enterocyte, ferrous iron is exported by ferroportin across the basolateral membrane of the enterocyte into the extracellular space, undergoes oxidation by the ferroxidase hephaestin, and is loaded as ferric iron onto the plasma iron carrier transferrin. The critical control point for iron absorption is the activity of ferroportin on the basolateral membrane of duodenal enterocytes, which is regulated by the systemic iron regulatory peptide hormone hepcidin (Fig. 3.2). Hepcidin secondarily affects apical iron uptake through DMT1 as well, in part because the blockade of iron export through ferroportin causes iron accumulation in enterocytes and modulates the activity of the oxygen- and iron-regulated transcription factor (hypoxia-inducible factor) HIF2α (Schwartz et al. 2019) and the RNA-binding regulators (iron regulatory proteins) IRP1 and IRP2. As iron accumulates in

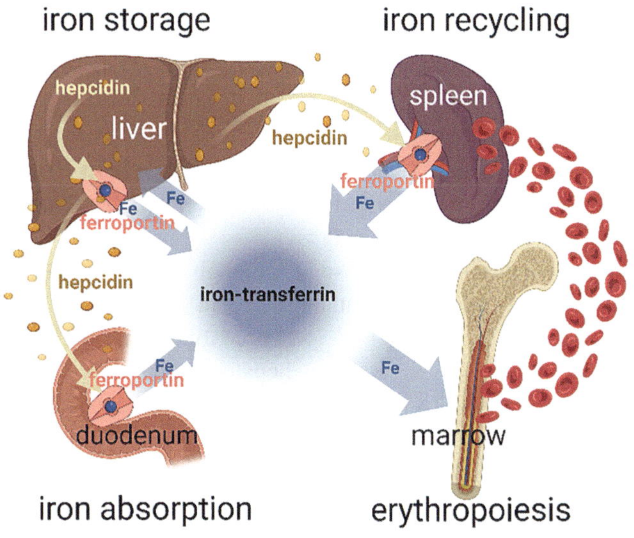

Fig. 3.2 The interaction of hepcidin and ferroportin controls the major flows of iron in the organism. The hormone hepcidin is produced in the liver and circulates in blood. Hepcidin controls the export of dietary iron by enterocyte ferroportin, the export of iron recycled from senescent erythrocytes by ferroportin in macrophages in the spleen, and the export of stored iron by ferroportin in hepatocytes and macrophages in the liver. Most of the iron from these sources flows to the marrow where it is largely consumed for the synthesis of hemoglobin in red blood cells

enterocytes whose iron export is inhibited by hepcidin, these mechanisms decrease the expression of apical DMT1 and its associated ferric reductase dCytB.

3.4 Iron Recycling

In healthy humans, the iron requirements of erythropoiesis and other metabolic processes are met largely by iron that is recycled from senescent erythrocytes in macrophages of the spleen and liver, with a small contribution of recycled iron from relatively iron-poor nonerythroid cells. Absorption of dietary iron accounts for about 5% of the iron turnover and normally balances out the losses incurred by intestinal and other epithelial shedding and through menstrual blood loss. Like the delivery of iron from intestinal iron absorption to blood plasma, the release of recycled ferrous iron from macrophages also takes place through membrane ferroportin (Fig. 3.2), assisted by a different but related plasma ferroxidase, ceruloplasmin, that oxidizes iron to its requisite ferric state for loading onto plasma transferrin.

Macrophage ferroportin is post-translationally regulated by hepcidin (Nemeth et al. 2004b; Knutson et al. 2005) so that in the presence of high concentrations of hepcidin, macrophage ferroportin is internalized or its function is blocked, and as a result, recycled iron is retained within macrophages. Intracellular iron in macrophages induces ferroportin expression by transcriptional and posttranscriptional mechanisms (Knutson et al. 2003; Delaby et al. 2005), which acts to counterbalance the effect of hepcidin during excessive iron accumulation in macrophages so that some iron is returned to the plasma compartment.

3.5 Regulation of Iron Compartments and Reserves

The average adult human contains 3–5 g of iron. Most of the iron, approximately 2/3, is in the hemoglobin of erythrocytes, and 0.3–1 g of iron is in the liver stored in ferritin, mainly in hepatocytes, and functions as an iron reserve. The amount of reserve iron is smaller and more variable in menstruating women. In contrast, the amount of iron in plasma transferrin is about a thousand times smaller (3–4 mg) than the total body iron content and represents a very dynamic compartment that turns over approximately every 3 h. It is remarkable that both the hepatic iron reserve and the serum iron concentration remain relatively stable provided that the iron losses from the body (1–2 mg/day) do not greatly increase and are matched by at least sufficient iron in the diet (normally only about 10% of dietary iron is absorbed). Once the setpoint for body iron stores is reached, iron absorption decreases, and stores do not increase further.

The tight regulation of both plasma iron concentration and body iron stores has been appreciated for at least the last three decades (Finch 1994). Remarkably, both are mediated by the hepcidin-ferroportin dyad, with plasma concentrations regulated predominantly by the rapid effect of hepcidin on macrophage ferroportin, and the hepatic iron stores regulated by the long-term effect of hepcidin on enterocyte ferroportin.

The mobilization of iron from hepatocyte stores also occurs via ferroportin (Zhang et al. 2012) and is subject to regulation by hepcidin (Ramey et al. 2010) by a similar mechanism as in enterocytes and macrophages.

3.6 Hepcidin Production Is Transcriptionally Regulated

Although small amounts of hepcidin are produced in other tissues (see Chap. 7), the main source of hepcidin for systemic regulation is hepatocytes (Zumerle et al. 2014). Hepcidin is transcriptionally regulated by two major signaling pathways (Fig. 3.3), the BMP pathway and the IL6 pathway. The BMP-BMP receptor-SMAD pathway (Fisher and Babitt 2022) mediates the influences of iron and erythropoiesis on hepcidin transcription. Iron stores modulate the production of BMP ligands, and erythropoietic activity generates the hormone erythroferrone

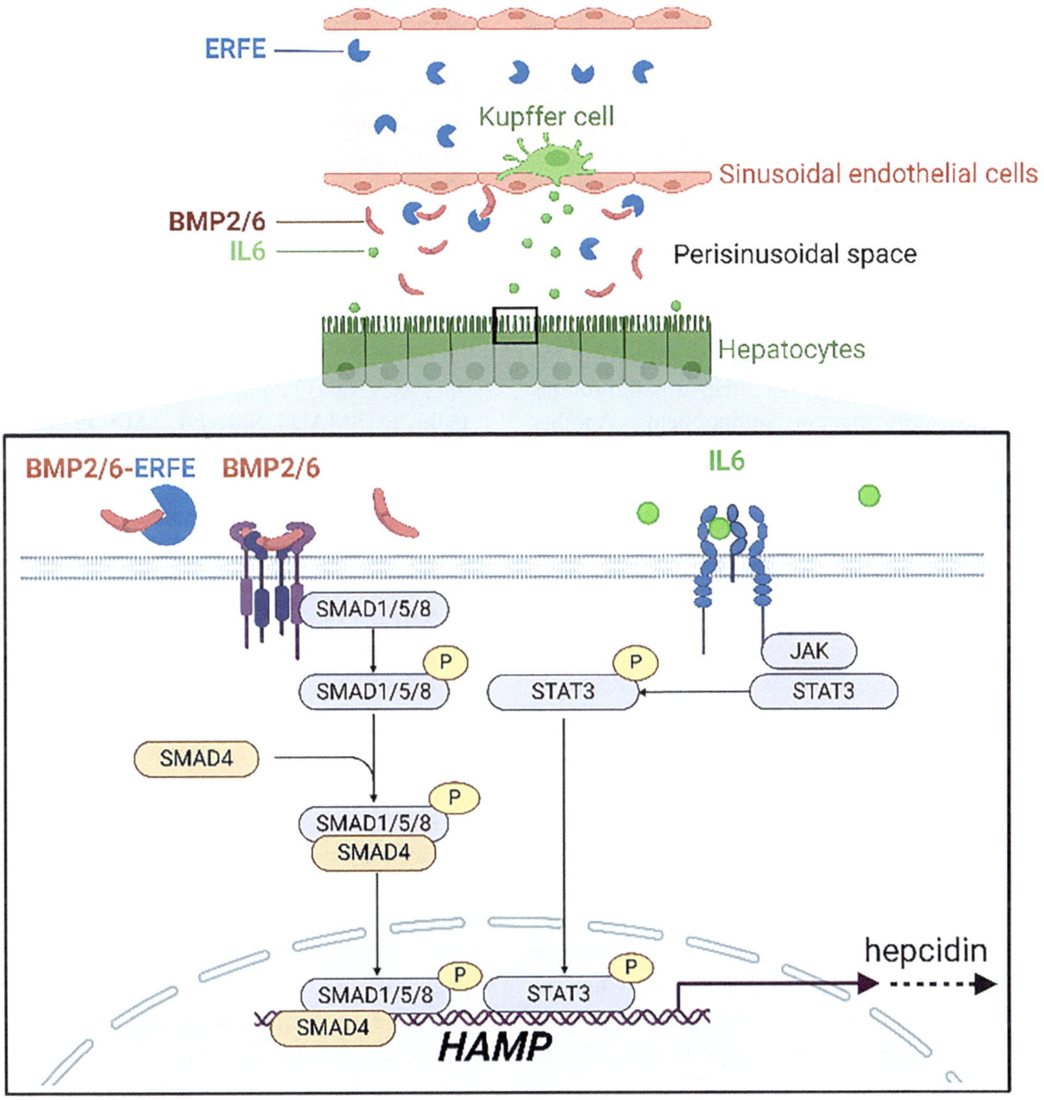

Fig. 3.3 Hepcidin synthesis in hepatocytes is transcriptionally controlled by BMPs produced in siunoidal endothelial cells, and by IL6 and other cytokines, all acting through their cognate hepatocyte receptors. The erythroid hormone erythroferrone (ERFE) binds BMPs and interferes with their ability to signal. The SMAD and JAK-STAT3 pathways converge on their respective binding sites in the hepcidin promoter and control the transcription of the *HAMP* gene

(ERFE), which binds BMPs and thereby inhibits BMP signaling to suppress hepcidin secretion by hepatocytes. The IL6-IL6 receptor-STAT3 pathway (Nemeth et al. 2004a; Verga Falzacappa et al. 2008; Pietrangelo et al. 2007; Wrighting and Andrews 2006) mediates the effects of inflammation on hepcidin transcription. Other cytokines also contribute to inflammatory signaling for hepcidin regulation, but their role is not as well defined.

3.7 Hepcidin Elimination

Hepcidin is a small peptide (2.7 kD) which is weakly bound to plasma proteins (Diepeveen et al. 2019). The peptide is small enough to be freely filtered in the glomerulus and eliminated in urine. Detailed measurements of its concentration in human urine and in laboratory mouse models suggest that it is mostly reabsorbed in the proximal tubule by a megalin-dependent mechanism (Peters et al. 2013) and degraded by the pathway that proteolyzes other small proteins and reabsorbs them as amino acids. Another mechanism for hepcidin elimination is ferroportin-mediated cellular endocytosis and proteolysis in lysosomes (Preza et al. 2013; Nemeth et al. 2004b), but it is not clear how much each mechanism contributes to the overall catabolism of hepcidin.

3.8 Regulation of Hepcidin Transcription by Iron

Our understanding of the regulation of hepcidin transcription by iron has been hindered by the difficulty of reconstituting the regulatory system in vitro. Most of the relevant information therefore came from genetic analysis of patients (Chaps. 9 and 10) and animal models (Chap. 8) with abnormalities of hepcidin regulation. These are usually detected as spontaneous iron overload with inappropriately low hepcidin or as unexplained iron deficiency/malabsorption of iron with inappropriately high hepcidin. The ability to selectively ablate candidate genes in specific organs or tissues has been particularly helpful in confirming the involvement of specific molecular regulators.

From these studies, the following principles emerged:

1. Two types of iron sensors appear to be involved in hepcidin regulation (Ramos et al. 2011; Corradini et al. 2011), one rapidly sensing the concentration of transferrin-bound iron in plasma (plasma iron regulator), the other sensing the slowly changing iron content of the liver (stores regulator).
2. The regulation of hepcidin transcription by iron requires the cooperation of two cell types: hepatocytes and sinusoidal endothelial cells (Canali et al. 2017b), as revealed by their cell-type-specific transgenic manipulation.
3. In hepatocytes, hepcidin deficiency can be caused by ablation or destructive mutation of HFE, TfR2, HJV, components of the BMP receptor (Alk2, Alk3, or combined BMPR2 and ACVR2A), components of the SMAD pathway (SMAD4, SMAD1, SMAD5, or to a lesser extent SMAD8), or the receptor scaffold protein neogenin.
4. Additional factors influencing BMP signaling and hepcidin transcription whose physiological role requires further clarification include FKBP12 (Colucci et al. 2017), FOXO1 (Xu et al. 2024), and proteoglycans associated with the BMP receptor (Asperti et al. 2019).
5. Ablation in hepatic sinusoidal endothelial cells of either BMP2 or BMP6 or both (Koch et al. 2017; Canali et al. 2017a; Xiao et al. 2020) causes severe impairment of hepcidin production by hepatocytes and severe iron overload. Additional systemic ablation of BMP2 by an antibody in BMP6 KO mice further suppressed hepcidin production (Canali et al. 2017a).
6. Secretion of BMP2 and BMP6 by hepatic sinusoidal endothelial cells is increased by iron overload in vivo (Canali et al. 2017a). Although BMP2 is expressed at a somewhat higher level than BMP6, its expression is less increased by iron overload. BMP2/6 heterodimers are more potent inducers of SMAD signaling than either BMP2 or BMP6 homodimers (Valera et al. 2010). Moreover, the requirement for both BMP2 and BMP6 for normal iron regulation may also involve the activity of proBMP2 in the processing of proBMP6 to its bioactive BMP6 form (Chauhan et al. 2024).
7. Hepcidin concentrations that are too high relative to plasma iron and hepatic iron content are clinically caused by homozygous or compound heterozygous mutations in the mem-

brane protease TMPRSS6 (also called MT2). A prominent target of this protease is hemojuvelin (Silvestri et al. 2008). However, the protease function is not required for the activity of TMPRSS6, arguing that the WT TMPRSS6 predominantly acts as a conformational inhibitor (Enns et al. 2020). Additional hepatocyte-specific mouse mutations that cause hepcidin increase relative to the iron status include homozygous ablation of the inhibitory SMAD7 (An et al. 2018).

Based on these findings, a model of hepcidin regulation by iron can be constructed (Fig. 3.4), admittedly with important remaining gaps:

1. The iron content of the liver stimulates the secretion of BMP2/6 by hepatic sinusoidal endothelial cells, but how endothelial cells sense liver iron content over its physiological range is not known.
2. The molecules TfR1, HFE, and TfR2 sense the concentration of iron bound to transferrin and modulate BMP signaling downstream of BMP-BMP receptor interaction, but the molecular details of this modulatory effect are not known.
3. Net BMP signaling via the SMAD pathway regulates hepcidin transcription via SMAD binding sites in the hepcidin promoter.

3.9 Iron Metabolism in Host Defense and Inflammation

Infection and inflammation have profound effects on iron homeostasis (see also Chap. 14). The production of myeloid cells by the marrow greatly increases while that of erythroid cells decreases. As a result, the consumption of iron for hemoglobin synthesis is decreased. Moreover, the lifespan of circulating erythrocytes is variably decreased

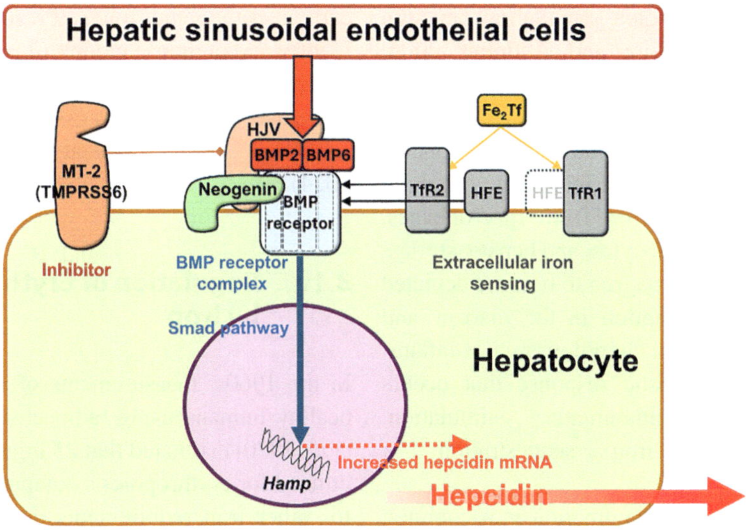

Fig. 3.4 Regulation of hepcidin transcription in hepatocytes by iron. Extracellular iron concentration is sensed predominantly as the differic form of transferrin Fe_2Tf by transferrin receptors TfR1 and TfR2. Binding of Fe_2Tf to TfR1 displaces HFE. HFE and TfR2 interact with the BMP receptor to stimulate BMP signaling and hepcidin transcription. TfR1 may also convey an intracellular iron signal. When intracellular iron is high, TfR1 expression is suppressed due to its unstable mRNA. In low-iron conditions TfR1 is induced transcriptionally and post-transcriptionally by iron-regulatory proteins (IRP1 and 2). The differing amounts of TfR1 modulate how much HFE available for signaling to the BMP receptor. HJV and neogenin are accessory molecules required for functional BMP receptor signaling. MT-2 (TMPRSS6) is a membrane protease that dampens BMP signaling. Hepatic sinusoidal endothelial cells secrete BMP2 and BMP6 in proportion to hepatic intracellular iron stores but the specific mechanisms that sense and signal intracelular iron stores are not understood

because of macrophage activation that increases erythrophagocytosis of senescent erythrocytes, and in some diseases also by antibody deposition on erythrocytes. Injury in inflamed or infected tissues results in the recycling of iron from necrotic cells. In the aggregate, these changes have the potential to decrease iron removal from plasma while increasing the entry of iron into plasma. This imbalance could result in iron accumulation in the plasma compartment that could fuel the proliferation of infectious microorganisms, for which iron is an essential and often limiting nutrient.

However, in a normal inflammatory response, this maladaptive scenario does not occur. Infection and inflammation dramatically increase hepcidin synthesis, acting through the effects of IL6 on IL6 receptors (Pigeon et al. 2001; Nemeth et al. 2003, 2004b) and to a lesser extent other cytokines on their cytokine receptors. The IL6 receptors signal through their canonical JAK-STAT pathways, resulting in the phosphorylation of STAT3 and its binding to specific binding sites in the hepcidin promoter, stimulating hepcidin transcription (Wrighting and Andrews 2006; Pietrangelo et al. 2007; Verga Falzacappa et al. 2007). Increased hepcidin concentrations in plasma act to block and proteolyze ferroportin, inhibiting the export of iron into plasma from all its sources: recycled iron from macrophages, dietary iron from enterocytes, and hepatic storage iron. This causes plasma iron to become depleted by continued consumption in the marrow and elsewhere, resulting in "hypoferremia of inflammation," a characteristic response that occurs within hours of inflammatory stimulation, accompanied by iron sequestration in macrophages.

Hypoferremia of inflammation is a common response to infection in animals and has an important but underappreciated host defense function. In vertebrates, the response is mediated by an inflammation-induced increase in hepcidin concentration, but other mechanisms exist in invertebrates (Iatsenko et al. 2020), suggesting that the response evolved multiple times. Hepcidin also has modest intrinsic antimicrobial properties as an antimicrobial peptide produced by the liver ("hep" + "cidin") (Park et al. 2001), akin to the inducibility of some antimicrobial peptides by inflammation, but the direct antimicrobial activity of hepcidin may not be functionally important. The documented role of hepcidin in host defense turns out to be closely connected to hepcidin's ability to lower the plasma concentration of iron. Mice lacking hepcidin lose control over iron efflux from iron-recycling macrophages and over dietary iron absorption, causing the concentration of iron in plasma to exceed the binding capacity of transferrin for iron and leading to the formation of non-transferrin-bound iron (NTBI). Unlike transferrin-bound iron, which is tightly complexed and inaccessible to most microbes, the NTBI form of iron is highly available to microbes and stimulates the rapid growth of several types of gram-negative bacteria, including *Vibrio vulnificus, Yersinia enterocolitica Klebsiella pneumoniae and E. coli* (Arezes et al. 2015; Stefanova et al. 2017; Michels et al. 2017; Stefanova et al. 2018) and certain pathogenic fungi (Petzer et al. 2019) in vitro and in mouse models of infection. Excess NTBI in hemochromatosis, iron-loading anemias, and liver failure is associated with similar infections in human patients (Ganz and Nemeth 2024).

3.10 Regulation of Erythropoiesis by Iron

In the 1960s, measurements of ferrokinetics in healthy humans using radioactive iron (FINCH et al. 1970) estimated that 25 mg/day of iron was utilized for erythropoiesis compared to 5 mg/day for other iron requirements, and that normally only about 3 mg of iron is in plasma transferrin. The utilization of iron for erythropoiesis can go up four to sixfold or more in chronic anemias when iron supply is adequate. Conversely, erythropoietic demand for iron can decrease severalfold if erythropoiesis is inhibited by disease or RBC transfusion. Not only do plasma iron flows vary with erythropoietic rates, but also erythropoiesis is strongly influenced by the availability of iron. After acute correction of iron deficiency with parenteral iron, the rate of erythropoiesis, as measured by the number of erythroblasts in the marrow, and erythrocyte output measured as reticulocytosis, can increase several-fold, indicating that iron deficiency profoundly constrains erythropoiesis (Kimura et al. 1986). The large and variable flows of iron through a relatively small transferrin compartment imply that iron homeostasis and erythropoiesis must regulate each other. Without mutual regulation, high erythropoietic rates have the potential to deplete the plasma iron compartment within hours, and low erythropoiesis rates would cause a rapid increase in plasma iron, saturating transferrin, and causing NTBI formation on the same time scale.

In the marrow, erythropoiesis is more sensitive to iron deficiency than the production of white cells and platelets (Bullock et al. 2010), and nearly all other organs and tissues are also more resistant to the effects of iron deficiency than erythropoiesis. When iron stores are depleted, this evolutionary feature likely blunts the effects of purely nutritional iron deficiency on other organs at the cost of the gradual development of iron deficiency anemia (Chap. 12). The erythron effectively functions as a backup iron reservoir for the rest of the organism. However, iron deficiency will develop rapidly after even a moderate blood loss, as the loss of each ml of packed erythrocytes means the loss of 1 mg of iron from the body.

The phase of erythrocyte development that is inhibited by iron deficiency is the cell division and differentiation of erythroblasts (Kimura et al. 1986), perhaps explaining why much greater reticulocytosis occurs in hemolytic anemias where iron is abundant than in bleeding-induced anemia, where iron deficiency is the norm. Through several incompletely understood mechanisms, iron deficiency inhibits multiple erythroid functions, including EPO receptor signaling (Khalil et al. 2018) and globin synthesis (Chen and Zhang 2022).

3.11 Regulation of Systemic Iron Metabolism by Erythropoiesis

It has long been known that erythropoiesis exerts a prominent role in the regulation of systemic iron metabolism (Finch 1994), but the mechanisms involved remained a mystery until recently. The existence of physiological erythroid regulators of iron homeostasis was anticipated from observations of greatly increased dietary iron absorption after erythropoietin administration, despite normal iron status. Anemias with ineffective erythropoiesis (Chap. 11) manifest increased iron absorption despite increased iron stores, pointing to the existence of pathological erythroid regulators of iron homeostasis. These signals were thought to originate in the marrow because in aplastic anemia, iron absorption was very low, despite very high levels of erythropoietin.

Later it became clear that the iron homeostatic response to hemorrhage or erythropoietin was largely mediated by factors that suppressed the production of hepcidin by the liver (Pak et al. 2006; Vokurka et al. 2006). The acute (8–24 h) suppression of hepcidin after erythropoietin or hemorrhage is mediated by erythroferrone (ERFE), a member of the C1qTnfα superfamily secreted by erythropoietin-stimulated erythroblasts (Kautz et al. 2014). The production of ERFE by erythroblasts is stimulated by EPO and its receptor via the JAK-STAT5 pathway and is proportional to the number of erythroblasts (Fig. 3.5). The dependence of ERFE production on both the number of erythroblasts and their degree of stimulation by EPO likely explains the suppression of hepcidin and non-transfusional iron overload in anemias with ineffective erythropoiesis, with their often massive expansion of erythroblast numbers (Olivera et al. 2023).

ERFE lowers hepcidin concentrations by binding to BMPs and interfering with their ability to signal through the BMP receptor to stimulate hepcidin transcription. Other mechanisms also contribute to hepcidin suppression during

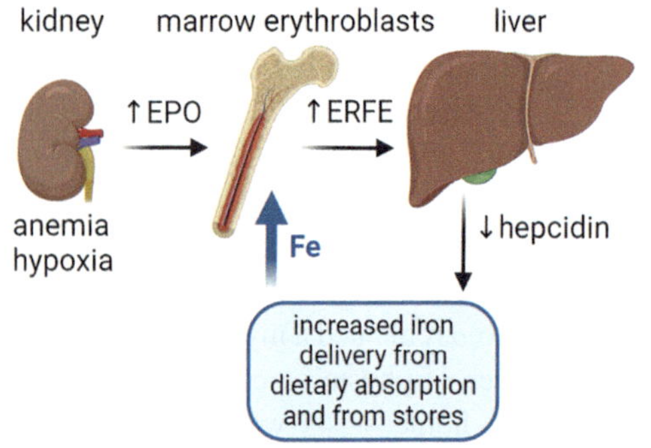

Fig. 3.5 ERFE inhibits hepcidin synthesis by hepatocytes and thereby helps mobilize iron for erythropoiesis. ERFE secretion by erythroblasts is stimulated by EPO in response to anemia and renal hypoxia

erythropoietic stimulation, including the increase in erythroblast transferrin receptor synthesis, which causes the depletion of the diferric form of transferrin from circulation (Mirciov et al. 2018), even though overall plasma iron levels decrease very little in the first 24 h in response to EPO or hemorrhage.

Anemia and hypoxia affect hepcidin synthesis chronically, but this response is less well understood. In a mouse model of bleeding-induced anemia, fibrinogen-like-1 (FGL1) is a hepcidin suppressor produced by the liver in response to anemia after the first 24 h following hemorrhage (Sardo et al. 2024). Its production is not responsive to erythropoietin but is directly stimulated by hepatic hypoxia. Like ERFE, FGL-1 suppresses hepcidin by binding to BMPs and interfering with BMP signaling that stimulates hepcidin transcription.

3.12 Conclusions

Systemic iron regulation by hepcidin and its receptor ferroportin serves three largely independent functions. First, it maintains a stable concentration of extracellular iron and safe amounts of storage iron so that individual tissues and cells can utilize iron according to their requirements. Second, it assures that erythropoiesis, the main consumer of iron in the body, can obtain sufficient iron for its variable hemoglobin synthesis requirements without compromising the supply of iron to other tissues. Third, it allows host defense to respond to infections by inhibiting the production of iron forms that are accessible to invading microorganisms. These functions are carried out by regulating the transcription of hepcidin in hepatocytes through signaling pathways activated by BMP receptors and cytokine receptors and their modulators. Many aspects of this complex system are not yet well understood and are an appropriate subject for further study.

References

An P, Wang H, Wu Q, Wang J, Xia Z, He X, Wang X, Chen Y, Min J, Wang F (2018) Smad7 deficiency decreases iron and haemoglobin through hepcidin up-regulation by multilayer compensatory mechanisms. J Cell Mol Med 22(6):3035–3044. https://doi.org/10.1111/jcmm.13546

Arezes J, Jung G, Gabayan V, Valore E, Ruchala P, Gulig PA, Ganz T, Nemeth E, Bulut Y (2015) Hepcidin-induced hypoferremia is a critical host defense mechanism against the siderophilic bacterium Vibrio vulnificus. Cell Host Microbe 17(1):47–57. https://doi.org/10.1016/j.chom.2014.12.001

Aschemeyer S, Qiao B, Stefanova D, Valore EV, Sek AC, Ruwe TA, Vieth KR, Jung G, Casu C, Rivella S, Jormakka M, Mackenzie B, Ganz T, Nemeth E (2018) Structure-function analysis of ferroportin defines the binding site and an alternative mechanism of action of hepcidin. Blood 131(8):899–910. https://doi.org/10.1182/blood-2017-05-786590

Asperti M, Denardo A, Gryzik M, Arosio P, Poli M (2019) The role of heparin, heparanase and heparan sulfates in hepcidin regulation. Vitam Horm 110:157–188. https://doi.org/10.1016/bs.vh.2019.01.008

Bernard C (1878) Phénomènes de la vie communs aux animaux et aux végétaux. J. B. Bailliere, Paris

Billesbolle CB, Azumaya CM, Kretsch RC, Powers AS, Gonen S, Schneider S, Arvedson T, Dror RO, Cheng Y, Manglik A (2020) Structure of hepcidin-bound ferroportin reveals iron homeostatic mechanisms. Nature 586(7831):807–811. https://doi.org/10.1038/s41586-020-2668-z

Bullock GC, Delehanty LL, Talbot AL, Gonias SL, Tong WH, Rouault TA, Dewar B, Macdonald JM, Chruma JJ, Goldfarb AN (2010) Iron control of erythroid development by a novel aconitase-associated regulatory pathway. Blood 116(1):97–108. blood-2009-10-251496 [pii];10.1182/blood-2009-10-251496 [doi]

Canali S, Wang CY, Zumbrennen-Bullough KB, Bayer A, Babitt JL (2017a) Bone morphogenetic protein 2 controls iron homeostasis in mice independent of Bmp6. Am J Hematol 92(11):1204–1213. https://doi.org/10.1002/ajh.24888

Canali S, Zumbrennen-Bullough KB, Core AB, Wang CY, Nairz M, Bouley R, Swirski FK, Babitt JL (2017b) Endothelial cells produce bone morphogenetic

protein 6 required for iron homeostasis in mice. Blood 129(4):405–414. https://doi.org/10.1182/blood-2016-06-721571
Chauhan P, Xue Y, Fisher AL, Kim H-S, Babitt JL, Christian JL (2024) The BMP2 prodomain promotes dimerization and cleavage of BMP6 homodimers and BMP2/6 heterodimers in vivo. bioRxiv 2024.2006.2019.599755. https://doi.org/10.1101/2024.06.19.599755
Chen JJ, Zhang S (2022) Translational control by heme-regulated eIF2alpha kinase during erythropoiesis. Curr Opin Hematol 29(3):103–111. https://doi.org/10.1097/MOH.0000000000000704
Colucci S, Pagani A, Pettinato M, Artuso I, Nai A, Camaschella C, Silvestri L (2017) The immunophilin FKBP12 inhibits hepcidin expression by binding the BMP type I receptor ALK2 in hepatocytes. Blood 130(19):2111–2120. https://doi.org/10.1182/blood-2017-04-780692
Corradini E, Meynard D, Wu Q, Chen S, Ventura P, Pietrangelo A, Babitt JL (2011) Serum and liver iron differently regulate the bone morphogenetic protein 6 (BMP6)-SMAD signaling pathway in mice. Hepatology 54(1):273–284. https://doi.org/10.1002/hep.24359
Delaby C, Pilard N, Goncalves AS, Beaumont C, Canonne-Hergaux F (2005) Presence of the iron exporter ferroportin at the plasma membrane of macrophages is enhanced by iron loading and down-regulated by hepcidin. Blood 106(12):3979–3984
Diepeveen LE, Laarakkers CM, Peters HPE, van Herwaarden AE, Groenewoud H, IntHout J, Wetzels JF, van Swelm RPL, Swinkels DW (2019) Unraveling hepcidin plasma protein binding: evidence from peritoneal equilibration testing. Pharmaceuticals (Basel) 12(3). https://doi.org/10.3390/ph12030123
Enns CA, Jue S, Zhang AS (2020) The ectodomain of matriptase-2 plays an important nonproteolytic role in suppressing hepcidin expression in mice. Blood 136(8):989–1001. https://doi.org/10.1182/blood.2020005222
Finch C (1994) Regulators of iron balance in humans. Blood 84(6):1697–1702
Finch CA, Deubelbeiss K, Cook JD, Eschbach JW, Barker LA, Funk DD, Marsaglia G, Hillman RS, Slichter S, Adamson JW, Ganzoni A, Giblett ER (1970) Ferrokinetics in man. Medicine 49(1):17–54
Fisher AL, Babitt JL (2022) Coordination of iron homeostasis by bone morphogenetic proteins: current understanding and unanswered questions. Dev Dyn 251(1):26–46. https://doi.org/10.1002/dvdy.372
Fuqua BK, Vulpe CD, Anderson GJ (2012) Intestinal iron absorption. J Trace ElemMedBiol 26(2–3):115–119. S0946-672X(12)00054-5 [pii];10.1016/j.jtemb.2012.03.015 [doi]
Ganz T, Nemeth E (2024) Hypoferremia of inflammation: innate host defense against infections. Blood Cells Mol Dis 104:102777. https://doi.org/10.1016/j.bcmd.2023.102777
Iatsenko I, Marra A, Boquete J-P, Peña J, Lemaitre B (2020) Iron sequestration by transferrin 1 mediates nutritional immunity in *Drosophila melanogaster*. Proc Natl Acad Sci 117(13):7317–7325. https://doi.org/10.1073/pnas.1914830117
Kautz L, Jung G, Valore EV, Rivella S, Nemeth E, Ganz T (2014) Identification of erythroferrone as an erythroid regulator of iron metabolism. Nat Genet 46(7):678–684. ng.2996 [pii];10.1038/ng.2996 [doi]
Khalil S, Delehanty L, Grado S, Holy M, White Z 3rd, Freeman K, Kurita R, Nakamura Y, Bullock G, Goldfarb A (2018) Iron modulation of erythropoiesis is associated with scribble-mediated control of the erythropoietin receptor. J Exp Med 215(2):661–679. https://doi.org/10.1084/jem.20170396
Kimura H, Finch CA, Adamson JW (1986) Hematopoiesis in the rat: quantitation of hematopoietic progenitors and the response to iron deficiency anemia. J Cell Physiol 126(2):298–306. https://doi.org/10.1002/jcp.1041260221
Knutson MD, Vafa MR, Haile DJ, Wessling-Resnick M (2003) Iron loading and erythrophagocytosis increase ferroportin 1 (FPN1) expression in J774 macrophages. Blood 102(12):4191–4197. https://doi.org/10.1182/blood-2003-04-1250
Knutson MD, Oukka M, Koss LM, Aydemir F, Wessling-Resnick M (2005) Iron release from macrophages after erythrophagocytosis is up-regulated by ferroportin 1 overexpression and down-regulated by hepcidin. Proc Natl Acad Sci USA 102(5):1324–1328. 0409409102 [pii];10.1073/pnas.0409409102 [doi]
Koch PS, Olsavszky V, Ulbrich F, Sticht C, Demory A, Leibing T, Henzler T, Meyer M, Zierow J, Schneider S, Breitkopf-Heinlein K, Gaitantzi H, Spencer-Dene B, Arnold B, Klapproth K, Schledzewski K, Goerdt S, Geraud C (2017) Angiocrine Bmp2 signaling in murine liver controls normal iron homeostasis. Blood 129(4):415–419. https://doi.org/10.1182/blood-2016-07-729822
McCord JM (1998) Iron, free radicals, and oxidative injury. Semin Hematol 35(1):5–12
Michels KR, Zhang Z, Bettina AM, Cagnina RE, Stefanova D, Burdick MD, Vaulont S, Nemeth E, Ganz T, Mehrad B (2017) Hepcidin-mediated iron sequestration protects against bacterial dissemination during pneumonia. JCI Insight 2(6):e92002. https://doi.org/10.1172/jci.insight.92002
Mirciov CSG, Wilkins SJ, Hung GCC, Helman SL, Anderson GJ, Frazer DM (2018) Circulating iron levels influence the regulation of hepcidin following stimulated erythropoiesis. Haematologica 103(10):1616–1626. https://doi.org/10.3324/haematol.2017.187245
Nemeth E, Valore EV, Territo M, Schiller G, Lichtenstein A, Ganz T (2003) Hepcidin, a putative mediator of anemia of inflammation, is a type II acute-phase protein. Blood 101(7):2461–2463
Nemeth E, Rivera S, Gabayan V, Keller C, Taudorf S, Pedersen BK, Ganz T (2004a) IL-6 mediates

hypoferremia of inflammation by inducing the synthesis of the iron regulatory hormone hepcidin. J Clin Invest 113(9):1271–1276. https://doi.org/10.1172/JCI20945

Nemeth E, Tuttle MS, Powelson J, Vaughn MB, Donovan A, Ward DM, Ganz T, Kaplan J (2004b) Hepcidin regulates cellular iron efflux by binding to ferroportin and inducing its internalization. Science 306(5704):2090–2093

Olivera J, Zhang V, Nemeth E, Ganz T (2023) Erythroferrone exacerbates iron overload and ineffective extramedullary erythropoiesis in a mouse model of beta-thalassemia. Blood Adv 7(14):3339–3349. https://doi.org/10.1182/bloodadvances.2022009307

Pak M, Lopez MA, Gabayan V, Ganz T, Rivera S (2006) Suppression of hepcidin during anemia requires erythropoietic activity. Blood 108(12):3730–3735. blood-2006-06-028787 [pii];10.1182/blood-2006-06-028787 [doi]

Park CH, Valore EV, Waring AJ, Ganz T (2001) Hepcidin, a urinary antimicrobial peptide synthesized in the liver. J Biol Chem 276(11):7806–7810

Peters HP, Laarakkers CM, Pickkers P, Masereeuw R, Boerman OC, Eek A, Cornelissen EA, Swinkels DW, Wetzels JF (2013) Tubular reabsorption and local production of urine hepcidin-25. BMC Nephrol 14:70. https://doi.org/10.1186/1471-2369-14-70

Petzer V, Wermke M, Tymoszuk P, Wolf D, Seifert M, Ovacin R, Berger S, Orth-Holler D, Loacker L, Weiss G, Haas H, Platzbecker U, Theurl I (2019) Enhanced labile plasma iron in hematopoietic stem cell transplanted patients promotes aspergillus outgrowth. Blood Adv 3(11):1695–1700. https://doi.org/10.1182/bloodadvances.2019000043

Pietrangelo A, Dierssen U, Valli L, Garuti C, Rump A, Corradini E, Ernst M, Klein C, Trautwein C (2007) STAT3 is required for IL-6-gp130-dependent activation of hepcidin in vivo. Gastroenterology 132(1):294–300

Pigeon C, Ilyin G, Courselaud B, Leroyer P, Turlin B, Brissot P, Loreal O (2001) A new mouse liver-specific gene, encoding a protein homologous to human antimicrobial peptide hepcidin, is overexpressed during iron overload. J Biol Chem 276(11):7811–7819

Preza GC, Pinon R, Ganz T, Nemeth E (2013) Cellular catabolism of the iron-regulatory peptide hormone hepcidin. PLoS One 8(3):e58934. https://doi.org/10.1371/journal.pone.0058934[doi];PONE-D-13-00071[pii]

Ramey G, Deschemin JC, Durel B, Canonne-Hergaux F, Nicolas G, Vaulont S (2010) Hepcidin targets ferroportin for degradation in hepatocytes. Haematologica 95(3):501–504. haematol.2009.014399 [pii];10.3324/haematol.2009.014399 [doi]

Ramm GA, Ruddell R (2005) Hepatotoxicity of iron overload: mechanisms of iron-induced hepatic fibrogenesis. Semin Liver Dis 04:433–449

Ramos E, Kautz L, Rodriguez R, Hansen M, Gabayan V, Ginzburg Y, Roth MP, Nemeth E, Ganz T (2011) Evidence for distinct pathways of hepcidin regulation by acute and chronic iron loading in mice. Hepatology 53(4):1333–1341. https://doi.org/10.1002/hep.24178

Sardo U, Perrier P, Cormier K, Sotin M, Personnaz J, Medjbeur T, Desquesnes A, Cannizzo L, Ruiz-Martinez M, Thevenin J, Billore B, Jung G, Abboud E, Peyssonnaux C, Nemeth E, Ginzburg YZ, Ganz T, Kautz L (2024) The hepatokine FGL1 regulates hepcidin and iron metabolism during anemia in mice by antagonizing BMP signaling. Blood 143(13):1282–1292. https://doi.org/10.1182/blood.2023022724

Schwartz AJ, Das NK, Ramakrishnan SK, Jain C, Jurkovic MT, Wu J, Nemeth E, Lakhal-Littleton S, Colacino JA, Shah YM (2019) Hepatic hepcidin/intestinal HIF-2alpha axis maintains iron absorption during iron deficiency and overload. J Clin Invest 129(1):336–348. https://doi.org/10.1172/JCI122359

Sharp PA (2010) Intestinal iron absorption: regulation by dietary & systemic factors. Int J Vitam Nutr Res 80(4–5):231–242. https://doi.org/10.1024/0300-9831/a000029

Shawki A, Engevik MA, Kim RS, Knight PB, Baik RA, Anthony SR, Worrell RT, Shull GE, Mackenzie B (2016) Intestinal brush-border Na+/H+ exchanger-3 drives H+-coupled iron absorption in the mouse. Am J Physiol Gastrointest Liver Physiol 311(3):G423–G430. https://doi.org/10.1152/ajpgi.00167.2016

Silvestri L, Pagani A, Nai A, De Domenico I, Kaplan J, Camaschella C (2008) The serine protease matriptase-2 (TMPRSS6) inhibits hepcidin activation by cleaving membrane hemojuvelin. Cell Metab 8(6):502–511. S1550-4131(08)00319-7 [pii];10.1016/j.cmet.2008.09.012 [doi]

Stefanova D, Raychev A, Arezes J, Ruchala P, Gabayan V, Skurnik M, Dillon BJ, Horwitz MA, Ganz T, Bulut Y, Nemeth E (2017) Endogenous hepcidin and its agonist mediate resistance to selected infections by clearing non-transferrin-bound iron. Blood 130(3):245–257. https://doi.org/10.1182/blood-2017-03-772715

Stefanova D, Raychev A, Deville J, Humphries R, Campeau S, Ruchala P, Nemeth E, Ganz T, Bulut Y (2018) Hepcidin protects against lethal Escherichia coli sepsis in mice inoculated with isolates from septic patients. Infect Immun 86(7). https://doi.org/10.1128/IAI.00253-18

Valera E, Isaacs MJ, Kawakami Y, Izpisúa Belmonte JC, Choe S (2010) BMP-2/6 heterodimer is more effective than BMP-2 or BMP-6 homodimers as inductor of differentiation of human embryonic stem cells. PLoS One 5(6):e11167. https://doi.org/10.1371/journal.pone.0011167

Verga Falzacappa MV, Vujic SM, Kessler R, Stolte J, Hentze MW, Muckenthaler MU (2007) STAT3 mediates hepatic hepcidin expression and its inflammatory stimulation. Blood 109(1):353–358

Verga Falzacappa MV, Casanovas G, Hentze MW, Muckenthaler MU (2008) A bone morphogenetic

protein (BMP)-responsive element in the hepcidin promoter controls HFE2-mediated hepatic hepcidin expression and its response to IL-6 in cultured cells. J Mol Med 86(5):531–540
Vokurka M, Krijt J, Sulc K, Necas E (2006) Hepcidin mRNA levels in mouse liver respond to inhibition of erythropoiesis. Physiol Res 55(6):667–674
Wrighting DM, Andrews NC (2006) Interleukin-6 induces hepcidin expression through STAT3. Blood 108(9):3204–3209
Xiao X, Dev S, Canali S, Bayer A, Xu Y, Agarwal A, Wang CY, Babitt JL (2020) Endothelial bone morphogenetic protein 2 (Bmp2) knockout exacerbates hemochromatosis in homeostatic iron regulator (Hfe) knockout mice but not Bmp6 knockout mice. Hepatology 72(2):642–655. https://doi.org/10.1002/hep.31048
Xu T, Zhang X, Zhao W, Shi J, Wan S, Zhang Y, Hao Y, Sun M, He J, Jiang L, Wang H, Gao H, Luo J, Luo Y, An P (2024) Foxo1 is an iron-responsive transcriptional factor regulating systemic iron homeostasis. Blood 144(12):1314–1328. https://doi.org/10.1182/blood.2024024293
Zhang Z, Zhang F, Guo X, An P, Tao Y, Wang F (2012) Ferroportin1 in hepatocytes and macrophages is required for the efficient mobilization of body iron stores in mice. Hepatology 56(3):961–971. https://doi.org/10.1002/hep.25746
Zumerle S, Mathieu JR, Delga S, Heinis M, Viatte L, Vaulont S, Peyssonnaux C (2014) Targeted disruption of hepcidin in the liver recapitulates the hemochromatotic phenotype. Blood 123(23):3646–3650. https://doi.org/10.1182/blood-2014-01-550467

The Labile Side of Iron in Health and Disease: A Narrative Review

4

Ioav Z. Cabantchik, Ola Karmi, and Rachel Nechushtai

Abstract

Labile iron is an exchangeable and redox-active form of Fe^{2+}/Fe^{3+} that is identified in cells under normal or pathological conditions, particularly with the aid of an increasing number of metal-sensitive probes, especially those carrying fluorescent-based sensors, some targetable to specific cell organelles, some useful for real-time measurements. Cell labile iron pools (LIPs) are associated with multiple ligands, predominantly as Fe^{2+}-GSH adducts, whose chemical identity remains to be firmly identified. Estimated cellular LIP levels are in the lower μMolar range, rising substantially in systemic iron overload, thus serving as targets for screening of drugs for chelation efficacy and protection from oxidative damage, including ferroptosis. Pathological LIP elevation in specific organs and/or cells results from mutations that affect some FeS-cluster (ISC) protein biogenesis, as in Wolfram syndrome-2 (WS2) patients, who lack NAF-1 (a member of the CISD-coded NEET protein family that carries labile ISCs). In cells from WS2 patients or shRNA NAF-1-repressed, there are elevated mitochondrial LIP and ROS together with reduced GSH levels that are accompanied by morphological and functional damage, most significantly correctable by applying anti-ferroptotic measures, including a conservative form of chelation in conjunction with an antioxidant, offering novel opportunities to treat the otherwise incurable WS2.

Keywords

Labile iron pool (LIP) · Non-transferrin bound iron (NTBI) · Iron homeostasis · Ferroptosis · Oxidative stress · Iron chelation therapy

4.1 Introduction

4.1.1 "Labile Iron" in Biological Systems

Iron is present in biological systems mostly bound to proteins, either directly via distinct ligands or as part of prosthetic groups such as heme or iron-sulfur clusters (ISCs). However, early physicochemical calculations have led authors to postulate the existence of transitory

I. Z. Cabantchik (✉) · O. Karmi · R. Nechushtai
The Alexander Silberman Institute of Life Sciences, The Hebrew University of Jerusalem, Jerusalem, Israel
e-mail: ioav.caban@mail.huji.ac.il

K. Pantopoulos (ed.), *Iron Metabolism in Human Health and Disease*, Advances in Experimental Medicine and Biology 1480, https://doi.org/10.1007/978-3-031-92033-2_4

pools of labile iron[1] that, in living entities, might play a physiological role (Greenberg and Wintrobe 1946; Jacobs 1976; Williams 1982; Egyed' and Saltman 1984). The term "labile iron" referred specifically to the biologically prevalent Fe^{2+}/Fe^{3+} pair bound to chemical structures with: (a) moderate affinity, allowing for an exchange with other metals (of the same or different kind) or vice versa, for an exchange of the same metal between different ligands or a facile transfer to other ligands (Cabantchik 2014; Koppenol and Hider 2019; Condeles et al. 2024); and (b) a catalytic activity associated with the redox potential of the pair that, in an O_2 environment, might prompt generation of reactive oxygen species (ROS) of potential toxicity (Rothman et al. 1992; Kruszewski 2003; Cabantchik et al. 2013b). However, redox-active forms of labile iron are also components of catalytic centers of key enzymes involved in multiple functions such as the synthesis of deoxynucleotides or demethylation of DNA and histones (Camarena et al. 2021).

The potentially toxic side of labile iron (Rothman et al. 1992; Cabantchik et al. 2013b; Koppenol and Hider 2019) (Chap. 1) was assumed to be normally handled by cells by keeping the labile iron pool (LIP) at relatively "safe" levels based on cell regulatory mechanisms (Zhang et al. 2014; Katsarou and Pantopoulos 2020; Camarena et al. 2021; Pezacki et al. 2024) (Chap. 2) aided by antioxidant tools that, either chemically or enzymatically, attenuate ROS levels. Moreover, cellular LIP was shown to serve as a metabolic sensor that triggers a cAMP-signaling path of histone and DNA demethylation (Camarena et al. 2021). However, deranged levels of LIP and ROS were implicated in the etiopathology of multiple human disorders, many of which have, in the last decade, become increasingly associated with ferroptosis, a form of programmed cell death prompted by labile iron and reduced levels of glutathione (GSH) that jointly lead to peroxidative lipid damage (Jiang et al. 2021; Dixon and Pratt 2023) (Chap. 5). In plasma, iron normally circulates safely bound to its carrier, transferrin (Tf), but might also appear in labile forms that are not bound to Tf, generically referred to by the apophatic term NTBI (non-Tf-bound iron) (Breuer et al. 2000; Garbowski et al. 2023). NTBI is detected in pathological conditions whereby Tf-binding capacity is surpassed due to systemic iron overload, such as in hereditary hemochromatosis or iatrogenic siderosis (of transfusional or iatrogenic origin). For further information, the reader is referred to recently published reviews (Breuer et al. 2000; Cabantchik et al. 2013b; Garbowski et al. 2023) and Chaps. 9, 10 and 11.

Experimental support for the concept of a cellular **LIP** that is comprised of loosely bound Fe^{2+} (often referred to as "free iron") was initially provided by studies of Fe^{2+} mobilization from erythroid cells using an ionophore capable of establishing a transmembrane ionic equilibrium of divalent metals while sparing protein-bound iron (Egyed' and Saltman 1984). The putative LIP was associated with a cytosolic iron pool of transitory or dynamic nature that purportedly serves as crossroads of cell iron metabolism in both anabolic and catabolic directions (Jacobs 1976). For mammalian cells, the cytosolic LIP was envisioned to be "fed" physiologically from plasma by the Tf/Tf receptor-mediated endocytosis machinery (Zhang et al. 2014; Katsarou and Pantopoulos 2020) or from within cells by degradation of iron proteins, later attributed to ferritinophagy or autophagy of iron proteins, primarily in conditions of iron deprivation. These processes of iron supply were shown to operate in a balanced or regulated manner, namely, vis-a-vis the mobilization of the metal needed for metalloprotein biogenesis, a process that takes place in either the cytosol or mitochondria, where most of the iron is incorporated into heme and ISC proteins (Zhang et al. 2014; Katsarou and Pantopoulos 2020).

[1]According to the Oxford Dictionary, **labile means** "liable to change; easily altered," and in chemistry, easily broken down or displaced. In the BioIron field, labile was used as an adjective to denote chemical forms of iron of a protean nature, i.e., that change once approached, like the elusive Loch Ness monster. RC Crichton. Iron Metabolism: From Molecular Mechanisms to Clinical Consequences. 4th ed. 2016

The cytosolic traffic of metabolic iron destined for integration into proteins has not been regarded as random or free-diffusional but rather mediated by dedicated chaperones of the poly-C-RNA binding protein family (PCPB1 and PCPB2) (Philpott et al. 2023) (Chap. 2). Analogous PCBP-mediated traffic has been proposed to be involved in facilitating: (a) the incorporation of "excess" cytosolic LIP into nascent ferritin shells (FT) whose expression is controlled by iron regulatory proteins 1 & 2 (IRP1 & IRP2) (Zhang et al. 2014, Katsarou and Pantopoulos 2020); and (b) the export of iron from specific cells to body fluids via the iron exporter ferroportin (Zhang et al. 2014). Whether or not PCBP associates with cellular LIP's Fe^{2+} directly and/or with a putative Fe^{2+}-X labile conjugate (e.g., X = GSH), assumed to be the main component of cytosolic LIP (Hider and Kong 2011; Cabantchik et al. 2013b; Hider and Kong 2013; Hider et al. 2021), has yet to be experimentally demonstrated in the actual live-cell setting (Brawley et al. 2023).

4.2 Attempting to Estimate LIP Levels and Reveal Its Chemical Identity(ies)

Various attempts were made to associate LIPs with potential Fe^{2+}/Fe^{3+} binders that populate cells, such as di- and tri-carboxylates and/or nucleoside-phosphates. Early studies were most likely flawed since they were based on the analysis of extracts from cells decompartmentalized by disruption and fractionation in aerobic conditions, thus leading to Fe^{2+} oxidation concomitantly with redistribution of the labile metal between potential metal ligands, particularly nucleotide-phosphates and carboxylates (Weaver and Pollack 1989; Brawley and Lindahl 2021; Brawley et al. 2023). Hitherto, the chemical identification of a predominant physiological LIP constituent remains in dispute, including the recently proposed GSH-Fe^{2+} adduct, a putative candidate that meets the above criteria of lability in quasi-physiological solutions (Hider and Kong 2011, 2013; Hider et al. 2021). However, that candidate has thus far failed to be detected in situ in cells or in cell extracts that ideally retain the environmentally sensitive chemical composition (Brawley et al. 2023).

Irrespective of LIP's true chemical identity(ies), numerous attempts were made to acquire estimates of LIP levels in intact cells and in organelles relevant to iron metabolism, such as the cytosol, mitochondria, and lysosomes/endosomes (e.g., Petrat et al. 1999 & 2001; Breuer et al. 2008; Cabantchik 2014; Ma et al. 2015; Abbate et al. 2016; Lv and Shang 2018; Zhu et al. 2018; Hirayama 2019, 2024; Hirayama et al. 2019; Silswal et al. 2023; Gao et al. 2020; Xie et al. 2023). Those attempts relied on chemical tools known to detect readily exchangeable iron (Fe^{2+} or Fe^{3+}), ideally complemented with others that monitor the catalytic properties of the metal ion in question (Domaille et al. 2008; Au-Yeung et al. 2013; Cabantchik et al. 2013a; Carter et al. 2014; Lindahl and Moore 2016; Aron et al. 2018; Hirayama 2024). Short of bona fide chemical standards, the applied procedures have, thus far, provided only estimates of LIP levels in whole cells or their organelles. Thus, even though LIP has been often reported in μMolar units, these values should be regarded as estimates that are particularly useful for comparative purposes, e.g., in assessing changes in LIP levels that are generated by physiological stimuli, pathological factors, or iatrogenic and pharmacological treatments. Hitherto, the real-time measurements of cellular LIPs were carried out with fluorescent probes applied to cells isolated from normal individuals or experimental animals, and mostly to cells grown in culture conditions. The cytosolic LIP values were estimated to: (a) span the high-nMolar to lower μMolar range in basal conditions; (b) rise in a time-dependent manner when exposed to transportable iron sources (Breuer et al. 1996, 2008; Niwa et al. 2014, 2018; Soleja and Mohsin 2024) and vice versa; and (c) fall when treated with permeable chelators (Zanninelli et al. 1997; Cabantchik et al. 2013b). Such measurements were instrumental for the screening of chelating agents with demonstrable ability to reduce the abnormal tissue LIPs associated with iron overload pathologies of systemic origin, as in hemochromatosis or transfusional siderosis

(Zanninelli et al. 1997; Breuer et al. 2000; Cabantchik 2014). Fluorescent iron sensors were also designed for differential targeting to specific cell compartments (i.e., organelles) (Kakhlon and Cabantchik 2002) by incorporating chemical moieties that confer potentiometric features (for accumulation in mitochondria), acidotropic accumulation in lysosomes, or targeting tags for the Golgi apparatus, as exemplified in Table 4.1. Those probes not only facilitated the detection of LIP in specific organelles but also the susceptibility of the latter to be scavenged by chelators in clinical practice (Zanninelli et al. 1997; Cabantchik et al. 2013b; Hirayama et al. 2019) or mobilized and transferred to other cell compartments or to the medium (Cabantchik et al. 2013a). Chelators with moderate iron-binding affinities, such as deferiprone (DFP), were shown to mobilize iron from iron-overloaded cells and transfer the metal to external apo-Tf, thus allowing for the development of a conservative chelation modality suitable for treating regional siderosis (Cabantchik et al. 2013a). Such a form of siderosis, whereby elevated LIPs appear only in specific tissues and/or in specific cell organelles, is found in the inherited diseases Friedreich ataxia and Wolfram syndrome-2 (WS2) or in neurodegenerative disorders such as Parkinson's disease or neurodegeneration with brain iron accumulation (Chap. 19). Many of these disorders are at least partially affected by ferroptosis, a form of cell death prompted by mismanagement of LIP and GSH (Jiang et al. 2021; Dixon and Pratt 2023). An apparently non-canonical modality of

Table 4.1 Selected probes in use for fluorescence tracing of labile iron in cells and their compartments

PROBE	λex→λem nm	Response mode	System	Reference
Calcein	486→517	on→off	Cells live	Espósito, Epsztejn et al. 2002, Condeles, da Silva et al 2024
FlCFe1 (modified calcein)	405 →490 /530	on→off (ratio)	Cells live	Carney, Kolanowski et al. 2018
CP665	458→475	on→off	Cells live	Ma, Abbate et al. 2015
Phen-Green-SK	507→532	on→off	Cells live	Petrat, Rauen et al. 1999
RPA	564→600	on→off	Mitochondria live	Petrat, de Groot et al. 2001
N-oxide Rhonox-1	555→575	off→on	lysosomes live, fixed	Halcrow, Kumar et al. 2022, Hirayama 2024
HMRhoNox-M	555→575	off→on	lysosomes live, fixed	Halcrow, Kumar et al. 2022, Hirayama 2024
CoNox-1	405→495	off→on	Cells live	Hirayama 2024
FluNox-1	488→530	off→on	Cells live	Hirayama 2024
SiRhoNox-1	630→660	off→on	Cells live	Xie, Jiang et al. 2023
FIP-1	495→545/517	on→off (ratio)	Cells live	Au-Yeung, Chan et al. 2013
Mito-RhFe	543→575	off→on	Mitochondria live	Zhu, Wang et al. 2018
Gol-SiRhoNox	655→665	off→on	Golgi	Hirayama, Inden et al 2019

The various probes denoted by their acronyms, as per the indicated references were used for tracing either Fe (2^+ or 3^+) as changes in fluorescence signals, either normal on- > off or off- > mode and where applicable, in ratiometric mode. Adittonal references to other probes for labile iron: Chen et al. (2006), Carter et al. (2014), Ma et al. (2015), Abbate et al. (2016), Ackerman et al. (2017), Gao et al. (2020), Xie et al. (2023), Wu et al. (2023), Che et al. (2024)

ferroptosis, traced to mitochondria, has recently been proposed to play a pivotal role in generating a deadly ferroptotic-like cascade (Li et al. 2023).

4.3 Tracing LIP Levels in Living Cells

The monitoring of LIPs in living cells has traditionally relied on fluorescent metal sensors that interact with components of the intracellular LIPs. The interaction with the labile metal should evoke a fast and measurable change in the signal which ideally should be reversible or preventable by the addition of an iron-specific, high-affinity chelator (Fig. 4.1). The latter was essential to ascertain that signal changes genuinely reflected the interaction of the probe with Fe^{2+} or Fe^{3+}, and not a photophysical or photochemical effect on the probe or an interaction with another metal. Thus, changes in the fluorescence signal ΔF (generally quenching) are interpreted as resulting from interaction with LIP components, provided those are demonstrably susceptible to quantitative reversal (or prevention), by competition with a high-affinity chelator.

Cellular LIP was initially traced in intact cells with metal-sensitive probes, primarily those endowed with a fluorescent tag attached to an Fe^{3+}- or Fe^{2+}-chelating moiety comprised of an array of coordinating groups such as hydroxamate, catecholate, hydroxypyridinone, or carboxylate. The binding of the labile iron is meant to elicit a commensurate and traceable change in the fluorescence signal (Espósito et al. 2002; Petrat et al. 2002; Ma et al. 2006; Cabantchik 2014). However, for live-cell measurements, the iron-sensing probe ought to be loadable into cells or particular organelles and stably retained therein for experimentation. Calceins, either as green fluorescein- or bluish-umbeliferyl-based fluorophores containing chelating ethylene diamino-carboxylate groups, were not originally designed for tracing labile iron. However, they were found to be highly sensitive to Fe^{3+} or Fe^{2+} in aerobic conditions irrespective of Ca^{2+} or Mg^{2+} and highly versatile for live-cell tracing of LIP without interfering with cell processes (Williams 1982).[2] A recent modification of calcein has led to the application of a novel fluorescent probe that is traceable in both ratiometric and turn-on modes, two features that confer reliability and sensitivity to biological monitoring of labile iron (Carney et al. 2018).

More recent developments are associated with the advent of signal off-on probes that, in the presence of LIP, either Fe^{2+} or Fe^{3+} in a given cell compartment, can trigger a local reaction that renders the probe itself (or a local protein) fluorescent or luminescent (Domaille et al. 2008; Lindahl and Holmes-Hampton 2011; Ackerman et al. 2017; Aron et al. 2018; Wu et al. 2023; Che et al. 2024; Silswal et al. 2023; Hirayama 2024). In addition, a genetically encoded probe for real-time monitoring and detection of iron has also been recently introduced (Soleja and Mohsin 2024). The advent of those novel analytical probes for in situ detection of labile iron in whole cells and in different cell compartments, some even in ratiometric mode (Table 4.1), has opened new opportunities not only for associating labile iron in the pathogenesis of various human disorders but also for developing LIP-targeted treatments aimed at changing the course of disease, as the following section exemplifies.

[2]The fluorescent metal sensor Calcein can be swiftly generated in the cytosol of living cells by loading it as the permeable (non-fluorescent) precursor calcein-acetomethoxy-ester that, once swiftly hydrolyzed by resident esterases, releases the polyanionic fluorescent metal sensor, thus facilitating the monitoring of LIP over time. The levels of cell LIP are revealed by scavenging the calcein-bound **iron** with a strong and permeable iron chelator. Verification that changes in fluorescence are associated with LIP and no leakage or bleaching of the probe or interactions with other metals is validated if the original fluorescence (before quenching) is regained following chelator treatment. The calcein-LIP methodology has allowed the estimation of LIP levels in basal conditions in healthy and pathological conditions such as in systemic or regional siderosis, often manifested as ferroptosis. Real-time changes in cytosolic LIP allow the monitoring of cell iron ingress upon exposure of cells to either TBI (transferrin-bound iron) or to NTBI species, but also the screening of iron chelators. Other fluorescent probes endowed with higher iron specificity and binding affinity have obvious advantages, provided they are proven to spare iron scavenging from resident proteins.

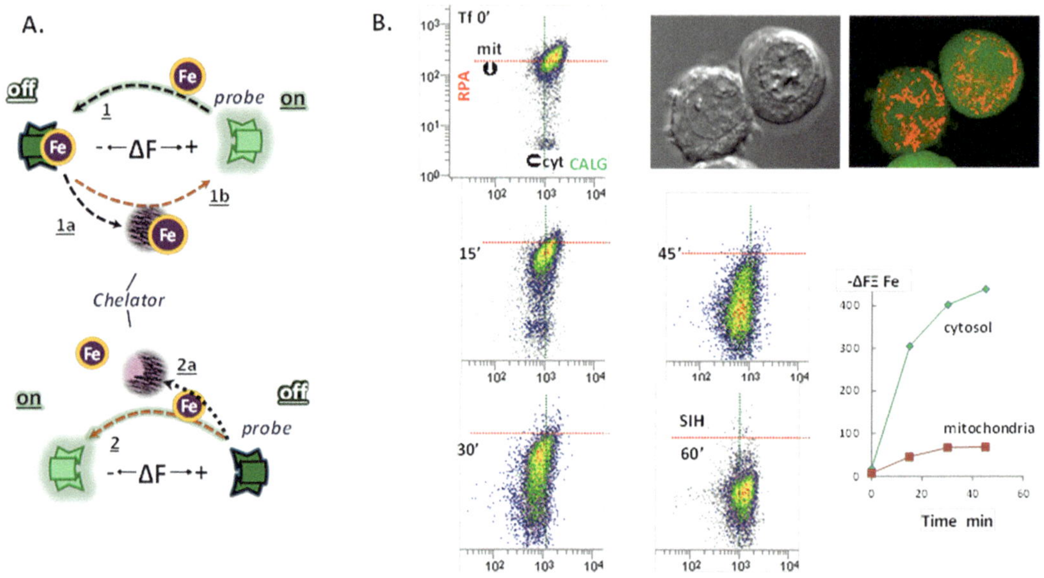

Fig. 4.1 Principles of detection of labile iron in biological systems using fluorescent "turn-on" and "turn-off" iron sensors (see also Table 4.1)(**a**) Metal detection mode: *Left-*. A fluorescent probe endowed with a metal chelating moiety (on) binds iron stoichiometrically undergoes a signal quenching (ΔF − = off) (1) while addition of a chelator to the quenched probe (scavenges the bound metal leading to restoration of the original signal—back "on"(1a)*Right*. A fluorogenic metal sensitive probe becomes fluorescent (ΔF+ = on) (2) either by binding labile iron stoichiometrically or via a chemical reaction triggered by labile iron directly or catalytically via intermediatorsPractical aspects of different approaches used for tracing iron in biological systems can be found in the Intern. BioIron Society website https://bioiron.org/meetings/past-meetings/2019-presentations.aspx under EMBL COURSE Techniques for Studying Iron in Health and Disease(**b**) Flow cytometry tracing of K562 cells (phase and fluoro-micrographs shown on top right) double labelled with the fluorescent calcein-green (CALG) (cytosol) and red RPA (mitochondria), following exposure to transferrin (Tf). The simultaneous ingress of iron into the two compartments was reflected as time-dependent changes in the fluorescence intensity **(F)** of the probes in both compartments, shown both as fluorescence intensity maps taken at different times following the addition of Tf (time zero), from where the graph of changes in fluorescence intensity **(ΔF)** vs. time was constructed. The cytosolic -permeant chelator SIH (salicyl-isonicotinoyl-hydrazone) was added to scavenge the CALG bound Fe (CALG-Fe) and restore the original fluorescence (adapted from Shvartsman and Cabantchik 2012)

4.4 Mitochondrial Labile Iron in Diseases of Deranged Fe-S Cluster (ISC) Protein Biogenesis: The Case of Wolfram Syndrome 2

A rise in mitochondrial labile iron and ensuing oxidative damage with both morphological changes and functional consequences have been recorded in cells from iron-overloaded patients and also in a variety of normal cells exposed to iron sources that gain access to cells by non-regulated pathways (Cabantchik et al. 2013b). More recent findings indicate that an analogous rise in mitochondrial LIP, with pathological consequences, can result from disruptions in the ISC protein biogenetic machinery (Lill and Freibert 2020; Maio and Rouault 2022), as found in human diseases such as Friedreich's ataxia (Chen et al. 2006; Che et al. 2024), X-linked sideroblastic anemia with ataxia, and a form of sideroblastic anemia associated with a deletion in the *GLRX5* gene (Ye et al. 2010). The rise in mitochondrial LIP is allegedly derived from an aborted transfer of the ISC to a missing or mutated apo-ISC acceptor, leading to cluster disintegration and the uncontrolled release of labile iron with potentially toxic consequences. However, the major pathological effects in these disorders are most likely linked to the cells' limited ability to cope with a scarcity in a gamut of life-essential ISC proteins, caused by ruptures in their biogenesis. Regarding mutations in the human *CISD2* gene implicated in WS2 pathology (Amr et al. 2007; Chen et al. 2009, 2010; Conlan et al. 2009; Rigoli and Di Bella 2012; Wu et al. 2012; Wiley et al. 2013; Urano 2016; Delprat et al. 2018; Delvecchio et al. 2021; Liao et al. 2021; Loncke et al. 2021; Shen et al. 2021), the prevalent ones are those that affect expression levels of the ISC protein NAF-1 (Nutrient-deprivation autophagy factor-1), a member of the NEET family of 2Fe-2S ISC proteins (Karmi et al. 2018; Nechushtai et al. 2020). Although those mutations apparently do not affect overall ISC protein biogenesis, the mere lack of expression of NAF-1 imparts oxidative damage, with structural and functional consequences for both mitochondrial and ER (Amr et al. 2007; Chen et al. 2009, 2010; Conlan et al. 2009; Rigoli and Di Bella 2012; Wu et al. 2012; Wiley et al. 2013; Mozzillo et al. 2014; Urano 2016; Delprat et al. 2018; Nechushtai et al. 2020; Delvecchio et al. 2021; Karmi et al. 2021a, 2022a; Loncke et al. 2021; Shen et al. 2021; Rosanio et al. 2022).

With the aim of elucidating the mode whereby a deficiency in NAF-1 leads to WS2, we inquired:

1. To what extent do the abnormalities that result from silencing *CISD2* expression of NAF-1 facilitate a rise in mitochondrial LIP and ensuing oxidative damage, analogously to the abnormalities caused by mutations in components of the ISC protein biosynthetic machinery (Nechushtai et al. 2020; Karmi et al. 2021a, 2022b). We initially suspected that apo-NAF-1 binding of the ISC is of a labile nature, in analogy with what has been obtained with the other members of the NEET family whose ISC binding domain is also comprised of three *Cys* and one *His* residues (Karmi et al. 2018, Nechushtai et al. 2020). On that basis, we also reckoned that the putative labile features of ISC binding might subserve physiologically a more facile 2Fe-2S ISC transfer to a receptive apoprotein acceptor, a property subsequently verified experimentally (Ferecatu et al. 2014; Lipper et al. 2015; Karmi et al. 2017; Nechushtai et al. 2020). However, the downside of facile transfer from a putative ISC donor, in the presence of insufficient acceptor, may be the premature release and disintegration of the cluster and a rise in mitochondrial LIP.
2. Whether the presumed rise in mitochondrial LIP of NAF-1 deficient cells results in oxidative damage and an ensuing drop in GSH levels and thereby, both combinedly, prompt a sequence of events characteristic of ferroptosis (Karmi et al. 2021a),
3. Whether the phenotypic changes in genetically manipulated cells faithfully recapitulate some of the basic features observed in skin fibroblasts derived from WS2 patients, which have no detectable levels of NAF-1 (mRNA or protein) (Karmi and Nechushtai 2019),
4. Whether the phenotypic changes in NAF-1 deficiency, including the putative ferroptosis, can be ameliorated by pharmacological measures suitable for clinical application on WS2 patients (Karmi et al. 2021a),

The set goals were of scientific and clinical relevance for several reasons:

Homozygous mutations in the *CISD2* gene, which codes mutated forms of NAF, have been identified in WS2 patients, a rare autosomal recessive disorder clinically characterized by a plethora of pathological changes and morbidities that lead to early death. Among the first verified patients were three Jordanian individuals (Amr et al. 2007), and subsequently, members of the Palestinian population in the Bethlehem area, where the disease mostly prevails. In those patients, the identified mutations seemingly preclude the expression of the mature NAF-1 protein, as found in skin fibroblasts isolated from various affected individuals (Karmi and Nechushtai 2019).

4.4.1 Phenotypic Consequences of NAF-1 Repression

Most studies aimed at understanding the biological consequences of NAF-1 repression were carried out either with established lines of cells grown in culture (Nechushtai et al. 2020; Karmi et al. 2021a) or with skin fibroblasts derived from local Palestinian WS2 patients and maintained in culture conditions (Karmi and Nechushtai 2019).

Biochemical Studies The shRNA-driven repression of NAF-1 (*CISD2*) in mammalian cells results in increased levels of mitochondrial LIP and increased ROS formation, both of which jointly implicated in the generation of mitochondrial damage of structural and functional nature (Fig. 4.2). The mitochondrial morphological changes comprise cristae deformations, increased fragmentation, and enhanced fission, while multiple changes are observed in reduced OXPHOS and ATP production, diminished lipid β-oxidation, and decreased mitochondrial membrane potential (Shen et al. 2021).

The fact that the joint application of a permeant iron chelator, DFP, and the GSH precursor N-acetyl cysteine (NAC) markedly reduces mitochondrial LIP and ROS formation and partially reverses the above parameters indicative of structural and functional damage causatively associates LPI and GSH as drivers of ferroptosis, let alone that similar results were obtained with the ferroptosis inhibitor ferrostatin (Karmi et al. 2021a). In line with the above changes resulting from the repression of NAF-1 is the concomitant downregulation of thioredoxin-interacting protein (TXNIP) expression, which, besides affecting a wide range of metabolic activities, promotes ferroptosis (Karmi et al. 2021a, b).

Physiological Studies A characteristic feature of WS2 is a progressive insulin deficiency. We also assessed the impact of repressing NAF-1 expression on both biochemical and physiological parameters in insulinoma-derived INS-1E pancreatic β-cells that were stably transfected with an NAF-1 shRNA. As in previous cell models, the INS-1E cells deficient in NAF-1 recapitulated the previously described abnormalities, i.e., (a) elevated levels of LIP, ROS, and indicators of lipid peroxidation, resulting in deformed mitochondria that interact abnormally with components of the endoplasmic reticulum; and (b) decreased levels of glutathione and glutathione peroxidase 4, implicating ferroptosis in the pathology of NAF-1 deficiency. However, more physiologically relevant was the finding that the suppression of NAF expression significantly impaired the ability of the INS1E cells to respond to glucose-stimulated insulin secretion in a manner sensitive to LIP and ROS. The fact that the combined DFP-NAC treatment mitigated the rise in LIP and ROS and significantly restored glucose-induced insulin secretion opens the road for clinical testing of the treatment in WS2 patients. The fact that those drugs are already in clinical use should facilitate their joint application in a modality that combines conservative chelation with DFP with antioxidant properties of NAC (Karmi et al. 2021a).[3] The *CISD2* gene product NAF-1 has been implicated in longevity by virtue of its proposed role in regulating cytosolic Ca^{2+} homeostasis (Shen et al. 2021).

4.5 Animal Studies with *CISD2* KO Mice

An early study using engineered *CISD2* KO mice, which showed no detectable expression of NAF-1, described animals with signs of premature aging and a shortened life span (Chen et al. 2009; Conlan et al. 2009; Shen et al. 2021; Karmi et al. 2022b). External features of the mutant mice included protruding eyes and ears, along with fur showing decreased hair pigmentation and density, accompanied by signs of follicular atrophy, epidermal hyperplasia, increased dermis thickness, and reduced subcutaneous fat and muscle mass. Degenerative signs were found in the optic nerve and muscle, including osteopenia and an age-dependent decline in metabolic markers, including a progressive change in diabetic indicators such as impaired glucose intolerance and decreased glucose-stimulated insulin secretion, a salient feature in WS2 patients. The biological role associated with NAF-1 involves Ca signaling and homeostasis, as those properties were disrupted in embryonic fibroblasts and in

[3] By conservative chelation, we refer to a treatment that does not lead to body iron loss, let alone to iron deficiency.

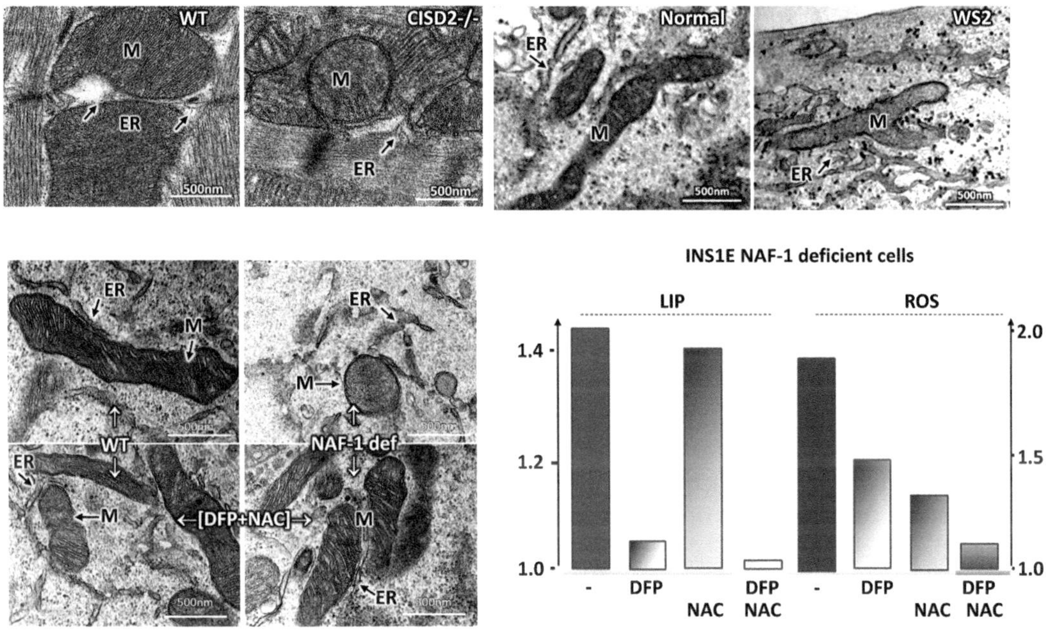

Fig. 4.2 Phenotypic features of NAF-1 deficient cells***Upper left***. Electron microscopy (EM) images of heart tissue isolated from wild type (**WT**, left) mice and ***CISD2(−/−)*** KO mice highlighting differences in **ER** and mitochondria **M** structures (adapted from Karmi et al. 2022b). ***Upper right***. EM images of skin fibroblasts isolated from a Palestinian WS2 patient indicating changes in in **ER** and mitochondria **M** structures (Karmi and Nechushtai 2019)***Lower left***. EM images from INS1E insulinoma cells, wild type (**WT**, left) = normal and **NAF-1-deficient** cells (right) with arrows indicating ER contacts with mitochondria M, before and after treatment with the iron chelator deferiprone (DFP) combined with N-acetyl-cysteine (NAC), as indicated elsewhere (Karmi et al. 2021a). ***Lower right***. Changes in the labile iron pool (LIP) and basal ROS formation of NAF-1 deficient INS1E cells relative to INS1E WT cells

cells isolated from skeletal and cardiac muscles of *CISD2* KO mice (Chen et al. 2009; Conlan et al. 2009; Shen et al. 2021). Those cells show depleted Ca stores in the ER and increased levels of Ca in mitochondria, leading to increased ratios of $NAD^+/NADH$ and ADP/ATP ratios and reduced levels of GSH (Shen et al. 2021). In an analogous *CISD2* KO mouse model (Karmi et al. 2022b), cardiomyocytes displayed both ER and mitochondrial abnormalities, but also deranged homeostasis of iron, which was manifested as high metal accumulation and high levels of expression of the ferritin H-subunit but also of TfR1.

Interesting to note is that mice deleted of the mitoNEET-encoding gene *CISD1* had abnormal movement capacity; this was probably linked to observed changes in the striatum that displayed reduced respiratory activity and ATP production and abnormal mitochondria structures with labile iron and increased ROS formation (Furihata et al. 2021).

4.5.1 Clinical Applications on WS2 Patients—Preliminary Results

WS-2 is a rare, fatal, multisystem neurodegenerative and β-cell degenerative disorder of monogenic autosomal recessive inheritance, caused by mutations in the *CISD2* gene. Although worldwide rare, a WS-2-linked mutation (*CISD2*: G109->C) was attributed to a founder effect in areas neighboring Jerusalem (but within the Palestinian Authority) that showed a carrier frequency rate of 1 in 40 (2.5%).[4] WFS-2 patients

[4]Reported by Prof. D. Zangen et al., Pediatric Endocrinology, Hadassah Medical Center, Mt. Scopus, Jerusalem, Israel (in preparation).

suffer from an early-onset insulin deficiency that progressively leads to juvenile diabetes mellitus, gastrointestinal ulcerations, abnormal platelet aggregation, sensorineural hearing loss, and optical nerve atrophy, often followed by episodes of psychotic behavior (Amr et al. 2007; Karmi and Nechushtai 2019).

A preliminary study with four patients subjected to a combined DFP + NAC daily therapy for a period of 3 months indicated significant improvements in the quality of life, frequency of hypoglycemic episodes, platelet aggregability, and HbA1c levels. However, the complexities inherent in the areas where the syndrome prevails have hampered both patients' access to qualified medical centers and the ability of medical teams to reach the affected patients.

4.5.2 A Working Model for the Source of Elevated Mitochondrial LIP and ROS in NAF-1 Deficiency

In the second part of the review, we covered pathological changes resulting from excessive mitochondrial accumulation of labile iron identified in cells deficient in the ISC protein NAF-1. The elevated LIP was found in the mitochondria of skin fibroblasts isolated from WS2 patients who have no detectable expression of the mutated *CISD2* gene that codes for NAF-1, as well as in mitochondria of cells with shRNA-mediated silencing of *CISD2* expression. The observed rise in mitochondrial LIP was tentatively attributed to a putative release of labile iron from the disintegration of ISCs due to the failure of the ISC normally present in NAF-1 to transfer the cluster to potential acceptors that seemingly interact with NAF-1 (Nechushtai et al. 2020). As the primary location of NAF-1 is in the ER outer membrane, the questions that arise are from which ISC donor does apo-NAF-1 normally gets its ISC, and to which acceptor does it transfer the ISC. Based on early observations, NAF-1 was found to interact with the two other ISC members of the same *CISD* gene-NEET protein family, directly with mitoNEET (Karmi et al. 2017), which seemingly gets its ISC from MiNT (Karmi et al. 2022a) (Fig. 4.3). The three NEET proteins share a similar ISC binding domain characterized by an intrinsic ability to transfer the ISC between them or to other ISC acceptors such as anamorsin, a protein required for ISC assembly in the CIA (Lipper et al. 2015), or mitoNEET to IRP1 (Ferecatu et al. 2014). Based on the cellular location of the three NEET proteins - MiNT inside mitochondria, mitoNEET on the outer mitochondrial membrane (OMM), and NAF-1 that resides mainly in the ER membrane that also interacts with mitochondria via the MAM, one can envision a traffic of ISC from within mitochondria towards NAF-1. However, in the absence of NAF-1, some ISC apparently goes astray and disintegrates, releasing labile iron and setting a sequela of oxidative events and ensuing damage. The fact that the structure-function abnormalities found in the NAF-1-deficient phenotype can be partly mitigated by reducing the levels of mitochondrial labile iron with a mild chelator such as DFP, but even further improved when combined with the GSH precursor NAC, indicates (Karmi et al. 2021a): (a) the involvement of deranged LIP and associated ROS in the etiopathology of NAF-1 deficiency, possibly associated with ferroptosis; and (b) the possibility of pharmacologically addressing the symptoms without affecting other iron-dependent functions, including those associated with ISC proteins, whose biogenesis is apparently not affected by NAF-1 deficiency. The precise role of NAF-1 in normal iron physiology, however, remains to be elucidated. Interestingly, note that most of the phenotypic changes evoked in cells by NAF-1 repression can be recapitulated by analogous repression of mitoNEET (Karmi et al. 2017). Moreover, re-expression of *CISD1* appears to protect mitochondria from a potential toxicity associated with a rise in labile iron in cells defective in ISC transfer between NEET proteins.

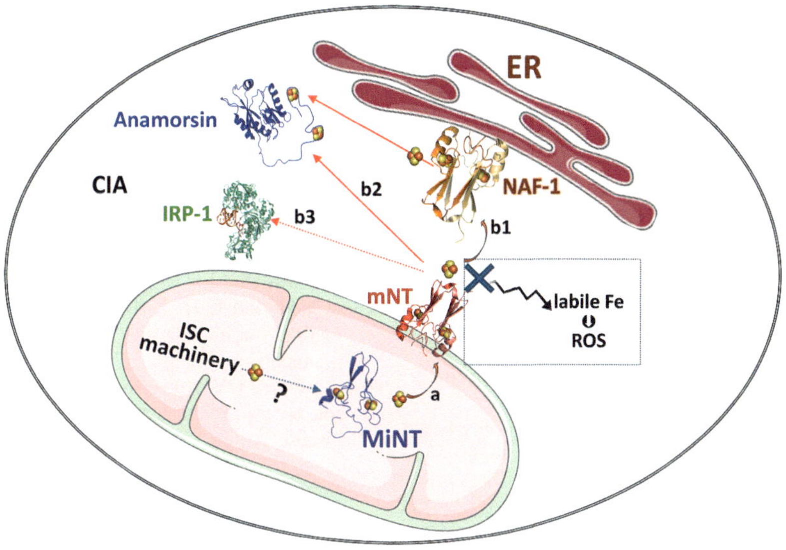

Fig. 4.3 A working model that attempts to explain the source of increased LIP and ROS in NAF-1-deficient cell-sThe model was based on identified cell sites of NEET proteins and their interactions in terms of ISC transfer abilities: (**a**) Adapted from **MiNT** in the mitochondrial matrix **M** to the mitochondrial outer membrane **mNT** (Karmi et al. 2022a); (**b**) Adapted from **mNT** to either (b1)- the ER-located **NAF-1** (Karmi et al. 2017))- that, in turn, might transfer the ISC to the CIA-machinery via anamorsin (Lipper et al. 2015) or (b2)- anamorsin or (**b3**)- IRP1 (Ferecatu et al. 2014). Failure to transfer the ISC to NAF-1, is assumed to lead to rise in LIP and ROS due to chemical decomposition of the ISC components and release of labile iron. (inset)

4.6 Conclusions

Although cellular LIPs have been tentatively associated with multiple ligands, predominantly as Fe^{2+}-GSH adducts, their chemical identity remains to be firmly established in bona fide native conditions. In most fluorescence-based studies, the normal cellular LIP levels are estimated to be in the low μMolar range, rising substantially under iron overload and in multiple pathological conditions, some of which are increasingly identified with ferroptosis. Monitoring of cytosolic and mitochondrial LIPs has facilitated the development of chelators with the requisite efficacy and accessibility to neutralize intracellular labile iron targets and ensuing ROS formation.

Among the diseases that show pathological elevation of LIPs are those with a component of local siderosis, classically found in cases of aborted biogenesis of ISC proteins. A special case of local siderosis and the ensuing oxidative damage is addressed in this review in relation to WS2 patients who lack NAF-1, an ISC protein of the NEET family encoded by a mutated *CISD2* gene. WS2 is a rare fatal disease with multi-manifestation that we found to prevail among Palestinians of the Bethlehem region. Based on experimentation with cell and animal models, as well as clinical specimens, we proposed a mechanistic model that attempts to explain how an aborted protein-protein ISC transfer, from the intramitochondrial *CISD3*-coded MiNT, via the OMM *CISD1*-coded mitoNEET (mNT), to the missing NAF-1, coded by a mutated *CISD2*, leads to ISC disintegration and a rise in LIP, promoting ROS formation and ensuing damage, partially attributed to ferroptosis. The fact that both structural and functional damages in NAF-1 deficiency can be significantly corrected by applying a conservative form of chelation in conjunction with antioxidant measures, both of which are already used independently in the clinic, offers a novel opportunity for the treatment of the hitherto incurable WS2.

References

Abbate V, Reelfs O, Kong X, Pourzand C, Hider RC (2016) Dual selective iron chelating probes with a potential to monitor mitochondrial labile iron pools. Chem Commun 52(4):784–787

Ackerman CM, Lee S, Chang CJ (2017) Analytical methods for imaging metals in biology: from transition metal metabolism to transition metal signaling. Anal Chem 89(1):22–41

Amr S, Heisey C, Zhang M, Xia X-J, Shows KH, Ajlouni K, Pandya A, Satin LS, El-Shanti H, Shiang R (2007) A homozygous mutation in a novel zinc-finger protein, ERIS, is responsible for Wolfram syndrome 2. Am J Hum Genet 81(4):673–683

Aron AT, Reeves AG, Chang CJ (2018) Activity-based sensing fluorescent probes for iron in biological systems. Curr Opin Chem Biol 43:113–118

Au-Yeung HY, Chan J, Chantarojsiri T, Chang CJ (2013) Molecular imaging of labile iron (II) pools in living cells with a turn-on fluorescent probe. J Am Chem Soc 135(40):15165–15173

Brawley HN, Lindahl PA (2021) Low-molecular-mass labile metal pools in Escherichia coli: advances using chromatography and mass spectrometry. JBIC J Biol Inorg Chem 26(4):479–494

Brawley HN, Kreinbrink AC, Hierholzer JD, Vali SW, Lindahl PA (2023) Labile iron pool of isolated Escherichia coli cytosol likely includes Fe-ATP and Fe-Citrate but not Fe-Glutathione or aqueous Fe. J Am Chem Soc 145(4):2104–2117

Breuer W, Epsztejn S, Ioav Cabantchik Z (1996) Dynamics of the cytosolic chelatable iron pool of K562 cells. FEBS Lett 382(3):304–308

Breuer W, Hershko C, Cabantchik Z (2000) The importance of non-transferrin bound iron in disorders of iron metabolism. Transfus Sci 23(3):185–192

Breuer W, Shvartsman M, Cabantchik ZI (2008) Intracellular labile iron. Int J Biochem Cell Biol 40(3):350–354

Cabantchik ZI (2014) Labile iron in cells and body fluids: physiology, pathology, and pharmacology. Front Pharmacol 5:45

Cabantchik ZI, Munnich A, Youdim MB, Devos D (2013a) Regional siderosis: a new challenge for iron chelation therapy. Front Pharmacol 4:167

Cabantchik ZI, Sohn YS, Breuer W, Espósito BP (2013b) The molecular and cellular basis of iron toxicity in iron overload disorders. Diagnostic and therapeutic approaches. Thalassemia Rep 3(s1):e3

Camarena V, Huff TC, Wang G (2021) Epigenomic regulation by labile iron. Free Radic Biol Med 170:44–49

Carney IJ, Kolanowski JL, Lim Z, Chekroun B, Torrisi AG, Hambley TW, New EJ (2018) A ratiometric iron probe enables investigation of iron distribution within tumour spheroids. Metallomics 10(4):553–556

Carter KP, Young AM, Palmer AE (2014) Fluorescent sensors for measuring metal ions in living systems. Chem Rev 114(8):4564–4601

Che Y, Yang J, Dong Z, Wang J, Yan X, Wang Y, Shuang S (2024) A sensitive "turn-on" Schiff-base fluorescent probe for the selective detection of Fe3+ and bio-imaging. Spectrochim Acta A Mol Biomol Spectrosc 308:123799

Chen J-L, Zhuo S-J, Wu Y-Q, Fang F, Li L, Zhu C-Q (2006) High selective determination iron (II) by its enhancement effect on the fluorescence of pyrene-tetramethylpiperidinyl (TEMPO) as a spin fluorescence probe. Spectrochim Acta A Mol Biomol Spectrosc 63(2):438–443

Chen Y-F, Kao C-H, Kirby R, Tsai T-F (2009) Cisd2 mediates mitochondrial integrity and life span in mammals. Autophagy 5(7):1043–1045

Chen YF, Wu CY, Kirby R, Kao CH, Tsai TF (2010) A role for the CISD2 gene in lifespan control and human disease. Ann N Y Acad Sci 1201(1):58–64

Condeles AL, da Silva GS, Hernandes MBB, Toledo Junior JC (2024) Insights on the endogenous labile iron pool binding properties. In: Biometals, pp 1–13

Conlan AR, Axelrod HL, Cohen AE, Abresch EC, Zuris J, Yee D, Nechushtai R, Jennings PA, Paddock ML (2009) Crystal structure of Miner1: the redox-active 2Fe-2S protein causative in Wolfram syndrome 2. J Mol Biol 392(1):143–153

Delprat B, Maurice T, Delettre C (2018) Wolfram syndrome: MAMs' connection? Cell Death Dis 9(3):364

Delvecchio M, Iacoviello M, Pantaleo A, Resta N (2021) Clinical spectrum associated with Wolfram syndrome type 1 and type 2: a review on genotype–phenotype correlations. Int J Environ Res Public Health 18(9):4796

Dixon SJ, Pratt DA (2023) Ferroptosis: a flexible constellation of related biochemical mechanisms. Mol Cell 83(7):1030–1042

Domaille DW, Que EL, Chang CJ (2008) Synthetic fluorescent sensors for studying the cell biology of metals. Nat Chem Biol 4(3):168–175

Egyed' A, Saltman P (1984) Iron is maintained as Fe (II) under aerobic conditions in erythroid cells. Biol Trace Elem Res 6:357–364

Espósito BP, Epsztejn S, Breuer W, Cabantchik ZI (2002) A review of fluorescence methods for assessing labile iron in cells and biological fluids. Anal Biochem 304(1):1–18

Ferecatu I, Gonçalves S, Golinelli-Cohen M-P, Clémancey M, Martelli A, Riquier S, Guittet E, Latour J-M, Puccio H, Drapier J-C (2014) The diabetes drug target MitoNEET governs a novel trafficking pathway to rebuild an Fe-S cluster into cytosolic aconitase/iron regulatory protein 1. J Biol Chem 289(41):28070–28086

Furihata T, Takada S, Kakutani N, Maekawa S, Tsuda M, Matsumoto J, Mizushima W, Fukushima A, Yokota T, Enzan N (2021) Cardiac-specific loss of mitoNEET expression is linked with age-related heart failure. Commun Biol 4(1):138

Gao J, Chen Y, Guo Z, He W (2020) Recent endeavors on molecular imaging for mapping metals in biology. Biophys Rep 6(5):159–178

Garbowski MW, Cabantchik I, Hershko C, Hider R, Porter JB (2023) The clinical relevance of detectable plasma iron species in iron overload states and subsequent to intravenous iron-carbohydrate administration. Am J Hematol 98(3):533–540

Greenberg GR, Wintrobe MM (1946) A labile iron pool. J Biol Chem 165(1):397

Halcrow PW, Kumar N, Afghah Z, Fischer JP, Khan N, Chen X, Meucci O, Geiger JD (2022) Heterogeneity of ferrous iron-containing endolysosomes and effects of endolysosome iron on endolysosome numbers, sizes, and localization patterns. J Neurochem 161(1):69–83

Hider RC, Kong XL (2011) Glutathione: a key component of the cytoplasmic labile iron pool. Biometals 24:1179–1187

Hider RC, Kong X (2013) Iron speciation in the cytosol: an overview. Dalton Trans 42(9):3220–3229

Hider R, Aviles MV, Chen YL, Latunde-Dada GO (2021) The role of GSH in intracellular iron trafficking. Int J Mol Sci 22(3):1278

Hirayama T (2019) Fluorescent probes for the detection of catalytic Fe (II) ion. Free Radic Biol Med 133:38–45

Hirayama Y (2024) Development and application of fluorescent probes for iron (II) ions and free heme. Seibutsubutsuri 64(1):12–16

Hirayama T, Inden M, Tsuboi H, Niwa M, Uchida Y, Naka Y, Hozumi I, Nagasawa H (2019) A Golgi-targeting fluorescent probe for labile Fe (ii) to reveal an abnormal cellular iron distribution induced by dysfunction of VPS35. Chem Sci 10(5):1514–1521

Jacobs A (1976) An intracellular transit iron pool. Ciba Found Symp(51):91–106

Jiang X, Stockwell BR, Conrad M (2021) Ferroptosis: mechanisms, biology and role in disease. Nat Rev Mol Cell Biol 22(4):266–282

Kakhlon O, Cabantchik ZI (2002) The labile iron pool: characterization, measurement, and participation in cellular processes. Free Radic Biol Med 33(8):1037–1046

Karmi O, Nechushtai R (2019) NEET proteins in Health and Disease, PhD thesis

Karmi O, Holt SH, Song L, Tamir S, Luo Y, Bai F, Adenwalla A, Darash-Yahana M, Sohn Y-S, Jennings PA (2017) Interactions between mitoNEET and NAF-1 in cells. PLoS One 12(4):e0175796

Karmi O, Marjault H-B, Pesce L, Carloni P, Onuchic JN, Jennings PA, Mittler R, Nechushtai R (2018) The unique fold and lability of the [2Fe-2S] clusters of NEET proteins mediate their key functions in health and disease. JBIC J Biol Inorg Chem 23(4):599–612

Karmi O, Sohn Y-S, Marjault H-B, Israeli T, Leibowitz G, Ioannidis K, Nahmias Y, Mittler R, Cabantchik IZ, Nechushtai R (2021a) A combined drug treatment that reduces mitochondrial iron and reactive oxygen levels recovers insulin secretion in NAF-1-deficient pancreatic cells. Antioxidants 10(8):1160

Karmi O, Sohn Y-S, Zandalinas SI, Rowland L, King SD, Nechushtai R, Mittler R (2021b) Disrupting CISD2 function in cancer cells primarily impacts mitochondrial labile iron levels and triggers TXNIP expression. Free Radic Biol Med 176:92–104

Karmi O, Marjault H-B, Bai F, Roy S, Sohn Y-S, Darash Yahana M, Morcos F, Ioannidis K, Nahmias Y, Jennings PA (2022a) A VDAC1-mediated NEET protein chain transfers [2Fe-2S] clusters between the mitochondria and the cytosol and impacts mitochondrial dynamics. Proc Natl Acad Sci 119(7):e2121491119

Karmi O, Rowland L, King SD, Manrique-Acevedo C, Cabantchik IZ, Nechushtai R, Mittler R (2022b) The [2Fe-2S] protein CISD2 plays a key role in preventing iron accumulation in cardiomyocytes. FEBS Lett 596(6):747–761

Katsarou A, Pantopoulos K (2020) Basics and principles of cellular and systemic iron homeostasis. Mol Asp Med 75:100866

Koppenol WH, Hider RH (2019) Iron and redox cycling. Do's and don'ts. Free Radic Biol Med 133:3–10

Kruszewski M (2003) Labile iron pool: the main determinant of cellular response to oxidative stress. Mutat Res 531(1–2):81–92

Li Y, Ouyang Q, Chen W, Liu K, Zhang B, Yao J, Zhang S, Ding J, Cong M, Xu A (2023) An iron-dependent form of non-canonical ferroptosis induced by labile iron. Sci China Life Sci 66(3):516–527

Liao H-Y, Liao B, Zhang H-H (2021) CISD2 plays a role in age-related diseases and cancer. Biomed Pharmacother 138:111472

Lill R, Freibert S-A (2020) Mechanisms of mitochondrial iron-sulfur protein biogenesis. Annu Rev Biochem 89(1):471–499

Lindahl PA, Holmes-Hampton GP (2011) Biophysical probes of iron metabolism in cells and organelles. Curr Opin Chem Biol 15(2):342–346

Lindahl PA, Moore MJ (2016) Labile low-molecular-mass metal complexes in mitochondria: trials and tribulations of a burgeoning field. Biochemistry 55(30):4140–4153

Lipper CH, Paddock ML, Onuchic JN, Mittler R, Nechushtai R, Jennings PA (2015) Cancer-related NEET proteins transfer 2Fe-2S clusters to anamorsin, a protein required for cytosolic iron-sulfur cluster biogenesis. PLoS One 10(10):e0139699

Loncke J, Vervliet T, Parys JB, Kaasik A, Bultynck G (2021) Uniting the divergent Wolfram syndrome–linked proteins WFS1 and CISD2 as modulators of Ca2+ signaling. Sci Signal 14(702):eabc6165

Lv H, Shang P (2018) The significance, trafficking and determination of labile iron in cytosol, mitochondria and lysosomes. Metallomics 10(7):899–916

Ma Y, de Groot H, Liu Z, Hider RC, Petrat F (2006) Chelation and determination of labile iron in primary hepatocytes by pyridinone fluorescent probes. Biochem J 395(1):49–55

Ma Y, Abbate V, Hider RC (2015) Iron-sensitive fluorescent probes: monitoring intracellular iron pools. Metallomics 7(2):212–222

Maio N, Rouault TA (2022) Mammalian iron sulfur cluster biogenesis: from assembly to delivery to recipient proteins with a focus on novel targets of the chaperone and co-chaperone proteins. IUBMB Life 74(7):684

Mozzillo E, Delvecchio M, Carella M, Grandone E, Palumbo P, Salina A, Aloi C, Buono P, Izzo A, D'Annunzio G (2014) A novel CISD2 intragenic deletion, optic neuropathy and platelet aggregation defect in Wolfram syndrome type 2. BMC Med Genet 15:1–7

Nechushtai R, Karmi O, Zuo K, Marjault H-B, Darash-Yahana M, Sohn Y-S, King SD, Zandalinas SI, Carloni P, Mittler R (2020) The balancing act of NEET proteins: iron, ROS, calcium and metabolism. Biochim Biophys Acta Mol Cell Res 1867(11):118805

Niwa M, Hirayama T, Okuda K, Nagasawa H (2014) A new class of high-contrast Fe(II) selective fluorescent probes based on spirocyclized scaffolds for visualization of intracellular labile iron delivered by transferrin. Org Biomol Chem 12(34):6590–6597

Niwa M, Hirayama T, Oomoto I, Wang DO, Nagasawa H (2018) Fe(II) ion release during Endocytotic uptake of iron visualized by a membrane-anchoring Fe(II) fluorescent probe. ACS Chem Biol 13(7):1853–1861

Petrat F, Rauen U, de Groot H (1999) Determination of the chelatable iron pool of isolated rat hepatocytes by digital fluorescence microscopy using the fluorescent probe, phen green SK. Hepatology 29(4):1171–1179

Petrat F, de Groot H, Rauen U (2001) Subcellular distribution of chelatable iron: a laser scanning microscopic study in isolated hepatocytes and liver endothelial cells. Biochem J 356(Pt 1):61–69

Petrat F, de Groot H, Sustmann R, Rauen U (2002) The chelatable iron pool in living cells: a methodically defined quantity. Biol Chem 383(3–4):489–502

Pezacki AT, Gonciarz RL, Okamura T, Matier CD, Torrente L, Cheng K, Miller SG, Ralle M, Ward NP, DeNicola GM, Renslo AR, Chang CJ (2024) A tandem activity-based sensing and labeling strategy reveals antioxidant response element regulation of labile iron pools. Proc Natl Acad Sci USA 121(28):e2401579121

Philpott CC, Protchenko O, Wang Y, Novoa-Aponte L, Leon-Torres A, Grounds S, Tietgens AJ (2023) Iron-tracking strategies: chaperones capture iron in the cytosolic labile iron pool. Front Mol Biosci 10:1127690

Rigoli L, Di Bella C (2012) Wolfram syndrome 1 and Wolfram syndrome 2. Curr Opin Pediatr 24(4):512–517

Rosanio FM, Di Candia F, Occhiati L, Fedi L, Malvone FP, Foschini DF, Franzese A, Mozzillo E (2022) Wolfram syndrome type 2: a systematic review of a not easily identifiable clinical spectrum. Int J Environ Res Public Health 19(2):835

Rothman RJ, Serroni A, Farber JL (1992) Cellular pool of transient ferric iron, chelatable by deferoxamine and distinct from ferritin, that is involved in oxidative cell injury. Mol Pharmacol 42(4):703–710

Shen Z-Q, Huang Y-L, Teng Y-C, Wang T-W, Kao C-H, Yeh C-H, Tsai T-F (2021) "CISD2 maintains cellular homeostasis." Biochimica et Biophysica Acta (BBA)-molecular. Cell Res 1868(4):118954

Shvartsman M, Cabantchik I (2012) PhD dissertation: mechanisms of iron trafficking to mitochondria of mammalian cells, stud-ied by live and real-time fluorescence. The Hebrew University, PhD Dissertation

Silswal A, Weslie P, Koner AL (2023) Bioimaging of labile lysosomal iron through naphthalimide-based fluorescent probe. Talanta 254:124147

Soleja N, Mohsin M (2024) A genetically encoded probe for monitoring and detection of iron in real-time. Sensors & Diagnostics

Urano F (2016) Wolfram syndrome: diagnosis, management, and treatment. Curr Diab Rep 16:1–8

Weaver J, Pollack S (1989) Low-M r iron isolated from Guinea pig reticulocytes as AMP-Fe and ATP-Fe complexes. Biochem J 261(3):787–792

Wiley SE, Andreyev AY, Divakaruni AS, Karisch R, Perkins G, Wall EA, van der Geer P, Chen YF, Tsai TF, Simon MI (2013) Wolfram syndrome protein, Miner1, regulates sulphydryl redox status, the unfolded protein response, and Ca2+ homeostasis. EMBO Mol Med 5(6):904–918

Williams R (1982) Free manganese (II) and iron (II) cations can act as intracellular cell controls. FEBS Lett 140(1):3–10

Wu C-Y, Chen Y-F, Wang C-H, Kao C-H, Zhuang H-W, Chen C-C, Chen L-K, Kirby R, Wei Y-H, Tsai S-F (2012) A persistent level of Cisd2 extends healthy lifespan and delays aging in mice. Hum Mol Genet 21(18):3956–3968

Wu R, Tang D, Kang R (2023) Fluorogenic probes for intracellular iron detection. Ferroptosis: methods and protocols. Springer, pp 1–8

Xie W, Jiang J, Shu D, Zhang Y, Yang S, Zhang K (2023) Recent progress in the rational design of biothiol-responsive fluorescent probes. Molecules 28(10):4252

Ye H, Jeong SY, Ghosh MC, Kovtunovych G, Silvestri L, Ortill D, Uchida N, Tisdale J, Camaschella C, T.A. Rouault TA. (2010) Glutaredoxin 5 deficiency causes sideroblastic anemia by specifically impairing heme biosynthesis and depleting cytosolic iron in human erythroblasts. J Clin Invest 120(5):1749–1761

Zanninelli G, Glickstein H, Breuer W, Milgram P, Brissot P, Hider R, Konijn A, Libman J, Shanzer A, Cabantchik ZI (1997) Chelation and mobilization of cellular iron by different classes of chelators. Mol Pharmacol 51(5):842–852

Zhang D-L, Ghosh MC, Rouault TA (2014) The physiological functions of iron regulatory proteins in iron homeostasis-an update. Front Pharmacol 5:124

Zhu C, Wang M, Qiu L, Hao S, Li K, Guo Z, He W (2018) A mitochondria-targeting fluorescent Fe3+ probe and its application in labile Fe3+ monitoring via imaging and flow cytometry. Dyes Pigments 157:328–333

Iron and Cell Death

5

Giorgia Federico, Marialuisa Moccia, and Francesca Carlomagno

Abstract

Iron is an essential micronutrient that supports cellular pathways such as DNA replication, metabolism, oxygen transport and oxidative phosphorylation via reduction-oxidation reactions. Nevertheless, in specific conditions, iron can be deleterious since it promotes different types of unprogrammed and programmed cell death, including apoptosis, necroptosis, and ferroptosis. In this chapter, we will address the role of iron in promoting cell death and the signaling pathways involved.

Keywords

Apoptosis · Ferroptosis · Necroptosis · NFR2 · ROS

G. Federico · F. Carlomagno (✉)
Dipartimento di Medicina Molecolare e Biotecnologie Mediche, Università Federico II, Napoli, Italy
e-mail: francesca.carlomagno@unina.it

M. Moccia
Istituto di Endocrinologia e Oncologia Sperimentale del CNR, Napoli, Italy

5.1 Introduction

Iron is one of the most abundant transition metals on earth, and it is an essential trace element in living organisms (Aisen et al. 2001), see also Chap. 1. Iron is necessary for life, being implicated in fundamental biological processes such as oxygen transport to tissues, cellular metabolism and respiration, DNA homeostasis (including DNA replication, DNA repair and epigenetic regulation), and RNA stability and translation (Coffey and Ganz 2017; Teh et al. 2024).

Iron is linked to around 400 cellular proteins and enzymes, either by directly binding the polypeptide chain (as in ribonucleotide reductase) or being part of prosthetic groups, such as heme (as in cytochromes or hemoglobin) or Fe-S clusters (as in DNA polymerases and helicases) (Wang and Pantopoulos 2011; Andreini et al. 2018). Protein-associated iron may play various roles: it can stabilize the protein in a specific conformation (as in the case of Iron Regulatory Proteins, IRPs), it can promote protein stabilization and correct folding (as in the case of some DNA polymerases) or it can be functional to catalysis (as in the case of ribonucleotide reductase) (Teh et al. 2024). Iron has been selected as a cofactor in several enzymatic reactions probably because of its ability to assume different atomic conformations (mainly ferrous Fe^{2+} or ferric Fe^{3+} in living organisms) being, therefore, able to both donate or acquire electrons in reduction-oxidation (redox) reactions (Johnson et al. 2024).

K. Pantopoulos (ed.), *Iron Metabolism in Human Health and Disease*, Advances in Experimental Medicine and Biology 1480, https://doi.org/10.1007/978-3-031-92033-2_5

In mammalian cells, free iron (Labile Iron Pool, LIP) represents 5% of total cellular iron and it is mainly represented by the highly reactive Fe^{2+} which, if in excess, may represent a danger (Kakhlon and Cabantchik 2002; Jomova and Valko 2011), see also Chap. 4. Indeed, iron might exert toxic effects on intracellular and extracellular components, namely nucleic acids, proteins, and lipids, through the generation of reactive oxidative species (ROS) or reactive lipid species (RLS) via Fenton/Haber-Weiss chemistry (see below) (Dixon and Stockwell 2014; Nakamura et al. 2019). Iron-mediated oxidative damage can elicit different types of cell death (including apoptosis, ferroptosis, and necroptosis) or cell senescence and is involved in the pathogenesis of a variety of diseases such as hemochromatosis, neurodegeneration, ischemia-reperfusion injury, and cancer. Here, we will summarize how iron can activate and sustain different types of programmed cell death via redox imbalance, with a focus on ferroptosis.

5.2 Iron-Mediated Generation of Reactive Oxygen or Lipid Species

ROS comprise a wide variety of molecules including superoxide (O2•–), peroxides (H_2O_2 and ROOH), and free radicals (HO• and RO•). They contain reduced oxygen and, although involved in regulating several normal cell functions, can be deleterious if produced in excess for their ability to damage macromolecules such as DNA, proteins, and lipids (Sies and Jones 2020). The most dangerous ROS, hydroxyl radicals (HO•), are generated via the Fenton Haber-Weiss reaction in the presence of hydrogen peroxide (H_2O_2), mainly produced by superoxide dismutase (SOD). The Fenton/Haber-Weiss reaction is dependent on Fe^{2+} which is converted in Fe^{3+} and then converted back to Fe^{2+} in the presence of superoxide (O2•–), produced by mitochondrial respiration (Kehrer 2000).

Reactive lipid species (RLS), including lipid peroxide (LOOH), lipid alkoxyl (LO•), and lipid peroxyl (LOO•) radicals, are also involved in multiple physiologic and pathologic processes, and iron participates in their production as well (Higdon et al. 2012). Indeed, iron can promote Fenton Haber-Weiss reactions to generate lipid peroxide radicals from lipid peroxide, which turn into cytotoxic aldehydes, like malondialdehyde (MDA) or 4-hydroxy-2-nonenal (HNE) (Pratt et al. 2011).

ROS and RLS can be also generated by enzymatic reactions catalyzed by iron-containing proteins such as NADPH oxidases (NOXs), nitric oxide synthases (NOSs), Cytochrome p450 enzymes, lipoxygenases (LOXs), xanthine oxidases, and mitochondrial electron transport chain proteins (Dixon and Stockwell 2014). In turn, ROS and RLS are inactivated by the action of detoxifying enzymes, including superoxide dismutase (SOD), catalase, glutathione peroxidases (GPXs), and thioredoxins (TRXs) (Valko et al. 2007). Upon redox imbalance, ROS can attack Fe-S clusters within proteins, inducing the release of iron atoms. This may trigger a vicious cycle that, by generating a progressive increase of LIP, favors the production of ROS and RLS leading to cell damage (Vernis et al. 2017).

By contributing to ROS and RLS production, iron excess furthers not only necrosis, through destruction of cellular components, but also different types of programmed cell death that can be initiated and sustained by the accumulation of such reactive species. These regulated forms of cell death foresee the activation of specific signaling pathways and molecular events that lead to distinctive cellular changes, such as DNA degradation and formation of apoptotic bodies (apoptosis), or dismantling of organelles and plasma membrane integrity via the propagation of phospholipid peroxidation (ferroptosis) (Fuchs and Steller 2015) (Fig. 5.1).

5.3 Iron in Apoptosis

Apoptosis is one of the best-known programmed cell death mechanisms exploited by cells under different physiologic and pathologic conditions; it relies on the activation of specific proteases called Caspases (Galluzzi et al. 2016b). Apoptosis is generally activated by internal beacons such as oxidative or endoplasmic reticulum stress, DNA

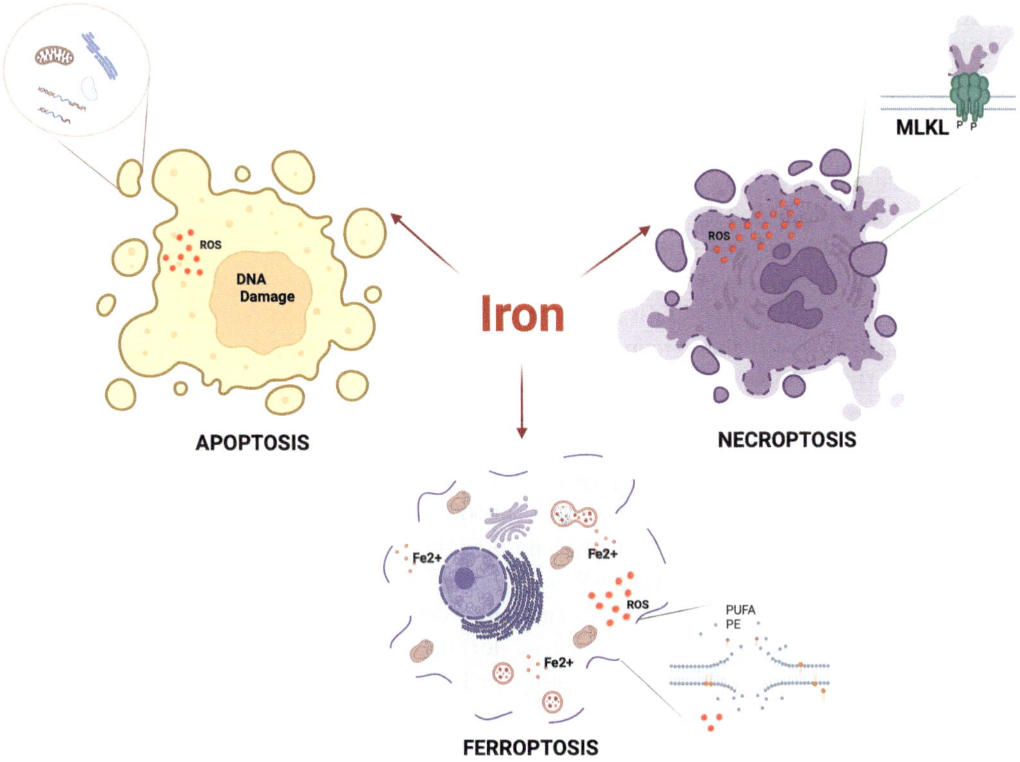

Fig. 5.1 Representative images of programmed cell death (apoptosis, necroptosis, and ferroptosis) that involve iron. *ROS* reactive oxygen species, *MLKL* mixed lineage kinase domain-like pseudokinase, *PUFA-PE* polyunsaturated fatty acid- phosphatidyl ethanolamine

damage, and mitochondrial dysfunction, that, through specific signaling pathways, promote the release of Cytochrome c from mitochondria and the formation of the apoptosome (intrinsic apoptosis) (Galluzzi et al. 2016a). In other conditions, apoptosis can be induced by autocrine or paracrine soluble factors, such as FASL (FS7-associated cell surface antigen ligand), and TNFα (tumor necrosis factor α), which act on specific death receptors, such as FAS and TNFR1, promoting, via defined signaling pathways, the activation of Caspases (extrinsic apoptosis).

Oxidative stress (often associated with conditions of iron overload) activates several signaling pathways, the best known being the ASK1(MAP3K5)-dependent one. In redox balance, ASK1 (Apoptosis signal-regulating kinase 1) is bound to dithiol oxidoreductase binding partners such as Thioredoxin (TRX), Glutaredoxin (GRX), and peroxiredoxin 1 (PRX1) that inhibit its activity. Upon oxidative stress, TRX, GRX, and PRX1 form intrapeptide disulfide bonds that result in conformational changes and detachment from ASK1. Free ASK1 undergoes autophosphorylation and recruits TRAF2, TRAF6, and ASK2 to form the ASK1 signalosome that, in turn, activates the MAPK (mitogen-activated protein kinases) p38/JNKs (c-Jun N-terminal kinases) via MKK3/6 and MKK4/7, respectively (Ichijo et al. 1997; Hattori et al. 2009; Shiizaki et al. 2013). P38 and JNKs are typical advocates of apoptosis. Thus, via direct phosphorylation, p38 activates both the pro-apoptotic BCL-2 like protein BIM, to induce Cytochrome c release from the mitochondria, and p53 that, as we will discuss later, is a key regulator of apoptosis (Grab and Rybniker 2019). JNKs, on the other hand, phosphorylate several pro-apoptotic BCL2-like proteins such as BIM, BAD, and BAX, contributing to their induction upon release from scaffold proteins. JNKs also phosphorylate the BCL2 anti-apoptotic protein, mediating its inactivation (Dhanasekaran and Reddy 2017). Iron is typically involved in p38/JNKs-mediated cell death. As an example, one of the most common triggers for apoptosis mediated by JNKs is treatment with H_2O_2, which uses iron-dependent Fenton Haber-Weiss chemistry to generate hydroxyl radicals that promote ASK1 activation, as discussed earlier.

Via generation of ROS, iron generates different types of DNA damage, including single- and double-strand breaks (SSBs and DSBs), as well as base oxidation with the production of 8-oxo-guanine (Linn 1998; Hahm et al. 2022). Interestingly, 8-oxoguanine can be generated during histone lysine demethylation by the activity of Lysine-specific histone demethylase 1 (LSD1), which produces a molecule of hydrogen peroxide which can be converted into oxygen radicals via iron-dependent Fenton Haber-Weiss chemistry (Forneris et al. 2008). DNA DSBs and DNA replication fork stalling (promoted by damaged bases) switch on the DNA damage response (DDR) by activating the ATM and ATR kinases, respectively. In turn, ATM and ATR phosphorylate and activate CHK1 and CHK2 kinases (Blackford and Jackson 2017). DDR results in p53 protein stabilization by reducing its interaction with the E3-ligase MDM2 that would otherwise promote p53 degradation by the proteasome (Jackson and Bartek 2009). As a transcription factor, p53 initially triggers cell cycle arrest by inducing the expression of the cyclin-dependent kinase inhibitor p21 and, if DNA damage persists, it promotes apoptosis by increasing the transcription of pro-apoptotic genes such as BAX, PUMA, NOXA, FASL, and FASR, or by directly interacting with and activating apoptosis-stimulating of p53 protein 1 and 2 (ASPP1 and ASPP2) (Roos et al. 2016). It is important to underline that, although genomic instability is often associated with impaired cell viability via activation of p53-dependent and -independent apoptosis, it can also promote neoplastic transformation (Torti and Torti 2020). Thus, one of the consequences of iron overload in the liver of patients affected by hereditary hemochromatosis is the development of hepatocellular carcinoma (Hentze et al. 2010).

Iron can also activate the apoptotic extrinsic pathway via modulation of the alternative splicing of the death receptor FAS. FAS is present in two different alternatively spliced isoforms: one is bound to the plasma membrane and triggers apoptosis upon binding to FASL; the second lacking exon 6, is soluble and, being released out of the cells, exerts anti-apoptotic effects (Ruberti et al. 1996). Changes in LIP caused by modulation of Ferritin levels can influence FAS alternative splicing. Ferritin is one of the main regulators of intracellular iron concentration (Chap. 2). It consists of an intracellular nanocage composed of 24 subunits of either Ferritin Heavy (FTH) or Ferritin Light (FTL) chains that shelter up to 4500 atoms of ferric iron (Arosio et al. 2017). Low iron conditions, triggered by iron chelation or by increasing Ferritin Light Chain (FTL) translation, via ACO1 (an FTL negative regulator) depletion, promote FAS alternative splicing and favor formation of the anti-apoptotic soluble form of the receptor. In contrast, iron overload, achieved by treatment with heme or by interfering with FTL expression, increases the abundance of transmembrane FAS, promoting apoptosis. Iron-dependent regulation of FAS alternative splicing is mediated by iron's ability to bind to the splicing factor SRSF7 that promotes the splicing of exon 6 from FAS mRNA (Tejedor et al. 2015).

In summary, iron overload can activate apoptotic intrinsic and extrinsic pathways by directly modulating pro-apoptotic pathways via ROS generation, inducing DNA damage, and regulating expression of FAS receptors on the plasma membrane.

5.4 Iron in Necroptosis

Being involved in ROS production and modulation of FAS, iron is also a crucial player in necroptosis, another type of programmed cell death. Necroptosis is distinguished from apoptosis in many aspects, including the lack of caspase activation and the triggering of an inflammatory response (Vanden Berghe et al. 2014; Linkermann and Green 2014). Nevertheless, necroptosis and apoptosis share similar features, including the involvement of death receptors such as TNFR1 and FAS, the increase of ROS levels that activate signaling pathways (e.g. JNK), as well as the activation of kinases that act in extrinsic apoptosis, such as RIPK1 and 3 (Receptor-interacting serine/threonine-protein kinase 1 and 3).

Necroptosis is typically initiated by increased intracellular ROS or death receptor activation that induces the formation of a microfilament-like complex, named necrosome, composed of RIPK1 and 3 kinases. RIPK1/3 auto- and trans-phosphorylate each other and recruit the pseudokinase Mixed Lineage Kinase Domain Like (MLKL). Once MLKL gets phosphorylated by RIPK1/3, it translocates within the lipid bilayer of internal organelles and plasma membrane leading to the expulsion of cellular contents into the extracellular space. Such substances function as damage associated molecular patterns (DAMPs) and induce inflammatory cell death (Wang and Pantopoulos 2011; Galluzzi et al. 2017).

Since iron is directly involved in ROS production and FAS regulation, both strong inducers of necroptosis, it is not surprising that iron triggers this type of programmed cell death. Nevertheless, there are also other ways in which iron is involved in necroptosis, mainly relying on a TNFα/Ferritin axis. Ferritin protects from necroptosis induced by TNFα, while the latter promotes iron release from ferritin to increase ROS production (Xie et al. 2005). Thus, TNFα induces proteasomal degradation of ferritin by activation of JNK1 (Antosiewicz et al. 2007). It must be pointed out that ferritin is not primarily degraded by the proteasome, but by a receptor-mediated autophagic pathway named ferritinophagy (see later). On the other hand, NF-κB (nuclear factor kappa-light-chain-enhancer of activated B cells) restrains ROS production by inducing FTH transcription, increasing iron sequestration in Ferritin, and preventing JNKs activation (Pham et al. 2004). In addition, TNFα activates NOX1 (NADPH oxidase 1) enhancing superoxide production to induce cell death (Kim et al. 2007).

5.5 Iron in Ferroptosis

5.5.1 General Aspects

Ferroptosis is a relatively novel type of programmed cell death that relies on the increase of intracellular iron concentration to promote a diffuse peroxidation of membrane polyunsatured fatty acids (PUFAs)-containing phospholipids, leading to organelle and plasma membrane disruption and cell death (Jiang et al. 2021) (Fig. 5.2). Iron is necessary to propagate peroxidation within cellular membranes via Fenton/Haber-Weiss reactions that convert PUFA-containing phospholipids in phospholipids with peroxidized polyunsaturated fatty acyl tail (Zhang 2024). Such an oxidative stress-driven non-apoptotic cell death was described for the first time in different experimental settings, either searching for compounds that could have a selective action on cancer cells or aiming at describing the lethal phenotype caused by genetic inactivation of GPX4 (Glutathione peroxidase 4) (Dolma et al. 2003; Yang and Stockwell 2008; Seiler et al. 2008). Since this novel form of programmed cell death was dependent on iron (indeed it could be suppressed by antioxidant agents such as vitamin E and by iron chelators such as deferoxamine, DFO), it was named ferroptosis (Dixon et al. 2012).

At ferroptosis onset, the site of initiation of lipid peroxidation is believed to be the endoplasmic reticulum, yet dysfunctional mitochondria (e.g. with alterations of electron transport chain or reduction of anti-ferroptosis molecules like DHODH or CoQ_{10}, as described below) or altered lysosomes (e.g. with accumulation of lipofuscin) can also contribute to ferroptosis initiation (Gao et al. 2019; Tian et al. 2021; von Krusenstiern et al. 2023). Once initiated, lipid peroxidation is propagated throughout the cell to different compartments, including the plasma membrane; this eventually results in the loss of cellular components and cell death. The intervention of the ESCRT III (endosomal sorting complexes required for transport) complex, able to repair cellular membranes, counteracts ferroptosis fir-

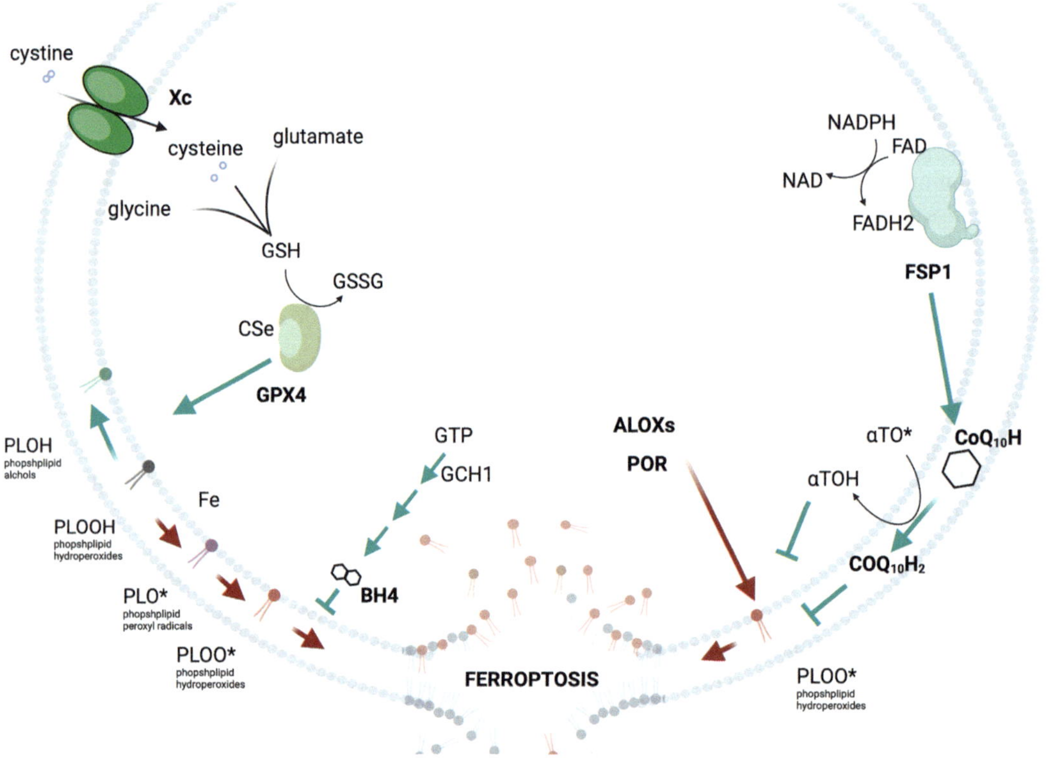

Fig. 5.2 Schematic representation of anti-ferroptosis cellular pathways. *GSH* reduced glutathione, *GSSG* oxidized glutathione, *GPX4* Glutathione Peroxidase 4, *CSe* SelenoCysteine, *GCH1* GTP cyclohydrolase 1, *BH4* tetrahydrobiopterin, *ALOX* Arachidonate Lipoxygenase, *POR* p450 Cytochrome, *αTO** oxidized α-tocopherol, *αTOH* reduced α-tocopherol, *COQ10H* quinone, *COQ10H2* hydroquinone, *FSP1* Ferroptosis suppressor protein 1

ing (Pedrera et al. 2021). It is still unknown how lipid peroxidation diffuses from one organelle to another so that, from a spotty alteration, diffuse damage is generated.

5.5.2 Pro- and Anti-ferroptosis Cellular Systems

Ferroptosis relies on a complex interplay between the iron state, lipid metabolism, and redox balance. Thus, multiple systems can influence the sensitivity of individual cells to ferroptosis (Fig. 5.2).

During ferroptosis, PUFA-containing phospholipid peroxidation relies on LIP increase. In the first instance, this is guaranteed by the intensification of iron uptake from the extracellular environment via the upregulation of Transferrin receptor 1 (TfR1). Accordingly, TfR1 is a valuable marker of ferroptosis (Stockwell 2022) (Table 5.1). In addition, mobilization of intracellular ferritin-bound iron storage via ferritin degradation by the autophagic pathway (ferritinophagy) is required to activate ferroptosis (Gao et al. 2016; Hou et al. 2016). Ferritinophagy is mediated by a cargo-receptor, NCOA4 (nuclear receptor coactivator 4), able to bind to FTH and ATG8-like proteins targeting ferritin to the autophagosome (Mancias et al. 2014). Besides direct lipid peroxidation, iron is also necessary for the enzymatic function of arachidonate lipoxygenases (ALOXs), that catalyze the deoxygenation of arachidonic acid generating hydroperoxidated lipids), and of Cytochrome p450 oxyreductase (POR), a membrane-bound enzyme able to transfer electrons from NADPH to Cytochrome p450 involved in lipid peroxidation during ferroptosis (Yang et al. 2016; Zou et al. 2020) (Fig. 5.2).

Table 5.1 Ferroptosis markers

Marker	Variation	Method of detection	Target
C-11 BODIPY	↑	IF (microscopy or FACS)	Lipid peroxidation
Lipid peroxide	↑	LS/MS	
MDA-adduct	↑	Biochemical assay	
HNE-adduct	↑	WB/IF/IHC	
NADPH	↓	Biochemical assay	FSP1 activity
GSH	↓	Biochemical assay	GPX4 activity
ROS	↑	DFCE assay	Redox unbalance
LIP	↑	Calcein assay	Intracellular iron levels
TfR1	↑	WB/IF/IHC	Extracellular iron uptake
Iron deposit	↑	Pearls blue tissue staining	Tissue iron accumulation
CHAC1	↑	mRNA rt-PCR	Ferroptosis-dependent gene regulation
PTGS2	↑	mRNA rt-PCR	
SLC7A11	↑	mRNA rt-PCR	
ACSL4	↑	mRNA rt-PCR	
NFR2 pathway	↓↑	mRNA rt-PCR	Antioxidant response

http://www.zhounan.org/ferrdb/current/

Sensitivity to ferroptosis is strongly influenced by the lipid composition of cellular membranes. Indeed, monounsaturated fatty acid (MUFA)-phospholipids are protective against ferroptosis, possibly because of their ability to counteract the propagation of lipid peroxidation. On the other hand, PUFA-phospholipids, especially those in which PUFAs are linked to phosphatidyl ethanolamine (PE) and that contain 20 or 22 carbons with 4 double bonds (20:4 and 22:4), are the most susceptible to peroxidation initiation and propagation, and therefore prone to trigger ferroptosis (Kagan et al. 2017). Along this line, enzymes that favor the incorporation of PUFAs into membrane phospholipids such as acyl coenzyme A synthase long chain family member 4 (ACSL4), which regulates PUFA levels, and lyso phosphatidyl choline acyl transferase 3 (LPCAT3), which promotes PUFA incorporation into phospholipids, display a pro-ferroptotic action (Dixon et al. 2015). In contrast, phospholipase iPLA2B that removes oxidized PUFA tails from phospholipids is protective against ferroptosis (Chen et al. 2021) (Fig. 5.2).

Cells are equipped with several lipid peroxidation detoxifying pathways to antagonize the onset of ferroptosis. One of the first described, and enabled the discovery of ferroptosis itself, is based on the GSH-dependent peroxidase GPX4 that can convert lipid peroxides into lipid alcohols, blocking ferroptosis onset (Seiler et al. 2008; Yang et al. 2014). Thus, GSH is considered one of the most important scavengers of lipid peroxidization and its intracellular levels profoundly influence susceptibility to ferroptosis. GSH levels are strongly dependent on the availability of cysteine, a building block for GSH synthesis. Indeed, cysteine deprivation by either removal from the cell medium or inhibiting the system Xc (SCL7A11), a cystine (oxidized dimer of cysteines)/glutamate antiporter, can induce ferroptosis (Dolma et al. 2003). Along this line, an important pro-ferroptotic modulator is the Multiple Drug Resistance 1 (MDR1) proteins (belonging to the ATP binding cassette subfamily B member 1/ABCB1 family) that, by exporting GSH outside cells, sensitizes them to ferroptosis (Cao et al. 2019). Similarly, the Cysteine dioxygenase type 1 (CDO1) also decreases GSH levels, by promoting cysteine oxidation, and is therefore considered an important pro-ferroptotic player (Hao et al. 2017) (Fig. 5.2).

A second system to detoxify lipid peroxides consists of CoQ10/ FSP1 (ferroptosis suppressor protein 1). CoQ10 is a vitamin-like component of mitochondrial membranes; it participates in electron transport but, by residing as well in other cellular membranes, it can also scavenge lipid peroxidization throughout the cell. Using NADPH, FSP1 is the enzyme that maintains CoQ10 in the active reduced state, thereby suppressing ferroptosis (Bersuker et al. 2019; Doll et al. 2019). Within mitochondria, CoQ10 can be also reduced by the dihydroorotate dehydrogenase enzyme, DHODH, that indeed displays a specific protection for these organelles against ferroptosis (Mao et al. 2021). FSP1 has been recently shown to also reduce vitamin K to its hydroquinone, which functions as a potent radical-trapping antioxidant, neutralizing lipid peroxidation (Mishima et al. 2022). This indicates that, besides its involvement in blood clotting, vitamin K has adjunctive features in mechanisms protecting against ferroptosis (Fig. 5.2).

The last lipid peroxide scavenging system described so far involves the GTP cyclohydrolase 1 (GCH1) and tetrahydrobiopterin (BH4). BH4 works similarly to CoQ10 in reducing oxidized lipids in membranes, and GCH1 maintains BH4 in the reduced active conformation. In addition, GCH1 can increase the amount of CoQ10 within cell membranes (Kraft et al. 2020; Soula et al. 2020) (Fig. 5.2).

The knowledge of all these systems acting in ferroptosis induction/protection mechanisms allows the definition of distinct molecular and morphologic ferroptotic markers. The main cellular markers of ferroptosis are listed are listed in Table 5.1. In addition, several small molecules, with the ability to induce ferroptosis by intercepting most of the above-mentioned pathways, have been described in the past years (Table 5.2). An extensive database listing ferroptosis regulators (both genes and substances) is available at: http://www.zhounan.org/ferrdb/current/

5.5.3 Signaling Pathways Regulating Ferroptosis

Ferroptosis is regulated by specific signaling pathways, that comprise cell differentiation, antioxidant responses and cellular metabolic state.

One of such pathways relies on the epithelial-mesenchymal transition (EMT), a physiologic process used during development that is also fostered in tissue repair and in cancer progression. In EMT, cells lose the expression of epithelial differentiation markers such as E-cadherin, reducing cell-cell adhesion (Yang et al. 2020). E-cadherin activates the NF2 (Merlin)-Hippo pathway that suppresses the transcription factor YAP, which is hyperactive in cells featuring EMT. YAP promotes the expression of TfR1 and ACSL4, increasing cell commitment to ferroptosis (Wu et al. 2019).

Using a genome-wide CRISPR/Cas9 screening, a different signaling pathway, involving PKCβII kinase and acting on ACSL4 as well, has been identified. Thus, PKCβII senses initial lipid peroxidation and amplifies it by phosphorylating ACSL4. Such phosphorylation enhances the synthesis of PUFA-containing lipids, sensitizing cells to ferroptosis (Zhang et al. 2022).

One of the most important antioxidant responses in cells is mediated by the NFR2/KEAP1 pathway. NFR2 (nuclear factor erythroid

Table 5.2 Ferroptosis-inducing drugs

Ferroptosis inducer	Mechanism of action	Reference
Erastin	Inhibition of system Xc	Dolma et al. (2003)
IKE	Inhibition of system Xc	Zhang et al. (2019)
Sorafenib	Inhibition of system Xc	Zheng et al. (2021)
RSL3	Inhibition of GPX4	Yang and Stockwell (2008)
FIN56	Inhibition of GPX4	Shimada et al. (2016)
ML210	Inhibition of GPX4	Weiwer et al. (2012)
FINO2	Inhibition of GPX4	Gaschler et al. (2018)
iFSP1	Inhibition of FSP1	Doll et al. (2019)
FSEN1	Inhibition of FSP1	Hendricks et al. (2023)
Buthionine sulfoximine	Inhibition of γGCS	Fruehauf et al. (1997)

2-related factor 2) is a basic leucine zipper (bZIP) transcription factor that is normally controlled by the KEAP1/CULLIN3 complex. KEAP1 functions as an adaptor that promotes NFR2 ubiquitination by CULLIN3 and its degradation by the proteosome. Under oxidative stress, specific KEAP1 cysteines are oxidized, causing the release, stabilization, and translocation of NFR2 into the nucleus, where it interacts with Maf (musculoaponeurotic fibrosarcoma oncogene homolog) proteins. NRF2/Maf complexes bind to antioxidant response elements (ARE) mapping in the promoter regions of genes encoding cytoprotective proteins (Ngo and Duennwald 2022). They include proteins that act as anti-ferroptotic factors, such as FTH1, to scavenge iron and reduce LIP, heme oxygenase 1 (HO-1), that catalyzes the breakdown of heme into the antioxidant biliverdin, NADPH quinone oxidoreductase 1 (NQO1), which catalyzes the reduction and detoxification of highly reactive quinones, and microsomal GSH S-transferase 1 (MGST1), that inhibits ALOX5 (Sun et al. 2016). In addition, NRF2 induces expression of the E3 ligase HERC2, which, in turn, ubiquitinates NCOA4, reducing its protein levels and restraining iron release from ferritin (Anandhan et al. 2023). In contrast, NCOA4 activity is promoted upon ferroptosis by the ATM (Ataxia telangiectasia mutated) kinase. Thus, oxidative stress activates ATM that phosphorylates NCOA4 to stimulate interaction with ferritin and activation of ferritinophagy (Guo et al. 2010; Wu et al. 2023). These data suggest that NCOA4 is a central cellular hub of anti- or pro-ferroptosis signals to decide cell fate.

mTOR (mammalian/mechanistic target of rapamycin) is another important regulator of ferroptosis. Activation of mTOR increases GPX4 expression through the Rag-mTORC1-4EBP signaling axis (Yan et al. 2021). In addition, mTOR induces sterol regulatory element-binding protein 1 (SREBP) that, in turn, activates stearoyl-CoA desaturase-1 (SCD1), promoting MUFA lipogenesis and protection from ferroptosis (Yi et al. 2020). Regulation of both GPX4 and SREBP results in protection from ferroptosis. Unexpectedly, during starvation, when mTOR is downregulated by the AMPK (AMP-dependent protein kinase)-dependent activation of the TSC1/TSC2 (Tuberous sclerosis complex 1 and 2) GTPase activating proteins, cells are protected from ferroptosis, because AMPK reduces PUFA synthesis restraining acetyl-CoA carboxylase (ACC) activity (Lee et al. 2020). In addition, an upstream mTOR activator, the serine/threonine kinase AKT protects from ferroptosis by inactivating GSK3β (glycogen synthase kinase 3β). Thus, GSK3β promotes ferroptosis by phosphorylating NRF2 and promoting its recognition by the β-TrCP E3 ligase and proteasome degradation (Cahuzac et al. 2023; Liu et al. 2023).

The observation that specific signaling pathways can dynamically promote or protect from ferroptosis is a strong indication that, similarly to apoptosis, this is an actively programmed cell death program that needs the orchestration of a platform of several biochemical events. However, ferroptosis and apoptosis are non-redundant cell responses; indeed, the non-enzymatic mechanism of ferroptosis typically occurs under conditions of oxidative stress, in which apoptosis might be difficult to be achieved because of its strict dependence on the action of several (lytic) enzymes.

5.6 Conclusions

Life without iron is not possible because iron is an essential component of many enzymes that are involved in cell metabolism and DNA synthesis. Nevertheless, there are different contests in which iron causes different types of programmed cell death such as apoptosis, necroptosis, and ferroptosis. Thus, both iron deficiency and iron overload are involved in the pathophysiology of human diseases. The possibility of modulating iron levels (either by supplementation or by chelation) or by targeting the proteins involved in these processes offers opportunities to treat such diseases.

Acknowledgement We thank Massimo Santoro, Mario Chiariello and Marco De Martino for critically revising the manuscript. This work was supported by Fondazione AIRC, Fondazione Telethon, and Italian Ministry of University and Research.

Conflict of Interest Disclosure The authors declare no conflict of interest.

References

Aisen P, Enns C, Wessling-Resnick M (2001) Chemistry and biology of eukaryotic iron metabolism. Int J Biochem Cell Biol 33(10):940–959. https://doi.org/10.1016/s1357-2725(01)00063-2

Anandhan A, Dodson M, Shakya A, Chen J, Liu P, Wei Y, Tan H, Wang Q, Jiang Z, Yang K, Garcia JG, Chambers SK, Chapman E, Ooi A, Yang-Hartwich Y, Stockwell BR, Zhang DD (2023) NRF2 controls iron homeostasis and ferroptosis through HERC2 and VAMP8. Sci Adv 9(5):eade9585. https://doi.org/10.1126/sciadv.ade9585

Andreini C, Putignano V, Rosato A, Banci L (2018) The human iron-proteome. Metallomics 10(9):1223–1231. https://doi.org/10.1039/c8mt00146d

Antosiewicz J, Ziolkowski W, Kaczor JJ, Herman-Antosiewicz A (2007) Tumor necrosis factor-alpha-induced reactive oxygen species formation is mediated by JNK1-dependent ferritin degradation and elevation of labile iron pool. Free Radic Biol Med 43(2):265–270. https://doi.org/10.1016/j.freeradbiomed.2007.04.023

Arosio P, Elia L, Poli M (2017) Ferritin, cellular iron storage and regulation. IUBMB Life 69(6):414–422. https://doi.org/10.1002/iub.1621

Bersuker K, Hendricks JM, Li Z, Magtanong L, Ford B, Tang PH, Roberts MA, Tong B, Maimone TJ, Zoncu R, Bassik MC, Nomura DK, Dixon SJ, Olzmann JA (2019) The CoQ oxidoreductase FSP1 acts parallel to GPX4 to inhibit ferroptosis. Nature 575(7784):688–692. https://doi.org/10.1038/s41586-019-1705-2

Blackford AN, Jackson SP (2017) ATM, ATR, and DNA-PK: the trinity at the heart of the DNA damage response. Mol Cell 66(6):801–817. https://doi.org/10.1016/j.molcel.2017.05.015

Cahuzac KM, Lubin A, Bosch K, Stokes N, Shoenfeld SM, Zhou R, Lemon H, Asara J, Parsons RE (2023) AKT activation because of PTEN loss upregulates xCT via GSK3beta/NRF2, leading to inhibition of ferroptosis in PTEN-mutant tumor cells. Cell Rep 42(5):112536. https://doi.org/10.1016/j.celrep.2023.112536

Cao JY, Poddar A, Magtanong L, Lumb JH, Mileur TR, Reid MA, Dovey CM, Wang J, Locasale JW, Stone E, Cole SPC, Carette JE, Dixon SJ (2019) A genome-wide haploid genetic screen identifies regulators of glutathione abundance and Ferroptosis sensitivity. Cell Rep 26(6):1544–1556. e1548. https://doi.org/10.1016/j.celrep.2019.01.043

Chen D, Chu B, Yang X, Liu Z, Jin Y, Kon N, Rabadan R, Jiang X, Stockwell BR, Gu W (2021) iPLA2beta-mediated lipid detoxification controls p53-driven ferroptosis independent of GPX4. Nat Commun 12(1):3644. https://doi.org/10.1038/s41467-021-23902-6

Coffey R, Ganz T (2017) Iron homeostasis: an anthropocentric perspective. J Biol Chem 292(31):12727–12734. https://doi.org/10.1074/jbc.R117.781823

Dhanasekaran DN, Reddy EP (2017) JNK-signaling: a multiplexing hub in programmed cell death. Genes Cancer 8(9–10):682–694. https://doi.org/10.18632/genesandcancer.155

Dixon SJ, Stockwell BR (2014) The role of iron and reactive oxygen species in cell death. Nat Chem Biol 10(1):9–17. https://doi.org/10.1038/nchembio.1416

Dixon SJ, Lemberg KM, Lamprecht MR, Skouta R, Zaitsev EM, Gleason CE, Patel DN, Bauer AJ, Cantley AM, Yang WS, Morrison B 3rd, Stockwell BR (2012) Ferroptosis: an iron-dependent form of nonapoptotic cell death. Cell 149(5):1060–1072. https://doi.org/10.1016/j.cell.2012.03.042

Dixon SJ, Winter GE, Musavi LS, Lee ED, Snijder B, Rebsamen M, Superti-Furga G, Stockwell BR (2015) Human haploid cell genetics reveals roles for lipid metabolism genes in nonapoptotic cell death. ACS Chem Biol 10(7):1604–1609. https://doi.org/10.1021/acschembio.5b00245

Doll S, Freitas FP, Shah R, Aldrovandi M, da Silva MC, Ingold I, Goya Grocin A, Xavier da Silva TN, Panzilius E, Scheel CH, Mourao A, Buday K, Sato M, Wanninger J, Vignane T, Mohana V, Rehberg M, Flatley A, Schepers A, Kurz A, White D, Sauer M, Sattler M, Tate EW, Schmitz W, Schulze A, O'Donnell V, Proneth B, Popowicz GM, Pratt DA, Angeli JPF, Conrad M (2019) FSP1 is a glutathione-independent ferroptosis suppressor. Nature 575(7784):693–698. https://doi.org/10.1038/s41586-019-1707-0

Dolma S, Lessnick SL, Hahn WC, Stockwell BR (2003) Identification of genotype-selective antitumor agents using synthetic lethal chemical screening in engineered human tumor cells. Cancer Cell 3(3):285–296. https://doi.org/10.1016/s1535-6108(03)00050-3

Forneris F, Binda C, Battaglioli E, Mattevi A (2008) LSD1: oxidative chemistry for multifaceted functions in chromatin regulation. Trends Biochem Sci 33(4):181–189. https://doi.org/10.1016/j.tibs.2008.01.003

Fruehauf JP, Zonis S, al-Bassam M, Kyshtoobayeva A, Dasgupta C, Milovanovic T, Parker RJ, Buzaid AC (1997) Selective and synergistic activity of L-S,R-buthionine sulfoximine on malignant melanoma is accompanied by decreased expression of glutathione-S-transferase. Pigment Cell Res 10(4):236–249. https://doi.org/10.1111/j.1600-0749.1997.tb00490.x

Fuchs Y, Steller H (2015) Live to die another way: modes of programmed cell death and the signals emanating from dying cells. Nat Rev Mol Cell Biol 16(6):329–344. https://doi.org/10.1038/nrm3999

Galluzzi L, Bravo-San Pedro JM, Kepp O, Kroemer G (2016a) Regulated cell death and adaptive stress

responses. Cell Mol Life Sci 73(11–12):2405–2410. https://doi.org/10.1007/s00018-016-2209-y
Galluzzi L, Lopez-Soto A, Kumar S, Kroemer G (2016b) Caspases connect cell-death signaling to organismal homeostasis. Immunity 44(2):221–231. https://doi.org/10.1016/j.immuni.2016.01.020
Galluzzi L, Kepp O, Chan FK, Kroemer G (2017) Necroptosis: mechanisms and relevance to disease. Annu Rev Pathol 12:103–130. https://doi.org/10.1146/annurev-pathol-052016-100247
Gao M, Monian P, Pan Q, Zhang W, Xiang J, Jiang X (2016) Ferroptosis is an autophagic cell death process. Cell Res 26(9):1021–1032. https://doi.org/10.1038/cr.2016.95
Gao M, Yi J, Zhu J, Minikes AM, Monian P, Thompson CB, Jiang X (2019) Role of mitochondria in ferroptosis. Mol Cell 73(2):354–363 e353. https://doi.org/10.1016/j.molcel.2018.10.042
Gaschler MM, Andia AA, Liu H, Csuka JM, Hurlocker B, Vaiana CA, Heindel DW, Zuckerman DS, Bos PH, Reznik E, Ye LF, Tyurina YY, Lin AJ, Shchepinov MS, Chan AY, Peguero-Pereira E, Fomich MA, Daniels JD, Bekish AV, Shmanai VV, Kagan VE, Mahal LK, Woerpel KA, Stockwell BR (2018) FINO(2) initiates ferroptosis through GPX4 inactivation and iron oxidation. Nat Chem Biol 14(5):507–515. https://doi.org/10.1038/s41589-018-0031-6
Grab J, Rybniker J (2019) The expanding role of p38 mitogen-activated protein kinase in programmed host cell death. Microbiol Insights 12:1178636119864594. https://doi.org/10.1177/1178636119864594
Guo Z, Kozlov S, Lavin MF, Person MD, Paull TT (2010) ATM activation by oxidative stress. Science 330(6003):517–521. https://doi.org/10.1126/science.1192912
Hahm JY, Park J, Jang ES, Chi SW (2022) 8-Oxoguanine: from oxidative damage to epigenetic and epitranscriptional modification. Exp Mol Med 54(10):1626–1642. https://doi.org/10.1038/s12276-022-00822-z
Hao S, Yu J, He W, Huang Q, Zhao Y, Liang B, Zhang S, Wen Z, Dong S, Rao J, Liao W, Shi M (2017) Cysteine dioxygenase 1 mediates erastin-induced ferroptosis in human gastric cancer cells. Neoplasia 19(12):1022–1032. https://doi.org/10.1016/j.neo.2017.10.005
Hattori K, Naguro I, Runchel C, Ichijo H (2009) The roles of ASK family proteins in stress responses and diseases. Cell Commun Signal 7:9. https://doi.org/10.1186/1478-811X-7-9
Hendricks JM, Doubravsky CE, Wehri E, Li Z, Roberts MA, Deol KK, Lange M, Lasheras-Otero I, Momper JD, Dixon SJ, Bersuker K, Schaletzky J, Olzmann JA (2023) Identification of structurally diverse FSP1 inhibitors that sensitize cancer cells to ferroptosis. Cell Chem Biol 30(9):1090–1103. e1097. https://doi.org/10.1016/j.chembiol.2023.04.007
Hentze MW, Muckenthaler MU, Galy B, Camaschella C (2010) Two to tango: regulation of mammalian iron metabolism. Cell 142(1):24–38. https://doi.org/10.1016/j.cell.2010.06.028
Higdon A, Diers AR, Oh JY, Landar A, Darley-Usmar VM (2012) Cell signalling by reactive lipid species: new concepts and molecular mechanisms. Biochem J 442(3):453–464. https://doi.org/10.1042/BJ20111752
Hou W, Xie Y, Song X, Sun X, Lotze MT, Zeh HJ 3rd, Kang R, Tang D (2016) Autophagy promotes ferroptosis by degradation of ferritin. Autophagy 12(8):1425–1428. https://doi.org/10.1080/15548627.2016.1187366
Ichijo H, Nishida E, Irie K, ten Dijke P, Saitoh M, Moriguchi T, Takagi M, Matsumoto K, Miyazono K, Gotoh Y (1997) Induction of apoptosis by ASK1, a mammalian MAPKKK that activates SAPK/JNK and p38 signaling pathways. Science 275(5296):90–94. https://doi.org/10.1126/science.275.5296.90
Jackson SP, Bartek J (2009) The DNA-damage response in human biology and disease. Nature 461(7267):1071–1078. https://doi.org/10.1038/nature08467
Jiang X, Stockwell BR, Conrad M (2021) Ferroptosis: mechanisms, biology and role in disease. Nat Rev Mol Cell Biol 22(4):266–282. https://doi.org/10.1038/s41580-020-00324-8
Johnson JE, Present TM, Valentine JS (2024) Iron: life's primeval transition metal. Proc Natl Acad Sci USA 121(38):e2318692121. https://doi.org/10.1073/pnas.2318692121
Jomova K, Valko M (2011) Advances in metal-induced oxidative stress and human disease. Toxicology 283(2–3):65–87. https://doi.org/10.1016/j.tox.2011.03.001
Kagan VE, Mao G, Qu F, Angeli JP, Doll S, Croix CS, Dar HH, Liu B, Tyurin VA, Ritov VB, Kapralov AA, Amoscato AA, Jiang J, Anthonymuthu T, Mohammadyani D, Yang Q, Proneth B, Klein-Seetharaman J, Watkins S, Bahar I, Greenberger J, Mallampalli RK, Stockwell BR, Tyurina YY, Conrad M, Bayir H (2017) Oxidized arachidonic and adrenic PEs navigate cells to ferroptosis. Nat Chem Biol 13(1):81–90. https://doi.org/10.1038/nchembio.2238
Kakhlon O, Cabantchik ZI (2002) The labile iron pool: characterization, measurement, and participation in cellular processes(1). Free Radic Biol Med 33(8):1037–1046. https://doi.org/10.1016/s0891-5849(02)01006-7
Kehrer JP (2000) The Haber-Weiss reaction and mechanisms of toxicity. Toxicology 149(1):43–50. https://doi.org/10.1016/s0300-483x(00)00231-6
Kim YS, Morgan MJ, Choksi S, Liu ZG (2007) TNF-induced activation of the Nox1 NADPH oxidase and its role in the induction of necrotic cell death. Mol Cell 26(5):675–687. https://doi.org/10.1016/j.molcel.2007.04.021
Kraft VAN, Bezjian CT, Pfeiffer S, Ringelstetter L, Muller C, Zandkarimi F, Merl-Pham J, Bao X, Anastasov N, Kossl J, Brandner S, Daniels JD, Schmitt-Kopplin P, Hauck SM, Stockwell BR, Hadian K, Schick JA (2020) GTP cyclohydrolase 1/tetrahydrobiopterin counteract ferroptosis through lipid remodeling. ACS Cent Sci 6(1):41–53. https://doi.org/10.1021/acscentsci.9b01063
Lee H, Zandkarimi F, Zhang Y, Meena JK, Kim J, Zhuang L, Tyagi S, Ma L, Westbrook TF, Steinberg GR, Nakada D, Stockwell BR, Gan B (2020) Energy-stress-mediated AMPK activation inhibits ferroptosis.

Nat Cell Biol 22(2):225–234. https://doi.org/10.1038/s41556-020-0461-8
Linkermann A, Green DR (2014) Necroptosis. N Engl J Med 370(5):455–465. https://doi.org/10.1056/NEJMra1310050
Linn S (1998) DNA damage by iron and hydrogen peroxide in vitro and in vivo. Drug Metab Rev 30(2):313–326. https://doi.org/10.3109/03602539808996315
Liu Y, Chou FJ, Lang F, Zhang M, Song H, Zhang W, Davis DL, Briceno NJ, Zhang Y, Cimino PJ, Zaghloul KA, Gilbert MR, Armstrong TS, Yang C (2023) Protein kinase B (PKB/AKT) protects IDH-mutated glioma from ferroptosis via Nrf2. Clin Cancer Res 29(7):1305–1316. https://doi.org/10.1158/1078-0432.CCR-22-3179
Mancias JD, Wang X, Gygi SP, Harper JW, Kimmelman AC (2014) Quantitative proteomics identifies NCOA4 as the cargo receptor mediating ferritinophagy. Nature 509(7498):105–109. https://doi.org/10.1038/nature13148
Mao C, Liu X, Zhang Y, Lei G, Yan Y, Lee H, Koppula P, Wu S, Zhuang L, Fang B, Poyurovsky MV, Olszewski K, Gan B (2021) DHODH-mediated ferroptosis defence is a targetable vulnerability in cancer. Nature 593(7860):586–590. https://doi.org/10.1038/s41586-021-03539-7
Mishima E, Ito J, Wu Z, Nakamura T, Wahida A, Doll S, Tonnus W, Nepachalovich P, Eggenhofer E, Aldrovandi M, Henkelmann B, Yamada KI, Wanninger J, Zilka O, Sato E, Feederle R, Hass D, Maida A, Mourao ASD, Linkermann A, Geissler EK, Nakagawa K, Abe T, Fedorova M, Proneth B, Pratt DA, Conrad M (2022) A non-canonical vitamin K cycle is a potent ferroptosis suppressor. Nature 608(7924):778–783. https://doi.org/10.1038/s41586-022-05022-3
Nakamura T, Naguro I, Ichijo H (2019) Iron homeostasis and iron-regulated ROS in cell death, senescence and human diseases. Biochim Biophys Acta Gen Subj 1863(9):1398–1409. https://doi.org/10.1016/j.bbagen.2019.06.010
Ngo V, Duennwald ML (2022) Nrf2 and oxidative stress: a general overview of mechanisms and implications in human disease. Antioxidants (Basel) 11(12). https://doi.org/10.3390/antiox11122345
Pedrera L, Espiritu RA, Ros U, Weber J, Schmitt A, Stroh J, Hailfinger S, von Karstedt S, Garcia-Saez AJ (2021) Ferroptotic pores induce Ca(2+) fluxes and ESCRT-III activation to modulate cell death kinetics. Cell Death Differ 28(5):1644–1657. https://doi.org/10.1038/s41418-020-00691-x
Pham CG, Bubici C, Zazzeroni F, Papa S, Jones J, Alvarez K, Jayawardena S, De Smaele E, Cong R, Beaumont C, Torti FM, Torti SV, Franzoso G (2004) Ferritin heavy chain upregulation by NF-kappaB inhibits TNFalpha-induced apoptosis by suppressing reactive oxygen species. Cell 119(4):529–542. https://doi.org/10.1016/j.cell.2004.10.017
Pratt DA, Tallman KA, Porter NA (2011) Free radical oxidation of polyunsaturated lipids: new mechanistic insights and the development of peroxyl radical clocks. Acc Chem Res 44(6):458–467. https://doi.org/10.1021/ar200024c
Roos WP, Thomas AD, Kaina B (2016) DNA damage and the balance between survival and death in cancer biology. Nat Rev Cancer 16(1):20–33. https://doi.org/10.1038/nrc.2015.2
Ruberti G, Cascino I, Papoff G, Eramo A (1996) Fas splicing variants and their effect on apoptosis. Adv Exp Med Biol 406:125–134. https://doi.org/10.1007/978-1-4899-0274-0_13
Seiler A, Schneider M, Forster H, Roth S, Wirth EK, Culmsee C, Plesnila N, Kremmer E, Radmark O, Wurst W, Bornkamm GW, Schweizer U, Conrad M (2008) Glutathione peroxidase 4 senses and translates oxidative stress into 12/15-lipoxygenase dependent- and AIF-mediated cell death. Cell Metab 8(3):237–248. https://doi.org/10.1016/j.cmet.2008.07.005
Shiizaki S, Naguro I, Ichijo H (2013) Activation mechanisms of ASK1 in response to various stresses and its significance in intracellular signaling. Adv Biol Regul 53(1):135–144. https://doi.org/10.1016/j.jbior.2012.09.006
Shimada K, Skouta R, Kaplan A, Yang WS, Hayano M, Dixon SJ, Brown LM, Valenzuela CA, Wolpaw AJ, Stockwell BR (2016) Global survey of cell death mechanisms reveals metabolic regulation of ferroptosis. Nat Chem Biol 12(7):497–503. https://doi.org/10.1038/nchembio.2079
Sies H, Jones DP (2020) Reactive oxygen species (ROS) as pleiotropic physiological signalling agents. Nat Rev Mol Cell Biol 21(7):363–383. https://doi.org/10.1038/s41580-020-0230-3
Soula M, Weber RA, Zilka O, Alwaseem H, La K, Yen F, Molina H, Garcia-Bermudez J, Pratt DA, Birsoy K (2020) Metabolic determinants of cancer cell sensitivity to canonical ferroptosis inducers. Nat Chem Biol 16(12):1351–1360. https://doi.org/10.1038/s41589-020-0613-y
Stockwell BR (2022) Ferroptosis turns 10: emerging mechanisms, physiological functions, and therapeutic applications. Cell 185(14):2401–2421. https://doi.org/10.1016/j.cell.2022.06.003
Sun X, Ou Z, Chen R, Niu X, Chen D, Kang R, Tang D (2016) Activation of the p62-Keap1-NRF2 pathway protects against ferroptosis in hepatocellular carcinoma cells. Hepatology 63(1):173–184. https://doi.org/10.1002/hep.28251
Teh MR, Armitage AE, Drakesmith H (2024) Why cells need iron: a compendium of iron utilisation. Trends Endocrinol Metab. https://doi.org/10.1016/j.tem.2024.04.015
Tejedor JR, Papasaikas P, Valcarcel J (2015) Genome-wide identification of Fas/CD95 alternative splicing regulators reveals links with iron homeostasis. Mol Cell 57(1):23–38. https://doi.org/10.1016/j.molcel.2014.10.029
Tian R, Abarientos A, Hong J, Hashemi SH, Yan R, Drager N, Leng K, Nalls MA, Singleton AB, Xu K, Faghri

F, Kampmann M (2021) Genome-wide CRISPRi/a screens in human neurons link lysosomal failure to ferroptosis. Nat Neurosci 24(7):1020–1034. https://doi.org/10.1038/s41593-021-00862-0

Torti SV, Torti FM (2020) Iron and cancer: 2020 vision. Cancer Res 80(24):5435–5448. https://doi.org/10.1158/0008-5472.CAN-20-2017

Valko M, Leibfritz D, Moncol J, Cronin MT, Mazur M, Telser J (2007) Free radicals and antioxidants in normal physiological functions and human disease. Int J Biochem Cell Biol 39(1):44–84. https://doi.org/10.1016/j.biocel.2006.07.001

Vanden Berghe T, Linkermann A, Jouan-Lanhouet S, Walczak H, Vandenabeele P (2014) Regulated necrosis: the expanding network of non-apoptotic cell death pathways. Nat Rev Mol Cell Biol 15(2):135–147. https://doi.org/10.1038/nrm3737

Vernis L, El Banna N, Baille D, Hatem E, Heneman A, Huang ME (2017) Fe-S clusters emerging as targets of therapeutic drugs. Oxidative Med Cell Longev 2017:3647657. https://doi.org/10.1155/2017/3647657

von Krusenstiern AN, Robson RN, Qian N, Qiu B, Hu F, Reznik E, Smith N, Zandkarimi F, Estes VM, Dupont M, Hirschhorn T, Shchepinov MS, Min W, Woerpel KA, Stockwell BR (2023) Identification of essential sites of lipid peroxidation in ferroptosis. Nat Chem Biol 19(6):719–730. https://doi.org/10.1038/s41589-022-01249-3

Wang J, Pantopoulos K (2011) Regulation of cellular iron metabolism. Biochem J 434(3):365–381. https://doi.org/10.1042/BJ20101825

Weiwer M, Bittker JA, Lewis TA, Shimada K, Yang WS, MacPherson L, Dandapani S, Palmer M, Stockwell BR, Schreiber SL, Munoz B (2012) Development of small-molecule probes that selectively kill cells induced to express mutant RAS. Bioorg Med Chem Lett 22(4):1822–1826. https://doi.org/10.1016/j.bmcl.2011.09.047

Wu J, Minikes AM, Gao M, Bian H, Li Y, Stockwell BR, Chen ZN, Jiang X (2019) Intercellular interaction dictates cancer cell ferroptosis via NF2-YAP signalling. Nature 572(7769):402–406. https://doi.org/10.1038/s41586-019-1426-6

Wu H, Liu Q, Shan X, Gao W, Chen Q (2023) ATM orchestrates ferritinophagy and ferroptosis by phosphorylating NCOA4. Autophagy 19(7):2062–2077. https://doi.org/10.1080/15548627.2023.2170960

Xie C, Zhang N, Zhou H, Li J, Li Q, Zarubin T, Lin SC, Han J (2005) Distinct roles of basal steady-state and induced H-ferritin in tumor necrosis factor-induced death in L929 cells. Mol Cell Biol 25(15):6673–6681. https://doi.org/10.1128/MCB.25.15.6673-6681.2005

Yan Y, Lei G, Gan B (2021) A mTORC1-mediated cyst(e)ine sensing mechanism governing GPX4 synthesis and ferroptosis. Mol Cell Oncol 8(3):1919006. https://doi.org/10.1080/23723556.2021.1919006

Yang WS, Stockwell BR (2008) Synthetic lethal screening identifies compounds activating iron-dependent, nonapoptotic cell death in oncogenic-RAS-harboring cancer cells. Chem Biol 15(3):234–245. https://doi.org/10.1016/j.chembiol.2008.02.010

Yang WS, SriRamaratnam R, Welsch ME, Shimada K, Skouta R, Viswanathan VS, Cheah JH, Clemons PA, Shamji AF, Clish CB, Brown LM, Girotti AW, Cornish VW, Schreiber SL, Stockwell BR (2014) Regulation of ferroptotic cancer cell death by GPX4. Cell 156(1–2):317–331. https://doi.org/10.1016/j.cell.2013.12.010

Yang WS, Kim KJ, Gaschler MM, Patel M, Shchepinov MS, Stockwell BR (2016) Peroxidation of polyunsaturated fatty acids by lipoxygenases drives ferroptosis. Proc Natl Acad Sci USA 113(34):E4966–E4975. https://doi.org/10.1073/pnas.1603244113

Yang J, Antin P, Berx G, Blanpain C, Brabletz T, Bronner M, Campbell K, Cano A, Casanova J, Christofori G, Dedhar S, Derynck R, Ford HL, Fuxe J, Garcia de Herreros A, Goodall GJ, Hadjantonakis AK, Huang RYJ, Kalcheim C, Kalluri R, Kang Y, Khew-Goodall Y, Levine H, Liu J, Longmore GD, Mani SA, Massague J, Mayor R, McClay D, Mostov KE, Newgreen DF, Nieto MA, Puisieux A, Runyan R, Savagner P, Stanger B, Stemmler MP, Takahashi Y, Takeichi M, Theveneau E, Thiery JP, Thompson EW, Weinberg RA, Williams ED, Xing J, Zhou BP, Sheng G, Association EMTI (2020) Guidelines and definitions for research on epithelial-mesenchymal transition. Nat Rev Mol Cell Biol 21(6):341–352. https://doi.org/10.1038/s41580-020-0237-9

Yi J, Zhu J, Wu J, Thompson CB, Jiang X (2020) Oncogenic activation of PI3K-AKT-mTOR signaling suppresses ferroptosis via SREBP-mediated lipogenesis. Proc Natl Acad Sci USA 117(49):31189–31197. https://doi.org/10.1073/pnas.2017152117

Zhang DD (2024) Ironing out the details of ferroptosis. Nat Cell Biol 26(9):1386–1393. https://doi.org/10.1038/s41556-024-01361-7

Zhang Y, Tan H, Daniels JD, Zandkarimi F, Liu H, Brown LM, Uchida K, O'Connor OA, Stockwell BR (2019) Imidazole ketone erastin induces ferroptosis and slows tumor growth in a mouse lymphoma model. Cell Chem Biol 26(5):623–633 e629. https://doi.org/10.1016/j.chembiol.2019.01.008

Zhang HL, Hu BX, Li ZL, Du T, Shan JL, Ye ZP, Peng XD, Li X, Huang Y, Zhu XY, Chen YH, Feng GK, Yang D, Deng R, Zhu XF (2022) PKCbetaII phosphorylates ACSL4 to amplify lipid peroxidation to induce ferroptosis. Nat Cell Biol 24(1):88–98. https://doi.org/10.1038/s41556-021-00818-3

Zheng J, Sato M, Mishima E, Sato H, Proneth B, Conrad M (2021) Sorafenib fails to trigger ferroptosis across a wide range of cancer cell lines. Cell Death Dis 12(7):698. https://doi.org/10.1038/s41419-021-03998-w

Zou Y, Li H, Graham ET, Deik AA, Eaton JK, Wang W, Sandoval-Gomez G, Clish CB, Doench JG, Schreiber SL (2020) Cytochrome P450 oxidoreductase contributes to phospholipid peroxidation in ferroptosis. Nat Chem Biol 16(3):302–309. https://doi.org/10.1038/s41589-020-0472-6

Dietary Iron Absorption: Biochemical and Nutritional Aspects

6

David M. Frazer, Gregory J. Anderson, and James F. Collins

Abstract

Iron is an essential nutrient for most organisms; however, it can also be toxic when present in excess. For this reason, life has evolved complex pathways to tightly control the amount of iron contained within body tissues. Unlike most other nutrients, mammals do not have a regulated excretory mechanism for iron and, therefore, must regulate body iron levels at the point of dietary iron absorption in the proximal small intestine. In this chapter, we will describe the molecules and pathways involved in this important process, including our current understanding of the mechanisms of both heme and non-heme iron absorption. The regulation of this process will also be discussed, with particular emphasis on local and systemic factors that affect how much iron is absorbed from the diet. We also highlight areas where our knowledge is incomplete and where more research is required to fully understand the molecular mechanisms responsible for this essential process.

Keywords

Divalent metal-ion transporter 1 · Ferroportin Heme iron · Hepcidin · Iron absorption

6.1 Introduction

Iron is an essential nutrient for most organisms, so the development of pathways to acquire iron from the environment was fundamental to the evolution of life (Chap. 1). In vertebrates, the main physiological way in which iron can enter the body is through intestinal absorption of iron from the diet (except for iron transferred to the fetus by the mother during pregnancy). As a result, mammals have developed sophisticated mechanisms that govern dietary iron acquisition and its subsequent transport, storage, and recycling. Importantly, these mechanisms adapt to variations in body iron demand to maintain iron at optimal levels to support many physiological functions and to avoid toxicity from excess iron. Basic aspects of intestinal iron absorption have been understood for decades, but it is only with the more recent application of molecular techniques that a detailed appreciation of the mechanisms of iron absorption has been developed.

D. M. Frazer · G. J. Anderson (✉)
QIMR Berghofer Medical Research Institute, Brisbane, QLD, Australia
e-mail: david.frazer@qimrberghofer.edu.au; greg.anderson@qimrberghofer.edu.au

J. F. Collins
Food Science & Human Nutrition Department, University of Florida, Gainesville, FL, USA
e-mail: jfcollins@ufl.edu

K. Pantopoulos (ed.), *Iron Metabolism in Human Health and Disease*, Advances in Experimental Medicine and Biology 1480, https://doi.org/10.1007/978-3-031-92033-2_6

6.2 Body Iron Requirements

Body iron requirements, and thus the amount of iron absorbed, vary throughout life in accordance with physiological needs. In the first 10–15 years of life, when the growth rate is high, iron demands and iron absorption are also high, but absorption declines as adulthood is reached and growth ceases. Importantly, the capacity of the human body to divest itself of iron is extremely limited, so the body's iron complement is largely determined by regulating the amount of dietary iron absorbed (Collins et al. 2018).

Over a lifetime, a typical human will absorb 25–50 g of iron from their diet. This amount is required to replace the obligatory (unregulated) losses that occur via the sloughing of cells, such as keratinocytes and intestinal enterocytes, small losses in bile and urine, and occasional pathological blood loss. Adult males absorb approximately 1.1 mg of iron each day (Hurrell and Egli 2010; Lu et al. 2019). Women have higher iron requirements as they lose iron through menstrual bleeding and pregnancy during their reproductive years, so their average daily iron requirement is ~1.4 mg. After menopause, the iron requirements of women decline to levels similar to those of men (0.9 mg/day).

6.3 Sources of Dietary Iron and Factors Influencing Its Absorption

In the diet, most iron is bound to proteins and is broadly considered to be present in either heme or non-heme forms (Collins et al. 2018), with the latter being particularly diverse biochemically. Heme iron is mainly derived from hemoglobin and myoglobin, which are abundant in meat products. Heme is a stable organic source of iron, and its absorption is largely unaffected by other dietary constituents. Non-heme iron, which is predominantly present in the ferric (Fe^{3+}) form, is also abundant in meat, but it is also the main form of iron in plant-derived foodstuffs. Ferric iron is highly insoluble and must be reduced to the more soluble ferrous (Fe^{2+}) form before it can be absorbed (Chap. 1). Because non-heme iron is not sequestered within a stable porphyrin ring, like heme iron, it is highly susceptible to interactions with both dietary constituents and endogenous factors, which can affect its absorption (Piskin et al. 2022) (Fig. 6.1). Small organic acids, such as ascorbic acid and citric acid, can keep iron in a soluble state and may promote its reduction to the ferrous form. The low pH of the stomach and proximal small intestine is also important for maintaining iron in a soluble and reduced state, and pathological conditions or pharmacological interventions associated with reduced gastric acid secretion can inhibit iron absorption (Skikne et al. 1981). For similar chemical reasons, acidification of the unstirred water layer on the apical surface of intestinal enterocytes via the action of a sodium-hydrogen exchanger (NHE), particularly NHE3, is likely to promote iron uptake by enterocytes (Shawki et al. 2016). Conversely, plant-derived substances such as phytates, tannins, and polyphenols can bind iron and reduce its bioavailability. Iron from plant-derived dietary components is typically less efficiently absorbed compared to iron in meat products. Moreover, iron absorption can be impaired by *Helicobacter pylori* infection and in celiac disease and disorders that affect intestinal motility (Collins et al. 2018).

6.4 Anatomical and Physical Influences on Iron Absorption

Most iron absorption occurs across the wall of the duodenum and proximal jejunum, although small amounts of iron can be absorbed by more distal parts of the small intestine and even the colon (Collins et al. 2018). Mature differentiated enterocytes of the mid- to upper-villi of the proximal small intestine are specifically adapted for the vectorial transport of iron and other nutrients (Conrad Jr. and Crosby 1963). The lower pH of the upper small intestine is a major contributor to the high iron absorption of this region, but the higher expression of iron transport proteins in the duodenal mucosa suggests that this part of the

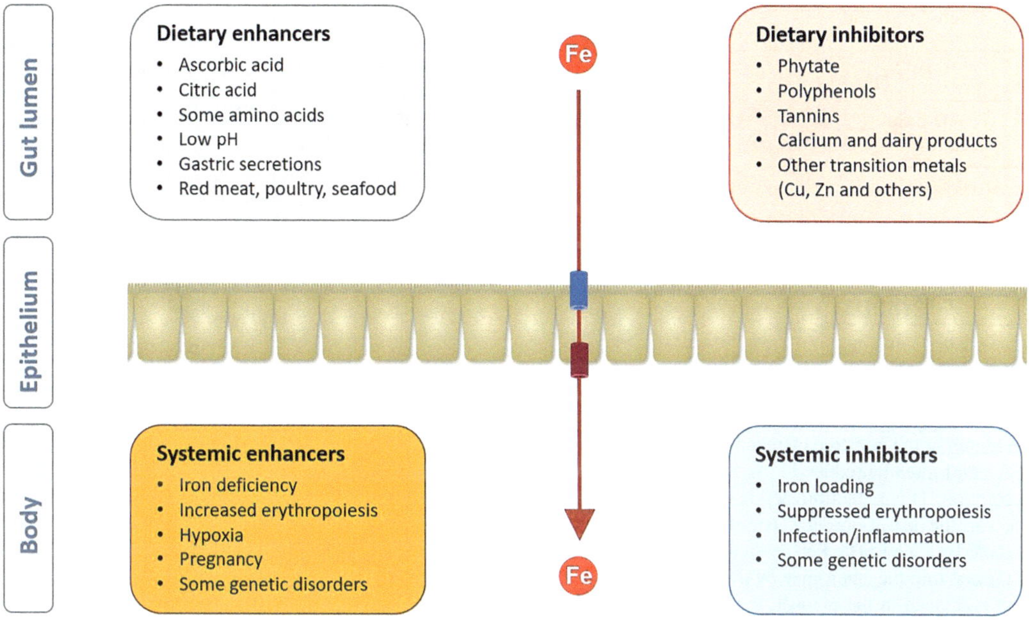

Fig. 6.1 Factors affecting iron absorption. The amount of iron absorbed from the diet is affected by numerous factors, which can either enhance or inhibit absorption. These encompass factors in the gut lumen that affect the bioavailability of iron, as well as systemic factors that reflect body iron requirements and various pathologies

small bowel is specifically optimized to maximize iron absorption (Frazer et al. 2001).

Absorption can be affected by a range of physical factors, including intestinal transit time and mucosal surface area (Collins et al. 2018). In celiac disease, decreased mucosal surface for nutrient absorption may impair iron absorption. Importantly, morphological adaptations during iron deficiency include an increase in villus size, the recruitment of a higher proportion of enterocytes on each villus for absorption, and the extension of absorption further down the gut. The combined effect of these changes is to provide a larger mucosal surface area, which maximizes iron transport.

6.5 Mechanisms of Non-heme Iron Absorption

Under normal physiological conditions, the epithelium of the small intestine forms a tight barrier to control the intake of nutrients and prevent the ingress of pathogens. The absorption of iron is generally considered to consist of three phases. In the first, iron in a suitable form in the intestinal lumen is transported across the apical or brush-border membrane (BBM) of enterocytes. Once within the enterocyte (phase two), iron can either be directed to the basolateral membrane (BLM) for subsequent export or sequestered within the cell in the iron storage protein ferritin. In the final phase, iron is moved across the BLM and makes its way to the circulatory system. The process of iron absorption can be very rapid, and radioactive iron administered into the intestinal lumen can appear in the blood within 15 seconds (Duthie 1964). The mechanism of absorption of non-heme iron is by far the best understood, but our current state of knowledge of the absorption of heme iron will also be considered.

6.5.1 Iron Transport Across the Enterocyte Apical Membrane—The Uptake Phase

Luminal non-heme iron is taken up across the enterocyte BBM via the transmembrane protein

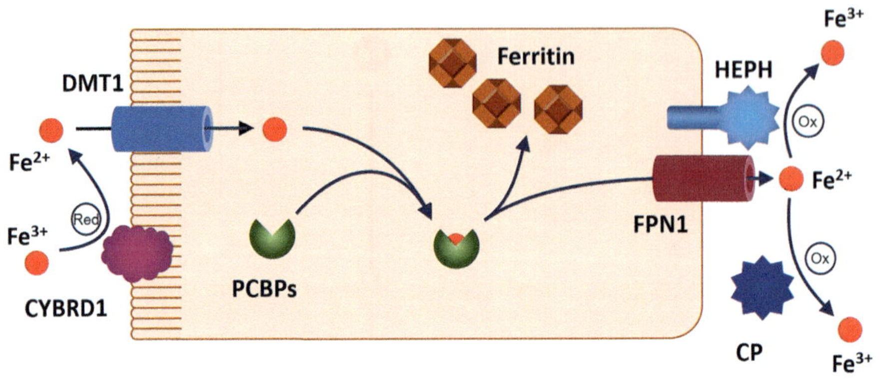

Fig. 6.2 Non-heme iron absorption. Non-heme iron in the lumen of the intestine is typically present as ferric iron (Fe^{3+}) and must be reduced to ferrous iron (Fe^{2+}) prior to absorption. This reduction step is carried out by enzymes such as the brush-border ferrireductase duodenal cytochrome B (CYBRD1). Fe^{2+} is then transported across the BBM and into the enterocyte by divalent metal-iron transporter 1. Once in the cytosol, the iron is likely bound to chaperones such as the poly(rC) binding proteins (PCBPs) before being transported to its ultimate destination. That destination may include being stored within the intracellular iron storage protein ferritin if the iron is not immediately required by the cell or the body. This iron can subsequently be released from ferritin, if needed, but is most likely lost from the body when the enterocyte is sloughed into the intestinal lumen at the end of its lifespan. The iron that is required by the body is delivered to the iron export protein ferroportin 1 (FPN1) on the BLM and is exported from the enterocyte. This process is facilitated by the ferroxidase hephaestin (HEPH), which oxidizes newly released Fe^{2+} to Fe^{3+}. This oxidation step may also be performed by the circulating protein ceruloplasmin (CP)

divalent metal-ion transporter 1 (DMT1) (Fleming et al. 1997; Gunshin et al. 1997; Mackenzie and Garrick 2005) (Fig. 6.2). This transporter is expressed in most body cells, where it plays an essential role in iron export from endosomes as part of the transferrin (TF) cycle (Chap. 2), but it plays a particularly important role in the gut in the absorption of dietary iron. DMT1 transports ferrous iron (Fe^{2+}), so any ferric iron (Fe^{3+}) in the gut lumen must first be reduced before it can be absorbed. While dietary constituents, such as ascorbic acid, and gastric factors can contribute to iron reduction in the lumen, enzymatic reduction is likely required. Duodenal cytochrome B (CYBRD1) is one brush-border ferric reductase that has been shown to contribute to iron absorption (McKie et al. 2001; Lane et al. 2015), but the effects of its deletion in mice are quite small (Gunshin et al. 2005b), suggesting that other (unidentified) reductases may exist.

DMT1 is a proton/ferrous iron symporter, which efficiently transports iron at a slightly acidic pH, so it is well adapted to operate in the upper small intestine. The required proton gradient is at least partially supplied by the apically expressed sodium-hydrogen exchanger NHE3 (Shawki et al. 2016). Although DMT1 is expressed widely throughout the gastrointestinal tract, it is most prominent in the enterocytes of the proximal small intestine, where most iron absorption occurs (Gunshin et al. 1997). Interestingly, some amino acids in the diet have been shown to enhance the trafficking of DMT1 to the BBM, thereby enhancing iron absorption (Woloshun et al. 2022), and luminal iron administration can lead to the internalization of DMT1 from the brush-border (Arita et al. 2022). As its name suggests, DMT1 is able to transport a variety of divalent metal ions; however, studies, particularly in mouse models, suggest that its main physiological substrate is iron (Shawki et al. 2012). Mouse models have also been used to demonstrate that DMT1 is the major brush-border protein facilitating the uptake of iron from the lumen of the intestine into enterocytes (Gunshin et al. 2005a). This appears to be true for humans as well, as mutations in *DMT1* lead to severe hypochromic, microcytic anemia in affected patients (Iolascon and De Falco 2009). These mutations are often associated with tissue iron loading, particularly in the liver, possibly due to residual DMT1 iron transport activity and a particularly strong erythropoietic drive to increase iron absorption.

6.5.2 The Handling of Iron Within Enterocytes—The Intracellular Phase

Iron transferred to the enterocyte cytoplasm likely exists bound to small molecular weight organic acids such as citrate or amino acids, or perhaps with low affinity to cytoplasmic proteins (Kruszewski 2003). However, at least some of this iron is bound by poly (rC) binding proteins (PCBP) 1 and 2, which act as iron chaperones (see Chap. 2). PCBP1 is able to deliver iron to the iron-storage protein ferritin (Shi et al. 2008). The PCBPs have also been shown to interact with the iron transporters DMT1 and ferroportin 1 (FPN1) (Yanatori et al. 2014; Yanatori et al. 2016; Wang et al. 2023). Interestingly, enterocytes in which PCBP1 has been deleted do not retain iron effectively, as iron efflux through FPN1 is increased (Wang et al. 2023). Consistent with this, the specific deletion of H-ferritin in the intestinal epithelium of mice leads to increased iron absorption (Vanoaica et al. 2010). These data suggest that the delivery of iron to, and its storage within, ferritin may play an active role in regulating the passage of iron across enterocytes, rather than simply being a passive sink for excess cellular iron.

While some of the iron stored as ferritin might subsequently be released by autophagy following the binding of nuclear receptor coactivator 4 (NCOA4) and then transferred to the portal circulation, much of this iron likely stays within the enterocyte and is lost when cells are sloughed off into the intestinal lumen. Mobilization of ferritin iron has been described as a regulated process, with the transcription factor hypoxia-inducible factor 2α (HIF2α) playing an important role in determining the fate of iron within enterocytes. HIF2α modulates the expression of NCOA4, thereby influencing the rate of iron release from ferritin (Das et al. 2022), as well as iron transport proteins. Intracellular iron released from ferritin may also modulate the expression of iron transporters through the iron-responsive element (IRE) /iron regulatory protein (IRP) system, as detailed below.

6.5.3 Iron Export Across the Enterocyte Basolateral Membrane—The Transfer Phase

FPN1 is the only known mammalian non-heme iron exporter and is expressed at particularly high levels in cells with high rates of iron flux, such as intestinal enterocytes (Drakesmith et al. 2015). FPN1 localizes to the BLM of enterocytes (Fig. 6.1) and has a similar gastrointestinal expression pattern to DMT1, with both transporters expressed at higher levels in the duodenum and proximal jejunum than in the more distal regions (McKie et al. 2000; Frazer et al. 2001).

The ablation of FPN1 in mice and zebrafish leads to embryonic lethality (Drakesmith et al. 2015), and mice in which FPN1 is deleted postnatally in the small intestine become severely iron deficient due to an inability to effectively absorb dietary iron (Donovan et al. 2005). These studies showed that FPN1 is responsible for almost all of the iron movement across the BLM and imply that, irrespective of the form in which iron crosses the BBM, its export from enterocytes is likely via this shared pathway.

Like DMT1, FPN1 uses ferrous iron as a substrate (Mitchell et al. 2014), but iron exported from enterocytes through FPN1 is ultimately bound in the ferric form to plasma TF for distribution in the bloodstream. Thus, the efficiency of FPN1-mediated iron export is greatly increased by the presence of an iron oxidase (a ferroxidase). For most body cells, the plasma-borne, copper-dependent ferroxidase ceruloplasmin performs this function (Hellman and Gitlin 2002). In the intestine, however, the major iron oxidase is a membrane-bound ceruloplasmin homolog called hephaestin (Vulpe et al. 1999; Helman et al. 2023), which, like FPN1, localizes to the enterocyte BLM (Fig. 6.1). While hephaestin increases the efficiency of enterocyte iron export (Doguer et al. 2017), it is not essential for this process, with the disruption of hephaestin in mice leading to a much less severe iron-deficiency anemia than that observed when FPN1 is knocked out (Fuqua et al. 2014). Ceruloplasmin can also contribute to intestinal iron absorption, particularly when body iron demands are high (Cherukuri et al. 2005), so it may enable at least some iron absorption to continue even in the absence of HEPH. It is also possible that other, yet unidentified, ferroxidases might exist (Ranganathan et al. 2012).

6.5.4 The Fate of Iron Released from Enterocytes

How iron is exported from enterocytes and moves to the circulation is incompletely understood. Newly absorbed iron must ultimately bind to TF for distribution in the bloodstream, but the rate of TF exchange across capillaries does not account for all iron absorbed (Morgan 1980). In addition, there is also no strong evidence that TF interacts with the BLM of mature enterocytes, nor that it interacts with HEPH (Hudson et al. 2008). Furthermore, in the rare condition hypotransferrinemia, where TF levels are negligible, iron absorption is not impaired (Bernstein 1987). These data suggest that not all newly absorbed iron binds to TF in the intestinal lamina propria, so it must transit to the circulation in either a low molecular weight chelated form or bound with low affinity to plasma proteins other than TF. A recent study has suggested a role for TF in the lamina propria in iron absorption, but this has yet to be verified (Sukhbaatar et al. 2023). Thus, the role of TF in dietary iron absorption requires further investigation.

6.6 Heme Iron Absorption

Heme is an important dietary iron source (Bezwoda et al. 1983; Carpenter and Mahoney 1992). It is probably absorbed intact by enterocytes, likely via receptor-mediated endocytosis followed by transport into the cytosol, but the receptor (or transporter) involved has not been identified (West and Oates 2008a) (Fig. 6.3). Heme-carrier protein 1 (HCP1) was identified as a putative low-affinity heme transporter (Shayeghi et al. 2005) but was subsequently shown to be a high-affinity folate transporter that likely has a very limited role in heme absorption in humans (Qiu et al. 2006). More recently, heme-regulated gene 1 (HRG1), a member of a family of heme transport proteins, was found to be expressed in the small intestine, where it could possibly contribute to heme iron absorption (Dutt et al. 2022).

Once within the enterocyte, heme is likely catabolized by heme oxygenase 1 and possibly heme oxygenase 2 in the endoplasmic reticulum, with the released iron entering the same intracellular pool as non-heme iron and subsequently exiting the enterocyte via FPN1 (Raffin et al. 1974; West and Oates 2008b) (Fig. 6.3). It also remains possible that some intact heme traverses the BLM. Two heme export proteins, feline leukemia virus, subgroup C, receptor (FLVCR) and ATP binding cassette subfamily G member 2 (JR blood group) (ABCG2), that have been detected in enterocytes may facilitate this process (Quigley et al. 2004; Krishnamurthy and Schuetz 2006); however, if such a pathway does exist, it is probably a quantitatively minor one. The mechanism governing heme iron absorption remains a major unresolved issue in the iron biology field.

6.7 Regulation of Iron Absorption

The amount of iron absorbed from the diet is determined by body iron requirements. It is increased in times of high iron demand and reduced when the body is iron replete (Collins et al. 2018). Under normal physiological conditions, the level of storage iron in the body and the adequacy of iron supply to the erythroid marrow are the systemic factors most often associated with changes in intestinal iron absorption. When body iron stores fall below the optimal level, and notably when iron stores are exhausted, intestinal

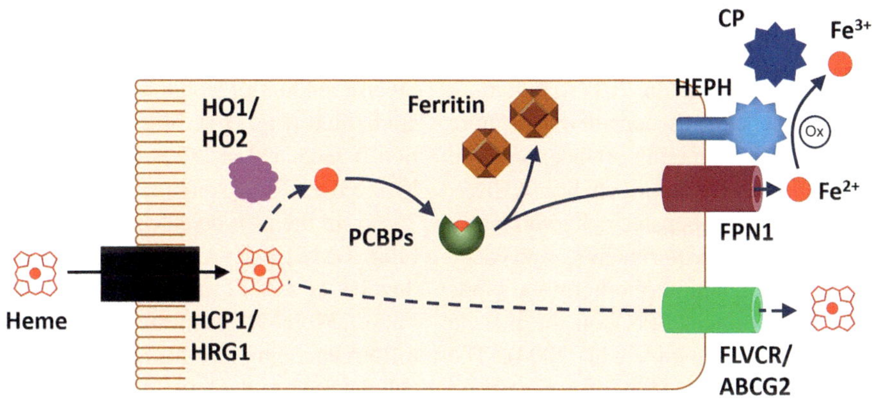

Fig. 6.3 Heme iron absorption. The mechanism of dietary heme iron absorption is poorly understood. Heme is thought to be absorbed intact across the BBM, a process possibly mediated by heme-regulated gene 1 (HRG1) or heme-carrier protein 1 (HCP1), although the latter is predominantly a folate transporter. Once inside the enterocyte, heme is likely broken down by heme oxygenase 1 and/or 2 (HO1/HO2), and the iron is released to join the intracellular iron pool. From here, iron can bind to poly(rC) binding proteins (PCBPs) and be delivered to ferritin for storage, or to ferroportin 1 (FPN1) for export from the cell followed by oxidation to Fe^{3+} by hephaestin (HEPH) or ceruloplasmin (CP). Some heme may be exported from the enterocyte intact, a process possibly facilitated by the heme transporters FLVCR, and ATP binding cassette subfamily G member 2 (JR blood group) (ABCG2)

iron absorption increases to make up for the shortfall (Sayers et al. 1994). If body iron levels are particularly low, the amount of iron bound by circulating TF can drop to a level at which the iron supply to critical tissues is compromised. Consistent with this, increased levels of *apo*-TF in the circulation are associated with enhanced iron absorption (Tsyplenkova et al. 2024). In contrast, when body iron stores are adequate or elevated, intestinal iron absorption is reduced.

Most body iron is found in hemoglobin within circulating erythrocytes, so it is not surprising that the iron demands of erythropoiesis influence iron absorption (Pasricha et al. 2016). An increase in the rate of erythropoiesis, e.g., following bleeding, often induces a corresponding increase in the amount of iron absorbed from the diet. In contrast, the suppression of erythropoiesis, as may occur following transfusion or irradiation, reduces iron absorption. In conditions such as β-thalassemia, the erythroid stimulus can be sufficiently strong that it overrides the suppressive effect of elevated iron stores, such that iron absorption remains elevated despite significant body iron load (Erlandson et al. 1962; Heinrich et al. 1973).

Hypoxia is also an important regulator of iron absorption (Mastrogiannaki et al. 2013). Absorption is increased under hypoxic conditions and returns to normal after the hypoxic stimulus is removed. Part of the effect of hypoxia can be ascribed to increased erythropoiesis, but hypoxia can also influence absorption through direct effects on iron transport and storage within enterocytes and via its effects on the body's iron homeostatic machinery in the liver. In each case, hypoxia directly or indirectly alters the expression of genes encoding iron transport proteins in the intestine (Mastrogiannaki et al. 2009; Shah et al. 2009; Peyssonnaux et al. 2007; Schwartz et al. 2019).

Absorption is also increased in pregnancy and in early postnatal life and is reduced by inflammation. These aspects of iron homeostasis are covered in detail in Chaps. 13 and 21, respectively.

6.7.1 Systemic Regulation of Iron Absorption

The liver-derived peptide hormone hepcidin plays a central role in the regulation of iron absorption and indeed in regulating iron homeostasis systemically (Collins et al. 2018; Ganz 2013; Sangkhae and Nemeth 2017) (Fig. 6.2). The biology of hepcidin is considered in more detail in Chap. 3, so its importance for iron absorption will only be described briefly here. Hepcidin production by hepatocytes is inversely related to body iron requirements, and once secreted into the circulation, the hormone binds to the iron export protein FPN1 on the plasma membrane of cells (Nemeth et al. 2004). The hepcidin-FPN1 complex is then internalized and degraded, thereby reducing the export of iron by target cells (Nemeth et al. 2004; Jiang et al. 2021). Hence, high circulating hepcidin levels mean low-iron absorption and vice versa.

Given its importance in controlling body iron intake, it is not surprising that hepcidin levels are low in various iron-loading disorders, and this can explain why affected individuals absorb excessive iron from their diet even though they may already have a high body iron load (Darshan et al. 2010). This is the case in the various forms of hemochromatosis that result from mutations in the genes encoding iron homeostatic regulator (*HFE*), transferrin receptor 2 (*TFR2*), hemojuvelin (*HJV*), and hepcidin itself (*HAMP*); see Chaps. 9 and 10. Certain mutations in FPN1 that make it unable to effectively respond to the action of hepcidin can also lead to increased absorption and body iron loading (Drakesmith et al. 2015). In contrast, mutations in the gene encoding transmembrane protease serine 6 (*TMPRSS6*), an upstream regulator of hepcidin, cause constitutively high hepcidin activity, which reduces iron absorption and results in profound anemia (known as iron refractory, iron-deficiency anemia [IRIDA]) in mice (Du et al. 2008) and in humans (Finberg et al. 2008); see also Chap. 12.

6.7.2 Local Regulation of Iron Absorption by Enterocytes

While systemic factors reflect how much iron the body needs to absorb and, via the hepcidin pathway, broadly determine the expression of key iron transport proteins in the gut, notably FPN1, these signals are integrated with enterocyte-specific regulatory mechanisms to determine the precise amount of iron that crosses the intestinal epithelium (Fig. 6.4). Much of this local regulation is mediated at the post-transcriptional level by the IRE/IRP system (see Chap. 2) (Galy et al. 2024). In the iron-depleted enterocyte, IRPs stabilize DMT1 mRNA, thereby increasing the production of DMT1 protein and facilitating iron import. Moreover, in most cells, translation of the mRNA encoding FPN1 is inhibited by IRPs when cellular iron is depleted, but this is inconsistent with increased iron absorption during iron deficiency. However, in enterocytes, a splice variant of FPN1 that does not contain an IRE predominates, and thus FPN1 levels remain relatively high when intracellular iron is low (Kuhn 2015; Zhang et al. 2009). The IRE/IRP system also regulates iron storage, with the synthesis of the iron storage protein ferritin being repressed when enterocyte iron levels are low.

Conversely, when enterocyte iron levels are high, a situation that might occur after iron-rich foods are consumed, or when systemic increases in hepcidin reduce FPN1 expression on the BLM, the IRE/IRP system downregulates the expression of DMT1 on the BBM to prevent the accumulation of potentially toxic iron levels and stimulates ferritin synthesis to enhance iron storage capacity. This mechanism can explain the phenomenon known as the "mucosal block," where a large dose of oral iron can inhibit the absorption of a subsequent smaller dose of iron (Frazer et al. 2003). It has also been demonstrated that high iron will lead to the removal of DMT1 from the BBM by endocytosis (Nunez 2010), thus adding another layer of regulatory complexity at the level of the enterocyte.

The hypoxia-inducible factors (HIF) family also contributes to the local regulation of iron absorption. Widely recognized as being regulated by oxygen levels, the HIFs are also stabilized in a low-iron environment, as prolyl hydroxylases, which degrade the HIFs under normoxic conditions, require iron for their activity (Shah et al. 2009). In the small intestine, the genes encoding DMT1, DCYTB, FPN1, and NCOA4 are all transcriptional targets of HIF2α, so under iron-

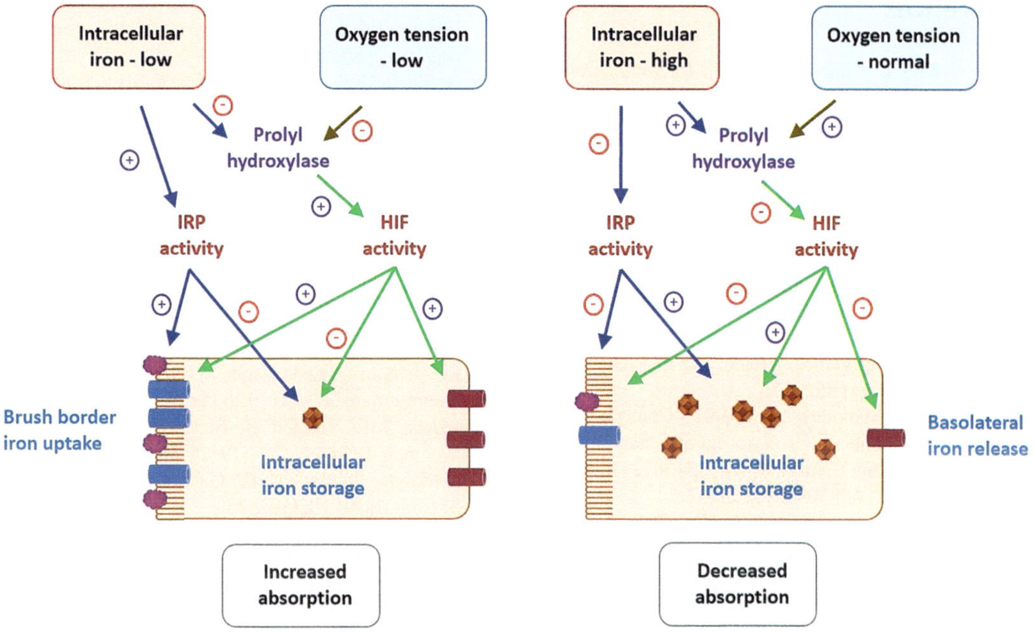

Fig. 6.4 Local regulation of iron absorption. Local factors that alter the amount of iron absorbed from the diet act at three points: the expression and localization of the brush-border iron uptake machinery; iron storage within ferritin; and the expression of the iron export protein ferroportin 1 (FPN1). Intracellular iron levels alter the binding activity of IRPs and affect the expression of DMT1 and ferritin, such that when iron levels are low, DMT1 is increased and ferritin is decreased. The opposite occurs when enterocyte iron levels are high. Intracellular iron can also affect the activity of prolyl hydroxylase and, subsequently, the HIFs. HIFs can regulate the expression of DMT1, duodenal cytochrome b (CYBRD1), FPN1, and nuclear receptor coactivator 4 (NCOA4), which stimulates ferritin iron release. Localized hypoxia can also affect HIF activity and, therefore, enterocyte iron absorption. Superimposed on these local regulatory mechanisms is the systemic regulation of iron absorption by circulating hepcidin

deficient/hypoxic conditions, their expression is increased (Mastrogiannaki et al. 2013). Consistent with the important role played by HIF2α in the gut, depleting HIF2α expression or activity in mice leads to major disturbances in iron absorption, including the inability of the animals to increase their absorption in response to iron deficiency or anemia (Yu et al. 2024).

6.8 Conclusions

The appropriate regulation of intestinal iron absorption is required for the health and well-being of all mammals, as this is the point at which body iron levels are controlled. While our understanding of the molecules and mechanisms involved in this process and how they are regulated has grown enormously in recent decades, there are still many areas in which our knowledge is, at best, rudimentary. One obvious limitation is our understanding of how heme iron is absorbed across the intestinal mucosa and whether it does so intact or is broken down within the enterocyte. Our knowledge of how non-heme iron traverses the enterocyte is also relatively poor, despite the discovery of the PCBPs as iron chaperones and their role in the intestine. Other emerging areas of importance include the effect of the microbiome on dietary iron uptake, with recent evidence suggesting that certain microbes can either enhance or inhibit iron uptake from the lumen of the gut (Das et al. 2020); see also Chap. 22. As many pathological changes in iron homeostasis are fundamentally due to inappropriate dietary iron absorption, it is important that this research continues, and we gain a greater knowledge of these processes to enable the development of better treatments for iron-related conditions.

References

Arita A, Kita I, Shinoda S (2022) Internalization and decrease of duodenal DMT1 involved in transient suppression of iron uptake in short-acting mucosal block. Biol Trace Elem Res 200(11):4795–4806. https://doi.org/10.1007/s12011-021-03053-x

Bernstein SE (1987) Hereditary hypotransferrinemia with hemosiderosis, a murine disorder resembling human atransferrinemia. J Lab Clin Med 110(6):690–705. doi:0022-2143(87)90467-7 [pii]

Bezwoda WR, Bothwell TH, Charlton RW, Torrance JD, MacPhail AP, Derman DP, Mayet F (1983) The relative dietary importance of haem and non-haem iron. S Afr Med J 64(14):552–556

Carpenter CE, Mahoney AW (1992) Contributions of heme and nonheme iron to human nutrition. Crit Rev Food Sci Nutr 31(4):333–367. https://doi.org/10.1080/10408399209527576

Cherukuri S, Potla R, Sarkar J, Nurko S, Harris ZL, Fox PL (2005) Unexpected role of ceruloplasmin in intestinal iron absorption. Cell Metab 2(5):309–319. https://doi.org/10.1016/j.cmet.2005.10.003

Collins JF, Flores SRL, Wang X, Anderson GJ (2018) Mechanisms and regulation of intestinal iron transport. In: Said HM (ed) Physiology of the gastrointestinal tract. Academic, pp 1451–1483. https://doi.org/10.1016/b978-0-12-809954-4.00060-8

Conrad ME Jr, Crosby WH (1963) Intestinal mucosal mechanisms controlling iron absorption. Blood 22:406–415

Darshan D, Frazer DM, Anderson GJ (2010) Molecular basis of iron-loading disorders. Expert Rev Mol Med 12:e36. https://doi.org/10.1017/S1462399410001687

Das NK, Schwartz AJ, Barthel G, Inohara N, Liu Q, Sankar A, Hill DR, Ma X, Lamberg O, Schnizlein MK, Arques JL, Spence JR, Nunez G, Patterson AD, Sun D, Young VB, Shah YM (2020) Microbial metabolite signaling is required for systemic iron homeostasis. Cell Metab 31(1):115–130 e116. https://doi.org/10.1016/j.cmet.2019.10.005

Das NK, Jain C, Sankar A, Schwartz AJ, Santana-Codina N, Solanki S, Zhang Z, Ma X, Parimi S, Rui L, Mancias JD, Shah YM (2022) Modulation of the HIF2alpha-NCOA4 axis in enterocytes attenuates iron loading in a mouse model of hemochromatosis. Blood 139(16):2547–2552. https://doi.org/10.1182/blood.2021013452

Doguer C, Ha JH, Gulec S, Vulpe CD, Anderson GJ, Collins JF (2017) Intestinal hephaestin potentiates iron absorption in weanling, adult, and pregnant mice under physiological conditions. Blood Adv 1(17):1335–1346. https://doi.org/10.1182/bloodadvances.2017008359

Donovan A, Lima CA, Pinkus JL, Pinkus GS, Zon LI, Robine S, Andrews NC (2005) The iron exporter ferroportin/Slc40a1 is essential for iron homeostasis. Cell Metab 1(3):191–200. https://doi.org/10.1016/j.cmet.2005.01.003

Drakesmith H, Nemeth E, Ganz T (2015) Ironing out Ferroportin. Cell Metab 22(5):777–787. https://doi.org/10.1016/j.cmet.2015.09.006

Du X, She E, Gelbart T, Truksa J, Lee P, Xia Y, Khovananth K, Mudd S, Mann N, Moresco EM, Beutler E, Beutler B (2008) The serine protease TMPRSS6 is required to sense iron deficiency. Science 320(5879):1088–1092. https://doi.org/10.1126/science.1157121

Duthie HL (1964) The relative importance of the duodenum in the intestinal absorption of iron. Br J Haematol 10:59–68. https://doi.org/10.1111/j.1365-2141.1964.tb00678.x

Dutt S, Hamza I, Bartnikas TB (2022) Molecular mechanisms of iron and heme metabolism. Annu Rev Nutr 42:311–335. https://doi.org/10.1146/annurev-nutr-062320-112625

Erlandson ME, Walden B, Stern G, Hilgartner MW, Wehman J, Smith CH (1962) Studies on congenital hemolytic syndromes, IV. Gastrointestinal absorption of iron. Blood 19:359–378

Finberg KE, Heeney MM, Campagna DR, Aydinok Y, Pearson HA, Hartman KR, Mayo MM, Samuel SM, Strouse JJ, Markianos K, Andrews NC, Fleming MD (2008) Mutations in TMPRSS6 cause iron-refractory iron deficiency anemia (IRIDA). Nat Genet 40(5):569–571. https://doi.org/10.1038/ng.130

Fleming MD, Trenor CC 3rd, Su MA, Foernzler D, Beier DR, Dietrich WF, Andrews NC (1997) Microcytic anaemia mice have a mutation in Nramp2, a candidate iron transporter gene. Nat Genet 16(4):383–386. https://doi.org/10.1038/ng0897-383

Frazer DM, Vulpe CD, McKie AT, Wilkins SJ, Trinder D, Cleghorn GJ, Anderson GJ (2001) Cloning and gastrointestinal expression of rat hephaestin: relationship to other iron transport proteins. Am J Physiol Gastrointest Liver Physiol 281(4):G931–G939. https://doi.org/10.1152/ajpgi.2001.281.4.G931

Frazer DM, Wilkins SJ, Becker EM, Murphy TL, Vulpe CD, McKie AT, Anderson GJ (2003) A rapid decrease in the expression of DMT1 and Dcytb but not Ireg1 or hephaestin explains the mucosal block phenomenon of iron absorption. Gut 52(3):340–346. https://doi.org/10.1136/gut.52.3.340

Fuqua BK, Lu Y, Darshan D, Frazer DM, Wilkins SJ, Wolkow N, Bell AG, Hsu J, Yu CC, Chen H, Dunaief JL, Anderson GJ, Vulpe CD (2014) The multicopper ferroxidase hephaestin enhances intestinal iron absorption in mice. PLoS One 9(6):e98792. https://doi.org/10.1371/journal.pone.0098792

Galy B, Conrad M, Muckenthaler M (2024) Mechanisms controlling cellular and systemic iron homeostasis. Nat Rev Mol Cell Biol 25(2):133–155. https://doi.org/10.1038/s41580-023-00648-1

Ganz T (2013) Systemic iron homeostasis. Physiol Rev 93(4):1721–1741. https://doi.org/10.1152/physrev.00008.2013

Gunshin H, Mackenzie B, Berger UV, Gunshin Y, Romero MF, Boron WF, Nussberger S, Gollan JL, Hediger MA (1997) Cloning and characterization of a mammalian proton-coupled metal-ion transporter. Nature 388(6641):482–488. https://doi.org/10.1038/41343

Gunshin H, Fujiwara Y, Custodio AO, Direnzo C, Robine S, Andrews NC (2005a) Slc11a2 is required for intestinal iron absorption and erythropoiesis but dispensable in placenta and liver. J Clin Invest 115(5):1258–1266. https://doi.org/10.1172/JCI24356

Gunshin H, Starr CN, Direnzo C, Fleming MD, Jin J, Greer EL, Sellers VM, Galica SM, Andrews NC (2005b) Cybrd1 (duodenal cytochrome b) is not necessary for dietary iron absorption in mice. Blood 106(8):2879–2883. https://doi.org/10.1182/blood-2005-02-0716

Heinrich HC, Gabbe EE, Oppitz KH, Whang DH, Bender-Gotze C, Schafer KH, Schroter W, Pfau AA (1973) Absorption of inorganic and food iron in children with heterozygous and homozygous beta-thalassemia. Z Kinderheilkd 115(1):1–22. https://doi.org/10.1007/BF00438987

Hellman NE, Gitlin JD (2002) Ceruloplasmin metabolism and function. Annu Rev Nutr 22:439–458. https://doi.org/10.1146/annurev.nutr.22.012502.114457

Helman SL, Zhou J, Fuqua BK, Lu Y, Collins JF, Chen H, Vulpe CD, Anderson GJ, Frazer DM (2023) The biology of mammalian multi-copper ferroxidases. Biometals 36(2):263–281. https://doi.org/10.1007/s10534-022-00370-z

Hudson DM, Krisinger MJ, Griffiths TA, Macgillivray RT (2008) Neither human hephaestin nor ceruloplasmin forms a stable complex with transferrin. J Cell Biochem 103(6):1849–1855. https://doi.org/10.1002/jcb.21566

Hurrell R, Egli I (2010) Iron bioavailability and dietary reference values. Am J Clin Nutr 91(5):1461S–1467S. https://doi.org/10.3945/ajcn.2010.28674F

Iolascon A, De Falco L (2009) Mutations in the gene encoding DMT1: clinical presentation and treatment. Semin Hematol 46(4):358–370. https://doi.org/10.1053/j.seminhematol.2009.06.005

Jiang L, Wang J, Wang K, Wang H, Wu Q, Yang C, Yu Y, Ni P, Zhong Y, Song Z, Xie E, Hu R, Min J, Wang F (2021) RNF217 regulates iron homeostasis through its E3 ubiquitin ligase activity by modulating ferroportin degradation. Blood 138(8):689–705. https://doi.org/10.1182/blood.2020008986

Krishnamurthy P, Schuetz JD (2006) Role of ABCG2/BCRP in biology and medicine. Annu Rev Pharmacol Toxicol 46:381–410. https://doi.org/10.1146/annurev.pharmtox.46.120604.141238

Kruszewski M (2003) Labile iron pool: the main determinant of cellular response to oxidative stress. Mutat Res 531(1–2):81–92. https://doi.org/10.1016/j.mrfmmm.2003.08.004

Kuhn LC (2015) Iron regulatory proteins and their role in controlling iron metabolism. Metallomics 7(2):232–243. https://doi.org/10.1039/c4mt00164h

Lane DJ, Bae DH, Merlot AM, Sahni S, Richardson DR (2015) Duodenal cytochrome b (DCYTB) in iron metabolism: an update on function and regulation. Nutrients 7(4):2274–2296. https://doi.org/10.3390/nu7042274

Lu J, Cai J, Ren T, Wu J, Mao D, Li W, Zhang Y, Piao J, Wang J, Yang L, Yang X, Ma Y (2019) Physiological requirements for iron in women of reproductive age assessed by the stable isotope tracer technique. Nutr Metabol 16:55. https://doi.org/10.1186/s12986-019-0384-1

Mackenzie B, Garrick MD (2005) Iron imports. II. Iron uptake at the apical membrane in the intestine. Am J Physiol Gastrointest Liver Physiol 289(6):G981–G986. https://doi.org/10.1152/ajpgi.00363.2005

Mastrogiannaki M, Matak P, Keith B, Simon MC, Vaulont S, Peyssonnaux C (2009) HIF-2alpha, but not HIF-1alpha, promotes iron absorption in mice. J Clin Invest 119(5):1159–1166. https://doi.org/10.1172/JCI38499

Mastrogiannaki M, Matak P, Peyssonnaux C (2013) The gut in iron homeostasis: role of HIF-2 under normal and pathological conditions. Blood 122(6):885–892. https://doi.org/10.1182/blood-2012-11-427765

McKie AT, Marciani P, Rolfs A, Brennan K, Wehr K, Barrow D, Miret S, Bomford A, Peters TJ, Farzaneh F, Hediger MA, Hentze MW, Simpson RJ (2000) A novel duodenal iron-regulated transporter, IREG1, implicated in the basolateral transfer of iron to the circulation. Mol Cell 5(2):299–309. https://doi.org/10.1016/s1097-2765(00)80425-6

McKie AT, Barrow D, Latunde-Dada GO, Rolfs A, Sager G, Mudaly E, Mudaly M, Richardson C, Barlow D, Bomford A, Peters TJ, Raja KB, Shirali S, Hediger MA, Farzaneh F, Simpson RJ (2001) An iron-regulated ferric reductase associated with the absorption of dietary iron. Science 291(5509):1755–1759. https://doi.org/10.1126/science.1057206

Mitchell CJ, Shawki A, Ganz T, Nemeth E, Mackenzie B (2014) Functional properties of human ferroportin, a cellular iron exporter reactive also with cobalt and zinc. Am J Physiol Cell Physiol 306(5):C450–C459. https://doi.org/10.1152/ajpcell.00348.2013

Morgan EH (1980) The role of plasma transferrin in iron absorption in the rat. Q J Exp Physiol Cogn Med Sci 65(3):239–252. https://doi.org/10.1113/expphysiol.1980.sp002510

Nemeth E, Tuttle MS, Powelson J, Vaughn MB, Donovan A, Ward DM, Ganz T, Kaplan J (2004) Hepcidin regulates cellular iron efflux by binding to ferroportin and inducing its internalization. Science 306(5704):2090–2093. https://doi.org/10.1126/science.1104742

Nunez MT (2010) Regulatory mechanisms of intestinal iron absorption-uncovering of a fast-response mechanism based on DMT1 and ferroportin endocytosis. Biofactors 36(2):88–97. https://doi.org/10.1002/biof.84

Pasricha SR, McHugh K, Drakesmith H (2016) Regulation of hepcidin by erythropoiesis: the story so far. Annu Rev Nutr 36:417–434. https://doi.org/10.1146/annurev-nutr-071715-050731

Peyssonnaux C, Zinkernagel AS, Schuepbach RA, Rankin E, Vaulont S, Haase VH, Nizet V, Johnson RS (2007)

Regulation of iron homeostasis by the hypoxia-inducible transcription factors (HIFs). J Clin Invest 117(7):1926–1932. https://doi.org/10.1172/JCI31370
Piskin E, Cianciosi D, Gulec S, Tomas M, Capanoglu E (2022) Iron absorption: factors, limitations, and improvement methods. ACS Omega 7(24):20441–20456. https://doi.org/10.1021/acsomega.2c01833
Qiu A, Jansen M, Sakaris A, Min SH, Chattopadhyay S, Tsai E, Sandoval C, Zhao R, Akabas MH, Goldman ID (2006) Identification of an intestinal folate transporter and the molecular basis for hereditary folate malabsorption. Cell 127(5):917–928. https://doi.org/10.1016/j.cell.2006.09.041
Quigley JG, Yang Z, Worthington MT, Phillips JD, Sabo KM, Sabath DE, Berg CL, Sassa S, Wood BL, Abkowitz JL (2004) Identification of a human heme exporter that is essential for erythropoiesis. Cell 118(6):757–766. https://doi.org/10.1016/j.cell.2004.08.014
Raffin SB, Woo CH, Roost KT, Price DC, Schmid R (1974) Intestinal absorption of hemoglobin iron-heme cleavage by mucosal heme oxygenase. J Clin Invest 54(6):1344–1352. https://doi.org/10.1172/JCI107881
Ranganathan PN, Lu Y, Fuqua BK, Collins JF (2012) Discovery of a cytosolic/soluble ferroxidase in rodent enterocytes. Proc Natl Acad Sci USA 109(9):3564–3569. https://doi.org/10.1073/pnas.1120833109
Sangkhae V, Nemeth E (2017) Regulation of the iron homeostatic hormone hepcidin. Adv Nutr 8(1):126–136. https://doi.org/10.3945/an.116.013961
Sayers MH, English G, Finch C (1994) Capacity of the store-regulator in maintaining iron balance. Am J Hematol 47(3):194–197. https://doi.org/10.1002/ajh.2830470309
Schwartz AJ, Das NK, Ramakrishnan SK, Jain C, Jurkovic MT, Wu J, Nemeth E, Lakhal-Littleton S, Colacino JA, Shah YM (2019) Hepatic hepcidin/intestinal HIF-2alpha axis maintains iron absorption during iron deficiency and overload. J Clin Invest 129(1):336–348. https://doi.org/10.1172/JCI122359
Shah YM, Matsubara T, Ito S, Yim SH, Gonzalez FJ (2009) Intestinal hypoxia-inducible transcription factors are essential for iron absorption following iron deficiency. Cell Metab 9(2):152–164. https://doi.org/10.1016/j.cmet.2008.12.012
Shawki A, Knight PB, Maliken BD, Niespodzany EJ, Mackenzie B (2012) H(+)-coupled divalent metal-ion transporter-1: functional properties, physiological roles and therapeutics. Curr Top Membr 70:169–214. https://doi.org/10.1016/B978-0-12-394316-3.00005-3
Shawki A, Engevik MA, Kim RS, Knight PB, Baik RA, Anthony SR, Worrell RT, Shull GE, Mackenzie B (2016) Intestinal brush-border Na+/H+ exchanger-3 drives H+-coupled iron absorption in the mouse. Am J Physiol Gastrointest Liver Physiol 311(3):G423–G430. https://doi.org/10.1152/ajpgi.00167.2016
Shayeghi M, Latunde-Dada GO, Oakhill JS, Laftah AH, Takeuchi K, Halliday N, Khan Y, Warley A, McCann FE, Hider RC, Frazer DM, Anderson GJ, Vulpe CD, Simpson RJ, McKie AT (2005) Identification of an intestinal heme transporter. Cell 122(5):789–801. https://doi.org/10.1016/j.cell.2005.06.025
Shi H, Bencze KZ, Stemmler TL, Philpott CC (2008) A cytosolic iron chaperone that delivers iron to ferritin. Science 320(5880):1207–1210. https://doi.org/10.1126/science.1157643
Skikne BS, Lynch SR, Cook JD (1981) Role of gastric acid in food iron absorption. Gastroenterology 81(6):1068–1071
Sukhbaatar N, Scholler M, Fritsch SD, Linke M, Horer S, Trager M, Mazic M, Forisch S, Gonzales K, Kahler JP, Binder C, Lassnig C, Strobl B, Muller M, Scheiber-Mojdehkar B, Gundacker C, Dabsch S, Kain R, Hengstschlager M, Verhelst SHL, Weiss G, Theurl I, Weichhart T (2023) Duodenal macrophages control dietary iron absorption via local degradation of transferrin. Blood 141(23):2878–2890. https://doi.org/10.1182/blood.2022016632
Tsyplenkova S, Charlebois E, Fillebeen C, Pantopoulos K (2024) Excess of circulating apotransferrin enhances dietary iron absorption in mice. Blood 144(1):117–121. https://doi.org/10.1182/blood.2023022916
Vanoaica L, Darshan D, Richman L, Schumann K, Kuhn LC (2010) Intestinal ferritin H is required for an accurate control of iron absorption. Cell Metab 12(3):273–282. https://doi.org/10.1016/j.cmet.2010.08.003
Vulpe CD, Kuo YM, Murphy TL, Cowley L, Askwith C, Libina N, Gitschier J, Anderson GJ (1999) Hephaestin, a ceruloplasmin homologue implicated in intestinal iron transport, is defective in the sla mouse. Nat Genet 21(2):195–199. https://doi.org/10.1038/5979
Wang Y, Protchenko O, Huber KD, Shakoury-Elizeh M, Ghosh MC, Philpott CC (2023) The iron chaperone poly(rC)-binding protein 1 regulates iron efflux through intestinal ferroportin in mice. Blood 142(19):1658–1671. https://doi.org/10.1182/blood.2023020504
West AR, Oates PS (2008a) Mechanisms of heme iron absorption: current questions and controversies. World J Gastroenterol 14(26):4101–4110. https://doi.org/10.3748/wjg.14.4101
West AR, Oates PS (2008b) Subcellular location of heme oxygenase 1 and 2 and divalent metal transporter 1 in relation to endocytotic markers during heme iron absorption. J Gastroenterol Hepatol 23(1):150–158. https://doi.org/10.1111/j.1440-1746.2007.05047.x
Woloshun RR, Yu Y, Xu X, Lee JK, Zhu S, Shine JS, Ebea P, Stevens BR, Vidyasagar S, Collins JF (2022) Four AAs increase DMT1 abundance in duodenal brush-border membrane vesicles and enhance iron absorption in iron-deprived mice. Blood Adv 6(10):3011–3021. https://doi.org/10.1182/bloodadvances.2021005111
Yanatori I, Yasui Y, Tabuchi M, Kishi F (2014) Chaperone protein involved in transmembrane transport of iron. Biochem J 462(1):25–37. https://doi.org/10.1042/BJ20140225

Yanatori I, Richardson DR, Imada K, Kishi F (2016) Iron export through the transporter ferroportin 1 is modulated by the iron chaperone PCBP2. J Biol Chem 291(33):17303–17318. https://doi.org/10.1074/jbc.M116.721936

Yu Y, Su Y, Yang S, Liu Y, Lin Z, Das NK, Wu Q, Zhou J, Sun S, Li X, Yue W, Shah YM, Min J, Wang F (2024) Activation of intestinal HIF2alpha ameliorates iron-refractory anemia. Adv Sci (Weinh) 11(12):e2307022. https://doi.org/10.1002/advs.202307022

Zhang DL, Hughes RM, Ollivierre-Wilson H, Ghosh MC, Rouault TA (2009) A ferroportin transcript that lacks an iron-responsive element enables duodenal and erythroid precursor cells to evade translational repression. Cell Metab 9(5):461–473. https://doi.org/10.1016/j.cmet.2009.03.006

7 Hepcidin and Tissue-Specific Iron Regulatory Networks

Samira Lakhal-Littleton and Carole Peyssonnaux

Abstract

Hepcidin is primarily secreted by the liver and functions as an endocrine hormone. However, a growing number of studies show that hepcidin can also be produced locally by other cells and organs, where it acts in an autocrine/paracrine manner to mediate important iron-dependent pathways. These pathways can operate under normal homeostatic conditions or become relevant in pathophysiological conditions (inflammation, infection, cancer, liver disease, myocardial infarction, etc.). This chapter will delve into the local roles of hepcidin, highlighting its unconventional functions in barrier maintenance, host defense, growth, tissue housekeeping, and injury repair.

Keywords

Hepcidin · Local iron regulation · Host defense · Barrier functions · Tissue housekeeping · Repair · Endocrine control · Cell division

S. Lakhal-Littleton (✉)
Department of Physiology, Anatomy & Genetics, University of Oxford, Oxford, UK
e-mail: samira.lakhal-littleton@dpag.ox.ac.uk

C. Peyssonnaux
Université Paris Cité, CNRS, INSERM, Institut Cochin, Paris, France

Laboratory of Excellence GR-Ex, Paris, France
e-mail: carole.peyssonnaux@inserm.fr

7.1 Extrahepatic Expression and Local Functions of Hepcidin

Hepcidin is the key hormone maintaining systemic iron homeostasis in the organism. This small, 25-amino-acid peptide, mainly produced by hepatocytes, controls plasma iron concentration by inhibiting intestinal iron absorption and iron recycling by macrophages. It prevents cellular iron efflux by binding to ferroportin (FPN), the only known exporter of cellular iron (Nemeth and Ganz 2021; see Chap. 3).

As suggested by its name, hepcidin is mainly produced in the liver, by hepatocytes (Pigeon et al. 2001; Park et al. 2001) in physiological conditions. However, pioneer experiments of subcutaneous pouch infection model in mice revealed hepcidin expression at the infection site, suggesting for the first time, that hepcidin could be directly expressed by cells other than hepatocytes, such as inflammatory cells recruited to the site of infection (Motley et al. 2004).

Building on this initial observation, the list of immune cells and non-hepatic tissues expressing hepcidin has been growing over the last two decades. Hepcidin-expressing immune cells include macrophages, neutrophils, lymphocytes,

K. Pantopoulos (ed.), *Iron Metabolism in Human Health and Disease*, Advances in Experimental Medicine and Biology 1480, https://doi.org/10.1007/978-3-031-92033-2_7

and dendritic cells (Peyssonnaux et al. 2006; Pinto et al. 2010; Bessman et al. 2020), while hepcidin-expressing tissues include the heart, lung, kidney, adipocytes, pancreas, brain, and retina (Kulaksiz et al. 2005, 2008; Bekri et al. 2006; Zechel et al. 2006; Merle et al. 2007; Gnana-Prakasam et al. 2008; Frazier et al. 2011a, b). Table 7.1 summarizes some of the cells that have been described to produce hepcidin and the proposed mode of action.

In order to discriminate the functions of extrahepatic hepcidin from that of systemic hepcidin, mouse models allowing tissue-specific gene deletion of hepcidin or harboring hepcidin-resistant ferroportin (FPN) variants have been generated (Zumerle et al. 2014; Lakhal-Littleton et al. 2015, 2016). Liver-specific hepcidin knockout (KO) mice have been shown to fully replicate the severe iron overload phenotype seen in total KO mice, harboring elevated plasma iron levels and extensive iron accumulation in the parenchyma. These findings confirm that hepatocytes serve as the primary reservoir for systemic hepcidin, with other tissues unable to compensate for its absence. This leaves open the question of the role of local hepcidin production in pathophysiological conditions.

A growing body of data suggests that locally produced hepcidin has unconventional functions in barrier maintenance, host defense, growth, tissue housekeeping, injury repair, and other mechanisms (outlined in Fig. 7.1).

7.2 Host Defense and Barrier Functions

Epithelial interfaces in the skin, airway, gastrointestinal tract, and renal tract are sites of exposure to pathogens and environmental agents, such as air pollutants, smoke, microbes, and microbial-derived toxins. Iron homeostasis at these sites plays an important role in maintaining epithelial integrity and barrier function and defending against pathogens.

In infectious conditions, local hepcidin production has been notably identified in myeloid cells (macrophages, neutrophils), lung epithelial cells, skin (keratinocytes), kidney (epithelial tubular cells) or stomach (gastric epithelial cells), with different functions/roles. Local production of hepcidin can be induced by inflammatory cytokines and by direct interactions with pathogens through pattern recognition receptors (PRRs), which stimulate hepcidin expression in response to microbial invasion. Depending on the context, the local actions of hepcidin at these sites can be beneficial or detrimental to the host. For instance, if infection or inflammation is sustained over a long period of time, locally produced hepcidin can contribute to iron-restricted anemia.

In macrophages, hepcidin production seems unaffected by iron status but is stimulated by pro-inflammatory signals like IL-6 or PAMPs. Bacteria, such as group A *Streptococcus* strains or *Pseudomonas aeruginosa*, can induce a 20- to 80-fold increase in hepcidin expression through a TLR-4–dependent pathway (Liu et al. 2005; Peyssonnaux et al. 2006). FPN expression, downregulated by autocrine hepcidin production, promotes iron sequestration within macrophages (Theurl et al. 2008), which subsequently limits extracellular bacterial infection. A growing set of data suggests that the hepcidin-FPN axis has various effects on the outcome of infections caused by intracellular bacteria (Michels et al. 2022). Interestingly, in infections with Leishmania, an obligate intracellular parasite of macrophages residing in the phagolysosomal compartment, parasites induce hepcidin-mediated proteasomal degradation of Nramp1, thereby enhancing phagolysosomal iron availability (Banerjee and Datta 2020). This specific strategy used by Leishmania allows its own intracellular growth.

In the airway, hepcidin and its target FPN are found both in alveolar macrophages and epithelial cells. In epithelial cells, hepcidin is induced by IFN-γ (Frazier et al. 2011a, b). In the contexts of bacterial pneumonia and sepsis-induced acute lung injury, epithelial hepcidin plays a dual role via local sequestration of iron: it mitigates bacterial growth and inhibits the phagocytic activity of macrophages (Yang et al. 2020; Chen et al. 2014). The net outcome of these effects is that epithelial hepcidin elevation improves barrier function at

Table 7.1 Extrahepatic hepcidin production in pathophysiological conditions and proposed mode(s) of action

Cells/organs	Hepcidin expression	Proposed mode(s) of action	References
Splenic macrophages Macrophages Peripheral blood mononuclear cells	Inflammation Bacterial infection (TLR agonists, group A *Streptococcus, Pseudomonas aeruginosa, Salmonella, Mycobacterium tuberculosis, Candida albicans, Influenza A virus*)	Autocrine/paracrine	Liu et al. (2005), Peyssonnaux et al. (2006), Armitage et al. (2011), Sow et al. (2011) and Theurl et al. (2008)
Alveolar macrophages Cardiac macrophages	*Streptococcus pneumonia* inflammation in infarcted heart		Nguyen et al. (2006), Perez et al. (2020) and Zlatanova et al. (2019)
Lymphocytes	Inflammation	Autocrine/paracrine	Pinto et al. (2010)
Conventional dendritic cells (intestine)	Colitis	Paracrine	Bessman et al. (2020)
Airway epithelial cells	Inflammation	Autocrine/paracrine	Frazier et al. (2011)
Cardiomyocytes	Basal level Inflammation Hypoxia	Cell autonomous	Lakhal-Littleton et al. (2015, 2016), Isoda et al. (2010), Simonis et al. (2010) and Merle et al. (2007)
Smooth muscle cells	Abdominal aortic aneurysm	Cell autonomous/paracrine	Loick et al. (2023)
Adipocytes	Inflammation	Autocrine/paracrine	Bekri et al. (2006) and Gotardo et al. (2013)
Pancreas (beta-cells of the islets of Langerhans)	Basal level Inflammation	Autocrine/paracrine	Kulaksiz et al. (2008)
Brain: Neurons Astrocytes Microglia	Basal level Inflammation	Autocrine/paracrine	Wang et al. (2008), Urrutia et al. (2013) and Qian et al. (2014)
Retina (Müller, photoreceptors, retinal pigment epithelium cells)	Inflammation	Autocrine/paracrine	Gnana-Prakasam et al. (2008)
Kidney (renal epithelial cells)	Basal level Infection (Uropathogenic *Escherichia coli*)	Autocrine/paracrine	Kulaksiz et al. (2005) and Houamel et al. (2016)
Colon (epithelial cells)	Colitis, cancer	Autocrine/paracrine	Gotardo et al. (2014) and Schwartz et al. (2021)
Stomach (parietal cells)	Infection (*Helicobacter pylori*)	Autocrine/paracrine	Schwarz et al. (2012)
Keratinocytes	Infection (group A *Streptococcus*) Inflammation (psoriasis)	Autocrine/paracrine	Malerba et al. (2020) and Abboud et al. (2024)

the airway epithelial interface (Lim et al. 2018). This barrier function is somewhat compromised by certain environmental stimuli, such as smoke. Indeed, cigarette smoke has been shown to impair lung hepcidin expression, resulting in elevated FPN expression in alveolar macrophages, which blunts their response to streptococcus pneumonia (Perez et al. 2020).

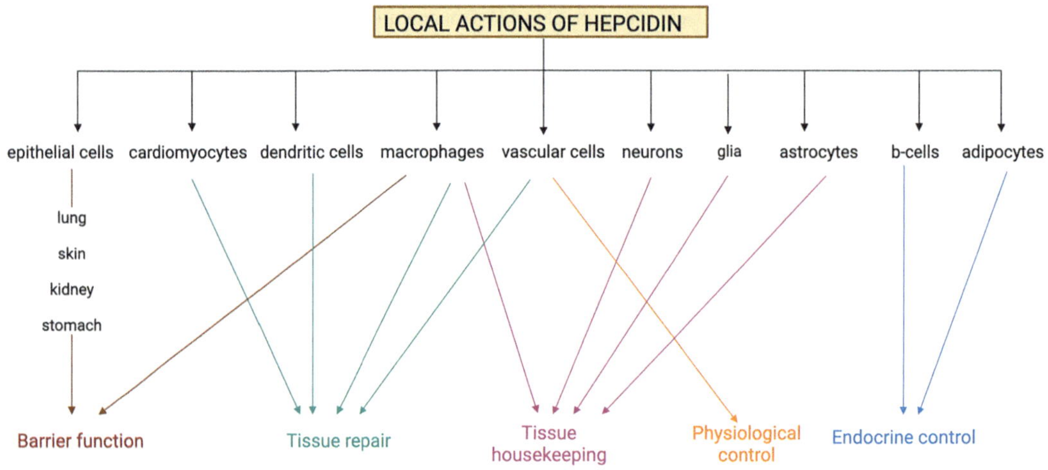

Fig. 7.1 Local actions of hepcidin

In the context of the skin, Malerba and colleagues demonstrated that hepcidin production is upregulated in the skin of patients with group A *Streptococcus* (GAS)-induced necrotizing fasciitis (NF). In a mouse model of GAS-induced NF, keratinocyte-specific hepcidin deficiency impaired the ability to restrict systemic infection spread from the initial tissue site. Surprisingly, this effect was linked to hepcidin's role in promoting CXCL1 chemokine production by keratinocytes, which drives neutrophil recruitment. Unlike CXCL1, hepcidin was resistant to degradation by major GAS proteases, allowing it to act as a reservoir to sustain CXCL1 levels in infected tissue. Furthermore, localized injection of synthetic hepcidin at the infection site effectively limited or prevented systemic GAS spread, highlighting the potential therapeutic value of hepcidin agonists in treating NF (Malerba et al. 2020).

Hepcidin levels have also been described to be higher in the lesions of multibacillary (MB) leprosy patients than in the skin biopsies of healthy control subjects (de Souza et al. 2012). Skin hepcidin levels directly correlated with the bacilloscopic index suggesting that hepcidin-mediated iron sequestration could facilitate *M. leprae* growth. However, in leprosy lesions, the majority of the bacilli exhibit a granular appearance, suggesting bacterial death. Therefore, a potential role for hepcidin in bacterial killing cannot be excluded.

Hepcidin has recently been shown to be highly expressed in the epidermis of psoriasis patients, particularly in the pustular type, a severe form of psoriasis resistant to treatments. In a mouse model of psoriasiform skin inflammation, keratinocyte-derived hepcidin was essential for iron retention in keratinocytes by promoting the degradation of FPN. Hepcidin was found to trigger psoriasiform skin inflammation by driving iron-induced epidermal cell proliferation and CXCL1-mediated neutrophil recruitment, two hallmark features of psoriatic lesions. Furthermore, transgenic mice with specific overexpression of hepcidin in keratinocytes developed iron accumulation, keratinocyte hyperproliferation, epidermal thickening (acanthosis), and neutrophil infiltration. Transcriptomic analysis of these keratinocytes revealed the activation of canonical pathways associated with human psoriasis, suggesting that hepcidin plays a causal role in the disease's pathogenesis. This study also uncovered a strong link between hepcidin and IL-36α as evidenced in particular by the highest hepcidin expression in pustular psoriasis patients with an IL36RN (interleukin 36 receptor antagonist) mutation. Taken together, these findings identify keratinocyte-derived hepcidin as a critical mediator in the development of psoriasiform skin inflammation, especially during the initiation and acute phases of the disease, and highlight its potential as a therapeutic target for psoriasis management (Abboud et al. 2024).

Finally, IL-36 has been linked to various inflammatory skin conditions beyond psoriasis, including hidradenitis suppurativa, atopic dermatitis, and allergic contact dermatitis. Investigating whether hepcidin also plays a role in these inflammatory skin disorders would be of interest. Altogether, skin hepcidin is an essential player of epidermis homeostasis, with an unanticipated role in neutrophil recruitment during diseases, which can be either beneficial, in the setting of cutaneous bacterial infection, or deleterious such as in psoriasis. This opens new avenues for therapeutic innovations, i.e., the use of local hepcidin agonists in cutaneous bacterial infection and hepcidin antagonists to alleviate the symptoms of cutaneous psoriasis.

Figure 7.2 outlines the functions of hepcidin in the skin.

In the gastric mucosa, *Helicobacter pylori* infection causes chronic inflammation which can potentially develop into gastric cancer. Several studies reported an increase in hepcidin levels in gastric parietal cells of the stomach infected with *Helicobacter pylori* (Schwarz et al. 2012; Kim et al. 2013) as well as in the lymph follicles of infected patients with nodular gastritis (Nishigaki et al. 2021). Unlike hepatic hepcidin, gastric hepcidin expression was downregulated in response to iron. In gastric cell lines, hepcidin was upregulated in response to IL-6 and bacterial infection. In humans, gastric hepcidin levels elevated during *Helicobacter pylori* infection normalized post-eradication. Hepcidin KO mice exhibited reduced acid secretion, linked to decreased Atp4a proton pump expression and subsequent gastric bacterial overgrowth. This study was pioneering in demonstrating hepcidin's role as a gastric factor crucial for acid secretion and combating Helicobacter infection.

In the context of the renal epithelium, the role of hepcidin has been explored during urinary tract infection. Using a urinary tract infection model, Houamel and colleagues found that hepcidin KO mice infected with Uropathogenic Escherichia coli (UPEC) exhibited greater bacterial loads and reduced renal inflammation compared to wild-type mice. Interestingly, Hepcidin KO mice also showed urine alkalization, linked to reduced Atp4a and Atp12a proton pump expression, as demonstrated in the stomach. In vitro studies demonstrated hepcidin's bacteriostatic effect against UPEC, which inhibited bacterial growth. Notably, UPEC repressed renal

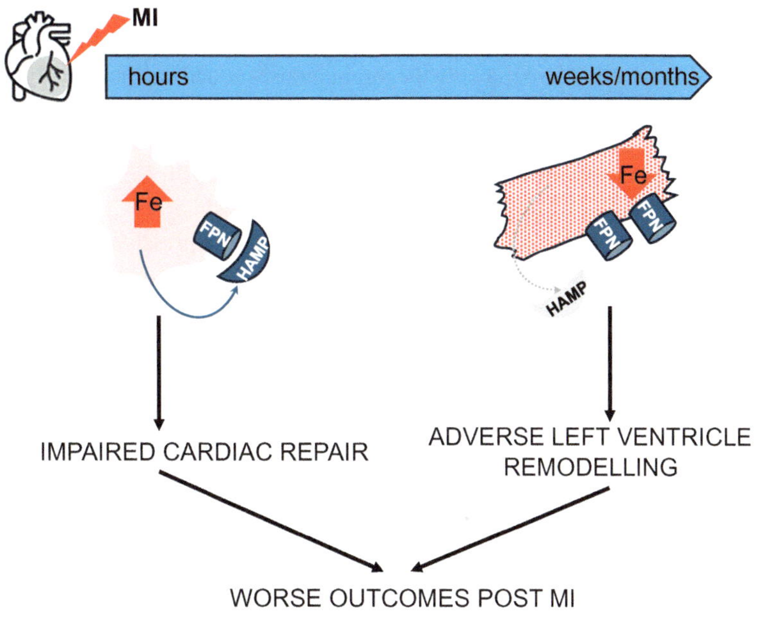

Fig. 7.2 Perturbations in cardiac hepcidin post MI determine outcomes post MI. In the acute stage post MI, raised hepcidin production by macrophages acts cell autonomously to promote injurious functions and impair repair functions. In the weeks and months post MI, cardiomyocyte hepcidin becomes suppressed, resulting in cardiomyocyte iron depletion, which promotes adverse LV remodeling

hepcidin in infected mice by reducing SMAD signaling, suggesting that bacteria attempt to counter the antibacterial function of hepcidin in early infection stages (Houamel et al. 2016).

Altogether, during inflammation, whether due to infection or other noninfectious agents, hepcidin can have pleiotropic roles: in nutritional immunity by regulating iron availability for pathogens, exerting immunomodulatory roles, or having direct or indirect antibacterial activity (through its antimicrobial properties or acid regulation, respectively).

7.3 Tissue Housekeeping and Repair

Tissue housekeeping is concerned with the appropriate handling of energy resources and potentially harmful substances such as misfolded proteins or reactive iron.

The histological sequelae observed in the brains of patients with neurodegenerative disorders illustrate the importance of tissue housekeeping. Iron is relevant in this context because abnormal iron deposits are a common denominator in neurodegenerative disorders, collectively termed neurodegeneration with brain iron accumulation (Chap. 19). Mis-management of brain iron content is increasingly being regarded as a contributor to the process of neurodegeneration. Astrocytes and microglia are the brain housekeepers that maintain brain ionic balance, and the integrity of the blood-brain barrier. They express both hepcidin and FPN. Astrocyte-derived hepcidin has been implicated in controlling iron trafficking at the blood-brain barrier, by regulating FPN in endothelial cells. This control appears to maintain normal brain iron levels and mitigate against β-amyloid-induced brain damage in mice (Zhang et al. 2020; Urrutia et al. 2017). On the other hand, astrocyte-derived hepcidin appears to have detrimental effects by promoting iron retention in neurons, which also express FPN. For instance, LPS-induced neuronal apoptosis appears to be mediated in part by the action of astrocyte hepcidin on neuronal FPN (You et al. 2022).

Sanfilippo syndrome is a pediatric neurodegenerative disease that shares many of the features of adult neurodegenerative diseases, including abnormal iron deposits. In a mouse model of Sanfilippo syndrome, hepcidin levels were seen to be elevated in microglia within the cerebral cortex and associated with lower brain FPN expression and greater iron retention (Puy et al. 2018). TLR4 signaling and STAT3 mediate induction of hepcidin in microglia, suggesting that local low-grade inflammation may contribute to abnormal iron deposits through the actions of glial or astrocyte-derived hepcidin. This process could also underpin increasing iron deposition within the ageing brain (Mezzanotte et al. 2022).

Tissue repair after injury involves a complex series of processes and interplay between resident and infiltrating immune cells, local fibroblasts, and the extracellular matrix. The net outcome of these repair processes can be adaptive or maladaptive. Hepcidin is elevated locally in the injured tissue, and this elevation is increasingly recognized as an important determinant of the outcome of tissue repair.

One example is the vascular injury sustained by endothelial and vascular cells lining the site of the thrombus in abdominal aortic aneurysm (AAA). An AAA is an abnormal dilatation of the abdominal aorta, defined as an aortic diameter of >3 cm. Left untreated, an AAA gradually expands and can potentially lead to a catastrophic rupture. Raised plasma levels of hepcidin have been reported in AAA patients (Martinez-Pinna et al. 2014; Loick et al. 2023). There is evidence that this rise is a spillover from raised hepcidin expression within smooth muscle cells lining the aneurysmal wall (Loick et al. 2023). In this setting, hepcidin acts cell-autonomously to contain thrombus-derived iron within smooth muscle cells. This containment triggers reparative autophagic pathways within smooth muscle cells, while also suppressing the local release of lipocalin 2, consequently limiting neutrophil infiltration into the aortic wall (Loick et al. 2023).

Kawasaki disease (KD) is a febrile coronary vasculitis that affects children and can result in complications such as coronary artery aneurysm.

Hepcidin gene polymorphisms and promoter hypomethylation are associated with susceptibility to KD, as well as worse clinical outcomes (Huang et al. 2017). This is accompanied by higher levels of hepcidin in the plasma and development of anemia (Yang et al. 2022). Another vascular injury setting where hepcidin is emerging as player is aneurysmal subarachnoid hemorrhage (aSAH), which significantly threaten the integrity of the BBB (Tan et al. 2006, Wei et al. 2024). Again, hepcidin gene variants and promoter methylation are associated with clinical outcomes of aSAH (Heinsberg et al. 2020). In a rat model of experimental SAH, hepcidin expression increased in the cerebral cortex and hippocampus, associated with higher brain water content and BBB permeability (Tan G et al. 2016). These lines of evidence give a clear indication that local iron regulation has an essential role in maintaining the integrity of the vascular wall and suggest that local overproduction of hepcidin, perhaps by smooth muscle or endothelial cells, could be either injurious or reparative, depending on the context (Naito and Ishihara 2022).

Tissue repair is also an important determinant of clinical outcomes after myocardial infarction (MI). In the healthy well-perfused heart, cardiomyocytes express both hepcidin and FPN. In this context, the cardiomyocyte hepcidin/FPN axis acts cell-autonomously to finetune intracellular iron levels and mitigate or buffer against fluctuations in plasma iron availability (Lakhal-Littleton et al. 2015, 2016). After MI, locally produced hepcidin, from the myocardium and infiltrating immune cells, and circulating hepcidin influence many of the stages of cardiac repair. In the acute stages of injury, hepcidin levels rise rapidly in the plasma, infarct, and remote myocardium (Van Breda et al. 2016; Isoda et al. 2010; Simonis et al. 2010). This is likely under the influence of IL-6, which has been shown to upregulate hepcidin in cardiac myocytes (Lakhal-Littleton et al. 2016). Furthermore, IL-6 receptor antagonism with tocilizumab inhibits the rise in plasma hepcidin in patients post MI (George et al. 2020).

Work in hepcidin knockout mice suggests that hepcidin elevation in heart-infiltrating macrophages compromises their reparative function. Interestingly, despite elevation of myocardial hepcidin itself in this context, it did not appear to influence macrophages in the same manner (Zlatanova et al. 2019). Clearly, more work is needed to address the interplay between the myocardial and macrophage hepcidin/FPN axes in the acute stages of myocardial injury. Later, in the process of cardiac remodeling post MI, myocardial hepcidin has been shown to become suppressed, resulting in elevated myocardial FPN, myocardial iron depletion and adverse left ventricular remodeling (Chung et al. 2023). Elevation of myocardial FPN and the ensuing myocardial iron depletion have been reported both in experimental murine model of MI and in biopsies from failing human hearts (Zhang et al. 2022; Melenovsky et al. 2017).

Figure 7.3 outlines the functions of hepcidin at different stages of myocardial repair.

Furthermore, in heart failure, lower plasma hepcidin was found be an independent predictor of mortality (Jankowska et al. 2014). Notably, modulation of myocardial iron and hepcidin appear to mediate some of the pleiotropic benefits of Sodium Glucose Transporter 2 inhibitors (SGLT2i). These drugs, previously designed for the management of glucose levels in diabetes, are now also part of guideline recommendations for the management of heart failure, because of their beneficial effects on clinical and functional outcomes, even in non-diabetic patients (Lakhal-Littleton and Cleland 2024). There is mounting evidence that these drugs have direct and profound effects on myocardial iron homeostasis (Schmidt et al. 2024). One study showed that SGLT2i prevent myocardial hepcidin elevation after ischemia-reperfusion injury, resulting in less adverse remodeling (Chen et al. 2023).

Hepcidin is also important for tissue repair in the gut. Bessman and colleagues revealed that hepcidin was essential for tissue repair in the mouse intestine following experimental injury. This function was independent of hepatocyte-derived hepcidin or systemic iron levels. Instead, they discovered that conventional dendritic cells (cDCs) serve as a source of hepcidin, induced by microbial stimulation in mice, prominently expressed in inflamed human intestines, and criti-

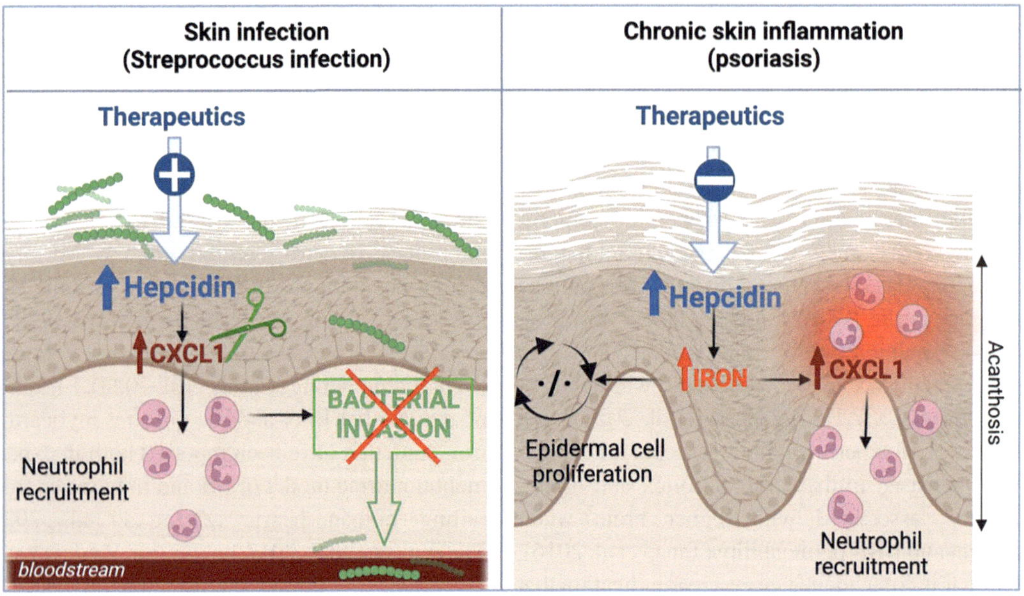

Fig. 7.3 Epidermal hepcidin in infectious and inflammatory diseases. Epidermal hepcidin has an unexpected function in neutrophil recruitment that can be either advantageous, as seen in cutaneous bacterial infections, or harmful, as in the case of psoriasis. These findings suggest potential new therapeutic approaches, such as the use of local hepcidin agonists for treating cutaneous bacterial infections and hepcidin antagonists to alleviate symptoms of psoriasis

cal for tissue repair. Hepcidin produced by cDCs acted on FPN-expressing phagocytes to promote local iron sequestration, which modulated the microbiota and, in turn, supported intestinal healing. These findings reveal a pathway in which cDC-derived hepcidin facilitates mucosal repair in the intestine via mechanisms of nutritional immunity (Bessman et al. 2020).

In trinitrobenzene sulfonic acid-induced colitis in rats, hepcidin was detected in the colon epithelium, and correlated with IL-6 production and STAT3 activation (Gotardo et al. 2014). While local iron deposition was also noted, further research is needed to determine whether colonic hepcidin represents a reparative or injurious factor in this context.

A recent study reported that in chronic atrophic gastritis (CAG) patients, elevated levels of hepcidin induced ferroptosis specifically in the stomach and impeded iron absorption in the duodenum. Expression of hepcidin during CAG was also mediated by the IL-6/STAT3 pathways (Zhao et al. 2023).

7.4 Physiological and Endocrine Control

Given the requirement for iron by a variety of metabolic and sensing proteins, it is perhaps not surprising that local functions of hepcidin can influence or even directly regulate the physiological and metabolic functions of certain tissues.

It has long been recognized that iron levels modify normal physiological responses to hypoxia, because the prolyl hydroxylases that sense oxygen and accordingly target HIF for degradation, also require iron as a co-factor. This means that tissue iron deficiency can induce a state of pseudohypoxia. If chronic, it can engender the same maladaptive sequelae as chronic hypoxia. This paradigm is exemplified in the context of pulmonary vascular response to hypoxia. Hypoxic pulmonary vasoconstriction (HPV) serves to optimize gas exchange by redirecting blood flow to the better-ventilated parts of the lungs. The HPV response requires endothelial and smooth muscle cells lining the pulmonary arteries to sense a drop in oxygen tension and upregulate the HIF pathway. HIF drives HPV through multiple targets, notably the vasoconstrictor endothelin 1. In healthy individuals, the magnitude of HPV, measured by a rise in pulmonary arterial pressure, is greater if they are iron-deficient (Frise et al. 2016). Furthermore, iron chelation enhances whereas iron supplementation dampens these responses (Bart et al. 2016; Smith et al. 2009). Hepcidin comes into play because pulmonary arterial smooth muscle cells (PASMCs) express both hepcidin and FPN and require this axis for autocrine regulation of intracellular iron levels. Loss of this control results in PASMC iron deficiency, sustained elevation of endothelin-1 expression, remodeling of the pulmonary arteries, and pulmonary arterial hypertension in mice (Lakhal-Littleton et al. 2019). In humans, PAH results from irreversible remodeling of the pulmonary arteries due to sustained vasoconstriction under the influence of chronic hypoxia. Thus, loss of hepcidin control in the pulmonary arteries can engender a sustained pseudo-hypoxic response and consequently progression to PAH (Lakhal-Littleton et al., 2019). Of note, iron deficiency is associated with worse outcomes in hereditary and idiopathic PAH (Soon et al. 2011; van Empel et al. 2014). Hereditary PAH is driven primarily by mutations in bone morphogenetic protein receptor 2 (BMPR2). It has been shown that PAH-causing BMPR2 mutations reduce hepcidin production in PASMCs, suggesting this could be one of the mechanistic links between mutation and disease (Lakhal-Littleton et al., 2019).

Another emerging aspect of local hepcidin function is in endocrine control. Clinical studies have reported that inappropriately elevated hepcidin levels are associated with many components of metabolic syndrome (Chap. 17), likely reflecting the actions of hepcidin in the pancreas and adipose tissue (Bek et al. 1992; Yook et al., 2021). In the pancreas, hepcidin localizes to beta-cells of the islets of Langerhans, where its expression is regulated by iron (Kulaksiz et al. 2008). Paradoxically, mice with mutations in matriptase 2, which have iron-refractory iron deficiency anemia due inappropriate elevation of hepcidin, also have improved glucose tolerance and insulin sensitivity (Folgueras et al., 2008). While FPN expression has yet to be demonstrated in β-cells, hephaestin is expressed (Maxel et al. 2019). There are a number of ways through which intracellular iron levels can modulate insulin secretion by β-cells. First, as iron is required in mitochondrial aconitase, its availability regulates the TCA cycle and glucose oxidation to generate ATP. The ATP/ADP ratio determines the opening of ATP-dependent potassium channels, membrane depolarization, and the opening of voltage-gated calcium channels that promote insulin secretion. Another mechanism is through modulation of HIF degradation. In β-cells, clock gene regulation of HIF appears to be important for circadian control of insulin release (Peek et al. 2017). Thus, excessive or inappropriately low hepcidin action at this site could perturb the circadian control of glucose levels by modulating HIF degradation.

Adipocytes also express FPN and hepcidin (Gotardo et al. 2013; Deschemin et al. 2023). This expression is not involved in the regulation of systemic iron status because adipocyte-specific FPN knockouts do not display overt changes in systemic iron levels (Britton et al. 2018). Instead, they regulate intracellular iron levels, which in turn positively or negatively influence the endocrine and energetic functions of the adipose tissue. Increased intracellular iron content within adipocytes decreases leptin production, thereby disrupting appetite control and energy homeostasis. Conversely, leptin receptor activation directly affects iron metabolism by increasing hepcidin expression. Sustained cytosolic aconitase activity, which would occur as a result of the autocrine action of hepcidin promotes adipogenic capacity. At the same time, adipose fat expansion in mice that are fed a high-fat diet is accompanied by increased hepcidin expression in adipose tissue (Gotardo 2013). Hepcidin deficiency in mice has been shown to impair white adipose tissue browning possibly due to a defect in de novo adipogenesis (Yook et al. 2021).

7.5 Growth, Development, and Cell Division

Iron is important for growth and development. Cells require iron to divide because iron is a cofactor for many of the proteins involved in DNA replication and repair (Lakhal-Littleton 2024). Additionally, iron is needed for the synthesis of many of the structural and functional components of tissues. For instance, iron is required for collagen synthesis in connective tissue, myelin synthesis in neuronal development, and hematopoietic stem cells (HSCs) that will later populate the bone marrow. It is therefore not surprising that the regulation of locally produced hepcidin can be used to secure the iron that tissues need for growth and development.

The clearest evidence for the role of hepcidin in supporting development is seen in the context of pregnancy (see also Chap. 21). Besides maternal hepcidin itself that is typically suppressed to increase maternal iron supply, hepcidin is also present in fetal liver, cord blood, and placenta (Principe et al. 2023).

In mice, fetuses harboring a paternally-inherited ubiquitous knock-in of the HAMP-resistant fpnC326Y, or a liver-specific knock-in of fpnC326Y or knockout of the *Hamp* gene, have reduced liver iron stores and hemoglobin, and markedly increased FPN in the liver. Thus, fetal liver HAMP operates cell-autonomous manner to increase fetal liver iron stores and support hepatic hematopoiesis (Kämmerer et al. 2020).

In the placenta, hepcidin is found in syncytiotrophoblasts and villous tissue, where it is proposed to regulate iron flux and, consequently, iron supply to meet the demands of fetal growth. Placental hepcidin expression is increased in response to nitric oxide (NO), and this has been suggested to contribute to fetal growth restriction in preeclampsia (Principe et al. 2023). Conversely, hepcidin gene expression is suppressed in the placentas of women with gestational diabetes, consistent with the notion that increased iron flux through the placenta contributes to excessive fetal growth in this context (Yang et al. 2016).

Because cell division has a high requirement for iron, cancers can also modulate hepcidin to secure iron for proliferation (see also Chap. 18). Elevated hepcidin levels are observed in various cancer types, restricting FPN-mediated iron export and promoting iron sequestration within cancer cells (Lin et al. 2023). A recent study showed that hepcidin is ectopically activated in colorectal cancer (CRC) (Schwartz et al. 2021). Mice lacking hepcidin specifically in the colon tumor epithelium exhibit significantly reduced tumor number, burden, and size in a sporadic CRC model compared to wild-type littermates. In contrast, the accumulation of intracellular iron due to the deletion of the iron exporter FPN worsens these tumor characteristics. Metabolomic analysis of patient-derived CRC tumor enteroids reveals that CRC prioritizes iron for nucleotide production, a phenomenon also observed in mouse CRC models. Mechanistically, the data suggest that iron chelation impairs mitochondrial function, leading to altered nucleotide synthesis, whereas the supplementation of nucleosides or aspartate partially rescues tumor growth in patient-derived enteroids and CRC cell lines treated with an iron chelator. Taken together, these findings indicate that ectopic hepcidin in the tumor epithelium creates a mechanism to sequester iron, supporting nucleotide synthesis and promoting proliferation in colorectal tumors. Analysis of primary tumors and axillary lymph nodes in breast cancer patients has revealed a significant iron reservoir in stromal inflammatory cells (Marques et al. 2016). Concurrently, breast cancer cells exhibit an "iron-utilization" phenotype, marked by increased expression of hepcidin and transferrin receptor 1 (Tfr1) and decreased expression of ferritin, while macrophages in primary tumors and metastatic lymph nodes display an "iron-donor" phenotype, characterized by heightened expression of FPN and ferritin (Marques et al. 2016). Hepcidin expression in macrophages and neutrophils could be further upregulated by proinflammatory cytokines of the tumor microenvironment (TME), inhibiting FPN on cancer cells as a paracrine factor, leading to increased intracellular iron content and promoting cancer cell proliferation.

7.6 Conclusions

Since its discovery at the turn of the century, our understanding of the pleiotropic functions of hepcidin, beyond systemic iron control, has grown. The local functions of hepcidin in various tissues are diverse, ranging from tissue homeostasis and repair to barrier function, as well as physiological and endocrine regulation. Sometimes, these functions are exerted in a healthy state to maintain adequate intracellular iron levels. In certain circumstances, however, local hepcidin dysregulation exacerbates underlying pathology. What remains unexplored is how the different environments in the tissues that express hepcidin dictate differences in mode of regulation, structure, and post-translational modifications, and how autocrine-acting and paracrine-acting hepcidins differ from circulating hepcidin. Ultimately, a firm understanding of these differences could enable targeted correction of tissue iron dys-homeostasis for therapeutic purposes.

References

Abboud E, Chrayteh D, Boussetta N et al (2024) Skin hepcidin initiates psoriasiform skin inflammation via Fe-driven hyperproliferation and neutrophil recruitment. Nat Commun 15:6718. https://doi.org/10.1038/s41467-024-50993-8

Armitage AE, Eddowes LA, Gileadi U, Cole S, Spottiswoode N, Selvakumar TA, Ho LP, Townsend AR, Drakesmith H (2011) Hepcidin regulation by innate immune and infectious stimuli. Blood 118(15):4129–4139. https://doi.org/10.1182/blood-2011-04-351957

Banerjee S, Datta R (2020) Leishmania infection triggers hepcidin-mediated proteasomal degradation of Nramp1 to increase phagolysosomal iron availability. Cell Microbiol 22:e13253. https://doi.org/10.1111/cmi.13253

Bart NK, Curtis MK, Cheng HY et al (2016) Elevation of iron storage in humans attenuates the pulmonary vascular response to hypoxia. J Appl Physiol 121(2):537–544. https://doi.org/10.1152/japplphysiol.00032.2016

Bek SG, Üstüner B, Eren N, Sentürk Z, Gönüllü BK (1992) The effect of hepcidin on components of metabolic syndrome in chronic kidney disease: a cross-sectional study. Rev Assoc Med Bras 66(8):1100–1107. https://doi.org/10.1590/1806-9282.66.8.1100

Bekri S, Gual P, Anty R et al (2006) Increased adipose tissue expression of hepcidin in severe obesity is independent from diabetes and NASH. Gastroenterology 131:788–796

Bessman NJ, Mathieu JRR, Renassia C et al (2020) Dendritic cell-derived hepcidin sequesters iron from the microbiota to promote mucosal healing. Science 368:186–189. https://doi.org/10.1126/science.aau6481

Britton L, Jaskowski LA, Bridle K et al (2018) Ferroportin expression in adipocytes does not contribute to iron homeostasis or metabolic responses to a high calorie diet. Cell Mol Gastroenterol Hepatol 5(3):319–331. Published 2018 Jan 9. https://doi.org/10.1016/j.jcmgh.2018.01.005

Chen QX, Song SW, Chen QH et al (2014) Silencing airway epithelial cell-derived hepcidin exacerbates sepsis induced acute lung injury. Crit Care 18(4):470. Published 2014 Aug 6. https://doi.org/10.1186/s13054-014-0470-8

Chen R, Zhang Y, Zhang H et al (2023) SGLT2 inhibitor dapagliflozin alleviates intramyocardial hemorrhage and adverse ventricular remodeling via suppressing hepcidin in myocardial ischemia-reperfusion injury. Eur J Pharmacol 950:175729. https://doi.org/10.1016/j.ejphar.2023.175729

Chung B, Wang Y, Thiel M et al (2023) Pre-emptive iron supplementation prevents myocardial iron deficiency and attenuates adverse remodelling after myocardial infarction. Cardiovasc Res 119(10):1969–1980. https://doi.org/10.1093/cvr/cvad092

de Souza VNB, de Malaspina TS, Campanelli AP et al (2012) Increased hepcidin expression in multibacillary leprosy. Mem Inst Oswaldo Cruz 107(Suppl 1):183–189. https://doi.org/10.1590/s0074-02762012000900026

Deschemin JC, Ransy C, Bouillaud F et al (2023) Hepcidin deficiency in mice impairs white adipose tissue browning possibly due to a defect in de novo adipogenesis. Sci Rep 13(1):12794. Published 2023 Aug 7. https://doi.org/10.1038/s41598-023-39305-0

Folgueras AR, de Lara FM, Pendás AM, Garabaya C, Rodríguez F, Astudillo A, Bernal T, Cabanillas R, López-Otín C, Velasco G. Membrane-bound serine protease matriptase-2 (Tmprss6) is an essential regulator of iron homeostasis. Blood. 2008 Sep 15;112(6):2539-45. https://doi.org/10.1182/blood-2008-04-149773. Epub 2008 Jun 3. PMID:18523150.

Frazier MD, Mamo LB, Ghio AJ, Turi JL (2011a) Hepcidin expression in human airway epithelial cells is regulated by interferon-γ. Respir Res 12:100. https://doi.org/10.1186/1465-9921-12-100

Frazier MD, Mamo LB, Ghio AJ, Turi JL (2011b) Hepcidin expression in human airway epithelial cells is regulated by interferon-γ. Respir Res 12(1):100. Published 2011 Aug 2. https://doi.org/10.1186/1465-9921-12-100

Frise MC, Cheng HY, Nickol AH et al (2016) Clinical iron deficiency disturbs normal human responses to hypoxia. J Clin Invest 126(6):2139–2150. https://doi.org/10.1172/JCI85715

Frazier MD, Mamo LB, Ghio AJ, Turi JL. Hepcidin expression in human airway epithelial cells is regulated by interferon-γ. Respir Res. 2011 Aug 2;12(1):100. https://doi.org/10.1186/1465-9921-12-100. PMID: 21810240; PMCID: PMC3160958.

George MJ, Kleveland O, Garcia-Hernandez J et al (2020) Novel insights into the effects of interleukin 6 antagonism in non-ST-segment-elevation myocardial infarction employing the SOMAscan proteomics platform. J Am Heart Assoc 9(12):e015628. https://doi.org/10.1161/JAHA.119.015628

Gnana-Prakasam JP, Martin PM, Mysona BA et al (2008) Hepcidin expression in mouse retina and its regulation via lipopolysaccharide/toll-like receptor-4 pathway independent of Hfe. Biochem J 411:79–88. https://doi.org/10.1042/BJ20071377

Gotardo ÉM, dos Santos AN, Miyashiro RA et al (2013) Mice that are fed a high-fat diet display increased hepcidin expression in adipose tissue. J Nutr Sci Vitaminol (Tokyo) 59(5):454–461. https://doi.org/10.3177/jnsv.59.454

Gotardo EM, Ribeiro Gde A, Clemente TR et al (2014) Hepcidin expression in colon during trinitrobenzene sulfonic acid-induced colitis in rats. World J Gastroenterol 20:4345–4352. https://doi.org/10.3748/wjg.v20.i15.4345

Heinsberg LW, Arockiaraj AI, Crago EA et al (2020) Genetic variability and trajectories of DNA methylation may support a role for HAMP in patient outcomes after aneurysmal subarachnoid hemorrhage. Neurocrit Care 32(2):550–563. https://doi.org/10.1007/s12028-019-00787-4

Houamel D, Ducrot N, Lefebvre T et al (2016) Hepcidin as a major component of renal antibacterial defenses against Uropathogenic Escherichia coli. J Am Soc Nephrol 27:835–846. https://doi.org/10.1681/ASN.2014101035

Huang YH, Yang KD, Hsu YW et al (2017) Correlation of HAMP gene polymorphisms and expression with the susceptibility and length of hospital stays in Taiwanese children with Kawasaki disease. Oncotarget 8(31):51859–51868. Published 2017 May 8. https://doi.org/10.18632/oncotarget.17700

Isoda M, Hanawa H, Watanabe R et al (2010) Expression of the peptide hormone hepcidin increases in cardiomyocytes under myocarditis and myocardial infarction. J Nutr Biochem 21(8):749–756. https://doi.org/10.1016/j.jnutbio.2009.04.009

Jankowska EA, Kasztura M, Sokolski M et al (2014) Iron deficiency defined as depleted iron stores accompanied by unmet cellular iron requirements identifies patients at the highest risk of death after an episode of acute heart failure. Eur Heart J 35(36):2468–2476. https://doi.org/10.1093/eurheartj/ehu235

Kämmerer L, Mohammad G, Wolna M, Robbin PA, Lakhal-Littleton S (2020) Fetal liver hepcidin secures iron stores in utero. Blood 136(13):1549–1557. https://doi.org/10.1182/blood.2019003907

Kim H-K, Jang E-C, Yeom J-O et al (2013) Serum prohepcidin levels are lower in patients with atrophic gastritis. Gastroenterol Res Pract 2013:201810. https://doi.org/10.1155/2013/201810

Kulaksiz H, Theilig F, Bachmann S et al (2005) The iron-regulatory peptide hormone hepcidin: expression and cellular localization in the mammalian kidney. J Endocrinol 184:361–370

Kulaksiz H, Fein E, Redecker P, Stremmel W, Adler G, Cetin Y (2008) Pancreatic beta-cells express hepcidin, an iron-uptake regulatory peptide. J Endocrinol 197(2):241–249. https://doi.org/10.1677/JOE-07-0528

Lakhal-Littleton S (2024) Anaemia, neurohormonal activation and myocardial iron depletion in heart failure: can this vicious circle be broken? Eur Heart J. (in press)

Lakhal-Littleton S, Cleland JGF (2024) Iron deficiency and supplementation in heart failure. Nat Rev Cardiol 21(7):463–486. https://doi.org/10.1038/s41569-024-00988-1

Lakhal-Littleton S, Wolna M, Carr CA et al (2015) Cardiac ferroportin regulates cellular iron homeostasis and is important for cardiac function. Proc Natl Acad Sci USA 112:3164–3169. https://doi.org/10.1073/pnas.1422373112

Lakhal-Littleton S, Wolna M, Chung YJ et al (2016) An essential cell-autonomous role for hepcidin in cardiac iron homeostasis. eLife 5:e19804. https://doi.org/10.7554/eLife.19804

Lakhal-Littleton S, Crosby A, Frise MC, Mohammad G, Carr CA, Loick PAM, Robbins PA. Intracellular iron deficiency in pulmonary arterial smooth muscle cells induces pulmonary arterial hypertension in mice. Proc Natl Acad Sci U S A. 2019 Jun 25;116(26):13122-13130. https://doi.org/10.1073/pnas.1822010116. Epub 2019 May 31. PMID:31152133; PMCID: PMC6600981

Lim D, Kim KS, Jeong JH et al (2018) The hepcidin-ferroportin axis controls the iron content of Salmonella-containing vacuoles in macrophages. Nat Commun 9(1):2091. Published 2018 May 29. https://doi.org/10.1038/s41467-018-04446-8

Lin F, Tuffour A, Hao G et al (2023) Distinctive modulation of hepcidin in cancer and its therapeutic relevance. Front Oncol 13:1141603. https://doi.org/10.3389/fonc.2023.1141603

Liu XB, Nguyen NB, Marquess KD et al (2005) Regulation of hepcidin and ferroportin expression by lipopolysaccharide in splenic macrophages. Blood Cells Mol Dis 35:47–56

Loick P, Mohammad GH, Cassimjee I et al (2023) Protective role for smooth muscle cell hepcidin in abdominal aortic aneurysm. Arterioscler Thromb Vasc Biol 43(5):713–725. https://doi.org/10.1161/ATVBAHA.123.319224

Malerba M, Louis S, Cuvellier S et al (2020) Epidermal hepcidin is required for neutrophil response to bacterial infection. J Clin Invest 130:329–334. https://doi.org/10.1172/JCI126645

Marques O, Porto G, Rêma A et al (2016) Local iron homeostasis in the breast ductal carcinoma microenvironment. BMC Cancer 16:187. https://doi.org/10.1186/s12885-016-2228-y

Martinez-Pinna R, Lindholt JS, Madrigal-Matute J et al (2014) From tissue iron retention to low systemic haemoglobin levels, new pathophysiological biomarkers of human abdominal aortic aneurysm. Thromb Haemost 112(1):87–95. https://doi.org/10.1160/TH13-08-0721

Maxel T, Smidt K, Petersen CC et al (2019) The zinc transporter Zip14 (SLC39a14) affects Beta-cell function: proteomics, gene expression, and insulin secretion studies in INS-1E cells. Sci Rep 9(1):8589. Published 2019 Jun 13. https://doi.org/10.1038/s41598-019-44954-1

Melenovsky V, Petrak J, Mracek T et al (2017) Myocardial iron content and mitochondrial function in human heart failure: a direct tissue analysis. Eur J Heart Fail 19(4):522–530. https://doi.org/10.1002/ejhf.640

Merle U, Fein E, Gehrke SG et al (2007) The iron regulatory peptide hepcidin is expressed in the heart and regulated by hypoxia and inflammation. Endocrinology 148:2663–2668. https://doi.org/10.1210/en.2006-1331

Merle U, Fein E, Gehrke SG, Stremmel W, Kulaksiz H. The iron regulatory peptide hepcidin is expressed in the heart and regulated by hypoxia and inflammation. Endocrinology. 2007 Jun;148(6):2663-8. https://doi.org/10.1210/en.2006-1331. Epub 2007 Mar 15. PMID: 17363462.

Mezzanotte M, Ammirata G, Boido M, Stanga S, Roetto A (2022) Activation of the Hepcidin-Ferroportin1 pathway in the brain and astrocytic-neuronal crosstalk to counteract iron dyshomeostasis during aging. Sci Rep 12(1):11724. Published 2022 July 9. https://doi.org/10.1038/s41598-022-15812-4

Michels K, Solomon AL, Scindia Y, Sordo Vieira L, Goddard Y, Whitten S, Vaulont S, Burdick MD, Atkinson C, Laubenbacher R, Mehrad B (2022) Aspergillus Utilizes Extracellular Heme as an Iron Source During Invasive Pneumonia, Driving Infection Severity. J Infect Dis. 16;225(10):1811-1821. https://doi.org/10.1093/infdis/jiac079. PMID:35267014; PMCID: PMC9113461.

Motley ST, Morrow BJ, Liu X et al (2004) Simultaneous analysis of host and pathogen interactions during an in vivo infection reveals local induction of host acute phase response proteins, a novel bacterial stress response, and evidence of a host-imposed metal ion limited environment. Cell Microbiol 6:849–865

Naito Y, Ishihara M (2022) Focus on cellular iron metabolism in aortic disease. Rev Cardiovasc Med 23(5):169. Published 2022 May 11. https://doi.org/10.31083/j.rcm2305169

Nemeth E, Ganz T (2021) Hepcidin-Ferroportin interaction controls systemic iron homeostasis. Int J Mol Sci 22:6493. https://doi.org/10.3390/ijms22126493

Nishigaki Y, Sato Y, Sato H et al (2021) Influence of *Helicobacter pylori* infection on Hepcidin expression in the gastric mucosa. Kurume Med J 68:107–113. https://doi.org/10.2739/kurumemedj.MS682011

Nguyen NB, Callaghan KD, Ghio AJ, Haile DJ, Yang F. (2006) Hepcidin expression and iron transport in alveolar macrophages. Am J Physiol Lung Cell Mol Physiol. 291(3):L417–25. https://doi.org/10.1152/ajplung.00484.2005. Epub 2006 Apr 28. PMID: 16648237.

Park CH, Valore EV, Waring AJ, Ganz T (2001) Hepcidin, a urinary antimicrobial peptide synthesized in the liver. J Biol Chem 276:7806–7810. https://doi.org/10.1074/jbc.M008922200

Peek CB, Levine DC, Cedernaes J et al (2017) Circadian clock interaction with HIF1α mediates oxygenic metabolism and anaerobic glycolysis in skeletal muscle. Cell Metab 25(1):86–92. https://doi.org/10.1016/j.cmet.2016.09.010

Perez E, Baker JR, Di Giandomenico S et al (2020) Hepcidin is essential for alveolar macrophage function and is disrupted by smoke in a murine chronic obstructive pulmonary disease model. J Immunol 205(9):2489–2498. https://doi.org/10.4049/jimmunol.190128

Peyssonnaux C, Zinkernagel AS, Datta V et al (2006) TLR4-dependent hepcidin expression by myeloid cells in response to bacterial pathogens. Blood 107:3727–3732. https://doi.org/10.1182/blood-2005-06-2259

Pigeon C, Ilyin G, Courselaud B et al (2001) A new mouse liver specific gene, encoding a protein homologous to human antimicrobial peptide hepcidin, is overexpressed during iron overload. J Biol Chem 276:7811–7819

Pinto JP, Dias V, Zoller H et al (2010) Hepcidin messenger RNA expression in human lymphocytes. Immunology 130:217–230. https://doi.org/10.1111/j.1365-2567.2009.03226.x

Principe P et al (2023) Nitric oxide affects Heme Oxygenase-1, Hepcidin, and transferrin receptor expression in the placenta. Int J Mol Sci 24,6:5887. https://doi.org/10.3390/ijms24065887

Puy V, Darwiche W, Trudel S et al (2018) Predominant role of microglia in brain iron retention in Sanfilippo syndrome, a pediatric neurodegenerative disease. Glia 66(8):1709–1723. https://doi.org/10.1002/glia.23335

Qian ZM, He X, Liang T, Wu KC, Yan YC, Lu LN, Yang G, Luo QQ, Yung WH, Ke Y. (2014) Lipopolysaccharides upregulate hepcidin in neuron via microglia and the IL-6/STAT3 signaling pathway. Mol Neurobiol. 50(3):811-20. https://doi.org/10.1007/s12035-014-8671-3. PMID: 24659348.

Schmidt K, Fuchs M, Weber N et al (2024) Single-nucleus RNA sequencing identifies cell-type-specific effects of sodium-glucose co-transporter 2 inhibitors in human myocardial slices. Eur Heart J 45(35):3292–3295. https://doi.org/10.1093/eurheartj/ehae472

Schwartz AJ, Goyert JW, Solanki S et al (2021) Hepcidin sequesters iron to sustain nucleotide metabolism and mitochondrial function in colorectal cancer epithelial cells. Nat Metab 3:969–982. https://doi.org/10.1038/s42255-021-00406-7

Schwarz P, Kübler JAM, Strnad P et al (2012) Hepcidin is localised in gastric parietal cells, regulates acid secretion and is induced by *Helicobacter pylori* infection. Gut 61:193–201. https://doi.org/10.1136/gut.2011.241208

Simonis G, Mueller K, Schwarz P et al (2010) The iron-regulatory peptide hepcidin is upregulated in the ischemic and in the remote myocardium after myocardial infarction. Peptides 31(9):1786–1790. https://doi.org/10.1016/j.peptides.2010.05.013

Smith TG, Talbot NP, Privat C et al (2009) Effects of iron supplementation and depletion on hypoxic pulmonary hypertension: two randomized controlled trials. JAMA 302(13):1444–1450. https://doi.org/10.1001/jama.2009.1404

Soon E, Treacy CM, Toshner MR et al (2011) Unexplained iron deficiency in idiopathic and heritable pulmonary arterial hypertension. Thorax 66(4):326–332. https://doi.org/10.1136/thx.2010.147272

Sow FB, Nandakumar S, Velu V, Kellar KL, Schlesinger LS, Amara RR, Lafuse WP, Shinnick TM, Sable SB. Mycobacterium tuberculosis components stimulate production of the antimicrobial peptide hepcidin. Tuberculosis (Edinb). 2011 Jul;91(4):314-21. https://doi.org/10.1016/j.tube.2011.03.003. Epub 2011 Apr 8. PMID: 21482189.

Tan G, Liu L, He Z, Sun J, Xing W, Sun X (2016) Role of hepcidin and its downstream proteins in early brain injury after experimental subarachnoid hemorrhage in rats. Mol Cell Biochem 418(1–2):31–38. https://doi.org/10.1007/s11010-016-2730-1

Theurl I, Theurl M, Seifert M et al (2008) Autocrine formation of hepcidin induces iron retention in human monocytes. Blood 111:2392–2399

Urrutia PJ, Hirsch EC, González-Billault C, Núñez MT (2017) Hepcidin attenuates amyloid beta-induced inflammatory and pro-oxidant responses in astrocytes and microglia. J Neurochem 142(1):140–152. https://doi.org/10.1111/jnc.14005

Urrutia P, Aguirre P, Esparza A, Tapia V, Mena NP, Arredondo M, González-Billault C, Núñez MT (2013) Inflammation alters the expression of DMT1, FPN1 and hepcidin, and it causes iron accumulation in central nervous system cells. J Neurochem. 126(4):541–9. https://doi.org/10.1111/jnc.12244. Epub 2013 Apr 3. PMID: 23506423.

van Breda GF, Bongartz LG, Zhuang W, van Swelm RP, Pertijs J, Braam B, Cramer MJ, Swinkels DW, Doevendans PA, Verhaar MC, Masereeuw R, Joles JA, Gaillard CA (2016) Cardiac Hepcidin Expression Associates with Injury Independent of Iron. Am J Nephrol. 44(5):368-378. https://doi.org/10.1159/000449419. Epub 2016 Oct 22. PMID: 27771699; PMCID: PMC5296884.

van Empel VP, Lee J, Williams TJ, Kaye DM (2014) Iron deficiency in patients with idiopathic pulmonary arterial hypertension. Heart Lung Circ 23(3):287–292. https://doi.org/10.1016/j.hlc.2013.08.007

Wang Q, Du F, Qian ZM, Ge XH, Zhu L, Yung WH, Yang L, Ke Y (2008) Lipopolysaccharide induces a significant increase in expression of iron regulatory hormone hepcidin in the cortex and substantia nigra in rat brain. Endocrinology. 149(8):3920-5. https://doi.org/10.1210/en.2007-1626. Epub 2008 May 1. PMID: 18450970; PMCID:PMC2488231.

Wei B, Liu W, Jin L et al (2024) Hepcidin depending on astrocytic NEO1 ameliorates blood-brain barrier dysfunction after subarachnoid hemorrhage. Cell Death Dis 15(8):569. Published 2024 Aug 7. https://doi.org/10.1038/s41419-024-06909-x

Yang A et al (2016) Expression of Hepcidin and Ferroportin in the placenta, and ferritin and transferrin receptor 1 levels in maternal and umbilical cord blood in pregnant women with and without gestational diabetes. Int J Environ Res Public Health 13,8:766. https://doi.org/10.3390/ijerph13080766

Yang Y, Zeng C, Yang S et al (2020) Airway epithelial Hepcidin coordinates lung macrophages and immunity against bacterial pneumonia. Shock 54(3):402–412. https://doi.org/10.1097/SHK.0000000000001471

Yang YL, Kuo HC, Chen KD et al (2022) Combination of hemoglobin-for-age Z-score and plasma Hepcidin identified as a novel predictor for Kawasaki disease. Children (Basel) 9(6):913. Published 2022 Jun 18. https://doi.org/10.3390/children9060913

Yook JS, You M, Kim J et al (2021) Essential role of systemic iron mobilization and redistribution for adaptive thermogenesis through HIF2-α/hepcidin axis. Proc Natl Acad Sci USA 118(40):e2109186118. https://doi.org/10.1073/pnas.2109186118

You L, Yu PP, Dong T et al (2022) Astrocyte-derived hepcidin controls iron traffic at the blood-brain-barrier via regulating ferroportin 1 of microvascular endothelial cells. Cell Death Dis 13(8):667. Published 2022 Aug 1. https://doi.org/10.1038/s41419-022-05043-w

Zechel S, Huber-Wittmer K, von Bohlen, Halbach O (2006) Distribution of the iron-regulating protein hepcidin in the murine central nervous system. J Neurosci Res 84:790–800

Zhang X, Gou YJ, Zhang Y et al (2020) Hepcidin overexpression in astrocytes alters brain iron metabolism and protects against amyloid-β induced brain damage in mice. Cell Death Discov 6(1):113. Published 2020 Oct 30. https://doi.org/10.1038/s41420-020-00346-3

Zhang H, Jamieson KL, Grenier J et al (2022) Myocardial iron deficiency and mitochondrial dysfunction in advanced heart failure in humans. J Am Heart Assoc 11(11):e022853. https://doi.org/10.1161/JAHA.121.022853

Zhao Y, Zhao J, Ma H et al (2023) High Hepcidin levels promote abnormal iron metabolism and ferroptosis in chronic atrophic gastritis. Biomedicines 11:2338. https://doi.org/10.3390/biomedicines11092338

Zlatanova I, Pinto C, Bonnin P et al (2019) Iron regulator Hepcidin impairs macrophage-dependent cardiac repair after injury. Circulation 139(12):1530–1547. https://doi.org/10.1161/CIRCULATIONAHA.118.034545

Zumerle S, Mathieu JRR, Delga S et al (2014) Targeted disruption of hepcidin in the liver recapitulates the hemochromatotic phenotype. Blood 123:3646–3650. https://doi.org/10.1182/blood-2014-01-550467

Control of Systemic Iron Homeostasis: Insights Gained from Studying Mouse Models

8

Sandro Altamura and Martina U. Muckenthaler

Abstract

Disorders of iron metabolism pose a global health challenge. Over the past two decades, novel regulators of systemic iron homeostasis have been identified that can be manipulated for therapeutic purposes. In this context, inbred and genetically modified mice have proven to be valuable tools enabling precise investigation of gene functions and testing of new pharmacological treatments that in some cases have reached the clinical stage. Here, we summarize the most recent discoveries on mechanisms regulating systemic iron homeostasis, with a specific focus on the role of murine models in advancing this research.

Keywords

Systemic iron homeostasis · Hepcidin-ferroportin axis · Murine models in iron regulation · BMP/SMAD signaling pathway · Iron metabolism disorders

S. Altamura · M. U. Muckenthaler (✉)
Center for Translational Biomedical Iron Research, Department of Pediatric Hematology, Oncology and Immunology, University of Heidelberg, Heidelberg, Germany

Molecular Medicine Partnership Unit, Heidelberg, Germany
e-mail: martina.muckenthaler@med.uni-heidelberg.de

8.1 Regulation of Systemic Iron Homeostasis: The Hepcidin/Ferroportin Axis

The regulation of systemic iron homeostasis ensures adequate dietary iron absorption while preventing toxic iron overload (Chap. 3). Central in this regulatory network is the liver-derived peptide hormone hepcidin and its target receptor ferroportin (Fpn). Hepcidin production is increased in response to elevated plasma and liver iron levels or inflammation/infection to limit plasma iron availability. From an evolutionary perspective, high hepcidin levels induced by inflammation limit iron for pathogen proliferation. Conversely, if the body requires more iron, such as during increased erythropoietic activity, hepcidin expression is reduced. Hepcidin achieves its regulatory function by binding to ferroportin, the only known iron exporter, which mediates iron release into the bloodstream from reticuloendothelial macrophages and duodenal enterocytes. Hepcidin inhibits Fpn activity by closing the iron channel or mediating its internalization and degradation.

8.1.1 Hepcidin

Hepcidin was initially identified as an acute-phase protein and a member of the beta-defensin family of antimicrobial peptides. Its pivotal role

K. Pantopoulos (ed.), *Iron Metabolism in Human Health and Disease*, Advances in Experimental Medicine and Biology 1480, https://doi.org/10.1007/978-3-031-92033-2_8

in controlling iron metabolism became apparent by a serendipitous event: Sophie Vaulont's team created *Usf2* knock-out mice and unintentionally deleted the previously unknown hepcidin gene (*Hamp*) located just 1 kilobase downstream of Usf2. They noted severe systemic iron overload (Nicolas et al. 2001). Since this groundbreaking discovery, various additional mouse models have been developed to confirm hepcidin's role in iron homeostasis. Constitutive hepcidin knock-out mice (Lesbordes-Brion et al. 2006) and transgenic mice overexpressing hepcidin (Lou et al. 2004) have provided direct evidence of hepcidin's essential function in maintaining systemic iron balance. More recently, a tamoxifen-inducible hepcidin knock-out mouse model has been generated. This model has been applied to study the immune role of hepcidin in the absence of an iron overload background (Armitage et al. 2016).

Of note, mice are unique among vertebrates in that they possess a second homologous hepcidin gene, *Hamp2*, which encodes a mature peptide with only 68% identity to its counterpart, hepcidin 1. Despite the high degree of homology with hepcidin 1, hepcidin 2 does not control iron homeostasis. Transgenic mice overexpressing *hamp2* show unaltered iron indices (Lou et al. 2004).

Since the discovery of hepcidin, significant progress has been made in understanding how its expression is controlled in response to changes in iron levels, inflammation, and hypoxia. It is now well-established that in the liver hepcidin is predominantly synthesized by hepatocytes, and that hepatocyte-derived hepcidin is the unique contributor to regulating systemic iron homeostasis. Indeed, conditional hepatocyte-specific deletion of *Hamp* mirrors the iron overload phenotype observed in constitutive knock-out mice (Zumerle et al. 2014). The therapeutic potential of hepcidin was tested in mouse models for hepcidin deficiency diseases, such as hemochromatosis, beta-thalassemia, and polycythemia vera, by applying mini-hepcidins or hepcidin analogs (Ramos et al. 2012; Casu et al. 2016, 2020); see also Chap. 26.

8.1.2 The BMP/SMAD Signaling Pathway Controls Hepcidin Expression in Hepatocytes

Hepcidin regulation by iron is complex and only partially understood, involving an intricate cross-talk among various hepatic cell types (Chap. 3). The current knowledge of this process is summarized in Fig. 8.1. Central to this process are Bone Morphogenetic Proteins (BMPs), secreted members of the transforming growth factor-β (TGF-β) superfamily. BMPs are produced at high levels by liver sinusoidal endothelial cells (LSECs) and act in a paracrine manner on hepatocytes to stimulate hepcidin transcription. Studies utilizing LSEC-specific knock-out models for *Bmp2* (Canali et al. 2017a; Koch et al. 2017) and *Bmp6* (Canali et al. 2017b) have highlighted the essential contribution of these two BMPs to hepcidin control. Upon LSEC-specific BMP deletion, mice show severe iron overload as a consequence of reduced hepcidin expression.

Under basal conditions, *Bmp2* is expressed at higher levels than *Bmp6*. However, while *Bmp2* is only modestly controlled by iron excess, *Bmp6* expression is substantially upregulated during iron overload. BMPs can act either as homodimers or heterodimers to activate their receptors. Mice with simultaneous deletion of *Bmp2* and *Bmp6* phenocopy *Bmp6* single knock-out mice, suggesting that BMP2 and BMP6 primarily operate as a heterodimer in hepcidin regulation (Xiao et al. 2020). Recently, the role of *Bmp5* in hepcidin regulation was investigated by analyzing mice carrying the *Bmp5* short-ear inactivating mutation (*Bmp5^se^*). These mice display a strong reduction in hepcidin expression only in the first weeks of life, without substantial alterations in transferrin saturation and hepatic iron content, indicating that BMP5 only has a minor role in the regulation of iron homeostasis (Xiao et al. 2023).

The mechanism(s) through which iron controls *Bmp6* expression are incompletely understood. Iron-induced oxidative stress has been proposed as a *Bmp6* inducer. Consistently, mice knock-out for the antioxidant surveillance-related transcription factor *Nrf2* fail to upregulate Bmp6 and hepcidin in iron overload conditions (Lim

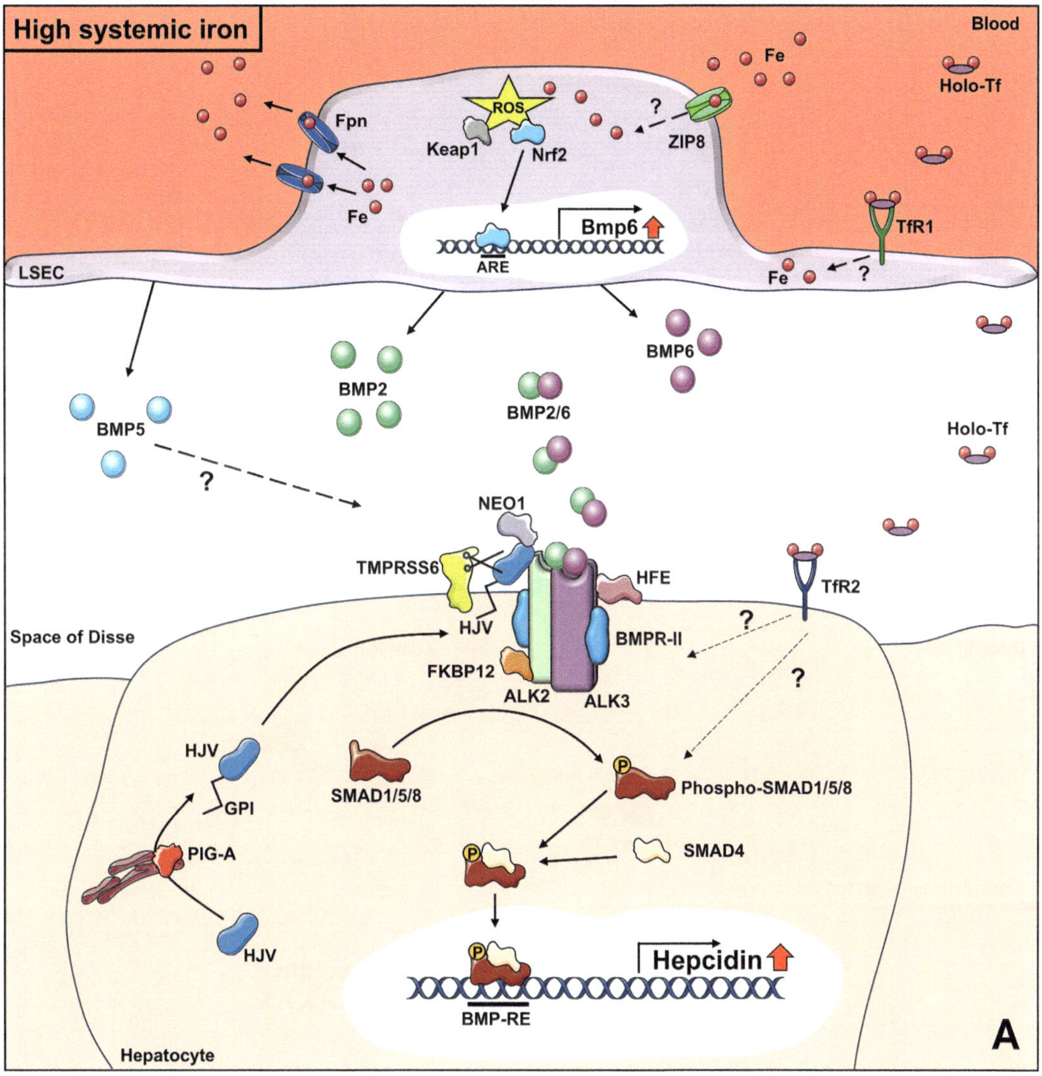

Fig. 8.1 Hepcidin regulation by systemic iron levels (**a**) Under conditions of high systemic iron, LSECs import transferrin-bound iron (holo-transferrin) and non-transferrin-bound iron (NTBI) through TfR1, ZIP8, and other yet unidentified mechanisms. The accumulation of intracellular iron leads to the generation of ROS, which promote the release of Nrf2 from its regulatory protein, Keap1. Once free, Nrf2 translocates to the nucleus and activates the transcription of *Bmp6*. The resulting BMP6 homodimers, BMP2 homodimers, and BMP2/BMP6 heterodimers interact with heterotetrameric BMP receptors on hepatocytes. These receptors consist of type I ALK2 and ALK3 receptors, type II BMPR-II receptors, the HJV co-receptor, and the regulatory proteins NEO1, FKBP12, HFE, and TMPRSS6. Upon activation, the receptor complex phosphorylates the regulatory R-SMADs (SMAD1/5/8), which then associate with SMAD4, translocate to the nucleus, and drive hepcidin expression, thereby reducing systemic iron levels. (**b**) In low systemic iron conditions, LSECs decrease BMP6 expression, thus reducing the activity of the BMP receptor on hepatocytes, resulting in lower R-SMAD phosphorylation and decreased hepcidin expression. This mechanism promotes higher Fpn-mediated iron export into the bloodstream. This figure was partially generated using Servier Medical Art (license CC BY 4.0)

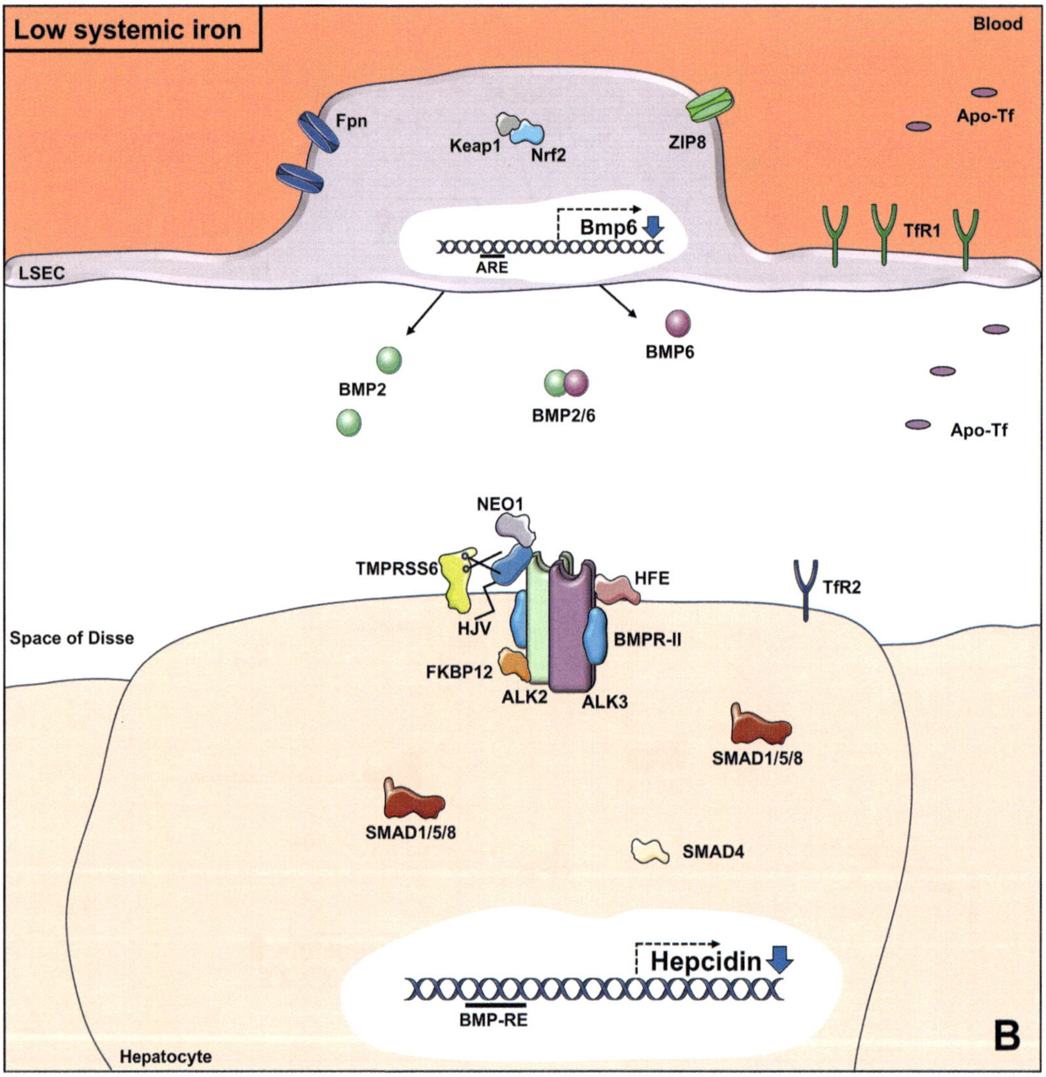

Fig. 8.1 (continued)

et al. 2019). However, murine disease models of hemochromatosis with severe iron overload (e.g., mice with HJV deletion and Fpn^{C326S} knock-in mice—see related chapters) increase *Bmp6* levels despite an iron-deficient molecular signature in LSECs (Rausa et al. 2015; Colucci et al. 2022a). This strongly suggests that additional mechanisms control *Bmp6* expression in LSECs, which are independent of intracellular iron accumulation.

An open question still is how LSECs import iron. The LSEC-specific deletion of murine TfR1, the receptor responsible for transferrin-bound iron uptake, does not affect *Bmp6* and *hepcidin* mRNA expression under basal conditions and shows only a minimal impact upon iron deficiency (Fisher et al. 2022; Charlebois et al. 2023). LSECs also express high levels of the non-transferrin-bound iron importer ZIP8. However, LSEC-specific *Zip8* knock-out mice show only a mild reduction in their iron-mediated Bmp6 response (Fisher et al. 2024a). Despite the fact that the mechanism is still unclear, LSECs can import non-transferrin-bound iron as well as ferritin-bound iron, both of which seem to increase *Bmp6* mRNA expression (Fisher et al. 2022; Feng et al. 2012).

LSECs further express Fpn and thus have the capacity to export iron. Similar to duodenal enterocytes and reticuloendothelial macrophages, Fpn expressed in LSECs is sensitive to hepcidin action, resulting in its internalization and degradation and subsequent iron retention. Mice lacking Fpn in LSECs accumulate iron, activate Nrf2, and increase Bmp6 and hepcidin mRNA levels (Fisher et al. 2024b). These findings indicate that Fpn modulation is involved in controlling BMP6 levels. However, additional regulatory mechanisms must exist to prevent a feedback loop, where LSEC iron accumulation induces BMP6 and hepcidin expression, which further acts on Fpn to enhance iron accumulation and perpetuate BMP6/hepcidin synthesis.

The BMPs secreted by LSECs bind to a hepatocytic heterotetrameric receptor complex, which consists of two type I (ALK2 and ALK3) (Steinbicker et al. 2011) and two type II receptors (ACVR2A and BMPR2) (Mayeur et al. 2014). Mouse models have been generated to understand the role of each receptor subunit in BMP-mediated signal transduction. Type II receptors exhibit functional redundancy (Mayeur et al. 2014), while ALK2 and ALK3 do not overlap in their roles. Both *Alk2* and *Alk3* knock-out mice show reduced hepcidin mRNA expression, but the effect is more pronounced in *Alk3*-deficient animals, consistent with more severe iron overload (Steinbicker et al. 2011). The combined deletion of *Alk2* and *Alk3* causes more severe iron overload compared to mice with only the *Alk3* ablation, highlighting the critical and partially distinct functions of these receptors in iron regulation (Traeger et al. 2018).

The binding of BMPs to the hepatocyte BMP receptor complex induces the constitutively active BMP type II receptors to transphosphorylate the glycine/serine-rich (GS) domain of the BMPR-type I receptors, which in turn phosphorylates the regulatory SMAD (R-SMAD) proteins SMAD1, SMAD5, and SMAD8. Once activated, the R-SMADs form a complex with SMAD4, causing their translocation into the nucleus (Mayeur et al. 2014). Here, the R-SMAD/SMAD4 heterodimers bind to BMP-responsive elements (BRE) to initiate the transcription of hepcidin (Casanovas et al. 2009; Verga Falzacappa et al. 2008). The pivotal role of SMAD4 in this signaling pathway has been confirmed through studies using hepatocyte-specific *Smad4* knock-out mice, which exhibit reduced hepcidin expression and iron overload (Wang et al. 2005). The role of each R-SMAD in controlling hepcidin expression has been explored by generating various conditional knock-out mouse models. In adult mice, it was proposed that the three R-SMADs exhibit functional redundancy, as single knock-outs do not impair hepcidin synthesis. However, the combined deletion of Smad1 and Smad5 causes reduced hepcidin transcription and elevated systemic iron levels. This phenotype is further aggravated upon concomitant knock-out of the three R-SMADs, resulting in dramatic hepcidin suppression and severe iron overload (Wang et al. 2017, 2019). In contrast, during the suckling stage, the roles of SMAD1 and SMAD5 are not redundant, suggesting distinct functions for these proteins in early development (Wang et al. 2017).

8.1.3 Proteins Activating Hepcidin Expression via the BMP/SMAD Pathway

Hemojuvelin (HJV), a GPI-anchored protein member of the repulsive guidance molecule (RGM) family, functions as a co-receptor of the BMP receptor complex. The role of HJV in hepcidin regulation has been extensively studied due to the clinical significance of HJV mutations in patients with a severe subtype of juvenile hemochromatosis (HC); see Chap. 10. Two independent research groups have developed constitutive *Hjv* knock-out mouse models, both of which accurately reproduce the clinical features observed in *HJV* HC patients (Huang et al. 2005; Niederkofler et al. 2005). The role of HJV in hepcidin regulation is restricted to hepatocytes, as evidenced by the characterization of hepatocyte-specific *Hjv* knock-out mice, which phenocopy the constitutive knock-out models (Gkouvatsos et al. 2011). The same study explored a possible iron regulatory role of HJV specifically in the skeletal muscle tissue, where HJV is dispensable for modifying systemic iron levels (Gkouvatsos et al. 2011). HJV only functions appropriately in the presence of neogenin (NEO1), a transmembrane protein that belongs to the immunoglobulin superfamily. How exactly neogenin affects HJV function is not yet understood. However, both constitutive and hepatocyte-specific knock-out mice for *neogenin* phenocopy HJV knock-out animals, suggesting that the two proteins work in concert to control hepcidin mRNA expression (Lee et al. 2010; Enns et al. 2021). The GPI anchor of HJV is essential for its function. Consequently, somatic mutations in Phosphatidylinositol glycan anchor biosynthesis class A (PIGA), a rate-limiting enzyme involved in the biosynthesis of the GPI anchors, cause a hemochromatosis-like phenotype (Muckenthaler et al. 2022). HJV-mediated activation of hepcidin mRNA expression requires both ALK2 and ALK3 (Dogan et al. 2024).

The type I receptor Alk3 undergoes E3 ubiquitination and proteasomal degradation, which reduces receptor availability on the cell surface, limiting hepatocyte responsiveness to BMP ligands. HFE, a ubiquitously expressed transmembrane protein that belongs to the family of MHC class I-like proteins, binds to ALK3 in hepatocytes, inhibiting its ubiquitination and degradation. As a result, ALK3 protein levels on the plasma membrane are increased, promoting mRNA expression of hepcidin (Wu et al. 2014). Mutations in *HFE* cause HC, the most common inherited autosomal recessive disorder in Caucasians, with nearly 80% of patients being homozygous for the p.C282Y HFE mutation (Chap. 9). A second, less common but still widespread HFE mutation is p.H63D. Given its prominent clinical relevance, the role of HFE in hepcidin regulation has been studied by generating various murine disease models, including knock-out (Herrmann et al. 2004; Zhou et al. 1998; Bahram et al. 1999; Levy et al. 1999) and knock-in mice for the p.C282Y and p.H63D mutations (Levy et al. 1999; Tomatsu et al. 2003). In all cases, the mouse models recapitulate the human disease, with inappropriately low hepcidin production and systemic iron overload.

MHC class I family members, like HFE, bind to β2-microglobulin (B2M) as an intrinsic part of their protein structures. Consistently, *B2M* knock-out mice exhibit a hemochromatosis-like phenotype (de Sousa et al. 1994). Given the ubiquitous expression of HFE, its role in various cell types critical for regulating iron homeostasis has been investigated by the generation of conditional mouse models. While duodenal- and LSEC-specific Hfe knock-out mice do not alter hepcidin synthesis (Vujic Spasic et al. 2007; Colucci et al. 2022b), hepatocyte-specific knock-out animals display all the hallmarks of the human disorder, phenocopying the constitutive knock-out model (Vujic Spasic et al. 2008). In macrophages, Hfe plays a minor role in iron regulation, which seems most apparent during the aging process (Vujic Spasic et al. 2008; Tangudu et al. 2021). It may be further involved in maintaining iron balance during infections (Tangudu et al. 2021).

In addition to mutations in HFE and HJV, inactivating mutations in the transferrin receptor 2 (TfR2) also cause HC (Chap. 10). Most of the patients show the p.Y250X nonsense mutation and reduced hepcidin production. Similarly, murine models with constitutive or hepatocyte-specific *Tfr2* deletions show iron overload due to decreased BMP/SMAD pathway activation and inappropriately low hepcidin levels (Wallace et al. 2005, 2007; Corradini et al. 2011). TfR2, a homolog of TfR1, exhibits a 27-times lower affinity for holo-transferrin compared to TfR1 and has been proposed to sense plasma iron levels to control hepcidin mRNA expression. TfR2 also exhibits regulatory roles outside hepatocytes. In osteoblasts, it directly binds BMP2, regulating osteoblastogenesis (Rauner et al. 2019). However, BMP2 injection in *Tfr2* knock-out and wild-type mice increases hepcidin levels to a similar extent (Pagani et al. 2020). TfR2 is further expressed in the erythron, where it controls red blood cell production by stabilizing the erythropoietin receptor (Nai et al. 2015; Rishi et al. 2016).

8.1.4 Proteins Inhibiting Hepcidin Expression via the BMP/SMAD Pathway

Several mechanisms prevent excessive activation of the BMP/SMAD pathway, counteracting high hepcidin expression and iron deficiency.

TMPRSS6 (also known as Matriptase-2) is one of the best-characterized hepcidin suppressors. TMPRSS6 is a transmembrane serine protease located on the hepatocytic plasma membrane. It cleaves the BMP co-receptor hemojuvelin (HJV), thereby limiting hepcidin expression. An additional study overexpressing TMPRSS6 in hepatoma cells demonstrated its ability to cleave other components of the BMP receptor, including ALK3, ACVR2A, and HFE (Enns et al. 2020). However, it remains uncertain whether such cleavage occurs under physiological stoichiometric conditions. Consistent with its role in fine-tuning hepcidin expression, its expression is itself controlled by the BMP/SMAD pathway through a negative feedback loop. Mutations in the TMPRSS6 gene are detected in patients with iron-refractory iron deficiency anemia (IRIDA), a rare genetic disorder characterized by inappropriately high hepcidin levels that causes severe iron deficiency, which is unresponsive to iron supplementation (Chap. 12). The critical role of TMPRSS6 in causing IRIDA was demonstrated in two mouse models: the *mask* mouse, which harbors a spontaneous mutation in TMPRSS6 (Du et al. 2008), and the *Tmprss6* knock-out mouse (Folgueras et al. 2008). Both models closely mimic the human disorder, exhibiting hallmark features such as elevated hepcidin expression, reduced plasma and tissue iron levels, and microcytic anemia. Given its biological role in controlling hepcidin levels, TMPRSS6 has emerged as a promising pharmacological target for restoring hepcidin expression in hepcidin deficiency diseases, such as HC or iron-loading anemias (e.g., thalassemia). As a proof-of-concept, double knock-out mice lacking both Tmprss6 and Hfe were generated, demonstrating that Tmprss6 ablation increased hepcidin expression and mitigated the iron overload phenotype of HC (Finberg et al. 2011). Likewise, Tmprss6 deletion ameliorated iron overload phenotypes in thalassemic mice and improved erythropoiesis (Nai et al. 2012). Therapeutic strategies targeting TMPRSS6 have been developed (Chap. 26) and tested in disease models of HC and beta-thalassemia. These include TMPRSS6 antisense oligonucleotides (Guo et al. 2013), TMPRSS6 RNAi delivered via lipid nanoparticles (Schmidt et al. 2013), and GalNAc-conjugated TMPRSS6 RNAi (Altamura et al. 2019; Vadolas et al. 2021). Common to all approaches are increased hepcidin levels and diminished iron overload, highlighting the therapeutic promise of TMPRSS6 inhibition.

An additional negative hepcidin regulator is the immunophilin FK506-binding protein 1A (FKBP12). FKBP12 is a peptidyl-prolyl-*cis-trans* cytosolic isomerase that binds ALK2. Modifying its interaction with components of the heterotetrameric BMP receptor complex causes attenuation of BMP/SMAD pathway activity. Consistently, the treatment of wild-type mice with the drug tacrolimus that causes the displacement of FKBP12 from ALK2 results in increased hepcidin expression (Pettinato et al. 2023).

The inhibitory SMADs (I-SMADs), SMAD6, and SMAD7 are intrinsic regulators of the BMP/SMAD pathway functioning as negative feedback regulators. Their expression is increased via the BMP/SMAD signaling cascade, allowing fine-tuning of the pathway's activity (Vujic Spasic et al. 2013; Mleczko-Sanecka et al. 2010). Consistent with their inhibitory role, overexpression of SMAD7 in mice reduces hepcidin expression, whereas *Smad7* knock-out mice increase hepcidin levels (An et al. 2018; Lai et al. 2018). Despite the established role of I-SMADs, therapeutic strategies to target I-SMAD expression have not yet been developed to increase hepcidin levels in hepcidin deficiency diseases.

8.1.5 Hepcidin Suppression in Response to Hypoxia

In humans, the daily production of approximately 200 billion erythrocytes requires about 20–25 mg of iron for incorporation into hemoglobin. The blood hormone erythroferrone (ERFE) assures that hepcidin expression is decreased following an increased iron demand for erythropoiesis (e.g., during hypoxia). ERFE is produced by erythroblasts upon erythropoietin (EPO) stimulation under hypoxic conditions (Kautz et al. 2014). Mechanistically, ERFE directly interacts with BMP2 and BMP6 to antagonize their binding to the BMP receptor. As a consequence, hepcidin transcription is decreased, and more iron is released from iron-recycling reticuloendothelial macrophages and duodenal enterocytes that take up dietary iron (Arezes et al. 2018).

The hepatokine fibrinogen-like 1 (FGL1) is an additional, recently identified suppressor of hepcidin in hypoxic conditions. FGL1 is expressed in hepatocytes and bone marrow, and its expression is directly regulated by hypoxia through two Hypoxia-Inducible Factor (HIF) Responsive Elements (HREs) located in its promoter. Similar to ERFE, FGL1 functions by directly binding to BMP6, thereby inhibiting the canonical BMP-SMAD signaling pathway that drives hepcidin transcription. Consistently, mice lacking FGL1 show elevated hepcidin levels under basal conditions and following hypoxic signaling induced by bleeding (Sardo et al. 2024). How the effects of FGL-1 and ERFE are coordinated requires further studies. The mechanism of action of ERFE and FGL1 in regulating hepcidin expression is summarized in Fig. 8.2.

8.1.6 Additional Mechanisms of Hepcidin Control

Heparan sulfate proteoglycans (HSPGs) are glycoproteins containing one or more covalently attached heparan sulfate (HS) chains. They are localized on the plasma membrane or secreted in the extracellular matrix. HSPGs can act as co-receptors for the binding of ligands, such as fibroblast growth factor (FGF), transforming growth factor (TGFβ), and BMPs, including BMP2 and BMP6 (Denardo et al. 2021). Conditional knock-out mice lacking the heparan sulfate biosynthetic gene N-deacetylase/N-sulfotransferase 1 (*Ndst1*) in hepatocytes show decreased BMP/SMAD pathway activation, reduced hepcidin levels, and increased serum and hepatic iron levels, thus linking HS, BMP signaling, and hepcidin expression (Poli et al. 2019).

Heparin is a highly sulfated soluble form of HS that is clinically used as an anticoagulant. Heparin administration to wild-type mice causes a marked reduction in hepcidin expression (Poli et al. 2011). Similar results were obtained with modified versions of heparin with diminished anticoagulant activity, such as Sevuparin (Asperti et al. 2024), Pentosan polysulfate (Asperti et al. 2020), and oversulfated heparins (Poli et al. 2014), opening the possibility to use these molecules in disorders characterized by inappropriately high hepcidin levels, including IRIDA and anemia of inflammation. Altogether, these results suggest that soluble HS may function as a BMP antagonist by sequestering BMP2 and BMP6 and preventing their interaction with the BMP receptor complex. Conversely, HSPGs located on the hepatocyte plasma membrane may play a crucial role in facilitating proper BMP receptor assembly (Kuo et al. 2010).

8.2 Ferroportin, the Hepcidin Receptor

In vertebrates, ferroportin (Fpn) is the sole iron exporter of ferrous iron into the bloodstream. Much of our understanding of the unique and non-redundant role of Fpn is derived from studies utilizing constitutive and conditional Fpn knock-out mouse models.

During embryogenesis, Fpn is essential for the iron transport from the mother to the fetus. Consistently, homozygous Fpn knock-out mice show embryonic lethality due to a failure to transport iron across the extraembryonic visceral endoderm (Donovan et al. 2005). Postnatally,

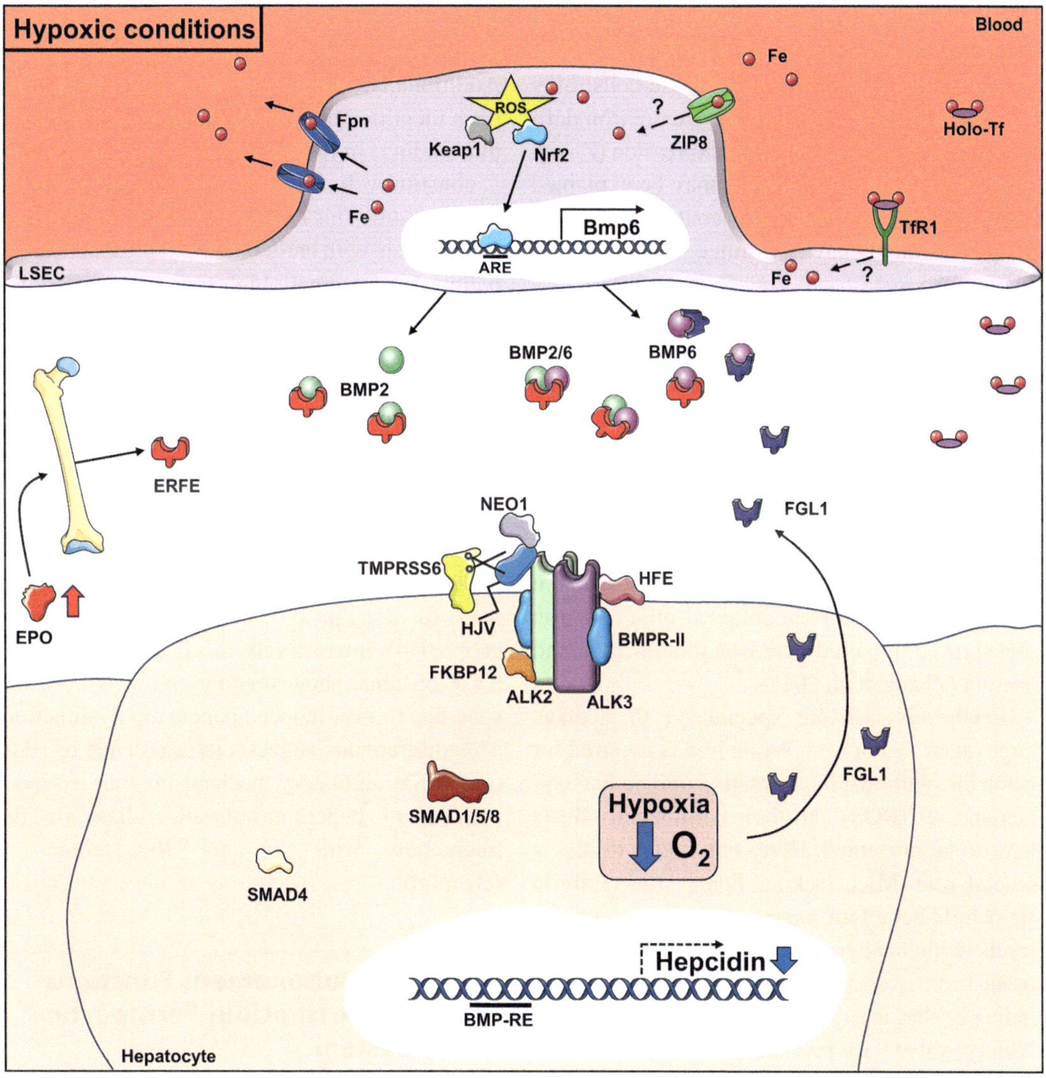

Fig. 8.2 Hepcidin suppression by increased erythropoiesis/hypoxia. Hypoxia or anemia stimulates the production of erythropoietin (EPO), which acts on erythroblasts in the bone marrow to induce erythroferrone (ERFE) synthesis. ERFE interacts with BMP2 homodimers, BMP6 homodimers, and BMP2/BMP6 heterodimers, preventing their binding to BMP receptors on hepatocytes. In addition, hypoxia in hepatocytes promotes the production of fibrinogen-like protein 1 (FGL1), which functions similarly to ERFE by binding to BMP6 and inhibiting its interaction with BMP receptors. These BMP antagonists reduce BMP receptor activation, leading to decreased R-SMAD phosphorylation and a subsequent reduction in hepcidin expression. As a result, iron flux into the bloodstream increases to support enhanced erythropoiesis. This figure was partially generated using Servier Medical Art (license CC BY 4.0)

Fpn plays a critical role in dietary iron export from duodenal enterocytes. This knowledge is derived from the characterization of a conditional intestinal Fpn knock-out model, which shows duodenal iron retention, anemia, and systemic iron deficiency, indicating impaired dietary iron export (Donovan et al. 2005).

Macrophage recycling is estimated to account for 95% of the daily iron required for erythropoiesis. To elucidate the role of Fpn in this process, two independent macrophage-specific Fpn knock-out mouse models have been generated (Zhang et al. 2011; Recalcati et al. 2019). The analysis of these mice yielded inconsistent results. While both models display iron retention in splenic macrophages and Kupffer cells, consistent with the lack of Fpn in these cells, only one of the two models shows systemic iron deficiency and decreased hepcidin expression (Zhang et al. 2011). This discrepancy may be explained by differences in dietary iron content and/or the unique feeding behavior of mice, which have ad libitum access to food. Thus, continuous dietary iron uptake may be sufficient to supply iron to the mouse.

The liver plays a critical role in maintaining iron homeostasis. It not only produces the iron regulatory hormone hepcidin, but it also serves as an iron storage organ to compensate for fluctuations in iron availability due to fasting or reduced dietary content. Consistently, Fpn plays an essential role in hepatocytes. Mice with Fpn ablation in hepatocytes that were maintained on a low iron diet showed impaired liver iron mobilization and anemia (Zhang et al. 2012).

Erythroid cells are specialized to manage large quantities of iron. While iron is required for heme biosynthesis, the formation of reactive oxygen species (ROS) and their detrimental effects have to be prevented. Here, Fpn seems to play a crucial role. Mice lacking Fpn in the erythron show mild hemolytic anemia, reduced serum iron levels with unaltered hepcidin expression, and tissue iron overload (Zhang et al. 2018). This indicates that in erythroid cells, Fpn serves as a "safety valve" to prevent pathological consequences that arise from excessive iron accumulation. In addition, these mice show that in iron deficiency the erythron becomes an organ that can supply iron to avoid the pathological consequences of excessive systemic iron deficiency (Zhang et al. 2018).

FPN mutations are detected in patients with altered iron metabolism. Overall, they can be divided into two phenotypic groups. Loss of function mutations of ferroportin cause "ferroportin disease" (Pietrangelo 2004); see also Chap. 10. This disorder is hallmarked by iron retention in iron-exporting organs (e.g., duodenum and spleen), high serum ferritin levels but low-to-normal transferrin saturation. The flatiron mouse (ffe), carrying the p.H32R Fpn mutation, is the best-characterized murine model to study ferroportin disease (Zohn et al. 2007). Additionally, gain-of-function mutations have been identified that render Fpn resistant to hepcidin binding. In response to these mutations, Fpn is constitutively active in exporting iron into the bloodstream. This causes a severe form of hemochromatosis with early onset that presents unique hallmarks: extremely high transferrin saturation and serum ferritin, and systemic iron overload despite the presence of elevated serum hepcidin. The Fpn p.C326S and p.C326Y mouse models recapitulate this human disorder (Altamura et al. 2014; Lakhal-Littleton et al. 2019).

Given the mechanism of action of the hepcidin-ferroportin regulatory circuitry, mice knock-out for hepcidin (Lesbordes-Brion et al. 2006) should in principle phenocopy knock-in mice for the Fpn p.C326S mutation (Altamura et al. 2014). Interestingly, this is not the case. Fpn p.C326S mice show wasting and a reduced life span due to iron-induced pancreatic dysfunction. This discrepancy suggests the existence of additional iron regulatory mechanisms that are independent of hepcidin but still reliant on the interaction with the p.C326 residue of ferroportin.

8.3 Cell-Autonomous Functions of the Hepcidin-Ferroportin System

Recent studies have revealed that the function of the hepcidin-ferroportin regulatory system extends beyond its primary task of controlling systemic iron balance by exerting a direct effect at a cell-autonomous level (see also Chap. 7). The first indication of the existence of such a mechanism came from the analysis of macrophages, which express both hepcidin and Fpn. Theurl and coworkers have shown that in this cell type, cell-intrinsic control of ferroportin by hepcidin can contribute to iron retention in anemia of inflammation (Theurl et al. 2008). Since then, due to the generation of tissue-specific Fpn p.C326Y knock-in and/or hepcidin knock-out animals, more and more evidence is available demonstrating the existence of cell-autonomous hepcidin/ferroportin iron regulation.

An example is cardiomyocytes, which need to fine-tune their iron content to assure sufficient iron supplies to sustain intensive mitochondrial oxidative phosphorylation while avoiding excessive iron accumulation and its pathological consequences. Cardiomyocytes produce low levels of hepcidin. Mice lacking cardiomyocyte hepcidin or expressing the Fpn p.C326Y hepcidin-resistant ferroportin allele show cardiac iron deficiency (Lakhal-Littleton et al. 2016). While cardiomyocyte-derived hepcidin is important during homeostasis, it also contributes to cardiac recovery following injury. Hepcidin ablation in cardiomyocytes inhibits efficient cardiac repair after acute myocardial infarction, highlighting the critical role of iron balance in the healing process (Zlatanova et al. 2019).

Ferroportin is also expressed in pulmonary arterial smooth muscle cells (PASMCs). Mice were developed that express the hepcidin-resistant FPN p.C326Y mutation specifically in PASMCs. These mice exhibit PASMCs intracellular iron deficiency, which is associated with the development of pulmonary arterial hypertension (Lakhal-Littleton et al. 2019). These findings further underscore the critical importance of the cell-intrinsic hepcidin-ferroportin axis in preventing disease.

The role of hepcidin in smooth muscular cells (SMCs) has been studied in the context of abdominal aortic aneurysm (AAA). Mice with SMC-specific deletion of hepcidin or knock-in of hepcidin-resistant Fpn p.C326Y had a more severe AAA phenotype, which is associated with minor iron retention in SMCs. These results indicate that hepcidin in SMCs plays a protective role in the setting of AAA (Loick et al. 2023).

Hepcidin is also expressed in keratinocytes, and its function has been studied in a model of necrotizing fasciitis. The generation of mice lacking hepcidin in this cell type revealed its importance in containing infection spread. Hepcidin facilitates the synthesis of the chemokine CXCL1, which in turn promotes neutrophil recruitment to the site of infection. While the exact mechanism underlying hepcidin-induced CXCL1 production is not fully understood, it involves an autocrine interaction with Fpn expressed on keratinocytes (Malerba et al. 2020).

8.4 Conclusions

The development of mouse models for critical players and regulators of the hepcidin/ferroportin regulatory axis contributed enormously to our understanding of mechanisms that control systemic iron homeostasis. These mechanisms involve the crosstalk of various organs and cell types that produce growth factors (e.g., BMPs) and hormones (EPO, ERFE, and FGL1). Such a degree of complexity can only be studied in mouse models and is difficult to reproduce in cellular assays or in silico.

Murine disease models of iron metabolism disorders, such as hemochromatosis, largely phenocopy the human disease and thus were instrumental to understand the mechanisms of how the mutations identified in patients cause alterations of systemic iron homeostasis. These unique tools also allow us to assess novel therapeutic approaches to prevent iron overload or deficiency. However, limitations exist. For example, mice are resistant to the pathological consequences of iron overload seen in humans (e.g., fibrosis, cirrhosis, liver cancer, diabetes). Therefore, the pathogenesis of such iron-induced pathologies is difficult to study in mice.

Of note, analyses of the same mouse model may yield different results in different laboratories. These inconsistencies can be explained by the iron content in mouse diets used by different research groups or by the genetic background mice are maintained. It was an important finding that the degree of tissue iron accumulation in mice can vary considerably depending on the genetic background they are maintained (Cavey et al. 2015). In addition, unlike humans, mice have ad libitum access to food that does not contain some components of the human diet, such as heme.

Finally, in some cell types, the Cre transgene used for the generation of tissue-specific knock-in and knock-out mice may exhibit toxicity.

Therefore, Cre controls should always be included in the experimental design and reported in publications.

Despite some limitations, murine models are and will always be an indispensable ally for unraveling the intricacies of systemic iron homeostasis regulation.

References

Altamura S, Kessler R, Grone HJ, Gretz N, Hentze MW, Galy B, Muckenthaler MU (2014) Resistance of ferroportin to hepcidin binding causes exocrine pancreatic failure and fatal iron overload. Cell Metab 20(2):359–367. https://doi.org/10.1016/j.cmet.2014.07.007

Altamura S, Schaeper U, Dames S, Loffler K, Eisermann M, Frauendorf C, Mudder K, Neves J, Muckenthaler MU (2019) SLN124, a GalNAc-siRNA conjugate targeting TMPRSS6, efficiently prevents iron overload in hereditary haemochromatosis type 1. Hema 3(6):e301. https://doi.org/10.1097/HS9.0000000000000301

An P, Wang H, Wu Q, Wang J, Xia Z, He X, Wang X, Chen Y, Min J, Wang F (2018) Smad7 deficiency decreases iron and haemoglobin through hepcidin up-regulation by multilayer compensatory mechanisms. J Cell Mol Med 22(6):3035–3044. https://doi.org/10.1111/jcmm.13546

Arezes J, Foy N, McHugh K, Sawant A, Quinkert D, Terraube V, Brinth A, Tam M, LaVallie ER, Taylor S, Armitage AE, Pasricha SR, Cunningham O, Lambert M, Draper SJ, Jasuja R, Drakesmith H (2018) Erythroferrone inhibits the induction of hepcidin by BMP6. Blood 132(14):1473–1477. https://doi.org/10.1182/blood-2018-06-857995

Armitage AE, Lim PJ, Frost JN, Pasricha SR, Soilleux EJ, Evans E, Morovat A, Santos A, Diaz R, Biggs D, Davies B, Gileadi U, Robbins PA, Lakhal-Littleton S, Drakesmith H (2016) Induced disruption of the iron-regulatory hormone Hepcidin inhibits acute inflammatory Hypoferraemia. J Innate Immun 8(5):517–528. https://doi.org/10.1159/000447713

Asperti M, Denardo A, Gryzik M, Castagna A, Girelli D, Naggi A, Arosio P, Poli M (2020) Pentosan polysulfate to control hepcidin expression in vitro and in vivo. Biochem Pharmacol 175:113867. https://doi.org/10.1016/j.bcp.2020.113867

Asperti M, Denardo A, Gryzik M, Persson KEM, Westerberg G, Ohd J, Poli M (2024) Sevuparin strongly reduces hepcidin expression in cells, mice, and healthy human volunteers. Hema 8(12):e70035. https://doi.org/10.1002/hem3.70035

Bahram S, Gilfillan S, Kuhn LC, Moret R, Schulze JB, Lebeau A, Schumann K (1999) Experimental hemochromatosis due to MHC class I HFE deficiency: immune status and iron metabolism. Proc Natl Acad Sci USA 96(23):13312–13317. https://doi.org/10.1073/pnas.96.23.13312

Canali S, Wang CY, Zumbrennen-Bullough KB, Bayer A, Babitt JL (2017a) Bone morphogenetic protein 2 controls iron homeostasis in mice independent of Bmp6. Am J Hematol 92(11):1204–1213. https://doi.org/10.1002/ajh.24888

Canali S, Zumbrennen-Bullough KB, Core AB, Wang CY, Nairz M, Bouley R, Swirski FK, Babitt JL (2017b) Endothelial cells produce bone morphogenetic protein 6 required for iron homeostasis in mice. Blood 129(4):405–414. https://doi.org/10.1182/blood-2016-06-721571

Casanovas G, Mleczko-Sanecka K, Altamura S, Hentze MW, Muckenthaler MU (2009) Bone morphogenetic protein (BMP)-responsive elements located in the proximal and distal hepcidin promoter are critical for its response to HJV/BMP/SMAD. J Mol Med (Berl) 87(5):471–480. https://doi.org/10.1007/s00109-009-0447-2

Casu C, Oikonomidou PR, Chen H, Nandi V, Ginzburg Y, Prasad P, Fleming RE, Shah YM, Valore EV, Nemeth E, Ganz T, MacDonald B, Rivella S (2016) Minihepcidin peptides as disease modifiers in mice affected by beta-thalassemia and polycythemia vera. Blood 128(2):265–276. https://doi.org/10.1182/blood-2015-10-676742

Casu C, Chessa R, Liu A, Gupta R, Drakesmith H, Fleming R, Ginzburg YZ, MacDonald B, Rivella S (2020) Minihepcidins improve ineffective erythropoiesis and splenomegaly in a new mouse model of adult beta-thalassemia major. Haematologica 105(7):1835–1844. https://doi.org/10.3324/haematol.2018.212589

Cavey T, Ropert M, de Tayrac M, Bardou-Jacquet E, Island ML, Leroyer P, Bendavid C, Brissot P, Loreal O (2015) Mouse genetic background impacts both on iron and non-iron metals parameters and on their relationships. Biometals 28(4):733–743. https://doi.org/10.1007/s10534-015-9862-8

Charlebois E, Fillebeen C, Presley J, Cagnone G, Lisi V, Lavallee VP, Joyal JS, Pantopoulos K (2023) Liver sinusoidal endothelial cells induce BMP6 expression in response to non-transferrin-bound iron. Blood 141(3):271–284. https://doi.org/10.1182/blood.2022016987

Colucci S, Altamura S, Marques O, Mudder K, Agarvas AR, Hentze MW, Muckenthaler MU (2022a) Iron-dependent BMP6 regulation in liver sinusoidal endothelial cells is instructed by hepatocyte-derived secretory signals. Hema 6(10):e773. https://doi.org/10.1097/HS9.0000000000000773

Colucci S, Mudder K, Muckenthaler MU, Altamura S (2022b) Hfe is highly expressed in liver sinusoidal endothelial cells but is not needed to maintain systemic iron homeostasis in vivo. Hema 6(1):e667. https://doi.org/10.1097/HS9.0000000000000667

Corradini E, Rozier M, Meynard D, Odhiambo A, Lin HY, Feng Q, Migas MC, Britton RS, Babitt JL, Fleming RE (2011) Iron regulation of hepcidin despite attenuated Smad1,5,8 signaling in mice without transferrin receptor 2 or Hfe. Gastroenterology 141(5):1907–1914. https://doi.org/10.1053/j.gastro.2011.06.077

de Sousa M, Reimao R, Lacerda R, Hugo P, Kaufmann SH, Porto G (1994) Iron overload in beta 2-microglobulin-deficient mice. Immunol Lett 39(2):105–111. https://doi.org/10.1016/0165-2478(94)90094-9

Denardo A, Elli S, Federici S, Asperti M, Gryzik M, Ruzzenenti P, Carmona F, Bergese P, Naggi A, Arosio P, Poli M (2021) BMP6 binding to heparin and heparan sulfate is mediated by N-terminal and C-terminal clustered basic residues. Biochim Biophys Acta Gen Subj 1865(2):129799. https://doi.org/10.1016/j.bbagen.2020.129799

Dogan DY, Urzica EI, Hornung I, Kastl P, Oguama D, Fette FM, Nguyen LH, Rosenbauer F, Zacharowski K, Klingmuller U, Gradhand E, von Knethen A, Popp R, Fleming I, Schrader L, Steinbicker AU (2024) Hemojuvelin-mediated hepcidin induction requires both bone morphogenetic protein type I receptors ALK2 and ALK3. Blood Adv 8(11):2870–2879. https://doi.org/10.1182/bloodadvances.2023012322

Donovan A, Lima CA, Pinkus JL, Pinkus GS, Zon LI, Robine S, Andrews NC (2005) The iron exporter ferroportin/Slc40a1 is essential for iron homeostasis. Cell Metab 1(3):191–200. https://doi.org/10.1016/j.cmet.2005.01.003

Du X, She E, Gelbart T, Truksa J, Lee P, Xia Y, Khovananth K, Mudd S, Mann N, Moresco EM, Beutler E, Beutler B (2008) The serine protease TMPRSS6 is required to sense iron deficiency. Science 320(5879):1088–1092. https://doi.org/10.1126/science.1157121

Enns CA, Jue S, Zhang AS (2020) The ectodomain of matriptase-2 plays an important nonproteolytic role in suppressing hepcidin expression in mice. Blood 136(8):989–1001. https://doi.org/10.1182/blood.2020005222

Enns CA, Jue S, Zhang AS (2021) Hepatocyte neogenin is required for hemojuvelin-mediated hepcidin expression and iron homeostasis in mice. Blood 138(6):486–499. https://doi.org/10.1182/blood.2020009485

Feng Q, Migas MC, Waheed A, Britton RS, Fleming RE (2012) Ferritin upregulates hepatic expression of bone morphogenetic protein 6 and hepcidin in mice. Am J Physiol Gastrointest Liver Physiol 302(12):G1397–G1404. https://doi.org/10.1152/ajpgi.00020.2012

Finberg KE, Whittlesey RL, Andrews NC (2011) Tmprss6 is a genetic modifier of the Hfe-hemochromatosis phenotype in mice. Blood 117(17):4590–4599. https://doi.org/10.1182/blood-2010-10-315507

Fisher AL, Wang CY, Xu Y, Joachim K, Xiao X, Phillips S, Moschetta GA, Alfaro-Magallanes VM, Babitt JL (2022) Functional role of endothelial transferrin receptor 1 in iron sensing and homeostasis. Am J Hematol 97(12):1548–1559. https://doi.org/10.1002/ajh.26716

Fisher AL, Phillips S, Wang CY, Paulo JA, Xiao X, Moschetta GA, Sridhar A, Mancias JD, Babitt JL (2024a) Endothelial ZIP8 plays a minor role in BMP6 regulation by iron in mice. Blood 143(23):2433–2437. https://doi.org/10.1182/blood.2023023385

Fisher AL, Phillips S, Wang CY, Paulo JA, Xiao X, Xu Y, Moschetta GA, Xue Y, Mancias JD, Babitt JL (2024b) The hepcidin-ferroportin axis modulates liver endothelial cell BMP expression to influence iron homeostasis in mice. Blood. https://doi.org/10.1182/blood.2024024795

Folgueras AR, de Lara FM, Pendas AM, Garabaya C, Rodriguez F, Astudillo A, Bernal T, Cabanillas R, Lopez-Otin C, Velasco G (2008) Membrane-bound serine protease matriptase-2 (Tmprss6) is an essential regulator of iron homeostasis. Blood 112(6):2539–2545. https://doi.org/10.1182/blood-2008-04-149773

Gkouvatsos K, Wagner J, Papanikolaou G, Sebastiani G, Pantopoulos K (2011) Conditional disruption of mouse HFE2 gene: maintenance of systemic iron homeostasis requires hepatic but not skeletal muscle hemojuvelin. Hepatology 54(5):1800–1807. https://doi.org/10.1002/hep.24547

Guo S, Casu C, Gardenghi S, Booten S, Aghajan M, Peralta R, Watt A, Freier S, Monia BP, Rivella S (2013) Reducing TMPRSS6 ameliorates hemochromatosis and beta-thalassemia in mice. J Clin Invest 123(4):1531–1541. https://doi.org/10.1172/JCI66969

Herrmann T, Muckenthaler M, van der Hoeven F, Brennan K, Gehrke SG, Hubert N, Sergi C, Grone HJ, Kaiser I, Gosch I, Volkmann M, Riedel HD, Hentze MW, Stewart AF, Stremmel W (2004) Iron overload in adult Hfe-deficient mice independent of changes in the steady-state expression of the duodenal iron transporters DMT1 and Ireg1/ferroportin. J Mol Med (Berl) 82(1):39–48. https://doi.org/10.1007/s00109-003-0508-x

Huang FW, Pinkus JL, Pinkus GS, Fleming MD, Andrews NC (2005) A mouse model of juvenile hemochromatosis. J Clin Invest 115(8):2187–2191. https://doi.org/10.1172/JCI25049

Kautz L, Jung G, Valore EV, Rivella S, Nemeth E, Ganz T (2014) Identification of erythroferrone as an erythroid regulator of iron metabolism. Nat Genet 46(7):678–684. https://doi.org/10.1038/ng.2996

Koch PS, Olsavszky V, Ulbrich F, Sticht C, Demory A, Leibing T, Henzler T, Meyer M, Zierow J, Schneider S, Breitkopf-Heinlein K, Gaitantzi H, Spencer-Dene B, Arnold B, Klapproth K, Schledzewski K, Goerdt S, Geraud C (2017) Angiocrine Bmp2 signaling in murine liver controls normal iron homeostasis. Blood 129(4):415–419. https://doi.org/10.1182/blood-2016-07-729822

Kuo WJ, Digman MA, Lander AD (2010) Heparan sulfate acts as a bone morphogenetic protein coreceptor by facilitating ligand-induced receptor hetero-oligomerization. Mol Biol Cell 21(22):4028–4041. https://doi.org/10.1091/mbc.E10-04-0348

Lai D, Teng F, Hammad S, Werle J, Maas T, Teufel A, Muckenthaler MU, Dooley S, Vujic Spasic M (2018) Hepatic Smad7 overexpression causes severe iron overload in mice. Blood 131(5):581–585. https://doi.org/10.1182/blood-2017-07-796797

Lakhal-Littleton S, Wolna M, Chung YJ, Christian HC, Heather LC, Brescia M, Ball V, Diaz R, Santos A, Biggs D, Clarke K, Davies B, Robbins PA (2016) An essential cell-autonomous role for hepcidin in cardiac iron homeostasis. elife 5. https://doi.org/10.7554/eLife.19804

Lakhal-Littleton S, Crosby A, Frise MC, Mohammad G, Carr CA, Loick PAM, Robbins PA (2019) Intracellular iron deficiency in pulmonary arterial smooth muscle cells induces pulmonary arterial hypertension in mice. Proc Natl Acad Sci USA 116(26):13122–13130. https://doi.org/10.1073/pnas.1822010116

Lee DH, Zhou LJ, Zhou Z, Xie JX, Jung JU, Liu Y, Xi CX, Mei L, Xiong WC (2010) Neogenin inhibits HJV secretion and regulates BMP-induced hepcidin expression and iron homeostasis. Blood 115(15):3136–3145. https://doi.org/10.1182/blood-2009-11-251199

Lesbordes-Brion JC, Viatte L, Bennoun M, Lou DQ, Ramey G, Houbron C, Hamard G, Kahn A, Vaulont S (2006) Targeted disruption of the hepcidin 1 gene results in severe hemochromatosis. Blood 108(4):1402–1405. https://doi.org/10.1182/blood-2006-02-003376

Levy JE, Montross LK, Cohen DE, Fleming MD, Andrews NC (1999) The C282Y mutation causing hereditary hemochromatosis does not produce a null allele. Blood 94(1):9–11

Lim PJ, Duarte TL, Arezes J, Garcia-Santos D, Hamdi A, Pasricha SR, Armitage AE, Mehta H, Wideman S, Santos AG, Santos-Goncalves A, Morovat A, Hughes JR, Soilleux E, Wang CY, Bayer AL, Klenerman P, Willberg CB, Hartley RC, Murphy MP, Babitt JL, Ponka P, Porto G, Drakesmith H (2019) Nrf2 controls iron homeostasis in haemochromatosis and thalassaemia via Bmp6 and hepcidin. Nat Metab 1(5):519–531. https://doi.org/10.1038/s42255-019-0063-6

Loick P, Mohammad GH, Cassimjee I, Chandrashekar A, Lapolla P, Carrington A, Vera-Aviles M, Handa A, Lee R, Lakhal-Littleton S (2023) Protective role for smooth muscle cell Hepcidin in abdominal aortic aneurysm. Arterioscler Thromb Vasc Biol 43(5):713–725. https://doi.org/10.1161/ATVBAHA.123.319224

Lou DQ, Nicolas G, Lesbordes JC, Viatte L, Grimber G, Szajnert MF, Kahn A, Vaulont S (2004) Functional differences between hepcidin 1 and 2 in transgenic mice. Blood 103(7):2816–2821. https://doi.org/10.1182/blood-2003-07-2524

Malerba M, Louis S, Cuvellier S, Shambat SM, Hua C, Gomart C, Fouet A, Ortonne N, Decousser JW, Zinkernagel AS, Mathieu JR, Peyssonnaux C (2020) Epidermal hepcidin is required for neutrophil response to bacterial infection. J Clin Invest 130(1):329–334. https://doi.org/10.1172/JCI126645

Mayeur C, Leyton PA, Kolodziej SA, Yu B, Bloch KD (2014) BMP type II receptors have redundant roles in the regulation of hepatic hepcidin gene expression and iron metabolism. Blood 124(13):2116–2123. https://doi.org/10.1182/blood-2014-04-572644

Mleczko-Sanecka K, Casanovas G, Ragab A, Breitkopf K, Muller A, Boutros M, Dooley S, Hentze MW, Muckenthaler MU (2010) SMAD7 controls iron metabolism as a potent inhibitor of hepcidin expression. Blood 115(13):2657–2665. https://doi.org/10.1182/blood-2009-09-238105

Muckenthaler L, Marques O, Colucci S, Kunz J, Fabrowski P, Bast T, Altamura S, Hochsmann B, Schrezenmeier H, Langlotz M, Richter-Pechanska P, Rausch T, Hofmeister-Mielke N, Gunkel N, Hentze MW, Kulozik AE, Muckenthaler MU (2022) Constitutional PIGA mutations cause a novel subtype of hemochromatosis in patients with neurologic dysfunction. Blood 139(9):1418–1422. https://doi.org/10.1182/blood.2021013519

Nai A, Pagani A, Mandelli G, Lidonnici MR, Silvestri L, Ferrari G, Camaschella C (2012) Deletion of TMPRSS6 attenuates the phenotype in a mouse model of beta-thalassemia. Blood 119(21):5021–5029. https://doi.org/10.1182/blood-2012-01-401885

Nai A, Lidonnici MR, Rausa M, Mandelli G, Pagani A, Silvestri L, Ferrari G, Camaschella C (2015) The second transferrin receptor regulates red blood cell production in mice. Blood 125(7):1170–1179. https://doi.org/10.1182/blood-2014-08-596254

Nicolas G, Bennoun M, Devaux I, Beaumont C, Grandchamp B, Kahn A, Vaulont S (2001) Lack of hepcidin gene expression and severe tissue iron overload in upstream stimulatory factor 2 (USF2) knockout mice. Proc Natl Acad Sci USA 98(15):8780–8785. https://doi.org/10.1073/pnas.151179498

Niederkofler V, Salie R, Arber S (2005) Hemojuvelin is essential for dietary iron sensing, and its mutation leads to severe iron overload. J Clin Invest 115(8):2180–2186. https://doi.org/10.1172/JCI25683

Pagani A, Pettinato M, Colucci S, Dulja A, Rauner M, Nai A, Camaschella C, Altamura S, Muckenthaler MU, Silvestri L (2020) Hemochromatosis proteins are dispensable for the acute hepcidin response to BMP2. Haematologica 105(10):e493. https://doi.org/10.3324/haematol.2019.241984

Pettinato M, Dulja A, Colucci S, Furiosi V, Fette F, Steinbicker AU, Muckenthaler MU, Nai A, Pagani A, Silvestri L (2023) FKBP12 inhibits hepcidin expression by modulating BMP receptors interaction and ligand responsiveness in hepatocytes. Am J Hematol 98(8):1223–1235. https://doi.org/10.1002/ajh.26961

Pietrangelo A (2004) The ferroportin disease. Blood Cells Mol Dis 32(1):131–138. https://doi.org/10.1016/j.bcmd.2003.08.003

Poli M, Girelli D, Campostrini N, Maccarinelli F, Finazzi D, Luscieti S, Nai A, Arosio P (2011) Heparin: a potent inhibitor of hepcidin expression in vitro and in vivo. Blood 117(3):997–1004. https://doi.org/10.1182/blood-2010-06-289082

Poli M, Asperti M, Ruzzenenti P, Mandelli L, Campostrini N, Martini G, Di Somma M, Maccarinelli F, Girelli D, Naggi A, Arosio P (2014) Oversulfated heparins with low anticoagulant activity are strong and fast inhibitors of hepcidin expression in vitro and in vivo. Biochem Pharmacol 92(3):467–475. https://doi.org/10.1016/j.bcp.2014.09.007

Poli M, Anower EKF, Asperti M, Ruzzenenti P, Gryzik M, Denardo A, Gordts P, Arosio P, Esko JD (2019) Hepatic heparan sulfate is a master regulator of hepcidin expression and iron homeostasis in human hepatocytes and mice. J Biol Chem 294(36):13292–13303. https://doi.org/10.1074/jbc.RA118.007213

Ramos E, Ruchala P, Goodnough JB, Kautz L, Preza GC, Nemeth E, Ganz T (2012) Minihepcidins prevent iron overload in a hepcidin-deficient mouse model of severe hemochromatosis. Blood 120(18):3829–3836. https://doi.org/10.1182/blood-2012-07-440743

Rauner M, Baschant U, Roetto A, Pellegrino RM, Rother S, Salbach-Hirsch J, Weidner H, Hintze V, Campbell G, Petzold A, Lemaitre R, Henry I, Bellido T, Theurl I, Altamura S, Colucci S, Muckenthaler MU, Schett G, Komla Ebri D, Bassett JHD, Williams GR, Platzbecker U, Hofbauer LC (2019) Transferrin receptor 2 controls bone mass and pathological bone formation via BMP and Wnt signaling. Nat Metab 1(1):111–124. https://doi.org/10.1038/s42255-018-0005-8

Rausa M, Pagani A, Nai A, Campanella A, Gilberti ME, Apostoli P, Camaschella C, Silvestri L (2015) Bmp6 expression in murine liver non parenchymal cells: a mechanism to control their high iron exporter activity and protect hepatocytes from iron overload? PLoS One 10(4):e0122696. https://doi.org/10.1371/journal.pone.0122696

Recalcati S, Gammella E, Buratti P, Doni A, Anselmo A, Locati M, Cairo G (2019) Macrophage ferroportin is essential for stromal cell proliferation in wound healing. Haematologica 104(1):47–58. https://doi.org/10.3324/haematol.2018.197517

Rishi G, Secondes ES, Wallace DF, Subramaniam VN (2016) Hematopoietic deletion of transferrin receptor 2 in mice leads to a block in erythroid differentiation during iron-deficient anemia. Am J Hematol 91(8):812–818. https://doi.org/10.1002/ajh.24417

Sardo U, Perrier P, Cormier K, Sotin M, Personnaz J, Medjbeur T, Desquesnes A, Cannizzo L, Ruiz-Martinez M, Thevenin J, Billore B, Jung G, Abboud E, Peyssonnaux C, Nemeth E, Ginzburg YZ, Ganz T, Kautz L (2024) The hepatokine FGL1 regulates hepcidin and iron metabolism during anemia in mice by antagonizing BMP signaling. Blood 143(13):1282–1292. https://doi.org/10.1182/blood.2023022724

Schmidt PJ, Toudjarska I, Sendamarai AK, Racie T, Milstein S, Bettencourt BR, Hettinger J, Bumcrot D, Fleming MD (2013) An RNAi therapeutic targeting Tmprss6 decreases iron overload in Hfe(−/−) mice and ameliorates anemia and iron overload in murine beta-thalassemia intermedia. Blood 121(7):1200–1208. https://doi.org/10.1182/blood-2012-09-453977

Steinbicker AU, Bartnikas TB, Lohmeyer LK, Leyton P, Mayeur C, Kao SM, Pappas AE, Peterson RT, Bloch DB, Yu PB, Fleming MD, Bloch KD (2011) Perturbation of hepcidin expression by BMP type I receptor deletion induces iron overload in mice. Blood 118(15):4224–4230. https://doi.org/10.1182/blood-2011-03-339952

Tangudu NK, Yilmaz D, Worle K, Gruber A, Colucci S, Leopold K, Muckenthaler MU, Vujic Spasic M (2021) Macrophage-HFE controls iron metabolism and immune responses in aged mice. Haematologica 106(1):259–263. https://doi.org/10.3324/haematol.2019.235630

Theurl I, Theurl M, Seifert M, Mair S, Nairz M, Rumpold H, Zoller H, Bellmann-Weiler R, Niederegger H, Talasz H, Weiss G (2008) Autocrine formation of hepcidin induces iron retention in human monocytes. Blood 111(4):2392–2399. https://doi.org/10.1182/blood-2007-05-090019

Tomatsu S, Orii KO, Fleming RE, Holden CC, Waheed A, Britton RS, Gutierrez MA, Velez-Castrillon S, Bacon BR, Sly WS (2003) Contribution of the H63D mutation in HFE to murine hereditary hemochromatosis. Proc Natl Acad Sci USA 100(26):15788–15793. https://doi.org/10.1073/pnas.2237037100

Traeger L, Gallitz I, Sekhri R, Baumer N, Kuhlmann T, Kemming C, Holtkamp M, Muller JC, Karst U, Canonne-Hergaux F, Muckenthaler MU, Bloch DB, Olschewski A, Bartnikas TB, Steinbicker AU (2018) ALK3 undergoes ligand-independent homodimerization and BMP-induced heterodimerization with ALK2. Free Radic Biol Med 129:127–137. https://doi.org/10.1016/j.freeradbiomed.2018.09.021

Vadolas J, Ng GZ, Kysenius K, Crouch PJ, Dames S, Eisermann M, Nualkaew T, Vilcassim S, Schaeper U, Grigoriadis G (2021) SLN124, a GalNac-siRNA targeting transmembrane serine protease 6, in combination with deferiprone therapy reduces ineffective erythropoiesis and hepatic iron-overload in a mouse model of beta-thalassaemia. Br J Haematol 194(1):200–210. https://doi.org/10.1111/bjh.17428

Verga Falzacappa MV, Casanovas G, Hentze MW, Muckenthaler MU (2008) A bone morphogenetic protein (BMP)-responsive element in the hepcidin promoter controls HFE2-mediated hepatic hepcidin expression and its response to IL-6 in cultured cells. J Mol Med (Berl) 86(5):531–540. https://doi.org/10.1007/s00109-008-0313-7

Vujic Spasic M, Kiss J, Herrmann T, Kessler R, Stolte J, Galy B, Rathkolb B, Wolf E, Stremmel W, Hentze MW, Muckenthaler MU (2007) Physiologic systemic iron metabolism in mice deficient for duodenal Hfe. Blood 109(10):4511–4517. https://doi.org/10.1182/blood-2006-07-036186

Vujic Spasic M, Kiss J, Herrmann T, Galy B, Martinache S, Stolte J, Grone HJ, Stremmel W, Hentze MW, Muckenthaler MU (2008) Hfe acts in hepatocytes to prevent hemochromatosis. Cell Metab 7(2):173–178. https://doi.org/10.1016/j.cmet.2007.11.014

Vujic Spasic M, Sparla R, Mleczko-Sanecka K, Migas MC, Breitkopf-Heinlein K, Dooley S, Vaulont S, Fleming RE, Muckenthaler MU (2013) Smad6 and Smad7 are co-regulated with hepcidin in mouse models of iron overload. Biochim Biophys Acta 1:76–84. https://doi.org/10.1016/j.bbadis.2012.08.013

Wallace DF, Summerville L, Lusby PE, Subramaniam VN (2005) First phenotypic description of transfer-

rin receptor 2 knockout mouse, and the role of hepcidin. Gut 54(7):980–986. https://doi.org/10.1136/gut.2004.062018
Wallace DF, Summerville L, Subramaniam VN (2007) Targeted disruption of the hepatic transferrin receptor 2 gene in mice leads to iron overload. Gastroenterology 132(1):301–310. https://doi.org/10.1053/j.gastro.2006.11.028
Wang RH, Li C, Xu X, Zheng Y, Xiao C, Zerfas P, Cooperman S, Eckhaus M, Rouault T, Mishra L, Deng CX (2005) A role of SMAD4 in iron metabolism through the positive regulation of hepcidin expression. Cell Metab 2(6):399–409. https://doi.org/10.1016/j.cmet.2005.10.010
Wang CY, Core AB, Canali S, Zumbrennen-Bullough KB, Ozer S, Umans L, Zwijsen A, Babitt JL (2017) Smad1/5 is required for erythropoietin-mediated suppression of hepcidin in mice. Blood 130(1):73–83. https://doi.org/10.1182/blood-2016-12-759423
Wang CY, Xiao X, Bayer A, Xu Y, Dev S, Canali S, Nair AV, Masia R, Babitt JL (2019) Ablation of hepatocyte Smad1, Smad5, and Smad8 causes severe tissue iron loading and liver fibrosis in mice. Hepatology 70(6):1986–2002. https://doi.org/10.1002/hep.30780
Wu XG, Wang Y, Wu Q, Cheng WH, Liu W, Zhao Y, Mayeur C, Schmidt PJ, Yu PB, Wang F, Xia Y (2014) HFE interacts with the BMP type I receptor ALK3 to regulate hepcidin expression. Blood 124(8):1335–1343. https://doi.org/10.1182/blood-2014-01-552281
Xiao X, Dev S, Canali S, Bayer A, Xu Y, Agarwal A, Wang CY, Babitt JL (2020) Endothelial bone morphogenetic protein 2 (Bmp2) knockout exacerbates hemochromatosis in homeostatic iron regulator (Hfe) knockout mice but not Bmp6 knockout mice. Hepatology 72(2):642–655. https://doi.org/10.1002/hep.31048
Xiao X, Xu Y, Moschetta GA, Yu Y, Fisher AL, Alfaro-Magallanes VM, McMillen S, Phillips S, Wang CY, Christian J, Babitt JL (2023) BMP5 contributes to hepcidin regulation and systemic iron homeostasis in mice. Blood 142(15):1312–1322. https://doi.org/10.1182/blood.2022019195
Zhang Z, Zhang F, An P, Guo X, Shen Y, Tao Y, Wu Q, Zhang Y, Yu Y, Ning B, Nie G, Knutson MD, Anderson GJ, Wang F (2011) Ferroportin1 deficiency in mouse macrophages impairs iron homeostasis and inflammatory responses. Blood 118(7):1912–1922. https://doi.org/10.1182/blood-2011-01-330324
Zhang Z, Zhang F, Guo X, An P, Tao Y, Wang F (2012) Ferroportin1 in hepatocytes and macrophages is required for the efficient mobilization of body iron stores in mice. Hepatology 56(3):961–971. https://doi.org/10.1002/hep.25746
Zhang DL, Ghosh MC, Ollivierre H, Li Y, Rouault TA (2018) Ferroportin deficiency in erythroid cells causes serum iron deficiency and promotes hemolysis due to oxidative stress. Blood 132(19):2078–2087. https://doi.org/10.1182/blood-2018-04-842997
Zhou XY, Tomatsu S, Fleming RE, Parkkila S, Waheed A, Jiang J, Fei Y, Brunt EM, Ruddy DA, Prass CE, Schatzman RC, O'Neill R, Britton RS, Bacon BR, Sly WS (1998) HFE gene knockout produces mouse model of hereditary hemochromatosis. Proc Natl Acad Sci USA 95(5):2492–2497. https://doi.org/10.1073/pnas.95.5.2492
Zlatanova I, Pinto C, Bonnin P, Mathieu JRR, Bakker W, Vilar J, Lemitre M, Voehringer D, Vaulont S, Peyssonnaux C, Silvestre JS (2019) Iron regulator Hepcidin impairs macrophage-dependent cardiac repair after injury. Circulation 139(12):1530–1547. https://doi.org/10.1161/CIRCULATIONAHA.118.034545
Zohn IE, De Domenico I, Pollock A, Ward DM, Goodman JF, Liang X, Sanchez AJ, Niswander L, Kaplan J (2007) The flatiron mutation in mouse ferroportin acts as a dominant negative to cause ferroportin disease. Blood 109(10):4174–4180. https://doi.org/10.1182/blood-2007-01-066068
Zumerle S, Mathieu JR, Delga S, Heinis M, Viatte L, Vaulont S, Peyssonnaux C (2014) Targeted disruption of hepcidin in the liver recapitulates the hemochromatotic phenotype. Blood 123(23):3646–3650. https://doi.org/10.1182/blood-2014-01-550467

Diagnosis and Treatment of HFE^{C282Y}-Linked Hemochromatosis

9

Paul C. Adams

Abstract

Genetic variants for hemochromatosis have been identified in human fossils that are over 4000 years old in Northern Ireland. Iron overload can lead to life-threatening complications, including liver fibrosis, cirrhosis, and hepatocellular carcinoma. In this review, the emphasis is on developments in the diagnosis and treatment of C282Y-linked hemochromatosis. There is a need for earlier diagnosis leading to earlier treatment to prevent morbidity and mortality.

Keywords

Hemochromatosis · Haemochromatosis · Iron overload

9.1 Introduction

Hemochromatosis is the most common genetic disease in persons of European ancestry (Adams et al. 2023; Adams and Ryan 2025). The prevalence of the typical genetic profile for hemochromatosis (*HFE*^{C282Y} homozygote, 845G> A) is 1 in 227 in North American patients of European ancestry in the Hemochromatosis and Iron Overload Screening (HEIRS) study (Adams et al. 2005) and 1 in 156 in northern England in the UK Biobank Study (Pilling et al. 2019). In Ireland, a study in which 1000 neonates were screened at birth for *HFE* variants revealed a prevalence of 1%, the highest rate in the world (Byrnes et al. 2001). Although the C282Y variant is relatively common in certain populations, the clinical penetrance is highly variable, and reports have varied widely depending on the definition of clinical penetrance used. More recent evidence from the UK Biobank has indicated that 1 in 5 males and 1 in 10 females will develop morbidity associated with hemochromatosis, changing the focus away from a condition previously believed to be without significant consequences (Pilling et al. 2019; Lucas et al. 2024a). Moreover, 1 in 10 males will develop severe liver disease unless identified early and treated (Grosse et al. 2018).

P. C. Adams (✉)
Western University, London, ON, Canada
e-mail: padams@uwo.ca

9.2 Diagnosis

The signs and symptoms of hemochromatosis are too non-specific to suggest a diagnosis of hemochromatosis. In C282Y-related hemochromatosis, the onset of clinical signs of iron overload may not occur until 40 s in men or 50 s in women, who may be protected by physiological blood loss through menstruation and pregnancy (Grosse et al. 2018; European Association for the Study

K. Pantopoulos (ed.), *Iron Metabolism in Human Health and Disease*, Advances in Experimental Medicine and Biology 1480, https://doi.org/10.1007/978-3-031-92033-2_9

of the Liver. Electronic address and European Association for the Study of the 2022). While the clinical penetrance and severity of iron-related organ damage are greater in men than women, as evidenced by higher risks of liver cirrhosis and liver cancer, reported symptoms may not be significantly different between the two genders (Pilling et al. 2019).

The most common non-specific symptoms include fatigue and arthralgias of the knuckles, affecting over half of individuals (Pilling et al. 2019). No symptoms are common in early hemochromatosis, but a family history of hemochromatosis may prompt testing in siblings who have the highest risk of being affected by this autosomal recessive condition. In most countries, the diagnosis of hemochromatosis is incidental, for instance, when a serum ferritin is ordered as a test for iron deficiency to investigate fatigue. Hemochromatosis may also be picked up as part of a blood investigation for fatigue, arthritis, liver disease, or diabetes. It may even be found incidentally on magnetic resonance imaging (MR) for another indication that demonstrates a "black liver" secondary to iron overload. In the largest population screening test to date (UK Biobank) using whole genomic testing, 451,270 participants in northern England were tested, and 1 in 156 were C282Y homozygotes (Pilling et al. 2019). In these homozygotes, a previous diagnosis of hemochromatosis had been made in only 12% of men and 3.4% of women (Table 9.1). The average age at diagnosis in a cohort of 554 referred patients was 58.5 years for women and 61 years for men (Fig. 9.1). This is strong evidence of a ***failure to diagnose*** hemochromatosis in clinical practice.

A paradox in hemochromatosis is that we miss the diagnosis in most cases (*missed diagnosis*), and we also overdiagnose hemochromatosis in patients with an elevation in serum ferritin who usually do not have hemochromatosis or iron overload (*misdiagnosis*). In the HEIRS study (Adams et al. 2005), participants from the United States and Canada were tested for serum ferritin, transferrin saturation, and the C282Y and H63D mutations of the *HFE* gene. All participants were notified of their results and offered counseling. There were 99,711 participants with 299 C282Y homozygotes reported in this multi-ethnic population. There were 12,993 participants (13%) with an elevated serum ferritin (>200 μg/L in women and >300 μg/L in men), 5997 participants (6%) with an elevated transferrin saturation (>45% in women and >50% in men). These results demonstrate that an elevated serum ferritin is extremely common; it is a non-specific test, and most patients with an elevated serum ferritin do not have hemochromatosis or iron overload (Fig. 9.2). The probability of being C282Y homozygous does increase with increasing ferritin levels for males and females (Lim et al. 2014). For these reasons, patients should not be told that they have hemochromatosis or "high iron levels" at their first sign of an elevated ferritin since most patients have elevations from much more common conditions, including metabolic syndrome, fatty liver, daily alcohol use, obesity (termed metabolic hyperferritinemia (Valenti et al. 2023)), and inflammation. There were no significant cases of non-*HFE* iron overload in the HEIRS study, and in that study, patients with an unexplained elevation in serum ferritin had extended genetic testing for rare iron overload genes (Barton et al. 2009). A new important observation from the HEIRS study was that the transferrin saturation test was not reproducible (Adams et al. 2007). The typical recommendations to warrant *HFE* testing are serum ferritin >200 μg/L in women and >300 μg/L in men, along with a transferrin saturation >45% (European Association for the Study of the Liver. Electronic address and European Association for the Study of the 2022).

As a condition that is readily identifiable and treatable, the argument for population screening in hemochromatosis remains pertinent today. When the C282Y genetic test became widely available in 1997, initial concerns were raised regarding the detection of C282Y homozygotes without iron overload, the potential genetic discrimination, and the cost of testing. In the HEIRS study, 12% of the newly identified male C282Y homozygotes and 43% of the newly identified female homozygotes had a normal serum ferritin (Adams et al. 2005). There were no cases of

Table 9.1 Inability to diagnose C282Y hemochromatosis in community practice

Study population	n	C282 homozygotes	Undiagnosed men	Undiagnosed women
UK Biobank (Pilling et al. 2019; Lucas et al. 2024b)	451,243	2890	88%	97%
All of us (Rao et al. 2025)	249,915	633	77%	84%
Geisinger community health (Savatt et al. 2023)	86,601	201	68%	75%

genetic discrimination reported in the first year of the study (Hall et al. 2007). The cost of genetic testing for a single mutation (C282Y), which can be done without PCR amplification (Invader assay), is reported to be as low as $USD 5. This may not be the charge in a commercial laboratory, but it is clear that the cost of C282Y testing is not prohibitive and in most cases is less than the cost of serum ferritin and transferrin saturation. There have been at least seven studies using genetic testing as a screening test for hemochromatosis (Adams et al. 2005; Asberg et al. 2001; Beutler et al. 2002; Nisselle et al. 2004; Savatt et al. 2023). The test is reproducible, inexpensive, widely available, and can be done on a saliva sample. A difficult issue has been defining the logistics of a screening program. Most economic analysis studies have shown cost-effectiveness with population screening for hemochromatosis (de Graaff et al. 2017; Adams and Valberg 1999), but have modeled targeted screening in groups such as those of northern European ancestry or only men, which is unappealing to public policymakers (Schmidtke 2022). Screening through a simple request for a saliva sample for hemochromatosis in a mail-back program may also not be a successful approach. In Northern Ireland, the

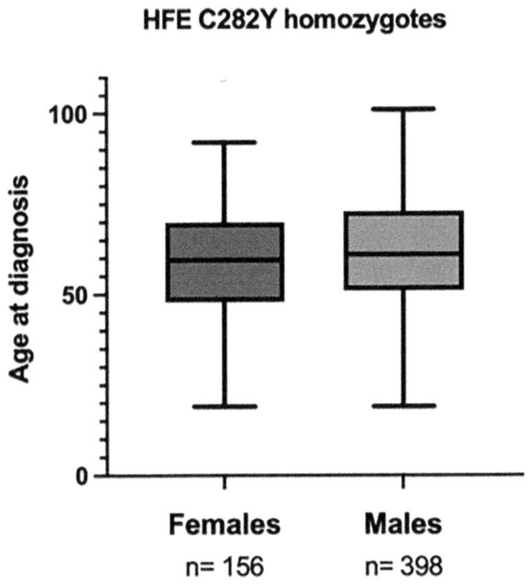

Fig. 9.1 Age at diagnosis in 554 referred C282Y patients from Canada and Ireland. (Figure is courtesy of John Ryan)

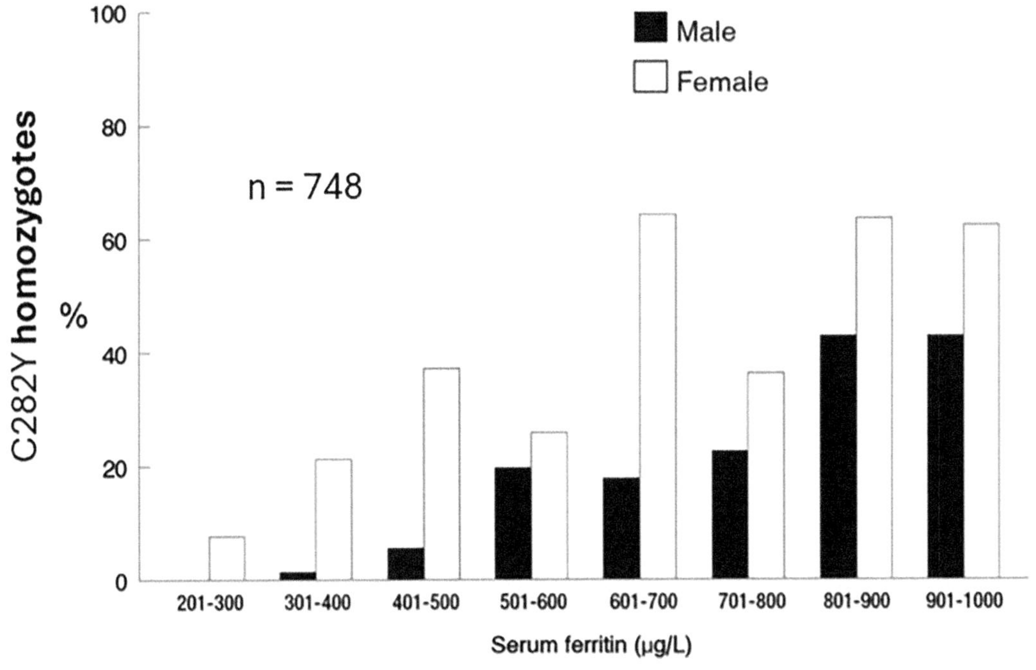

Fig. 9.2 The probability (%) of having C282Y hemochromatosis in 748 Caucasian participants with elevated serum ferritin (>200 μg/L women and >300 μg/L men) and transferrin saturation (>45% women and >50% men)

patient support organization mailed 36,000 people information about free genetic tests for hemochromatosis with instructions, and only 360 requested the test (Mahase and Iacobucci 2023). The acceptance rate of the genetic test to screen for hemochromatosis in the first study using the genetic test for screening in 2000 in Canada was 97%, but these were blood donors with a recruiter on site (Adams et al. 2000). Whole genome sequencing may offer an additional mechanism by which C282Y homozygosity/hemochromatosis could be detected, were it to become more commonly used in clinical practice; indeed, hemochromatosis has been a Tier 1 genetic result since 2021, in which notification of the result to the patient is recommended (Miller et al. 2023). The addition of C282Y genetic testing to an existing panel of other genetic diseases (hitched to another wagon) could improve the logistics of any screening program and remove the need to focus on Caucasian men since the other genetic diseases tested will likely affect diverse populations.

A new diagnosis of hemochromatosis in a C282Y homozygote should include a family history, and siblings and parents should be instructed to request an *HFE* (C282Y) genetic test. Adult children may also be tested, or alternatively, the partner/parent of the children may be tested, and if negative for the C282Y variant, then the children do not need testing in the future.

Upon diagnosis, the initial serum ferritin for a C282Y homozygote with hemochromatosis remains the best non-invasive marker of iron burden, the likelihood of iron-related complications, and the expected number of phlebotomies required to achieve "de-ironing." Serum ferritin is also a useful indicator of liver disease risk. For instance, those with a serum ferritin >1000 μg/L are much more likely to have severe liver fibrosis than those who do not. A small French study of 77 C282Y homozygotes found advanced (F3/F4) fibrosis on liver biopsy in 1 in 5 patients with a ferritin >1000 μg/L and/or increased liver transaminases (Legros et al. 2015).

All hemochromatosis patients should be screened for liver fibrosis, as liver disease will affect approximately 20% of adult male C282Y homozygotes (Lucas et al. 2024a). Typically a baseline abdominal ultrasound is performed to assess for concomitant hepatic steatosis, as well as for evidence of cirrhosis, portal hypertension, and liver cancer. MR liver can also provide information on steatosis, focal lesions, and fibrosis through MR elastography.

9.3 Hemochromatosis—Treatment

The management of hemochromatosis utilizes the medieval treatment of blood removal to decrease body iron stores. To focus treatment on those in need, it is essential to have a correct diagnosis that includes C282Y genetic testing and, in non-C282Y homozygotes, proof of iron overload (Table 9.2). It has been demonstrated in France in a multi-center study that most patients

Table 9.2 Methods to determine the presence of iron overload

Method	
Serum ferritin	Widely available. This test was originally developed for iron deficiency and is very non-specific for iron overload. Most patients with elevated ferritin do *not* have iron overload or hemochromatosis.
Transferrin saturation	This test has wide biological variability which is problematic for a diagnostic test (Adams et al. 2007). You can have iron overload with a normal transferrin saturation.
Liver biopsy	This is a definitive test but is invasive and is not commonly used now in C282Y homozygotes because of other methods.
MR liver	This is currently the best non-invasive test for iron overload and can provide additional information on liver fibrosis and steatosis. Availability and cost vary between countries (Hernando et al. 2014).
Trial of phlebotomy	This is commonly done but uses important resources at ambulatory therapy centers for patients who usually do not have iron overload (Le Gac et al. 2022).

receiving venesections for presumed iron overload are not C282Y homozygotes and do not have iron overload (Le Gac et al. 2022). The early hypothesis that many of these cases would have genetic mutations in other iron genes (see Chap. 10) has not been demonstrated, and the most common diagnosis is the metabolic syndrome without iron overload.

Once a diagnosis has been established, treatment should not be delayed, given the numerous reported benefits associated with the normalization of iron stores, not limited to an improvement in fatigue, the reversal of liver fibrosis, and a reduced risk of diabetes, liver cirrhosis, liver cancer, and death (Bardou-Jacquet et al. 2020; Prabhu et al. 2020). Moreover, treatment should be initiated as a matter of urgency in symptomatic patients or in those with significant iron overload (e.g., serum ferritin >1000 ug/L). Treatment should also not be withheld from patients with mildly increased iron stores, as patients with moderate elevations in serum ferritin (e.g., 300–500) have been shown to benefit from venesection, with a reported reduction in cardiovascular disease and extrahepatic malignancies (Bardou-Jacquet et al. 2020). Ultimately, patients should be encouraged to attend voluntary blood donation once de-ironed, as transfusion services now accept blood from hemochromatosis patients in many countries, with a positive effect on the blood donor pool (Cabrera et al. 2022).

9.4 C282Y Homozygote

A C282Y homozygote with elevated serum ferritin can be assumed to have iron overload and should be offered phlebotomy treatment. Patients are typically treated by weekly 500 ml phlebotomy by an experienced team until the serum ferritin is in the low normal range. Many studies suggest a target serum ferritin of 50 μg/L. This number was arbitrarily chosen since it was anticipated that some patients would not return for follow-up and allowed possibly for a slow rise of serum ferritin over time. A patient survey from France suggested that aggressive phlebotomy therapy to also decrease the transferrin saturation would reduce symptoms in some patients (Bardou-Jacquet et al. 2017). This strategy places patients on the verge of iron deficiency, requires more intensive follow-up by the treating physician, and is difficult to implement in routine practice.

There can be non-expressing C282Y homozygotes (normal ferritin, transferrin saturation, no symptoms) discovered through family or population screening studies. Adult patients with normal ferritin at the time of diagnosis are unlikely to develop iron overload later in life and can have periodic screening of ferritin every few years (Yamashita and Adams 2003). The side effects of phlebotomy at an experienced treatment center are very low. The most common are local effects around the needle insertion, including pain, redness, and inability to find a good vein (Hicken et al. 2003). Syncope and fatigue can occur. Volume depletion can be minimized by the use of a salt-containing sports drink.

In those with established cirrhosis, 6-monthly ultrasound and alpha feto-protein (AFP) levels for hepatocellular carcinoma (HCC) screening are mandatory, given a significantly increased risk. This is the most devastating and life-threatening complication of hemochromatosis, typically affecting men. Data from the UK Biobank has revealed a hazard ratio of 10.5 [95% CI, 6.6–16.7] for hepatic malignancies in C282Y homozygous males (Atkins et al. 2020). Moreover, the type of HCC seen in hemochromatosis tends to be more poorly differentiated and associated with worse outcomes (Di Capua et al. 2024). In addition, hemochromatosis is a known cause of non-cirrhotic HCC, and the merits of HCC screening can be considered in those with liver fibrosis, a family history, or concomitant risks for progressive liver disease. Early detection of HCC can lead to successful treatments with liver transplantation, liver resection, focal radiotherapy, or tumor ablation therapies.

Not all of the symptoms of hemochromatosis respond to phlebotomy therapy. Arthritis of the hands and ankles and the need for joint replacement can progress despite phlebotomy and can develop years after an asymptomatic C282Y homozygote has been iron depleted by phlebotomy (Banfield et al. 2023; Adams et al. 2021). It can be difficult to distinguish hemochromatosis arthropathy from osteoarthritis seen in the aging population. Treatment at the present time is symptomatic treatment with anti-inflammatory medications and joint replacement in severe cases. Joint replacement in hemochromatosis is more common in men than women.

Diabetes was once considered a major symptom of hemochromatosis when hemochromatosis was called "Bronze Diabetes," but larger population studies like HEIRS could not demonstrate an increase in diabetes compared to a control population. Early studies of severe hemochromatosis with autopsies demonstrated iron in the pancreas, but metabolic studies have suggested that insulin resistance, often associated with liver disease and obesity, is the most likely cause of diabetes in hemochromatosis (Barton and Acton 2017).

Cardiac dysfunction is only seen in severe cases of hemochromatosis, and cardiac function and dyspnea improve with iron depletion.

Skin pigmentation, which is related to melanin, not iron in the skin, improves with iron depletion.

Endocrine dysfunction is possible but uncommon in most hemochromatosis patients.

There are some neurological symptoms reported like tremors and brain iron has been seen using newer MR methodology (Sharma et al. 2022). Brain fog and fatigue are common patient-reported symptoms that often do not improve with iron depletion. In a cirrhotic patient, subclinical hepatic encephalopathy should also be considered since treatments are available.

9.5 Other *HFE* Genotypes

This group can include compound heterozygotes (C282Y/H63D) and H63D homozygotes (H63D/H63D). Large population-based studies have shown that individuals with these genotypes do not experience an increase in mortality or morbidity and can be reassured. Mild elevations in serum ferritin have been demonstrated in some individuals but are almost always associated with other risk factors like metabolic syndrome, obesity with fatty liver, or daily alcohol consumption (Lucas et al. 2024b). Mild biochemical iron overload can be seen in C282Y/H63D compound heterozygotes in approximately 10% of cases and in 1% of H63D homozygotes (Hasan et al. 2022). Furthermore, 10% of C282Y heterozygotes can demonstrate mild increases in serum ferritin or transferrin saturation, but not of clinical significance (Adams et al. 2023). Although venesection has not been proven to be of benefit in such patients, the most pragmatic and reasonable approach to normalizing iron stores is to promote voluntary blood donation and treat other metabolic risk factors.

9.6 Erythrocytapheresis

This method is a type of automated red blood cell exchange. It is a pump-driven system similar to plasmapheresis, and it has been shown to be an efficient method of removing iron while maintaining blood volume. Each procedure can remove up to 800 ml and remove 2.3 times the amount of iron compared to a 500 ml phlebotomy. The procedure has more supply costs per treatment, but fewer treatments are required. It has been used in several European centers (Rombout-Sestrienkova et al. 2016) and can be used with acceptable safety in experienced centers. It is the only treatment for iron overload that has been studied in a randomized participant-blinded study (Ong et al. 2017). In this project, mildly affected C282Y homozygotes were compared between those with iron removed and a placebo group that had the apheresis treatment without iron removal. Many measures were unchanged between the two groups, and there was a small improvement in fatigue.

9.7 Dietary Therapies

Patients are often very interested in dietary therapies, and this view is reinforced by many Internet patient support groups. It is not possible to have an iron-free diet. An iron-reduced diet has been studied and has had a minor effect in reducing serum ferritin. Iron supplementation of common foods was a popular marketing tool in the 1950s, but the removal of these supplements in Scandinavia has been studied with small effects on population health (Olsson et al. 1997). English tea with every meal can reduce iron absorption. There are several studies on the use of oral iron-binding compounds that can reduce iron absorption (Buerkli et al. 2022; Qian et al. 2017). Inhibition of gastric acid by proton pump inhibitors can also reduce iron absorption but not enough to manage a hemochromatosis patient without phlebotomy (Vanclooster et al. 2017). Weight reduction diets, including possibly GLP1 Medications, may be helpful since obesity and high waist-to-hip circumference ratios are associated with a higher mortality in C282Y homozygotes (Lucas et al. 2024b).

9.8 Therapies That May Reduce the Need for Maintenance Therapy

In our current understanding of the pathogenesis of iron overload in hemochromatosis, hepcidin dysregulation plays a central role. For this reason, synthetic mini-hepcidin was developed as a parenteral therapy for hemochromatosis. However, it became apparent in animal models that hepcidin therapy did not mobilize excess liver iron but did decrease intestinal iron absorption (Katsarou and Pantopoulos 2018). Newer concepts of therapy have considered an initial period of phlebotomy to iron depletion followed by parenteral hepcidin injections to decrease the need for maintenance phlebotomy (Kowdley et al. 2023; Swinkels and Matlung 2023). There are several potential flaws in the rationale for combination therapy. It assumes that patients or providers are dissatisfied with phlebotomy therapy. This is incorrect, and many patients enjoy phlebotomy treatment, including the psychosocial aspects of meeting other patients and families with hemochromatosis. There are many patients that will not require maintenance therapy. This is particularly true in young women who present with serum ferritin <1000 μg/L. In 1991, a treatment study that did not offer maintenance therapy found that 52% did not have significant iron re-accumulation over a 4-year observation period (Adams et al. 1993). Any study of an expensive biological hepcidin therapy would need to be compared to a phlebotomy group, and a cost analysis will always favor phlebotomy since a blood donation can be considered a benefit rather than a cost. A study group of no phlebotomy, if ferritin does not rise above the reference range, would help define the need for maintenance therapy.

9.9 Chelation Therapy

Deferasirox is the only oral iron chelator that has been studied in the treatment of hemochromatosis (Phatak et al. 2010). It is an oral therapy that has been approved for secondary iron overload associated with anemia. It has not been approved for the treatment of hemochromatosis.

9.10 The Future Is Now

The UK Biobank Study has redefined the mortality and morbidity of C282Y homozygotes discovered through population testing. The C282Y genetic test from whole genomic testing could be used to identify early disease and initiate phlebotomy therapy to prevent cirrhosis and hepatocellular carcinoma. It is unclear at this time whether whole genomic testing will become an accepted and affordable component of the patient's medical record.

The advances in our understanding of the pathophysiology of iron overload mean that new therapies may finally be on the horizon. In particular, overcoming hepcidin deficiency, which underpins hemochromatosis (see Chap. 3) either through supplementation or the modulation of its regulators, is a logical approach. While promising, the need for parenteral administration and projected costs may limit their use to exceptional patients. Hepcidin agonists do not appear to reduce liver iron concentration and would therefore not be effective in the de-ironing phase of treatment.

Table 9.3 Future Gene therapy for hemochromatosis

Promising features
Common disease
Single mutation (C282Y)
Intravenous infusion
Good nanoparticle uptake with CRISPR/cas by the liver in animal studies
C282Y mutation is ideally suited for adenosine base editing, so less non-specific gene cutting
No viral vectors
No chemotherapy
Areas of concern
Hemochromatosis still seems rare to policymakers.
Are the patients sick enough to warrant gene therapy?
What is the ideal stage of disease to treat?
Are there off-target effects?
Who will pay for this type of treatment?
How would you know the gene therapy worked?
Will existing stored iron be mobilized by gene therapy?
How often do you need an infusion?
Are the IRB ethics issues insurmountable?

Gene editing by CRISPR, base editing, and RNA editing has the potential to cure hemochromatosis, but the methods will likely be tried first in more life-threatening diseases.

The UK Biobank and other whole genomic studies (Our Future Health, All of Us) represent a rich resource from which crucial clinical questions may be answered. Having origins traced back to more than 4000 years ago, the care for individuals with hemochromatosis must now evolve into the modern era of precision medicine.

9.11 Conclusions

The diagnosis of hemochromatosis has evolved from the use of non-specific iron blood tests (serum ferritin) and tests with unreliable reproducibility (transferrin saturation) to a simple genetic test (C282Y test) and advanced imaging (MR liver, elastography).

The treatment of hemochromatosis by phlebotomy has stood the test of time and is a very safe and cost-effective treatment. Biological therapies with mini-hepcidins, or gene editing using CRISPR technology (Table 9.3), are on the horizon but must be tested against phlebotomy.

References

Adams P, Ryan J (2025) Diagnosis and treatment of C282Y linked hemochromatosis. Clin Gastro Hep. (in press) PMID: 39889898 DOI: 10.1016/j.cgh.2024.10.041

Adams PC, Valberg LS (1999) Screening blood donors for hereditary hemochromatosis: decision analysis model comparing genotyping to phenotyping. Am J Gastroenterol 94(6):1593–1600. https://doi.org/10.1111/j.1572-0241.1999.1120_f.x

Adams PC, Kertesz AE, Valberg LS (1993) Rate of iron reaccumulation following iron depletion in hereditary hemochromatosis. Implications for venesection therapy. J Clin Gastroenterol 16(3):207–210. https://doi.org/10.1097/00004836-199304000-00008

Adams PC, Kertesz AE, McLaren CE, Barr R, Bamford A, Chakrabarti S (2000) Population screening for hemochromatosis: a comparison of unbound iron-binding capacity, transferrin saturation, and C282Y genotyping in 5,211 voluntary blood donors. Hepatology 31(5):1160–1164. https://doi.org/10.1053/he.2000.6984

Adams PC, Reboussin DM, Barton JC, McLaren CE, Eckfeldt JH, McLaren GD, Dawkins FW, Acton RT, Harris EL, Gordeuk VR, Leiendecker-Foster C, Speechley M, Snively BM, Holup JL, Thomson E, Sholinsky P (2005) Hemochromatosis and iron-overload screening in a racially diverse population. N Engl J Med 352(17):1769–1778. https://doi.org/10.1056/NEJMoa041534

Adams PC, Reboussin DM, Press RD, Barton JC, Acton RT, Moses GC, Leiendecker-Foster C, McLaren GD, Dawkins FW, Gordeuk VR, Lovato L, Eckfeldt JH (2007) Biological variability of transferrin saturation and unsaturated iron-binding capacity. Am J Med 120(11):e991–e997. https://doi.org/10.1016/j.amjmed.2007.02.027

Adams PC, Richard L, Weir M, Speechley M (2021) Survival and development of health conditions after iron depletion therapy in C282Y-linked hemochromatosis patients. Can Liver J 4(4):381–390. https://doi.org/10.3138/canlivj-2021-0016

Adams PC, Jeffrey G, Ryan J (2023) Haemochromatosis. Lancet 401(10390):1811–1821. https://doi.org/10.1016/S0140-6736(23)00287-8

Asberg A, Hveem K, Thorstensen K, Ellekjter E, Kannelonning K, Fjosne U, Halvorsen TB, Smethurst HB, Sagen E, Bjerve KS (2001) Screening for hemochromatosis: high prevalence and low morbidity in an unselected population of 65,238 persons. Scand J Gastroenterol 36(10):1108–1115. https://doi.org/10.1080/003655201750422747

Atkins JL, Pilling LC, Masoli JAH, Kuo CL, Shearman JD, Adams PC, Melzer D (2020) Association of hemochromatosis HFE p.C282Y homozygosity with hepatic malignancy. JAMA 324(20):2048–2057. https://doi.org/10.1001/jama.2020.21566

Banfield LR, Knapp KM, Pilling LC, Melzer D, Atkins JL (2023) Hemochromatosis genetic variants and musculoskeletal outcomes: 11.5-year follow-up in the UK biobank cohort study. JBMR Plus 7(10):e10794. https://doi.org/10.1002/jbm4.10794

Bardou-Jacquet E, Laine F, Guggenbuhl P, Morcet J, Jezequel C, Guyader D, Moirand R, Deugnier Y (2017) Worse outcomes of patients with HFE hemochromatosis with persistent increases in transferrin saturation during maintenance therapy. Clin Gastroenterol Hepatol 15(10):1620–1627. https://doi.org/10.1016/j.cgh.2016.12.039

Bardou-Jacquet E, Morandeau E, Anderson GJ, Ramm GA, Ramm LE, Morcet J, Bouzille G, Dixon J, Clouston AD, Laine F, Turlin B, Powell LW, Deugnier YM (2020) Regression of fibrosis stage with treatment reduces long-term risk of liver cancer in patients with hemochromatosis caused by mutation in HFE. Clin Gastroenterol Hepatol 18(8):1851–1857. https://doi.org/10.1016/j.cgh.2019.10.010

Barton JC, Acton RT (2017) Diabetes in HFE hemochromatosis. J Diabetes Res 2017:9826930. https://doi.org/10.1155/2017/9826930

Barton JC, Lafreniere SA, Leiendecker-Foster C, Li H, Acton RT, Press RD, Eckfeldt JH (2009) HFE, SLC40A1, HAMP, HJV, TFR2, and FTL mutations detected by denaturing high-performance liquid chromatography after iron phenotyping and HFE C282Y and H63D genotyping in 785 HEIRS study participants. Am J Hematol 84(11):710–714. https://doi.org/10.1002/ajh.21524

Beutler E, Felitti VJ, Koziol JA, Ho NJ, Gelbart T (2002) Penetrance of 845G--> A (C282Y) HFE hereditary haemochromatosis mutation in the USA. Lancet 359(9302):211–218. https://doi.org/10.1016/S0140-6736(02)07447-0

Buerkli S, Salvioni L, Koller N, Zeder C, Teles MJ, Porto G, Habermann JH, Dubach IL, Vallelian F, Frey BM, Moretti D, Baumgartner J, Zimmermann MB (2022) The effect of a natural polyphenol supplement on iron absorption in adults with hereditary hemochromatosis. Eur J Nutr 61(6):2967–2977. https://doi.org/10.1007/s00394-022-02829-8

Byrnes V, Ryan E, Barrett S, Kenny P, Mayne P, Crowe J (2001) Genetic hemochromatosis, a Celtic disease: is it now time for population screening? Genet Test 5(2):127–130. https://doi.org/10.1089/109065701753145583

Cabrera E, Crespo G, VanWagner LB (2022) Diagnosis and management of hereditary hemochromatosis. JAMA 328(18):1862–1863. https://doi.org/10.1001/jama.2022.17727

de Graaff B, Neil A, Si L, Yee KC, Sanderson K, Gurrin L, Palmer AJ (2017) Cost-effectiveness of different population screening strategies for hereditary haemochromatosis in Australia. Appl Health Econ Health Policy 15(4):521–534. https://doi.org/10.1007/s40258-016-0297-3

Di Capua DM, Shanahan W, Bourke M, Ramlaul N, Appel J, Canney A, Docherty NG, McGrath E, Ring E, Jones F, Boyle M, McCormack J, Gallagher T, Hoti E, Nolan N, Ryan JD, Houlihan DD, Fabre A (2024) Tumour stemness and poor clinical outcomes in haemochromatosis patients with hepatocellular carcinoma. J Clin Pathol 77(10):669–675. https://doi.org/10.1136/jcp-2022-208679

European Association for the Study of the Liver. Electronic address eee, European Association for the Study of the L (2022) EASL clinical practice guidelines on haemochromatosis. J Hepatol 77(2):479–502. https://doi.org/10.1016/j.jhep.2022.03.033

Grosse SD, Gurrin LC, Bertalli NA, Allen KJ (2018) Clinical penetrance in hereditary hemochromatosis: estimates of the cumulative incidence of severe liver disease among HFE C282Y homozygotes. Genet Med 20(4):383–389. https://doi.org/10.1038/gim.2017.121

Hall MA, Barton JC, Adams PC, McLaren CE, Reiss JA, Castro O, Ruggiero A, Acton RT, Power TE, Bent TC (2007) Genetic screening for iron overload: no evidence of discrimination at 1 year. J Fam Pract 56(10):829–834

Hasan SMM, Farrell J, Borgaonkar M (2022) C282Y/H63D compound heterozygosity is a low penetrance genotype for iron overload-related disease. J Can Assoc Gastroenterol 5(5):240–247. https://doi.org/10.1093/jcag/gwac025

Hernando D, Levin YS, Sirlin CB, Reeder SB (2014) Quantification of liver iron with MRI: state of the art and remaining challenges. J Magn Reson Imaging 40(5):1003–1021. https://doi.org/10.1002/jmri.24584

Hicken BL, Tucker DC, Barton JC (2003) Patient compliance with phlebotomy therapy for iron overload associated with hemochromatosis. Am J Gastroenterol 98(9):2072–2077. https://doi.org/10.1111/j.1572-0241.2003.07292.x

Katsarou A, Pantopoulos K (2018) Hepcidin therapeutics. Pharmaceuticals (Basel) 11(4). https://doi.org/10.3390/ph11040127

Kowdley KV, Modi NB, Peltekian K, Vierling JM, Ferris C, Valone FH, Gupta S (2023) Rusfertide for the treatment of iron overload in HFE-related haemochromatosis: an open-label, multicentre, proof-of-concept phase 2 trial. Lancet Gastroenterol Hepatol 8(12):1118–1128. https://doi.org/10.1016/S2468-1253(23)00250-9

Le Gac G, Scotet V, Gourlaouen I, L'Hostis C, Merour MC, Karim Z, Deugnier Y, Bardou-Jacquet E, Lefebvre T, Assari S, Ferec C (2022) Prevalence of HFE-related haemochromatosis and secondary causes of hyperferritinaemia and their association with iron overload in 1059 French patients treated by venesection. Aliment Pharmacol Ther 55(8):1016–1027. https://doi.org/10.1111/apt.16775

Legros L, Bardou-Jacquet E, Latournerie M, Guillygomarc'h A, Turlin B, Le Lan C, Desille Y, Laine F, Moirand R, Brissot P, Deugnier Y, Guyader

D (2015) Non-invasive assessment of liver fibrosis in C282Y homozygous HFE hemochromatosis. Liver Int 35(6):1731–1738. https://doi.org/10.1111/liv.12762

Lim A, Speechley M, Adams PC (2014) Predicting C282Y homozygote genotype for hemochromatosis using serum ferritin and transferrin saturation values from 44,809 participants of the HEIRS study. Can J Gastroenterol Hepatol 28(9):502–504. https://doi.org/10.1155/2014/698352

Lucas MR, Atkins JL, Pilling LC, Shearman JD, Melzer D (2024a) HFE genotypes, haemochromatosis diagnosis and clinical outcomes at age 80 years: a prospective cohort study in the UK Biobank. BMJ Open 14(3):e081926. https://doi.org/10.1136/bmjopen-2023-081926

Lucas MR, Pilling LC, Atkins JL, Melzer D (2024b) Incidence of liver complications with hemochromatosis associated HFE p.C282Y homozygosity: the role of central adiposity. Hepatology. https://doi.org/10.1097/HEP.0000000000001056

Mahase E, Iacobucci G (2023) Genetic test "screening campaign" may be causing unnecessary alarm, experts warn. BMJ 381:1264. https://doi.org/10.1136/bmj.p1264

Miller DT, Lee K, Abul-Husn NS, Amendola LM, Brothers K, Chung WK, Gollob MH, Gordon AS, Harrison SM, Hershberger RE, Klein TE, Richards CS, Stewart DR, Martin CL, documents@acmg.net ASFWGEa (2023) ACMG SF v3.2 list for reporting of secondary findings in clinical exome and genome sequencing: a policy statement of the American College of Medical Genetics and Genomics (ACMG). Genet Med 25(8):100866. https://doi.org/10.1016/j.gim.2023.100866

Nisselle AE, Delatycki MB, Collins V, Metcalfe S, Aitken MA, du Sart D, Halliday J, Macciocca I, Wakefield A, Hill V, Gason A, Warner B, Calabro V, Williamson R, Allen KJ (2004) Implementation of HaemScreen, a workplace-based genetic screening program for hemochromatosis. Clin Genet 65(5):358–367. https://doi.org/10.1111/j.1399-0004.2004.0239.x

Olsson KS, Vaisanen M, Konar J, Bruce A (1997) The effect of withdrawal of food iron fortification in Sweden as studied with phlebotomy in subjects with genetic hemochromatosis. Eur J Clin Nutr 51(11):782–786. https://doi.org/10.1038/sj.ejcn.1600488

Ong SY, Gurrin LC, Dolling L, Dixon J, Nicoll AJ, Wolthuizen M, Wood EM, Anderson GJ, Ramm GA, Allen KJ, Olynyk JK, Crawford D, Ramm LE, Gow P, Durrant S, Powell LW, Delatycki MB (2017) Reduction of body iron in HFE-related haemochromatosis and moderate iron overload (Mi-Iron): a multicentre, participant-blinded, randomised controlled trial. Lancet Haematol 4(12):e607–e614. https://doi.org/10.1016/S2352-3026(17)30214-4

Phatak P, Brissot P, Wurster M, Adams PC, Bonkovsky HL, Gross J, Malfertheiner P, McLaren GD, Niederau C, Piperno A, Powell LW, Russo MW, Stoelzel U, Stremmel W, Griffel L, Lynch N, Zhang Y, Pietrangelo A (2010) A phase 1/2, dose-escalation trial of deferasirox for the treatment of iron overload in HFE-related hereditary hemochromatosis. Hepatology 52(5):1671–1779. https://doi.org/10.1002/hep.23879

Pilling LC, Tamosauskaite J, Jones G, Wood AR, Jones L, Kuo CL, Kuchel GA, Ferrucci L, Melzer D (2019) Common conditions associated with hereditary haemochromatosis genetic variants: cohort study in UK Biobank. BMJ 364:k5222. https://doi.org/10.1136/bmj.k5222

Prabhu A, Cargill T, Roberts N, Ryan JD (2020) Systematic review of the clinical outcomes of iron reduction in hereditary hemochromatosis. Hepatology 72(4):1469–1482. https://doi.org/10.1002/hep.31405

Qian J, Sullivan BP, Peterson SJ, Berkland C (2017) Nonabsorbable iron binding polymers prevent dietary iron absorption for the treatment of iron overload. ACS Macro Lett 6(4):350–353. https://doi.org/10.1021/acsmacrolett.6b00945

Rao ND, Moonesinghe R, Shi L, Adams PC, Jarvik GP, Kowdley KV, Schieve LA, Grosse SD, Dotson WD, Khoury MJ (2025) Associations of HFE genotypes with hemochromatosis related phenotypes in the all of us research program. Genet Med Open. DOI: 10.1016/j.gimo.2024.101959

Rombout-Sestrienkova E, Winkens B, Essers BA, Nieman FH, Noord PA, Janssen MC, van Deursen CT, Boonen A, Reuser-Kaasenbrood EP, Heeremans J, van Kraaij M, Masclee A, Koek GH (2016) Erythrocytapheresis versus phlebotomy in the maintenance treatment of HFE hemochromatosis patients: results from a randomized crossover trial. Transfusion 56(1):261–270. https://doi.org/10.1111/trf.13328

Savatt JM, Johns A, Schwartz MLB, McDonald WS, Salvati ZM, Oritz NM, Masnick M, Hatchell K, Hao J, Buchanan AH, Williams MS (2023) Testing and Management of Iron Overload after Genetic Screening-Identified Hemochromatosis. JAMA Netw Open 6(10):e2338995. https://doi.org/10.1001/jamanetworkopen.2023.38995

Schmidtke J (2022) Twenty-five years of contemplating genotype-based hereditary hemochromatosis population screening. Genes (Basel) 13(9). https://doi.org/10.3390/genes13091622

Sharma S, Sethi SK, Reese D, Gharabaghi S, Yerramsetty KK, Palutla VK, Chen Y, Haacke EM, Jog MS (2022) Brain iron deposition and movement disorders in hereditary haemochromatosis without liver failure: a cross-sectional study. Eur J Neurol 29(5):1417–1426. https://doi.org/10.1111/ene.15242

Swinkels DW, Matlung HL (2023) Does phlebotomy in people with haemochromatosis become history? Lancet Gastroenterol Hepatol 8(12):1063–1064. https://doi.org/10.1016/S2468-1253(23)00283-2

Valenti L, Corradini E, Adams LA, Aigner E, Alqahtani S, Arrese M, Bardou-Jacquet E, Bugianesi E, Fernandez-Real JM, Girelli D, Hagstrom H, Henninger B, Kowdley K, Ligabue G, McClain D, Laine F, Miyanishi K, Muckenthaler MU, Pagani A, Pedrotti P, Pietrangelo A, Prati D, Ryan JD, Silvestri L, Spearman CW, Stal P, Tsochatzis EA, Vinchi F, Zheng MH,

Zoller H (2023) Consensus statement on the definition and classification of metabolic hyperferritinaemia. Nat Rev Endocrinol 19(5):299–310. https://doi.org/10.1038/s41574-023-00807-6

Vanclooster A, van Deursen C, Jaspers R, Cassiman D, Koek G (2017) Proton pump inhibitors decrease phlebotomy need in HFE hemochromatosis: double-blind randomized placebo-controlled trial. Gastroenterology 153(3):678–680 e672. https://doi.org/10.1053/j.gastro.2017.06.006

Yamashita C, Adams PC (2003) Natural history of the C282Y homozygote for the hemochromatosis gene (HFE) with a normal serum ferritin level. Clin Gastroenterol Hepatol 1(5):388–391. https://doi.org/10.1053/s1542-3565(03)00187-3

10 Diagnosis and Management of Non-*HFE* Hemochromatosis, Ferroportin Disease, and Rare Hereditary Iron-Loading Disorders

Antonello Pietrangelo

Abstract

Beyond the classic HFE-hemochromatosis, several genetic iron-loading disorders arise from mutations in genes regulating iron homeostasis, such as *TFR2*, *HAMP*, *HJV*, and the *SLC40A1* (ferroportin) gene, as well as those involved in iron transport and mitochondrial function. These disorders lead to systemic or localized iron overload, resulting in complications such as liver disease, cardiomyopathy, endocrine dysfunction, and neurodegeneration. This chapter reviews the genetic basis, clinical presentations, and therapeutic approaches, emphasizing the importance of early diagnosis and intervention to mitigate organ damage and improve outcomes.

Keywords

Non-HFE Hemochromatosis · Ferroportin disease · Juvenile hemochromatosis · Mitochondrial iron genes · Iron overload

A. Pietrangelo (✉)
Chief Internal Medicine Unit and Center for Hemochromatosis, Genomic Medicine & Rare Diseases, Italy, Modena
e-mail: pietra@unimore.it

10.1 Introduction

The discovery of hepcidin, a liver-derived peptide hormone (Pigeon et al. 2001; Park et al. 2001), has been a cornerstone in advancing our understanding of iron homeostasis. The regulatory hepcidin-ferroportin axis has profoundly reshaped our comprehension of iron-related disorders, particularly hereditary diseases of iron overload and deficiency (see Chaps. 9 and 12). Most proteins involved in hereditary iron overload diseases are linked to the hepcidin-ferroportin axis. Table 10.1 summarizes the key proteins involved in iron metabolism and the mechanisms by which their defective function leads to iron overload, as discussed in the following sections.

These diseases arise from mutations that directly or indirectly disrupt hepcidin production, ferroportin function, or the signaling pathways regulating this axis. Such disruptions lead to dysregulated iron absorption and systemic iron overload. The significance of this pathway is underscored by the wide variety of genetic defects that impair its function. For instance, in non-HFE hemochromatosis, loss-of-function mutations in genes such as *TFR2*, *HAMP*, and *HJV* impair hepcidin production; gain-of-function mutations in *SLC40A1* reduce responsiveness of the gene product ferroportin (FPN) to hepcidin.

K. Pantopoulos (ed.), *Iron Metabolism in Human Health and Disease*, Advances in Experimental Medicine and Biology 1480, https://doi.org/10.1007/978-3-031-92033-2_10

Table 10.1 Proteins involved in iron metabolism and mechanisms of iron overload

Gene/Protein	Role	Mechanism for causing iron overload
HAMP (Hepcidin)	Key hormone controlling systemic iron homeostasis by degrading ferroportin	Loss-of-function mutations result in unregulated ferroportin activity and iron absorption, causing rapid and severe systemic iron overload
SLC40A1 (Ferroportin)	Iron exporter regulated by hepcidin	Gain-of-function mutations lead to systemic iron overload due to ferroportin resistance to hepcidin
		Loss-of-function mutations lead to intracellular iron retention and anemia (ferroportin disease)
HFE (high Fe-iron)	Regulates hepcidin expression in response to iron levels	Mutations impair hepcidin regulation, increasing iron absorption and systemic overload
TfR2 (transferrin receptor 2)	Regulates hepcidin in response to iron levels	Loss-of-function mutations reduce hepcidin production, increasing iron absorption and systemic overload
HJV (Hemojuvelin)	Co-regulator of hepcidin production	Mutations reduce hepcidin activation, resulting in unregulated ferroportin activity and iron absorption, and causing rapid and severe systemic iron overload
CP (Ceruloplasmin)	Oxidizes Fe^{2+} to Fe^{3+} for transferrin binding; supports iron export	Deficiency causes impaired iron export, leading to tissue iron accumulation
TF (transferrin)	Iron transporter in plasma, delivering iron to cells	Dysfunction leads to inadequate iron delivery to cells and disrupted homeostasis
Mitochondrial iron genes	FXN (Frataxin) involved in mitochondrial iron storage and Fe-S cluster biogenesis	Deficiency leads to mitochondrial iron accumulation and oxidative damage
	GLRX5 and PIH1D3 are essential for Fe-S cluster assembly in mitochondria, critical for erythroid iron balance	Mutations lead to ineffective erythropoiesis, mitochondrial iron accumulation, and systemic overload
	ALAS2 key enzyme in erythroid heme biosynthesis	
SLC11A2 (DMT1)	Iron transporter involved in dietary iron absorption and cellular uptake	Deficiency reduces cellular iron uptake, leading to anemia and compensatory overload
FTL (ferritin light chain)	Stores iron safely within cells to prevent oxidative damage	Dysfunction leads to cellular iron mismanagement and oxidative damage

Other non-HFE genetic iron-loading disorders are linked to hepcidin resistance, such as in ferroportin disease (FD) (Pietrangelo 2004a), or to reduced hepcidin levels caused by ineffective erythropoiesis, as seen in genetic iron-loading anemias (Chap. 11). These anemias are characterized by low hepcidin levels, excessive iron absorption, and secondary iron overload (Camaschella and Nai 2016).

In addition to impaired hepcidin regulation, excess iron accumulation—often associated with varying degrees of mild to severe anemia—can result from increased tissue iron release, as observed in aceruloplasminemia (Miyajima et al. 1987), impaired iron handling, such as in DMT1 deficiency (Mims et al. 2005), and hypotransferrinemia/atransferrinemia (Dabboubi et al. 2020).

Furthermore, iron overload may arise from local (cellular or subcellular) deregulation of iron homeostasis, as seen in mitochondrial iron disorders such as Friedreich's ataxia (Llorens et al. 2019) and neuroferritinopathy (Curtis et al. 2001).

Figure 10.1 illustrates a diagnostic algorithm for suspecting and diagnosing the vari-

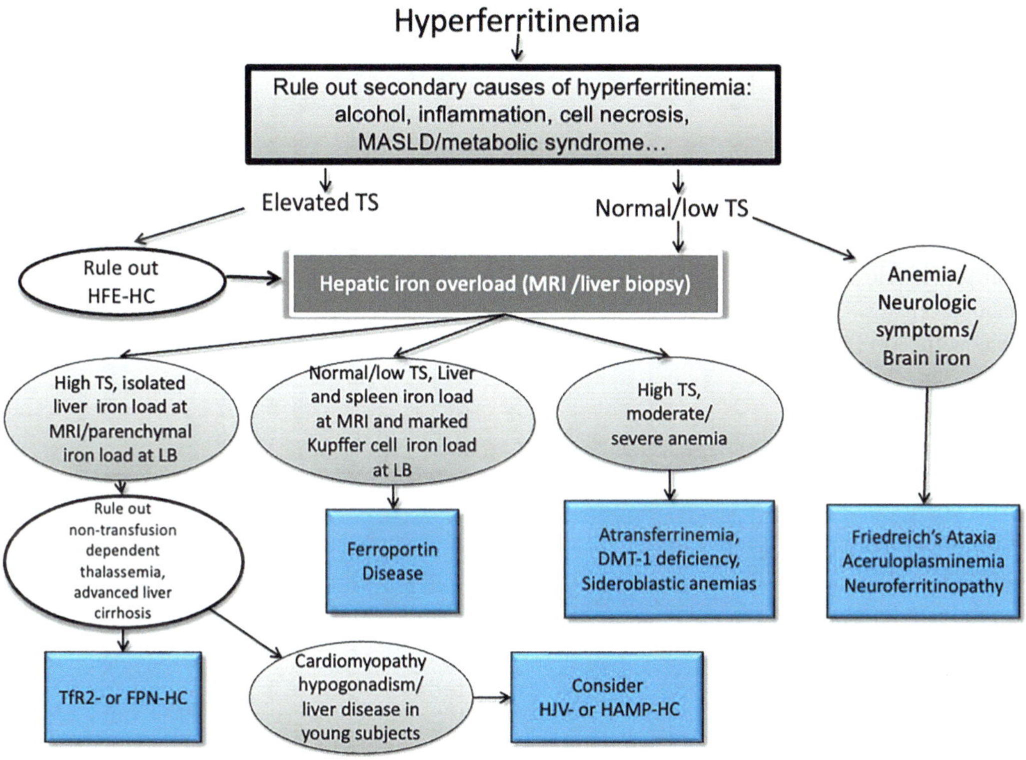

Fig. 10.1 Algorithm for diagnosing non-HFE hemochromatosis, FD, and rare genetic iron-loading conditions **Stepwise diagnostic algorithm for distinguishing various forms of non-HFE hemochromatosis, starting with the presence of hyperferritinemia. The algorithm incorporates TS levels and liver iron accumulation assessed via MRI or liver biopsy (LB) to guide differential diagnosis. Specific patterns of liver and spleen iron deposition direct the suspicion toward conditions such as TfR2- or FPN-related hemochromatosis, juvenile forms like HJV- or HAMP-HC, and the FD. Overlapping conditions, such as moderate to severe anemia, neurologic symptoms, and brain iron, may suggest alternative diagnoses (see text for details)**

ous forms of non-HFE iron disorders reported below, starting with the biochemical iron parameters (serum ferritin and transferrin saturation) and tissue iron status.

10.2 Non-HFE Hemochromatosis

Beyond hereditary/genetic hemochromatosis (HC)—the classic disorder caused by mutations in the HFE gene, which serves as the prototype and by far the most prevalent form (Feder et al. 1996) (see Chap. 9)—other rare disorders resulting from mutations in genes encoding transferrin receptor 2 (TfR2) (Camaschella et al. 2000), hepcidin (HAMP) (Roetto et al. 2003), hemojuvelin (HJV) (Papanikolaou et al. 2004), or ferroportin (FPN) (Njajou et al. 2001) can also present with a hemochromatosis phenotype.

All non-HFE HC forms share the same underlying pathophysiologic mechanism, involving progressive expansion of the circulatory iron pool, tissue iron accumulation, and subsequent organ damage and disease (Table 10.1). The specific phenotype of non-HFE HC is largely determined by the rate and extent of circulatory and tissue iron overload, which depends on the role the defective protein plays in hepcidin regulation (Table 10.2).

Table 10.2 Key epidemiologic, genetic, and clinical features of Non-HFE Hemochromatosis, Ferroportin Disease, and Hereditary Iron Overload Disorders

Disorder	Affected gene (location)	Epidemiology (estimated frequency)	Genetics	Clinical onset (decade)	Main clinical manifestation	Clinical course
Non-HFE-hemochromatosis	*TfR2*/7q22	Affects Caucasians and non-Caucasians; rare	Autosomal recessive	Second to fourth decade	Liver disease, cardiomyopathy	Moderate to severe
	SLC40A1/2q32	Rare, worldwide	Autosomal dominant	Third to fourth decade	Liver disease, diabetes, cardiomyopathy	Moderate to severe
	HJV/1p21	Affects Caucasians and non-Caucasians; rare	Autosomal recessive	Second to third decade	Hypogonadism, liver disease, cardiomyopathy	Severe
	HAMP/19q13.1	Affects Caucasians and non-Caucasians; rare	Autosomal Recessive	Second to third decade	Hypogonadism, liver disease, cardiomyopathy	Severe
Ferroportin disease	*SLC40A1*/2q32	Affects Caucasians and non-Caucasians; rare	Autosomal dominant	Any decade	Liver disease, marginal anemia	Mild
Aceruloplasminemia	*CP*/3q23-q25	Extremely rare, affects Caucasians and non-Caucasians	Autosomal recessive	Second to third decade	Multiorgan iron overload (primarily the liver) neurologic manifestations, anemia	Severe
A(hypo) transferrinemia	*TF*/3q21	Extremely rare, affects Caucasians and non-Caucasians	Autosomal recessive	First to second decade	Anemia, non-transferrin-bound iron (NTBI)-mediated iron overload)	Severe
Mitochondrial iron gene disorders	*FXN*/9q21	Worldwide, rare	Autosomal recessive	Second decade	Ataxia, cardiomyopathy, diabetes	Progressive
	GLRX5/14q32, *PIH1D3*/Xq28, *ALAS2*/Xp11.21	*GLRX5*,*PIH1D3*, extremely rare, *ALAS2* less rare; male-only for *PIH1D3* and *ALAS2*	Autosomal recessive (*GLRX5*), X-linked recessive (*PIH1D3*, *ALAS2*)	First to second decade	Sideroblastic anemia, microcytic anemia, mitochondrial iron accumulation, ineffective erythropoiesis	Variable, progressive
Neuroferritinopathy	*FTL*/19q13.3	Very rare; primarily in families of northern European descent	Autosomal dominant	Third to fourth decade	Movement disorders, cognitive decline	Progressive
DMT-1 deficiency	*SLC11A2*/12q13	Extremely rare	Autosomal recessive	First decade	Microcytic anemia, growth retardation	Variable
HMOX-1 deficiency	*HMOX1*/22q12.3	Extremely rare; two reported cases	Autosomal recessive	First to second decade	Growth retardation, anemia, iron overload in liver and kidney, vascular injury, hemolysis	Severe; early death reported

Loss of key regulators of hepcidin expression, such as hemojuvelin (HJV) or hepcidin (HAMP), itself results in a rapid and massive influx of iron into the bloodstream. This process leads to early-onset organ damage, typical of juvenile forms of hemochromatosis.

Loss of an ancillary regulator of hepcidin, such as in patients with mutations in transferrin receptor 2 (TfR2), leads to a less severe phenotype, with symptoms typically presenting in the teenage years or young adulthood.

Ferroportin-associated HC manifests biochemically at any age and typically presents clinically with a mild phenotype.

See Table 10.1. Unlike HFE-HC, non-HFE HC forms are not restricted to individuals of Northern European ancestry.

Common features of non-HFE HC often overlap with those of HFE-HC and include:

- An autosomal recessive inheritance pattern (except forms associated with gain-of-function mutations in FPN, which are autosomal dominant).
- Early and progressive elevation of plasma iron levels, characterized by increasing transferrin saturation.
- Progressive parenchymal iron deposition, with potentially severe damage to the liver, endocrine glands, heart, and joints.
- Preservation of erythropoiesis, allowing for an optimal response to therapeutic phlebotomy (Pietrangelo 2004b).

10.3 Suspecting and Diagnosing Non-HFE Hemochromatosis

10.3.1 TfR2-Associated HC

Common in regions with higher rates of consanguinity (e.g., Southern Italy) but also reported globally, TfR2-HC cases have also been documented in France, Portugal, Spain, and among individuals of mixed European descent (Camaschella et al. 2000). Isolated cases have been reported in non-European populations, such as Taiwan and Japan, highlighting its global presence.

Patients with TfR2-associated hemochromatosis typically experience an earlier onset of symptoms compared to *HFE*-related forms, with manifestations often appearing between the second and third decades of life (Le Gac et al. 2004). Clinically, this form of hemochromatosis presents an "intermediate" phenotype—more severe than *HFE*-hemochromatosis but less aggressive than juvenile hemochromatosis (see below). Among the affected organs, the liver is particularly vulnerable, with significant iron deposition and associated complications.

10.3.2 Ferroportin-Associated HC (FPN-HC)

The first clinical description of an autosomal dominant form of classic hemochromatosis was reported in Melanesian kindred in 1990 (Eason et al. 1990). This condition was later, in 2003, linked to a ferroportin gain-of-function mutation (Arden et al. 2003).

In FPN-HC, excessive iron export into the bloodstream results in iron overload and subsequent tissue iron deposition (Njajou et al. 2001).

FPN-HC closely resembles other forms of hemochromatosis but must be carefully differentiated from FD, which results from loss-of-function mutations in ferroportin (Pietrangelo 2004a) (see next section).

Unlike FD, patients with FPN-HC exhibit clinical and biochemical features similar to other forms of hemochromatosis, including elevated ferritin levels and high transferrin saturation, leading to progressive parenchymal iron deposition (Wallace et al. 2004) (see Fig. 10.1).

Studies have also reported heterozygous missense mutations in the BMP6 gene, a crucial modulator of hepcidin expression through the BMP-SMAD signaling pathway, among patients with unexplained late-onset non-HFE hemochromatosis (Daher et al. 2016).

10.3.3 Juvenile Hemochromatosis (J-HC)

J-HC, characterized by a strong genetic or congenital component, typically manifests early during adolescence or young adulthood, presenting with a more severe phenotype (Camaschella et al. 2002).

Most cases are associated with pathogenic mutations in the *HJV* gene, particularly the p. Gly320Val variant (Papanikolaou et al. 2004). Less commonly, mutations in the *HAMP* gene (Roetto et al. 2003) are implicated.

The rapid accumulation of iron in J-HC can lead to symptoms appearing as early as the first or second decade of life, resulting in severe complications, especially in the heart and endocrine system (Sandhu et al. 2018).

Common manifestations include (Fig. 10.1): cardiomyopathy, which progresses to severe heart failure, often life-threatening before age 30; hypogonadism, manifesting as amenorrhea in young women and delayed puberty or infertility in men; liver disease, with fibrosis or cirrhosis; and metabolic disorders, such as diabetes or glucose intolerance. Additional symptoms include joint pain and skin pigmentation.

Although J-HC typically presents early and severely, rare cases of delayed onset have been reported (Hamdi-Roze et al. 2019).

The severe manifestations of J-HC compared to HFE-HC are attributed to the critical role of HJV in hepcidin production and the heart and endocrine glands' greater susceptibility to iron toxicity than the liver.

Moreover, rare cases have been reported in patients with combined mutations of HFE and TfR2 (digenic hemochromatosis), which present with the same severe phenotype as J-HC (Pietrangelo et al. 2005). Recently, pathogenic mutations in the *PIGA* gene were found to cause J-HC by severely impairing the glycosylphosphatidylinositol (GPI) anchorage of hemojuvelin (Muckenthaler et al. 2022; Steinbicker 2022).

10.3.4 Treatment

Phlebotomy remains the cornerstone of therapy in non-HFE HC, as it is in HFE-HC. Given the severity of J-HC, timely diagnosis is crucial to prevent irreversible organ damage, and early and intensive treatment is essential to improve outcomes and prevent complications. However, in severe cases where phlebotomy alone is insufficient, chelation therapy using combinations of iron-chelating agents may be necessary (Fabio et al. 2007; Cooray et al. 2018; Zinecker et al. 2024).

10.4 Ferroportin Disease (FD)

FD is a relatively common hereditary iron overload disorder and a leading cause of hereditary hyperferritinemia. It results from pathogenic, often missense, mutations in the *SLC40A1* gene (encoding ferroportin). The FD phenotype was first described in 1999 in a large Italian family (Pietrangelo et al. 1999). FD is characterized by predominant iron accumulation in reticuloendothelial cells, elevated ferritin levels with normal to low transferrin saturation (TS), and a tendency toward hypochromic anemia (Pietrangelo 2017) (Fig. 10.1). The disorder stems from loss-of-function mutations in ferroportin, which impair its ability to export iron effectively, particularly in cells with high iron turnover, such as macrophages.

10.4.1 Genetic Features

FD differs from other hereditary iron overload disorders (e.g., those caused by mutations in *HFE*, *TfR2*, *HJV*, or *HAMP*) by its autosomal dominant inheritance pattern. Numerous *SLC40A1* mutations associated with FD have been identified globally, showing that FD crosses ethnic boundaries. Common mutations such as *Val192del*, *A77D*, and *G80S* have been found in diverse populations, suggesting that these variants arose independently in isolated groups worldwide (Pietrangelo 2017).

Global analysis of *SLC40A1* variants, including those associated with both FD and FPN1-associated hemochromatosis, has revealed relatively high allele frequencies, with a predicted pathogenic genotype carrier rate approaching that of *HFE*-hemochromatosis (HFE-HC), highlighting the potential underdiagnosis of FD (Wallace and Subramaniam 2016). The high frequency of *FPN1* variants in Africans, particularly the *Q248H* variant, contributes significantly to these estimates (Barton et al. 2003; Gordeuk et al. 2004). Knowing that a strong connection exists between iron and pathogens (Ullah and Lang 2023), as well as iron and host responses to malaria (Wideman et al. 2023), it has been proposed that the *Q248H* variant might have undergone positive selection in malaria-endemic areas due to a potential protective effect (Zhang et al. 2018). However, recent studies have challenged this idea, showing no protection conferred by the *Q248H* mutation against malaria or bacteremia (Manake et al. 2021; Muriuki et al. 2019; Bahati et al. 2022).

10.4.2 Pathogenesis

Recent studies, such as those by Billesbølle et al. (Billesbolle et al. 2020) and Pan et al. (2020), have investigated the structural and membrane organization of FPN and its interaction with hepcidin. However, a conclusive pathogenic model for FD that would explain the functional effect of different FPN variants and FD pathogenesis has yet to be defined. According to one model (Sabelli et al. 2017), FPN1 monomers are able to reach the cell surface and adequately export iron in cells with lower iron demand, such as enterocytes. However, in cells like macrophages with high iron turnover, FPN1 is unable to reach the plasma membrane sufficiently, likely due to bottlenecks in degradation or endocytic cycling pathways. This model aligns with FD's clinical picture, where iron accumulates early in hepatic Kupffer cells, TS remains normal, and TS levels can drop significantly in young women post-menarche or after intensive phlebotomy, when higher iron demands for erythropoiesis increase the need for FPN1 activity (Pietrangelo 2004a) (Table 10.2).

10.4.3 Suspecting and Diagnosing FD

FD should be considered in cases of unexplained familial hyperferritinemia and low-to-normal TS, non-parenchymal cell iron accumulation at liver biopsy, or a characteristic iron magnetic resonance imaging (MRI) pattern, particularly in young adults (Fig. 10.2). In fact, hyperferritinemia in FD often appears early in life, and MRI can reveal the characteristic iron accumulation in the spleen, spine, and liver, known as the "SSL triad" (Pietrangelo et al. 2006).

Clinical expressivity tends to be mild with variable symptoms, including hypochromic anemia often seen in young menstruating women (Pietrangelo 2004a; Pietrangelo et al. 1999; Le Lan et al. 2011). Due to the lack of prospective studies, it's presently unclear whether the extensive iron loading of macrophages of the liver and the spleen might adversely affect the course and outcome of chronic liver disease, immune response during infections, and cancer. In the original family with FD, one individual developed liver cancer after stopping phlebotomy while carrying an occult HBV infection, and two siblings experienced progressive liver fibrosis, suggesting that iron might play an additional pathogenic role in FD (Corradini et al. 2007).

10.4.4 Treatment

While data on the natural progression of FD and standardized treatment guidelines are limited, iron-depletion therapy is commonly recommended. While iron chelation may be considered in select cases (Unal et al. 2015), venesection remains the primary treatment. However, its tolerance varies; low TS and anemia may develop

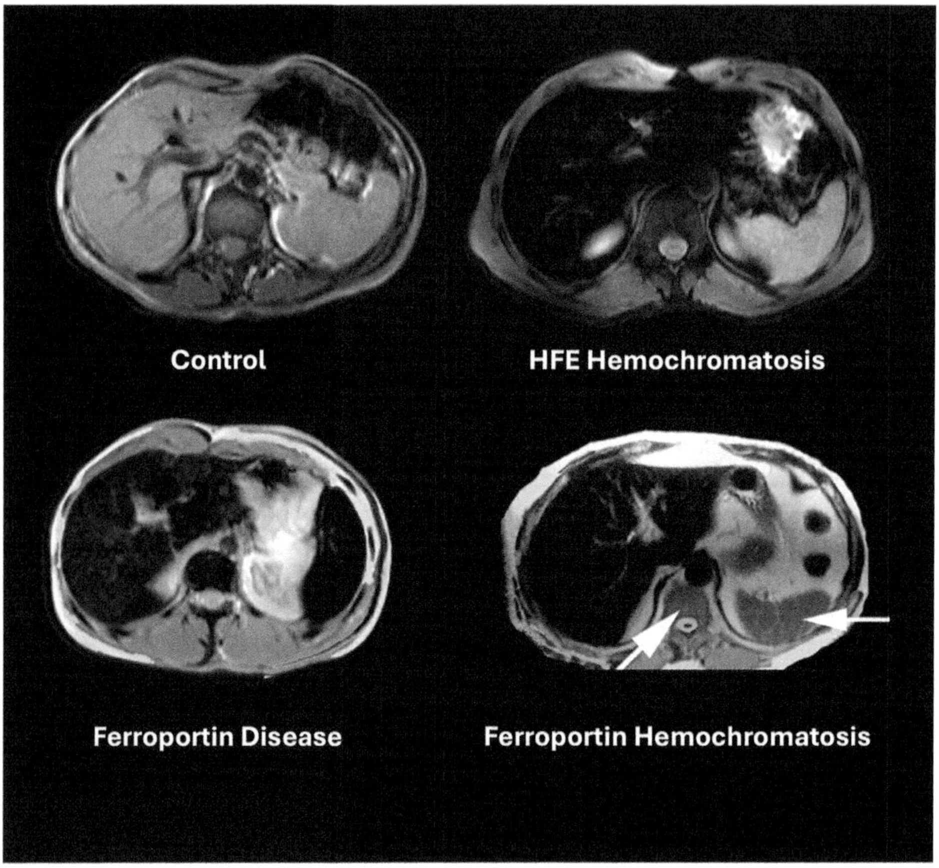

Fig. 10.2 *T2* Magnetic resonance imaging (MRI) for the diagnosis of HFE, non-HFE hemochromatosis, and FD. T2-weighted GRE sequences were utilized to detect iron accumulation. Compared to a control subject, the patient with HFE-hemochromatosis exhibits marked hepatic iron accumulation. The patient with FD, carrying a loss-of-function FPN mutation, shows iron overload in the liver, spine, and spleen due to large iron deposits in non-parenchymal liver cells, bone, and spleen macrophages. In contrast, the patient with FPN-hemochromatosis, carrying a gain-of-function FPN mutation, presents an MRI pattern similar to that of the HFE-hemochromatosis subject, with iron sparing in the spine and spleen (arrows).

during phlebotomy, while ferritin levels paradoxically remain elevated (Pietrangelo 2004a). Therefore, TS levels should be closely monitored during treatment, and the target ferritin level—unlike in HFE-HC—should not be overly restrictive to avoid anemia.

10.5 Aceruloplasminemia

Aceruloplasminemia (AC) is a rare autosomal recessive disorder caused by mutations in the ceruloplasmin (*CP*) gene; see also Chap. 19. CP is essential for iron efflux from cells by promoting iron release to plasma transferrin (Tf) and stabilizing membrane ferroportin (Miyajima et al. 2003; Kono et al. 2010). The absence or deficiency of CP activity leads to iron overload in multiple organs—including the liver, pancreas, heart, thyroid, and central nervous system—and restricts iron availability for red blood cell production (Kono et al. 2010). Though first identified in Japan (Miyajima 1993), AC has been reported globally and shows variable clinical expressivity (Vroegindeweij et al. 2015; Vila Cuenca et al. 2020). Laboratory findings typically include absent or reduced serum CP, hyperferritinemia, low-to-normal TS, micro/normocytic anemia, and isolated microcytosis (Vila Cuenca et al. 2020). Major clinical features include diabetes mellitus, retinal degeneration, and a range of neurological symptoms (Kono 2013).

Iron accumulation primarily affects the liver (especially hepatocytes) and the brain, while the spleen remains largely unaffected, despite the role of macrophage ferroportin in iron recycling (Table 10.2). Notably, severe iron overload in the liver does not commonly result in fibrosis. Neurodegeneration may be exacerbated by iron deficiency within neurons due to limited iron release from astrocytes (Kono 2013; Miyajima 2015). This cell-specific damage from restricted iron supply may also occur in other organs (Piperno and Alessio 2018). AC is a progressive and potentially fatal disease; hence, treatment is essential, though therapeutic strategies are not standardized. Early initiation of iron chelation therapy seems to postpone the onset of neurological manifestations, and iron chelation therapy in combination with phlebotomy seems beneficial in non-anemic AC patients (Vroegindeweij et al. 2015; Miyake et al. 2020).

10.6 Hypo/Atransferrinemia

Congenital hypo/atransferrinemia is an extremely rare autosomal recessive disorder resulting from mutations in the transferrin (TF) gene. Worldwide, with fewer than 20 cases reported in the medical literature to date (not all genetically confirmed) (Dabboubi et al. 2020). Inefficient delivery of iron to the bone marrow through Tf causes severe microcytic hypochromic anemia from birth, while intestinal iron absorption is either unaffected or further stimulated by the anemia. Iron overload can develop in organs like the liver, myocardium, pancreas, and thyroid due to non-transferrin-bound iron (NTBI) transport mechanisms. Laboratory tests reveal low or undetectable levels of fully saturated Tf and high serum ferritin (Dabboubi et al. 2020) (Table 10.2). Patients may require blood transfusions and benefit from combined therapy with fresh frozen plasma (to increase the Tf pool) and chelation therapy (Piperno et al. 2020). Clinical severity can vary, as a total absence of Tf seems incompatible with life. Disease severity may relate to the specific mutations involved or other genetic/acquired modifiers.

10.7 Miscellaneous Genetic Disorders with Systemic or Local Iron Deposition

10.7.1 Mitochondrial Iron Gene Disorders

Friedreich ataxia, the most common inherited ataxia, involves progressive degeneration of the central and peripheral nervous systems and various non-neural tissues, such as the heart and pancreas (Camaschella 1998). It is caused by a deficiency in frataxin, a mitochondrial protein involved in iron-sulfur cluster formation. This deficiency disrupts mitochondrial iron balance and leads to iron accumulation, especially in cardiomyocytes and other organs. While the exact disease mechanisms remain unclear, the role of iron in disease progression is still being explored (Llorens et al. 2019) (Table 10.2).

Mutations in genes involved in Fe-S cluster biogenesis or heme synthesis, such as *GLRX5*, *PIH1D3*, and *ALAS2*, can lead to mitochondrial iron accumulation and systemic dysregulation (Table 10.2). GLRX5 mutations disrupt Fe-S cluster assembly, causing sideroblastic anemia, while ALAS2 mutations impair heme biosynthesis in X-linked sideroblastic anemia, both resulting in compensatory iron absorption and systemic overload. These mechanisms further illustrate the critical role of hepcidin suppression and mitochondrial dysfunction in hereditary anemias with iron dysregulation (Rouault 2013; Fleming and Ponka 2012).

10.7.2 Neuroferritinopathy

Neuroferritinopathy (NF) is a rare, autosomal dominant, adult-onset disorder caused by mutations in the L-ferritin (*FTL*) gene (Curtis et al. 2001); see also Chap. 19. It is predominantly reported in certain families or clusters, such as in the UK and other European populations. As part of the NBIA (neurodegeneration with brain iron accumulation) disorders, which also include aceruloplasminemia, NF is associated with focal iron buildup in the brain, especially in the basal ganglia. Only NF and AC in this group have links to genes directly involved in iron metabolism (Cozzi et al. 2021). NF pathology features aggregates of ferritin and iron in the brain, as well as other tissues like skin, liver, kidney, and muscle. Patients typically exhibit low-to-normal serum ferritin and neurological symptoms, including movement disorders. Oxidative damage from free iron, due to compromised ferritin storage, is believed to be central to NF pathogenesis (Cozzi et al. 2021), with non-neural tissues appearing relatively unaffected.

10.7.3 DMT1 Deficiency

DMT1 deficiency, a rare autosomal recessive disorder, results from mutations in the *SLC11A2* gene encoding DMT1, which is responsible for absorbing inorganic iron from enterocytes and processing Tf-bound iron in acidified endosomes (Iolascon and De Falco 2009). Patients present with hypochromic microcytic anemia of variable severity, similar to findings in rodent disease models (Iolascon and De Falco 2009). They may also have elevated serum iron, TS, mildly increased serum ferritin, and hepatic iron overload, often appearing at a young age (Mims et al. 2005; Iolascon and De Falco 2009). Iron overload may be further contributed by increased absorption of heme iron, a pathway unaffected by the genetic defect (Mims et al. 2005). Patients may require blood transfusions, but iron chelation is not usually effective. Treatment with erythropoietin may offer some benefit (Pospisilova et al. 2006).

10.7.4 HMOX1 Deficiency

Heme oxygenase 1 (HMOX1) deficiency has been reported in two children who exhibited growth retardation, severe anemia, hyperferritinemia, hepatic and renal iron overload, intravascular hemolysis, vascular injury, and early death (Kawashima et al. 2002; Radhakrishnan et al. 2011). Iron deposition in tissues seems secondary to impaired iron recycling, intravascular hemolysis, and heme accumulation.

10.8 Conclusions

Non-HFE hemochromatosis, FD, and related genetic iron-loading disorders provide a diverse spectrum of pathophysiological mechanisms that disrupt iron homeostasis. These conditions, arising from mutations in key genes such as *TFR2*, *HAMP*, *HJV*, *SLC40A1*, and others involved in mitochondrial function, result in systemic or localized iron overload with significant clinical implications. Early recognition and accurate diagnosis, supported by advanced genetic and biochemical tools, are essential to mitigate organ damage and improve patient outcomes. While therapeutic phlebotomy remains the cornerstone treatment, adjunct therapies like iron chelation have become crucial for severe or refractory cases. Continued research into the molecular underpinnings of these disorders will be vital for advancing diagnostic precision and therapeutic strategies and, ultimately, improving the management and prognosis of affected individuals.

References

Arden KE, Wallace DF, Dixon JL, Summerville L, Searle JW, Anderson GJ, Ramm GA, Powell LW, Subramaniam VN (2003) A novel mutation in ferroportin1 is associated with haemochromatosis in a Solomon Islands patient. Gut 52(8):1215–1217

Bahati YL, Delanghe J, Balaluka GB, Vandepoele K, Cirhuza JC, Kishabongo AS, Philippe J (2022) Ferroportin Q248H mutation was not found to be protective against malaria and anemia in children under 5 years living in South Kivu/Democratic Republic of Congo, an endemic area of P lasmodium infection. Heliyon 8(9):e10460. https://doi.org/10.1016/j.heliyon.2022.e10460

Barton JC, Acton RT, Rivers CA, Bertoli LF, Gelbart T, West C, Beutler E (2003) Genotypic and phenotypic heterogeneity of African Americans with primary iron overload. Blood Cells Mol Dis 31(3):310–319

Billesbolle CB, Azumaya CM, Kretsch RC, Powers AS, Gonen S, Schneider S, Arvedson T, Dror RO, Cheng Y, Manglik A (2020) Structure of hepcidin-bound ferroportin reveals iron homeostatic mechanisms. Nature 586(7831):807–811. https://doi.org/10.1038/s41586-020-2668-z

Camaschella C (1998) Juvenile haemochromatosis. Baillieres Clin Gastroenterol 12(2):227–235

Camaschella C, Nai A (2016) Ineffective erythropoiesis and regulation of iron status in iron loading anaemias. Br J Haematol 172(4):512–523. https://doi.org/10.1111/bjh.13820

Camaschella C, Roetto A, Cali A, De Gobbi M, Garozzo G, Carella M, Majorano N, Totaro A, Gasparini P (2000) The gene TFR2 is mutated in a new type of haemochromatosis mapping to 7q22. Nat Genet 25(1):14–15

Camaschella C, Roetto A, De Gobbi M (2002) Juvenile hemochromatosis. Semin Hematol 39(4):242–248. https://doi.org/10.1053/shem.2002.35635

Cooray SD, Heerasing NM, Selkrig LA, Subramaniam VN, Hamblin PS, McDonald CJ, McLean CA, McNamara E, Leet AS, Roberts SK (2018) Reversal of end-stage heart failure in juvenile hemochromatosis with iron chelation therapy: a case report. J Med Case Rep 12(1):18. https://doi.org/10.1186/s13256-017-1526-6

Corradini E, Ferrara F, Pollicino T, Vegetti A, Abbati GL, Losi L, Raimondo G, Pietrangelo A (2007) Disease progression and liver cancer in the ferroportin disease. Gut 56(7):1030–1032. https://doi.org/10.1136/gut.2007.122549

Cozzi A, Santambrogio P, Ripamonti M, Rovida E, Levi S (2021) Pathogenic mechanism and modeling of neuroferritinopathy. Cell Mol Life Sci 78(7):3355–3367. https://doi.org/10.1007/s00018-020-03747-w

Curtis AR, Fey C, Morris CM, Bindoff LA, Ince PG, Chinnery PF, Coulthard A, Jackson MJ, Jackson AP, McHale DP, Hay D, Barker WA, Markham AF, Bates D, Curtis A, Burn J (2001) Mutation in the gene encoding ferritin light polypeptide causes dominant adult-onset basal ganglia disease. Nat Genet 28(4):350–354

Dabboubi R, Amri Y, Yahyaoui S, Mahjoub R, Sahli CA, Sahli C, Hadj Fredj S, Bibi A, Sammoud A, Messaoud T (2020) A new case of congenital atransferrinemia with a novel splice site mutation: c.293-63del. Eur J Med Genet 63(5):103874. https://doi.org/10.1016/j.ejmg.2020.103874

Daher R, Kannengiesser C, Houamel D, Lefebvre T, Bardou-Jacquet E, Ducrot N, de Kerguenec C, Jouanolle AM, Robreau AM, Oudin C, Le Gac G, Moulouel B, Loustaud-Ratti V, Bedossa P, Valla D, Gouya L, Beaumont C, Brissot P, Puy H, Karim Z, Tchernitchko D (2016) Heterozygous mutations in BMP6 pro-peptide Lead to inappropriate Hepcidin synthesis and moderate iron overload in humans. Gastroenterology 150(3):672–683. https://doi.org/10.1053/j.gastro.2015.10.049

Eason RJ, Adams PC, Aston CE, Searle J (1990) Familial iron overload with possible autosomal dominant inheritance. AustNZJMed 20:226–230

Fabio G, Minonzio F, Delbini P, Bianchi A, Cappellini MD (2007) Reversal of cardiac complications by deferiprone and deferoxamine combination therapy in a patient affected by a severe type of juvenile hemochromatosis (JH). Blood 109(1):362–364. https://doi.org/10.1182/blood-2006-04-016949

Feder JN, Gnirke A, Thomas W, Tsuchihashi Z, Ruddy DA, Basava A, Dormishian F, Domingo R Jr, Ellis MC, Fullan A, Hinton LM, Jones NL, Kimmel BE, Kronmal GS, Lauer P, Lee VK, Loeb DB, Mapa FA, McClelland E, Meyer NC, Mintier GA, Moeller N, Moore T, Morikang E, Prass CE, Quintana L, Starnes SM, Schatzman RC, Brunke KJ, Drayna DT, Risch NJ, Bacon BR, Wolff RK (1996) A novel MHC class I-like gene is mutated in patients with hereditary haemochromatosis. Nat Genet 13(4):399–408. https://doi.org/10.1038/ng0896-399

Fleming RE, Ponka P (2012) Iron overload in human disease. N Engl J Med 366(4):348–359. https://doi.org/10.1056/NEJMra1004967

Gordeuk VR, Caleffi A, Corradini E, Ferrara F, Jones RA, Castro O, Onyekwere O, Kittles R, Pignatti E, Montosi G, Garuti C, Gangaidzo IT, Gomo ZAR, Moyo VM, Rouault TA, MacPhail P, Pietrangelo A (2004) Iron overload in Africans and African-Americans and a common mutation in the SCL40A1 (ferroportin 1) gene (vol 31, pg 299, 2003). Blood Cells Mol Dis 33(1):94–94. https://doi.org/10.1016/j.bcmd.2004.04.006

Hamdi-Roze H, Ben Ali Z, Ropert M, Detivaud L, Aggoune S, Simon D, Pelletier G, Deugnier Y, David V, Bardou-Jacquet E (2019) Variable expressivity of HJV related hemochromatosis: "Juvenile" hemochromatosis? Blood Cells Mol Dis 74:30–33. https://doi.org/10.1016/j.bcmd.2018.10.006

Iolascon A, De Falco L (2009) Mutations in the gene encoding DMT1: clinical presentation and treatment. Semin Hematol 46(4):358–370. https://doi.org/10.1053/j.seminhematol.2009.06.005

Kawashima A, Oda Y, Yachie A, Koizumi S, Nakanishi I (2002) Heme oxygenase-1 deficiency: the first autopsy case. Hum Pathol 33(1):125–130. https://doi.org/10.1053/hupa.2002.30217

Kono S (2013) Aceruloplasminemia: an update. Int Rev Neurobiol 110:125–151. https://doi.org/10.1016/B978-0-12-410502-7.00007-7

Kono S, Yoshida K, Tomosugi N, Terada T, Hamaya Y, Kanaoka S, Miyajima H (2010) Biological effects of mutant ceruloplasmin on hepcidin-mediated internalization of ferroportin. Biochim Biophys Acta 1802(11):968–975. https://doi.org/10.1016/j.bbadis.2010.07.011

Le Gac G, Mons F, Jacolot S, Scotet V, Ferec C, Frebourg T (2004) Early onset hereditary hemochromatosis resulting from a novel TFR2 gene nonsense mutation (R105X) in two siblings of north French descent. Br J Haematol 125(5):674–678

Le Lan C, Mosser A, Ropert M, Detivaud L, Loustaud-Ratti V, Vital-Durand D, Roget L, Bardou-Jacquet E, Turlin B, David V, Loreal O, Deugnier Y, Brissot P, Jouanolle AM (2011) Sex and acquired cofactors determine phenotypes of ferroportin disease. Gastroenterology 140(4):1199–1207. e1192. https://doi.org/10.1053/j.gastro.2010.12.049

Llorens JV, Soriano S, Calap-Quintana P, Gonzalez-Cabo P, Molto MD (2019) The role of iron in Friedreich's ataxia: insights from studies in human tissues and cellular and animal models. Front Neurosci 13:75. https://doi.org/10.3389/fnins.2019.00075

Manake MG, Ramatlho P, Ntereke TD, Tawe L, Bango ZA, Quaye IK, Paganotti GM, Kasvosve I (2021) Similar Ferroportin Q248H polymorphism prevalence in patients with plasmodium falciparum malaria and control subjects in the low-endemic setting of Botswana. Clin Chim Acta 523:77–80. https://doi.org/10.1016/j.cca.2021.09.009

Mims MP, Guan Y, Pospisilova D, Priwitzerova M, Indrak K, Ponka P, Divoky V, Prchal JT (2005) Identification of a human mutation of DMT1 in a patient with microcytic anemia and iron overload. Blood 105(3):1337–1342. https://doi.org/10.1182/blood-2004-07-2966

Miyajima H (1993) Aceruloplasminemia. In: Pagon RA, Bird TD, Dolan CR, Stephens K (eds) Gene Reviews, Seattle

Miyajima H (2015) Aceruloplasminemia. Neuropathology 35(1):83–90. https://doi.org/10.1111/neup.12149

Miyajima H, Nishimura Y, Mizoguchi K, Sakamoto M, Shimizu T, Honda N (1987) Familial apoceruloplasmin deficiency associated with blepharospasm and retinal degeneration. Neurology 37(5):761–767

Miyajima H, Takahashi Y, Kono S (2003) Aceruloplasminemia, an inherited disorder of iron metabolism. Biometals 16(1):205–213

Miyake Z, Nakamagoe K, Yoshida K, Kondo T, Tamaoka A (2020) Deferasirox might be effective for microcytic anemia and neurological symptoms associated with Aceruloplasminemia: a case report and review of the literature. Intern Med 59(14):1755–1761. https://doi.org/10.2169/internalmedicine.4178-19

Muckenthaler L, Marques O, Colucci S, Kunz J, Fabrowski P, Bast T, Altamura S, Hochsmann B, Schrezenmeier H, Langlotz M, Richter-Pechanska P, Rausch T, Hofmeister-Mielke N, Gunkel N, Hentze MW, Kulozik AE, Muckenthaler MU (2022) Constitutional PIGA mutations cause a novel subtype of hemochromatosis in patients with neurologic dysfunction. Blood 139(9):1418–1422. https://doi.org/10.1182/blood.2021013519

Muriuki JM, Mentzer AJ, Band G, Gilchrist JJ, Carstensen T, Lule SA, Goheen MM, Joof F, Kimita W, Mogire R, Cutland CL, Diarra A, Rautanen A, Pomilla C, Gurdasani D, Rockett K, Mturi N, Ndungu FM, Scott JAG, Sirima SB, Morovat A, Prentice AM, Madhi SA, Webb EL, Elliott AM, Bejon P, Sandhu MS, Hill AVS, Kwiatkowski DP, Williams TN, Cerami C, Atkinson SH (2019) The ferroportin Q248H mutation protects from anemia, but not malaria or bacteremia. Sci Adv 5(9):eaaw0109. https://doi.org/10.1126/sciadv.aaw0109

Njajou OT, Vaessen N, Joosse M, Berghuis B, van Dongen JW, Breuning MH, Snijders PJ, Rutten WP, Sandkuijl LA, Oostra BA, van Duijn CM, Heutink P (2001) A mutation in SLC11A3 is associated with autosomal dominant hemochromatosis. Nat Genet 28(3):213–214

Pan Y, Ren Z, Gao S, Shen J, Wang L, Xu Z, Yu Y, Bachina P, Zhang H, Fan X, Laganowsky A, Yan N, Zhou M (2020) Structural basis of ion transport and inhibition in ferroportin. Nat Commun 11(1):5686. https://doi.org/10.1038/s41467-020-19458-6

Papanikolaou G, Samuels ME, Ludwig EH, MacDonald ML, Franchini PL, Dube MP, Andres L, MacFarlane J, Sakellaropoulos N, Politou M, Nemeth E, Thompson J, Risler JK, Zaborowska C, Babakaiff R, Radomski CC, Pape TD, Davidas O, Christakis J, Brissot P, Lockitch G, Ganz T, Hayden MR, Goldberg YP (2004) Mutations in HFE2 cause iron overload in chromosome 1q-linked juvenile hemochromatosis. Nat Genet 36(1):77–82

Park CH, Valore EV, Waring AJ, Ganz T (2001) Hepcidin, a urinary antimicrobial peptide synthesized in the liver. J Biol Chem 276(11):7806–7810

Pietrangelo A (2004a) The ferroportin disease. Blood Cells Mol Dis 32(1):131–138

Pietrangelo A (2004b) Non-HFE hemochromatosis. Hepatology 39(1):21–29

Pietrangelo A (2017) The ferroportin disease: pathogenesis, diagnosis and treatment. Haematologica 102(12):1972–1984. https://doi.org/10.3324/haematol.2017.170720

Pietrangelo A, Montosi G, Totaro A, Garuti C, Conte D, Cassanelli S, Fraquelli M, Sardini C, Vasta F, Gasparini P (1999) Hereditary hemochromatosis in adults without pathogenic mutations in the hemochromatosis gene. N Engl J Med 341(10):725–732. https://doi.org/10.1056/NEJM199909023411003

Pietrangelo A, Caleffi A, Henrion J, Ferrara F, Corradini E, Kulaksiz H, Stremmel W, Andreone P, Garuti C (2005) Juvenile hemochromatosis associated with pathogenic mutations of adult hemochromatosis genes. Gastroenterology 128(2):470–479

Pietrangelo A, Corradini E, Ferrara F, Vegetti A, De Jong G, Luca Abbati G, Paolo Arcuri P, Martinelli S, Cerofolini E (2006) Magnetic resonance imaging to identify classic and nonclassic forms of ferroportin disease. Blood Cells Mol Dis 37(3):192–196

Pigeon C, Ilyin G, Courselaud B, Leroyer P, Turlin B, Brissot P, Loreal O (2001) A new mouse liver-specific gene, encoding a protein homologous to human antimicrobial peptide hepcidin, is overexpressed during iron overload. J Biol Chem 276(11):7811–7819

Piperno A, Alessio M (2018) Aceruloplasminemia: waiting for an efficient therapy. Front Neurosci 12:903. https://doi.org/10.3389/fnins.2018.00903

Piperno A, Pelucchi S, Mariani R (2020) Inherited iron overload disorders. Transl Gastroenterol Hepatol 5:25. https://doi.org/10.21037/tgh.2019.11.15

Pospisilova D, Mims MP, Nemeth E, Ganz T, Prchal JT (2006) DMT1 mutation: response of anemia to darbepoetin administration and implications for iron homeostasis. Blood 108(1):404–405. https://doi.org/10.1182/blood-2006-02-003962

Radhakrishnan N, Yadav SP, Sachdeva A, Pruthi PK, Sawhney S, Piplani T, Wada T, Yachie A (2011) Human heme oxygenase-1 deficiency presenting with hemolysis, nephritis, and asplenia. J Pediatr

Hematol Oncol 33(1):74–78. https://doi.org/10.1097/MPH.0b013e3181fd2aae

Roetto A, Papanikolaou G, Politou M, Alberti F, Girelli D, Christakis J, Loukopoulos D, Camaschella C (2003) Mutant antimicrobial peptide hepcidin is associated with severe juvenile hemochromatosis. Nat Genet 33(1):21–22

Rouault TA (2013) Iron metabolism in the CNS: implications for neurodegenerative diseases. Nat Rev Neuroscience14(8):551–564. https://doi.org/10.1038/nrn3453

Sabelli M, Montosi G, Garuti C, Caleffi A, Oliveto S, Biffo S, Pietrangelo A (2017) Human macrophage ferroportin biology and the basis for the ferroportin disease. Hepatology 65(5):1512–1525. https://doi.org/10.1002/hep.29007

Sandhu K, Flintoff K, Chatfield MD, Dixon JL, Ramm LE, Ramm GA, Powell LW, Subramaniam VN, Wallace DF (2018) Phenotypic analysis of hemochromatosis subtypes reveals variations in severity of iron overload and clinical disease. Blood 132(1):101–110. https://doi.org/10.1182/blood-2018-02-830562

Steinbicker AU (2022) PIGA mutations (can) cause juvenile hemochromatosis. Blood 139(9):1273–1275. https://doi.org/10.1182/blood.2021014935

Ullah I, Lang M (2023) Key players in the regulation of iron homeostasis at the host-pathogen interface. Front Immunol 14:1279826. https://doi.org/10.3389/fimmu.2023.1279826

Unal S, Piperno A, Gumruk F (2015) Iron chelation with deferasirox in a patient with de-novo ferroportin mutation. J Trace Elem Med Biol 30:1–3. https://doi.org/10.1016/j.jtemb.2015.01.002

Vila Cuenca M, Marchi G, Barque A, Esteban-Jurado C, Marchetto A, Giorgetti A, Chelban V, Houlden H, Wood NW, Piubelli C, Dorigatti Borges M, Martins de Albuquerque D, Yotsumoto Fertrin K, Jove-Buxeda E, Sanchez-Delgado J, Baena-Diez N, Burnyte B, Utkus A, Busti F, Kaubrys G, Suku E, Kowalczyk K, Karaszewski B, Porter JB, Pollard S, Eleftheriou P, Bignell P, Girelli D, Sanchez M (2020) Genetic and clinical heterogeneity in thirteen new cases with Aceruloplasminemia. Atypical anemia as a clue for an early diagnosis. Int J Mol Sci 21(7). https://doi.org/10.3390/ijms21072374

Vroegindeweij LH, van der Beek EH, Boon AJ, Hoogendoorn M, Kievit JA, Wilson JH, Langendonk JG (2015) Aceruloplasminemia presents as type 1 diabetes in non-obese adults: a detailed case series. Diabet Med 32(8):993–1000. https://doi.org/10.1111/dme.12712

Wallace DF, Subramaniam VN (2016) The global prevalence of HFE and non-HFE hemochromatosis estimated from analysis of next-generation sequencing data. Genet Med 18(6):618–626. https://doi.org/10.1038/gim.2015.140

Wallace DF, Clark RM, Harley HA, Subramaniam VN (2004) Autosomal dominant iron overload due to a novel mutation of ferroportin1 associated with parenchymal iron loading and cirrhosis. J Hepatol 40(4):710–713

Wideman SK, Frost JN, Richter FC, Naylor C, Lopes JM, Viveiros N, Teh MR, Preston AE, White N, Yusuf S, Draper SJ, Armitage AE, Duarte TL, Drakesmith H (2023) Cellular iron governs the host response to malaria. PLoS Pathog 19(10):e1011679. https://doi.org/10.1371/journal.ppat.1011679

Zhang DL, Wu J, Shah BN, Greutelaers KC, Ghosh MC, Ollivierre H, Su XZ, Thuma PE, Bedu-Addo G, Mockenhaupt FP, Gordeuk VR, Rouault TA (2018) Erythrocytic ferroportin reduces intracellular iron accumulation, hemolysis, and malaria risk. Science 359(6383):1520–1523. https://doi.org/10.1126/science.aal2022

Zinecker KA, Sliwinska A, Wilson LM, Zahr R (2024) Juvenile hemochromatosis connecting cardiac arrest and hypogonadotropic hypogonadism in a young woman. JCEM Case Rep 2(11):luae196. https://doi.org/10.1210/jcemcr/luae196

11 Iron-Loading Anemias

Maayan V. Levy and Yelena Z. Ginzburg

Abstract

Erythropoiesis is a process by which red blood cells (RBCs) are produced in the adult bone marrow. It involves the differentiation of erythroid progenitor cells into mature RBCs, which transport oxygen from the lungs to all cells in the body. Erythropoiesis is a complex process with a nuanced crosstalk of regulation involving hormones, cytokines, and growth factors. In recent years, multiple sources of evidence have increased our understanding of the molecular mechanisms that coordinate erythropoiesis to enable the daily production of approximately 200 billion RBCs. These advances have shed light on the pathophysiology in a variety of diseases, i.e., iron-loading anemias, and are paving the way for novel therapeutic strategies in preclinical and clinical development to treat such dyserythropoietic disorders. This chapter elucidates our current understanding of iron-loading anemias in general and specifically in β-thalassemia and myelodysplastic syndrome (MDS), describes the current cutting-edge understanding in pathophysiology, and delineates what novel therapies are currently being developed to target these disorders.

Keywords

Ineffective erythropoiesis · Iron overload · Erythroferrone (ERFE) · β-thalassemia · Myelodysplastic syndromes (MDS)

11.1 Normal Erythropoiesis

Erythropoiesis is the process by which erythroid progenitors differentiate into erythroblasts, progressing through several rounds of cell division while synthesizing hemoglobin, the protein that binds oxygen in RBCs (Caulier and Sankaran 2022). This complex orchestrated process involves the coordination of multiple pathways in which proliferation, differentiation, cell death, enucleation, and cellular microenvironments with different cytokines, hormones, and growth factors predominantly in the bone marrow and, when necessary, potentially also in the liver and spleen. Morphologically, erythroblasts undergo nuclear and cellular condensation, gradually reducing in size while increasing their cytoplasmic volume, to ultimately enucleate and become reticulocytes (Fig. 11.1). The output of erythropoiesis is regulated by a number of cytokines and growth factors, including erythropoietin (EPO), stem cell factor (SCF), and inter-

M. V. Levy · Y. Z. Ginzburg (✉)
The Tisch Cancer Institute, Division of Hematology and Medical Oncology, Tisch Cancer Center, Icahn School of Medicine at Mount Sinai, New York, NY, USA
e-mail: yelena.ginzburg@mssm.edu

K. Pantopoulos (ed.), *Iron Metabolism in Human Health and Disease*, Advances in Experimental Medicine and Biology 1480, https://doi.org/10.1007/978-3-031-92033-2_11

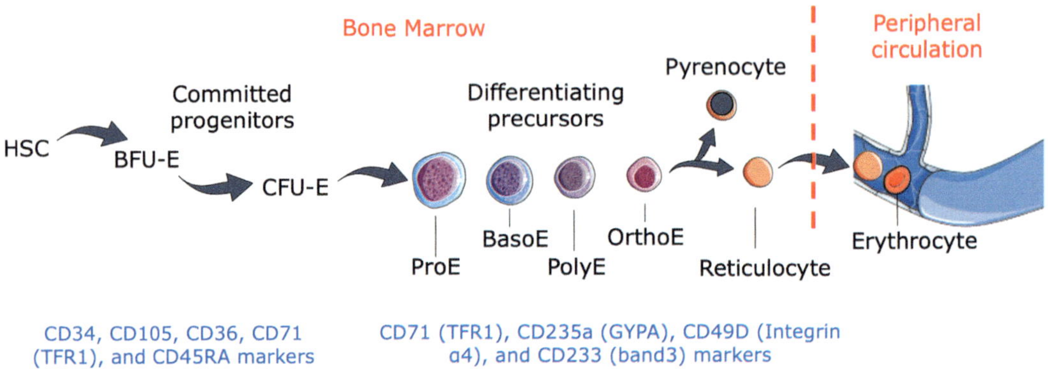

Fig. 11.1 Definitive Erythropoiesis. Definitive erythropoiesis in the adult organism is derived from hematopoietic stem cells (HSC) with progressive movement of cells through three compartments: progenitors, erythroblast precursors, and erythrocytes. Erythroid progenitors (BFU-E, CFU-E) are defined by their capacity to form colonies of maturing erythroid cells in vitro. Erythroid precursors are defined as proerythroblasts (ProE), basophilic erythroblasts (BasoE), polychromatophilic erythroblasts (PolyE), and orthochromatic erythroblasts (OrthoE) based on morphology and progressive change in the expression of surface markers. OrthoE enucleates to form a pyrenocyte that contains the condensed nucleus and a reticulocyte that goes on to mature into an erythrocyte in the peripheral circulation. Erythroblast precursors undergo differentiation in contact with a macrophage within the EBI

leukin-3 (IL-3), among others. EPO, synthesized primarily by the kidney, binds EPO receptor on erythroid progenitors (and other cells in the same microenvironment, e.g., macrophages and osteoblasts), leading to downstream signaling that enables a balance between erythroblast proliferation, differentiation, and survival. EPO is especially crucial for early erythroid progenitors BFU-E and especially CFU-E, leading to activation of JAK2/STAT3/STAT5 signaling in response to EPO:EPO receptor binding to induce expression of erythroid genes, enhancing the survival of erythroid lineage progenitors (Shih et al. 2018). In addition, transcription factors, such as GATA and KLF, also play a crucial role in erythropoiesis. Furthermore, erythropoiesis in the bone marrow is also influenced by their microenvironment and the variety of cells in the erythroblast niche, including macrophages, osteoblasts, endothelial cells, and stromal cells, as well as extracellular matrix components (Zivot et al. 2018). These adjacent cells in the niche influence erythroblasts through the production of cytokines and growth factors, as a source of oxygen and other nutrients, as well as direct cell-to-cell interaction with erythroid progenitors.

Finally, iron is an essential nutrient and a component of an abundance of proteins and enzymes that carry out vital biochemical functions in all cells. Systemic regulation of iron metabolism, by hepcidin (Chap. 3), has a significant impact on the efficiency of erythropoiesis in several ways, predominantly by regulating the availability of iron for erythropoiesis. Most of the iron in the body is contained in hemoglobin within RBCs such that erythroblasts utilize more than 80% of serum iron to synthesize the hemoglobin needed for the high pace of daily RBC production at steady state. For this reason, mutual crosstalk between iron homeostasis and erythropoiesis dominates the regulation of each, with recent evidence demonstrating that iron availability influences responsiveness to EPO. First, the expression of the iron importer transferrin receptor (TFR1, also known as TFRC), which initiates iron uptake, and the second receptor of transferrin (TFR2) involved in iron sensing to modulate EPO responsiveness in erythroblasts are both triggered as a consequence of EPO:EPO receptor binding. Iron is thus essential both for hemoglobin synthesis and differentiation during terminal erythropoiesis. In addition, erythropoiesis also regulates iron homeostasis as elucidated by the

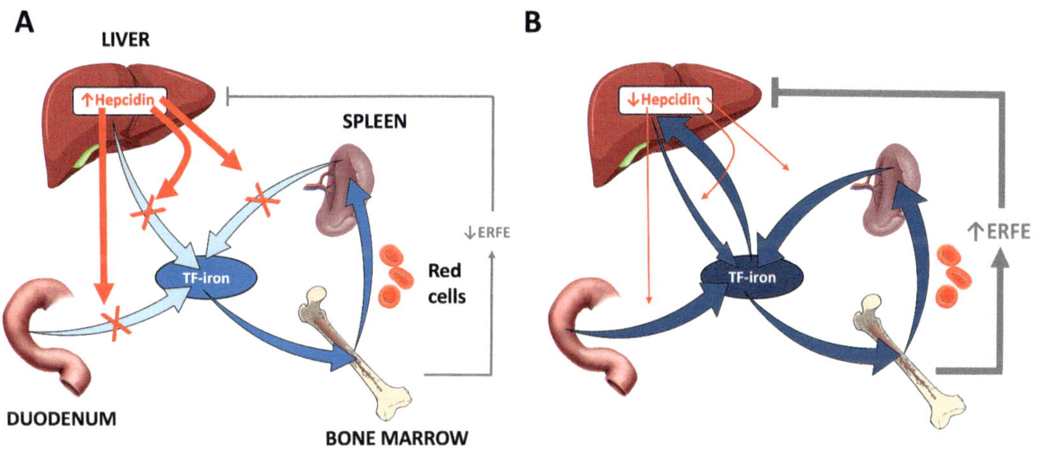

Fig. 11.2 Hepcidin and erythroferrone are central to the crosstalk regulation between erythropoiesis and iron metabolism. (**a**) Systemically, hepcidin is a negative regulator of iron flows such that increased hepcidin synthesis (which mainly occurs in the liver) leads to hypoferremia by decreasing iron absorption in the duodenum, iron recycling from splenic macrophages, and iron release from hepatocyte stores into the circulation where it is bound to transferrin (TF) and available for delivery to all cells. Small amounts of erythroferrone (ERFE) are produced in the bone marrow to enable coordination of iron supply with erythropoietic demand. (**b**) In iron-loading diseases (e.g., β-thalassemia and myelodysplastic syndrome), elevated ERFE leads to hepcidin suppression that supports an increase in iron absorption in the duodenum, iron recycling from splenic macrophages, and iron release from hepatocyte stores into the circulation where it may exceed the TF-Fe binding capacity and result in iron overload with parenchymal iron deposition

erythroblast production of erythroferrone (ERFE), which modulates the hepcidin:ferroportin axis to coordinate iron availability with erythropoietic demand (Fig. 11.2a). Disruption of any of these factors can lead to abnormal erythropoiesis, resulting in erythrocytosis, anemia, or abnormal morphological function.

11.2 Ineffective Erythropoiesis

Iron-loading anemias are a category of disparate conditions characterized by ineffective erythropoiesis. Ineffective erythropoiesis can be defined as the diminished production of enucleated RBCs despite an increase in the number of erythroid precursors (Fig. 11.3). Molecular characteristics of ineffective erythropoiesis remain incompletely understood. However, several conditions shed light on what unifies and defines them as a group and, in the absence of more nuanced mechanistic definitions, provide an opportunity to discuss how these disorders lead to iron loading while other similar disorders do not. The pathophysiology of iron-loading anemias is driven by an expanded population of erythroid progenitors and precursors, whose maturation is impaired. This blockade leads to elevated ERFE production, which suppresses hepcidin (Kautz et al. 2014), thereby promoting enhanced iron absorption and subsequent parenchymal iron loading (Fig. 11.2b). Hypoxia also plays a role in increasing the expression of genes responsible for iron absorption in the duodenum (Anderson et al. 2013). From a clinical perspective, when the quantity and/or the ability of enucleated RBCs to transport oxygen declines below a certain level, patients require RBC transfusions for survival, further exacerbating iron loading (Kattamis et al. 2022; Ginzburg and Rivella 2011). Taken together, in such diseases of concurrent iron overload and ineffective erythropoiesis (e.g., β-thalassemia, some cases of myelodysplastic syndromes (MDS), and dyserythropoietic anemias), lower than expected hepcidin expression, insufficiently elevated relative to increased iron stores, results in iron overload. Given the physi-

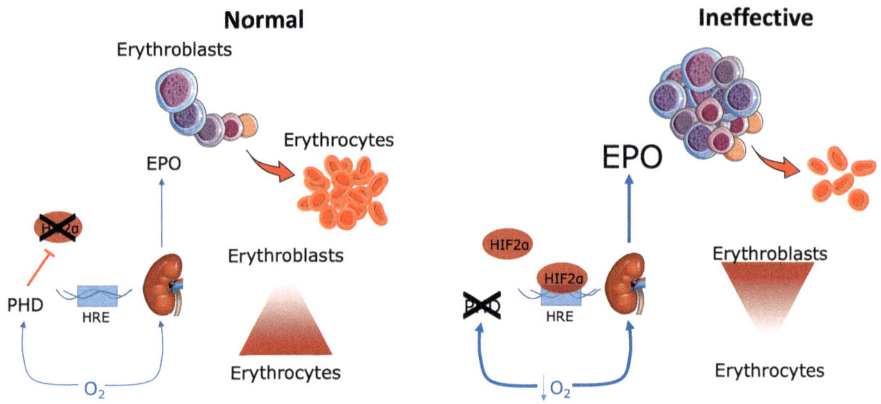

Fig. 11.3 Overview of ineffective erythropoiesis. Under normal conditions, small numbers of differentiating erythroblasts are needed to efficiently differentiate and enucleate to reticulocytes and ultimately mature RBCs (erythrocytes). Hypoxia-inducible factor-2α (HIF2α) is made constitutively by cells but is rapidly ubiquitinated and degraded in proteasomes after hydroxylation by the iron- and oxygen-dependent prolyl hydroxylases (PHDs). When oxygen levels decrease and limit the activity of PHDs, HIF2α levels rise, upregulating the transcription of HIF2α target genes, e.g., erythropoietin (EPO). In conditions associated with ineffective erythropoiesis (e.g., β-thalassemia, myelodysplastic syndrome and others), decreased O_2 delivery to peripheral tissues stimulates HIF2α bind at hypoxia response elements (HRE) which leads to increased EPO production in the kidney and enhances survival and proliferation of erythroblasts with a concurrent block in erythroblast differentiation, resulting in the accumulation of immature erythroblasts, preventing efficient production of erythrocytes, and consequent in anemia

ological regulation of hepcidin by ERFE in response to episodic blood loss and stress erythropoiesis necessary to return to steady state and prevent chronic anemia, ineffective erythropoiesis may be a type of chronic stress erythropoiesis. Alternatively, additional mechanisms that distinguish ineffective erythropoiesis from chronic stress erythropoiesis have been delineated (Fig. 11.4).

11.3 β-Thalassemia

Ineffective erythropoiesis has been the subject of intensive investigation in β-thalassemia, a relatively common monogenic inherited disorder, which manifests with an expanded number of erythroid precursors and their reduced ability to differentiate. This leads to blockade at the polychromatophilic erythroblasts, an increase in polychromatophilic erythroblast death (Arlet et al. 2016), and reduced survival of enucleated RBCs (Fibach and Dana 2019). β-thalassemia is highly prevalent in Africa, around the Mediterranean Sea, and in Southeast Asia, and as a result of population movement and migration, is now globally identified (Shah et al. 2019). In heterozygous carriers, β-thalassemia is associated with asymptomatic disease or mild anemia. Those affected in homozygosity are often symptomatic with either transfusion dependent (TDT, thalassemia major) or transfusion independent (NTDT, thalassemia intermedia) forms (Origa 2017). TDT patients experience a lifelong requirement for regularly scheduled RBC transfusions and iron chelation therapy that negatively impacts their quality of life (Scalone et al. 2008) (Fig. 11.5).

Genetically, point mutations or deletions of the β-globin gene cluster, located on the chromosome 11 p15.3 locus, reduce or eliminate β-globin synthesis (Taher et al. 2018), leading to the relative excess of α-globin chains (Kattamis et al. 2022; Ginzburg and Rivella 2011). Excess α-globin alters erythropoiesis in several ways. First, in a complex with free heme molecules, excess α-globin chains form hemichromes, which are the main source of oxidative stress and cell death (Kattamis et al. 2022). A scavenger protein, i.e., α-hemoglobin-stabilizing protein (AHSP), in

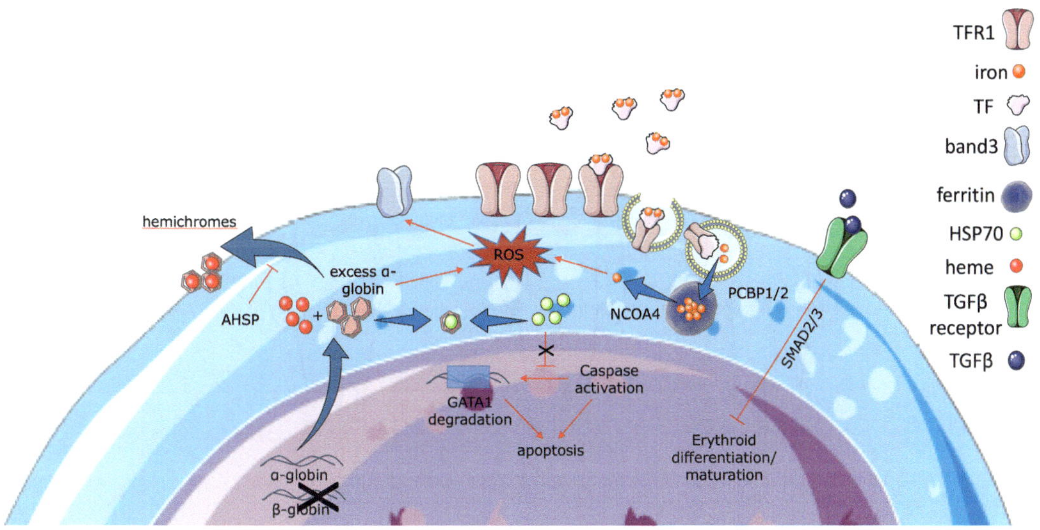

Fig. 11.4 Current mechanistic understanding of iron dysregulation causing ineffective erythropoiesis. Iron uptake in erythroblasts occurs predominantly via iron-TF binding to TFR1, leading to the formation of a clathrin-coated pit which undergoes acidification, leading to release of iron from TF to enable iron transfer across the pit membrane. PCBP1 and 2 are iron chaperones which traffic iron to ferritin and NCOA4 is an iron chaperone which traffics iron from ferritin and to the mitochondria for hemoglobin synthesis in erythroblasts. NCOA4-mediated ferritinophagy is necessary for the generation of iron-dependent ROS, implicated in ferroptosis, a lipid peroxidation-mediated type of cell death, and correlates with ineffective erythropoiesis. In addition, ROS is implicated in erythroblast death by denaturing band3, a critical membrane component, leading to poor erythroblast deformability and consequent clearance. Furthermore, in β-thalassemia, excess α-globin and heme form hemichromes, complexes known to damage erythroblast membranes; the quantity of excess α-globin in β-thalassemia overwhelms AHSP capacity to prevent hemichrome formation in discordant α:β-globin production by scavenging excess α-globin. Excess α-globin also complexes with HSP70 and prevents its translocation into the nucleus where it would otherwise protect GATA1 from caspase-mediated cleavage, leading to apoptosis. Finally, TGFβ binding to its receptor triggers SMAD2/3 signaling which inhibits erythroblast differentiation and maturation. AHSP = α hemoglobin stabilizing protein; HSP70 = heat shock protein 70; NCOA4 = nuclear co-activator 4; PCBP1/2 = Poly (rC)-binding protein 1 and 2; ROS = reactive oxygen species; TF = transferrin; TFR1 = transferrin receptor 1; TGFβ = transforming growth factor β)

normal erythroblasts, can prevent cell membrane damage caused by small amounts of α-globin precipitation (Han et al. 2022), but in typical quantities is not sufficient to prevent cell membrane damage from the significant excess hemichrome precipitation observed in β-thalassemia (Che Yaacob et al. 2020). In addition, excess α-globin chains reduce the stability of GATA1, the main erythroid transcription factor in erythroblasts (Dong et al. 2020), interfering with their survival and maturation (Arlet et al. 2014) (Fig. 11.4). Specifically, during terminal erythropoiesis, heat shock protein 70 (HSP70) is translocated from the cytoplasm into the nucleus, serving as a nuclear chaperone to prevent GATA1 cleavage by Caspase-3 (Ribeil et al. 2007; Arlet et al. 2014). However, in β-thalassemia, excess α-globin leads to HSP70 sequestration within the cytoplasm, compromising its protective effect on GATA1 and ultimately impeding erythroblast maturation (Guillem et al. 2020). Finally, the abnormal and reduced number of RBCs in circulation leads to hypoxia-mediated EPO production, which in turn exacerbates further the expanded number of immature erythroid precursors (Libani et al. 2008) (Fig. 11.3).

Anemia, tissue hypoxia, elevated erythropoietin, and erythroid expansion with consequently increased ERFE expression act in concert to suppress the production of hepcidin in the liver and to increase iron absorption, a process that results

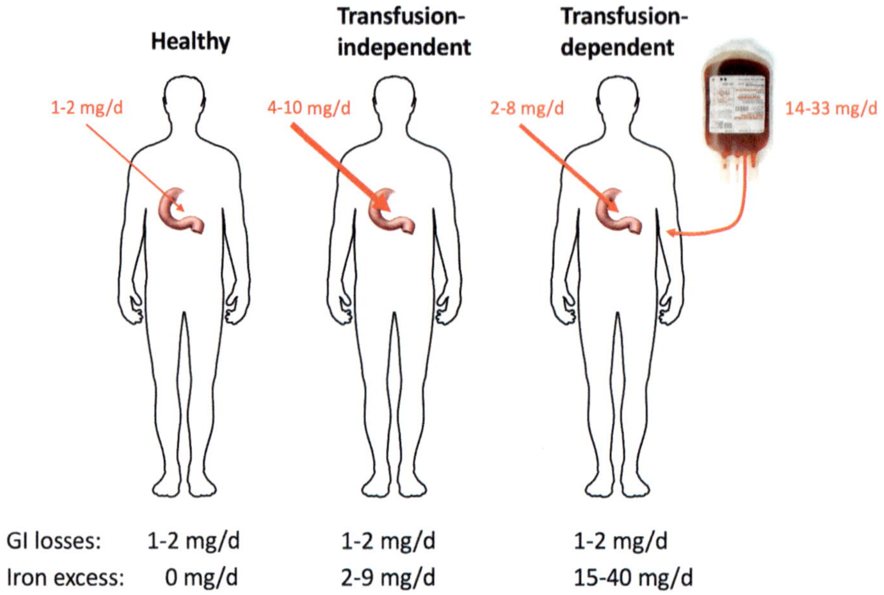

Fig. 11.5 Causes of iron overload in iron-loading anemias like β-thalassemia. Significantly increased iron absorption in the gastrointestinal (GI) tract results from insufficiently elevated hepcidin in transfusion-independent patients with iron-loading anemias and leads to parenchymal iron deposition and systemic iron overload. Iron absorption is also increased in transfusion-dependent patients with iron-loading anemias, but it is the large iron load from red blood cell transfusion that predominates in causing the more severe and/or earlier onset iron overload observed in transfusion dependent (vs. transfusion-independent) patients

in iron overload, even in the absence of RBC transfusions, and is theorized to exacerbate anemia (Camaschella and Nai 2016; Kautz et al. 2015). As a consequence, β-thalassemia patients are mainly treated with RBC transfusion to provide symptomatic management of anemia and suppress ineffective erythropoiesis, and iron chelation therapy to counter resultant iron overload (Origa 2017); see also Chap. 23. Possible risks of chronic RBC transfusion include allo-immunization, transfusion reactions, blood-borne infection, and iron overload (Cappellini et al. 2018). Although in severely affected patients, organ iron overload develops because of frequent RBC transfusions, in individuals with NTDT, the mechanisms responsible for increased iron absorption (i.e., insufficiently elevated hepcidin expression), along with chronic hemolysis, lead to progressive tissue iron deposition and toxicity, requiring even in these cases, the use of iron chelators to prevent significant morbidity and mortality (Kattamis et al. 2022; Musallam et al. 2021).

Furthermore, despite the availability of several FDA-approved iron chelators, many patients still experience complications and premature mortality related to iron overload (Shah et al. 2019), while frequent side effects require careful monitoring and can affect adherence (Cappellini et al. 2018). The financial costs of RBC transfusion and iron chelation are high, limiting availability to more affluent countries and wealthier individuals in lower income countries (Liem et al. 2011; Weidlich et al. 2016; McQuilten et al. 2019). Finally, in patients with severe TDT, a potentially curative treatment, hematopoietic stem cell transplantation, is possible for a subset of younger patients with an available compatible donor, who can tolerate the significant side effects of myeloablation required as part of the transplant, can manage the high cost, and have access to a treatment center with appropriate expertise (Cappellini et al. 2018).

Consequently, despite significant advancement in our understanding of the pathophysiology of dyserythropoiesis in β-thalassemia, the need for novel targeted therapies persists to further increase both the lifespan and health span of these chronically affected individuals.

11.4 Myelodysplastic Syndrome

Beyond the inherited forms of iron-loading anemias, i.e., β-thalassemia, MDS is an acquired form of ineffective erythropoiesis associated with iron overload. MDS is a heterogeneous group of clonal bone marrow stem cell disorders; several subtypes of MDS are characterized by ineffective erythropoiesis leading to blood cytopenias and increased incidence of transformation to acute myeloid leukemia (AML) (Haferlach 2019). The heterogeneity of MDS and its outcomes necessitates individual risk stratification to determine prognosis and suitable treatments for individual patients (Steensma 2018; Fenaux et al. 2021). The most commonly used risk stratification system, the revised International Prognostic Scoring System, delineates patients into (1) two lowest risk groups and part of the intermediate group described collectively as "lower risk MDS" and (2) the remaining "higher risk MDS" (Greenberg et al. 2012). The goal of patient care in those with higher risk MDS is primarily to prevent progression to AML, while in those with lower risk MDS, where the risk of progression to AML is much lower, the primary goal is typically treatment of anemia and ineffective erythropoiesis (Fenaux et al. 2021).

Approximately 80% of cases present with anemia-related symptoms, predominantly severe fatigue, with 30–50% requiring only regular RBC transfusions, to alleviate anemia and its associated symptoms (Dayyani et al. 2010; Kröger 2019). RBC transfusions, however, are the main cause of progressive iron overload and consequent end-organ damage in transfusion-dependent MDS patients (Oliva et al. 2010). However, the risk-benefit ratio of treating iron overload in MDS patients remains controversial and even then is only applicable to low-risk MDS patients with a long median survival (e.g., 10 years). Furthermore, RBC transfusion dependence and iron overload correlate strongly with decreased survival in MDS patients (Malcovati et al. 2005, 2006; Garcia-Manero et al. 2008; Malcovati et al. 2011).

The pathogenesis of ineffective erythropoiesis in MDS is also heterogeneous. Similar to that observed in β-thalassemia, ineffective erythropoiesis in MDS results from defects in cell signaling and dysregulation of important erythropoiesis-relevant growth factors (Santini 2015; Bulycheva et al. 2015), defective nuclear localization of HSP70 (Frisan et al. 2012), and changes in specific erythropoiesis regulators (i.e., GATA1, TGFβ, SMAD2/3) (Parisi et al. 2021; Verma et al. 2020; Maratheftis et al. 2007; Camaschella and Nai 2016) (Fig. 11.4). Furthermore, in vitro experiments demonstrate that excess iron inhibits erythroid lineage differentiation in both murine and human hematopoietic progenitors, exhibiting dysplastic changes with increased intracellular reactive oxygen species (ROS), decreased expression of anti-apoptotic genes, and DNA damage, triggering apoptosis, and worsening disease in MDS (Pan et al. 1999; Camaschella et al. 2007; Fibach and Rachmilewitz 2012; Taoka et al. 2012; Hartmann et al. 2013; Park et al. 2020; Claessens et al. 2005). In addition, alterations in the bone marrow microenvironment, including MDS-specific modification in mesenchymal stromal cell function and inflammasome-initiated pyroptosis have been proposed to increase malignant clone fitness at the expense of normal hematopoietic stem and precursor cells (Basiorka et al. 2016; Teodorescu et al. 2020; Medyouf et al. 2014).

Finally, specific disease-modifying mutations in the hematopoietic stem cell compartment, including those in epigenetic regulators (i.e., TET2 and ASXL1) and spliceosome genes (e.g., splicing factor 3B subunit 1 [SF3B1]) (Heuser et al. 2018), exhibit iron dependency and specific iron-mediated pathophysiology. Recent reports target the iron-dependent enzyme *TET2*, which is often mutated in MDS (20–25% of cases), demonstrating that *TET2* mutant hematopoietic cells are particularly dependent on iron for their competitive growth advantage (Loke et al. 2024). In addition, a specific subtype of low-risk MDS, namely MDS with ringed sideroblasts, occurs in

20% of all MDS patients. *SF3B1* mutations in hematopoietic stem and progenitor cells are a hallmark of this disease, inducing aberrant splicing of genes involved in heme biosynthesis and mitochondrial iron transport, leading to the abnormal deposition of iron in erythroblasts, and resulting in dysfunctional hemoglobin synthesis and the formation of ringed sideroblasts (Visconte et al. 2015; Dolatshad et al. 2015; Dolatshad et al. 2016; Shiozawa et al. 2018; Clough et al. 2022). Although preclinical data in mouse models predicted a therapeutic effect of splicing inhibition, a recent phase I clinical trial did not yield significant clinical improvement (Lee et al. 2016; Steensma et al. 2021). Growing evidence suggests that the heme-regulated EIF2AK1 kinase pathway affects erythropoiesis in health and disease. For example, EIF2AK1 effector DDIT3 is overexpressed in MDS hematopoietic stem and progenitor cells (Berastegui et al. 2022); DDIT3 overexpression delays erythroid differentiation in CD34-positive cells from MDS patients, and EIF2AK1 inhibition increases the expression of both mitochondrial heme biosynthesis enzymes and iron transporters, reversing *SF3B1* mutation-induced arrest of erythroid differentiation in vitro (Adema et al. 2022).

Taken together, these varied mechanisms result in an accumulation of immature erythroblasts with concurrently reduced numbers of mature RBCs, a classic characteristic of ineffective erythropoiesis. Furthermore, it remains unclear whether the mechanisms of dyserythropoiesis in MDS align with the broader pattern of ineffective erythropoiesis seen in iron-loading anemias, or if distinct mechanistic differences indicate fundamental variations in disease pathophysiology.

11.5 Iron Overload and Potential for Ferroptosis-Mediated Cell Death

Management of the anemia associated with these diseases often leads to chronic RBC transfusion requirements (Germing et al. 2019; Shah et al. 2019). Although they alleviate the symptoms of anemia, RBC transfusions not only carry a small risk of both acute and chronic infections or immune reactions (discussed in Chap. 24) but are also associated with iron overload, which can lead to end-organ damage (e.g., liver or heart disease) as well as correlate with poor outcomes and shortened survival. In fact, RBC transfusions are the main cause of progressive iron overload and consequent end-organ damage in TDT and transfusion-dependent MDS patients (Oliva et al. 2010). For example, in vitro experiments demonstrate that excess iron inhibits erythroid lineage differentiation in both murine and human hematopoietic progenitors, exhibiting dysplastic changes with increased intracellular ROS, decreased expression of anti-apoptotic genes, and DNA damage, triggering apoptosis, and worsening disease in MDS (Pan et al. 1999; Camaschella et al. 2007; Fibach and Rachmilewitz 2012; Taoka et al. 2012; Hartmann et al. 2013).

In addition to the iron loading from RBC transfusions, ineffective erythropoiesis in iron-loading anemias results in increased expression of ERFE and possibly other pathological regulators of iron metabolism, dampening hepcidin responsiveness to iron (Fig. 11.2b). The largest body of data on the crosstalk between erythropoiesis and iron metabolism comes from work done in β-thalassemia, but similar results are also evidenced IN MDS mouse models and patients (Santini et al. 2011; Gattermann 2018; An et al. 2023). Serum hepcidin concentration in β-thalassemic patients is increased in correlation with hemoglobin and increases in response to RBC transfusion, suggesting that enhanced erythropoietic activity induces hepcidin suppression. Recent evidence supports that ERFE is a major erythroid regulator of hepcidin, which in β-thalassemia overrides the regulation of hepcidin by iron and leads to the exacerbation of iron overload (Ganz et al. 2017). Preclinical evidence shows increased ERFE levels in mouse models of β-thalassemia, along with relatively increased hepcidin expression. However, iron overload was only partially mitigated in β-thalassemic/ERFE knockout mice compared to β-thalassemic mice (Kautz et al. 2015). The contribution of additional hepcidin regulators is currently being explored in preclinical studies (Sardo et al. 2024).

To understand the uniqueness of MDS with ringed sideroblasts, prior evidence indicates that these patients develop iron overload even before the initiation of RBC transfusion therapy (Gattermann 2005), likely due to hepcidin suppression that is particularly pronounced compared to other MDS subtypes (Gu et al. 2013; Santini et al. 2011). Interestingly, mutations in the splicing factor SF3B1 gene may also lead to increased production of an ERFE variant, causing particularly robust suppression of hepcidin and increased iron loading in MDS with ring sideroblasts (Bondu et al. 2019). Furthermore, *HFE* gene polymorphisms that predispose to iron overload are detected in up to 21% of MDS with ringed sideroblasts, significantly higher than in other MDS subtypes (Nearman et al. 2007; Valent et al. 2008). A more complete understanding of how ringed sideroblast formation in MDS exacerbates ineffective erythropoiesis, as well as the specific role of dysregulated iron metabolism in disease pathophysiology, is currently lacking.

Under physiological conditions, hepcidin is induced in response to high iron stores to prevent continued iron absorption and exacerbation of iron loading. However, in iron-loading anemias, insufficient hepcidin expression, relative to systemic iron levels, is implicated as the cause of iron overload. Thus, suppressed hepcidin results in further increased iron absorption, ultimately exceeding transferrin iron-carrying capacity and leading to the formation of non-transferrin bound iron (NTBI), which is not available for erythropoiesis and deposits in the parenchyma of non-hematopoietic tissues. NTBI is transported into cells by iron transporters such as ZIP14 (Knutson 2019; Jenkitkasemwong et al. 2015; Coffey and Knutson 2016) in hepatocytes and pancreatic islet cells. How cardiomyocytes take up NTBI is still an unresolved question, but L-type and T-type calcium channels likely contribute (Knutson 2019). The liver functions as the primary clearance organ for NTBI (Craven et al. 1987), leading to liver toxicity and consequent fibrosis, which correlates with liver iron concentration (Chin et al. 2019). Cardiac toxicity in β-thalassemia also strongly correlates with NTBI concentrations (Piga et al. 2009), thought to occur in the setting of high rates of NTBI generation, overwhelming or exceeding liver NTBI uptake. Cardiac toxicity is uncommon in conditions where iron accumulation is slow and NTBI concentrations are relatively low (e.g., NTDT and non-transfusion-dependent MDS patients) compared with those conditions where iron accumulation is rapid and NTBI concentrations are high (e.g., TDT and transfusion-dependent MDS patients) (Origa et al. 2008). Excessive intracellular iron that is unchaperoned by carrier proteins is known to catalyze chemical modifications of proteins, lipids, and nucleic acids (Ramm and Ruddell 2005), causing initially reversible functional deficits but eventually progressive fibrosis, organ failure, and carcinogenesis. Finally, NTBI is highly bioavailable to gram-negative bacteria and other pathogenic microbes that cause frequent infections in iron-overloaded patients, including those with thalassemias (Stefanova et al. 2017, 2018; Wang et al. 2003; Vento et al. 2006).

Recently, evidence of a newly elucidated form of cell death, ferroptosis, has been described (see Chap. 5). Canonical ferroptosis induction is a process requiring three elements considered its hallmarks—lipid peroxidation, presence of iron-mediated redox activity, and mitigation or exacerbation by manipulating steps in this pathway (Brigelius-Flohé and Maiorino 2013; Yang et al. 2016; Dixon and Stockwell 2019). Due to their high iron requirements, erythroblasts are particularly vulnerable to potential ferroptosis-mediated cell death, providing a rationale for evaluating ferroptosis-related physiology and/or pathophysiology in dyserythropoietic disorders. Furthermore, increased ROS have been identified in erythroblasts from patient samples and mouse models of iron-loading anemias (Lin et al. 2024). However, although a potential role for ferroptosis in the pathophysiology of diseases involving fibroblasts or hematopoietic stem cells, among other cellular lineages, has been proposed, no evidence is currently available regarding ferroptosis in dysregulated erythropoiesis. Ferroptosis has been proposed to contribute to hematopoietic stem and progenitor cell death due to iron overload in MDS (Lv et al. 2020; Teodorescu et al. 2020). Specifically, preclinical data suggest that (a) iron overload further aggravates oxidative stress in MDS patient bone marrow samples; (b) MDS erythroblasts are particularly sensitive to oxidative DNA damage and the accompanying bone marrow genomic instability and mutagenesis (Saigo et al. 2011; Chan et al. 2008, 2021a, b; Gattermann 2018); and (c)_in an MDS mouse model, elevated erythroblast ROS is reversed in response to iron chelation without impacting classical apoptosis (An et al. 2023). Taken together, the pathophysiology of dyserythropoiesis in at least some iron-loading anemias may involve ferroptosis. However, additional experimentation is needed to elucidate the extent of the contribution of ferroptosis in this process.

11.6 Iron Chelation Therapy

Three iron-chelating drugs have been approved for the treatment of iron overload: deferiprone, deferoxamine, and deferasirox (Chap. 23), with compelling evidence that these agents lower parenchymal iron concentration to reverse or prevent complications of iron overload, often improving event-free survival (EFS). However, although iron chelation is considered standard of care in β–thalassemia, the risk-benefit ratio of treating iron overload in MDS patients remains controversial despite the strong correlation between RBC transfusion dependence, iron overload, and decreased survival in MDS patients (Malcovati et al. 2005, 2006, 2011; Garcia-Manero et al. 2008). The TELESTO trial was designed and executed to test whether iron chelation provides clinical benefit in lower risk MDS patients with iron overload (Angelucci et al. 2020). The data demonstrated increased EFS in deferasirox-treated MDS patients without clear improvement in hemoglobin, reduction of RBC transfusion burden, or effect on overall survival. Deferasirox and most other commercially available iron chelators have a higher iron binding affinity relative to transferrin (Sohn et al. 2008). Only deferiprone's iron-chelating effect enables iron transfer from parenchymal cells to increase transferrin saturation (Schmidt et al. 2015; Casu et al. 2016) and may provide an opportunity to reap additional benefits if shown to ameliorate ineffective erythropoiesis in addition to reversing or preventing iron overload in MDS patients as in MDS mice (An et al. 2023).

11.7 Alternative Therapeutic Approaches

11.7.1 β-Thalassemia

Novel drugs and genetic approaches are recently approved or are in the process of being translated to improve the quality of life and even cure patients with β-thalassemia (and others). Although a full description of these therapeutics is beyond the scope of the current chapter, these drugs can be broadly classified based on their mechanism of action as shown in Table 11.1. Among these, only hematopoietic stem cell transplant and gene therapy provide curative approaches. However, the risks remain high, and alternative therapeutic options are welcome for those patients who are ineligible for a cure.

Luspatercept is a modified activin receptor IIB ligand trap, a member of the transforming growth factor-β (TGF-β) superfamily. Luspatercept modulates TGFβ signaling by acting as a trap for TGFβ ligands, such as GDF11, activin, and some BMPs, to restore late-stage erythropoiesis via inhibition of SMAD2/3 signaling (Suragani et al. 2014; Guerra et al. 2019). Specifically, Luspatercept has been found to increase nuclear GATA1 levels in a β-thalassemia mouse model, but its precise mechanism of action in patients remains an active area of research (Martinez et al. 2020). This agent was approved in 2019 for TDT based on results from the phase 3 BELIEVE trial, which demonstrated a reduction in RBC transfusion burden in Luspatercept vs. placebo-treated TDT patients (Cappellini et al. 2020). Similarly, the phase 2 BEYOND trial in NTDT patients revealed an increase in hemoglobin concentration in Luspatercept- vs. placebo-treated NTDT patients (Taher et al. 2022).

Table 11.1 Available therapeutic options and those currently in development for several iron-loading anemias

β-thalassemia		MDS	
Mechanism	Agent	Mechanism	Agent
Anemia management	RBC transfusion + iron chelation	Anemia management	RBC transfusion + iron chelation
Cure for inherited germline β-globin mutations	HSCT, gene therapy	Mitigation of defective EPO-mediated signaling	ESAs
Mitigation of metabolic defect leading to shortened survival of erythroblasts and RBCs	Mitapivat	Disease modification for acquired somatic HSC mutations	HSCT, HMAs
Mitigating blocked erythroblast maturation, aberrant TGFβ signaling	Luspatercept	Modulating enhanced erythroblast proliferation	Lenalidomide
Modulating ineffective erythropoiesis by reversing hepcidin suppression	Hepcidin agonists, transferrin	Mitigating blocked erythroblast maturation, aberrant TGFβ signaling	Luspatercept

MDS myelodysplastic syndrome, *RBC* red blood cells, *HSCT* hematopoietic stem cell transplant, *EPO* erythropoietin, *ESAs* erythropoietin stimulating agents, *HMAs* hypomethylating agents, *TGFβ* transforming growth factor beta

An alternative approach targets the increased metabolic demands on erythroblasts resulting from aggregates of excess globin chains; their clearance is a metabolically demanding process, yet ATP supply is reduced in thalassemic erythroblasts, in part contributing to anemia and cell death (Kuo et al. 2022; Chakraborty et al. 2012). Mitapivat, an activator of the RBC pyruvate kinase (PK) enzyme that is involved in ATP production, reduces hemolysis and leads to increased hemoglobin in patients with PK deficiency (Al-Samkari et al. 2022). On this basis, a phase 2 trial in NTDT patients was initiated with a robust increase in hemoglobin [93] and was followed by two phase 3 trials, ENERGIZE-T (NCT04770779) in TDT patients and ENERGIZE (NCT04770753) in NTDT patients. The currently reported data show robust effects of Mitapivat (Cappellini et al. 2024).

Drugs that act on iron metabolism (e.g., hepcidin-mimetics and transferrin) could also improve RBC quality and survival by limiting erythroid iron intake and hemichrome formation, as shown in mouse models of β-thalassemia (Li et al. 2010; Guo et al. 2013); see also Chap. 26. Manipulating the ERFE:hepcidin:FPN1 axis to increase hepcidin or block ERFE or FPN1 could help limit intestinal iron absorption and sequester iron in macrophages and hepatocytes to reverse or prevent iron overload. However, even if these drugs individually fail to do so in β-thalassemia patients, their use could reduce iron absorption in combination with iron chelators if the patient is already iron overloaded. In addition, they could prevent iron from accumulating in combination with drugs that improve RBC quality and production (e.g., luspatercept, mitapivat).

Furthermore, multiple lines of evidence support the beneficial effects of iron restriction on ineffective erythropoiesis, which could reverse splenomegaly and decrease RBC transfusion requirements by reducing RBC destruction in the spleen or improve erythroid differentiation and enucleation in β-thalassemia (Gardenghi et al. 2010; Li et al. 2010, 2017; Gelderman et al. 2015). Thus, in addition to preventing or reversing iron overload, modulating the ERFE:hepcidin:FPN1 axis may provide significant iron restriction within the erythroid compartment and enable more effective erythropoiesis in iron-loading anemias. Currently, the translation of hepcidin agonist agents and transferrin for TDT and NTDT patients is on hold due to insufficiently robust results in early phase clinical trials, although some agents may still be conceptually supporting translation (Pilo and Angelucci 2024).

11.7.2 MDS

Therapies currently available for the treatment of MDS include growth factor support, iron chelation, lenalidomide, hypomethylating agents

(HMAs, i.e., azacitidine and decitabine), intensive chemotherapy such as that used to treat AML, and allogeneic hematopoietic stem cell transplantation (HSCT) (Garcia-Manero et al. 2020; Cappellini et al. 2023). Due to the heterogeneity in MDS, therapy depends on the underlying risk of progression in individual MDS patients. Lower risk MDS patients are generally managed with growth factors and typical erythropoietic stimulating agents (ESAs), but only 30–60% of patients respond to ESAs (Steensma 2018). Lenalidomide is recommended as first-line therapy in MDS patients with 5q- who are transition-dependent; fewer options are available for low-risk MDS patients without 5q-, relegating them to more standard approaches, including RBC transfusion, iron chelation, and ESAs.

Eligible MDS patients with higher risk disease undergo HSCT, the only potentially curative option in MDS, while ineligible MDS patients typically receive HMAs as first-line therapy. HMAs have played a pivotal role in treating MDS in recent decades, providing sustained hematological responses (e.g., improving cytopenias, lowering blast counts), improved quality of life, and delayed progression to AML. Furthermore, the approval of oral HMAs has increased the convenience of these often elderly, patients. However, a vast majority of patients will experience treatment failure within 2 years, providing evidence that the disease-modifying effects of HMAs are insufficient to impact the long-term consequences of MDS.

In addition, a novel therapy has recently been approved for MDS. Luspatercept was approved in 2020 for lower risk RBC transfusion-dependent MDS patients with ring sideroblasts who are not responsive to an ESA based on the phase III MEDALIST trial, demonstrating that Luspatercept resulted in substantive RBC transfusion independence in treated patients vs. placebo (Fenaux et al. 2020; Farrukh et al. 2022). Luspatercept is currently also being investigated in ESA-naïve patients with anemia due to lower risk MDS in the phase 3 COMMANDS trial (NCT03682536). Taken together, these recent findings provide supporting evidence for dysregulated iron trafficking in the pathophysiology of ineffective erythropoiesis in MDS.

Finally, a variety of novel agents are currently under investigation in low-risk MDS patients, the delineation of which is beyond the scope of this chapter and is likely to be outdated given the activity in this area of translational development (Cappellini et al. 2023).

11.8 Perspectives and Future Directions

Taken together, a great deal is currently known about the physiology of erythropoiesis and iron metabolism as well as the pathophysiology of iron-loading anemias. Despite this, significant unknowns remain, both regarding the mechanisms of normal function as well as their disordered regulation in disease; we thus include here an incomplete list to guide the next generation of targeted inquiry along these lines:

- A mechanistic molecular understanding of ineffective erythropoiesis and a more complete delineation of how EPO responsiveness is modulated in both physiological and pathological conditions remain elusive. This direction of investigation could yield novel therapeutic development for a variety of diseases associated with anemia, e.g., anemia of chronic inflammation and renal failure.
- Whether and how EPOR expression outside of erythroid lineage cells regulates the crosstalk between iron metabolism and erythropoiesis remains to be more completely elucidated, e.g., in osteoblasts and erythroblastic island (EBI) macrophages.
- Determining whether iron-mediated pathophysiology in iron-loading anemias is central to dysregulated erythropoiesis or a common complication thereof would also significantly advance our understanding and provide novel therapeutic targets more broadly. Furthermore, understanding if cellular iron trafficking is central to disease pathophysiology even in the absence of overly iron-dependent epigenetic modifications (e.g., TET2, ASXL1, IDH1/2,

SF3B1 mutations in MDS) may increase the generalizability of targeting the regulation of iron metabolism for therapeutic purposes in iron-loading anemias.

- A novel mechanism for cell death, ferroptosis, was relatively recently discovered, an iron-mediated mechanism leading to lipid peroxidation and cell death that is being targeted as a therapeutic approach in various cancers. Understanding more definitively the role of ferroptosis as it pertains to iron trafficking in erythroblasts may shed light on iron-loading anemias associated with ineffective erythropoiesis (i.e., β-thalassemia and MDS).
- Elucidating the mechanism(s) leading to the robust effect of hepcidin agonists and mimetics in β-thalassemia mouse models yet insufficiently potent activity in patients during early phase clinical trials with multiple agents.

References

Adema V, Ma F, Kanagal-Shamanna R et al (2022) Targeting the EIF2AK1 signaling pathway rescues red blood cell production in SF3B1-mutant myelodysplastic syndromes with rings sideroblasts. Blood Cancer Discov 3(6):554–567. https://doi.org/10.1158/2653-3230.BCD-21-0220

Al-Samkari H, Galactéros F, Glenthøj A et al (2022) Mitapivat versus placebo for pyruvate kinase deficiency. N Engl J Med 386(15):1432–1442. https://doi.org/10.1056/NEJMoa2116634

An W, Feola M, Levy M et al (2023) Iron chelation improves ineffective erythropoiesis and iron overload in myelodysplastic syndrome mice. eLife 12:e83103. https://doi.org/10.7554/eLife.83103

Anderson ER, Taylor M, Xue X et al (2013) Intestinal HIF2α promotes tissue-iron accumulation in disorders of iron overload with anemia. Proc Natl Acad Sci USA 110(50):E4922–E4930. https://doi.org/10.1073/pnas.1314197110

Angelucci E, Li J, Greenberg P et al (2020) Iron chelation in transfusion-dependent patients with low- to intermediate-1-risk myelodysplastic syndromes: a randomized trial. Ann Intern Med 172(8):513–522. https://doi.org/10.7326/M19-0916

Arlet J, Ribeil J, Guillem F et al (2014) HSP70 sequestration by free α-globin promotes ineffective erythropoiesis in β-thalassaemia. Nature 514:242–246. https://doi.org/10.1038/nature3614

Arlet J, Dussiot M, Moura IC et al (2016) Novel players in β-thalassemia dyserythropoiesis and new therapeutic strategies. Curr Opin Hematol 23(3):181–188. https://doi.org/10.1097/MOH.0000000000000231

Basiorka AA, McGraw KL, Eksioglu EA et al (2016) The NLRP3 inflammasome functions as a driver of the myelodysplastic syndrome phenotype. Blood 128(25):2960–2975. https://doi.org/10.1182/blood-2016-07-730556

Berastegui N, Ainciburu M, Romero JP et al (2022) The transcription factor DDIT3 is a potential driver of dyserythropoiesis in myelodysplastic syndromes. Nat Commun 13(1):7619. https://doi.org/10.1038/s41467-022-35192-7

Bondu S, Alary AS, Lefèvre C et al (2019) A variant erythroferrone disrupts iron homeostasis in SF3B1-mutated myelodysplastic syndrome. Sci Transl Med 11(500):eaav5467. https://doi.org/10.1126/scitranslmed.aav5467

Brigelius-Flohé R, Maiorino M (2013) Glutathione peroxidases. Biochim Biophys Acta 1830(5):3289–3303. https://doi.org/10.1016/j.bbagen.2012.11.020

Bulycheva E, Rauner M, Medyouf H et al (2015) Myelodysplasia is in the niche: novel concepts and emerging therapies. Leukemia 29(2):259–268. https://doi.org/10.1038/leu.2014.325

Camaschella C, Nai A (2016) Ineffective erythropoiesis and regulation of iron status in iron loading anaemias. Br J Haematol 172(4):512–523. https://doi.org/10.1111/bjh.13820

Camaschella C, Campanella A, Falco LD et al (2007) The human counterpart of zebrafish shiraz shows sideroblastic-like microcytic anemia and iron overload. Blood 110(4):1353–1358. https://doi.org/10.1182/blood-2007-02-072520

Cappellini MD, Porter JB, Viprakasit V et al (2018) A paradigm shift on beta-thalassemia treatment: how will we manage this old disease with new therapies? Blood Rev 32(4):300–311. https://doi.org/10.1016/j.blre.2018.02.001

Cappellini MD, Viprakasit V, Taher AT et al (2020) A phase 3 trial of luspatercept in patients with transfusion-dependent β-thalassemia. N Engl J Med 382(13):1219–1231. https://doi.org/10.1056/NEJMoa1910182

Cappellini MD, Taher AT, Verma A et al (2023) Erythropoiesis in lower-risk myelodysplastic syndromes and beta-thalassemia. Blood Rev 59:101039. https://doi.org/10.1016/j.blre.2022.101039

Cappellini MD, Sheth S, Taher AT et al (2024) ENERGIZE-T: a global, phase 3, double-blind, randomized, placebo-controlled study of mitapivat in adults with transfusion-dependent alpha- or beta-thalassemia. Blood 144(Supplement 1):409. https://doi.org/10.1182/blood-2024-200867

Casu C, Oikonomidou PR, Chen H et al (2016) Minihepcidin peptides as disease modifiers in mice affected by β-thalassemia and polycythemia vera. Blood 128(2):265–276. https://doi.org/10.1182/blood-2015-10-676742

Caulier AL, Sankaran VG (2022) Molecular and cellular mechanisms that regulate human erythropoiesis.

Blood 139(16):2450–2459. https://doi.org/10.1182/blood.2021011044
Chakraborty I, Mishra R, Gachhui R et al (2012) Distortion of β-globin chain of hemoglobin alters the pathway of erythrocytic glucose metabolism through band 3 protein. Arch Med Res 43(2):112–116. https://doi.org/10.1016/j.arcmed.2012.02.003
Chan LSA, Buckstein R, Reis MD et al (2008) Iron overload and haematopoiesis in MDS: does blood transfusion promote progression to AML? Blood 112(11):2685. https://doi.org/10.1182/blood-V112.11.2685.2685
Chan LSA, Gu LC, Leitch HA et al (2021a) Intracellular ROS profile in hematopoietic progenitors of MDS patients: association with blast count and iron overload. Hematology 26(1):88–95. https://doi.org/10.1080/16078454.2020.1870373
Chan LSA, Gu LC, Wells RA (2021b) The effects of secondary iron overload and iron chelation on a radiation-induced acute myeloid leukemia mouse model. BMC Cancer 21(1):509. https://doi.org/10.1186/s12885-021-08259-9
Che Yaacob NS, Islam MA, Alsaleh H et al (2020) Alpha-hemoglobin-stabilizing protein (AHSP): a modulatory factor in β-thalassemia. Int J Hematol 111:352–359. https://doi.org/10.1007/s12185-091-02806-8
Chin J, Powell LW, Ramm LE et al (2019) Utility of hepatic or total body iron burden in the assessment of advanced hepatic fibrosis in HFE hemochromatosis. Sci Rep 9(1):20234. https://doi.org/10.1038/s41598-019-56732-0
Claessens Y, Park S, Dubart-Kupperschmitt A et al (2005) Rescue of early-stage myelodysplastic syndrome-deriving erythroid precursors by the ectopic expression of a dominant-negative form of FADD. Blood 105(10):4035–4042. https://doi.org/10.1182/blood-2004-08-3166
Clough CA, Pangallo J, Sarchi M et al (2022) Coordinated missplicing of TMEM14C and ABCB7 causes ring sideroblast formation in SF3B1-mutant myelodysplastic syndrome. Blood 139(13):2038–2049. https://doi.org/10.1182/blood.2021012652
Coffey R, Knutson MD (2016) The plasma membrane metal-ion transporter ZIP14 contributes to non-transferrin-bound iron uptake by human β-cells. Am J Physiol Cell Physiol 312(2):C169–C175. https://doi.org/10.1152/ajpcell.0016.2016
Craven CM, Alexander J, Eldridge M et al (1987) Tissue distribution and clearance kinetics of non-transferring-bound iron in the hypotransferrinemic mouse: a rodent model for hemochromatosis. PNAS 84(10):3457–3461. https://doi.org/10.1073/pnas.84.10.3457
Dayyani F, Conley AP, Strom SS et al (2010) Cause of death in patients with lower-risk myelodysplastic syndrome. Cancer 116(9):2174–2179. https://doi.org/10.1002/cncr.24984
Dixon SJ, Stockwell BR (2019) The hallmarks of ferroptosis. Annu Rev Cancer Biol 3:35–54. https://doi.org/10.1146/annurev-cancerbio-030518-055844
Dolatshad H, Pellagatti A, Fernandez-Mercado M et al (2015) Disruption of SF3B1 results in deregulated expression and splicing of key genes and pathways in myelodysplastic syndrome hematopoietic stem and progenitor cells. Leukemia 29(5):1092–1103. https://doi.org/10.1038/leu.2014.331
Dolatshad H, Pellagatti A, Liberante FG et al (2016) Cryptic splicing events in the iron transporter ABCB7 and other key target genes in SFB1-mutant myelodysplastic syndromes. Leukemia 30(12):2322–2331. https://doi.org/10.1038/leu.2016.149
Dong X, Zhao K, Zheng W et al (2020) EDAG mediates Hsp70 nuclear localization in erythroblasts and rescues dyserythropoiesis in myelodysplastic syndrome. FASEB 34(6):8416–8427. https://doi.org/10.1096/fj.201902946R
Farrukh F, Chetram D, Al-Kali A et al (2022) Real-world experience with luspatercept and predictors of response in myelodysplastic syndromes with ring sideroblasts. Am J Hematol 97(6):E210–E214. https://doi.org/10.1002/ajh.26533
Fenaux P, Platzbecker U, Mufti GJ et al (2020) Luspatercept in patients with lower-risk myelodysplastic syndromes. N Engl J Med 382(2):140–151. https://doi.org/10.1056/NEJMoa1908892
Fenaux P, Haase D, Santini V et al (2021) Myelodysplastic syndromes: ESMO clinical practice guidelines for diagnosis, treatment and follow-up. Ann Oncol 32(2):142–156. https://doi.org/10.1016/j.annonc.2020.11022
Fibach E, Dana M (2019) Oxidative stress in β-thalassemia. Mol Diagnos Ther 23:245–261. https://doi.org/10.1007/s40291-018-0373-5
Fibach E, Rachmilewitz EA (2012) Selective toxicity towards myelodysplastic hematopoietic progenitors—another rationale for iron chelation in MDS. Leuk Res 36(8):962–963. https://doi.org/10.1016/j.leukres.2012.04.030
Frisan E, Vandekerckhove J, de Thonel A et al (2012) Defective nuclear localization of Hsp70 is associated with dyserythropoiesis and GATA-1 cleavage in myelodysplastic syndromes. Blood 119(6):1532–1542. https://doi.org/10.1182/blood-2011-03-343475
Ganz T, Jung G, Naeim A et al (2017) Immunoassay for human serum erythroferrone. Blood 130(10):1243–1246. https://doi.org/10.1182/blood-2017-04-777987
Garcia-Manero G, Shan J, Faderl S et al (2008) A prognostic score for patients with lower risk myelodysplastic syndrome. Leukemia 22(3):538–543. https://doi.org/10.1038/sj.leu.2405070
Garcia-Manero G, Chien KS, Montalban-Bravo G (2020) Myelodysplastic syndromes: 2021 update on diagnosis, risk stratification and management. Am J Hematol 95(11):1399–1420. https://doi.org/10.1002/ajh.25950
Gardenghi S, Ramos P, Marongui MR et al (2010) Hepcidin as a therapeutic tool to limit iron overload and improve anemia in β-thalassemic mice. J Clin Invest 120(12):4466–4477. https://doi.org/10.1172/JCI41717

Gattermann N (2005) Clinical consequences of iron overload in myelodysplastic syndromes and treatment with chelators. Hematol Oncol Clin North Am 19(1):13–17

Gattermann N (2018) Iron overload in myelodysplastic syndromes (MDS). Int J Hematol 107(1):55–63. https://doi.org/10.1007/s12185-017-2367-1

Gelderman MP, Baek JH, Yalamanoglu A et al (2015) Reversal of hemochromatosis by apotransferrin in non-transfused and transfused Hbbth3/+ (heterozygous B1/B2 globin gene deletion) mice. Haematologica 100(5):611–622. https://doi.org/10.3324/haematol.2014.117325

Germing U, Oliva EN, Hiwase D et al (2019) Treatment of anemia in transfusion dependent and non-transfusion-dependent lower-risk MDS: current and emerging strategies. Hema 3(6):e314. https://doi.org/10.1097/HS9.0000000000000314

Ginzburg Y, Rivella S (2011) β-Thalassemia: a model for elucidating the dynamic regulation of ineffective erythropoiesis and iron metabolism. Blood 118(16):4321–4330. https://doi.org/10.1182/blood-2011-03-283614

Greenberg PL, Tuechler H, Schanz J et al (2012) Revised international prognostic scoring system for myelodysplastic syndromes. Blood 120(12):2454–2465. https://doi.org/10.1182/blood-2012-03-420489

Gu S, Song X, Zhao Y et al (2013) The evaluation of iron overload through hepcidin level and its related factors in myelodysplastic syndromes. Hematology 18(5):286–294. https://doi.org/10.1179/1607845412Y.0000000064

Guerra A, Oikonomidou PR, Sinha S et al (2019) Lack of Gdf11 does not improve anemia or prevent the activity of RAP-536 in a mouse model of β-thalassemia. Blood 134(6):568–572. https://doi.org/10.1182/blood.2019001057

Guillem F, Dussiot M, Colin E et al (2020) XPO1 regulates erythroid differentiation and is a new target for the treatment of β-thalassemia. Haematologica 105(9):2240–2249. https://doi.org/10.3324/haematol.2018.210054

Guo S, Casu C, Gardenghi S et al (2013) Reducing TMPRSS6 ameliorates hemochromatosis and β-thalassemia in mice. J Clin Invest 123(4):1531–1541. https://doi.org/10.1172/JCI66969

Haferlach T (2019) The molecular pathology of myelodysplastic syndrome. Pathobiology 86(1):24–29. https://doi.org/10.1159/000488712

Han G, Cao C, Yang X et al (2022) Nrf2 expands the intracellular pool of the chaperone AHSP in a cellular model of β-thalassemia. Redox Biol 50:102239. https://doi.org/10.1016/j.redox.2022.102239

Hartmann J, Braulke F, Sinzig U et al (2013) Iron overload impairs proliferation of erythroid progenitors cells (BFU-E) from patients with myelodysplastic syndromes. Leuk Res 37(3):327–332. https://doi.org/10.1016/j.leukres.2012.11.005

Heuser M, Yun H, Thol F (2018) Epigenetics in myelodysplastic syndromes. Semin Cancer Biol 51:170–179. j.semcancer.2017.07.009

Jenkitkasemwong S, Wang CY, Coffey R et al (2015) SLC39A14 is required for the development of hepatocellular iron overload in murine models of hereditary hemochromatosis. Cell Metab 22(1):138–150. https://doi.org/10.1016/j.cmet.2015.05.002

Kattamis A, Kwiatkowski JL, Aydinok T (2022) Thalassaemia. Lancet 399(10343):2310–2324. https://doi.org/10.1016/S0140-6736(22)00536-0

Kautz L, Jung G, Valor EV et al (2014) Identification of erythroferrone as an erythroid regulator of iron metabolism. Nat Genet 46(7):678–687. https://doi.org/10.1038/ng.2996

Kautz L, Jung G, Du X et al (2015) Erythroferrone contributes to hepcidin suppression and iron overload in a mouse model of β-thalassemia. Blood 126(17):2031–2037. https://doi.org/10.1182/blood-2015-07-658419

Knutson MD (2019) Non-transferrin-bound iron transporters. Free Radic Biol Med 133:101–111. https://doi.org/10.1016/j.freeradbiomed.2018.10.413

Kröger N (2019) Induction, bridging, or straight ahead: the ongoing dilemma of allografting in advanced myelodysplastic syndrome. BBMT 25(8):E247–E249. https://doi.org/10.1016/j.bbmt.2019.06.016

Kuo KHM, Layton DM, Lal A et al (2022) Safety and efficacy of mitapivat, an oral pyruvate kinase activator, in adults with non-transfusion dependent α-thalassaemia or β-thalassaemia: an open-label, multicentre, phase 2 study. Lancet 400(10351):493–501. https://doi.org/10.1016/S0140-6736(22)01337-X

Lee SC, Dvinge H, Kim E et al (2016) Modulation of splicing catalysis for therapeutic targeting of leukemia with mutation in genes encoding spliceosomal proteins. Nat Med 22(6):672–678. https://doi.org/10.1038/nm.4097

Li H, Rybicki AC, Suzuka SM et al (2010) Transferrin therapy ameliorates disease in beta-thalassemic mice. Nat Med 16(2):177–182. https://doi.org/10.1038/nm.2073

Li H, Choesang T, Bao W et al (2017) Decreasing TfR1 expression reverses anemia and hepcidin suppression in β-thalassemic mice. Blood 129(11):1514–1526. https://doi.org/10.1182/blood-2016-09-742387

Libani IV, Guy EC, Melchiori L et al (2008) Decreased differentiation of erythroid cells exacerbates ineffective erythropoiesis in β-thalassemia. Blood 112(3):875–885. https://doi.org/10.1182/blood-2007-12/126938

Liem RI, Gilgour B, Pelligra SA et al (2011) The impact of thalassemia on southeast Asian and Asian Indian families in the United States: a qualitative study. Ethn Dis 21(3):361–369

Lin S, Zheng Y, Chen M et al (2024) The interactions between ineffective erythropoiesis and ferroptosis in β-thalassemia. Front Physiol 15:1346173. https://doi.org/10.3389/fphys.2024.1346173

Loke JCT, Kim GG, Boileau M et al (2024) In vivo barcoded CRISPR-Cas9 screen identifies Ncoa4-mediated ferritinophagy as a dependence in Tet2-deficient haemopoiesis. Blood 144(Supplement 1):188. https://doi.org/10.1182/blood-2024-200955

Lv Q, Niu H, Yue L et al (2020) Abnormal ferroptosis in myelodysplastic syndrome. Front Oncol 10:1656. https://doi.org/10.3389/fonc.2020.01656

Malcovati L, Della Porta MG, Pascutto C et al (2005) Prognostic factors and life expectancy in myelodysplastic syndromes classified according to WHO criteria: a basis for clinical decision making. J Clin Oncol 23(30):7594–7603. https://doi.org/10.1200/JCO.2005.01.7038

Malcovati L, Della Porta MG, Cazzola M (2006) Predicting survival and leukemic evolution in patients with myelodysplastic syndrome. Haematologica 91(12):1588–1590

Malcovati L, Della Porta MG, Strupp C et al (2011) Impact of the degree of anemia on the outcome of patients with myelodysplastic syndrome and its integration into the WHO classification-based Prognostic Scoring System (WPSS). Haematologica 96(11):1433–1440. https://doi.org/10.3324/haematol.2011.044602

Maratheftis CI, Bolaraki PE, Voulgarelis M (2007) GATA-1 transcription factor is upregulated in bone marrow hematopoietic progenitor CD34(+) and erythroid CD71(+) cells in myelodysplastic syndromes. Am J Hematol 82(10):887–892. https://doi.org/10.1002/ajh.20993

Martinez PA, Li R, Ramanathan HN et al (2020) Smad2/3-pathway ligand trap luspatercept enhances erythroid differentiation in murine β-thalassaemia by increasing GATA-1 availability. J Cell Mol Med 24(11):6162–6177. https://doi.org/10.1111/jcmm.15243

McQuilten ZK, Higgins AM, Burns K et al (2019) The cost of blood: a study of the total cost of red blood cell transfusion in patients with β-thalassemia using time-driven activity-based costing. Transfusion 59(11):3386–3395. https://doi.org/10.1111/trf.15558

Medyouf H, Mossner M, Jann J et al (2014) Myelodysplastic cells in patients reprogram mesenchymal stromal cells to establish a transplantable stem cell niche disease unit. Cell Stem Cell 14(6):824–837. https://doi.org/10.1016/j.stem.2014.02.014

Musallam KM, Cappellini MD, Viprakasit V et al (2021) Revisiting the non-transfusion-dependent (NTDT) vs. transfusion-dependent (TDT) thalassemia classification 10 years later. Am J Hematol 96(2):E54–E56. https://doi.org/10.1002/ajh.26056

Nearman ZP, Szpurka H, Serio B et al (2007) Hemochromatosis-associated gene mutation in patients with myelodysplastic syndromes with refractory anemia with ringed sideroblasts. Am J Hematol 82(12):1076–1079. https://doi.org/10.1002/ajh.20995

Oliva EN, Ronco F, Marino A et al (2010) Iron chelation therapy associated with improvement of hematopoiesis in transfusion-dependent patients. Transfusion 50(7):1568–1570. https://doi.org/10.1111/j.1537-2995.2010.02617.x

Origa R (2017) β-Thalassemia. Genet Med 19:609–619. https://doi.org/10.1038/gim.2016.173

Origa R, Barella S, Argiolas GM et al (2008) No evidence of cardiac iron in 20 never- or minimally-transfused patients with thalassemia intermedia. Haematologica 93(7):1095–1096. https://doi.org/10.3324/haematol.12484

Pan Z, Voehringer DW, Meyn RE (1999) Analysis of redox regulation of cytochrome c-induced apoptosis in a cell-free system. Cell Death Differ 6(7):683–688. https://doi.org/10.1038/sj.cdd.4400544

Parisi S, Finelli C, Fazio A et al (2021) Clinical and molecular insights in erythropoiesis regulation of signal transduction pathways in myelodysplastic syndromes and β-thalassemia. Int J Mol Sci 22(2):827. https://doi.org/10.3390/ijms22020827

Park S, Kelaidi C, Meunier M et al (2020) The prognostic value of serum erythropoietin in patients with lower-risk myelodysplastic syndromes: a review of the literature and expert opinion. Ann Hematol 99(1):7–19. https://doi.org/10.1007/s00277-019-033799-4

Piga A, Longo F, Duca L et al (2009) High nontransferrin bound iron levels and heart disease in thalassemia major. Am J Hematol 84(1):29–33. https://doi.org/10.1002/ajh.21317

Pilo F, Angelucci E (2024) Vamifeport: monography of the first oral ferroportin inhibitor. J Clin Med 13(18):5524. https://doi.org/10.3390/jcm13185524

Ramm GA, Ruddell RG (2005) Hepatotoxicity of iron overload: mechanisms of iron-induced hepatic fibrogenesis. Semin Liver Dis 25(4):433–449. https://doi.org/10.1055/s-2005-923315

Ribeil J, Zermati Y, Vandekerckhove J et al (2007) Hsp70 regulates erythropoiesis by preventing caspase-3-mediated cleavage of GATA1. Nature 445:102–105. https://doi.org/10.1038/nature05378

Saigo K, Takenokuchi M, Hiramatsu Y et al (2011) Oxidative stress levels in myelodysplastic syndrome patients: their relationship to serum ferritin and haemoglobin values. J Int Med Res 39(5):1941–1945. https://doi.org/10.1177/147323001103900539

Santini V (2015) Anemia as the main manifestation of myelodysplastic syndromes. Semin Hematol 52(4):348–356. https://doi.org/10.1053/j.seminhaematol.2015.06.002

Santini V, Girelli D, Sanna A et al (2011) Hepcidin levels and their determinants in different types of myelodysplastic syndromes. PLoS One 6(8):e23109. https://doi.org/10.1371/journal.pone.0023109

Sardo U, Perrier P, Cormier K et al (2024) The hepatokine FGL1 regulates hepcidin and iron metabolism during anemia in mice by antagonizing BMP signaling. Blood 143(13):1282–1292. https://doi.org/10.1182/blood.2023022724

Scalone L, Mantovani LG, Krol M et al (2008) Costs, quality of life, treatment satisfaction and compliance in patients with β-thalassemia major undergoing iron chelation therapy: the ITHACA study. Curr Med Res Opin 24(7):1905–1917. https://doi.org/10.1185/03007990802160834

Schmidt PJ, Racie T, Westerman M et al (2015) Combination therapy with a Tmprss6 RNAi-

therapeutic and the oral iron chelator deferiprone additively diminishes secondary iron overload in a mouse model of β-thalassemia intermedia. Am J Hematol 90(4):310–313. https://doi.org/10.1002/ajh.23934
Shah FT, Sayani F, Trompeter S et al (2019) Challenges of blood transfusions in β-thalassemia. Blood Rev 37:100588. https://doi.org/10.1016/j.blre.2019.100588
Shih H, Wu C, Lin S (2018) Physiology and pathophysiology of renal erythropoietin-producing cells. J Formos Med Assoc 117(11):955–963. https://doi.org/10.1016/j.jfma.2018.03.017
Shiozawa Y, Malcovati L, Gallì A et al (2018) Aberrant splicing and defective mRNA production induced by somatic spliceosome mutations in myelodysplasia. Nat Commun 9(1):3649. https://doi.org/10.1038/s41467-018-06063-x
Sohn YS, Breuer W, Munnich A et al (2008) Redistribution of accumulated cell iron: a modality of chelation with therapeutic implications. Blood 111(3):1690–1699. https://doi.org/10.1182/blood-2007-07-102335
Steensma DP (2018) Myelodysplastic syndromes current treatment algorithm 2018. Blood Cancer J 8:47. https://doi.org/10.1038/s41408-018-0085-4
Steensma DP, Wermke M, Klimet V et al (2021) Phase I first-inhuman dose escalation study of the oral SF3B1 modulator H3B-8800 in myeloid neoplasms. Leukemia 35(12):3542–3550. https://doi.org/10.1038/s41375-021-01328-9
Stefanova D, Raychev A, Arezes J et al (2017) Endogenous hepcidin and its agonist mediate resistance to selected infections by clearing non-transferrin-bound iron. Blood 130(3):245–257. https://doi.org/10.1182/blood-2017-03-772715
Stefanova D, Raychev A, Deville J et al (2018) Hepcidin protects against lethal Escherichia coli sepsis in mice inoculated with isolates from septic patients. Infect Immun 86(7):e00253–e00318. https://doi.org/10.1128/IAI.00253-18
Suragani RNVS, Cadena SM, Cawley SM et al (2014) Transforming growth factor-β superfamily ligand trap ACE-536 corrects anemia by promoting late-stage erythropoiesis. Nat Med 20(4):408–414. https://doi.org/10.1038/nm.3512
Taher AT, Weatherall DJ, Dappellini MD (2018) Thalassaemia. Lancet 391:155–167. https://doi.org/10.1016/S0140-6736(17)31822-6
Taher AT, Cappellini MD, Kattamis A et al (2022) Luspatercept for the treatment of anaemia in non-transfusion-dependent β-thalassaemia (BEYOND): a phase 2, randomised, double-blind, multicentre, placebo-controlled trial. Lancet Haematol 9(10):e733–e744. https://doi.org/10.1016/S2352-3026(22)00208-3
Taoka K, Kumano K, Nakamura F et al (2012) The effect of iron overload and chelation on erythroid differentiation. Int J Hematol 95(2):149–159. https://doi.org/10.1007/s12185-011-0988-3
Teodorescu P, Pasca S, Dima D et al (2020) Targeting the microenvironment in MDS: the final frontier. Front Pharmacol 11:1044. https://doi.org/10.3389/fphar.2020.01044
Valent P, Krieger O, Stauder R et al (2008) Iron overload in myelodysplastic syndromes (MDS)—diagnosis, management, and response criteria: a proposal of the Austrian MDS platform. Eur J Clin Investig 38(3):143–149. https://doi.org/10.1111/j.1365-2362.2007.01915.x
Vento S, Cainelli F, Cesario F (2006) Infections and thalassaemia. Lancet Infect Dis 6(4):226–233. https://doi.org/10.1016/S1473-3099(06)70437-6
Verma A, Suragani RN, Aluri S et al (2020) Biological basis for efficacy of activin receptor ligand traps in myelodysplastic syndromes. J Clin Invest 130(2):582–589. https://doi.org/10.1172/JCI133678
Visconte V, Avishai N, Mahfouz R et al (2015) Distinct iron architecture in SF3B1-mutant myelodysplastic syndrome patients is linked to an SLC25A37 splice variant with a retained iron intron. Leukemia 29(1):188–195. https://doi.org/10.1038/leu.2014.1070
Wang SC, Lin KH, Chern JP et al (2003) Severe bacterial infection in transfusion-dependent patients with thalassemia major. Clin Infect Dis 37(7):984–988. https://doi.org/10.1086/378062
Weidlich D, Kefalas, Guest JF (2016) Healthcare costs and outcomes of managing β-thalassemia. Transfusion 56(5):1038–1045. https://doi.org/10.1111/trf.13513
Yang WS, Kim KJ, Gaschler MM et al (2016) Peroxidation of polyunsaturated fatty acids by lipoxygenases drives ferroptosis. Proc Natl Acad Sci USA 113(34):E4966–E4975. https://doi.org/10.1073/pnas.1603244113
Zivot A, Lipton JM, Narla A et al (2018) Erythropoiesis: insights into pathophysiology and treatments in 2017. Mol Med 24(1):11. https://doi.org/10.1186/s10020-018-0011-z

12 Iron Deficiency Anemia

Xiuqi Li and Karin E. Finberg

Abstract

Iron deficiency anemia (IDA) is estimated to affect over 1.2 billion people worldwide and is particularly common in children and reproductive-age women in low- and middle-income countries. Iron deficiency (ID) without anemia has an even higher prevalence. Absolute ID is caused by increased iron requirements, inadequate iron intake, impaired iron absorption, and chronic blood loss. ID triggers multiple adaptive physiological mechanisms, including modifications in cellular iron handling orchestrated by iron regulatory proteins, as well as alterations in systemic iron handling secondary to suppression of the iron regulatory hormone hepcidin. The laboratory diagnosis of ID is based on biochemical parameters and may be complicated during inflammatory conditions. ID is treated with oral or parenteral iron supplements, depending upon the severity of ID and its underlying cause(s). An improved understanding of systemic iron regulation has begun to inform oral iron dosing regimens and the selection of oral versus intravenous iron formulations for the management of ID.

Keywords

Iron deficiency anemia · Iron deficiency · Iron metabolism · Micronutrient · Dietary iron · Iron regulatory proteins · Hepcidin · Iron-refractory iron deficiency anemia · Iron supplementation

12.1 Introduction

Iron is an essential micronutrient that participates in a variety of cellular processes (Teh et al. 2024). Iron deficiency (ID) may be either absolute or relative. Absolute ID is characterized by the depletion of total body iron, particularly iron in macrophage and hepatocyte stores (Fig. 12.1) (Camaschella 2019). Because the majority of the body's daily iron requirement supports hemoglobin synthesis during erythropoiesis, iron deficiency anemia (IDA) is the most widely appreciated manifestation of ID. However, because iron plays key roles in physiological processes beyond erythropoiesis, such as myoglobin synthesis and mitochondrial function in skeletal and cardiac muscle, ID in the absence of anemia is clinically significant.

X. Li
Department of Cell Biology, Harvard Medical School, Boston, MA, USA

K. E. Finberg (✉)
Department of Pathology, Yale School of Medicine, New Haven, CT, USA
e-mail: karin.finberg@yale.edu

K. Pantopoulos (ed.), *Iron Metabolism in Human Health and Disease*, Advances in Experimental Medicine and Biology 1480, https://doi.org/10.1007/978-3-031-92033-2_12

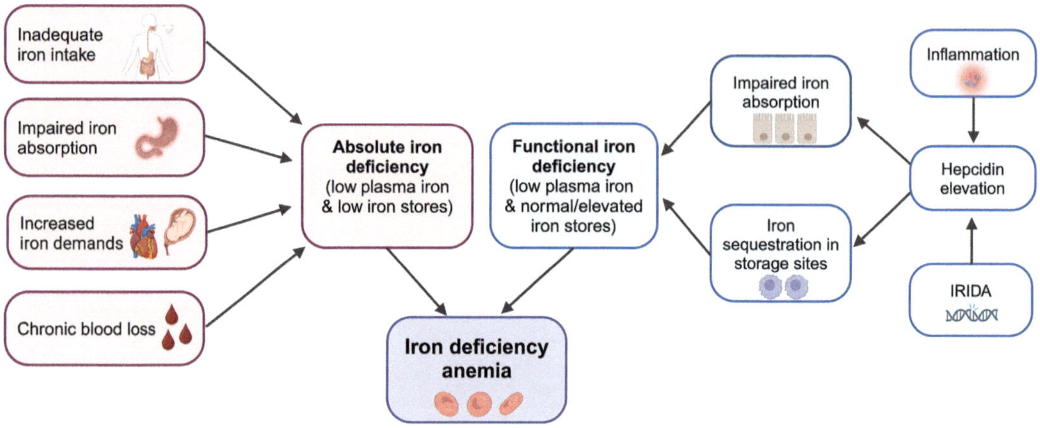

Fig. 12.1 Comparison of absolute iron deficiency versus functional iron deficiency. Absolute ID is characterized by the depletion of total body iron, especially iron that is stored in macrophages and hepatocytes, and may be caused by reduced iron intake, impaired intestinal iron absorption, increased systemic iron demands, and chronic blood loss. By contrast, functional ID occurs in clinical states where iron availability for erythropoiesis is restricted even though body iron stores are normal or elevated. Functional ID occurs in inflammatory states where cytokines induce the production of the key iron regulatory hormone hepcidin. In the congenital disorder Iron-Refractory Iron Deficiency Anemia (IRIDA), the underlying genetic defect results in inappropriately elevated levels of hepcidin, which impair both dietary iron absorption and utilization of iron stores

By contrast, functional ID occurs when iron availability for erythropoiesis is restricted, despite body iron stores being preserved or even elevated. Functional ID occurs in patients with inflammation, malignancy, and infection, where cytokines induce the production of the key iron regulatory hormone hepcidin (Fig. 12.1); see also Chap. 13. In practice, absolute and functional ID may coexist in a variety of clinical conditions.

12.2 Epidemiology of Iron Deficiency Anemia

According to recent estimates, anemia affects 24% of the world's population (1.92 billion people), with 66% of these cases attributed to ID (GBD 2021 Anaemia Collaborators 2023). Dietary ID is the leading cause of anemia across all ages in both sexes worldwide and the eighth leading cause of years lived with disability globally (GBD 2021 Diseases and Injuries Collaborators 2024). The global prevalence of ID without anemia is unknown but has been suggested to be at least twice that of IDA (Camaschella 2019). The prevalence of ID is highest in preschool children and in women of reproductive age in low- and middle-income countries. Indeed, the World Health Organization estimates that globally approximately 42% of anemia cases in children and 50% of anemia cases in women are amenable to iron supplementation (World Health Organization 2015). ID also remains a health concern in high-income countries. For example, in the United States, the prevalence of ID and IDA among children 1–5 years old is 7% and 1%, respectively (Gupta et al. 2016), and the prevalence of ID and IDA among pregnant females ages 12–21 years old is 38% and 6%, respectively (Weyand et al. 2023). Absolute ID is estimated to affect 14% of adults overall in the United States (Tawfik et al. 2024).

12.3 Pathophysiology of Iron Deficiency

ID induces adaptive mechanisms that operate at the cellular and systemic levels. In most cell types, ID promotes binding of the iron regulatory proteins IRP1 and IRP2 to iron regulatory elements (IREs) in untranslated regions (UTR) of

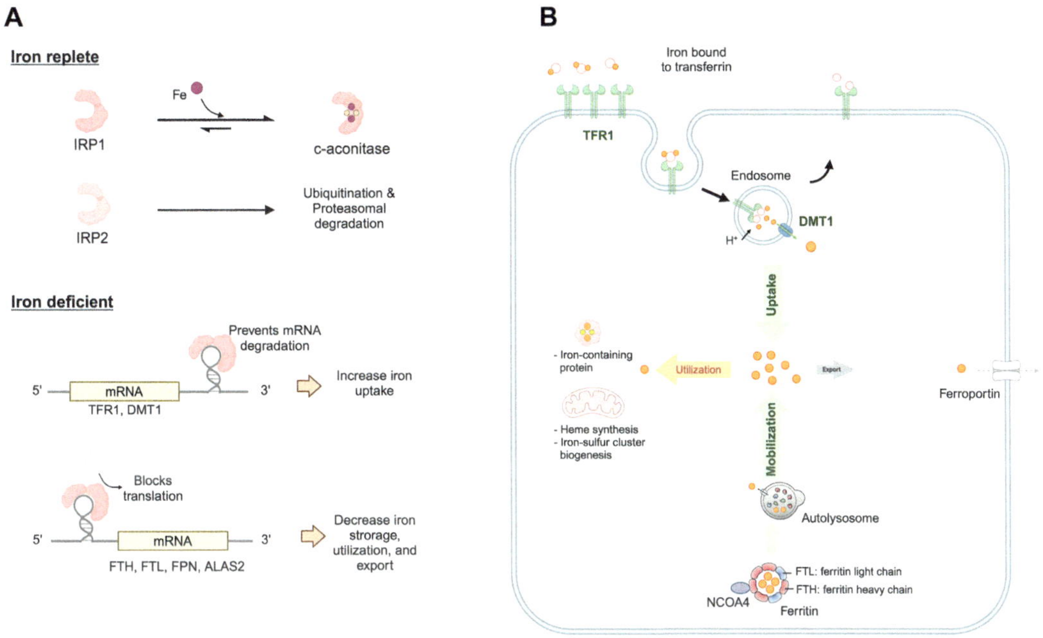

Fig. 12.2 Cellular adaptive responses to iron deficiency. (**a**) Iron regulatory protein function in iron-replete versus iron-deficient states. In the iron-replete state, IRP1 functions as a cytoplasmic aconitase (c-aconitase), and IRP2 undergoes ubiquitination and degradation. By contrast, under conditions of ID, IRP1 and IRP2 bind to iron regulatory elements in UTRs of target mRNAs. Messenger RNAs encoding TFR1 and DMT1 are stabilized by binding of IRPs to their 3′ UTRs, increasing translation of the respective proteins. Concurrently, the translation of mRNAs encoding the iron exporter ferroportin (FPN) and the L and H subunits of ferritin (FTL, FTH) is repressed by IRP binding to 5′ UTR IREs. In erythroid precursor cells, IRPs also regulate the translation of ALAS2, thereby coupling heme synthesis to iron availability. (**b**) Cellular adaptations in ID attempt to preserve iron availability for key cellular functions. During ID, increased expression of TFR1 and DMT1, mediated by IRPs, facilitates increased cellular uptake of transferrin-bound iron through receptor-mediated endocytosis. Decreased expression of the ferroportin and ferritin subunit proteins, also mediated by IRPs, reduces cellular capacity for iron export and iron storage. Additionally, iron is mobilized from ferritin stores through an autophagic process in which the cargo receptor NCOA4 delivers ferritin to the lysosome. Collectively, these mechanisms aim to maintain the supply of iron for key cellular functions

target messenger RNAs (mRNAs) to increase iron uptake and decrease iron storage and export (Fig. 12.2a) (Maio et al. 2021); see also Chap. 2. Binding to 3′ UTR IREs stabilizes mRNAs encoding transferrin receptor 1 (TFR1) and divalent metal transporter 1 (DMT1), while binding to 5′ UTR IREs represses translation of the mRNAs encoding the iron exporter ferroportin and the L and H subunits of ferritin. During ID, ferritin is also degraded by autophagy in a process in which the cargo receptor NCOA4 delivers ferritin to the lysosome for degradation and recovery of iron to meet cellular iron demands (Fig. 12.2b) (Santana-Codina and Mancias 2018).

ID also modulates systemic iron regulation through the hepcidin/ferroportin axis (Fig. 12.3); see also Chap. 3. During systemic ID, production of the key ligand bone morphogenetic protein 6 (BMP6) by liver sinusoidal endothelial cells is reduced, which in turn lowers BMP/SMAD protein signaling for hepcidin production in hepatocytes (Fisher and Babitt 2022). During IDA, hepcidin is also suppressed because erythropoietin (EPO) stimulates erythroblasts to release erythroferrone, a ligand trap for BMP6 (Arezes et al. 2018). Levels of the TMPRSS6 protein, an inhibitor of BMP/SMAD signaling, also increase during ID (Zhang et al. 2011). Because hepcidin

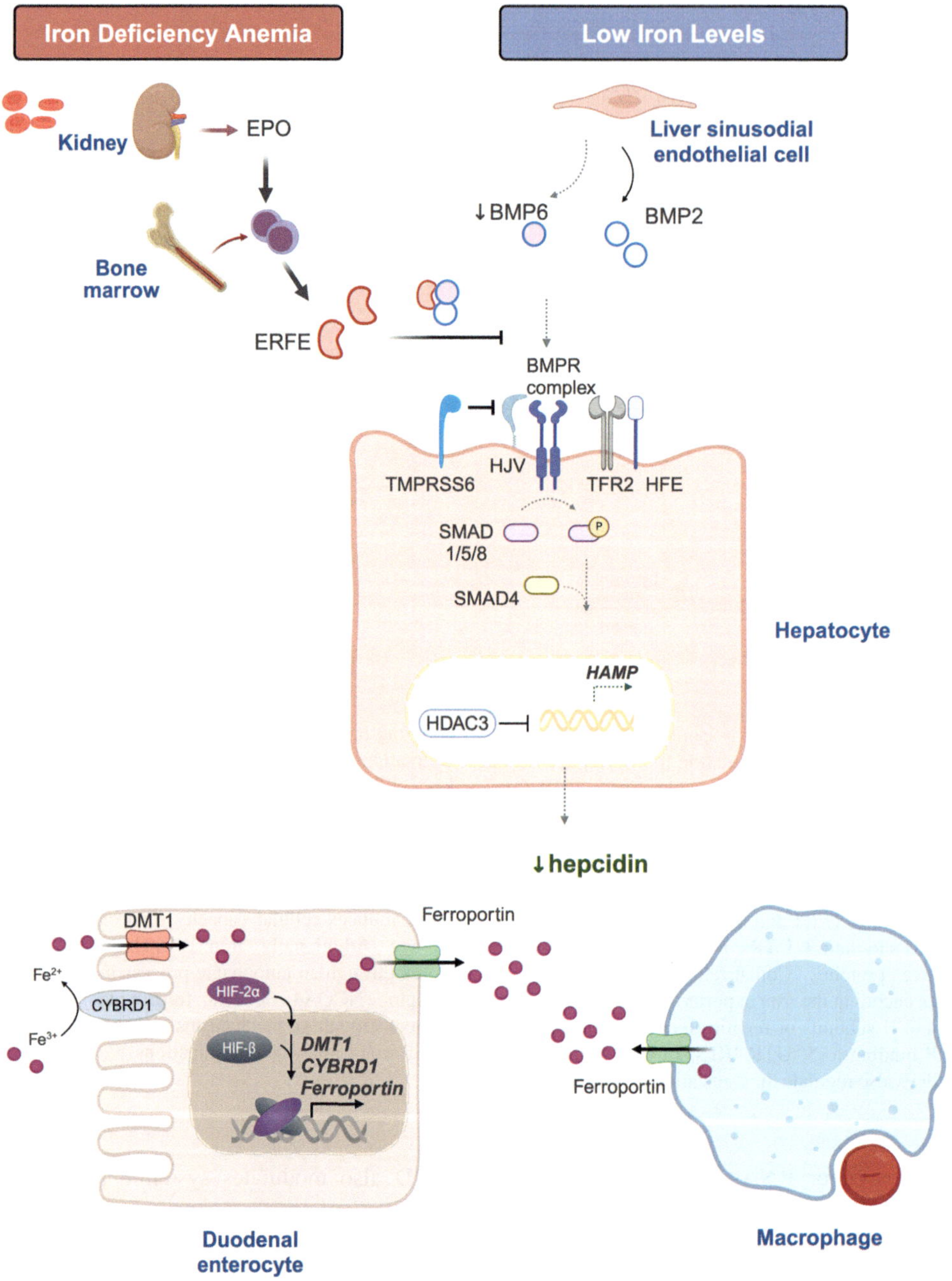

Fig. 12.3 Iron deficiency and iron deficiency anemia modulate the hepcidin/ferroportin axis. During ID and IDA, hepcidin production by hepatocytes is reduced through several mechanisms. During systemic ID, sinusoidal endothelial cells in the liver lower their production of the BMP6 ligand (top right); this in turn dampens BMP/SMAD protein signaling in hepatocytes (middle), thereby reducing transcription of the hepcidin gene (*HAMP*). Protein levels of TMPRSS6, a negative regulator of BMP/SMAD signaling in hepatocytes, also increase in response to ID. During IDA, increased renal production of EPO promotes erythroblast expression of ERFE (top left); by sequestering BMPs, ERFE functions as a negative regulator of hepcidin production. Hepcidin expression is also downregulated epigenetically by HDAC3 in ID. Because hepcidin downregulates the activity of the cellular iron exporter ferroportin, which is expressed on the basolateral membrane of duodenal enterocytes (bottom left) and the plasma membrane of macrophages (bottom right), reducing hepcidin production by hepatocytes is an adaptive response that attempts to preserve plasma iron levels by promoting both absorption of dietary iron and release of iron from reticuloendothelial macrophage stores. During IDA, duodenal iron uptake is also enhanced by local stabilization of HIF-2α, which upregulates transcription of the genes encoding ferroportin as well as the apical iron transport machinery (DMT1, CYBRD1)

downregulates the activity of the cellular iron exporter ferroportin by inducing its degradation and also by occluding ferroportin, hepcidin suppression enhances iron release from reticuloendothelial macrophage stores and iron uptake from duodenal enterocytes (Nemeth and Ganz 2023). Hepcidin suppression during ID also involves epigenetic regulation mediated by histone deacetylase 3 (HDAC3) (Pasricha et al. 2017). During IDA, duodenal iron uptake is enhanced by local stabilization of hypoxia-inducible factor 2α (HIF-2α), which upregulates transcription of the genes encoding iron transport machinery at the enterocyte basolateral (ferroportin) and apical (DMT1 and the ferric reductase cytochrome b reductase 1 (CYBRD1)) plasma membranes (Solanki and Shah 2024).

ID induces multiple compensatory mechanisms to coordinate heme and globin synthesis with the reduced iron availability. Translation of aminolevulinic acid synthetase 2 (ALAS2), the first and rate-limiting enzyme of heme synthesis in erythroid cells, is blocked by IRP1 binding to the 5′ UTR IRE in the *ALAS2* mRNA (Fig. 12.2a) (Maio et al. 2021). Heme deficiency, in turn, prevents eukaryotic translation initiation factor 2 alpha kinase 1 (also known as heme-regulated inhibitor) from activating the synthesis of α- and β-globin chains (Chen and Zhang 2022). Because the HIF-2α mRNA contains a 5′ UTR IRE, ID attenuates renal EPO production, which may preserve iron availability for other tissues (Maio et al. 2021). ID also downregulates plasma membrane levels of transferrin receptor 2 (TFR2), a homolog of TFR1 that binds iron-bound transferrin but does not play a primary role in cellular iron uptake. TFR2 functions as an EPO receptor (EPOR) binding partner in erythroid cells, and through mechanisms that are not yet fully understood, TFR2 downregulation enhances EPO-EPOR signaling, permitting erythropoiesis to proceed in ID despite inappropriately low EPO levels (Silvestri and Nai 2021). In the low hepcidin state of ID, mature red blood cells can export iron via ferroportin, buffering plasma iron levels (Zhang et al. 2018). ID also acts earlier in hematopoiesis to bias the commitment of megakaryocytic-erythroid progenitor cells towards the megakaryocyte lineage (Xavier-Ferrucio et al. 2019), which may conserve iron for other bodily functions.

12.4 Etiologies of Iron Deficiency

Causes of absolute ID may be grouped into the following major categories based on the general mechanism leading to negative iron balance: (1) increased iron requirements, (2) inadequate iron intake, (3) impaired iron absorption, and (4) chronic blood loss (Fig. 12.1). The likelihood that a particular mechanism is responsible for ID is influenced by age, sex, and geographical region, and in some cases, multiple etiologies may co-exist in an individual patient (Cappellini et al. 2020).

Increased Iron Requirements Iron needs increase during normal stages of the lifespan and in response to physiological stressors. Accordingly, national recommendations for daily iron intake vary by age, sex, and during pregnancy. Iron demands for body growth and the accompanying expansion of the erythroid mass increase during childhood and adolescence, particularly during growth spurts (Institute of Medicine 2001). In females of reproductive age, iron demands increase to counteract menstrual losses and in pregnancy to support the growing fetus and the increasing maternal red cell mass, primarily in the second and third trimesters (Sangkhae et al. 2023). Given the physiological demand for iron to support oxygen delivery and mitochondrial function, ID is common in athletes (Peeling and McKay 2023); additionally, dietary iron absorption may be negatively modulated in athletes due to exercise-induced hepcidin elevation (Larsuphrom and Latunde-Dada 2021). Individuals undergoing treatment with erythropoiesis-stimulating agents require additional iron to support the acute expansion of their erythroid mass (Del Vecchio and Minutolo 2021).

Inadequate Iron Intake Iron intake may be insufficient because the absolute iron content of the diet is low or because the consumed iron has low bioavailability. In low- and middle-income countries, malnutrition contributes to a deficiency of iron and other micronutrients (Brittenham et al. 2023). Additionally, the plant-based diets in these geographic regions provide predominantly non-heme iron, which is absorbed less efficiently than heme iron found in animal products (Chap. 6). The bioavailability of ingested iron may be reduced due to chelation by phytates (non-digestible organophosphorus compounds found in plant seeds, nuts, legumes, and grains) and by polyphenols (a large family of phenol compounds found in vegetables, cereals, spices, red wine, tea, coffee, and cocoa). Calcium decreases the absorption of both non-heme and heme iron through mechanisms that remain to be clarified (Piskin et al. 2022). In high-income countries, those consuming vegan or vegetarian diets are at increased risk of ID (Neufingerl and Eilander 2023).

Impaired Iron Absorption Dietary non-heme iron must be reduced from the ferric (Fe^{3+}) to the ferrous (Fe^{2+}) state to enable absorption by duodenal enterocytes. Accordingly, factors that decrease the production of gastric acid or modify the absorptive epithelium can promote ID. Gastric acid production may be reduced by medications, such as proton pump inhibitors and histamine receptor blockers (Vinnakota and Brett 2019), and in atrophic gastritis, where chronic inflammation due to *H. pylori* infection or autoimmunity causes loss of gastric glands (Shah et al. 2021). Absorption is also reduced in conditions that cause duodenal villous atrophy, and IDA is the most common extra-intestinal manifestation of gluten-sensitive enteropathy (celiac disease) (Talarico et al. 2021). Furthermore, the capacity for iron absorption is reduced after gastrointestinal surgeries, such as gastrectomy, duodenal bypass, and bariatric procedures (Kaberi-Otarod et al. 2024; Teixeira Farinha et al. 2023). Extremely rarely, impaired iron absorption may be secondary to a constitutional genetic defect that results in elevated circulating hepcidin levels (discussed below).

Chronic Blood Loss Conditions associated with chronic blood loss, which may be overt or occult, increase the risk of ID. A variety of benign or malignant anatomic lesions of the gastrointestinal or genitourinary tracts can lead to chronic blood loss (Willis and Tewelde 2019; Stein et al. 2016). Heavy menstrual bleeding is a major contributor to ID in women of childbearing age (Munro et al. 2023). Inherited bleeding disorders increase the risk of ID and predispose to heavy menstrual bleeding (VanderMeulen and Sholzberg 2018). Bleeding from the gastrointestinal tract may also be due to *H. pylori* infection or from the use of medications that either cause mucosal injury (e.g., non-steroidal anti-inflammatory drugs, salicylates, corticosteroids) (Tai and McAlindon 2021) or that promote systemic bleeding (e.g., anticoagulants, antiplatelet drugs) (Samuelson Bannow 2020). In low- and middle-income countries, intestinal parasitic infections (e.g., roundworms, hookworms, whipworms) may contribute to blood loss (Ellwanger et al. 2022). Blood donation, which removes approximately 250 mg of iron from the body, increases the risk of ID in healthy recurrent donors (Spencer and Mast 2022). In exceptional cases, ID may result from self-inflicted bloodletting (Karamanou and Androutsos 2010).

12.5 Genetic Contributions to Iron Deficiency

Extremely rarely, IDA may be congenital due to a primary inherited defect of iron metabolism. Iron-Refractory Iron Deficiency Anemia (IRIDA) is a rare, autosomal recessive disorder caused by loss-of-function variants in *TMPRSS6* (*transmembrane serine protease 6*) (Heeney and Finberg 2014). Patients with IRIDA exhibit hepcidin levels that are inappropriately elevated relative to their body iron stores (Fig. 12.1), leading to hypochromic, microcytic anemia that is refractory to oral iron and only partially responsive to parenteral iron. *TMPRSS6* encodes a hepatic protease that interacts with the BMP co-receptor hemojuvelin, downregulates SMAD1/5/8 signaling, and suppresses hepcidin transcription (Fig. 12.3); however, the exact mechanism by which TMPRSS6 regulates hepcidin expression is not yet fully understood (Silvestri et al. 2008; Enns et al. 2020; Finberg et al. 2010). The IRIDA phenotype has also been observed in a single individual with compound heterozygous variants in *TMPRSS6* and *ACVR1* (*activin A receptor type 1*; the gene encoding the BMP type I receptor also known as ALK2) (Pagani et al. 2017).

While IRIDA-causing *TMPRSS6* variants are rare, common genetic variation in *TMPRSS6* impacts iron status in the general population. Multiple genome-wide association studies (GWAS), which have been conducted mainly in European and Asian populations, have correlated common single nucleotide polymorphisms (SNPs) in *TMPRSS6* with mild changes in hemoglobin, erythrocyte mean corpuscular volume, erythrocyte mean corpuscular hemoglobin, serum iron, and transferrin saturation levels (Soranzo et al. 2009; Chambers et al. 2009; Benyamin et al. 2009; Ganesh et al. 2009). One of the most strongly associated TMPRSS6 SNPs in these GWAS, rs855791, encodes a valine (V) to alanine (A) substitution at amino acid 736 in the catalytic domain that has been found to modulate hepcidin transcription in vitro. Accordingly, VV homozygotes showed higher serum hepcidin and lower serum iron levels than AA homozygotes in an Italian population study (Nai et al. 2011). The rs855791 SNP has also been correlated with hematological and iron responses after phlebotomy in healthy blood donors. In a study of first-time blood donors in the United States, females of VV genotype showed a rapid drop in hemoglobin and ferritin when compared to those of VA or AA genotype (Mast et al. 2020). The V allele has also been associated with lower iron stores in male Danish blood donors (Sorensen et al. 2016). In addition to *TMPRSS6*, several novel loci (*DUOX2*, *F5*, *SLC11A2*) have been associated with biomarkers of ID in a recent meta-analysis of multiple GWAS (Bell et al. 2021).

12.6 Clinical Features and Consequences of Iron Deficiency

Clinical symptoms of ID are non-specific and may be attributed to reduced oxygen delivery to tissues and reduced activity of iron-containing enzymes (Fig. 12.4). Individuals with ID may report lethargy, fatigue, dizziness, shortness of breath, chest pain, rapid heartbeat, impaired concentration, tinnitus, and headache (Pasricha et al. 2021). Pica, the compulsive ingestion of non-nutritive substances such as ice (pagophagia), soil/clay (geophagia), and uncooked rice (rhizophagia), is strongly associated with ID (Ganesan and Vasauskas 2023). Those with chronic ID may exhibit a range of findings on physical exam, including pallor, dry skin, alopecia, brittle nails, atrophic glossitis (loss of tongue papillae), angular cheilitis (inflammation of corners of the mouth), and koilonychia (spoon-shaped nails) (Wong and Chu 2021). ID also promotes restless leg syndrome (Allen et al. 2013). If accompanied by severe anemia, ID may cause hemodynamic instability.

Chronic ID is associated with a variety of adverse outcomes. IDA during pregnancy increases the risk of low birth weight, preterm birth, postpartum hemorrhage, stillbirth, and neonatal death (Obianeli et al. 2024). Because iron is required for multiple critical processes in the developing brain, including myelination, ATP production, monoamine neurotransmitter synthesis, and epigenetic gene regulation, inadequate fetal iron loading during gestation is associated with long-term negative effects on neurodevelopment. (Georgieff 2023). ID has been shown to impair adaptive immunity in preclinical studies and is associated with reduced efficacy of childhood vaccines (Drakesmith et al. 2021). ID also reduces exercise capacity and athletic performance, which is due not only to a reduction in hemoglobin but also to changes in cellular metabolism in myocytes (Kardasis et al. 2023).

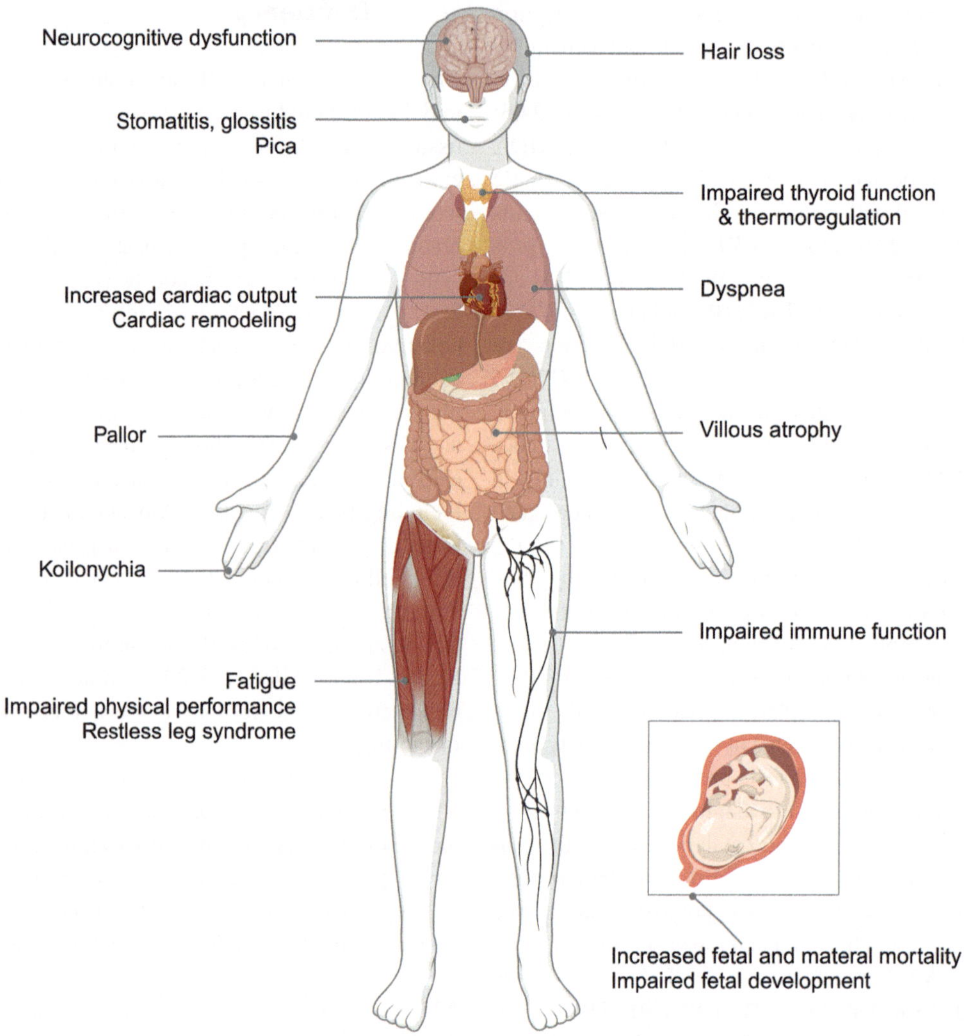

Fig. 12.4 Clinical signs and symptoms of iron deficiency. Iron deficiency may result in a diverse spectrum of clinical signs and symptoms involving a range of organ systems. Some of these signs and symptoms (e.g., pallor, dyspnea, fatigue) are shared with other forms of anemia, and in IDA may be attributed to low blood hemoglobin levels and the resulting reduction in oxygen delivery to tissues. Additional signs and symptoms reflect impairment of other biological processes that rely on other key iron-containing proteins, such as DNA/RNA synthesis, and mitochondrial, neuronal, and immune cell function

12.7 Laboratory and Clinical Investigations in Iron Deficiency

Laboratory tests for ID assess the three main iron compartments in the body: iron stores, transport iron, and functional iron (Pfeiffer and Looker 2017). The depletion of each compartment results in a different stage of ID. Short-term variation in physiological iron requirements is met through mobilization of iron from hepatocyte and reticuloendothelial macrophage stores. The first stage of ID is characterized by a reduction in body iron stores. The second stage is reached when circulating iron levels can no longer be maintained due to depletion of body iron stores. The third stage occurs when the iron supply to erythroblasts is inadequate to maintain blood hemoglobin concentrations. Microscopic examination of bone marrow aspirates stained with Perls Prussian blue is considered the gold standard method for assessing iron stores. However, bone marrow biopsy is rarely needed for the diagnosis of uncomplicated cases of ID and is avoided due to the invasive nature of the procedure. Clinical assessment for ID usually employs biomarkers in the blood (Fletcher et al. 2022); see also Chap. 25.

Storage iron is typically assessed by measurement of serum (or plasma) levels of ferritin, an intracellular iron storage protein consisting of H and L subunits is secreted from macrophages (Truman-Rosentsvit et al. 2018). Circulating ferritin levels, which consist mainly of the L subunit, are measured by immunological methods. While serum ferritin reflects reticuloendothelial iron stores in otherwise apparently healthy individuals, the interpretation of serum ferritin levels is complicated by the fact that ferritin is an acute phase reactant that is nonspecifically elevated in conditions such as inflammation, infection, and malignancy. The World Health Organization recommends the use of different serum ferritin cutoff values for defining ID in apparently healthy individuals (<12 μg/L for infants and children under 5 years; <15 μg/L for age ≥ 5 years) versus individuals with inflammation or infection (<30 μg/L for infants and children under 5 years; <70 μg/L for age ≥ 5 years) (World Health Organization 2020). However, the use of higher cutoffs may increase diagnostic sensitivity, and there is ongoing debate about the cutoff to use for clinical decision-making (Truong et al. 2024; Garcia-Casal et al. 2021).

Serum iron is a measure of ferric iron bound to the carrier protein transferrin and does not assess the iron component of hemoglobin in circulating red blood cells. Acidification of serum allows ferric iron to be released from transferrin, reduced to ferrous iron, and detected by a chromogenic iron-chelating agent. Transferrin saturation (TSAT), which represents the percentage of binding sites on all transferrin molecules that are occupied by iron, provides information about the adequacy of iron supply. TSAT is calculated as the ratio of serum iron to transferrin. Transferrin synthesis by the liver is upregulated during ID, while transferrin is also a negative acute phase reactant that decreases during inflammation. Transferrin levels may be measured directly by immunological methods or indirectly as the total iron binding capacity (TIBC). The TIBC is defined as the total amount of iron required to fully saturate all iron binding sites on transferrin. TIBC can be measured directly by adding sufficient ferric iron to fully saturate all unbound sites on transferrin, removing excess unbound iron, and measuring iron concentration in the remaining sample. Alternatively, TIBC can be calculated indirectly by summing the serum iron concentration and the unsaturated iron binding capacity (UIBC); the latter is determined by adding a predetermined amount of ferric iron to serum at near neutral pH, quantifying the excess unbound iron with a chromogen, and calculating the difference between the added and excess iron (Elsayed et al. 2016).

A truncated monomer of TFR1, known as soluble transferrin receptor (sTFR) is shed from erythroblasts into the circulation. Because iron regulatory proteins increase TFR1 expression in erythroblasts during ID, ID is associated with elevations of sTFR (Harms and Kaiser 2015). However, sTFR elevation is not specific for ID because sTFR levels are also increased in states of erythroid hyperplasia, including thalassemia.

Applications of sTFR-derived indices, such as the sTFR/log ferritin index, may have utility in identifying ID in individuals with anemia complicated by coexisting inflammatory disease (Shin et al. 2015). The utility of sTFR as a marker of ID is presently hindered by a lack of standardization of commercial assays (Lyle et al. 2023).

For diagnosis of IDA, a complete blood count will confirm anemia if blood hemoglobin concentration is below the reference range. Classically, ID is associated with hypochromic, microcytic red cells on a peripheral blood smear, reflecting impaired hemoglobinization; more pronounced morphological changes (e.g., elliptocytes and pencil cells) arise when IDA is severe. Red cell indices may show low mean corpuscular volume, mean corpuscular hemoglobin, and mean corpuscular hemoglobin concentration; however, these reductions are not specific to IDA and may occur in other conditions, most notably thalassemia (Fletcher et al. 2022).

Modern automated hematology analyzers provide additional parameters relevant to the assessment of ID. Because of the lifespan of mature erythrocytes, the percentage of hypochromic red blood cells may inform iron availability for erythropoiesis during the preceding 2–3 months (Amir et al. 2019). In contrast, because reticulocytes circulate for only a short period before becoming mature erythrocytes, measurement of the hemoglobin content of reticulocytes reflects iron availability for erythropoiesis in the past 3–4 days and can suggest ID before other distinguishing changes in red cell indices (Ogawa et al. 2020). Due to different technologies employed on different hematology analyzers, this measurement may be reported as CHr (reticulocyte hemoglobin content) and RET-He (reticulocyte hemoglobin equivalent). In individuals without thalassemia, a higher percentage of hypochromic red blood cells or a low mean reticulocyte hemoglobin content suggests inadequate availability of iron for erythropoiesis.

Because hepcidin is an acute phase reactant, hepcidin measurements may be clinically useful in the differentiation of absolute IDA from the anemia of inflammation (Ganz et al. 2008). Hepcidin measurements may also have utility in distinguishing IRIDA associated with *TMPRSS6* mutations (Heeney et al. 2018). Hepcidin can be measured in serum or urine by immunoassays and mass spectrometry assays that are currently available in the research setting; however, interpretation is currently complicated by the lack of international standardization (Diepeveen et al. 2019).

12.8 Treatment of Iron Deficiency and Iron Deficiency Anemia

In ID, the treatment goal is to replenish body iron stores and, if anemia is present, to restore normal hemoglobin levels through the administration of oral or parenteral iron supplements (Chap. 24). The use of red blood cell transfusions should be restricted to individuals with IDA and cardiovascular compromise (Ning and Zeller 2019).

Oral iron therapy is the mainstay of treatment in cases of ID uncomplicated by impaired absorption (Ning and Zeller 2019). Ferrous iron salts, such as ferrous sulfate, ferrous gluconate, and ferrous fumarate, are standard first-line therapies in the outpatient setting due to their low cost and excellent safety profiles. Consumption of oral iron with either vitamin C or orange juice can enhance iron bioavailability because ascorbic acid limits the oxidation of ferrous iron to ferric iron in the stomach, reduces ferric to ferrous iron to allow uptake by DMT1, and forms chelates with ferrous or ferric iron at acidic pH that remain soluble in the alkaline duodenal lumen (von Siebenthal et al. 2023). Historically, oral iron therapy has often been administered as elemental iron doses up to 100–200 mg per day given as several divided doses (Pasricha et al. 2021). However, recent studies that quantified fractional iron absorption and considered the hepcidin-inducing effects of oral iron dosing have found that morning doses of 60–120 mg iron as a ferrous salt given with ascorbic acid on alternate days may be an optimal oral dosing regimen, at least in young women with ID and mild IDA (Stoffel et al. 2020). Because only a fraction of the oral iron dose is absorbed by enterocytes, excess iron remains in the digestive tract, which can trigger gastrointestinal side effects (e.g., constipation, diarrhea, nausea, vomiting, abdominal pain) that may limit tolerability and alter the diversity and composition of the microbiome (Seyoum et al. 2021). Oral iron supplements, especially in liquid forms, may also cause discoloration of the teeth and a metallic taste in the mouth.

Alternative oral iron formulations have been developed, some of which are thought to be absorbed from the intestinal lumen by pathways distinct from ferrous iron salts (Ebea-Ugwuanyi et al. 2024). These include iron-polysaccharide complexes, iron complexed with milk proteins (iron protein succinylate), ferric citrate, heme-iron polypeptides, lipophilic iron-chelators (e.g., ferric maltol), and liposomal/nanoparticle encapsulated iron (Pantopoulos 2024). Ferric polysaccharide complexes are claimed to have fewer side effects than ferrous salts but may be less effective because iron liberation from the carbohydrate shell may not occur until distal segments of the small intestine. Ferric citrate, a phosphate binder used to control hyperphosphatemia in patients with dialysis-dependent chronic kidney disease, may also be used to treat IDA in patients with chronic kidney disease who are not dialysis-dependent. Heme-iron polypeptides are expected to be absorbed in a manner similar to dietary heme. Newer encapsulated forms of iron (e.g., liposomal ferric pyrophosphate and sucrosomial iron) are believed to be absorbed independently of DMT1 via paracellular and transcellular pathways, theoretically providing advantages for the treatment of anemia in high-hepcidin states.

Intravenous iron formulations are used for the treatment of ID caused by a variety of clinical conditions in which oral iron is inadequate or ineffective in repleting body iron stores. These include situations of ongoing blood loss where iron repletion cannot be adequately achieved with oral formulations (e.g., heavy menstrual bleeding, hereditary hemorrhagic telangiectasia), during the second- and third-trimester of pregnancy due to the high iron requirements of the growing fetus, following gastrointestinal surgeries that reduce the capacity for oral iron absorption, in inflammatory bowel disease where oral iron may exacerbate the disease, and in comorbid conditions associated with hepcidin elevation (e.g., chronic kidney disease, cancer, and chemotherapy-induced anemia) (Van Doren et al. 2024). In surgical patients, intravenous iron may be employed when rapid correction of IDA in the pre-operative period is desired (Kalra and Auron 2024). Intravenous iron repletion has been reported to improve symptoms and quality of life in patients with heart failure and concurrent ID, but controversies remain regarding its administration in this population (Cheema et al. 2024; Lakhal-Littleton and Cleland 2024); see also Chap. 15.

Administration of the early intravenous iron formulations in the 1950s was associated with severe toxicity due to labile iron release. Newer intravenous formulations, including ferric derisomaltose, ferric carboxymaltose, and ferumoxytol, feature tighter binding of elemental iron in the iron-carbohydrate complex, reducing the risk of labile-iron-mediated toxicity (Blumenstein et al. 2021). By providing for slower iron release, these newer formulations allow for the correction of the total iron deficit in one or two rapid infusions (Van Doren et al. 2024). Following infusion, the iron-carbohydrate complexes are phagocytosed by reticuloendothelial macrophages, which degrade the carbohydrate component and enable the iron to be either stored intracellularly as ferritin or exported into the circulation for utilization. Certain intravenous iron formulations, particularly ferric carboxymaltose, have been associated with treatment-emergent hypophosphatemia, which can result in debilitating myopathy, osteomalacia, and fractures (Schaefer et al. 2022).

In addition to treating the ID, the etiology of ID should be identified and addressed whenever possible. Further clinical investigations are largely guided by detailed clinical history and physical exams, taking into consideration the age-specific causes of ID. Evaluations may, for example, include upper and lower gastrointestinal endoscopy, serological testing for celiac disease, and non-invasive testing for *H. pylori* infection (Ko et al. 2020). Importantly, ID may serve as a marker of an occult gastrointestinal malignancy. Indeed, among men and postmenopausal women in a population-based cohort study in the United States, the proportion diagnosed with a gastrointestinal malignancy after 2 years of follow-up was 31 times greater in those with IDA and 5 times greater in those with ID without anemia, compared to those with normal hemoglobin and TSAT levels (Ioannou et al. 2002).

12.9 Conclusions

Recent insights into fundamental mechanisms of cellular and systemic iron regulation have enhanced our understanding of the pathophysiology of ID and IDA. However, opportunities remain to improve our ability to diagnose ID, particularly functional ID, through the identification and validation of novel biomarkers. Advances in iron metabolism may inform the optimization of oral iron dosing and clarify clinical indications for high-dose intravenous iron therapy, facilitating more personalized therapeutic approaches for ID.

Acknowledgements Figures were prepared with BioRender.

References

Allen RP, Auerbach S, Bahrain H, Auerbach M, Earley CJ (2013) The prevalence and impact of restless legs syndrome on patients with iron deficiency anemia. Am J Hematol 88(4):261–264. https://doi.org/10.1002/ajh.23397

Amir N, Md Noor S, Subbiah I, Osman M, Seman Z (2019) Percentage of hypochromic red cells as a potential screening test to evaluate iron status in blood donors. Int J Lab Hematol 41(3):418–423. https://doi.org/10.1111/ijlh.13009

Arezes J, Foy N, McHugh K, Sawant A, Quinkert D, Terraube V, Brinth A, Tam M, LaVallie ER, Taylor S, Armitage AE, Pasricha SR, Cunningham O, Lambert M, Draper SJ, Jasuja R, Drakesmith H (2018) Erythroferrone inhibits the induction of hepcidin by BMP6. Blood 132(14):1473–1477. https://doi.org/10.1182/blood-2018-06-857995

Bell S, Rigas AS, Magnusson MK, Ferkingstad E, Allara E, Bjornsdottir G, Ramond A, Sorensen E, Halldorsson GH, Paul DS, Burgdorf KS, Eggertsson HP, Howson JMM, Thorner LW, Kristmundsdottir S, Astle WJ, Erikstrup C, Sigurdsson JK, Vuckovic D, Dinh KM, Tragante V, Surendran P, Pedersen OB, Vidarsson B, Jiang T, Paarup HM, Onundarson PT, Akbari P, Nielsen KR, Lund SH, Juliusson K, Magnusson MI, Frigge ML, Oddsson A, Olafsson I, Kaptoge S, Hjalgrim H, Runarsson G, Wood AM, Jonsdottir I, Hansen TF, Sigurdardottir O, Stefansson H, Rye D, Consortium DG, Peters JE, Westergaard D, Holm H, Soranzo N, Banasik K, Thorleifsson G, Ouwehand WH, Thorsteinsdottir U, Roberts DJ, Sulem P, Butterworth AS, Gudbjartsson DF, Danesh J, Brunak S, Di Angelantonio E, Ullum H, Stefansson K (2021) A genome-wide meta-analysis yields 46 new loci associating with biomarkers of iron homeostasis. Commun Biol 4(1):156. https://doi.org/10.1038/s42003-020-01575-z

Benyamin B, Ferreira MA, Willemsen G, Gordon S, Middelberg RP, McEvoy BP, Hottenga JJ, Henders AK, Campbell MJ, Wallace L, Frazer IH, Heath AC, de Geus EJ, Nyholt DR, Visscher PM, Penninx BW, Boomsma DI, Martin NG, Montgomery GW, Whitfield JB (2009) Common variants in TMPRSS6 are associated with iron status and erythrocyte volume. Nat Genet 41(11):1173–1175. https://doi.org/10.1038/ng.456

Blumenstein I, Shanbhag S, Langguth P, Kalra PA, Zoller H, Lim W (2021) Newer formulations of intravenous iron: a review of their chemistry and key safety aspects—hypersensitivity, hypophosphatemia, and cardiovascular safety. Expert Opin Drug Saf 20(7):757–769. https://doi.org/10.1080/14740338.2021.1912010

Brittenham GM, Moir-Meyer G, Abuga KM, Datta-Mitra A, Cerami C, Green R, Pasricha SR, Atkinson SH (2023) Biology of anemia: a public health perspective. J Nutr 153(Suppl 1):S7–S28. https://doi.org/10.1016/j.tjnut.2023.07.018

Camaschella C (2019) Iron deficiency. Blood 133(1):30–39. https://doi.org/10.1182/blood-2018-05-815944

Cappellini MD, Musallam KM, Taher AT (2020) Iron deficiency anaemia revisited. J Intern Med 287(2):153–170. https://doi.org/10.1111/joim.13004

Chambers JC, Zhang W, Li Y, Sehmi J, Wass MN, Zabaneh D, Hoggart C, Bayele H, McCarthy MI, Peltonen L, Freimer NB, Srai SK, Maxwell PH, Sternberg MJ, Ruokonen A, Abecasis G, Jarvelin MR, Scott J, Elliott P, Kooner JS (2009) Genome-wide association study identifies variants in TMPRSS6 associated with hemoglobin levels. Nat Genet 41(11):1170–1172. https://doi.org/10.1038/ng.462

Cheema B, Chokshi A, Orimoloye O, Ardehali H (2024) Intravenous iron repletion for patients with heart failure and iron deficiency: JACC state-of-the-art review. J Am Coll Cardiol 83(25):2674–2689. https://doi.org/10.1016/j.jacc.2024.03.431

Chen JJ, Zhang S (2022) Translational control by heme-regulated eIF2alpha kinase during erythropoiesis. Curr Opin Hematol 29(3):103–111. https://doi.org/10.1097/MOH.0000000000000704

Del Vecchio L, Minutolo R (2021) ESA, iron therapy and new drugs: are there new perspectives in the treatment of Anaemia? J Clin Med 10(4). https://doi.org/10.3390/jcm10040839

Diepeveen LE, Laarakkers CMM, Martos G, Pawlak ME, Uguz FF, Verberne K, van Swelm RPL, Klaver S, de Haan AFJ, Pitts KR, Bansal SS, Abbas IM, Fillet M, Lefebvre T, Geurts-Moespot AJ, Girelli D, Castagna A, Herkert M, Itkonen O, Olbina G, Tomosugi N, Westerman ME, Delatour V, Weykamp CW, Swinkels DW (2019) Provisional standardization of hepcidin assays: creating a traceability chain with a primary reference material, candidate reference method and a commutable secondary reference material. Clin Chem Lab Med 57(6):864–872. https://doi.org/10.1515/cclm-2018-0783

Drakesmith H, Pasricha SR, Cabantchik I, Hershko C, Weiss G, Girelli D, Stoffel N, Muckenthaler MU, Nemeth E, Camaschella C, Klenerman P, Zimmermann MB (2021) Vaccine efficacy and iron deficiency: an intertwined pair? Lancet Haematol 8(9):e666–e669. https://doi.org/10.1016/S2352-3026(21)00201-5

Ebea-Ugwuanyi PO, Vidyasagar S, Connor JR, Frazer DM, Knutson MD, Collins JF (2024) Oral iron therapy: current concepts and future prospects for improving efficacy and outcomes. Br J Haematol 204(3):759–773. https://doi.org/10.1111/bjh.19268

Ellwanger JH, Ziliotto M, Kulmann-Leal B, Chies JAB (2022) Iron deficiency and soil-transmitted helminth infection: classic and neglected connections. Parasitol Res 121(12):3381–3392. https://doi.org/10.1007/s00436-022-07697-z

Elsayed ME, Sharif MU, Stack AG (2016) Transferrin saturation: a body iron biomarker. Adv Clin Chem 75:71–97. https://doi.org/10.1016/bs.acc.2016.03.002

Enns CA, Jue S, Zhang AS (2020) The ectodomain of matriptase-2 plays an important nonproteolytic role in suppressing hepcidin expression in mice. Blood 136(8):989–1001. https://doi.org/10.1182/blood.2020005222

Finberg KE, Whittlesey RL, Fleming MD, Andrews NC (2010) Down-regulation of Bmp/Smad signaling by Tmprss6 is required for maintenance of systemic iron homeostasis. Blood 115(18):3817–3826. https://doi.org/10.1182/blood-2009-05-224808

Fisher AL, Babitt JL (2022) Coordination of iron homeostasis by bone morphogenetic proteins: current understanding and unanswered questions. Dev Dyn 251(1):26–46. https://doi.org/10.1002/dvdy.372

Fletcher A, Forbes A, Svenson N, Wayne Thomas D, Paper ABSfHGP (2022) Guideline for the laboratory diagnosis of iron deficiency in adults (excluding pregnancy) and children. Br J Haematol 196(3):523–529. https://doi.org/10.1111/bjh.17900

Ganesan PR, Vasauskas AA (2023) The association between pica and iron-deficiency anemia: a scoping review. Cureus 15(4):e37904. https://doi.org/10.7759/cureus.37904

Ganesh SK, Zakai NA, van Rooij FJ, Soranzo N, Smith AV, Nalls MA, Chen MH, Kottgen A, Glazer NL, Dehghan A, Kuhnel B, Aspelund T, Yang Q, Tanaka T, Jaffe A, Bis JC, Verwoert GC, Teumer A, Fox CS, Guralnik JM, Ehret GB, Rice K, Felix JF, Rendon A, Eiriksdottir G, Levy D, Patel KV, Boerwinkle E, Rotter JI, Hofman A, Sambrook JG, Hernandez DG, Zheng G, Bandinelli S, Singleton AB, Coresh J, Lumley T, Uitterlinden AG, Vangils JM, Launer LJ, Cupples LA, Oostra BA, Zwaginga JJ, Ouwehand WH, Thein SL, Meisinger C, Deloukas P, Nauck M, Spector TD, Gieger C, Gudnason V, van Duijn CM, Psaty BM, Ferrucci L, Chakravarti A, Greinacher A, O'Donnell CJ, Witteman JC, Furth S, Cushman M, Harris TB, Lin JP (2009) Multiple loci influence erythrocyte phenotypes in the CHARGE Consortium. Nat Genet 41(11):1191–1198. https://doi.org/10.1038/ng.466

Ganz T, Olbina G, Girelli D, Nemeth E, Westerman M (2008) Immunoassay for human serum hepcidin. Blood 112(10):4292–4297. https://doi.org/10.1182/blood-2008-02-139915

Garcia-Casal MN, Pasricha SR, Martinez RX, Lopez-Perez L, Pena-Rosas JP (2021) Serum or plasma ferritin concentration as an index of iron deficiency and overload. Cochrane Database Syst Rev 5(5):CD011817. https://doi.org/10.1002/14651858.CD011817.pub2

GBD 2021 Anaemia Collaborators (2023) Prevalence, years lived with disability, and trends in anaemia burden by severity and cause, 1990-2021: findings from the global burden of disease study 2021. Lancet Haematol 10(9):e713–e734. https://doi.org/10.1016/S2352-3026(23)00160-6

GBD 2021 Diseases and Injuries Collaborators (2024) Global incidence, prevalence, years lived with disability (YLDs), disability-adjusted life-years (DALYs), and healthy life expectancy (HALE) for 371 diseases and injuries in 204 countries and territories and 811 subnational locations, 1990–2021: a systematic analysis for the global burden of disease study 2021. Lancet 403(10440):2133–2161. https://doi.org/10.1016/S0140-6736(24)00757-8

Georgieff MK (2023) Maternal gestational iron status and infant haematological and neurodevelopmental outcomes. BJOG 130(Suppl 3):92–98. https://doi.org/10.1111/1471-0528.17612

Gupta PM, Perrine CG, Mei Z, Scanlon KS (2016) Iron, anemia, and iron deficiency anemia among young children in the United States. Nutrients 8(6). https://doi.org/10.3390/nu8060330

Harms K, Kaiser T (2015) Beyond soluble transferrin receptor: old challenges and new horizons. Best Pract Res Clin Endocrinol Metab 29(5):799–810. https://doi.org/10.1016/j.beem.2015.09.003

Heeney MM, Finberg KE (2014) Iron-refractory iron deficiency anemia (IRIDA). Hematol Oncol Clin North Am 28(4):637–652. https://doi.org/10.1016/j.hoc.2014.04.009

Heeney MM, Guo D, De Falco L, Campagna DR, Olbina G, Kao PP, Schmitz-Abe K, Rahimov F, Gutschow P, Westerman K, Ostland V, Jackson T, Klaassen RJ,

Markianos K, Finberg KE, Iolascon A, Westerman M, London WB, Fleming MD (2018) Normalizing hepcidin predicts TMPRSS6 mutation status in patients with chronic iron deficiency. Blood 132(4):448–452. https://doi.org/10.1182/blood-2017-03-773028

Institute of Medicine (2001) Institute of Medicine (US) Panel on Micronutrients. Dietary Reference Intakes for Vitamin A, Vitamin K, Arsenic, Boron, Chromium, Copper, Iodine, Iron, Manganese, Molybdenum, Nickel, Silicon, Vanadium, and Zinc. The National Academies Press, Washington, DC

Ioannou GN, Rockey DC, Bryson CL, Weiss NS (2002) Iron deficiency and gastrointestinal malignancy: a population-based cohort study. Am J Med 113(4):276–280. https://doi.org/10.1016/s0002-9343(02)01214-7

Kaberi-Otarod J, Still CD, Wood GC, Benotti PN (2024) Iron treatment in patients with iron deficiency before and after metabolic and bariatric surgery: a narrative review. Nutrients 16(19). https://doi.org/10.3390/nu16193350

Kalra SK, Auron M (2024) Anemia and transfusion medicine. Med Clin North Am 108(6):1065–1085. https://doi.org/10.1016/j.mcna.2024.04.002

Karamanou M, Androutsos G (2010) Lasthenie de Ferjol syndrome: a rare disease with fascinating history. Intern Med J 40(5):381–382. https://doi.org/10.1111/j.1445-5994.2009.02154.x

Kardasis W, Naquin ER, Garg R, Arun T, Gopianand JS, Karmakar E, Gnana-Prakasam JP (2023) The IRONy in athletic performance. Nutrients 15(23). https://doi.org/10.3390/nu15234945

Ko CW, Siddique SM, Patel A, Harris A, Sultan S, Altayar O, Falck-Ytter Y (2020) AGA clinical practice guidelines on the gastrointestinal evaluation of iron deficiency anemia. Gastroenterology 159(3):1085–1094. https://doi.org/10.1053/j.gastro.2020.06.046

Lakhal-Littleton S, Cleland JGF (2024) Iron deficiency and supplementation in heart failure. Nat Rev Cardiol 21(7):463–486. https://doi.org/10.1038/s41569-024-00988-1

Larsuphrom P, Latunde-Dada GO (2021) Association of Serum Hepcidin Levels with aerobic and resistance exercise: a systematic review. Nutrients 13(2). https://doi.org/10.3390/nu13020393

Lyle AN, Budd JR, Kennerley VM, Smith BN, Danilenko U, Pfeiffer CM, Vesper HW (2023) Assessment of WHO 07/202 reference material and human serum pools for commutability and for the potential to reduce variability among soluble transferrin receptor assays. Clin Chem Lab Med 61(10):1719–1729. https://doi.org/10.1515/cclm-2022-1198

Maio N, Zhang DL, Ghosh MC, Jain A, SantaMaria AM, Rouault TA (2021) Mechanisms of cellular iron sensing, regulation of erythropoiesis and mitochondrial iron utilization. Semin Hematol 58(3):161–174. https://doi.org/10.1053/j.seminhematol.2021.06.001

Mast AE, Langer JC, Guo Y, Bialkowski W, Spencer BR, Lee TH, Kiss J, Cable RG, Brambilla D, Busch MP, Page GP, Nhlbi Recipient Epidemiology Donor Evaluation Study I (2020) Genetic and behavioral modification of hemoglobin and iron status among first-time and high-intensity blood donors. Transfusion 60(4):747–758. https://doi.org/10.1111/trf.15743

Munro MG, Mast AE, Powers JM, Kouides PA, O'Brien SH, Richards T, Lavin M, Levy BS (2023) The relationship between heavy menstrual bleeding, iron deficiency, and iron deficiency anemia. Am J Obstet Gynecol 229(1):1–9. https://doi.org/10.1016/j.ajog.2023.01.017

Nai A, Pagani A, Silvestri L, Campostrini N, Corbella M, Girelli D, Traglia M, Toniolo D, Camaschella C (2011) TMPRSS6 rs855791 modulates hepcidin transcription in vitro and serum hepcidin levels in normal individuals. Blood 118(16):4459–4462. https://doi.org/10.1182/blood-2011-06-364034

Nemeth E, Ganz T (2023) Hepcidin and iron in health and disease. Annu Rev Med 74:261–277. https://doi.org/10.1146/annurev-med-043021-032816

Neufingerl N, Eilander A (2023) Nutrient intake and status in children and adolescents consuming plant-based diets compared to meat-eaters: a systematic review. Nutrients 15(20). https://doi.org/10.3390/nu15204341

Ning S, Zeller MP (2019) Management of iron deficiency. Hematology Am Soc Hematol Educ Program 1:315–322. https://doi.org/10.1182/hematology.2019000034

Obianeli C, Afifi K, Stanworth S, Churchill D (2024) Iron deficiency Anaemia in pregnancy: a narrative review from a clinical perspective. Diagnostics (Basel) 14(20). https://doi.org/10.3390/diagnostics14202306

Ogawa C, Tsuchiya K, Maeda K (2020) Reticulocyte hemoglobin content. Clin Chim Acta 504:138–145. https://doi.org/10.1016/j.cca.2020.01.032

Pagani A, Colucci S, Bocciardi R, Bertamino M, Dufour C, Ravazzolo R, Silvestri L, Camaschella C (2017) A new form of IRIDA due to combined heterozygous mutations of TMPRSS6 and ACVR1A encoding the BMP receptor ALK2. Blood 129(25):3392–3395. https://doi.org/10.1182/blood-2017-03-773481

Pantopoulos K (2024) Oral iron supplementation: new formulations, old questions. Haematologica 109(9):2790–2801. https://doi.org/10.3324/haematol.2024.284967

Pasricha SR, Lim PJ, Duarte TL, Casu C, Oosterhuis D, Mleczko-Sanecka K, Suciu M, Da Silva AR, Al-Hourani K, Arezes J, McHugh K, Gooding S, Frost JN, Wray K, Santos A, Porto G, Repapi E, Gray N, Draper SJ, Ashley N, Soilleux E, Olinga P, Muckenthaler MU, Hughes JR, Rivella S, Milne TA, Armitage AE, Drakesmith H (2017) Hepcidin is regulated by promoter-associated histone acetylation and HDAC3. Nat Commun 8(1):403. https://doi.org/10.1038/s41467-017-00500-z

Pasricha SR, Tye-Din J, Muckenthaler MU, Swinkels DW (2021) Iron deficiency. Lancet 397(10270):233–248. https://doi.org/10.1016/S0140-6736(20)32594-0

Peeling P, McKay A (2023) Iron regulation and absorption in athletes: contemporary thinking and recommendations. Curr Opin Clin Nutr Metab

Care 26(6):551–556. https://doi.org/10.1097/MCO.0000000000000966
Pfeiffer CM, Looker AC (2017) Laboratory methodologies for indicators of iron status: strengths, limitations, and analytical challenges. Am J Clin Nutr 106(Suppl 6):1606S–1614S. https://doi.org/10.3945/ajcn.117.155887
Piskin E, Cianciosi D, Gulec S, Tomas M, Capanoglu E (2022) Iron absorption: factors, limitations, and improvement methods. ACS Omega 7(24):20441–20456. https://doi.org/10.1021/acsomega.2c01833
Samuelson Bannow B (2020) Management of heavy menstrual bleeding on anticoagulation. Hematology Am Soc Hematol Educ Program 2020(1):533–537. https://doi.org/10.1182/hematology.2020000138
Sangkhae V, Fisher AL, Ganz T, Nemeth E (2023) Iron homeostasis during pregnancy: maternal, placental, and fetal regulatory mechanisms. Annu Rev Nutr 43:279–300. https://doi.org/10.1146/annurev-nutr-061021-030404
Santana-Codina N, Mancias JD (2018) The role of NCOA4-mediated ferritinophagy in health and disease. Pharmaceuticals (Basel) 11(4). https://doi.org/10.3390/ph11040114
Schaefer B, Tobiasch M, Wagner S, Glodny B, Tilg H, Wolf M, Zoller H (2022) Hypophosphatemia after intravenous iron therapy: comprehensive review of clinical findings and recommendations for management. Bone 154:116202. https://doi.org/10.1016/j.bone.2021.116202
Seyoum Y, Baye K, Humblot C (2021) Iron homeostasis in host and gut bacteria—a complex interrelationship. Gut Microbes 13(1):1–19. https://doi.org/10.1080/19490976.2021.1874855
Shah SC, Piazuelo MB, Kuipers EJ, Li D (2021) AGA clinical practice update on the diagnosis and management of atrophic gastritis: expert review. Gastroenterology 161(4):1325–1332. e1327. https://doi.org/10.1053/j.gastro.2021.06.078
Shin DH, Kim HS, Park MJ, Suh IB, Shin KS (2015) Utility of access soluble transferrin receptor (sTfR) and sTfR/log ferritin index in diagnosing iron deficiency anemia. Ann Clin Lab Sci 45(4):396–402
Silvestri L, Nai A (2021) Iron and erythropoiesis: a mutual alliance. Semin Hematol 58(3):145–152. https://doi.org/10.1053/j.seminhematol.2021.05.002
Silvestri L, Pagani A, Nai A, De Domenico I, Kaplan J, Camaschella C (2008) The serine protease matriptase-2 (TMPRSS6) inhibits hepcidin activation by cleaving membrane hemojuvelin. Cell Metab 8(6):502–511. https://doi.org/10.1016/j.cmet.2008.09.012
Solanki S, Shah YM (2024) Hypoxia-induced signaling in gut and liver pathobiology. Annu Rev Pathol 19:291–317. https://doi.org/10.1146/annurev-pathmechdis-051122-094743
Soranzo N, Spector TD, Mangino M, Kuhnel B, Rendon A, Teumer A, Willenborg C, Wright B, Chen L, Li M, Salo P, Voight BF, Burns P, Laskowski RA, Xue Y, Menzel S, Altshuler D, Bradley JR, Bumpstead S, Burnett MS, Devaney J, Doring A, Elosua R, Epstein SE, Erber W, Falchi M, Garner SF, Ghori MJ, Goodall AH, Gwilliam R, Hakonarson HH, Hall AS, Hammond N, Hengstenberg C, Illig T, Konig IR, Knouff CW, McPherson R, Melander O, Mooser V, Nauck M, Nieminen MS, O'Donnell CJ, Peltonen L, Potter SC, Prokisch H, Rader DJ, Rice CM, Roberts R, Salomaa V, Sambrook J, Schreiber S, Schunkert H, Schwartz SM, Serbanovic-Canic J, Sinisalo J, Siscovick DS, Stark K, Surakka I, Stephens J, Thompson JR, Volker U, Volzke H, Watkins NA, Wells GA, Wichmann HE, Van Heel DA, Tyler-Smith C, Thein SL, Kathiresan S, Perola M, Reilly MP, Stewart AF, Erdmann J, Samani NJ, Meisinger C, Greinacher A, Deloukas P, Ouwehand WH, Gieger C (2009) A genome-wide meta-analysis identifies 22 loci associated with eight hematological parameters in the HaemGen consortium. Nat Genet 41(11):1182–1190. https://doi.org/10.1038/ng.467
Sorensen E, Rigas AS, Thorner LW, Burgdorf KS, Pedersen OB, Petersen MS, Hjalgrim H, Erikstrup C, Ullum H (2016) Genetic factors influencing ferritin levels in 14,126 blood donors: results from the Danish blood donor study. Transfusion 56(3):622–627. https://doi.org/10.1111/trf.13397
Spencer BR, Mast AE (2022) Iron status of blood donors. Curr Opin Hematol 29(6):310–316. https://doi.org/10.1097/MOH.0000000000000733
Stein J, Connor S, Virgin G, Ong DE, Pereyra L (2016) Anemia and iron deficiency in gastrointestinal and liver conditions. World J Gastroenterol 22(35):7908–7925. https://doi.org/10.3748/wjg.v22.i35.7908
Stoffel NU, von Siebenthal HK, Moretti D, Zimmermann MB (2020) Oral iron supplementation in iron-deficient women: how much and how often? Mol Asp Med 75:100865. https://doi.org/10.1016/j.mam.2020.100865
Tai FWD, McAlindon ME (2021) Non-steroidal anti-inflammatory drugs and the gastrointestinal tract. Clin Med (Lond) 21(2):131–134. https://doi.org/10.7861/clinmed.2021-0039
Talarico V, Giancotti L, Mazza GA, Miniero R, Bertini M (2021) Iron deficiency anemia in celiac disease. Nutrients 13(5). https://doi.org/10.3390/nu13051695
Tawfik YMK, Billingsley H, Bhatt AS, Aboelsaad I, Al-Khezi OS, Lutsey PL, Buckley LF (2024) Absolute and functional iron deficiency in the US, 2017–2020. JAMA Netw Open 7(9):e2433126. https://doi.org/10.1001/jamanetworkopen.2024.33126
Teh MR, Armitage AE, Drakesmith H (2024) Why cells need iron: a compendium of iron utilisation. Trends Endocrinol Metab. https://doi.org/10.1016/j.tem.2024.04.015
Teixeira Farinha H, Bouriez D, Grimaud T, Rotariu AM, Collet D, Mantziari S, Gronnier C (2023) Gastrointestinal disorders and micronutrient deficiencies following oncologic esophagectomy and gastrectomy. Cancers (Basel) 15(14). https://doi.org/10.3390/cancers15143554
Truman-Rosentsvit M, Berenbaum D, Spektor L, Cohen LA, Belizowsky-Moshe S, Lifshitz L, Ma J, Li W,

Kesselman E, Abutbul-Ionita I, Danino D, Gutierrez L, Li H, Li K, Lou H, Regoni M, Poli M, Glaser F, Rouault TA, Meyron-Holtz EG (2018) Ferritin is secreted via 2 distinct nonclassical vesicular pathways. Blood 131(3):342–352. https://doi.org/10.1182/blood-2017-02-768580

Truong J, Naveed K, Beriault D, Lightfoot D, Fralick M, Sholzberg M (2024) The origin of ferritin reference intervals: a systematic review. Lancet Haematol 11(7):e530–e539. https://doi.org/10.1016/S2352-3026(24)00103-0

Van Doren L, Steinheiser M, Boykin K, Taylor KJ, Menendez M, Auerbach M (2024) Expert consensus guidelines: intravenous iron uses, formulations, administration, and management of reactions. Am J Hematol 99(7):1338–1348. https://doi.org/10.1002/ajh.27220

VanderMeulen H, Sholzberg M (2018) Iron deficiency and anemia in patients with inherited bleeding disorders. Transfus Apher Sci 57(6):735–738. https://doi.org/10.1016/j.transci.2018.10.015

Vinnakota RD, Brett AS (2019) Iron deficiency anemia associated with acid-modifying medications: two cases and literature review. Am J Med Sci 357(2):160–163. https://doi.org/10.1016/j.amjms.2018.10.014

von Siebenthal HK, Moretti D, Zimmermann MB, Stoffel NU (2023) Effect of dietary factors and time of day on iron absorption from oral iron supplements in iron deficient women. Am J Hematol 98(9):1356–1363. https://doi.org/10.1002/ajh.26987

Weyand AC, Chaitoff A, Freed GL, Sholzberg M, Choi SW, McGann PT (2023) Prevalence of iron deficiency and iron-deficiency anemia in US females aged 12-21 years, 2003-2020. JAMA 329(24):2191–2193. https://doi.org/10.1001/jama.2023.8020

Willis GC, Tewelde SZ (2019) The approach to the patient with hematuria. Emerg Med Clin North Am 37(4):755–769. https://doi.org/10.1016/j.emc.2019.07.011

Wong CY, Chu DH (2021) Cutaneous signs of nutritional disorders. Int J Womens Dermatol 7(5Part A):647–652. https://doi.org/10.1016/j.ijwd.2021.09.003

World Health Organization (2015) The global prevalence of anaemia in 2011. World Health Organization, Geneva

World Health Organization (2020) WHO guideline on use of ferritin concentrations to assess iron status in individuals and populations. World Health Organization, Geneva

Xavier-Ferrucio J, Scanlon V, Li X, Zhang PX, Lozovatsky L, Ayala-Lopez N, Tebaldi T, Halene S, Cao C, Fleming MD, Finberg KE, Krause DS (2019) Low iron promotes megakaryocytic commitment of megakaryocytic-erythroid progenitors in humans and mice. Blood 134(18):1547–1557. https://doi.org/10.1182/blood.2019002039

Zhang AS, Anderson SA, Wang J, Yang F, DeMaster K, Ahmed R, Nizzi CP, Eisenstein RS, Tsukamoto H, Enns CA (2011) Suppression of hepatic hepcidin expression in response to acute iron deprivation is associated with an increase of matriptase-2 protein. Blood 117(5):1687–1699. https://doi.org/10.1182/blood-2010-06-287292

Zhang DL, Ghosh MC, Ollivierre H, Li Y, Rouault TA (2018) Ferroportin deficiency in erythroid cells causes serum iron deficiency and promotes hemolysis due to oxidative stress. Blood 132(19):2078–2087. https://doi.org/10.1182/blood-2018-04-842997

Anemia of Inflammation

13

Lukas Lanser and Günter Weiss

Abstract

Anemia of inflammation (AI), also known as anemia of chronic disease, is the most common anemia in hospitalized patients and considered to be the second most common anemia worldwide after iron deficiency anemia (IDA). The hallmark of AI is iron restriction within macrophages of the mononuclear phagocyte system (MPS) resulting in hypoferremia and hyperferritinemia together with suppression of erythropoiesis and shortened erythrocyte lifespan. Symptoms are comparable to other anemia entities and often related to the underlying disease. Patients usually present with normocytic, normochromic, hypoproliferative, mild-to-moderate anemia, reduced circulating iron levels (transferrin saturation), and increased stored iron (serum ferritin). However, AI is often associated with true iron deficiency on the basis of inflammatory diseases and blood losses of different reasons, which is why the correct identification of these patients, and their iron needs is a diagnostic challenge. Treatment of the underlying disease that causes immune activation is the primary therapeutic approach for AI which normally results in its resolution over time. Concomitant pathologies and factors contributing to the AI severity should be considered and, when feasible, specifically corrected. Iron supplementation is the first-line therapy for AI+IDA patients, while intravenously applied iron is trapped in macrophages of the MPS during advanced inflammation in patients with solely AI, whereas orally supplemented iron is not properly absorbed. Effectiveness of erythropoiesis-stimulating agents is limited in AI due to inflammation-mediated suppression of erythropoietin (Epo) signaling and impaired erythroid cell proliferation and differentiation, while red blood cell transfusion should primarily be used in life-threatening anemia. Clinical studies on hypoxia-inducible factor prolyl hydroxylase inhibitors seem promising although concerns about their safety and efficacy in AI arose within recent years. New treatment strategies aim to modify the hepcidin-ferroportin axis, yet clinical trials are still outstanding.

Keywords

Hepcidin-ferroportin axis · Iron sequestration · Erythropoiesis suppression · Inflammatory cytokines · Functional iron deficiency

L. Lanser · G. Weiss (✉)
Department of Internal Medicine II (Infectiology, Immunology, Rheumatology, Pneumology), Medical University of Innsbruck, Innsbruck, Austria
e-mail: guenter.weiss@i-med.ac.at

K. Pantopoulos (ed.), *Iron Metabolism in Human Health and Disease*, Advances in Experimental Medicine and Biology 1480, https://doi.org/10.1007/978-3-031-92033-2_13

13.1 Introduction

Patients with acute or chronic inflammatory disorders frequently develop anemia, termed anemia of inflammation (AI) (Weiss et al. 2019). AI was initially linked to chronic infections, cancer, and autoimmune disorders including rheumatoid arthritis, inflammatory bowel disease (IBD), systemic lupus erythematosus, or vasculitis, thus also termed anemia of chronic disease (ACD). Meanwhile, we have learned that inflammation is involved in the pathophysiology of several other chronic diseases, which are often accompanied by the merge of inflammatory anemia (Fig. 13.1) (Weiss and Schett 2013).

Anemia of critical illness can occur within hours of sepsis, major traumas, or surgery and is mainly caused by shortened erythrocyte lifespan and stimulated erythrophagocytosis, besides hemodilution and diagnostic phlebotomy (Hayden et al. 2012). Cancer-related anemia occurring before any antineoplastic treatment is multifactorial and related to immune activation-driven effects on iron homeostasis and erythropoiesis alongside bone marrow invasion by tumor cells and additional nutritional and vitamin deficiencies (Gaspar et al. 2015). Immune activation including inflammaging in immunosenescence also contributes to anemia at older age besides bone marrow failure syndromes, chronic kidney disease (CKD), and nutritional deficiencies (Stauder et al. 2018). Anemia in CKD is not only caused by reduced Epo production and bone marrow hyporesponsiveness, uremia-related inhibition of erythropoiesis, and vitamin deficiencies but also triggered by inflammation specifically in dialysis patients (Marques et al. 2022; Nakanishi et al. 2019). Anemia in patients with pulmonary hypertension or chronic obstructive pulmonary disease (COPD) is strongly related to

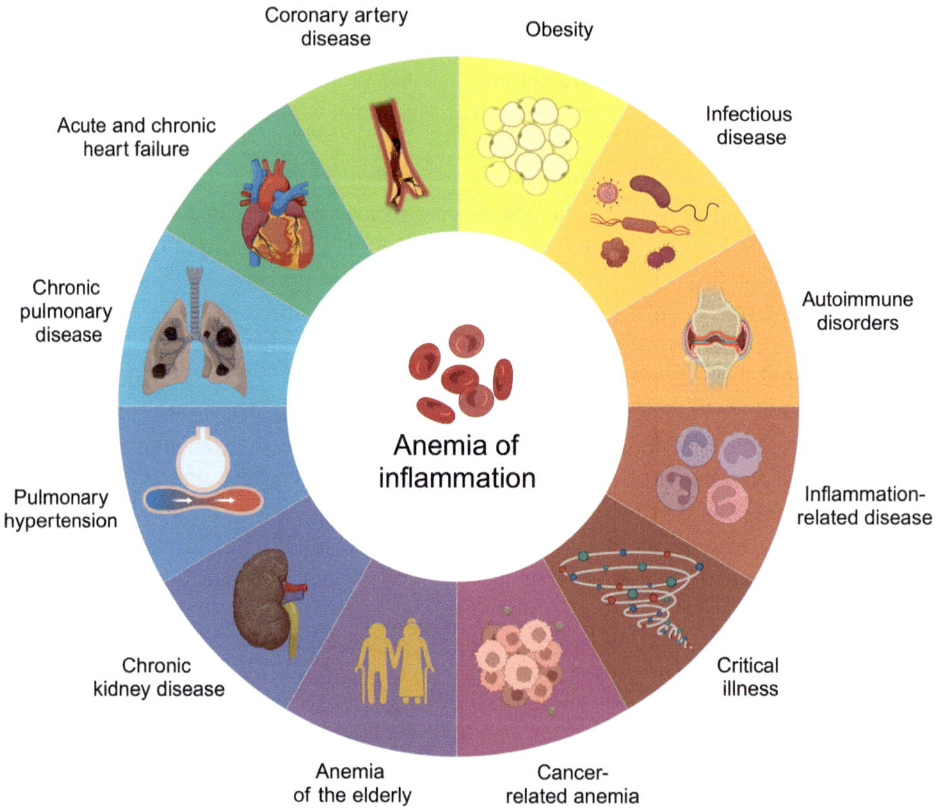

Fig. 13.1 Conditions and disorders that are related to anemia of inflammation

inflammatory-related alterations of iron homeostasis and erythropoiesis, and both iron deficiency (ID) and anemia impact on disease progression and prognosis (Sonnweber et al. 2018; Alisamir et al. 2022). AI, often in combination with true or absolute iron deficiency anemia (IDA; see Chap. 12), frequently occurs in patients with heart failure (HF) negatively affecting cardiovascular performance, disease progression, and mortality (Anand and Gupta 2018). Anemia and immune activation are also linked to disease progression and adverse outcomes in patients with coronary artery disease (CAD) (Lanser et al. 2021a). Finally, low-grade inflammation in obesity is linked to reduced inflammation-related iron availability for erythropoiesis (Purdy and Shatzel 2021).

According to estimates, AI and combined anemia with inflammation-related contributions account for up to 42% of anemias in the general population, thus suggested being the second most common anemia worldwide. It is even considered being the most frequent anemia in hospitalized patients (Weiss et al. 2019; Steinbicker and Muckenthaler 2013). AI is related to the severity and progression of the underlying disease. However, it often remains unclear whether and to which extent AI contributes to the pathology and course of the underlying disease, or whether it is solely a reflection of the degree of inflammation and/or the severity of the underlying disease.

13.2 Pathophysiology

AI development is primarily related to reduced iron availability, suppression of erythropoiesis, and shortened erythrocyte lifespan (Fig. 13.2). The relative contribution of each of these mechanisms to the pathogenesis of anemia might differ between individuals depending on the underlying disease, genetic and comorbid preconditions including body iron storage, erythropoietic capacity, responsiveness of hypoxia-related Epo production, type and severity of immune activation, or phagocytosis resistance of erythrocytes (Weiss et al. 2019).

13.2.1 Reduced Iron Availability

Immune activation following invasion of microbes or emergence of auto- or tumor antigens and subsequent release of pro-inflammatory cytokines by activated immune cells affects cellular and systemic iron homeostasis. Specifically, inflammation leads to iron accumulation and restriction by macrophages of the mononuclear phagocyte system (MPS) along with suppressed duodenal iron absorption. This decreases serum iron concentration and reduces iron availability for erythroid progenitor cells in the bone marrow, also described as functional ID (Weiss et al. 2019).

Iron withdrawal from the circulation is considered as an efficient defense strategy of the human body against circulating pathogens, known as nutritional immunity (Chap. 14). Iron is essential for most microbes, which require it for proliferation and pathogenicity. Thus, restriction of iron accessibility inhibits microbial growth (Nairz and Weiss 2020). On the other hand, iron restriction with consecutive hypoferremia and hyperferritinemia has distinctive consequences for erythropoiesis and metabolic processes in patients with immune activation. Over 90% of the daily iron requirements for hemoglobin (Hb) synthesis are provided by macrophages that recycle iron from senescent erythrocytes (Muckenthaler et al. 2017).

Inflammation-related changes of iron metabolism are particularly mediated by hepcidin (Chap. 3), which is the primary regulator of iron homeostasis and produced by hepatocytes. Hepcidin binds to the only known transmembrane iron exporter ferroportin (Fpn1) subsequently causing its internalization and lysosomal degradation (Nemeth et al. 2004). Consequently, iron in macrophages that was recycled from senescent erythrocytes, along with iron stored in hepatocytes and dietary iron absorbed in the duodenum, cannot be released into the circulation (Ganz and Nemeth 2012). There is evidence that high concentrations of hepcidin can occlude Fpn1 in addition, thus directly inhibiting iron export (Aschemeyer et al. 2018). Inflammation-related hepcidin expression in hepatocytes is primarily mediated by interleu-

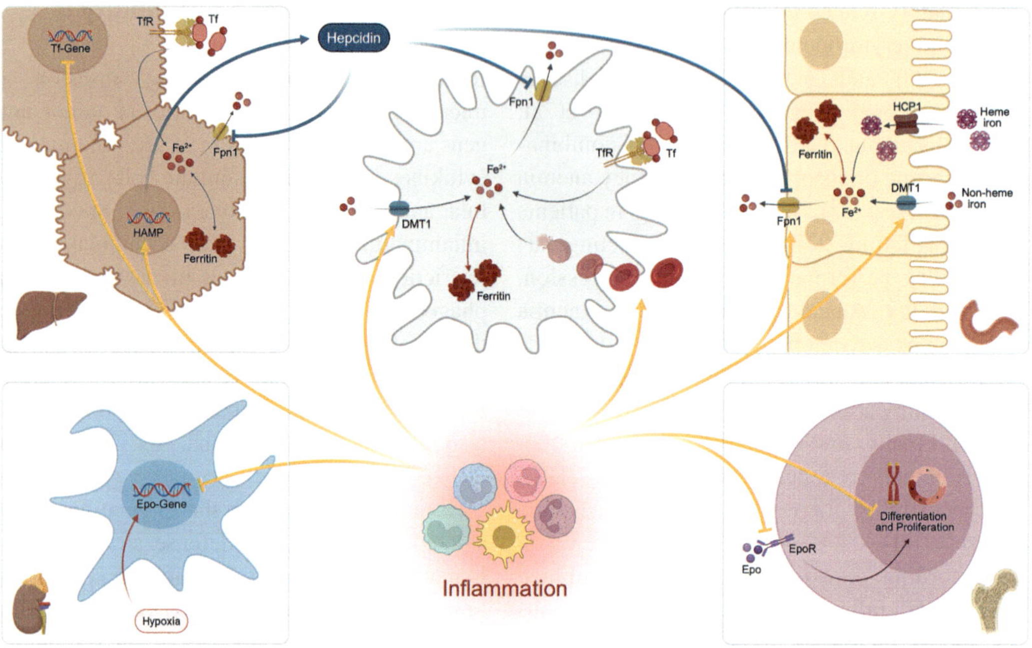

Fig. 13.2 Pathophysiological mechanisms contributing to the development of anemia of inflammation. *Tf* transferrin, *TfR* transferrin receptor, *Fpn1* ferroportin, *DMT1* divalent metal transporter 1, *HCP1* heme carrier protein 1, *Epo* erythropoietin, *EpoR* erythropoietin receptor

kin (IL)-6 and to a lesser extent by IL-22 via the JAK/STAT3 signaling pathway (Ganz and Nemeth 2012). IL-1β and activin B also stimulate hepcidin production, although their importance for inflammation-mediated hepcidin upregulation is not entirely clear yet (Armitage et al. 2011). Hepcidin is additionally produced by other cells including macrophages upon stimulation by IL-6 and lipopolysaccharide (LPS), then acting via autocrine signaling on Fpn1 (Theurl et al. 2008). Hepcidin production by intestinal dendritic cells is suggested to protect the local mucosa by reducing iron accessibility for the intestinal microbiome, thus promoting mucosal healing in IBD (Bessman et al. 2020).

Various receptors and proteins involved in iron metabolism are also directly affected by pro-inflammatory cytokines. Interferon gamma (IFN-γ), tumor necrosis factor alpha (TNF-α), and LPS upregulate transcription of divalent metal transporter 1 (DMT1) and block Fpn1 transcription subsequently increasing iron uptake and reducing iron release from MPS-resident macrophages (Ludwiczek et al. 2003; Drakesmith et al. 2015). The pro-inflammatory cytokines IL-1β, IL-6, and TNF-α, as well as the anti-inflammatory cytokines IL-4, IL-10, and IL-13, further stimulate transferrin-iron uptake along with increased synthesis of the iron storage protein ferritin via several mechanisms (Tilg et al. 2002; Weiss et al. 1997; Miller et al. 1991). Retained iron and heme again stimulate Fpn1 transcription in macrophages, thus partially counteracting the inflammation-related iron restriction (Delaby et al. 2008). This might explain why anemia in patients with AI is not as severe as it can be in patients with IDA (Camaschella 2015).

Preclinical studies suggest that erythrophagocytosis and iron recycling primarily take place in resident hepatic and not splenic macrophages when erythrocyte destruction is increased (Theurl et al. 2016). On the contrary, recent evidence indicates that during acute inflammation, iron restriction primarily occurs in the spleen with measurable increase in tissue iron content, while it remains unclear how iron is redistributed during chronic immune activation (Lanser et al. 2023).

13.2.2 Impaired Erythroid Cell Proliferation

Erythroid cell proliferation is impaired upon immune activation due to limited iron availability for erythroid cells, and decreased production and biological activity of Epo in response to inflammation (Nairz et al. 2012). Epo levels are lower than expected for anemia severity in AI. This might be related to suppression of hypoxia-mediated Epo stimulation or radical-mediated damage of Epo-producing renal cells by pro-inflammatory cytokines like IL-1β and TNF-α (Nairz et al. 2012).

Epo activates the JAK/STAT signaling pathway by binding to the erythropoietin receptor (EpoR). This Epo-mediated signaling is impaired and negatively related to serum IL-1β and IL-6 levels in AI although the EpoR number appears to be normal suggesting inflammation-related EpoR hyporesponsiveness (Jelkmann 2011; Okonko et al. 2013). Reduced erythroid iron availability in AI downregulates EpoR expression by interfering with iron sensing of erythroid cells mediated via transferrin receptor (TfR)-2 (Khalil et al. 2018). Epo and hypoxia usually inhibit hepcidin production which is why reduced Epo availability and activity in AI reverse these hepcidin-suppressing effects, thus intensifying macrophage iron retention and reduced iron availability for erythropoiesis (Latour et al. 2017a).

13.2.3 Shortened Erythrocyte Lifespan

Inflammation causes a shortened erythrocyte lifespan primarily due to enhanced erythrophagocytosis by macrophages resident in the spleen and liver. Besides macrophage activation, increased erythrophagocytosis is related to damage of erythrocytes by inflammation-derived radicals and fibrin deposition in the microvasculature together with antibody and complement deposition on erythrocytes (Mitlyng et al. 2006; Moldawer et al. 1989). In addition, pro-inflammatory cytokines, especially IFN-γ, induce apoptosis of erythroid cells via ceramide- and radical-mediated signaling pathways and reduce erythrocyte lifespan besides downregulation of EpoR expression on erythroid cells and inhibition of erythroid progenitor cell differentiation (Dallalio and Means Jr. 2003; Taniguchi et al. 1997; Libregts et al. 2011). Although shortened erythrocyte lifespan usually contributes slightly to AI development under chronic inflammatory conditions, it appears to be the most important mechanism in patients with acute infections, severe sepsis, or other disease with intensive and acute inflammatory activity and high circulating cytokine levels. Anemia found in these patients within hours or few days after disease onset is attributed to massive erythrophagocytosis, pooling of erythrocytes in the spleen, hemolysis, and hemodilution since iron restriction and impaired erythropoiesis are not immediately leading to anemia (Docherty et al. 2018).

13.2.4 Coexisting Factors Modifying Severity of AI

Simultaneous depletion of iron stores, known as true or absolute ID, can contribute to the severity of AI. Absolute ID can originate from reduced bioavailability and absorption of dietary iron attributable to dietary habits, genetic polymorphisms, or *Helicobacter pylori* infection. It can also originate from bleeding caused by menstruation, gastrointestinal blood loss (including disease-associated blood losses, medication-related subclinical bleeding following nonsteroidal anti-inflammatory drugs, heparin, or antiplatelet therapy), or repeated diagnostic phlebotomy or blood losses as a consequence of therapies such as dialysis. Erythropoiesis can be further impaired by vitamin deficiency (vitamin B12, folic acid), kidney dysfunction (reduced Epo formation, uremic toxins), chronic infections (parvovirus B19, CMV, EBV, HIV, HCV), medications (DMARDs, ACE inhibitors, proton-pump inhibitors [PPIs], neuroleptics, radio-/chemotherapy), thyroid hormones, or bone marrow infiltration by malignancies and infections (Leishmania, Plasmodium). Anemia can be intensified by concomitant hemolysis in autoimmune disease (immunoglobulin- or complement-mediated), infectious disease (Plasmodium, Babesia, parvovirus B19, EBV, CMV), and malignancies (immunoglobulin- or complement-mediated), and following vitamin deficiencies (vitamin B12, folic acid). Hepcidin synthesis or circulating levels can be further stimulated (vitamin D deficiency, genetic polymorphisms, obesity, Epo deficiency, kidney dysfunction) or suppressed (heparin therapy, testosterone, estrogen) at the same time (Fig. 13.3) (Weiss et al. 2019; Girelli et al. 2016).

13.3 Diagnostic Evaluation

Anemia in patients with activated immune system often is complex and multifactorial with coexisting factors contributing to anemia (Fig. 13.3). AI components are not always clearly distinguishable, making the diagnosis of AI challenging in clinical practice (Fig. 13.4).

Symptoms are comparable to other anemia entities and often related to the underlying disease. Patients present with weakness, fatigue, worsened exercise capacity, reduced QoL, and difficulties in concentrating with worsened learning and memory capacity. Although our knowledge of AI pathophysiology steadily increased over the past years, there is still no single specific biomarker for defining AI in patients (Weiss et al. 2019).

13.3.1 Basic Laboratory Assessment and Iron Status

Diagnostic evaluation starts with suspecting the diagnosis in patients with a disease related to acute or chronic immune activation. Occurrence of clinical or biochemical evidence of inflammation (elevation of erythrocyte sedimentation rate, high-sensitivity C-reactive protein, or serum IL-6) is obligatory for diagnosis of AI. Anemia (Hb <130 g/L in men, <120 g/L in women) is usually mild (Hb ≥110 g/L)-to-moderate (Hb 80–109 g/L) normochromic, normocytic, and hypoproliferative (reticulocyte count frequently <25,000/mL), thus clearly separating from the hypochromic and microcytic IDA (Camaschella 2015; Theurl et al. 2006).

Characteristic alterations of iron metabolism help to identify AI (Weiss et al. 2019). Circulating iron concentration and the transferrin saturation (TfS) are reduced, while levels of the iron storage protein ferritin are normal or elevated (Weiss and Goodnough 2005). Transferrin (Tf) synthesis is suppressed following inflammation; thus, TfS might appear normal, consequently leading to misdiagnosis of patients with pathophysiological AI (Theurl et al. 2006). Low circulating iron concentrations are also found in IDA; therefore, serum ferritin helps to discriminate between AI and IDA (Camaschella 2015). Elevated serum ferritin in AI is related to upregulation by various inflammatory cytokines (ferritin is also an acute-phase protein), tissue injury with cellular damage, and active secretion of iron-retaining macrophages (Dignass et al. 2018). This limits its usefulness in measuring body iron stores under inflammatory conditions although low serum ferritin concentrations in an inflammatory state suggest empty iron stores.

13.3.2 Inconsistency Thresholds for AI Definition

A widely used threshold for transferrin saturation (TfS) to define reduced circulating iron concentrations is 20%. Serum ferritin ;30 μg/L is related to absolute ID identifying IDA patients, while AI patients usually present with serum ferritin 100 μg/L and those with serum ferritin 30–100 μg/L are suggested to have combined AI+IDA (Weiss et al. 2019; Camaschella 2015). These thresholds are also recommended by the European Crohn's and Colitis Organization (ECCO) and an international consensus state-

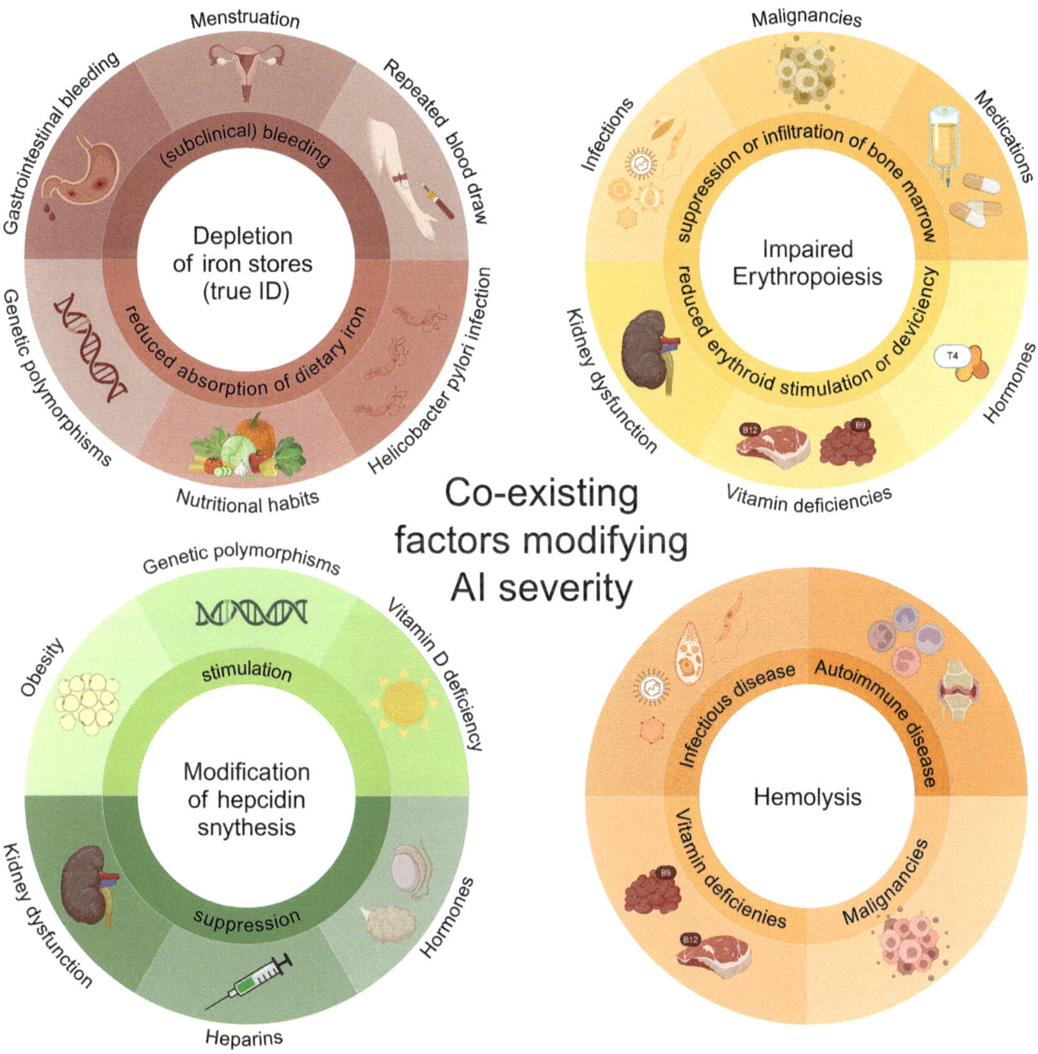

Fig. 13.3 Coexisting factors and disease that are modifying severity of anemia of inflammation

ment on perioperative management (Muñoz et al. 2017; Dignass et al. 2015).

Recently, the British Society of Gastroenterology (BSG) and American Gastroenterological Association (AGA) expanded the serum ferritin cutoffs, thus increasing sensitivity while reducing specificity for patients with empty iron stores. They recommend serum ferritin <45 μg/L for diagnosing absolute ID in clinical practice and suggest that absolute ID is unlikely under inflammatory conditions when serum ferritin >150 μg/L, defining a range of 45–150 μg/L where a combination of both absolute and functional ID (combined ID) might be considered (Snook et al. 2021; Ko et al. 2020). Increase in the serum ferritin cutoffs is supported by a recent study showing spleen iron accumulation when serum ferritin rises over 200 μg/L in patients with inflammatory diseases (Lanser et al. 2023).

The Kidney Disease Improving Global Outcomes (KDIGO) guidelines even recommend a serum ferritin <100 μg/L (<200 μg/L in end-stage kidney disease) to identify patients with absolute ID and suggest that bone marrow iron stores are normal in most patients with serum fer-

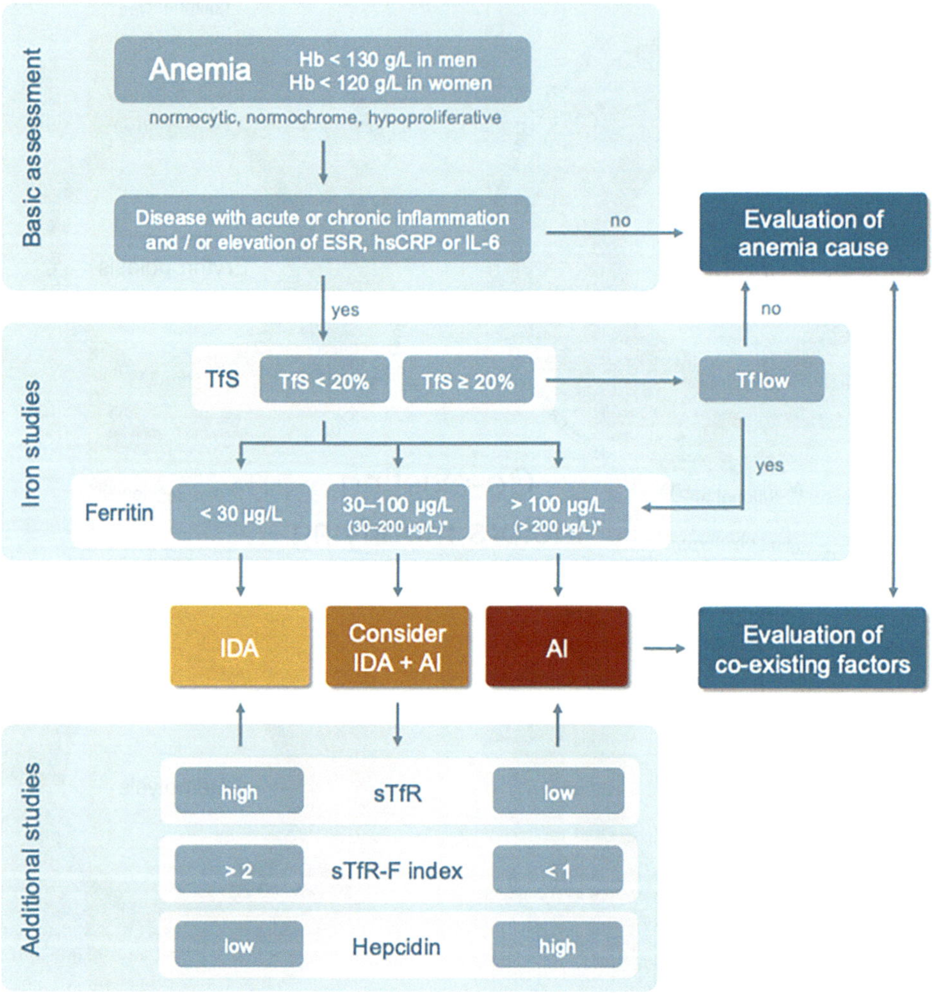

Fig. 13.4 Diagnostic evaluation of anemia in patients with inflammation. Evaluation of anemia cause or coexisting factors (see Fig. 13.3) includes testing for hemolysis (haptoglobin, lactate dehydrogenase, bilirubin), decreased kidney function (serum creatinine, estimated glomerular filtration rate), liver disease (serum transaminases), vitamin deficiencies (folic acid, vitamin B12, vitamin D3), and endocrine disorders (thyroid-stimulating hormone, cortisol), as well as age- and sex-appropriate cancer screening. In patients with IDA, further evaluation regarding the origin of absolute ID should be initiated. When heavy menstrual bleeding, repeated blood loss, and nutritional habits could be excluded, further evaluation including gastro- and colonoscopy should be performed.*Spleen iron accumulation above serum ferritin levels >200 µg/L. *Hb* hemoglobin, *ESR* erythrocyte sedimentation rate, *hsCRP* high-sensitivity C-reactive protein, *IL-6* interleukin-6, *TfS* transferrin saturation, *Tf* transferrin, *IDA* iron deficiency anemia, *AI* anemia of inflammation, *sTfR* soluble transferrin receptor, *sTfR-F index* sTfR/log(ferritin) index

ritin ≥300 µg/L, defining a range of 100–299 µg/L (respectively 200–299 µg/L) where combined ID might be considered (Babitt et al. 2021; Mcmurray et al. 2012). These cutoffs were also assumed for clinical trials and guidelines by the European Society of Cardiology (ESC), American College of Cardiology (ACC), and American Heart Association (AHA) (McDonagh et al. 2021; Heidenreich et al. 2022). However, these cutoffs were arbitrarily set without any clinical evidence from prospective randomized controlled studies.

13.3.3 Additional Laboratory Assessment in Selected Cases

The main challenge is identifying AI patients with simultaneously emptied iron stores (AI+IDA) since these patients need further diagnostic evaluation concerning potential source of bleeding (i.e., gastrointestinal bleeding, iatrogenic blood loss due to blood sampling, or other procedures), malnutrition, or increased iron requirements (i.e., pregnancies, growing children) with consequently different treatments compared to patients with pure AI (Valente de Souza et al. 2020).

Additional iron biomarkers could help to better identify patients with AI+IDA. These biomarkers strongly depend on laboratory equipment and are thus not suitable for general recommendations in clinical routine. The soluble transferrin receptor (sTfR) reflects iron requirements of TfR-expressing cells (primarily erythroid progenitor cells) and has been shown to be a reliable biomarker for detection of empty iron stores in anemic patients. Lack of assay standardization and common reference ranges limit the use of sTfR with decision threshold in clinical routine. Since sTfR could also be affected by inflammation, the sTfR-F index (sTfR/log(ferritin) ratio) was established to provide a better diagnostic accuracy in differentiating AI±IDA. Patients with a sTfR-F index ;2 are suggested having AI+IDA, while those with a sTfR-F index ;1 are suggested having "pure" AI, defining a clinically relevant gray area (Skikne et al. 2011; Weiss 2015).

The usefulness of hepcidin as a biomarker was extensively studied related to its central role in AI pathophysiology. Hepcidin is high in AI and low in IDA, whereas AI+IDA is related to lower hepcidin levels compared to AI, suggesting that ID-related hepcidin suppression might dominate over inflammation-related hepcidin stimulation (Ganz et al. 2008; Theurl et al. 2009). Finally, a reduced reticulocyte Hb content indicates iron-deficient erythropoiesis independent of inflammation (Mast et al. 2008). In addition, red blood cell indices may also help in discrimination between AI and AI+IDA, because MCV and MCH are normal in AI, whereas they can be reduced in AI+IDA (Theurl et al. 2006).

13.4 Management of Patients with AI

Treatment of the underlying disease that causes immune activation is the best therapy for AI usually resulting in its resolution over time. Concomitant pathologies contributing to the AI severity should be considered (Fig. 13.3) and when feasible specifically corrected. When the underlying disease could not be sufficiently treated, or AI persists after its treatment, other therapeutic options should be considered. Despite the growing literature on pathophysiology of AI, we are still lacking a specific made-to-measure therapy for patients with AI (Weiss et al. 2019).

Treatment considerations must take the origin of AI and anemia severity into account. As described above, AI with iron restriction is an efficient mechanism of host defense (Nairz and Weiss 2020). Iron affects also immune functions by modulating differentiation, proliferation, and antimicrobial effector mechanisms of immune cells (Soares and Weiss 2015). Specific AI treatment in patients with active infection (e.g., with iron supplementation) could promote microbial growth while impairing microbial immune response (Macdougall et al. 2016). Especially in high-burden countries for infectious disease, mild anemia ±ID might be even beneficial, while uncritical iron supplementation increases morbidity and mortality from infections (Nairz and Weiss 2020). Therefore, patients receiving iron replacement therapy or anemia correction must be carefully selected regarding their potential benefit.

Iron supplementation is the first-line therapy in AI+IDA patients, while supplemented iron is trapped in macrophages of the MPS during advanced inflammation in AI patients. New treatment strategies aiming to release restricted iron in AI gained increased interest in the last years. Inhibition of prolyl hydroxylase inhibitors (PHIs), thus stabilizing hypoxia-inducible factors (HIFs), and modification of hepcidin function and synthesis (as a master regulator of iron metabolism) seem promising from a pathophysiological perspective (Fig. 13.5).

13.4.1 Iron Supplementation

Iron that is supplemented under inflammatory conditions accumulates in macrophages because of hepcidin-related iron retention consequently being inaccessible for erythropoiesis (De Souza et al. 2023). Tissue iron retention might even promote radical formation and cellular injury, thus enhancing inflammation (Koskenkorva-Frank et al. 2013). On the contrary, iron supplementation in AI+IDA leads to hematopoietic response since ID-related hepcidin downregulation overwhelms its inflammation-related upregulation (De Souza et al. 2023; Prentice et al. 2012).

Iron can be administered orally or intravenously (Chap. 24). Inflammation impairs absorption of orally supplemented iron due to hepcidin-related Fpn1 internalization (Girelli et al. 2018). Iron that is orally supplemented but

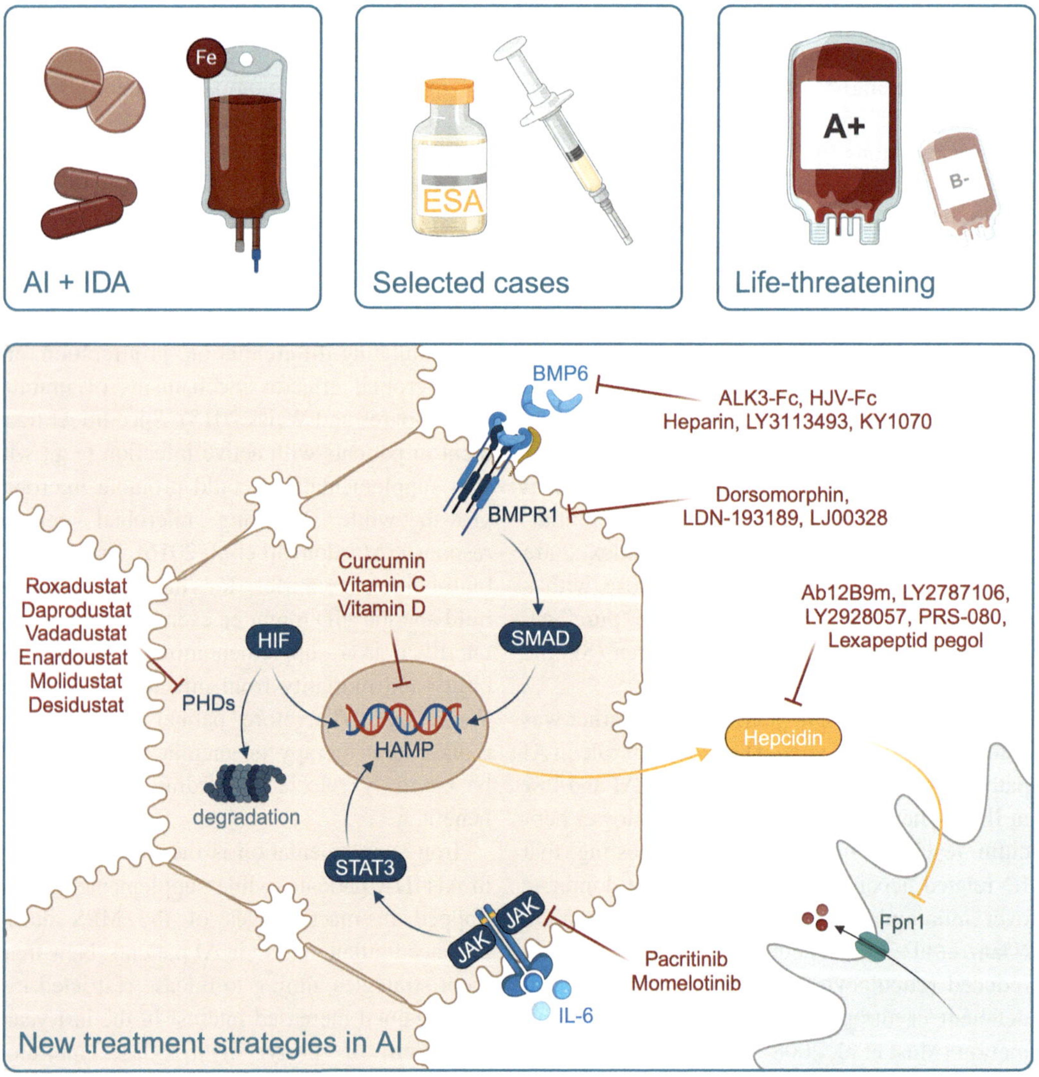

Fig. 13.5 Management of patients with AI

not absorbed leads to more side effects like nausea, constipation, and abdominal pain, thus reducing compliance in patients (Nielsen et al. 2018). Parenteral iron supplementation brings the opportunity of more rapid iron repletion, previously given only to patients not adequately responding or developing side effects to oral iron because of the very rare but potentially severe anaphylactic reactions (Nielsen et al. 2018). Nowadays, new intravenous iron formulas are more safe and established in treatment of AI+IDA and absolute ID in nonanemic patients with chronic (inflammatory) disease (Petzer et al. 2018). Low serum hepcidin concentrations might help to identify patients who benefit not only from oral but also from intravenous iron supplementation (Prentice et al. 2012).

Several clinical trials compared the effectiveness of oral and intravenous iron supplementation in inflammatory diseases. Although a meta-analysis suggested a better hematopoietic response in anemic IBD patients receiving intravenous vs. oral iron with fewer gastrointestinal side effects (Bonovas et al. 2016), oral iron supplementation might be similarly effective in IBD patients with low disease activity and absolute ID (Reinisch et al. 2013). Both oral and intravenous iron affect the gut microbiome and fecal metabolite signature, thus potentially also affecting the underlying disease (Nielsen et al. 2018; Lee et al. 2017). Clinical trials investigating iron treatment in HF showed that intravenous but not oral iron improved exercise capacity and re-hospitalization rate, while Hb levels increased independent of administration (von Haehling et al. 2019). Intravenous but not oral iron supplementation resulted in hematopoietic response with decreased blood transfusion rate and increased QoL in anemic cancer patients. However, the effects of such intervention on the clinical course and outcome of the underlying disease remain elusive so far (Busti et al. 2018). Iron absorption in tuberculosis-associated AI was shown to be negligible, while the metal was well absorbed by anemic patients after successful tuberculosis treatment (Cercamondi et al. 2021).

Bioavailability of oral iron could also be affected by other noninflammatory factors that should be considered. PPI intake, milk products, or tea decreases iron availability, while vitamin C and overnight fasting increase iron uptake (Aspuru et al. 2011). Moreover, oral iron should be taken only once daily or on alternate days since more frequent iron intake increases hepcidin production (negative feedback mechanism loop), thus reducing subsequent iron absorption (Stoffel et al. 2017). When administering intravenous iron, different physicochemical properties of intravenous iron formulations should be kept in mind. Ferric carboxymaltose and to a lesser extent ferric derisomaltose further upregulate fibroblast growth factor 23 (FGF23) expression (anyway increased in inflammation), thus favoring hypophosphatemia with hyperparathyroidism eventually leading to bone fractures, muscular weakness, and in rare cases respiratory failure (Zoller et al. 2017). FGF23 also stimulates cytokine and hepcidin production, suppresses Epo secretion, and enhances erythrocyte apoptosis, thereby exacerbating AI pathogenesis (Czaya and Faul 2019).

13.4.2 Erythropoiesis-Stimulating Agents and Red Blood Cell Transfusions

Erythropoiesis-stimulating agents (ESAs) were used for treatment of AI, especially in CKD and cancer patients for many years (Weiss and Goodnough 2005), but studies indicated a higher risk for thromboembolic complications, hemorrhage, hypertension, and tumor progression with higher adverse outcome in patients with CKD or cancer suggesting a more restrictive use of ESAs (Lanser et al. 2021b). Effectiveness of ESAs in AI patients is limited anyway due to suppressed Epo-mediated signaling and impaired erythroid cell proliferation and differentiation by inflammatory cytokines (Caiado et al. 2021) and FGF23-related inhibition of EpoR expression (Coe et al. 2014).

Red blood cell transfusions are essential for rapid correction of Hb levels in patients with life-threatening anemia including patients with acute bleeding and chemotherapy-induced anemia

(Weiss and Goodnough 2005). Otherwise, restrictive red blood cell transfusion management is recommended owing to lack of evidence in improving outcome and the increased risk of side effects including allergic or hemolytic reactions and acute lung injury (Goodnough and Murphy 2014).

13.4.3 Inhibition of Prolyl Hydroxylase Inhibitors

The HIF pathway represents an oxygen-sensing mechanism where HIFs stimulate Epo production, intestinal iron absorption, and mobilization of stored iron in response to hypoxia. Under normoxic conditions, HIFs are degraded by prolyl hydroxylases (PHs). HIF-PHIs (roxadustat, daprodustat, vadadustat, enarodustat, molidustat, desidustat) are a new class of orally administered medications that stabilize HIFs under normoxia (Nakanishi and Kuragano 2024). They were shown to mobilize stored iron (by increasing transferrin and transferrin-bound iron, and decreasing ferritin and hepcidin) and to increase Hb in anemic CKD patients. Although clinical trials excluded patients with chronic inflammation, elevated CRP levels did not impair the efficacy of HIF-PHIs. Few trials even demonstrated a better Hb response in patients with elevated CRP receiving roxadustat when compared to ESA (Stoumpos et al. 2024). Roxadustat and daprodustat were also shown to decrease LDL cholesterol although the consequences for cardiovascular endpoints are still unclear (Stoumpos et al. 2024). A recent meta-analysis involving 26,478 patients with CKD revealed no increased major adverse cardiovascular event risk or all-cause mortality compared to placebo. HIF-PHIs further had no higher risk for HF, atrial fibrillation, thromboembolic complications, or the composite kidney outcome compared to ESA, while they were related to an increased risk of thromboembolic complications when compared to placebo in nondialysis-dependent CKD (Ha et al. 2024). HIF stabilization affects not only erythropoiesis and iron metabolism but also the expression of numerous genes involved in biological functions including mitochondrial metabolism, and anti-osteoporotic, anti-inflammatory, and anti-carcinogenesis effects (Nakanishi and Kuragano 2024).

13.4.4 Modifying the Hepcidin-Ferroportin Axis

Novel targeted therapies for AI and other iron-related disorders are extensively discussed in Chap. 26. In brief, hepcidin function can be suppressed by its neutralization via hepcidin antibodies. Clinical trials with the fully humanized monoclonal antibody LY2787106, lexaptepid pegol, and Anticalin PRS-080 showed an increase in serum iron concentration and TfS (Vadhan-Raj et al. 2017; Boyce et al. 2016; Renders et al. 2019). Function of hepcidin can also be impaired by blocking its effector binding site Fpn1. The monoclonal antibody LY2928057 was shown to block the interaction of hepcidin with Fpn1, thus allowing iron efflux with a consequent increase in serum iron and TfS with a slower Hb drop (Sheetz et al. 2019).

Hepcidin expression can be suppressed by targeting the BMP/SMAD pathway, which plays an important role in the iron- and inflammation-related hepcidin expression. Preclinical studies have shown that BMPR1 inhibitors (dorsomorphin, LDN-193189, LJ00328), anti-BMP6 antibodies (LY3113493, KY1070), the human recombinant BMP ligand antagonist ALK3-Fc, and HJV-Fc decrease hepcidin expression in AI with subsequent mobilization of stored iron and recovering of anemia (Sheetz et al. 2019; Sun et al. 2013; Mayeur et al. 2015; Theurl et al. 2011, 2014; Steinbicker et al. 2011; Petzer et al. 2020). The JAK inhibitors momelotinib and pacritinib, approved for myelofibrosis treatment, also inhibit BMPR1, thus suppressing hepcidin expression and improving anemia (Duminuco et al. 2023). Preclinical studies of AI also demonstrated an anemia improvement following momelotinib treatment (Asshoff et al. 2017). Commercial heparins can suppress hepcidin expression by binding and sequestering BMP6 with the unfavorable effect of anticoagulation which is why modified heparins with reduced anticoagulation but increased hepcidin-suppressing activity were developed (Asperti et al. 2019).

Clinical trials investigating inhibition of the BMP/SMAD signaling pathway in AI patients are lacking by now. Actually, the BMP/SMAD signaling pathway also plays an important role in cardiovascular function, bone homeostasis, and cancer progression, thus limiting the therapeutic approach of BMP/SMAD signaling inhibitors for specific disease like AI (Gomez-Puerto et al. 2019). Loss of BMP6 or HJV function even promotes extrahepatic iron loading following suppression of SMAD-related hepcidin expression in the liver (Latour et al. 2017b). Therefore, pharmacological hepcidin modification must be carefully evaluated.

13.4.5 Nutritional Considerations

Besides the importance of folic acid and vitamin B12 for erythropoiesis, other vitamins also affect inflammation-related changes in iron metabolism. Vitamin D inhibits hepcidin expression with an increase of Hb levels by binding to vitamin D response elements located in the promoter region of hepcidin. Vitamin D signaling also has anti-inflammatory effects, thus probably positively affecting the course of the underlying disease. Vitamin C supports intestinal iron uptake, directly inhibits hepcidin expression, and increases EpoR expression. Moreover, curcumin reduces pro-inflammatory cytokine production while suppressing hepcidin expression. Finally, lactoferrin stimulates production of anti-infectious and anti-tumor cytokines by changing the macrophage phenotype from pro- to anti-inflammatory cytokines and consequently decreases IL-6 and hepcidin expression while increasing Fpn1 expression (Lanser et al. 2021b).

References

Alisamir M, Ebrahimi M, Rahim F (2022) Anemia in chronic obstructive pulmonary disease: a systematic review. Respir Investig 60(4):510–521. https://doi.org/10.1016/j.resinv.2022.03.006

Anand IS, Gupta P (2018) Anemia and iron deficiency in heart failure: current concepts and emerging therapies. Circulation 138(1):80–98. https://doi.org/10.1161/circulationaha.118.030099

Armitage AE, Eddowes LA, Gileadi U, Cole S, Spottiswoode N, Selvakumar TA, Ho LP, Townsend AR, Drakesmith H (2011) Hepcidin regulation by innate immune and infectious stimuli. Blood 118(15):4129–4139. https://doi.org/10.1182/blood-2011-04-351957

Aschemeyer S, Qiao B, Stefanova D, Valore EV, Sek AC, Ruwe TA, Vieth KR, Jung G, Casu C, Rivella S, Jormakka M, Mackenzie B, Ganz T, Nemeth E (2018) Structure-function analysis of ferroportin defines the binding site and an alternative mechanism of action of hepcidin. Blood 131(8):899–910. https://doi.org/10.1182/blood-2017-05-786590

Asperti M, Denardo A, Gryzik M, Arosio P, Poli M (2019) Chapter eight – The role of heparin, heparanase and heparan sulfates in hepcidin regulation. In: Litwack G (ed) Vitamins and hormones, vol 110. Academic Press, pp 157–188. https://doi.org/10.1016/bs.vh.2019.01.008

Aspuru K, Villa C, Bermejo F, Herrero P, López SG (2011) Optimal management of iron deficiency anemia due to poor dietary intake. Int J Gen Med 4:741–750. https://doi.org/10.2147/ijgm.S17788

Asshoff M, Petzer V, Warr MR, Haschka D, Tymoszuk P, Demetz E, Seifert M, Posch W, Nairz M, Maciejewski P, Fowles P, Burns CJ, Smith G, Wagner K-U, Weiss G, Whitney JA, Theurl I (2017) Momelotinib inhibits ACVR1/ALK2, decreases hepcidin production, and ameliorates anemia of chronic disease in rodents. Blood 129(13):1823–1830. https://doi.org/10.1182/blood-2016-09-740092

Babitt JL, Eisenga MF, Haase VH, Kshirsagar AV, Levin A, Locatelli F, Małyszko J, Swinkels DW, Tarng DC, Cheung M, Jadoul M, Winkelmayer WC, Drüeke TB (2021) Controversies in optimal anemia management: conclusions from a kidney disease: improving global outcomes (KDIGO) conference. Kidney Int 99(6):1280–1295. https://doi.org/10.1016/j.kint.2021.03.020

Bessman NJ, Mathieu JRR, Renassia C, Zhou L, Fung TC, Fernandez KC, Austin C, Moeller JB, Zumerle S, Louis S, Vaulont S, Ajami NJ, Sokol H, Putzel GG, Arvedson T, Sockolow RE, Lakhal-Littleton S, Cloonan SM, Arora M, Peyssonnaux C, Sonnenberg GF (2020) Dendritic cell-derived hepcidin sequesters iron from the microbiota to promote mucosal healing. Science 368(6487):186–189. https://doi.org/10.1126/science.aau6481

Bonovas S, Fiorino G, Allocca M, Lytras T, Tsantes A, Peyrin-Biroulet L, Danese S (2016) Intravenous versus oral iron for the treatment of anemia in inflammatory bowel disease: a systematic review and meta-analysis of randomized controlled trials. Medicine (Baltimore) 95(2):e2308. https://doi.org/10.1097/md.0000000000002308

Boyce M, Warrington S, Cortezi B, Zöllner S, Vauléon S, Swinkels DW, Summo L, Schwoebel F, Riecke K (2016) Safety, pharmacokinetics and pharmacodynamics of the anti-hepcidin Spiegelmer lexaptepid pegol in healthy subjects. Br J Pharmacol 173(10):1580–1588. https://doi.org/10.1111/bph.13433

Busti F, Marchi G, Ugolini S, Castagna A, Girelli D (2018) Anemia and iron deficiency in cancer patients: role of iron replacement therapy. Pharmaceuticals 11(4):94

Caiado F, Pietras EM, Manz MG (2021) Inflammation as a regulator of hematopoietic stem cell function in disease, aging, and clonal selection. J Exp Med 218(7). https://doi.org/10.1084/jem.20201541

Camaschella C (2015) Iron-deficiency anemia. N Engl J Med 372(19):1832–1843. https://doi.org/10.1056/NEJMra1401038

Cercamondi CI, Stoffel NU, Moretti D, Zoller T, Swinkels DW, Zeder C, Mhimibra F, Hella J, Fenner L, Zimmermann MB (2021) Iron homeostasis during anemia of inflammation: a prospective study of patients with tuberculosis. Blood 138(15):1293–1303. https://doi.org/10.1182/blood.2020010562

Coe LM, Madathil SV, Casu C, Lanske B, Rivella S, Sitara D (2014) FGF-23 is a negative regulator of prenatal and postnatal erythropoiesis. J Biol Chem 289(14):9795–9810. https://doi.org/10.1074/jbc.M113.527150

Czaya B, Faul C (2019) The role of fibroblast growth factor 23 in inflammation and anemia. Int J Mol Sci 20(17):4195. https://doi.org/10.3390/ijms20174195

Dallalio G, Means RT Jr (2003) Effects of oxidative stress on human erythroid colony formation: modulation by gamma-interferon. J Lab Clin Med 141(6):395–400. https://doi.org/10.1016/s0022-2143(03)00041-6

De Souza LV, Hoffmann A, Fischer C, Petzer V, Asshoff M, Theurl I, Tymoszuk P, Seifert M, Brigo N, Hilbe R, Demetz E, Von Raffay L, Berger S, Barros-Pinkelnig M, Weiss G (2023) Comparative analysis of oral and intravenous iron therapy in rat models of inflammatory anemia and iron deficiency. Haematologica 108(1):135–149. https://doi.org/10.3324/haematol.2022.281149

Delaby C, Pilard N, Puy H, Canonne-Hergaux F (2008) Sequential regulation of ferroportin expression after erythrophagocytosis in murine macrophages: early mRNA induction by haem, followed by iron-dependent protein expression. Biochem J 411(1):123–131. https://doi.org/10.1042/bj20071474

Dignass AU, Gasche C, Bettenworth D, Birgegård G, Danese S, Gisbert JP, Gomollon F, Iqbal T, Katsanos K, Koutroubakis I, Magro F, Savoye G, Stein J, Vavricka S, Crohn's tE, Organisation C (2015) European consensus on the diagnosis and management of iron deficiency and anaemia in inflammatory bowel diseases. J Crohn's Colitis 9(3):211–222. https://doi.org/10.1093/ecco-jcc/jju009

Dignass A, Farrag K, Stein J (2018) Limitations of serum ferritin in diagnosing iron deficiency in inflammatory conditions. Int J Chronic Dis 2018(1):9394060. https://doi.org/10.1155/2018/9394060

Docherty AB, Turgeon AF, Walsh TS (2018) Best practice in critical care: anaemia in acute and critical illness. Transfus Med 28(2):181–189. https://doi.org/10.1111/tme.12505

Drakesmith H, Nemeth E, Ganz T (2015) Ironing out Ferroportin. Cell Metab 22(5):777–787. https://doi.org/10.1016/j.cmet.2015.09.006

Duminuco A, Chifotides HT, Giallongo S, Giallongo C, Tibullo D, Palumbo GA (2023) ACVR1: a novel therapeutic target to treat anemia in myelofibrosis. Cancers (Basel) 16(1). https://doi.org/10.3390/cancers16010154

Ganz T, Nemeth E (2012) Hepcidin and iron homeostasis. Biochim Biophys Acta 1823(9):1434–1443. https://doi.org/10.1016/j.bbamcr.2012.01.014

Ganz T, Olbina G, Girelli D, Nemeth E, Westerman M (2008) Immunoassay for human serum hepcidin. Blood 112(10):4292–4297. https://doi.org/10.1182/blood-2008-02-139915

Gaspar BL, Sharma P, Das R (2015) Anemia in malignancies: pathogenetic and diagnostic considerations. Hematology 20(1):18–25. https://doi.org/10.1179/1607845414Y.0000000161

Girelli D, Nemeth E, Swinkels DW (2016) Hepcidin in the diagnosis of iron disorders. Blood 127(23):2809–2813. https://doi.org/10.1182/blood-2015-12-639112

Girelli D, Ugolini S, Busti F, Marchi G, Castagna A (2018) Modern iron replacement therapy: clinical and pathophysiological insights. Int J Hematol 107(1):16–30. https://doi.org/10.1007/s12185-017-2373-3

Gomez-Puerto MC, Iyengar PV, García de Vinuesa A, ten Dijke P, Sanchez-Duffhues G (2019) Bone morphogenetic protein receptor signal transduction in human disease. J Pathol 247(1):9–20. https://doi.org/10.1002/path.5170

Goodnough LT, Murphy MF (2014) Do liberal blood transfusions cause more harm than good? BMJ (Clinical research ed) 349:g6897. https://doi.org/10.1136/bmj.g6897

Ha JT, Hiremath S, Jun M, Green SC, Wheeler DC, Coyne DW, Perkovic V, Badve SV (2024) Hypoxia-inducible factor prolyl hydroxylase inhibitors in kidney disease. NEJM Evid 3(9):EVIDoa2300189. https://doi.org/10.1056/EVIDoa2300189

Hayden SJ, Albert TJ, Watkins TR, Swenson ER (2012) Anemia in critical illness: insights into etiology, consequences, and management. Am J Respir Crit Care Med 185(10):1049–1057. https://doi.org/10.1164/rccm.201110-1915CI

Heidenreich PA, Bozkurt B, Aguilar D, Allen LA, Byun JJ, Colvin MM, Deswal A, Drazner MH, Dunlay SM, Evers LR, Fang JC, Fedson SE, Fonarow GC, Hayek SS, Hernandez AF, Khazanie P, Kittleson MM, Lee CS, Link MS, Milano CA, Nnacheta LC, Sandhu AT, Stevenson LW, Vardeny O, Vest AR, Yancy CW (2022) 2022 AHA/ACC/HFSA guideline for the Management of Heart Failure: a report of the American College of Cardiology/American Heart Association joint committee on clinical practice guidelines. Circulation 145(18):e895–e1032. https://doi.org/10.1161/CIR.0000000000001063

Jelkmann W (2011) Regulation of erythropoietin production. J Physiol 589(Pt 6):1251–1258. https://doi.org/10.1113/jphysiol.2010.195057

Khalil S, Delehanty L, Grado S, Holy M, White Z 3rd, Freeman K, Kurita R, Nakamura Y, Bullock G, Goldfarb A (2018) Iron modulation of erythropoiesis is associated with scribble-mediated control of the erythropoietin receptor. J Exp Med 215(2):661–679. https://doi.org/10.1084/jem.20170396

Ko CW, Siddique SM, Patel A, Harris A, Sultan S, Altayar O, Falck-Ytter Y (2020) AGA clinical practice guidelines on the gastrointestinal evaluation of iron deficiency anemia. Gastroenterology 159(3):1085–1094. https://doi.org/10.1053/j.gastro.2020.06.046

Koskenkorva-Frank TS, Weiss G, Koppenol WH, Burckhardt S (2013) The complex interplay of iron metabolism, reactive oxygen species, and reactive nitrogen species: insights into the potential of various iron therapies to induce oxidative and nitrosative stress. Free Radic Biol Med 65:1174–1194. https://doi.org/10.1016/j.freeradbiomed.2013.09.001

Lanser L, Fuchs D, Scharnagl H, Grammer T, Kleber ME, März W, Weiss G, Kurz K (2021a) Anemia of chronic disease in patients with cardiovascular disease. Front Cardiovasc Med 8(939). https://doi.org/10.3389/fcvm.2021.666638

Lanser L, Fuchs D, Kurz K, Weiss G (2021b) Physiology and inflammation driven pathophysiology of iron homeostasis—mechanistic insights into anemia of inflammation and its treatment. Nutrients 13(11):3732

Lanser L, Plaikner M, Schroll A, Burkert FR, Seiwald S, Fauser J, Petzer V, Bellmann-Weiler R, Fritsche G, Tancevski I, Duftner C, Pircher A, Seeber A, Zoller H, Kremser C, Henninger B, Weiss G (2023) Tissue iron distribution in patients with anemia of inflammation: results of a pilot study. Am J Hematol. https://doi.org/10.1002/ajh.26909

Latour C, Wlodarczyk MF, Jung G, Gineste A, Blanchard N, Ganz T, Roth MP, Coppin H, Kautz L (2017a) Erythroferrone contributes to hepcidin repression in a mouse model of malarial anemia. Haematologica 102(1):60–68. https://doi.org/10.3324/haematol.2016.150227

Latour C, Besson-Fournier C, Gourbeyre O, Meynard D, Roth M-P, Coppin H (2017b) Deletion of BMP6 worsens the phenotype of HJV-deficient mice and attenuates hepcidin levels reached after LPS challenge. Blood 130(21):2339–2343. https://doi.org/10.1182/blood-2017-07-795658

Lee T, Clavel T, Smirnov K, Schmidt A, Lagkouvardos I, Walker A, Lucio M, Michalke B, Schmitt-Kopplin P, Fedorak R, Haller D (2017) Oral versus intravenous iron replacement therapy distinctly alters the gut microbiota and metabolome in patients with IBD. Gut 66(5):863–871. https://doi.org/10.1136/gutjnl-2015-309940

Libregts SF, Gutiérrez L, de Bruin AM, Wensveen FM, Papadopoulos P, van Ijcken W, Ozgür Z, Philipsen S, Nolte MA (2011) Chronic IFN-γ production in mice induces anemia by reducing erythrocyte life span and inhibiting erythropoiesis through an IRF-1/PU.1 axis. Blood 118(9):2578–2588. https://doi.org/10.1182/blood-2010-10-315218

Ludwiczek S, Aigner E, Theurl I, Weiss GN (2003) Cytokine-mediated regulation of iron transport in human monocytic cells. Blood 101(10):4148–4154. https://doi.org/10.1182/blood-2002-08-2459

Macdougall IC, Bircher AJ, Eckardt KU, Obrador GT, Pollock CA, Stenvinkel P, Swinkels DW, Wanner C, Weiss G, Chertow GM (2016) Iron management in chronic kidney disease: conclusions from a "kidney disease: improving global outcomes" (KDIGO) controversies conference. Kidney Int 89(1):28–39. https://doi.org/10.1016/j.kint.2015.10.002

Marques O, Weiss G, Muckenthaler MU (2022) The role of iron in chronic inflammatory diseases: from mechanisms to treatment options in anemia of inflammation. Blood 140(19):2011–2023. https://doi.org/10.1182/blood.2021013472

Mast AE, Blinder MA, Dietzen DJ (2008) Reticulocyte hemoglobin content. Am J Hematol 83(4):307–310. https://doi.org/10.1002/ajh.21090

Mayeur C, Kolodziej SA, Wang A, Xu X, Lee A, Yu PB, Shen J, Bloch KD, Bloch DB (2015) Oral administration of a bone morphogenetic protein type I receptor inhibitor prevents the development of anemia of inflammation. Haematologica 100(2):e68–e71. https://doi.org/10.3324/haematol.2014.111484

McDonagh TA, Metra M, Adamo M, Gardner RS, Baumbach A, Böhm M, Burri H, Butler J, Čelutkienė J, Chioncel O, Cleland JGF, Coats AJS, Crespo-Leiro MG, Farmakis D, Gilard M, Heymans S, Hoes AW, Jaarsma T, Jankowska EA, Lainscak M, CSP L, Lyon AR, McMurray JJV, Mebazaa A, Mindham R, Muneretto C, Francesco Piepoli M, Price S, Rosano GMC, Ruschitzka F, Kathrine Skibelund A, Group ESD (2021) 2021 ESC guidelines for the diagnosis and treatment of acute and chronic heart failure: developed by the task force for the diagnosis and treatment of acute and chronic heart failure of the European Society of Cardiology (ESC) with the special contribution of the heart failure association (HFA) of the ESC. Eur Heart J 42(36):3599–3726. https://doi.org/10.1093/eurheartj/ehab368

Mcmurray J, Parfrey PS, Adamson JW, Aljama P, Berns JS, Bohlius J, Drüeke TB, Finkelstein FO, Fishbane S, Ganz T, Macdougall IC, McDonald RA, McMahon LP, Obrador GT, Strippoli GFM, Weiss G, Więcek A (2012) KDIGO clinical practice guideline for anemia in chronic kidney disease

Miller LL, Miller SC, Torti SV, Tsuji Y, Torti FM (1991) Iron-independent induction of ferritin H chain by tumor necrosis factor. Proc Natl Acad Sci 88(11):4946–4950. https://doi.org/10.1073/pnas.88.11.4946

Mitlyng BL, Singh JA, Furne JK, Ruddy J, Levitt MD (2006) Use of breath carbon monoxide measurements to assess erythrocyte survival in subjects with chronic diseases. Am J Hematol 81(6):432–438. https://doi.org/10.1002/ajh.20644

Moldawer LL, Marano MA, Wei H, Fong Y, Silen ML, Kuo G, Manogue KR, Vlassara H, Cohen H, Cerami A, Lowry SF (1989) Cachectin/tumor necrosis factor-α alters red blood cell kinetics and induces

anemia in vivo. FASEB J 3(5):1637–1643. https://doi.org/10.1096/fasebj.3.5.2784116
Muckenthaler MU, Rivella S, Hentze MW, Galy B (2017) A red carpet for iron metabolism. Cell 168(3):344–361. https://doi.org/10.1016/j.cell.2016.12.034
Muñoz M, Acheson AG, Auerbach M, Besser M, Habler O, Kehlet H, Liumbruno GM, Lasocki S, Meybohm P, Rao Baikady R, Richards T, Shander A, So-Osman C, Spahn DR, Klein AA (2017) International consensus statement on the peri-operative management of anaemia and iron deficiency. Anaesthesia 72(2):233–247. https://doi.org/10.1111/anae.13773
Nairz M, Weiss G (2020) Iron in infection and immunity. Mol Asp Med 75:100864. https://doi.org/10.1016/j.mam.2020.100864
Nairz M, Sonnweber T, Schroll A, Theurl I, Weiss G (2012) The pleiotropic effects of erythropoietin in infection and inflammation. Microbes Infect 14(3):238–246. https://doi.org/10.1016/j.micinf.2011.10.005
Nakanishi T, Kuragano T (2024) Growing concerns about using hypoxia-inducible factor prolyl hydroxylase inhibitors for the treatment of renal anemia. Clin Kidney J 17:3. https://doi.org/10.1093/ckj/sfae051
Nakanishi T, Kimura T, Kuragano T (2019) The Hepcidin-anemia Axis: pathogenesis of anemia in chronic kidney disease. Contrib Nephrol 198:124–134. https://doi.org/10.1159/000496636
Nemeth E, Tuttle MS, Powelson J, Vaughn MB, Donovan A, Ward DM, Ganz T, Kaplan J (2004) Hepcidin regulates cellular iron efflux by binding to ferroportin and inducing its internalization. Science 306(5704):2090–2093. https://doi.org/10.1126/science.1104742
Nielsen OH, Soendergaard C, Vikner ME, Weiss G (2018) Rational management of iron-deficiency anaemia in inflammatory bowel disease. Nutrients 10(1):82
Okonko DO, Marley SB, Anker SD, Poole-Wilson PA, Gordon MY (2013) Erythropoietin resistance contributes to anaemia in chronic heart failure and relates to aberrant JAK-STAT signal transduction. Int J Cardiol 164(3):359–364. https://doi.org/10.1016/j.ijcard.2011.07.045
Petzer V, Theurl I, Weiss G (2018) Established and emerging concepts to treat imbalances of iron homeostasis in inflammatory diseases. Pharmaceuticals (Basel) 11(4):135. https://doi.org/10.3390/ph11040135
Petzer V, Tymoszuk P, Asshoff M, Carvalho J, Papworth J, Deantonio C, Bayliss L, Wake MS, Seifert M, Brigo N, Valente de Souza L, Hilbe R, Grubwieser P, Demetz E, Dichtl S, Volani C, Berger S, Böhm F, Hoffmann A, Pfeifhofer-Obermair C, von Raffay L, Sopper S, Arndt S, Bosserhoff A, Kautz L, Perrier P, Nairz M, Wolf D, Weiss G, Germaschewski V, Theurl I (2020) A fully human anti-BMP6 antibody reduces the need for erythropoietin in rodent models of the anemia of chronic disease. Blood 136(9):1080–1090. https://doi.org/10.1182/blood.2019004653
Prentice AM, Doherty CP, Abrams SA, Cox SE, Atkinson SH, Verhoef H, Armitage AE, Drakesmith H (2012) Hepcidin is the major predictor of erythrocyte iron incorporation in anemic African children. Blood 119(8):1922–1928. https://doi.org/10.1182/blood-2011-11-391219
Purdy JC, Shatzel JJ (2021) The hematologic consequences of obesity. Eur J Haematol 106(3):306–319. https://doi.org/10.1111/ejh.13560
Reinisch W, Staun M, Tandon RK, Altorjay I, Thillainayagam AV, Gratzer C, Nijhawan S, Thomsen LL (2013) A randomized, open-label, non-inferiority study of intravenous iron isomaltoside 1,000 (Monofer) compared with oral iron for treatment of anemia in IBD (PROCEED). Am J Gastroenterol 108(12):1877–1888. https://doi.org/10.1038/ajg.2013.335
Renders L, Budde K, Rosenberger C, van Swelm R, Swinkels D, Dellanna F, Feuerer W, Wen M, Erley C, Bader B, Sommerer C, Schaier M, Meurer K, Matis L (2019) First-in-human phase I studies of PRS-080#22, a hepcidin antagonist, in healthy volunteers and patients with chronic kidney disease undergoing hemodialysis. PLoS One 14(3):e0212023. https://doi.org/10.1371/journal.pone.0212023
Sheetz M, Barrington P, Callies S, Berg PH, McColm J, Marbury T, Decker B, Dyas GL, Truhlar SME, Benschop R, Leung D, Berg J, Witcher DR (2019) Targeting the hepcidin–ferroportin pathway in anaemia of chronic kidney disease. Br J Clin Pharmacol 85(5):935–948. https://doi.org/10.1111/bcp.13877
Skikne BS, Punnonen K, Caldron PH, Bennett MT, Rehu M, Gasior GH, Chamberlin JS, Sullivan LA, Bray KR, Southwick PC (2011) Improved differential diagnosis of anemia of chronic disease and iron deficiency anemia: a prospective multicenter evaluation of soluble transferrin receptor and the sTfR/log ferritin index. Am J Hematol 86(11):923–927. https://doi.org/10.1002/ajh.22108
Snook J, Bhala N, Beales ILP, Cannings D, Kightley C, Logan RP, Pritchard DM, Sidhu R, Surgenor S, Thomas W, Verma AM, Goddard AF (2021) British Society of Gastroenterology guidelines for the management of iron deficiency anaemia in adults. Gut 70(11):2030–2051. https://doi.org/10.1136/gutjnl-2021-325210
Soares MP, Weiss G (2015) The iron age of host–microbe interactions. EMBO Rep 16(11):1482–1500. https://doi.org/10.15252/embr.201540558
Sonnweber T, Nairz M, Theurl I, Petzer V, Tymoszuk P, Haschka D, Rieger E, Kaessmann B, Deri M, Watzinger K, Steringer-Mascherbauer R, Tancevski I, Weiss G, Löffler-Ragg J (2018) The crucial impact of iron deficiency definition for the course of precapillary pulmonary hypertension. PLoS One 13(8):e0203396. https://doi.org/10.1371/journal.pone.0203396
Stauder R, Valent P, Theurl I (2018) Anemia at older age: etiologies, clinical implications, and management. Blood 131(5):505–514. https://doi.org/10.1182/blood-2017-07-746446
Steinbicker AU, Muckenthaler MU (2013) Out of balance--systemic iron homeostasis in iron-related disorders.

Nutrients 5(8):3034–3061. https://doi.org/10.3390/nu5083034

Steinbicker AU, Sachidanandan C, Vonner AJ, Yusuf RZ, Deng DY, Lai CS, Rauwerdink KM, Winn JC, Saez B, Cook CM, Szekely BA, Roy CN, Seehra JS, Cuny GD, Scadden DT, Peterson RT, Bloch KD, Yu PB (2011) Inhibition of bone morphogenetic protein signaling attenuates anemia associated with inflammation. Blood 117(18):4915–4923. https://doi.org/10.1182/blood-2010-10-313064

Stoffel NU, Cercamondi CI, Brittenham G, Zeder C, Geurts-Moespot AJ, Swinkels DW, Moretti D, Zimmermann MB (2017) Iron absorption from oral iron supplements given on consecutive versus alternate days and as single morning doses versus twice-daily split dosing in iron-depleted women: two open-label, randomised controlled trials. Lancet Haematol 4(11):e524–e533. https://doi.org/10.1016/s2352-3026(17)30182-5

Stoumpos S, Crowe K, Sarafidis P, Barratt J, Bolignano D, Del Vecchio L, Małyszko J, Więcek A, Ortiz A, Cozzolino M (2024) Hypoxia-inducible factor prolyl hydroxylase inhibitors for anaemia in chronic kidney disease: a clinical practice document by the European renal best practice board of the European renal association. Nephrol Dialysis Transplant 39(10):1710–1730. https://doi.org/10.1093/ndt/gfae075

Sun CC, Vaja V, Chen S, Theurl I, Stepanek A, Brown DE, Cappellini MD, Weiss G, Hong CC, Lin HY, Babitt JL (2013) A hepcidin lowering agent mobilizes iron for incorporation into red blood cells in an adenine-induced kidney disease model of anemia in rats. Nephrol Dial Transplant 28(7):1733–1743. https://doi.org/10.1093/ndt/gfs584

Taniguchi S, Dai CH, Price JO, Krantz SB (1997) Interferon gamma downregulates stem cell factor and erythropoietin receptors but not insulin-like growth factor-I receptors in human erythroid colony-forming cells. Blood 90(6):2244–2252

Theurl I, Mattle V, Seifert M, Mariani M, Marth C, Weiss G (2006) Dysregulated monocyte iron homeostasis and erythropoietin formation in patients with anemia of chronic disease. Blood 107(10):4142–4148. https://doi.org/10.1182/blood-2005-08-3364

Theurl I, Theurl M, Seifert M, Mair S, Nairz M, Rumpold H, Zoller H, Bellmann-Weiler R, Niederegger H, Talasz H, Weiss G (2008) Autocrine formation of hepcidin induces iron retention in human monocytes. Blood 111(4):2392–2399. https://doi.org/10.1182/blood-2007-05-090019

Theurl I, Aigner E, Theurl M, Nairz M, Seifert M, Schroll A, Sonnweber T, Eberwein L, Witcher DR, Murphy AT, Wroblewski VJ, Wurz E, Datz C, Weiss G (2009) Regulation of iron homeostasis in anemia of chronic disease and iron deficiency anemia: diagnostic and therapeutic implications. Blood 113(21):5277–5286. https://doi.org/10.1182/blood-2008-12-195651

Theurl I, Schroll A, Sonnweber T, Nairz M, Theurl M, Willenbacher W, Eller K, Wolf D, Seifert M, Sun CC, Babitt JL, Hong CC, Menhall T, Gearing P, Lin HY, Weiss G (2011) Pharmacologic inhibition of hepcidin expression reverses anemia of chronic inflammation in rats. Blood 118(18):4977–4984. https://doi.org/10.1182/blood-2011-03-345066

Theurl M, Nairz M, Schroll A, Sonnweber T, Asshoff M, Haschka D, Seifert M, Willenbacher W, Wilflingseder D, Posch W, Murphy AT, Witcher DR, Theurl I, Weiss G (2014) Hepcidin as a predictive factor and therapeutic target in erythropoiesis-stimulating agent treatment for anemia of chronic disease in rats. Haematologica 99(9):1516–1524. https://doi.org/10.3324/haematol.2013.099481

Theurl I, Hilgendorf I, Nairz M, Tymoszuk P, Haschka D, Asshoff M, He S, Gerhardt LM, Holderried TA, Seifert M, Sopper S, Fenn AM, Anzai A, Rattik S, McAlpine C, Theurl M, Wieghofer P, Iwamoto Y, Weber GF, Harder NK, Chousterman BG, Arvedson TL, McKee M, Wang F, Lutz OM, Rezoagli E, Babitt JL, Berra L, Prinz M, Nahrendorf M, Weiss G, Weissleder R, Lin HY, Swirski FK (2016) On-demand erythrocyte disposal and iron recycling requires transient macrophages in the liver. Nat Med 22(8):945–951. https://doi.org/10.1038/nm.4146

Tilg H, Ulmer H, Kaser A, Weiss G (2002) Role of IL-10 for induction of anemia during inflammation. J Immunol 169(4):2204–2209

Vadhan-Raj S, Abonour R, Goldman JW, Smith DA, Slapak CA, Ilaria RL Jr, Tiu RV, Wang X, Callies S, Cox J, Tuttle JL, Lau Y-K, Roeland EJ (2017) A first-in-human phase 1 study of a hepcidin monoclonal antibody, LY2787106, in cancer-associated anemia. J Hematol Oncol 10(1):73–73. https://doi.org/10.1186/s13045-017-0427-x

Valente de Souza L, Hoffmann A, Weiss G (2020) Impact of bacterial infections on erythropoiesis. Expert Rev Anti Infect Ther:1–15. https://doi.org/10.1080/14787210.2021.1841636

von Haehling S, Ebner N, Evertz R, Ponikowski P, Anker SD (2019) Iron deficiency in heart failure: an overview. JACC. Heart Failure 7(1):36–46. https://doi.org/10.1016/j.jchf.2018.07.015

Weiss G (2015) Anemia of chronic disorders: new diagnostic tools and new treatment strategies. Semin Hematol 52(4):313–320. https://doi.org/10.1053/j.seminhematol.2015.07.004

Weiss G, Goodnough LT (2005) Anemia of chronic disease. N Engl J Med 352(10):1011–1023. https://doi.org/10.1056/NEJMra041809

Weiss G, Schett G (2013) Anaemia in inflammatory rheumatic diseases. Nat Rev Rheumatol 9(4):205–215. https://doi.org/10.1038/nrrheum.2012.183

Weiss G, Bogdan C, Hentze MW (1997) Pathways for the regulation of macrophage iron metabolism by the anti-inflammatory cytokines IL-4 and IL-13. J Immunol 158(1):420–425

Weiss G, Ganz T, Goodnough LT (2019) Anemia of inflammation. Blood 133(1):40–50. https://doi.org/10.1182/blood-2018-06-856500

Zoller H, Schaefer B, Glodny B (2017) Iron-induced hypophosphatemia: an emerging complication. Curr Opin Nephrol Hypertens. https://doi.org/10.1097/MNH.0000000000000329

14 Iron, Hepcidin, and Immunity

Daisy Crisell, Nick Talbot, and Hal Drakesmith

Abstract

Infection involves the utilization of host nutritional resources by the pathogen. Genetic, experimental, and clinical evidence supports the critical role of iron as a focus for both pathogen attack and host defense. The molecular basis of how mammals regulate iron transport to maintain homeostasis, and in response to infection, has been well characterized since the discovery of the hormone hepcidin. The multiple activities of iron in host innate and adaptive immune cell function have also been studied in increasing depth. Recently these research strands have been converging, showing how host regulation of iron affects not only classical "nutritional immunity" as a defense against pathogens but also influences the development and magnitude of immune responses. These findings are discussed in the context of different types of infectious microorganisms, against a background of host disease states (including respiratory disorders), and in relation to iron supplementation, vaccination, and pathological immune responses.

Keywords

Hepcidin · Infection · Immune responses · Hypoferremia · Vaccination

D. Crisell
Clinical Medical School, University of Oxford, Oxford, UK

Exeter College, University of Oxford, Oxford, UK

N. Talbot
Department of Physiology, Anatomy and Genetics, University of Oxford, Oxford, UK

Oxford Centre for Respiratory Medicine, Oxford University Hospitals NHS Foundation Trust, Oxford, UK

Nuffield Department of Medicine, University of Oxford, Oxford, UK

H. Drakesmith (✉)
MRC Translational Immune Discovery Unit, MRC Weatherall Institute of Molecular Medicine, Radcliffe Department of Medicine, John Radcliffe Hospital, University of Oxford, Oxford, UK
e-mail: alexander.drakesmith@imm.ox.ac.uk

14.1 Hepcidin in Infection

The iron regulatory hormone hepcidin (Chap. 3) is induced by inflammatory responses to infection, predominantly by the cytokine IL-6 (Nemeth et al. 2004). This is an innate immune response and part of "nutritional immunity": the action of inflammatory mediators on the expression and activity of transporters and controllers of metabolites and nutrients. The rapid upregulation of hepcidin can cause a state of profound hypoferremia, in which serum iron concentrations decline by four-fifths so that transferrin saturation can drop to 5% or less (Darton et al. 2015; Rivera et al. 2005). This redistribution of iron has

K. Pantopoulos (ed.), *Iron Metabolism in Human Health and Disease*, Advances in Experimental Medicine and Biology 1480, https://doi.org/10.1007/978-3-031-92033-2_14

consequences for the course of infections, but the effects vary considerably depending on the type of pathogen (Drakesmith and Prentice 2012). For example, the hypoferremic response decreases the availability of iron to blood-borne siderophilic pathogens. The importance of this response is illustrated by the consequences of deregulated hepcidin synthesis for susceptibility to extracellular bacteria.

14.2 Hepcidin and Extracellular Bacterial Infections

The iron overloading disorder hemochromatosis (Chaps. 9 and 10) is characterized by hepcidin insufficiency and leads to chronic iron over-accumulation, which if excessive enough, causes pathology via iron-catalyzed chemical tissue damage (Nemeth and Ganz 2022). However, the increase in iron also predisposes to certain infections. In a particularly striking case, a laboratory-acquired and usually safe attenuated strain of *Yersinia pestis*, in which a high-pathogenicity island encoding iron-uptake machinery was deleted (Carniel 2001), was fatal to an exposed researcher who had underlying hereditary hemochromatosis that was diagnosed post-mortem (Frank et al. 2011). *Vibrio vulnificus* is a siderophilic extracellular bacterium that usually causes few clinical symptoms in healthy individuals but can lead to fatal rapid sepsis in patients with hemochromatosis (Horseman and Surani 2011). Pre-clinical mechanistic experiments support the causal protective role of hepcidin. Compared to wild-type, hepcidin knock-out mice are unable to generate sufficient hypoferremia in response to *V. vulnificus* infection, the infection being fatal within 24 hours; but this can be rescued by administration of a hepcidin agonist to control the availability of excess (non-transferrin-bound) iron that stimulates bacterial proliferation (Arezes et al. 2015). Hepcidin agonists can also protect against other species of extracellular bacteria through the same mechanism (Stefanova et al. 2017). Many bacteria (*Neisseria* and *Shigella* species) can also directly acquire transferrin-bound iron through bacterially encoded transferrin-capturing machinery (Barber and Elde 2014); potentially a hepcidin-mediated decrease in transferrin-bound iron concentrations may also have a protective effect by delaying growth of such transferrin-targeting bacteria, but this has not been explored fully.

14.3 Hepcidin and *Salmonella* Infection

The protective effect of hepcidin-mediated hypoferremia against iron-demanding extracellular bacteria is clear. However, the balance of hepcidin-induced hypoferremia and an immune response to intracellular bacterial infection is often more complex.

The hepcidin-ferroportin axis is a key mediator of *Salmonella* infection and is well illustrated in the case of hemochromatosis. The allelic frequency of mutations in the HFE gene, causing hereditary hemochromatosis, are relatively high in people of northern European descent (Nairz et al. 2009); protection against *Salmonella*, or other intracellular bacteria, infection may provide an evolutionary explanation for this phenomenon. Hemochromatosis is characterized by a partial or complete hepcidin deficiency and subsequent macrophage iron depletion. Studies in $Hfe^{-/-}$ and $Hfe^{+/-}$ mouse models infected with *S. Typhimurium* demonstrate improved survival compared to wild type (Nairz et al. 2009, 2017). This protection was associated with lower iron availability in the macrophages of these mice and the antimicrobial activity of macrophages through factors such as lipocalin-2 (Lcn2) (Nairz et al. 2009). Therefore, the reduced activity of the hepcidin-ferroportin axis, leading to macrophage iron depletion, in models of hemochromatosis allows better control of *Salmonella* infection.

In the case of normal hepcidin-ferroportin axis activity, macrophage iron sequestration is detrimental during *Salmonella* infection. Nairz et al. (2017) showed no significant difference in hepcidin expression in response to *Salmonella* infection in both Hfe−/− and wild-type mice (Nairz et al. 2017), suggesting determinants of survival of Hfe−/− mice occur at the level of the infected macrophage. Iron sequestration in macrophages has been shown to affect intracellular immune pathways, reducing interferon gamma activity (Oexle et al. 2003), and impairing nitric oxide formation and TNF-alpha production in *Salmonella*-infected macrophages (Nairz et al. 2007). A dual RNA sequencing analysis of dendritic cell (DC) infection by different *Salmonella* strains found that, whilst infection with all strains upregulated transferrin receptor mRNA in DCs, transferrin receptor protein expression was significantly higher in *S. Typhi*-infected cells, reflecting an enhanced ability to drive DC iron acquisition (Aulicino et al. 2022). Furthermore, pre-treatment of DCs with iron citrate reduced recoverable *S. Typhimurium* after infection, but not *S. Typhi* (Aulicino et al. 2022). Compared to the immune dysfunction seen in macrophages, this suggests that DCs can continue antimicrobial responses in an iron-rich setting. These results also indicate strain-dependent differences in the ability of *Salmonella* to circumvent host iron homeostasis and immunity.

The key mechanism by which macrophages mount an antimicrobial response against *Salmonella* infection is the upregulation of ferroportin (Nairz et al. 2007), tipping the hepcidin-ferroportin axis toward iron efflux. Nairz et al. (2013) showed that nitric oxide (NO) signaling plays a key role in this; NO generated by inducible NO synthase 2 (NOS2) in response to *Salmonella* infection permits binding of nuclear factor erythroid 2-related factor (Nrf2) to the ferroportin gene promoter, upregulating its expression and promoting iron efflux (Nairz et al. 2013). Mirroring this, infection of *Nos2*$^{-/-}$ mice with *S. typhimurium* led to increased iron accumulation, reduced IFN-gamma expression, and impaired pathogen control compared to wild type; these measures were restored on administration of an iron chelator or hyperexpression of ferroportin (Nairz et al. 2013). Therefore, intracellular NO signaling is a key modulator of the hepcidin-ferroportin axis and cellular iron homeostasis during *Salmonella* infection, aiding the infected cell in subverting the systemic actions of hepcidin during infection.

Further intracellular host factors, including Lcn2 that binds the *Salmonella* siderophore enterobactin, also provide protection. *Hfe*$^{-/-}$*Lcn2*$^{-/-}$ and *Hfe*$^{+/+}$*Lcn2*$^{-/-}$ mice show an increased bacterial load compared to *Hfe*$^{-/-}$*Lcn2*$^{+/+}$ mice (Nairz et al. 2007), indicating that the protective effect of a locally-disrupted hepcidin-ferroportin axis is mediated through the activity of Lcn2. Therefore, weighting of the hepcidin-ferroportin axis toward macrophage iron efflux promotes downstream intracellular protective mechanisms against *Salmonella* infection.

Despite the host balancing systemic and cellular mechanisms to mediate an effective immune response, *Salmonella* is able to modulate this; in addition to producing siderophores, Salmonella targets the hepcidin-ferroportin axis. Kim et al. elucidated a pathway in mice by which *S. typhimurium* infection stimulated IL-6 production and subsequent expression of the STAT3 transcription factor, which activates the expression of nuclear receptor estrogen-related receptor gamma (ERR-gamma). ERR-gamma then binds to the Hamp promoter, stimulating hepcidin production and promoting hypoferremia and macrophage iron sequestration. This was supported by a range of evidence, including induction of hepatic hepcidin and ERR-gamma expression in infected mice and identification of conserved binding sites in ERR-gamma and Hamp promoter regions (Kim et al. 2014). Importantly, the use of the ERR-gamma reverse agonist GSK5182 attenuated the virulence of *Salmonella* in infected mice and led to reduced serum hepcidin and increased ferroportin expression (Kim et al. 2014), indicating the importance of ERR-gamma to the bacterium in disturbing host iron homeostasis. This emphasizes the central role of the hepcidin-ferroportin axis and reveals potential therapeutic targets for *Salmonella* infection.

Finally, effects of *Salmonella*-induced iron homeostasis disruption on the wider immune system, including skewing of T helper cells toward Th-2 polarization (Nairz et al. 2014), have been suggested. Interestingly, Aulicino et al. showed that *S. Typhi*-infected DCs expressed higher levels of STAT6 and IL4R (Aulicino et al. 2022), indicative of a type 2 immune response and consistent with other RNA sequencing studies demonstrating M2 macrophage polarization on viable salmonella infection (Saliba et al. 2016). Given the central role of iron in immunometabolism, the modulation of iron homeostasis by *Salmonella* is likely to affect the development of an immune response.

14.4 Tuberculosis, Hepcidin, and Anemia

Iron deficiency and anemia are associated with increased mortality to *Mycobacterium tuberculosis* (Mtb) infection (Isanaka et al. 2012), while iron loading has been shown to increase susceptibility to Mtb in murine and human studies (Schaible et al. 2002; Gangaidzo et al. 2001). In contrast, studies in human macrophages have shown that hemochromatosis is protective against Mtb infection (Olakanmi et al. 2007). Therefore, an effective hepcidin-induced hypoferremic response is a critical determinant of clinical outcomes to Mtb infection.

Cercamondi et al. conducted a prospective study of 18 Tanzanian adults with tuberculosis, monitoring iron homeostasis and inflammation before, during, and after intensive treatment, and on completion of treatment (Cercamondi et al. 2021). At baseline, 11 of the participants were experiencing anemia of inflammation (see Chap. 13), a hallmark of Mtb infection, reducing to one participant on treatment completion; concurrently, hepcidin, IL-6, and erythroferrone were notably elevated, while serum iron and transferrin saturation were low. By week 2 of treatment, hepcidin and IL-6 levels were significantly reduced, correlating with an increase in serum iron and reticulocytosis; erythroferrone levels did not significantly decrease until week 8 of treatment (Cercamondi et al. 2021). This indicates a reduction in inflammation permitting resolution of anemia; the authors proposed that the maintenance of erythroferrone levels aided this resolution. Notably, erythroferrone became more strongly negatively correlated with hepcidin, which it suppresses, throughout the course of treatment (Cercamondi et al. 2021), suggesting a correction of iron homeostasis disturbance. Overall, this demonstrates an Mtb-mediated disturbance of iron homeostasis in which the host is unable to compensate for; treatment aids normalization of iron homeostasis and permits resolution the anemia of inflammation associated with Mtb infection.

14.5 Iron, Inflammation, and Respiratory Diseases

In addition to tuberculosis, iron dysregulation is apparent in other inflammatory and infectious pulmonary disorders. Iron deficiency is very common in patients with respiratory disease. This has been well characterized in patients with idiopathic pulmonary arterial hypertension, in whom it has been attributed by some authors to elevated hepcidin levels (Rhodes et al. 2011a; Ruiter et al. 2011; Soon et al. 2011), and where there may be a specific link to iron signaling through the BMPR2 pathway (Rhodes et al. 2011b). However, iron deficiency is also seen in patients with chronic airways disease. In a cohort of patients with chronic obstructive pulmonary disease (COPD), non-anemic iron deficiency was shown to be present in 18% of patients, compared with just 5% in a matched control population (Viethen et al. 2014), and in patients with cystic fibrosis, the prevalence of iron deficiency is reported to be up to 60% (Pond et al. 1996; Reid et al. 2002; Smith 1964).

Given these observations, an important clinical issue is whether iron deficiency should be treated in the setting of chronic respiratory disease. On the one hand, In patients with heart failure or pulmonary hypertension, correction of non-anemic iron deficiency leads to improvements in functional status, exercise capacity and wellbeing (Anker et al. 2009; Jankowska et al. 2016; Kramer et al. 2021; Viethen et al. 2014), and there is growing evidence that iron supplementation (Chap. 24) can reduce pulmonary artery pressure in patients exposed to hypoxemia, regardless of baseline iron status (Frise et al. 2016; Smith et al. 2009). These potential benefits would be highly clinically relevant if replicated in patients with respiratory disease, for whom treatment options are often limited (Patel et al. 2019; Ramakrishnan et al. 2018).

On the other hand, there is longstanding concern about the potential for iron supplementation to increase the risk of infection, particularly in the context of known or possible colonization of the lung with iron-requiring organisms such as *Pseudomonas aeruginosa* or non-tuberculous mycobacteria. A major feature of the innate immune response to respiratory infection is the restriction of iron availability, and the growth of *P. aeruginosa* is known to be highly dependent on its ability to scavenge environmental iron (Konings et al. 2013; Reid et al. 2007). Indeed, there has been interest in iron chelators as an antibacterial strategy in cystic fibrosis, using both synthetic chelators and naturally occurring compounds such as lactoferrin (Smith et al. 2013). More generally, there have been anecdotal reports of worsening respiratory infection following iron supplementation in patients with cystic fibrosis (Hoo and Wildman 2012), elevated serum iron has been suggested as a poor prognostic factor in patients with critical illness (Tacke et al. 2016), and a recent large meta-analysis identified an association between infection and intravenous iron administration (Shah et al. 2021).

To date, these theoretical concerns appear not to be borne out by clinical data in the respiratory setting. In patients with cystic fibrosis given oral iron supplementation, there was no evidence of increased clinical infection and no change in the sputum microbiome (Gifford et al. 2014), and several small clinical trials of intravenous iron in patients with critical illness or COPD, who are at high risk of opportunistic respiratory infection, have shown no increase in such infections, compared with placebo (Litton et al. 2016; Pieracci et al. 2014; Santer et al. 2020; Shah et al. 2022). Given the balance of risks and benefits in relation to iron supplementation in patients with respiratory disease, and the resultant clinical equipoise, there is a need for further high-quality clinical studies, underpinned by mechanistic analysis of the effects of iron on several aspects of relevant physiology—including the immune response to infections.

14.6 Hepcidin and Innate Immunity

The role of innate immune responses altering iron trafficking has been well studied. In addition to the induction of hepcidin by IL-6, other regulators of iron trafficking are influenced by inflammation. For example, the secretion of hemopexin and haptoglobin, which capture heme and hemoglobin respectively, is increased by inflammation to protect against possible harmful effects of hemolysis (Smith and McCulloh 2015; Vinchi et al. 2016). Ferroportin is strongly downregulated at the transcriptional level by inflammation, which contributes to the hypoferremic response and macrophage iron accumulation (Guida et al. 2015). Transferrin is a negative acute phase protein, which may impair iron acquisition by bacteria that capture transferrin-iron complexes.

There is also considerable interest in how iron affects the activity of innate immune cells. Much work has focused on how iron accumulation influences macrophage polarization and metabolism (Pereira et al. 2019) and how inflammation controls macrophage iron handling (Alam et al. 2020) and has been reviewed elsewhere (Soares and Hamza 2016; Behmoaras 2021; Soares and Weiss 2015). Macrophage iron accumulation and dysfunction are observed in several disease states, including pulmonary disorders as mentioned above. Recent work indicates that iron trafficking alters the metabolism of macrophages, with consequences for their differentiation; and that the effects of heme iron and non-heme iron can be different (Cui et al. 2022). More work is required to link these observations to the pathophysiology of disease states.

One example of a key role for hepcidin in regulating pathology is in intestinal tissues. Healing of experimentally induced gastrointestinal wounds in mice required hepcidin produced by a subset of dendritic cells (cDCs). The hepcidin acted locally on ferroportin-expressing phagocytic cells causing them to sequester iron and regulate microbiota, enabling tissue repair (Bessman et al. 2020). Such tissue-specific immunomodulatory effects of hepcidin may be more widespread and deserve further investigation.

14.7 Neutrophils, Hepcidin, and Iron Regulatory Proteins

Like macrophages, neutrophils require iron to generate the oxidative burst that is used as part of microbicidal defense. Neutrophil granules are especially enriched in iron-interacting proteins. Furthermore, neutrophils are produced in large numbers, in humans around 10^{11} per day, but higher than this during infection (Cosgrove et al. 2021). The high rate of production and the enrichment of iron-interacting proteins together yielded a prediction that neutrophil generation during hematopoiesis would require around 100-fold more iron per day than other leukocytes, and around 10–20-fold less than erythrocyte production (Frost et al. 2022). Neutrophils have a considerably shorter lifespan than red blood cells, suggesting that iron restriction might impact rapidly on neutrophil numbers.

This was borne out in two studies, one using hepcidin agonists to control plasma iron availability for steady-state and inflammatory hematopoiesis (Frost et al. 2022) and a second that disrupted cellular iron homeostasis during hematopoiesis by deletion of iron regulatory proteins (IRP) 1 and 2, which together control iron acquisition and iron storage by cells (Bonadonna et al. 2022). Both studies found that the generation of neutrophils was particularly affected by suboptimal iron availability or acquisition while other hematopoietic lineages were relatively unaffected, notably including monocyte production, which was comparatively unscathed by effects of iron deprivation (either by hepcidin agonist or IRP deletion). Steady-state and inflammation-induced production of neutrophils was suppressed by iron restriction so that neutrophil differentiation programs were not completed. Neutrophil functions were also compromised, with neutrophils from mice treated with hepcidin agonists displaying reduced ability to phagocytose fluorescent *E. coli*, produce cytokines CCL2 and TNF-alpha, and kill *S. aureus* in ex vivo assays. In contrast, monocyte cytokine production ex vivo was not influenced by exposure in vivo to hypoferremic conditions, consistent with previous observations of unaltered NO_2 production by LPS-stimulated monocytes derived from individuals with iron deficiency. Therefore, the availability of iron for hematopoiesis, which is controlled at a systemic level by hepcidin (Frost et al. 2022) and at a cell-intrinsic level by IRP1 and IRP2 (Bonadonna et al. 2022), affects the monocyte:neutrophil ratio and innate immune functionality. This impacts our understanding pathology of inflammatory disorders, including chronic infection, when dysfunctional neutrophils can cause tissue damage. The findings also have relevance for the clinical use of granulopoiesis-mobilizing agents, employed post-chemotherapy in patients at risk of infection and iron deficiency.

Other aspects of neutrophil biology are influenced by hepcidin and iron. Iron overload due to hemochromatosis or secondary to iron supplementation can cause dysfunctional neutrophils with pathological consequences, for example, in the context of *Yersinia pseudotuberculosis* infection (Das et al. 2022). Interestingly, and with some conceptual parallels to the immunomodulatory role of dendritic cell-derived hepcidin described above (Bessman et al. 2020), it was found that keratinocyte-derived hepcidin is required for controlling neutrophil recruitment to the skin via CXCL1, and this is important for defense against group A Streptococcus infection (Malerba et al. 2020).

14.8 Hepcidin and Adaptive Immunity

An unexpected sensitivity of adaptive immune responses to iron availability has been recently uncovered. A strong indication of this came from the characterization of a rare mutation in the *TFRC* gene encoding transferrin receptor (Jabara et al. 2015). The mutation causes an amino acid substitution in the internalization motif of the cytoplasmic tail of the transferrin receptor, so that its ability to endocytose its cargo of transferrin-bound iron is reduced by around 50%. Rather than causing anemia, the predominant phenotype of children homozygous for this mutation was life-threatening immunodeficiency.

Neutropenia was observed in all the children, consistent with the iron requirements for neutropoiesis described above (Aljohani et al. 2020). However, in addition, T-cells and B-cells from affected children showed severe proliferation defects that were rescuable in vitro by the addition of non-transferrin-bound iron (Jabara et al. 2015). This indicated that the demands of activated proliferating antigen-specific lymphocytes for iron cannot be met by suboptimal iron acquisition, as a result of the mutated transferrin receptor.

After activation of T-cells and B-cells during an adaptive immune response, the expression of the transferrin receptor is increased by many orders of magnitude (Howden et al. 2019), consistent with an enhanced demand for iron that must be satisfied by supply from the plasma. Physiologically, the amount of iron in plasma bound to transferrin is controlled by hepcidin. Therefore, the effect of increased hepcidin and decreased iron availability on adaptive immune responses was tested (Frost et al. 2021). Mice on a low iron diet and injected daily with mini-hepcidin following immunization have reduced markers of an immunological response, including antigen-specific CD8+ T-cells, germinal center B-cells, plasma cells, T-follicular helper cells, and antigen-specific IgG, compared to mice on a normal diet receiving a placebo injection (Frost et al. 2021).

Similarly, the T- and B-cell response to influenza virus was suppressed by injection of hepcidin agonists, leading to failure to control viral infection and more severe lung damage and disease morbidity (Frost et al. 2021). This observation of decreased immunity to a viral infection caused by hepcidin agonists was reproduced with a murine Friend's virus infection model, in which the protective NK cell response was impaired by hypoferremia (Littwitz-Salomon et al. 2021). These deleterious effects of hepcidin and iron restriction on adaptive immune responses contrast with the protective effects of hepcidin agonists in the context of extracellular bacterial infections (Arezes et al. 2015; Stefanova et al. 2017) as mentioned earlier. Therefore, the beneficial innate immune response that decreases iron availability to some siderophilic pathogens appears to come at a price of withholding iron from adaptive immune responses against viral infections (see Fig. 14.1).

Examination of the proteome of activated T-cells indicated an upregulation of many iron-interacting proteins that modulate important aspects of biochemical pathways, suggesting that the remodeling of T-cell metabolism after activation is likely to be iron-dependent, consistent with the powerful inhibitory effects of iron deprivation (Teh et al. 2021). Furthermore, clinical studies have demonstrated that iron deficiency (ID) is associated with reduced T-cell function (Thibault et al. 1993) and lower levels of antibodies to *Plasmodium falciparum* (Bundi et al. 2021) and several microbial antigens (Brussow et al. 1995). Consistent with this, in humans carrying mutations in TMPRSS6 and who have increased hepcidin and very low serum iron, a reduced concentration of some vaccine-inducible antibodies was observed, indicating a potential interaction between iron deficiency and suboptimal vaccine responses (Frost et al. 2021).

Pre-clinical evidence has supported this. One of the first pre-clinical studies investigating ID and vaccine responses examined rats receiving an attenuated influenza virus vaccine; rats with severe iron deficiency anemia (IDA), but not with moderate ID, had a reduced antibody response to vaccination (Dhur et al. 1990). Piglets are naturally born with IDA and are dependent on iron supplements for normal growth (Abbas et al. 2022); 2 weeks after vaccination against *Mycoplasma hyopneumoniae*, compared to those on no supplements, piglets receiving iron to correct hemoglobin levels had significantly higher antibody titers (Frost et al. 2021). These studies indicated that iron deficiency might affect vaccine performance in humans.

14.9 Studies of Iron Status and Vaccination in Humans

Vaccines protect immunologically vulnerable members of society, reducing the risk of serious disease; important target groups for vaccination

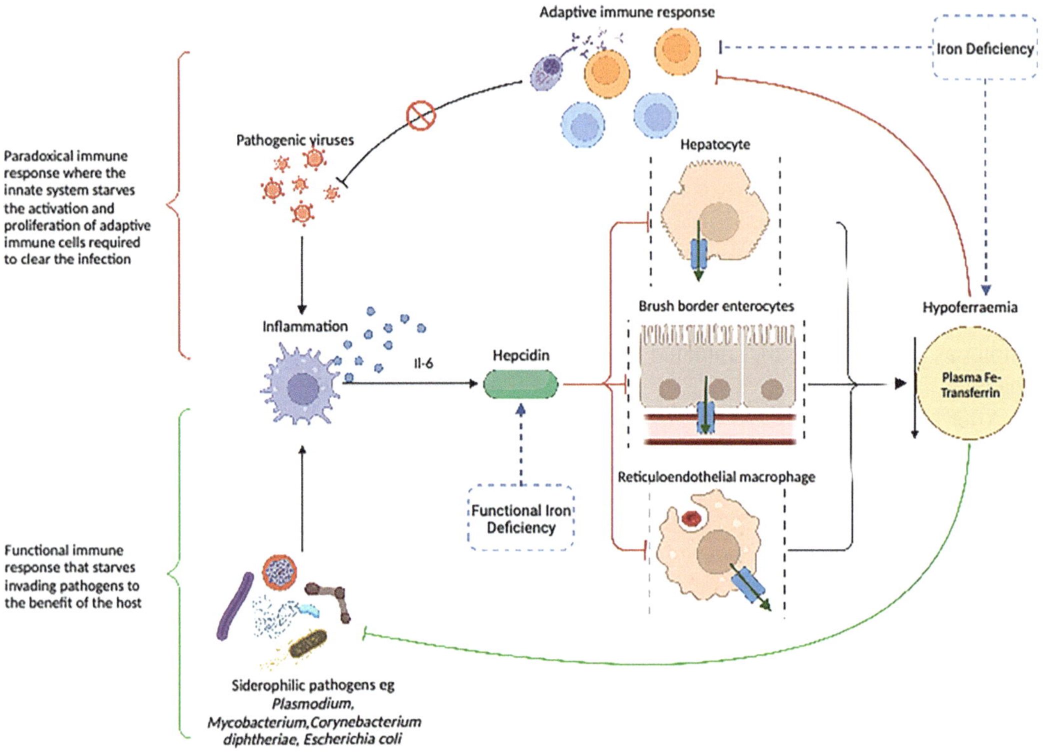

Fig. 14.1 Innate immune system-induced hypoferremia may inhibit the adaptive immune system. IL-6 expression resulting from inflammation promotes hepcidin synthesis, causing hypoferremia. In the case of siderophilic pathogen infection, this is beneficial, starving these pathogens of iron vital for their survival. However, hypoferremia also limits the activation and proliferation of lymphocytes, impairing the adaptive immune response and subsequent control of non-siderophilic pathogens. ID interacts with this process, promoting hypoferremia and further limiting the adaptive immune response

programs include newborns, infants, and the elderly. The efficacy and immunogenicity of several vaccines have been noted to vary by geographical location and age (Drakesmith et al. 2021), for example, reduced efficacy of rotavirus vaccines in infants in low- and middle-income countries (LMICs) (Burnett et al. 2020), an effectiveness of below 75% of the measles vaccine in sub-Saharan Africa (Stoffel et al. 2020), and a remarkable observation of 0% efficacy of an Influenza vaccine in infants in Senegal (Victor et al. 2016). In 5–17 month-old children, the RTS,S malaria vaccine has efficacy of ~55% for 18 months but which waned rapidly; a booster dose given at 18 months resulted in 36% efficacy over 3–4 years (Rts 2015). In the elderly, immunological markers of response to coronavirus and influenza vaccination are suboptimally altered (McElhaney et al. 2020; Collier et al. 2021). The elderly, and infants in LMICs especially, are at risk of iron deficiency. Moreover, children in LMICs experience higher rates of vaccine-preventable infection as shown in the 2019 Global Burden of Disease study (IHME 2020). Consequently, there are several instances of overlap between ID and reduced efficacy and immunogenicity of vaccines. Low iron availability could influence the T-cell and antibody response to immunization, but if neutrophil production and function are also impaired by iron restriction as indicated above (Bonadonna et al. 2022; Frost et al. 2022), this could alter the ability of antibodies to mediate protection via opsonization. This neutrophil-antibody interaction may play a role in the limited efficacy of RTS,S (Feng et al. 2022).

There have been several clinical trials investigating iron status or supplementation and vaccine response (Table 14.1). One major limitation is the small sample size in some of these studies, reducing the reliability of the data. For example, Crogan et al. studied only eight patients across two nursing homes who received both iron supplements and an influenza vaccine yet concluded that supplementation had no benefit (Crogan et al. 2005). Fulop et al. investigated 23 subjects separated into two groups based on iron status; the results showed that those with lower serum iron and hemoglobin had an insufficient response to influenza vaccination (Fülöp et al. 1999). The designs and aims of these studies were different and the results have opposing implications.

Several studies do not confirm that ID is corrected by the administered iron supplementation, nor have a "no supplementation" group; this may be due to specific governmental recommendations, for example, for pregnant women. Therefore, while these studies can compare supplements, they do not confirm an association between the correction of ID and improved vaccine response as a result of receiving supplementation. Isanaka et al. and Okala et al. investigated supplementation in pregnant mothers and subsequent vaccine responses in infants; Isanaka et al. found that supplementation type did not affect rotavirus vaccine responses (Isanaka et al. 2021), while Okala et al. found that diphtheria-tetanus-pertussis vaccine responses were improved by certain supplements (Okala et al. 2019). Specifically, iron-containing micronutrient powders were more effective than iron-alone supplements, potentially reflecting the extent of malnutrition in some populations or the promotion of non-hem iron absorption by vitamin C in the duodenum (Dev and Babitt 2017). These studies also illustrate that the potential benefits of iron supplementation prior to vaccination are likely to be vaccine-specific.

Stoffel et al. provide the most recent and specific evidence of the effects of ID and iron supplementation on pediatric vaccine responses, demonstrating that iron status affects the immunogenicity of several vaccines, and iron supplementation improves the primary response to the measles vaccine (Stoffel et al. 2020). This supports the correction of ID prior to vaccination. In contrast, Liu et al. show that hemodialysis patients receiving intravenous iron (IV-Fe) had significantly lower antibody titers in response to Hepatitis B vaccination (Liu et al. 2009), while Eiselt et al. (2016) suggest that iron status does not affect influenza vaccine response in these patients (Eiselt et al. 2016). Eiselt et al. (2010) also describe complex iron metabolism in hemodialysis patients, which may complicate the interplay between iron status, immune response, and iron supplementation (Eiselt et al. 2010). The contrast in these results could be due to the method of iron supplementation, vaccine type, or patient cohort. Nevertheless, this illustrates that general conclusions about the relationship between iron status or supplementation and vaccine response are difficult to draw from a single study.

14.10 Can Iron Parameters Predict Vaccine Response?

A measure of iron status that predicts vaccine response could identify people who may benefit from iron supplementation prior to vaccination. Studies in mice have suggested that serum iron availability is the major determinant of vaccine response, with tissue iron stores having relatively little effect (Frost et al. 2021). While yet to be confirmed in humans, this is consistent with some trials in Table 14.1 in which low serum iron, low serum transferrin, and anemia are associated with reduced vaccine response. These measures reflect the iron available to cells, such as activated and proliferating lymphocytes, and so are likely to correlate with vaccine response.

ID and IDA are usually diagnosed using serum ferritin and transferrin saturation (TSAT) (Fig. 14.1); however, these markers can vary significantly due to daily fluctuations or inflammation (Wish 2006), so may not be able to discriminate between chronic ID and acute functional ID that will resolve without intervention. Measures of iron status, such as serum transferrin receptor and serum iron, may aid the identification

Table 14.1 Clinical trials investigating iron status or supplementation and vaccine response

Citation	Vaccine	Study population	Parameters	Summary of findings
Stoffel et al. (2020)—randomized controlled trial	Monovalent measles at age 9 months (mo) and 18 mo	127 Kenyan infants, recruited at age 7.5 mo and given micronutrient powder (with 5 mg Fe ($n = 85$) or without Fe ($n = 42$)). End of intervention at 11.5 mo and follow up at 4.5 years	ID defined as serum ferritin <12 μg/mL (or serum transferrin receptor >8.3 μg/mL at 24 weeks (wks)) and anemia defined as Hb <11 g/dL for ages 24wks, 12 and 18 mo. IDA was occurrence of anemia with additional ID. Protective response defined as anti-measles IgG ≥0.12 IU/mL (international units/mL)	Iron supplementation improved Hb and serum ferritin parameters compared to the no supplementation group. The iron group had increased anti-measles IgG at 11.5 mo (primary vaccine response, $p = 0.0415$), but not at 4.5 years (secondary vaccine response, $p = 0.6800$) compared to the control group At age 11.5 mo, the control group had lower avidity IgG ($p = 0.0425$)
Stoffel et al. (2020)—birth cohort study	3-valent oral polio (OPV), diphtheria-tetanus- whole cell pertussis (DTP), hemophilus influenzae type b (Hib), 10-valent pneumococcal conjugate vaccine, and monovalent measles vaccine	303 Kenyan infants, all received vitamin A supplementation at 24 wks as per Kenyan government guidelines	ID, anemia, IDA, and protective anti-measles response are defined as above Protective response to vaccines in serum is defined as: Anti-OPV IgG ≥ 0.6, 2.1, and 0.7 IU/mL for S1, 2 and 3 Anti-diphtheria IgG ≥ 0.1 IU/mL Anti-tetanus IgG > 0.5 IU/mL Anti-Hib IgG > 1.0 μg/mL Anti-pneumococcal polysaccharide IgG ≥ 0.35 μg/mL There is no established protective anti-pertussis IgG concentration	IDA became apparent at ages 24 wks and 12mo; >90% of the cohort was anemic, with most cases due to iron depletion Hb at the time of vaccination was the strongest positive predictor of: Anti-diphtheria ($p = 0.0071$ and 0.0182) and anti-pertussis ($p = 0.0339$ and 0.0360) IgG at 24wks and 18mo Anti-pneumococcus 19 IgG at 18mo ($p = 0.0129$) Anemia and serum transferrin receptor at the time of vaccination were the strongest negative predictors of seroconversion against Diphtheria ($p = 0.0484$, $p = 0.0439$) at 18 mo Pneumococcus 19 at ($p = 0.0199$, $p = 0.0327$) 18 mo

(continued)

Table 14.1 (continued)

Citation	Vaccine	Study population	Parameters	Summary of findings
Isanaka et al. (2021)	Oral rotavirus (ORV) at age 6, 10, and 14 mo	Pregnant women in Niger were given either a lipid-based nutrient supplement, multiple micronutrient supplement, or iron-folic acid prior to birth After birth, 1525 infants were either given 3 doses of ORV ($n = 794$) or placebo ($n = 731$)	Primary immunogenicity endpoint in infant sera (collected pre-dose 1, and 28 days post-dose 3) was seroconversion defined as ≥3 fold increase in anti-rotavirus IgA Women with Hb < 7 g/dL were excluded from the trial	Seroconversion in response to vaccination did not differ by supplementation group The immune response varied based on previous exposure to rotavirus
Eiselt et al. (2016)	Influenza vaccine (A/Brisbane/59/2007 (H1N1), A/Brisbane/10/2007 (H3N2) and B/Brisbane/60/2008)	133 hemodialysis patients and 40 healthy controls Hemodialysis patients received weekly doses of IV-Fe; this was stopped 2 weeks prior to the start of the study	Seroconversion is defined as ≥4 fold increase in inhibitory antibody titer following vaccination	Hb, Il-6, transferrin, iron, hepcidin, and ferritin measures were all significantly lower in the patient group; C-reactive protein measure was significantly higher in the patient group Seroconversion rates were significantly higher in the control group (55.8–84.2% vs 35.6–52.5%) In the patient group, for the H1N1, H3N2, and B strains, older age ($p = 0.02$, 0.0001, 0.002), and vaccination in the previous season (0.0001, 0.0001, 0.01) were predictors of decreased seroconversion in both groups. There were no significant associations between serum iron ($p = 0.99$, 0.18, 0.16) or Hb ($p = 0.11$, 0.85, 0.07) and seroconversion

(continued)

Table 14.1 (continued)

Citation	Vaccine	Study population	Parameters	Summary of findings
Crogan et al. (2005)	Influenza vaccine—Strain not indicated	Two groups of nursing home residents in the USA aged over 65 years; each group contained 7 people. Each group received oral iron supplements for 30 days. Subsequently, 4 received the vaccine, and 3 received a placebo injection in each group	A sufficient immune response was defined as a fourfold increase in antigen-specific antibody titer	Serum transferrin significantly increased from day 1 to day 30 of iron supplementation in both groups There was no significant difference in antibody titer after vaccination for both vaccinated groups against influenza A ($p = 0.0920$, $p = 0.1030$) or influenza B ($p = 1.000$, $p = 1.000$). There was no change in measured antibody titers in the placebo group
Bagchi et al. (1980)	Diphtheria and tetanus toxoids	14 children with IDA and 24 normal controls (age range 2–10 years)	Anemia was defined as Hb ≤ 11.0 g/dL	There was no significant difference between the antigen-specific IgG, IgM, or IgA titers of the IDA and control groups. The IDA group had a significantly lower T-lymphocyte count than the control group, but there was no significant difference in B-lymphocyte count
MacDougall and Jacbos (1978)	Diphtheria toxoid, and typhoid vaccine	5 ID children and 4 healthy children received diphtheria toxoid; 8 ID children and 5 healthy children received the typhoid vaccine. The trial took place in South Africa	Response to vaccination was measured 14 days after administration. Protective level of anti-diphtheria toxin is defined as 0.04–0.5 U/mL	None of the ID children receiving diphtheria toxoid produced a measurable increase in anti-diphtheria toxin, but 3 of the control children did (the fourth contracted measles) 3 of the ID children receiving the typhoid vaccine showed no antibody response, while the other 5 showed a response. In the control group, 4 children developed a response (the fifth contracted chickenpox)

(continued)

Table 14.1 (continued)

Citation	Vaccine	Study population	Parameters	Summary of findings
Okala et al. (2019)	DTP vaccination	Pregnant women aged 18–45 years received either an iron-folic acid (Fe-Fol), multiple micronutrients (MMN), protein energy (PE), or PE + MMN supplement until delivery Infants were administered DTP vaccine at 8, 12, and 16 wks; blood samples at 12 ($n = 710$) and 24 ($n = 662$) wks The trial took place in rural Gambia	A sufficient vaccine response was defined as an antigen-specific antibody titer >0.1 IU/mL for diphtheria, and > 0.5 EU/mL (Elisa units/mL) for pertussis. The pertussis cut-off is arbitrary. Women with severe anemia (Hb < 7 g/dL) were excluded from the trial. All supplements contained 60 mg iron	For anti-diphtheria antibody titers, MMN and MMN + PE were associated with higher titers than Fe-Fol ($p < 0.001$) For anti-tetanus antibody titers, MMN, PE, and MMN + PE groups were associated with higher titers than Fe-Fol ($p < 0.001$) For anti-pertussis antibody titers, Fe-Fol, MMN, and MNN + PE groups all had higher titers than the PE group ($p < 0.001$), but no significant difference between themselves otherwise
Fülöp et al. (1999)	Influenza vaccine—1994–1995 preparation (A/Shangdong/09/93, A/Texas/36/91, B/Panama/45/90)	23 elderly patients in a nursing home in Canada	A fourfold increase in antibody titer against any of the antigens was considered seroconversion	11 patients responded to vaccination, 12 did not. In addition to three other parameters, hemoglobin and serum iron were significantly lower in the non-responders ($p = 0.03$, $p = 0.004$)
Liu et al. (2009)	Hepatitis B vaccination at 0, 1, 2, and 6 months	83 end-stage renal disease patients, including 51 on hemodialysis and 32 who had peritoneal dialysis Patients either received recombinant human erythropoietin (rHuEPO) ($n = 50$) or rHuEPO and IV-Fe ($n = 33$)	Anti-hepatitis B-antigen (anti-HBsAg) antibodies at 1 month after completing the vaccination program was used as a marker of immunologic responsiveness	Patients in the rHuEPO group had a significantly higher antiHBsAg antibody titers than patients in the rHuEPO and IV-Fe group ($p < 0.001$)

(continued)

Table 14.1 (continued)

Citation	Vaccine	Study population	Parameters	Summary of findings
Eiselt et al. (2010)	Influenza vaccine—2008–2009 preparation (A/Brisbane/59/2007 (H1N1), A/Brisbane/10/2007 (H3N2), and B/Florida/4/2006)	40 hemodialysis patients and 46 controls without renal disease Hemodialysis patients received IV-Fe, which was stopped 2 weeks prior to the start of the study to prevent false readouts of iron status	Seroconversion defined as fourfold increase in antigen-specific antibody titer	31 of the patients responded, 9 did not Responders received a significantly higher dose of IV-Fe in the 12 weeks before vaccination ($p = 0.0290$) Ferritin was significantly lower in responders ($p = 0.021$), but there was no significant difference in iron ($p = 0.2000$), hepcidin ($p = 0.9700$), transferrin ($p = 0.1900$), transferrin receptor ($p = 0.8200$), TSAT ($p = 0.1200$) or Hb ($p = 0.9600$)

of those with chronic ID requiring such intervention. Serum transferrin receptor reflects the demand for iron; a decrease in iron availability causes an increase in expression of the transferrin receptor, independent of inflammation status (Wish 2006; World Health Organisation 2014). However, a standardized serum transferrin receptor test and threshold have not been determined (Pasricha et al. 2021); standardizing this test and combining it with commonly used measures of iron availability could aid discrimination between chronic and acute ID. Subsequently, a threshold of iron availability should be determined for those with chronic ID, below which vaccine response is significantly affected, and iron supplementation should be given.

14.11 Conclusions

While it has long been known that host-mediated control of iron availability can influence the growth of infectious pathogens (Schade and Caroline 1944; Cartwright and Lauritsen 1946; Schade and Caroline 1946), the complex molecular mechanisms underlying the regulation of iron by immunity, and vice-versa, are still being revealed. The iron regulatory hormone hepcidin is of ancient antimicrobial lineage, and its deployment during inflammation limits the supply of extracellular iron to siderophilic pathogens. However, recent findings indicate that hepcidin acts both systemically (hepatocyte-derived) and locally (eg skin, intestinal mucosa) and regulates pathogen growth and tissue homeostasis (Chaps. 3 and 7). Iron is also required for hematopoiesis, including for emergency myelopoiesis during infections, and by activated antigen-specific lymphocytes proliferating to mount an adaptive immune response. Because immune cells are involved across so many disease states, and because iron imbalance due to iron deficiency, iron overload, and chronic inflammation is so common, the interactions of hepcidin, iron, and immunity are likely to be important in a general sense for health and wellbeing. Many critical unknowns are ranging from molecular mechanistic to conceptual, clinical, and logistical, which need to be addressed. An overarching goal must be to use our increasing knowledge of how iron influences immunity to develop powerful new methods to manipulate immune responses therapeutically.

Acknowledgments We thank many members of the Drakesmith laboratory and our collaborators for the discussions over several years that have shaped our thinking. We thank the Medical Research Council UK for funding (MCU_12010/10).

References

Abbas M, Hayirli Z, Drakesmith H, Andrews SC, Lewis MC (2022) Effects of iron deficiency and iron supplementation at the host-microbiota interface: could a piglet model unravel complexities of the underlying mechanisms? Front Nutr 9:927754. https://doi.org/10.3389/fnut.2022.927754

Alam Z, Devalaraja S, Li M, To TKJ, Folkert IW, Mitchell-Velasquez E, Dang MT, Young P, Wilbur CJ, Silverman MA, Li X, Chen YH, Hernandez PT, Bhattacharyya A, Bhattacharya M, Levine MH, Haldar M (2020) Counter regulation of spic by NF-kappaB and STAT signaling controls inflammation and iron metabolism in macrophages. Cell Rep 31(13):107825. https://doi.org/10.1016/j.celrep.2020.107825

Aljohani AH, Al-Mousa H, Arnaout R, Al-Dhekri H, Mohammed R, Alsum Z, Nicolas-Jilwan M, Alrogi F, Al-Muhsen S, Alazami AM, Al-Saud B (2020) Clinical and immunological characterization of combined immunodeficiency due to TFRC mutation in eight patients. J Clin Immunol 40(8):1103–1110. https://doi.org/10.1007/s10875-020-00851-1

Anker SD, Comin-Colet J, Filippatos G, Willenheimer R, Dickstein K, Drexler H, Luscher TF, Bart B, Banasiak W, Niegowska J, Kirwan BA, Mori C, von Eisenhart RB, Pocock SJ, Poole-Wilson PA, Ponikowski P, Investigators F-HT (2009) Ferric carboxymaltose in patients with heart failure and iron deficiency. N Engl J Med 361(25):2436–2448. https://doi.org/10.1056/NEJMoa0908355

Arezes J, Jung G, Gabayan V, Valore E, Ruchala P, Gulig Paul A, Ganz T, Nemeth E, Bulut Y (2015) Hepcidin-induced Hypoferremia is a critical host defense mechanism against the Siderophilic bacterium Vibrio vulnificus. Cell Host Microbe 17(1):47–57. https://doi.org/10.1016/j.chom.2014.12.001

Aulicino A, Antanaviciute A, Frost J, Sousa Geros A, Mellado E, Attar M, Jagielowicz M, Hublitz P, Sinz J, Preciado-Llanes L, Napolitani G, Bowden R, Koohy H, Drakesmith H, Simmons A (2022) Dual RNA sequencing reveals dendritic cell reprogramming in response to typhoidal salmonella invasion. Commun Biol 5(1):111–111. https://doi.org/10.1038/s42003-022-03038-z

Bagchi K, Mohanram M, Reddy V (1980) Humoral immune response in children with iron-deficiency anaemia. Br Med J 280(6226):1249–1251. https://doi.org/10.1136/bmj.280.6226.1249

Barber MF, Elde NC (2014) Escape from bacterial iron piracy through rapid evolution of transferrin. Science 346(6215):1362–1366. https://doi.org/10.1126/science.1259329

Behmoaras J (2021) The versatile biochemistry of iron in macrophage effector functions. FEBS J 288(24):6972–6989. https://doi.org/10.1111/febs.15682

Bessman NJ, Mathieu JRR, Renassia C, Zhou L, Fung TC, Fernandez KC, Austin C, Moeller JB, Zumerle S, Louis S, Vaulont S, Ajami NJ, Sokol H, Putzel GG, Arvedson T, Sockolow RE, Lakhal-Littleton S, Cloonan SM, Arora M, Peyssonnaux C, Sonnenberg GF (2020) Dendritic cell-derived hepcidin sequesters iron from the microbiota to promote mucosal healing. Science 368(6487):186–189. https://doi.org/10.1126/science.aau6481

Bonadonna M, Altamura S, Tybl E, Palais G, Qatato M, Polycarpou-Schwarz M, Schneider M, Kalk C, Rudiger W, Ertl A, Anstee N, Bogeska R, Helm D, Milsom MD, Galy B (2022) Iron regulatory protein (IRP)-mediated iron homeostasis is critical for neutrophil development and differentiation in the bone marrow. Sci Adv 8(40):eabq4469. https://doi.org/10.1126/sciadv.abq4469

Brussow H, Sidoti J, Dirren H, Freire WB (1995) Effect of malnutrition in Ecuadorian children on titers of serum antibodies to various microbial antigens. Clin Diagn Lab Immunol 2(1):62–68. https://doi.org/10.1128/cdli.2.1.62-68.1995

Bundi CK, Nalwoga A, Lubyayi L, Muriuki JM, Mogire RM, Opi H, Mentzer AJ, Mugyenyi CK, Mwacharo J, Webb EL, Bejon P, Williams TN, Gikunju JK, Beeson JG, Elliott AM, Ndungu FM, Atkinson SH (2021) Iron deficiency is associated with reduced levels of plasmodium falciparum-specific antibodies in African children. Clin Infect Dis 73(1):43–49. https://doi.org/10.1093/cid/ciaa728

Burnett E, Parashar UD, Tate JE (2020) Real-world effectiveness of rotavirus vaccines, 2006–19: a literature review and meta-analysis. Lancet Glob Health 8(9):e1195–e1202. https://doi.org/10.1016/S2214-109X(20)30262-X

Carniel E (2001) The Yersinia high-pathogenicity Island: an iron-uptake Island. Microbes Infect 3(7):561–569. https://doi.org/10.1016/s1286-4579(01)01412-5

Cartwright GE, Lauritsen MA (1946) The anemia of infection; the experimental production of hypoferremia and anemia in dogs. J Clin Invest 25:81–86. https://doi.org/10.1172/JCI101691

Cercamondi CI, Stoffel NU, Moretti D, Zoller T, Swinkels DW, Zeder C, Mhimibra F, Hella J, Fenner L, Zimmermann MB (2021) Iron homeostasis during anemia of inflammation: a prospective study of patients with tuberculosis. Blood 138(15):1293–1303. https://doi.org/10.1182/blood.2020010562

Collier DA, Ferreira IATM, Kotagiri P, Datir RP, Lim EY, Touizer E, Meng B, Abdullahi A, Collaboration C-NBC, Elmer A, Kingston N, Graves B, Le Gresley E, Caputo D, Bergamaschi L, KGC S, Bradley JR, Ceron-Gutierrez L, Cortes-Acevedo P, Barcenas-Morales G, Linterman MA, McCoy LE, Davis C, Thomson E, Lyons PA, McKinney E, Doffinger R, Wills M, Gupta RK (2021) Age-related immune response heterogeneity to SARS-CoV-2 vaccine BNT162b2. https://doi.org/10.17863/CAM.75333

Cosgrove J, Hustin LSP, de Boer RJ, Perie L (2021) Hematopoiesis in numbers. Trends Immunol 42(12):1100–1112. https://doi.org/10.1016/j.it.2021.10.006

Crogan NL, Velasquez D, Gagan MJ (2005) Testing the feasibility and initial effects of iron and vitamin C to enhance nursing home residents' immune status following an influenza vaccine. Geriatr Nurs 26(3):188–194. https://doi.org/10.1016/j.gerinurse.2005.03.007

Cui Y, Gutierrez S, Ariai S, Öberg L, Thörn K, Gehrmann U, Cloonan SM, Naessens T, Olsson H (2022) Non-heme iron overload impairs monocyte to macrophage differentiation via mitochondrial oxidative stress. Front Immunol 13. https://doi.org/10.3389/fimmu.2022.998059

Darton TC, Blohmke CJ, Giannoulatou E, Waddington CS, Jones C, Sturges P, Webster C, Drakesmith H, Pollard AJ, Armitage AE (2015) Rapidly escalating Hepcidin and associated serum iron starvation are features of the acute response to typhoid infection in humans. PLoS Negl Trop Dis 9(9):e0004029. https://doi.org/10.1371/journal.pntd.0004029

Das S, Saqib M, Meng RC, Chittur SV, Guan Z, Wan F, Sun W (2022) Hemochromatosis drives acute lethal intestinal responses to hyperyersiniabactin-producing Yersinia pseudotuberculosis. Proc Natl Acad Sci USA 119(2). https://doi.org/10.1073/pnas.2110166119

Dev S, Babitt JL (2017) Overview of iron metabolism in health and disease. Hemodial Int 21(S1):S6–S20. https://doi.org/10.1111/hdi.12542

Dhur A, Galan P, Hannoun C, Huot K, Hercberg S (1990) Effects of iron deficiency upon the antibody response to influenza virus in rats. J Nutr Biochem 1(12):629–634. https://doi.org/10.1016/0955-2863(90)90021-C

Drakesmith H, Prentice AM (2012) Hepcidin and the iron-infection axis. Science 338(6108):768–772. https://doi.org/10.1126/science.1224577

Drakesmith H, Pasricha S-R, Cabantchik I, Hershko C, Weiss G, Girelli D, Stoffel N, Muckenthaler MU, Nemeth E, Camaschella C, Klenerman P, Zimmermann MB (2021) Vaccine efficacy and iron deficiency: an intertwined pair? Lancet Haematol 8(9):e666–e669. https://doi.org/10.1016/S2352-3026(21)00201-5

Eiselt J, Kielberger L, Sedláčková T, Racek J, Pazdiora P (2010) High ferritin, but not Hepcidin, is associated with a poor immune response to an influenza vaccine in hemodialysis patients. Nephron Clin Pract 115(2):c147–c153. https://doi.org/10.1159/000312878

Eiselt J, Kielberger L, Rajdl D, Racek J, Pazdiora P, Malánová L (2016) Previous vaccination and age are more important predictors of immune response to influenza vaccine than inflammation and iron status in dialysis patients. Kidney Blood Press Res 41(2):139–147. https://doi.org/10.1159/000443416

Feng G, Kurtovic L, Agius PA, Aitken EH, Sacarlal J, Wines BD, Hogarth PM, Rogerson SJ, Fowkes FJI, Dobano C, Beeson JG (2022) Induction, decay, and determinants of functional antibodies following vaccination with the RTS,S malaria vaccine in young children. BMC Med 20(1):289. https://doi.org/10.1186/s12916-022-02466-2

Frank KM, Schneewind O, Shieh WJ (2011) Investigation of a researcher's death due to septicemic plague. N Engl J Med 364(26):2563–2564. https://doi.org/10.1056/NEJMc1010939

Frise MC, Cheng HY, Nickol AH, Curtis MK, Pollard KA, Roberts DJ, Ratcliffe PJ, Dorrington KL, Robbins PA (2016) Clinical iron deficiency disturbs normal human responses to hypoxia. J Clin Invest 126(6):2139–2150. https://doi.org/10.1172/JCI85715

Frost JN, Tan TK, Abbas M, Wideman SK, Bonadonna M, Stoffel NU, Wray K, Kronsteiner B, Smits G, Campagna DR, Duarte TL, Lopes JM, Shah A, Armitage AE, Arezes J, Lim PJ, Preston AE, Ahern D, Teh M, Naylor C, Salio M, Gileadi U, Andrews SC, Dunachie SJ, Zimmermann MB, van der Klis FRM, Cerundolo V, Bannard O, Draper SJ, Townsend ARM, Galy B, Fleming MD, Lewis MC, Drakesmith H (2021) Hepcidin-mediated Hypoferremia disrupts immune responses to vaccination and infection. Med (N Y) 2(2):164–179.e112. https://doi.org/10.1016/j.medj.2020.10.004

Frost JN, Wideman SK, Preston AE, Teh MR, Ai Z, Wang L, Cross A, White N, Yazicioglu Y, Bonadonna M, Clarke AJ, Armitage AE, Galy B, Udalova IA, Drakesmith H (2022) Plasma iron controls neutrophil production and function. Sci Adv 8(40):eabq5384. https://doi.org/10.1126/sciadv.abq5384

Fülöp T, Wagner JR, Khalil A, Weber J, Trottier L, Payette H (1999) Relationship between the response to influenza vaccination and the nutritional status in institutionalized elderly subjects. J Gerontol Ser A Biomed Sci Med Sci 54(2):M59–M64. https://doi.org/10.1093/gerona/54.2.M59

Gangaidzo IT, Moyo VM, Mvundura E, Aggrey G, Murphree NL, Khumalo H, Saungweme T, Kasvosve I, Gomo ZAR, Rouault T, Boelaert JR, Gordeuk VR (2001) Association of pulmonary tuberculosis with increased dietary iron. J Infect Dis 184(7):936–939. https://doi.org/10.1086/323203

Gifford AH, Alexandru DM, Li Z, Dorman DB, Moulton LA, Price KE, Hampton TH, Sogin ML, Zuckerman JB, Parker HW, Stanton BA, O'Toole GA (2014) Iron supplementation does not worsen respiratory health or alter the sputum microbiome in cystic fibrosis. J Cyst Fibros 13(3):311–318. https://doi.org/10.1016/j.jcf.2013.11.004

Guida C, Altamura S, Klein FA, Galy B, Boutros M, Ulmer AJ, Hentze MW, Muckenthaler MU (2015) A novel inflammatory pathway mediating rapid hepcidin-independent hypoferremia. Blood 125(14):2265–2275. https://doi.org/10.1182/blood-2014-08-595256

Hoo ZH, Wildman MJ (2012) Intravenous iron among cystic fibrosis patients. J Cyst Fibros 11(6):560–562. https://doi.org/10.1016/j.jcf.2012.05.002

Horseman MA, Surani S (2011) A comprehensive review of Vibrio vulnificus: an important cause of severe sepsis and skin and soft-tissue infection. Int J Infect Dis 15(3):e157–e166. https://doi.org/10.1016/j.ijid.2010.11.003

Howden AJM, Hukelmann JL, Brenes A, Spinelli L, Sinclair LV, Lamond AI, Cantrell DA (2019) Quantitative analysis of T cell proteomes and environmental sensors during T cell differentiation. Nat Immunol 20(11):1542–1554. https://doi.org/10.1038/s41590-019-0495-x

(IHME) IfHMaE (2020) GBD compare data visualisation. IHME, University of Washington. https://www.healthdata.org/data-visualization/gbd-compare. Accessed 25 Mar 2022

Isanaka S, Mugusi F, Urassa W, Willett WC, Bosch RJ, Villamor E, Spiegelman D, Duggan C, Fawzi WW (2012) Iron deficiency and anemia predict mortality in patients with tuberculosis. J Nutr 142(2):350–357. https://doi.org/10.3945/jn.111.144287

Isanaka S, Garba S, Plikaytis B, McNeal MM, Guindo O, Langendorf C, Adehossi E, Ciglenecki I, Grais RF (2021) Immunogenicity of an oral rotavirus vaccine administered with prenatal nutritional support in Niger: a cluster randomized clinical trial. PLoS Med 18(8):e1003720–e1003720. https://doi.org/10.1371/journal.pmed.1003720

Jabara HH, Boyden SE, Chou J, Ramesh N, Massaad MJ, Benson H, Bainter W, Fraulino D, Rahimov F, Sieff C, Liu Z-J, Alshemmari SH, Al-Ramadi BK, Al-Dhekri H, Arnaout R, Abu-Shukair M, Vatsayan A, Silver E, Ahuja S, Davies EG, Sola-Visner M, Ohsumi TK, Andrews NC, Notarangelo LD, Fleming MD, Al-Herz W, Kunkel LM, Geha RS (2015) A missense mutation in TFRC, encoding transferrin receptor 1, causes combined immunodeficiency. Nat Genet 48(1):74–78. https://doi.org/10.1038/ng.3465

Jankowska EA, Tkaczyszyn M, Suchocki T, Drozd M, von Haehling S, Doehner W, Banasiak W, Filippatos G, Anker SD, Ponikowski P (2016) Effects of intravenous iron therapy in iron-deficient patients with systolic heart failure: a meta-analysis of randomized controlled trials. Eur J Heart Fail 18(7):786–795. https://doi.org/10.1002/ejhf.473

Kim D-K, Jeong J-H, Lee J-M, Kim KS, Park S-H, Kim YD, Koh M, Shin M, Jung YS, Kim H-S, Lee T-H, Oh B-C, Kim JI, Park HT, Jeong W-I, Lee C-H, Park SB, Min J-J, Jung S-I, Choi S-Y, Choy HE, Choi H-S (2014) Inverse agonist of estrogen-related receptor γ controls salmonella typhimurium infection by modulating host iron homeostasis. Nat Med 20(4):419–424. https://doi.org/10.1038/nm.3483

Konings AF, Martin LW, Sharples KJ, Roddam LF, Latham R, Reid DW, Lamont IL (2013) Pseudomonas aeruginosa uses multiple pathways to acquire iron during chronic infection in cystic fibrosis lungs. Infect Immun 81(8):2697–2704. https://doi.org/10.1128/IAI.00418-13

Kramer T, Wissmuller M, Natsina K, Gerhardt F, Ten Freyhaus H, Dumitrescu D, Viethen T, Hellmich M, Baldus S, Rosenkranz S (2021) Ferric carboxymaltose in patients with pulmonary arterial hypertension and iron deficiency: a long-term study. J Cachexia Sarcopenia Muscle 12(6):1501–1512. https://doi.org/10.1002/jcsm.12764

Litton E, Baker S, Erber WN, Farmer S, Ferrier J, French C, Gummer J, Hawkins D, Higgins A, Hofmann A, De Keulenaer B, McMorrow J, Olynyk JK, Richards T, Towler S, Trengove R, Webb S, Australian, New Zealand Intensive Care Society Clinical Trials G (2016) Intravenous iron or placebo for anaemia in intensive care: the IRONMAN multicentre randomized blinded trial : a randomized trial of IV iron in critical illness. Intensive Care Med 42(11):1715–1722. https://doi.org/10.1007/s00134-016-4465-6

Littwitz-Salomon E, Moreira D, Frost JN, Choi C, Liou KT, Ahern DK, O'Shaughnessy S, Wagner B, Biron CA, Drakesmith H, Dittmer U, Finlay DK (2021) Metabolic requirements of NK cells during the acute response against retroviral infection. Nat Commun 12(1):5376. https://doi.org/10.1038/s41467-021-25715-z

Liu JH, Liu YL, Lin HH, Yang YF, Kuo HL, Lin PW, Huang CC (2009) Intravenous iron attenuates postvaccination anti-HBsAg titres after quadruple hepatitis B vaccination in dialysis patients with erythropoietin therapy. Int J Clin Pract 63(3):387–393. https://doi.org/10.1111/j.1742-1241.2008.01732.x

MacDougall LG, Jacbos MR (1978) The immune response in iron-deficient children—isohaemagglutinin titres and antibody response to immunization. S Afr Med J 53(11):405–407

Malerba M, Louis S, Cuvellier S, Shambat SM, Hua C, Gomart C, Fouet A, Ortonne N, Decousser JW, Zinkernagel AS, Mathieu JR, Peyssonnaux C (2020) Epidermal hepcidin is required for neutrophil response to bacterial infection. J Clin Invest 130(1):329–334. https://doi.org/10.1172/JCI126645

McElhaney JE, Verschoor CP, Andrew MK, Haynes L, Kuchel GA, Pawelec G (2020) The immune response to influenza in older humans: beyond immune senescence. Immun Ageing 17(1):10–10. https://doi.org/10.1186/s12979-020-00181-1

Nairz M, Theurl I, Ludwiczek S, Theurl M, Mair SM, Fritsche G, Weiss G (2007) The co-ordinated regulation of iron homeostasis in murine macrophages limits the availability of iron for intracellular salmonella typhimurium. Cell Microbiol 9(9):2126–2140. https://doi.org/10.1111/j.1462-5822.2007.00942.x

Nairz M, Theurl I, Schroll A, Theurl M, Fritsche G, Lindner E, Seifert M, Crouch M-LV, Hantke K, Akira S, Fang FC, Weiss G (2009) Absence of functional Hfe protects mice from invasive salmonella enterica Serovar Typhimurium infection via induction of lipocalin-2. Blood 114(17):3642–3651. https://doi.org/10.1182/blood-2009-05-223354

Nairz M, Schleicher U, Schroll A, Sonnweber T, Theurl I, Ludwiczek S, Talasz H, Brandacher G, Moser PL, Muckenthaler MU, Fang FC, Bogdan C, Weiss G (2013) Nitric oxide-mediated regulation of ferroportin-1 controls macrophage iron homeostasis and immune function in salmonella infection. J Exp Med 210(5):855–873. https://doi.org/10.1084/jem.20121946

Nairz M, Haschka D, Demetz E, Weiss G (2014) Iron at the interface of immunity and infection. Front Pharmacol 5:152–152. https://doi.org/10.3389/fphar.2014.00152

Nairz M, Schroll A, Haschka D, Dichtl S, Tymoszuk P, Demetz E, Moser P, Haas H, Fang FC, Theurl I, Weiss G (2017) Genetic and dietary iron overload differentially affect the course of salmonella Typhimurium infection. Front Cell Infect Microbiol 7:110–110. https://doi.org/10.3389/fcimb.2017.00110

Nemeth E, Ganz T (2022) Hepcidin and iron in health and disease. Annu Rev Med. https://doi.org/10.1146/annurev-med-043021-032816

Nemeth E, Rivera S, Gabayan V, Keller C, Taudorf S, Pedersen BK, Ganz T (2004) IL-6 mediates hypoferremia of inflammation by inducing the synthesis of the iron regulatory hormone hepcidin. J Clin Invest 113(9):1271–1276. https://doi.org/10.1172/JCI200420945

Oexle H, Kaser A, Most J, Bellmann-Weiler R, Werner ER, Werner-Felmayer G, Weiss G (2003) Pathways for the regulation of interferon-γ-inducible genes by iron in human monocytic cells. J Leukoc Biol 74(2):287–294. https://doi.org/10.1189/jlb.0802420

Okala SG, Darboe MK, Sosseh F, Sonko B, Faye-Joof T, Prentice AM, Moore SE (2019) Impact of nutritional supplementation during pregnancy on antibody responses to diphtheria-tetanus-pertussis vaccination in infants: a randomised trial in The Gambia. PLoS Med 16(8):–e1002854. https://doi.org/10.1371/journal.pmed.1002854

Olakanmi O, Schlesinger LS, Britigan BE (2007) Hereditary hemochromatosis results in decreased iron acquisition and growth by mycobacterium tuberculosis within human macrophages. J Leukoc Biol 81(1):195–204. https://doi.org/10.1189/jlb.0606405

Pasricha SR, Tye-Din J, Muckenthaler MU, Swinkels DW (2021) Iron deficiency. Lancet 397(10270):233–248. https://doi.org/10.1016/S0140-6736(20)32594-0

Patel MS, McKie E, Steiner MC, Pascoe SJ, Polkey MI (2019) Anaemia and iron dysregulation: untapped therapeutic targets in chronic lung disease? BMJ Open Respir Res 6(1):e000454. https://doi.org/10.1136/bmjresp-2019-000454

Pereira M, Chen TD, Buang N, Olona A, Ko JH, Prendecki M, Costa ASH, Nikitopoulou E, Tronci L, Pusey CD, Cook HT, McAdoo SP, Frezza C, Behmoaras J (2019) Acute iron deprivation reprograms human macrophage metabolism and reduces inflammation in vivo. Cell Rep 28:498–511.e495. https://doi.org/10.1016/j.celrep.2019.06.039

Pieracci FM, Stovall RT, Jaouen B, Rodil M, Cappa A, Burlew CC, Holena DN, Maier R, Berry S, Jurkovich J, Moore EE (2014) A multicenter, randomized clinical trial of IV iron supplementation for anemia of traumatic critical illness*. Crit Care Med 42(9):2048–2057. https://doi.org/10.1097/CCM.0000000000000408

Pond MN, Morton AM, Conway SP (1996) Functional iron deficiency in adults with cystic fibrosis. Respir Med 90(7):409–413

Ramakrishnan L, Pedersen SL, Toe QK, Quinlan GJ, Wort SJ (2018) Pulmonary arterial hypertension: iron matters. Front Physiol 9:641. https://doi.org/10.3389/fphys.2018.00641

Reid DW, Withers NJ, Francis L, Wilson JW, Kotsimbos TC (2002) Iron deficiency in cystic fibrosis: relationship to lung disease severity and chronic Pseudomonas aeruginosa infection. Chest 121(1):48–54

Reid DW, Carroll V, O'May C, Champion A, Kirov SM (2007) Increased airway iron as a potential factor in the persistence of Pseudomonas aeruginosa infection in cystic fibrosis. Eur Respir J 30(2):286–292. https://doi.org/10.1183/09031936.00154006

Rhodes CJ, Howard LS, Busbridge M, Ashby D, Kondili E, Gibbs JS, Wharton J, Wilkins MR (2011a) Iron deficiency and raised hepcidin in idiopathic pulmonary arterial hypertension: clinical prevalence, outcomes, and mechanistic insights. J Am Coll Cardiol 58(3):300–309. https://doi.org/10.1016/j.jacc.2011.02.057

Rhodes CJ, Wharton J, Howard L, Gibbs JS, Vonk-Noordegraaf A, Wilkins MR (2011b) Iron deficiency in pulmonary arterial hypertension: a potential therapeutic target. Eur Respir J 38(6):1453–1460. https://doi.org/10.1183/09031936.00037711

Rivera S, Nemeth E, Gabayan V, Lopez MA, Farshidi D, Ganz T (2005) Synthetic hepcidin causes rapid dose-dependent hypoferremia and is concentrated in ferroportin-containing organs. Blood 106(6):2196–2199. https://doi.org/10.1182/blood-2005-04-1766

Rts SCTP (2015) Efficacy and safety of RTS,S/AS01 malaria vaccine with or without a booster dose in infants and children in Africa: final results of a phase 3, individually randomised, controlled trial. Lancet 386(9988):31–45. https://doi.org/10.1016/S0140-6736(15)60721-8

Ruiter G, Lankhorst S, Boonstra A, Postmus PE, Zweegman S, Westerhof N, van der Laarse WJ, Vonk-Noordegraaf A (2011) Iron deficiency is common in idiopathic pulmonary arterial hypertension. Eur Respir J 37(6):1386–1391. https://doi.org/10.1183/09031936.00100510

Saliba A-E, Li L, Westermann AJ, Appenzeller S, Stapels DAC, Schulte LN, Helaine S, Vogel J (2016) Single-cell RNA-seq ties macrophage polarization to growth rate of intracellular salmonella. Nat Microbiol 2(2):16206–16206. https://doi.org/10.1038/nmicrobiol.2016.206

Santer P, McGahey A, Frise MC, Petousi N, Talbot NP, Baskerville R, Bafadhel M, Nickol AH, Robbins PA (2020) Intravenous iron and chronic obstructive pulmonary disease: a randomised controlled trial. BMJ Open Respir Res 7(1). https://doi.org/10.1136/bmjresp-2020-000577

Schade AL, Caroline L (1944) Raw hen egg White and the role of iron in growth inhibition of Shigella Dysenteriae, staphylococcus aureus, Escherichia Coli and Saccharomyces Cerevisiae. Science 100(2584):14–15. https://doi.org/10.1126/science.100.2584.14

Schade AL, Caroline L (1946) An iron-binding component in human blood plasma. Science 104(2702):340–341. https://doi.org/10.1126/science.104.2702.340

Schaible UE, Collins HL, Priem F, Kaufmann SHE (2002) Correction of the iron overload defect in β-2-microglobulin knockout mice by lactoferrin abolishes their increased susceptibility to tuberculosis. J Exp Med 196(11):1507–1513. https://doi.org/10.1084/jem.20020897

Shah AA, Donovan K, Seeley C, Dickson EA, Palmer AJR, Doree C, Brunskill S, Reid J, Acheson AG, Sugavanam A, Litton E, Stanworth SJ (2021) Risk of infection associated with administration of intravenous Iron: a systematic review and meta-analysis. JAMA Netw Open 4(11):e2133935. https://doi.org/10.1001/jamanetworkopen.2021.33935

Shah A, Chester-Jones M, Dutton SJ, Marian IR, Barber VS, Griffith DM, Singleton J, Wray K, James T, Drakesmith H, Robbins PA, Frise MC, Young JD, Walsh TS, McKechnie SR, Stanworth SJ, Investigators I (2022) Intravenous iron to treat anaemia following critical care: a multicentre feasibility randomised trial. Br J Anaesth 128(2):272–282. https://doi.org/10.1016/j.bja.2021.11.010

Smith RS (1964) Iron absorption in cystic fibrosis. Br Med J 1(5383):608–609

Smith A, McCulloh RJ (2015) Hemopexin and haptoglobin: allies against heme toxicity from hemoglobin not contenders. Front Physiol 6:187. https://doi.org/10.3389/fphys.2015.00187

Smith TG, Talbot NP, Privat C, Rivera-Ch M, Nickol AH, Ratcliffe PJ, Dorrington KL, Leon-Velarde F, Robbins PA (2009) Effects of iron supplementation and depletion on hypoxic pulmonary hypertension: two randomized controlled trials. JAMA 302(13):1444–1450. https://doi.org/10.1001/jama.2009.1404

Smith DJ, Lamont IL, Anderson GJ, Reid DW (2013) Targeting iron uptake to control Pseudomonas aeruginosa infections in cystic fibrosis. Eur Respir J 42(6):1723–1736. https://doi.org/10.1183/09031936.00124012

Soares MP, Hamza I (2016) Macrophages and iron metabolism. Immunity 44(3):492–504. https://doi.org/10.1016/j.immuni.2016.02.016

Soares MP, Weiss G (2015) The iron age of host-microbe interactions. EMBO Rep 16(11):1482–1500. https://doi.org/10.15252/embr.201540558

Soon E, Treacy CM, Toshner MR, MacKenzie-Ross R, Manglam V, Busbridge M, Sinclair-McGarvie M, Arnold J, Sheares KK, Morrell NW, Pepke-Zaba J (2011) Unexplained iron deficiency in idiopathic and heritable pulmonary arterial hypertension. Thorax 66(4):326–332. https://doi.org/10.1136/thx.2010.147272

Stefanova D, Raychev A, Arezes J, Ruchala P, Gabayan V, Skurnik M, Dillon BJ, Horwitz MA, Ganz T, Bulut Y, Nemeth E (2017) Endogenous hepcidin and its agonist mediate resistance to selected infections by clearing non-transferrin-bound iron. Blood 130(3):245–257. https://doi.org/10.1182/blood-2017-03-772715

Stoffel NU, Uyoga MA, Mutuku FM, Frost JN, Mwasi E, Paganini D, van der Klis FRM, Malhotra IJ, LaBeaud AD, Ricci C, Karanja S, Drakesmith H, King CH, Zimmermann MB (2020) Iron deficiency anemia at time of vaccination predicts decreased vaccine response and iron supplementation at time of vaccination increases humoral vaccine response: a birth cohort study and a randomized trial follow-up study in Kenyan infants. Front Immunol 11:1313–1313. https://doi.org/10.3389/fimmu.2020.01313

Tacke F, Nuraldeen R, Koch A, Strathmann K, Hutschenreuter G, Trautwein C, Strnad P (2016) Iron parameters determine the prognosis of critically ill patients. Crit Care Med 44(6):1049–1058. https://doi.org/10.1097/CCM.0000000000001607

Teh MR, Frost JN, Armitage AE, Drakesmith H (2021) Analysis of iron and iron-interacting protein dynamics during T-cell activation. Front Immunol 12:714613. https://doi.org/10.3389/fimmu.2021.714613

Thibault H, Galan P, Selz F, Preziosi P, Olivier C, Badoual J, Hercberg S (1993) The immune response in iron-deficient young children: effect of iron supplementation on cell-mediated immunity. Eur J Pediatr 152(2):120–124. https://doi.org/10.1007/BF02072487

Victor JC, Lewis KD, Diallo A, Niang MN, Diarra B, Dia N, Ortiz JR, Widdowson MA, Feser J, Hoagland R, Emery SL, Lafond KE, Neuzil KM (2016) Efficacy of a Russian-backbone live attenuated influenza vaccine among children in Senegal: a randomised, double-blind, placebo-controlled trial. Lancet Glob Health 4(12):e955–e965. https://doi.org/10.1016/S2214-109X(16)30201-7

Viethen T, Gerhardt F, Dumitrescu D, Knoop-Busch S, ten Freyhaus H, Rudolph TK, Baldus S, Rosenkranz S (2014) Ferric carboxymaltose improves exercise capacity and quality of life in patients with pulmonary arterial hypertension and iron deficiency: a pilot study. Int J Cardiol 175(2):233–239. https://doi.org/10.1016/j.ijcard.2014.04.233

Vinchi F, Costa da Silva M, Ingoglia G, Petrillo S, Brinkman N, Zuercher A, Cerwenka A, Tolosano E, Muckenthaler MU (2016) Hemopexin therapy reverts heme-induced proinflammatory phenotypic switching of macrophages in a mouse model of sickle cell disease. Blood 127(4):473–486. https://doi.org/10.1182/blood-2015-08-663245

Wish JB (2006) Assessing iron status: beyond serum ferritin and transferrin saturation. Clin J Am Soc Nephrol 1(Suppl 1):S4–S8. https://doi.org/10.2215/CJN.01490506

World Health Organisation (2014) Serum transferrin receptor levels for the assessment of iron status and iron deficiency in populations. World Health Organisation. https://apps.who.int/iris/bitstream/handle/10665/133707/WHO_NMH_NHD_EPG_14.6_eng.pdfjsessionid=59614CEDE74D43AB3A2B809A9CD6FD34?sequence=1. Accessed 23 Mar 2022

15 Iron Metabolism in Cardiovascular Disease

Anthony Kalousdian, Mariam Ardehali, and Hossein Ardehali

Abstract

Iron is pivotal in energy-demanding cells such as cardiomyocytes, supporting its critical functions such as electron transfer and energy production. Systemically, iron is mainly absorbed through the gut or through red blood cell turnover by macrophages. This involves the complex interplay of transporters and regulatory proteins such as transferrin, hepcidin, and ferroportin. Intracellularly, iron is regulated by iron regulatory proteins and mRNA-binding proteins such as tristetraprolin, maintaining iron balance within cardiomyocytes. However, iron dysregulation can manifest as iron deficiency or iron overload, each leading to distinct cardiovascular diseases, particularly heart failure, anthracycline-induced cardiotoxicity, hemochromatosis, and Friedreich's ataxia. Recent clinical trials have investigated the efficacy of intravenous iron supplementation in patients with heart failure and certain criteria for iron deficiency. These trials have examined the efficacy of these therapies in improving heart failure symptoms, hospitalizations, and mortality. Moreover, emerging biomarkers like soluble transferrin-receptors are showing promise in refining the diagnosis of iron deficiency anemia. Here, we discuss the role of iron dysregulation in cardiovascular diseases, therapeutic strategies in heart failure, and future perspectives.

Keywords

Iron dysregulation · Heart failure (HF) · Iron supplementation · Transferrin receptor 1 (TFR1) · Mitochondrial dysfunction

15.1 Introduction

Iron is the most abundant element on earth, and it is an essential cofactor for all life forms (Chap. 1). Due to iron's unique properties as an electron acceptor or donor, it plays an essential role in cells that are heavily reliant on energy production, such as cardiomyocytes (Andrews 1999). Intracellular iron is mainly stored as ferritin in the cytoplasm or mitochondria, or in iron-sulfur-containing proteins (Ghafourian et al. 2020; Sawicki et al. 2023). Iron is also a vital component of heme, which is present in hemoglobin (Hb) and allows the transfer of oxygen to cells (Sawicki et al. 2015), and in heme-containing proteins that are important for energy production in the electron transport chain. A small pool of free iron is also present intracellularly and is tightly regulated since this labile iron pool can generate reactive oxygen species and oxidative

A. Kalousdian · M. Ardehali · H. Ardehali (✉)
Department of Medicine, Northwestern University, Chicago, IL, USA
e-mail: h-ardehali@northwestern.edu

K. Pantopoulos (ed.), *Iron Metabolism in Human Health and Disease*, Advances in Experimental Medicine and Biology 1480, https://doi.org/10.1007/978-3-031-92033-2_15

stress, which is damaging to human cells (Jomova and Valko 2011).

Over the past few decades, the role of iron in the cardiovascular system has been studied extensively. In this chapter, we highlight systemic and cellular iron regulation, mechanisms involving iron homeostasis and cardiovascular diseases, and the role of iron supplementation in patients with heart failure (HF).

15.2 Systemic Iron Regulation

Every day, about 20–25 mg of iron is present in the serum at any given time (Ganz 2013). The main source of serum iron is through the turnover of damaged and senescent cells, mostly red blood cells, by macrophages. Iron is also absorbed through the digestive tract via duodenal enterocytes, which absorb an average of 1–2 mg of iron (Green et al. 1968); details on iron absorption are provided in Chap. 6. This absorption occurs in the form of free iron or heme. The mechanism of heme absorption through the duodenal enterocyte is not well understood and may involve heme carrier protein 1 (HCP-1) transporters (Przybyszewska and Zekanowska 2014; Le Blanc et al. 2012). After heme enters the enterocyte, heme oxygenase-1 and heme oxygenase-2 facilitate heme degradation and the release of iron. Inorganic iron can also be absorbed through enterocytes using divalent metal transporter 1 (DMT1) transporters after reduction from ferric to ferrous form, facilitated by the duodenal cytochrome b reductase 1 (DCYTB) (Latunde-Dada et al. 2008; McKie et al. 2001; Tandy et al. 2000). Subsequently, iron can either be stored intracellularly in enterocytes by binding to ferritin or be transported to the systemic circulation through ferroportin (FPN-1) (Abboud and Haile 2000; McKie et al. 2000). Once released into the circulation, iron binds to transferrin (Tf) for transport. The iron-bound Tf can be endocytosed by cells through the transferrin receptor-1 or -2 surface proteins (TFR1/2) (Hémadi et al. 2004). TFR1 is crucial for iron uptake in most cells, while TFR2 is mostly abundant in the liver and acts as a sensor for serum iron and hepcidin production (Ganz 2013; Hentze et al. 2010) (Fig. 15.1).

In cardiomyocytes, TFR1 and TFR2 play a crucial role in maintaining cardiac homeostasis. Cardiomyocyte-specific TFR1 knockout mice develop cardiomyopathy secondary to mitochondrial dysfunction and eventually die (Xu et al. 2015). TFR2 protein has two isoforms: α and β (Kawabata et al. 1999). While TFR2α is a transmembrane protein, the β isoform is an intracellular protein since its amino acid sequence lacks transmembrane portions (Ikuta et al. 2010). Also, the TFR2β isoform plays a crucial role in ferritin transcription in cardiomyocytes. Knockdown of TFR2β in mice protects the heart from ischemia/reperfusion injury due to the upregulation of antioxidant enzymes and kinases that play a role in cardioprotection and cell survival (Roetto et al. 2018). Additionally, non-transferrin-bound iron is imported into the cardiomyocyte through the divalent metal transporter (DMT), T-type or L-type calcium channels, or zinc transporters (Lupu et al. 2019; Tsushima et al. 1999; Huang et al. 2000). Iron can subsequently be stored in cardiomyocytes as ferritin, generate Fe/S or heme clusters, or remain as free iron in the cytosol (Abboud and Haile 2000).

The liver plays a crucial role in iron storage and homeostasis. Under conditions when serum iron is abundant, it can store iron and upregulate the production of hepcidin, a peptide that regulates systemic iron levels (Ganz 2013); see Chap. 3. Hepcidin, once released into circulation, binds to FPN-1, leading to its internalization and degradation, and ultimately downregulation of iron efflux from macrophages, enterocytes, and hepatocytes into the circulation, resulting in reduced iron absorption and release from storage sites (Fig. 15.2) (Nemeth and Ganz 2021; Nemeth et al. 2004). Conversely, in conditions of iron deficiency (ID), the liver ceases to produce hepcidin, and low hepcidin levels allow iron absorption through enterocytes continuously through increased FPN-1 transcription and stabilization (due to lack of its degradation). Additionally, increased FPN-1 leads to iron efflux from macrophages.

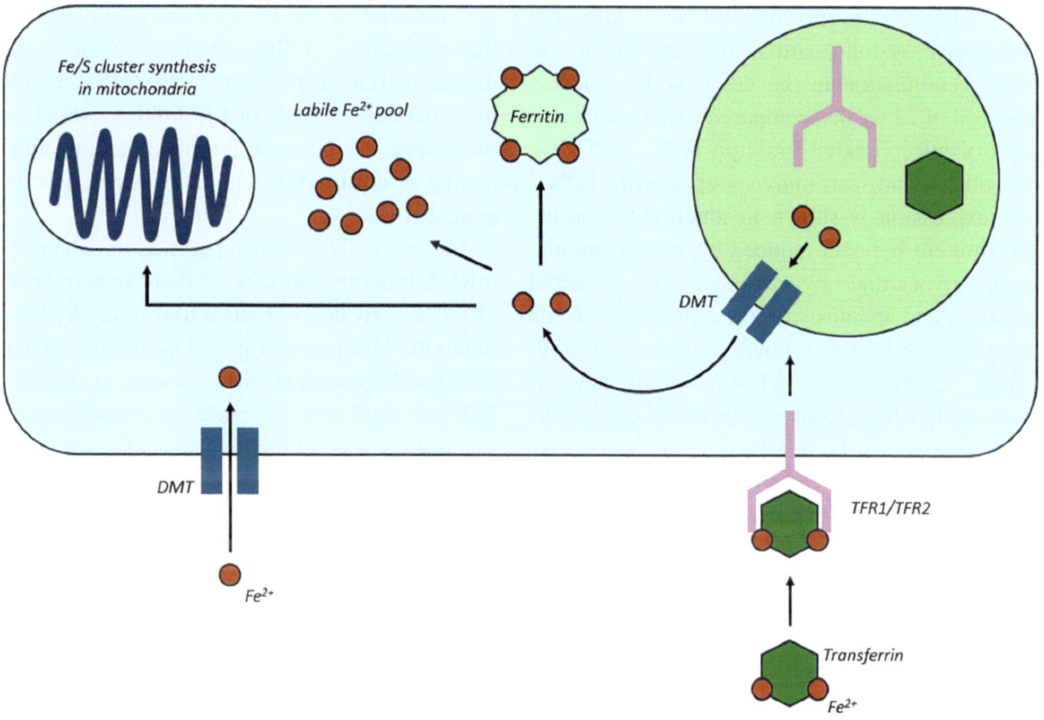

Fig. 15.1 Iron, bound to transferrin, binds to transferrin-receptors 1 or 2 (TFR1/TFR2) on the cardiomyocyte membrane and is endocytosed. From the endolysosome, iron is released into the cell as free iron in the cytosol, bound to ferritin, or to the mitochondria as Fe/S clusters. Non-transferrin-bound iron is imported through the cell through the DMT

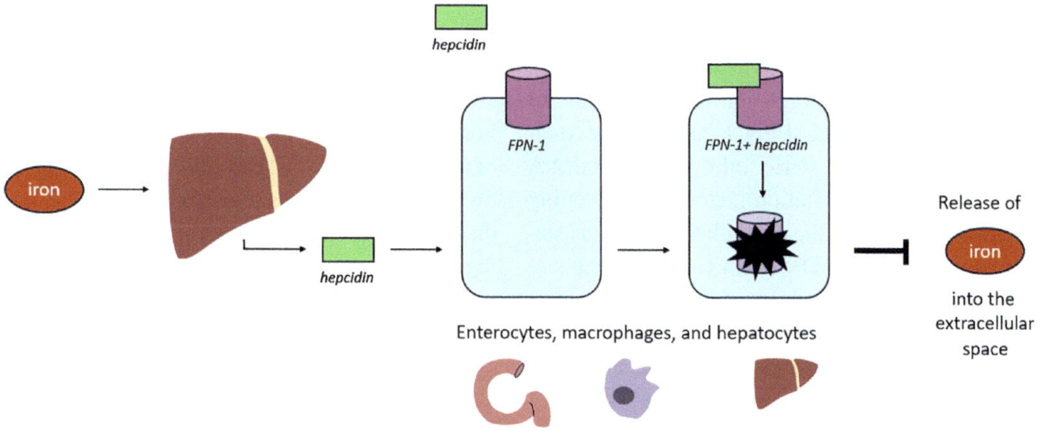

Fig. 15.2 Iron triggers the production of hepcidin in hepatocytes, which ultimately downregulates iron release into the extracellular space. Hepcidin binds to ferroportin (FPN-1), which promotes the receptor's internalization and degradation within the cell. This leads to the ultimate downregulation of the release of iron from enterocytes, macrophages, and hepatocytes into circulation

Cardiomyocyte-specific FPN-1 knockout mice have systolic cardiac dysfunction due to iron accumulation in the heart, with a median survival of 22 weeks, compared to the unaffected control mice (Lakhal-Littleton et al. 2015). On the other hand, cardiomyocytes-specific FPN-1 overexpression is shown to ameliorate chronic intermittent hypoxia induced by iron accumulation, myocardial hypertrophy, mitochondrial damage, and apoptosis (Song et al. 2024). When systemic iron levels are low, endogenous or exogenous erythropoietin (EPO) production is increased, which leads to hormone erythroferrone production by erythroid precursor cells. EFRE suppresses liver production of hepcidin to maintain the bone marrow's iron supply for erythropoiesis (Kautz et al. 2014).

15.3 Cellular Iron Regulation

Cellular iron homeostasis is regulated by iron regulatory proteins (IRPs); see Chap. 2. These mRNA-binding proteins regulate cellular iron homeostasis by binding to iron-responsive elements (IREs) in target mRNAs and altering gene expression (Pantopoulos 2004). IRPs can bind to 3′-untranslated regions of certain transcripts such as TFR1 and increase their expression through mRNA stabilization (Miyazawa et al. 2018). They also bind to the 5′-untraslated regions of other transcripts like FPN-1 and reduce their translation through steric inhibition of mRNA interaction with translation factors. In the setting of low cellular iron, these regulatory proteins stabilize the TFR1 and DMT1 mRNA transcripts to allow influx of iron while inhibiting iron efflux through inhibition of FPN-1 and ferritin mRNA translation (Paterek et al. 2019; Zhao et al. 2006). Cardiomyocyte-specific knockouts of both IRP1 and IRP2 did not lead to a phenotype at baseline but failed to show increased LV systolic function after the dobutamine test (Haddad et al. 2017). After myocardial infarction (MI), IRP-targeted mice also developed more severe LV dysfunction with increased mortality. Mechanistically, the activity of iron-sulfur cluster-containing complex I of the mitochondrial electron transport chain was reduced in the affected mice, highlighting the importance of IRPs in iron regulation and mitochondrial respiration. Single IRP mutant mice with either IRP1 or IRP2 did not show any phenotype, such as cardiac dysfunction, under resting or during dobutamine infusions (Haddad et al. 2017).

In severe ID, another pathway involving an mRNA-binding protein called tristetraprolin (TTP) is activated to reduce iron influx by binding to the TFR1 transcript and promoting its degradation (Bayeva et al. 2012; Sawicki et al. 2018). This mechanism is in place to ensure that the scarce iron is available to the vital organs (like the brain and the heart) in severe ID. To conserve iron for essential proteins, TTP degrades nonessential mRNA transcripts encoding for iron-containing proteins in the electron transport chain. Global TTP knockout mice have systemic inflammation since TNFα is one of its targets, and its deletion leads to increased TNFα levels. However, cardiac-specific TTP knockout mice develop cardiac dysfunction in the state of ID (Sato et al. 2018). Normally, TTP is induced by ID and leads to the degradation of mRNAs coding for mitochondrial Fe/S-cluster-containing proteins. This post-transcriptional regulation is crucial for maintaining mitochondrial integrity under conditions such as ID. The loss of TTP in cardiomyocytes does not decrease these Fe/S-cluster-containing proteins under ID, leading to non-functional complex III, electron leakage, and oxidative damage. These findings highlight the importance of TTP as a critical regulator of cardiac response to ID through post-transcriptional regulation.

Our group recently identified a novel pathway for iron sensing at the cellular level. Iron is crucial for histone demethylation, and its availability alters chromatin in response to iron abundance in the cellular environment (Shapiro et al. 2023). By using iron, the KDM3B protein demethylates lysine in histones, and low cellular levels of iron can result in reduced activity of this protein. In cases of prolonged ID, the anti-anabolic effect of ID is primarily mediated through KDM3B-directed changes in H3K9me^2 in enhancer regions of genes required to facilitate mTORC1 activity.

This would result in an inhibition of mTORC1 and reduced anabolic processes in the setting of ID. Thus, KDM3B acts as an iron sensor, and in the setting of ID, its lower activity leads to an inhibition of mTORC1 activity through increased histone methylation at the promoter of genes needed for mTORC1 activity. This phenomenon is seen in mice with systemic ID and in human-derived tumor cells treated with iron chelators.

15.4 Iron Deficiency in CVD

ID can be due to a depletion of total body iron stores (absolute ID) or because of restricted access to systemic iron despite adequate iron stores (functional ID); see Chap. 13. In absolute ID, blood loss or nutritional deficiencies are the primary culprits, whereas functional ID generally occurs in the setting of inflammation due to elevated hepcidin levels. Hepcidin subsequently downregulates FPN-1, decreasing iron efflux from macrophages, enterocytes, and hepatocytes. Thus, despite adequate levels of total body iron stores, iron is sequestered intracellularly and not released into circulation.

The gold standard for the diagnosis of absolute ID is by showing reduced iron staining on bone marrow biopsy (Chap. 25). However, in clinical practice, ID is generally diagnosed by measuring certain serum indices such as serum iron, ferritin, transferrin, and transferrin saturation (TSAT) (Lopez et al. 2016). The interpretation of these indices is challenging, as ferritin is an acute phase reactant and fluctuates in states of malignancy, inflammation, or infection (Wang et al. 2010). Moreover, systemic iron is not necessarily an accurate surrogate of myocardial iron levels, as there is not always a correlation between systemic iron levels and iron levels in cardiomyocytes (Paterek et al. 2021).

It is proposed that patients with HF may suffer from functional ID, and thus the classical criteria for the diagnosis of absolute ID should not be utilized in clinical practice. Instead, the current guidelines from the American Heart Association (AHA), American College of Cardiology (ACC), and European Society for Cardiology (ESC) recommend the following criteria for the diagnosis of ID: ferritin <100 μg/L or ferritin 100–300 μg/L with a TSAT <20% (McDonagh et al. 2021; Heidenreich et al. 2022). This definition originated from guidelines used in patients with chronic kidney disease (CKD), and it was subsequently adopted in clinical trials of intravenous (IV) iron supplementation in HFrEF (Graham et al. 2022; Anker et al. 2009).

To investigate the correlation between these adopted criteria and the gold standard of bone marrow staining in diagnosing ID, Grote Beverborg et al. (2018) performed bone marrow aspiration with iron staining in 42 patients with HFrEF undergoing coronary bypass surgery. The AHA/ACC/ESC criteria had a positive predictive value of 66.7%, revealing that one-third of the HFrEF patients diagnosed with ID had a normal BM stain. Moreover, many patients with HFrEF and hypoferritinemia (ferritin <100 μg/mL and a TSAT of >20%), though labeled as iron deficient by the AHA and ESC guidelines of ID, were found to have sufficient iron levels on BM staining. A TSAT ≤19.8% or serum iron ≤13 μmol/L was more specific and sensitive than the currently used ferritin-TSAT cutoffs for the correct diagnosis of ID. In the same study, all-cause mortality was assessed in patients with HFrEF in an outpatient setting, and it was demonstrated that the definition of either TSAT ≤19.8% or serum iron ≤13 μmol/L was associated independently with death, while ferritin was not.

Novel biomarkers such as soluble TfR1 (sTfR) are being assessed as a marker for accurate diagnosis of ID in patients with CVD. Elevated sTfR levels are associated with reduced BM iron staining and lower Hb levels. In patients with HFrEF, a sTfR cut-off of 1.25 mg/L had the best accuracy for the diagnosis of ID based on BM iron stores with a specificity of 100% and sensitivity of 84% (Sierpinski et al. 2021). An sTfR ≥1.41 mg/L was also optimal in predicting mortality in this patient population. The adjusted 3-year survival rates were 63 ± 6% (95% CI) for patients with HF and sTfR ≥ 1.41 mg/L compared to 80 ± 4% (95% CI) in the same patient population with a sTfR < 1.41 mg/L ($Z = 4.76$, $P < 0.001$). To further investigate the prognostic value of sTfR levels in non-anemic HF patients, another study demonstrated that tissue ID (defined as sTfR > 1.65 mg/L) was associated with all-cause death, elevated troponin and NT-proBNP levels, systemic inflammation, and impaired erythropoiesis (Ramos-Polo et al. 2024). Another marker used for diagnosing ID is serum hepcidin. In patients with a recent acute HF exacerbation, low levels of hepcidin (<14.5 ng/mL) and elevated sTfR were strongly predictive of all-cause mortality at 1 year, while ID defined as ferritin <100 μg/L or TSAT <20% was not predictive (Jankowska et al. 2014). Although hepcidin levels are generally low in ID-associated anemia (IDA), these levels might be detectable in inflammatory, neoplastic, or infectious processes (Girelli et al. 2016). Thus, the use of hepcidin has posed a challenge in routine clinical practice. Should hepcidin or sTfR be available in clinical practice, they will likely become more accurate predictors of ID than the current standard.

15.5 Dysregulation of Iron in CVD

15.5.1 HF with Reduced Ejection Fraction

HFrEF has been thought to be a chronic inflammatory condition; as such, it is associated with modest elevation in the levels of cytokines such as tumor necrosis factor alpha (TNFα), interleukin-1 (IL-1) and interleukin-6 (IL-6) (Mehta and Pothineni 2016; Briasoulis et al. 2016; Shirazi et al. 2017). Although these inflammatory cytokines can cause elevated hepcidin levels early in HF, recent studies have shown that the levels of hepcidin decline with clinical severity (Jankowska et al. 2013). Thus, the notion that patients with HFrEF suffer from functional ID is not supported by the fact that hepcidin is not ultimately increased in these patients. Thus, there is no definitive evidence to support a link between HFrEF and widespread functional ID. This suggests that the inflammatory landscape of HFrEF might be distinct from other chronic inflammatory conditions such as CKD or cancer (Sawicki and Ardehali 2021). Other postulated mechanisms of ID in HF include absolute ID due to reduced dietary iron absorption from bowel edema, concomitant proton pump inhibitor use, and an increase in gastrointestinal losses of iron due to anticoagulant and antiplatelet use.

At the cellular level, myocardial iron is dysregulated in advanced HFrEF. Some studies have shown that myocardial iron levels are reduced in the mitochondria of explanted human hearts with advanced HF (Leszek et al. 2012; Melenovsky et al. 2017), while other studies have shown increased mitochondrial iron and cellular heme/iron levels in failing hearts (Khechaduri et al. 2013; Sawicki et al. 2015). As mentioned before, systemic iron levels do not always correlate with cellular, and particularly myocardial, iron levels. Thus, systemic ID can occur despite normal myocardial iron levels and mitochondrial iron overload in the heart may occur independent of the systemic iron levels.

15.5.2 HF with Preserved Ejection Fraction (HFpEF)

Due to the heterogeneity of HFpEF phenotypes, the significance of systemic and cellular ID in these patients is unclear. In a systematic review, ID was associated with worse exercise capacity and functional outcomes in HFpEF patients, but mortality and hospitalization were not affected (Bekfani et al. 2019; Núñez et al. 2016). While ferritin was not associated with increased all-cause mortality in patients with HFpEF, specific biomarkers associated with systemic iron levels, such as sTfR > 1.76 mg/L, were correlated with increased mortality (Fitzsimons et al. 2022). In another study, ID was associated with worse quality of life (using the Minnesota Living with HF Questionnaire) and worse functional capacity (using the 6MWT) (Alcaide-Aldeano et al. 2020). Moreover, further analysis of the patients in this study suggested that these results also varied based on left ventricular ejection fraction (LVEF). For patients with HFpEF and an LVEF of 50–60%, ID was associated with worse 6MWT. While for patients with HFpEF and an LVEF > 60%, ID was associated with worse quality of life. Overall, given the unclear association between ID and HFpEF, more research is needed to study the role of iron dysregulation in this disorder.

15.5.3 Anthracycline-Induced Cardiomyopathy

Since the 1960s, anthracyclines have been used in treating acute leukemias and solid cancers (Di Marco et al. 1981; Tan et al. 1967). Anthracyclines have been long associated with cardiomyopathy, with iron playing a crucial role in the pathogenesis of this condition (Bristow et al. 1981). One in every 10 patients receiving doxorubicin will develop a reduction in LVEF by completion of 1 year of doxorubicin therapy (Curigliano et al. 2020; Smith et al. 2010). Current guidelines recommend baseline echocardiography, surveillance echocardiography during treatment, and follow-up imaging at 1 and 5 years after treatment completion (Members et al. 2022). Anthracyclines tend to accumulate in the mitochondria due to their high affinity to cardiolipin, which is present on the inner mitochondrial membrane. In the presence of iron, this anthracycline accumulation eventually leads to the overproduction of superoxide anions and highly toxic hydroxyl radicals. Our group has shown that doxorubicin downregulates the expression of ATP-binding cassette protein-B8 (ABCB8), a protein involved in mitochondrial iron export, leading to iron accumulation in the mitochondria (Ichikawa et al. 2014). Overexpression of ABCB8 or using an iron chelator that reduces mitochondrial iron is protective against doxorubicin-induced cardiotoxicity, implicating a role for mitochondrial iron in the pathogenesis of this disorder. In other studies, animals fed iron-rich diets showed significantly higher doxorubicin cardiotoxicity, further highlighting the role of iron in anthracycline-induced myocardial toxicity (Panjrath et al. 2007).

15.5.4 Myocarditis

Iron dysregulation has been shown to play a crucial role in myocardial inflammation. In patients with acute myocarditis, low serum iron and low TSAT levels have been associated with greater in-hospital neurohumoral activation and persistent LV dysfunction (Franczuk et al. 2023). In these patients, iron metabolism is dysregulated, and it takes 6 weeks to normalize after the acute episode (Bi et al. 2021).

Additionally, the mechanism of iron dysregulation in viral-induced myocarditis is through the accumulation of intracellular iron, which leads to the production of reactive oxygen species (ROS) and subsequently apoptosis (Ursu et al. 2014).

15.6 Iron Overload

When the body's capacity to store iron is exceeded, it can lead to iron overload (Radford-Smith et al. 2018). The retained iron is subsequently deposited in different organs, causing a constellation of symptoms (Richardson et al. 2018).

15.6.1 Primary Iron Overload

15.6.1.1 Hemochromatosis

Hemochromatosis is a genetic condition that leads to iron overload and deposition in different organs (Chaps. 9 and 10). It is a genetic disorder caused by mutations in the homeostatic iron regulator (*HFE*) gene. HFE is a cell surface protein on hepatocytes and negatively regulates iron uptake by competing with transferrin for TFR1 binding. As iron-bound transferrin increases in the circulation, it displaces HFE from TFR1 and allows endocytosis of iron into cells (Piperno et al. 2020). In a retrospective study, Udani et al. showed that 27.8% of patients with hemochromatosis had a higher incidence of CVD such as arrhythmias and HF (Udani et al. 2021). Iron-mediated toxicity, such as cellular damage and oxidative stress, affects iron channels responsible for electrical conduction and propagation, contributing to the development of arrhythmias (Oudit et al. 2006; Siri-Angkul et al. 2018).

Hemochromatosis also causes dilated cardiomyopathy, secondary to oxidative stress, iron-mediated toxicity, inflammation, and fibrosis (Sweet et al. 2018). Excess-free iron leads to impaired contraction and excitation coupling due to the dysfunction of the sarcoendoplasmic reticulum calcium-ATPase 2 and sodium-calcium exchanger. Dysfunction of these transporters increases intracellular calcium and sodium, leading to reduced contractile forces in cardiomyocytes (Oudit et al. 2006).

15.6.1.2 Friedrich's Ataxia

Friedrich's Ataxia (FA) is an autosomal recessive disorder caused by mutations in the mitochondrial frataxin protein with a trinucleotide repeat expansion of GAA within the gene. This manifests clinically in childhood with spinocerebellar ataxia, cardiac hypertrophy, hearing difficulties, diabetes, and liver dysfunction (Smith and Kosman 2020; Norrish et al. 2022). In this condition, deficiency of frataxin leads to impairment of Fe/S cluster formation, which negatively regulates mitochondrial enzymes, leading to dysfunctional mitochondrial respiration, increased oxidative damage, and mitochondrial iron overload (Böttinger et al. 2018; Petit et al. 2021). In a retrospective longitudinal study, children with FA had a high prevalence of hypertrophic cardiomyopathy (HCM) and atrial arrhythmias (Norrish et al. 2022). In terms of therapy, the use of idebenone, a coenzyme Q10 analog, has yielded mixed results in any improvement in cardiac function; the clinical implication of this treatment has not yet been determined (Hanson et al. 2019). Cardiac-specific frataxin knockout mouse models showed progressive and severe cardiomyopathy. Treatment of these mice with a vector that restored frataxin protein levels normalized systolic function, slowed histologic fibrosis, and restored mitochondrial function (Perdomini et al. 2014). This therapy established a preclinical proof of concept for potential use in humans in the future.

15.6.2 Secondary Iron Overload

Secondary causes of iron overload include iatrogenic (dietary or parenteral iron supplementation), repeated transfusion of packed RBCs, and hematologic diseases such as thalassemia, sickle cell disease, and myelodysplastic syndrome (Taher et al. 2021; Franke et al. 2020; Rasel et al. 2024).

15.6.2.1 Iron Overload Cardiomyopathy and Arrhythmias

Iron overload is associated with atrial fibrillation and heart block in murine models and humans (Gordan et al. 2018). Accumulation of iron also induces oxidative stress and results in mitochondrial impairment, leading to systolic and diastolic dysfunction of the heart (Shen et al. 2024). Patients with chronic iron overload have a high prevalence of cardiomyopathy and ventricular arrhythmias, leading to increased mortality in this population.

15.6.2.2 Valvular Disease

Iron promotes oxidative stress, which in turn leads to valvular calcification (Neven et al. 2011). Through interleaflet hemorrhages, iron can accumulate in aortic valves (Laguna-Fernandez et al. 2016). Heme and inorganic iron colocalize within valvular interstitial cells and induce macrophage-mediated mineralization of these cells (Xu et al. 2021). Valvular interstitial cells can take up iron in an inflammatory environment and actively contribute to ECM remodeling and calcification (Laguna-Fernandez et al. 2016).

15.6.2.3 Coronary Artery Disease (CAD) and Atherosclerosis

Iron accumulation has been shown to accelerate the formation of atherosclerotic plaques through several mechanisms. When intravascular iron levels exceed the capacity of TFR1, non-transferrin-bound iron promotes atherosclerosis through ROS generation, pro-inflammatory cell recruitment, and apoptosis (Vinchi 2021). Non-transferrin-bound iron through the circulation can also cause ROS-mediated endothelial cell injury (Madsen et al. 2023). All these processes can stimulate the progression of atherosclerosis. In patients with CAD, iron chelation improves endothelial function through NO-dependent vasodilation (Duffy et al. 2001). ROS also promotes LOX-1 expression on endothelial cells, causing autophagy and mitochondrial DNA damage (Ding et al. 2014). In a case-control study, iron imbalance, characterized by either low iron or high ferritin levels, was associated with a higher risk of CAD (Guo et al. 2022).

These findings prompted studies investigating the role of iron chelation in patients with stable coronary artery disease. In the TACT (Trial to Assess Chelation Therapy) trial, patients who had suffered from MI in at least the past 6 months were randomized to either divalent anion chelator ethylenediamine tetraacetic acid (EDTA) or placebo (Lamas et al. 2013). Iron chelation was associated with a reduction in the composite outcome, including mortality, MI, stroke, coronary artery revascularization, and hospitalization due to anginal pain. However, in contrast to the TACT1 trial, results from the TACT2 trial did not show any change in the primary endpoint, including all-cause mortality, MI, stroke, coronary revascularization, or hospitalization for unstable angina after 48 months between the chelation group and placebo group (Lamas et al. 2024). This difference in the results of the two studies was attributed to possible changes in the amount of lead in the environment, which was significantly less in the TACT2 trial. These findings warrant further studies on the role of iron chelation and cardiovascular diseases.

15.7 Iron Supplementation in HF

In the past decade, the role of IV iron supplementation in patients with HFrEF has been studied in several trials (Table 15.1). Details on IV iron therapy are provided in Chap. 24. In this section, we review these major trials, the potential risks of IV iron supplementation, and the role of oral iron supplementation in patients with HF.

15.8 Clinical Trials of IV Iron Supplementation and HF

The role of IV iron supplementation in patients with HF has been evaluated in three large randomized clinical trials (RCTs). These trials included patients with ID defined as serum ferritin <100 ng/mL or TSAT <20% with a ferritin between 100 and 300 ng/mL. The first three trials (FAIR-HF, CONFIRM-HF, and EFFECT-HF) did not evaluate major outcomes and only relied on symptomatic improvements, while AFFIRM-AHF, IRONMAN, and HEARTFID assessed hard outcomes with this therapy.

15.8.1 FAIR-HF Trial (2008)

The first major clinical trial to investigate the role of IV iron in patients with HF was the Ferinject Assessment in Patients with Iron Deficiency and Chronic HF (FAIR-HF) in 2008 (Anker et al. 2009). This trial randomized 459 patients with HFrEF (EF ≤40% and NYHA Class II or EF ≤45% and NYHA Class III) with ID to either IV ferric carboxymaltose (FCM) or placebo. Iron supplementation was by weekly FCM or placebo injections until iron indices normalized, followed by maintenance injections every 4 weeks. After 24 weeks, those in the treatment group showed an improvement in patient global assessment (PGA) and NYHA class. 47% of the patients in the treatment arm ended up with an NYHA functional class I or II compared to 30% in the placebo group, along with an improvement in quality of life and exercise capacity.

15.8.2 CONFIRM-HF Trial (2015)

The second major trial of IV supplementation was Ferric Carboxymaltose Evaluation on Performance in patients with Iron Deficiency in Combination with Chronic HF (CONFIRM-HF) (Ponikowski et al. 2015), which randomized 304 patients with HFrEF (EF ≤45%, NYHA Class II and III) and ID. After 52 weeks of treatment with either IV FCM or placebo, the treatment arm showed improvements in NYHA functional class, quality of life, and the 6-min walk test.

15.8.3 EFFECT-HF Trial (2017)

The third major trial was Effect of Ferric Carboxymaltose on Exercise Capacity in Patients with Iron Deficiency and Chronic HF (EFFECT-HF) (van Veldhuisen et al. 2017). This

Table 15.1 Major clinical trials of intravenous iron in heart failure

Study	ID definition	Inclusion criteria	Intervention + Follow-up	Primary endpoint(s)
FAIR-HF	Ferritin <100 μg/L or 100–299 μg/L with Tsat <20%	NYHA class II-III HF, LVEF ≤40% for NYHA Class II or ≤45% for NYHA Class III, presence of ID	IV FCM 200 mg weekly until repleted, then every 4 weeks. Follow-up: 24 weeks	Self-reported Patient Global Assessment and NYHA functional class
CONFIRM-HF	Ferritin <100 ng/mL or 100–300 ng/mL with Tsat <20%	NYHA class II-III HF, LVEF ≤45%, presence of ID, capable of completing 6MWT, elevated natriuretic peptides, Hb <15 g/dL	IV FCM 500 mg-2000 mg during treatment phase and 500 mg at weeks 12, 24, and 36 if still had ID Follow-up: 52 weeks	Change in 6MWT
EFFECT-HF	Ferritin <100 ng/mL or 100–300 ng/mL with Tsat <20%	≥18 years, NYHA class II-III HF, optimal GDMT for ≥4 weeks, LVEF ≤45%, BNP >100 pg/mL or NT-proBNP >400 pg/mL, reproducible exercise limitation, presence of ID	IV FCM 500–1000 mg at day 0, week 6, and week 12 if still had ID Follow-up: 24 weeks	Change in peak VO_2
AFFIRM-AHF	Ferritin <100 μg/L or 100–299 μg/L with Tsat <20%	≥18 years, hospitalized for acute HF, LVEF <50%, presence of ID	FCM at day 0, week 6, week 12, and week 24 if still had ID Follow-up: 52 weeks	CV death + HF hospitalizations
IRONMAN	Ferritin <100 μg/L or Tsat <20%	≥18 years, symptomatic HF, presence of ID, LVEF ≤45%, hospitalized for acute HF or elevated BNP or NT-proBNP	Ferric derisomaltose 20 mg/kg if ≤50 kg or Hb <10 g/dL, or 1000 mg if 50–70 kg and Hb ≥10 g/dL. Given at day 0, week 4, and every 4 months after if ferritin <100 μg/L or ≤400 μg/L with Tsat <25% Follow-up: 2.7 years	CV death + HF hospitalizations
HEART-FID	Ferritin <100 ng/mL or 100–300 ng/mL with Tsat <20%	≥18 years with HF, LVEF ≤40%, hospitalization for HF or elevated BNP, presence of ID	IV FCM week 1 and week 2, then every 6 months based on Hb and presence of ID Follow-up: 1.9 years	Death, HF hospitalizations, change in 6MWT

FAIR-HF Ferric carboxymaltose in patients with heart failure and iron deficiency, *CONFIRM-HF* beneficial effects of long-term intravenous iron therapy with ferric carboxymaltose in patients with symptomatic heart failure and iron deficiency, *EFFECT-HF* effect of ferric carboxymaltose on exercise capacity in patients with chronic heart failure and iron deficiency, *AFFIRM-AHF* Ferric carboxymaltose for iron deficiency at discharge after acute heart failure, *IRONMAN* Intravenous iron derisomaltose in patients with heart failure and iron deficiency in the UK, *HEART-FID* Ferric carboxymaltose in heart failure and iron deficiency, *ID* iron deficiency, *Tsat* transferrin saturation, *NYHA* New York Heart Association, *LVEF* left ventricular ejection fraction, *Hb* hemoglobin, *HF* heart failure, *BNP* B-type natriuretic peptide, *NT-proBNP* N-terminal pro-B-type natriuretic peptide, *IV* intravenous, *FCM* ferric carboxymaltose, *6MWT* 6-min walk test, *CV* cardiovascular

trial randomized 174 patients with HFrEF (EF ≤45%, NYHA Class II and III, elevated BNP) to either IV FCM or standard of care. At week 24, peak VO_2 decreased in the standard of care group but was unchanged in the treatment arm. Since the primary analysis was done by imputing a value of 0 in peak VO_2 for 4 patients who died in the standard of care arm, reanalysis showed no significant difference between the treatment arm and the standard of care arm.

15.8.4 AFFIRM-AHF Trial (2020)

AFFIRM-AHF (Study to Compare Ferric Carboxymaltose With Placebo in Patients With Acute HF and Iron Deficiency), a large placebo-controlled trial, randomized 1132 hospitalized patients for acute HF (EF < 50% and ID) to either FCM or placebo for up to 24 weeks (Ponikowski et al. 2020). After 52 weeks, there was no significant difference in the combined endpoint of total cardiovascular hospitalizations and cardiovascular death. Total HF hospitalizations and the combined secondary endpoint of first HF hospitalization and cardiovascular death were reduced in the treatment arm compared to the placebo arm. However, this combined endpoint was mainly driven by patients with a ferritin <100 ng/mL at baseline (RR 0.68 (0.51–0.92)) and ischemic HF etiology (RR 0.6 (0.43–0.84)).

15.8.5 IRONMAN Trial (2022)

IRONMAN (Intravenous Iron Treatment in Patients With HF and Iron Deficiency), a prospective open-label RCT, randomized 1137 patients with HFrEF (EF ≤45%, elevated BNP, and hospitalized patients due to HF for the past 6 months) and ID to either IV ferric derisomaltose or usual care (Kalra et al. 2022). The primary endpoint (composite of recurrent HF hospitalization and cardiovascular death) was not different between the two groups, but in the treatment arm, there was a reduction in the secondary endpoint of cardiovascular death or hospital admission due to stroke, MI, or HF.

15.8.6 HEART-FID Trial (2023)

HEART-FID (Randomized Placebo-Controlled Trial of Ferric Carboxymaltose as Treatment for HF With Iron Deficiency) is the most recent RCT of FCM use in patients with HFrEF (EF ≤40% and hospitalization for HF in the past 12 months or elevated BNP). In this study, ID was defined by the same criteria used in FAIR-HF and a Hb of 9.0–13.5 g/dL in women or 9.0–15 g/dL in men. 3065 patients were randomized to either IV FCM or placebo (Mentz et al. 2023). Death occurred in 8.6% of patients in the treatment arm compared to 10.3% in the placebo arm. Moreover, there were slightly less hospitalizations in the FCM arm (297 hospitalizations) compared to the placebo arm (332 hospitalizations). 6MWT was increased by 8 m in the treatment arm compared to 4 m in the placebo arm. There was no difference in the composite endpoint (hierarchical composite of death, hospitalizations for HF, or 6-min walk distance) between both arms. There were also no differences between the FCM arm and the placebo arm in the main secondary outcome of the trial, which was a composite score of cardiovascular death or first hospitalization for HF for the duration of follow-up.

In summary, these trials suggest that IV iron supplementation in patients with HFrEF and ID might improve patient symptoms and decrease HF hospitalizations. However, based on the large clinical trials conducted so far, there is no evidence that IV iron treatment results in any improvement in overall mortality or hard outcomes (except for possible HF hospitalization). Thus, the clinical decision to support IV iron supplementation in these patients should include a risk-benefit assessment before initiating those therapies.

15.9 Side Effects of IV Iron

IV iron supplementation delivers large amounts of iron to the circulation, bypassing the regulatory mechanisms and absorption through the gut (Chap. 24). This huge amount of iron delivery can overwhelm the body's capacity to carry iron through transferrin, leading to a rapid rise in non-transferrin bound iron and reactive labile iron pools intracellularly. This excess iron can cause damage to endothelial cells and tissues due to free radical production and cell death (Stamenkovic et al. 2019; Fang et al. 2019; Ghafourian et al. 2020; Sawicki et al. 2023; Brissot et al. 2012). In healthy human volunteers, IV iron supplementation led to increases in NTBI and was associated with endothelial cell dysfunction measured through flow-mediated vasodilation (Rooyakkers et al. 2002; Zheng et al. 2006). Conversely, iron chelation has been shown to reduce endothelial dysfunction in CAD (Duffy et al. 2001). Data from a prospective observational study with 12 patients showed that NTBI derived from IV iron (FCM) is rapidly and directly delivered to the myocardium via NTBI transporters and remains labile and potentially toxic for weeks (Vera-Aviles et al. 2024). Interestingly, high myocardial iron levels were sustained in the FCM-treated patients after 1 year, despite recurrence of ID (Piechnik et al. 2024).

IV iron use has also been linked to elevated fibroblast growth factor (FGF) 23 levels (Hardy and Vandemergel 2015; Schaefer et al. 2022). This factor is produced by bone cells and is involved in phosphate homeostasis. Elevated levels of FGF23 reduce phosphate absorption in the kidney, resulting in hypophosphatemia, bone resorption, and osteomalacia. The association between hypophosphatemia and elevated FGF23 levels has been primarily linked to the use of FCM versus other IV iron formulations (Wolf et al. 2018). In patients with CKD, elevated FGF23 level is associated with all-cause mortality, left ventricular hypertrophy, HF, and myocardial fibrosis (Ghafourian et al. 2020).

In animal models, the use of IV iron in excessive amounts increases the development of neoplasms where iron deposits (Weinberg 1999). In patients with hemochromatosis or chronic blood transfusion recipients, elevated iron levels were associated with hepatic carcinoma and non-Hodgkin's lymphoma, further highlighting the role of iron in the development and progression of malignancy (Cerhan et al. 2008; Edgren et al. 2008; Kowdley 2004).

The role of IV iron and the risk of infection is another potential side effect of this therapy. In a large systematic review and meta-analysis, IV iron was associated with an increased risk of infection when compared with oral or no iron (RR: 1.16; 95% CI, 1.03–1.29) (Shah et al. 2021). Conversely, some studies did not show any negative outcomes with IV iron supplementation in the setting of acute bacterial infection (Centanni et al. 2024; Macdougall et al. 2020). Further studies are needed to investigate the risk of IV iron use and bacterial infections.

15.10 Oral Iron Therapy

Oral iron supplementation is an appealing alternative to IV iron due to its low cost, availability, and safety profile (Chap. 24). This route of delivery is under the control of the body's mechanisms to titrate serum iron levels, where absorption is high when iron stores are low and vice versa. Oral preparations of iron differ in their side effect and efficacy profile. Some of the most common side effects of oral iron include gastrointestinal intolerability, such as nausea, bloating, constipation, and a metallic aftertaste.

15.10.1 Oral Iron Supplementation and HF

To this date, there have been no studies comparing oral iron to IV iron in patients with ID and HF. In the Iron Supplementation in HF Patients With Anemia (IRON-HF) study (Beck-da-Silva et al. 2013), patients with HFrEF were randomized to either IV iron sucrose, oral ferrous sulfate, or placebo. However, the study was not completed due to lack of recruitment and funding. In this limited follow-up, a change in peak VO2 was observed in the IV iron arm compared to the placebo arm. However, this effect was not reported in the group with oral iron supplementation compared to placebo.

The Effect of Oral Iron Repletion on Exercise Capacity in Patients With HF With Reduced Ejection Fraction and Iron Deficiency (IRONOUT HF) study was a double-blinded, placebo-controlled RCT where patients with HFrEF (LVEF $< 40\%$) and ID (ferritin <100 or TSAT $<20\%$) were randomized to either receiving oral iron polysaccharide 150 mg twice daily

($n = 111$) or placebo ($n = 114$) (Lewis et al. 2017). After 16 weeks, no significant decrease was seen in the study's primary endpoint (change in peak O_2 uptake) or secondary endpoints (change in natriuretic peptide levels, 6-min walk test, or quality of life score) in both arms. However, patients with low hepcidin displayed improvement in their iron indices, indicating that patients with true ID can benefit from oral iron therapy. Additionally, it is important to note that IV iron has not been shown to improve peak VO2 either, so lack of improvement with oral iron does not demonstrate its inferiority to IV iron.

IRON5 (The Effect of Short-Term Oral Iron Supplementation in Systolic HF Patients Suffering From Iron Deficiency Anemia) was a small study conducted in an outpatient setting in Indonesia, where patients with HFrEF (LVEF<45%; NYHA Class II or III), anemia (Hb <13 g/dL in men and <12 g/dL in women), ID by the FAIR-HF criteria, and an estimated glomerular filtration rate (eGFR) > 30 mL/min/1.73 m^2 were treated with either ferrous sulfate 200 mg 3 times per day or placebo. After 12 weeks, the patients in the treatment arm reported improvement in NYHA functional class and an increase in the 6-min walk test. No significant differences in EF, mortality, or HF hospitalizations were noted in this study. This study prompted further investigation into the use of sucrosomial iron, a new oral iron form with comparable improvement in iron indexes and Hb to IV iron while maintaining a relatively better GI side effect profile among different forms of oral iron (Karavidas et al. 2021). The ongoing randomized-controlled trials comparing sucrosomial iron with IV FCM include PREFER-HF (NCT03833336) in patients with HFpEF and IVOFER-HF (EudraCT 2017-005053-37) in patients with HFrEF. Recent studies have suggested that lower daily doses and spacing out of elemental iron are more efficacious than the traditional multi-daily high-dose oral iron (Ning and Zeller 2019). The every-other-day dosing has also been associated with a better GI side effect profile than the traditional dosing (Moretti et al. 2015).

15.11 Conclusions

Iron is crucial for many biological processes. After decades of research, precise iron regulation has been shown to be essential for normal physiology and cardiovascular function. The cardiovascular system utilizes several proteins to import iron intracellularly while only relying on FPN1 to export iron, making the heart susceptible to iron overload. Despite earlier trials showing reduced risk for rehospitalization and promising improvement in quality of life with IV iron supplementation in patients with HFrEF, this approach has failed to show improvement in cardiovascular outcomes. Also, due to the side effects of IV iron, its use in clinical settings should be limited to patients who do not tolerate oral iron therapy or only special cases of ID. More research is needed to explore the use of iron in patients with HFpEF, in addition to the role of oral iron supplementation in symptomatic improvement and cardiovascular outcomes in patients with HF.

References

Abboud S, Haile DJ (2000) A novel mammalian iron-regulated protein involved in intracellular iron metabolism. J Biol Chem 275(26):19906–19912. https://doi.org/10.1074/jbc.M000713200

Alcaide-Aldeano A, Garay A, Alcoberro L, Jiménez-Marrero S, Yun S, Tajes M, García-Romero E, Díez-López C, González-Costello J, Mateus-Porta G, Cainzos-Achirica M, Enjuanes C, Comín-Colet J, Moliner P (2020) Iron deficiency: impact on functional capacity and quality of life in heart failure with preserved ejection fraction. J Clin Med 9(4). https://doi.org/10.3390/jcm9041199

Andrews NC (1999) Disorders of iron metabolism. N Engl J Med 341(26):1986–1995. https://doi.org/10.1056/nejm199912233412607

Anker SD, Colet JC, Filippatos G, Willenheimer R, Dickstein K, Drexler H, Lüscher TF, Bart B, Banasiak W, Niegowska J, Kirwan B-A, Mori C, Rothe BVE, Pocock SJ, Poole-Wilson PA, Ponikowski P (2009) Ferric carboxymaltose in patients with heart failure and iron deficiency. N Engl J Med 361(25):2436–2448. https://doi.org/10.1056/NEJMoa0908355

Bayeva M, Khechaduri A, Puig S, Chang HC, Patial S, Blackshear PJ, Ardehali H (2012) mTOR regulates

cellular iron homeostasis through tristetraprolin. Cell Metab 16(5):645–657. https://doi.org/10.1016/j.cmet.2012.10.001
Beck-da-Silva L, Piardi D, Soder S, Rohde LE, Pereira-Barretto AC, de Albuquerque D, Bocchi E, Vilas-Boas F, Moura LZ, Montera MW, Rassi S, Clausell N (2013) IRON-HF study: a randomized trial to assess the effects of iron in heart failure patients with anemia. Int J Cardiol 168(4):3439–3442. https://doi.org/10.1016/j.ijcard.2013.04.181
Bekfani T, Pellicori P, Morris D, Ebner N, Valentova M, Sandek A, Doehner W, Cleland JG, Lainscak M, Schulze PC, Anker SD, von Haehling S (2019) Iron deficiency in patients with heart failure with preserved ejection fraction and its association with reduced exercise capacity, muscle strength and quality of life. Clin Res Cardiol 108(2):203–211. https://doi.org/10.1007/s00392-018-1344-x
Bi Y, Ajoolabady A, Demillard LJ, Yu W, Hilaire ML, Zhang Y, Ren J (2021) Dysregulation of iron metabolism in cardiovascular diseases: from iron deficiency to iron overload. Biochem Pharmacol 190:114661. https://doi.org/10.1016/j.bcp.2021.114661
Böttinger L, Mårtensson CU, Song J, Zufall N, Wiedemann N, Becker T (2018) Respiratory chain supercomplexes associate with the cysteine desulfurase complex of the iron-sulfur cluster assembly machinery. Mol Biol Cell 29(7):776–785. https://doi.org/10.1091/mbc.E17-09-0555
Briasoulis A, Androulakis E, Christophides T, Tousoulis D (2016) The role of inflammation and cell death in the pathogenesis, progression and treatment of heart failure. Heart Fail Rev 21(2):169–176. https://doi.org/10.1007/s10741-016-9533-z
Brissot P, Ropert M, Le Lan C, Loreal O (2012) Non-transferrin bound iron: a key role in iron overload and iron toxicity. Biochim Biophys Acta 1820 3:403–410. https://doi.org/10.1016/j.bbagen.2011.07.014
Bristow MR, Mason JW, Billingham ME, Daniels JR (1981) Dose-effect and structure-function relationships in doxorubicin cardiomyopathy. Am Heart J 102(4):709–718. https://doi.org/10.1016/0002-8703(81)90096-x
Centanni N, Hammond J, Carver J, Craig W, Nichols S (2024) Intravenous iron in patients hospitalized with bacterial infections: utilization and outcomes. J Maine Med Cent 6(2). https://doi.org/10.46804/2641-2225.1176
Cerhan JR, Engels EA, Cozen W, Davis S, Severson RK, Morton LM, Gridley G, Hartge P, Linet M (2008) Blood transfusion, anesthesia, surgery and risk of non-Hodgkin lymphoma in a population-based case–control study. Int J Cancer 123(4):888–894. https://doi.org/10.1002/ijc.23561
Curigliano G, Lenihan D, Fradley M, Ganatra S, Barac A, Blaes A, Herrmann J, Porter C, Lyon AR, Lancellotti P, Patel A, DeCara J, Mitchell J, Harrison E, Moslehi J, Witteles R, Calabro MG, Orecchia R, de Azambuja E, Zamorano JL, Krone R, Iakobishvili Z, Carver J, Armenian S, Ky B, Cardinale D, Cipolla CM, Dent S, Jordan K (2020) Management of cardiac disease in cancer patients throughout oncological treatment: ESMO consensus recommendations. Ann Oncol 31(2):171–190. https://doi.org/10.1016/j.annonc.2019.10.023
Di Marco A, Cassinelli G, Arcamone F (1981) The discovery of daunorubicin. Cancer Treat Rep 65(Suppl 4):3–8
Ding Z, Liu S, Wang X, Dai Y, Khaidakov M, Romeo F, Mehta JL (2014) LOX-1, oxidant stress, mtDNA damage, autophagy, and immune response in atherosclerosis. Can J Physiol Pharmacol 92(7):524–530. https://doi.org/10.1139/cjpp-2013-0420
Duffy SJ, Biegelsen ES, Holbrook M, Russell JD, Gokce N, Keaney JF Jr, Vita JA (2001) Iron chelation improves endothelial function in patients with coronary artery disease. Circulation 103(23):2799–2804. https://doi.org/10.1161/01.cir.103.23.2799
Edgren G, Reilly M, Hjalgrim H, Tran TN, Rostgaard K, Adami J, Titlestad K, Shanwell A, Melbye M, Nyrén O (2008) Donation frequency, iron loss, and risk of cancer among blood donors. JNCI J Natl Cancer Inst 100(8):572–579. https://doi.org/10.1093/jnci/djn084
Fang X, Wang H, Han D, Xie E, Yang X, Wei J, Gu S, Gao F, Zhu N, Yin X, Cheng Q, Zhang P, Dai W, Chen J, Yang F, Yang HT, Linkermann A, Gu W, Min J, Wang F (2019) Ferroptosis as a target for protection against cardiomyopathy. Proc Natl Acad Sci USA 116(7):2672–2680. https://doi.org/10.1073/pnas.1821022116
Fitzsimons S, Poppe KK, Choi Y, Devlin G, Lund M, Lam CSP, Troughton R, Richards AM, Doughty RN (2022) Relationship between soluble transferrin receptor and clinical outcomes in patients with heart failure according to ejection fraction phenotype: the New Zealand people study. J Card Fail 28(8):1255–1263. https://doi.org/10.1016/j.cardfail.2021.12.018
Franczuk P, Tkaczyszyn M, Kosiorek A, Kulej-Łyko K, Kobak KA, Kasztura M, Sołtowska A, Jaroch J, Ponikowski P, Jankowska EA (2023) Iron status and short-term recovery after non-severe acute myocarditis: a prospective observational study. Biomedicine 11(8):2136
Franke GN, Kubasch AS, Cross M, Vucinic V, Platzbecker U (2020) Iron overload and its impact on outcome of patients with hematological diseases. Mol Asp Med 75:100868. https://doi.org/10.1016/j.mam.2020.100868
Ganz T (2013) Systemic iron homeostasis. Physiol Rev 93(4):1721–1741. https://doi.org/10.1152/physrev.00008.201393/4/1721
Ghafourian K, Shapiro JS, Goodman L, Ardehali H (2020) Iron and heart failure: diagnosis, therapies, and future directions. JACC: Basic Transl Sci 5(3):300–313. https://doi.org/10.1016/j.jacbts.2019.08.009
Girelli D, Nemeth E, Swinkels DW (2016) Hepcidin in the diagnosis of iron disorders. Blood 127(23):2809–2813. https://doi.org/10.1182/blood-2015-12-639112
Gordan R, Wongjaikam S, Gwathmey JK, Chattipakorn N, Chattipakorn SC, Xie LH (2018) Involvement of cyto-

solic and mitochondrial iron in iron overload cardiomyopathy: an update. Heart Fail Rev 23(5):801–816. https://doi.org/10.1007/s10741-018-9700-5
Graham FJ, Masini G, Pellicori P, Cleland JGF, Greenlaw N, Friday J, Kazmi S, Clark AL (2022) Natural history and prognostic significance of iron deficiency and anaemia in ambulatory patients with chronic heart failure. Eur J Heart Fail 24(5):807–817. https://doi.org/10.1002/ejhf.2251
Green R, Charlton R, Seftel H, Bothwell T, Mayet F, Adams B, Finch C, Layrisse M (1968) Body iron excretion in man: a collaborative study. Am J Med 45(3):336–353. https://doi.org/10.1016/0002-9343(68)90069-7
Grote Beverborg N, Klip IT, Meijers WC, Voors AA, Vegter EL, van der Wal HH, Swinkels DW, van Pelt J, Mulder AB, Bulstra SK, Vellenga E, Mariani MA, de Boer RA, van Veldhuisen DJ, van der Meer P (2018) Definition of iron deficiency based on the gold standard of bone marrow iron staining in heart failure patients. Circ Heart Fail 11(2):e004519. https://doi.org/10.1161/CIRCHEARTFAILURE.117.004519
Guo S, Mao X, Li X, Ouyang H (2022) Association between iron status and incident coronary artery disease: a population based-cohort study. Sci Rep 12(1):17490. https://doi.org/10.1038/s41598-022-22275-0
Haddad S, Wang Y, Galy B, Korf-Klingebiel M, Hirsch V, Baru AM, Rostami F, Reboll MR, Heineke J, Flogel U, Groos S, Renner A, Toischer K, Zimmermann F, Engeli S, Jordan J, Bauersachs J, Hentze MW, Wollert KC, Kempf T (2017) Iron-regulatory proteins secure iron availability in cardiomyocytes to prevent heart failure. Eur Heart J 38(5):362–372. https://doi.org/10.1093/eurheartj/ehw333
Hanson E, Sheldon M, Pacheco B, Alkubeysi M, Raizada V (2019) Heart disease in Friedreich's ataxia. World J Cardiol 11(1):1–12. https://doi.org/10.4330/wjc.v11.i1.1
Hardy S, Vandemergel X (2015) Intravenous iron administration and hypophosphatemia in clinical practice. Int J Rheumatol 2015:468675. https://doi.org/10.1155/2015/468675
Heidenreich PA, Bozkurt B, Aguilar D, Allen LA, Byun JJ, Colvin MM, Deswal A, Drazner MH, Dunlay SM, Evers LR, Fang JC, Fedson SE, Fonarow GC, Hayek SS, Hernandez AF, Khazanie P, Kittleson MM, Lee CS, Link MS, Milano CA, Nnacheta LC, Sandhu AT, Stevenson LW, Vardeny O, Vest AR, Yancy CW (2022) 2022 AHA/ACC/HFSA guideline for the management of heart failure. J Am Coll Cardiol 79(17):e263–e421. https://doi.org/10.1016/j.jacc.2021.12.012
Hémadi M, Kahn PH, Miquel G, El Hage Chahine JM (2004) Transferrin's mechanism of interaction with receptor 1. Biochemistry 43(6):1736–1745. https://doi.org/10.1021/bi030142g
Hentze MW, Muckenthaler MU, Galy B, Camaschella C (2010) Two to tango: regulation of Mammalian iron metabolism. Cell 142(1):24–38. https://doi.org/10.1016/j.cell.2010.06.028
Huang B, Qin D, Deng L, Boutjdir M, El-Sherif N (2000) Reexpression of T-type Ca2+ channel gene and current in post-infarction remodeled rat left ventricle. Cardiovasc Res 46(3):442–449. https://doi.org/10.1016/s0008-6363(00)00017-1
Ichikawa Y, Ghanefar M, Bayeva M, Wu R, Khechaduri A, Naga Prasad SV, Mutharasan RK, Naik TJ, Ardehali H (2014) Cardiotoxicity of doxorubicin is mediated through mitochondrial iron accumulation. J Clin Invest 124(2):617–630. https://doi.org/10.1172/jci72931
Ikuta K, Yersin A, Ikai A, Aisen P, Kohgo Y (2010) Characterization of the interaction between diferric transferrin and transferrin receptor 2 by functional assays and atomic force microscopy. J Mol Biol 397(2):375–384. https://doi.org/10.1016/j.jmb.2010.01.026
Jankowska EA, Malyszko J, Ardehali H, Koc-Zorawska E, Banasiak W, von Haehling S, Macdougall IC, Weiss G, McMurray JJ, Anker SD, Gheorghiade M, Ponikowski P (2013) Iron status in patients with chronic heart failure. Eur Heart J 34(11):827–834. https://doi.org/10.1093/eurheartj/ehs377
Jankowska EA, Kasztura M, Sokolski M, Bronisz M, Nawrocka S, Oleśkowska-Florek W, Zymliński R, Biegus J, Siwołowski P, Banasiak W, Anker SD, Filippatos G, Cleland JG, Ponikowski P (2014) Iron deficiency defined as depleted iron stores accompanied by unmet cellular iron requirements identifies patients at the highest risk of death after an episode of acute heart failure. Eur Heart J 35(36):2468–2476. https://doi.org/10.1093/eurheartj/ehu235
Jomova K, Valko M (2011) Advances in metal-induced oxidative stress and human disease. Toxicology 283(2–3):65–87. https://doi.org/10.1016/j.tox.2011.03.001
Kalra PR, Cleland JGF, Petrie MC, Thomson EA, Kalra PA, Squire IB, Ahmed FZ, Al-Mohammad A, Cowburn PJ, Foley PWX, Graham FJ, Japp AG, Lane RE, Lang NN, Ludman AJ, Macdougall IC, Pellicori P, Ray R, Robertson M, Seed A, Ford I, Kalra PR, Cleland JGF, Petrie MC, Thomson EA, Kalra PA, Squire IB, Ahmed FZ, Al-Mohammad A, Cowburn PJ, Foley PWX, Graham FJ, Japp AG, Lane RE, Lang NN, Ludman AJ, Macdougall IC, Pellicori P, Ray R, Robertson M, Seed A, Ford I, Boon N, Amoils S, Chapman C, Diness TG, McMurray J, Mindham R, Sandu P, Strom CC, Travers M, Wilcox R, Struthers A, Mark P, Weir C, Cowan E, Turner C, Austin R, Rogers P, Chandrasekaran B, Fraile E, Kyeremeh L, McGregor L, Osmanska J, Meyer B, Ahmad F, Fisher J, Summersgill C, Adeniji K, Chinnadurai R, Massimo L, Hardman C, Sykes D, Frank S, Smith S, Anwar M, Whittington B, Sookhoo V, Lyons S, Middle J, Housley K, Clark A, Bulemfu J, Critoph C, Chong V, Wood S, Szwejkowski B, Lang C, Duff J, MacDonald S, Schiff R, Donnelly P, Nageh T, Kunhunny S, Gardner R, McAdam M, McPherson E, Banerjee P, Sear E, Edwards N, Glover J, Murphy

C, Cooke J, Spencer C, Francis M, Matthews I, McKie H, Marshall A, Large J, Stratford J, Clifford P, Boos C, Keeling P, Hughes D, Wong A, Jones D, James A, Williams R, Leslie S, Finlayson J, Hannah A, Campbell P, Walsh J, Quinn J, Piper S, Patale S, Gupta P, Sim V, Knibbs L, Lyons K, Dixon L, Petrie C, Wong Y-k, Labinjoh C, Duckett S, Massey I, Savage H, Matias S, Ramirez J, Manisty C, Hussain I, Sankaranarayanan R, Davis G, McClure S, Baxter J, Wicks E, Sobolewska J, Murphy J, Elzayat A, Cooke A, Wright J, Williams S, Muthumala A, Chaggar P, Webber S, Ellis G, Welch M, Bulugahapitiya S, Jackson T, Pakrashi T, Bakhai A, Krishnamurthy V, Gamma R, Ellery S, Jenkins G, Thomas G, Nightingale A, Greenlaw N, Wetherall K, Clarke R, Graham C, Kean S, Stevenson A, Wilson R, Boyle S, McHugh J, Hall L, Woollard J, Brunton C, Dinnett E, Reid A, Howe S, Nicholls J, Cunnington A, Douglas E, Fegen M, Jones M, McGowan S, Ross B, Surtees P, Stuart D (2022) Intravenous ferric derisomaltose in patients with heart failure and iron deficiency in the UK (IRONMAN): an investigator-initiated, prospective, randomised, open-label, blinded-endpoint trial. Lancet 400(10369):2199–2209. https://doi.org/10.1016/S0140-6736(22)02083-9

Karavidas A, Troganis E, Lazaros G, Balta D, Karavidas IN, Polyzogopoulou E, Parissis J, Farmakis D (2021) Oral sucrosomial iron improves exercise capacity and quality of life in heart failure with reduced ejection fraction and iron deficiency: a non-randomized, open-label, proof-of-concept study. Eur J Heart Fail 23(4):593–597. https://doi.org/10.1002/ejhf.2092

Kautz L, Jung G, Valore EV, Rivella S, Nemeth E, Ganz T (2014) Identification of erythroferrone as an erythroid regulator of iron metabolism. Nat Genet 46(7):678–684. https://doi.org/10.1038/ng.2996

Kawabata H, Yang R, Hirama T, Vuong PT, Kawano S, Gombart AF, Koeffler HP (1999) Molecular cloning of transferrin receptor 2. A new member of the transferrin receptor-like family. J Biol Chem 274(30):20826–20832. https://doi.org/10.1074/jbc.274.30.20826

Khechaduri A, Bayeva M, Chang HC, Ardehali H (2013) Heme levels are increased in human failing hearts. J Am Coll Cardiol 61(18):1884–1893. https://doi.org/10.1016/j.jacc.2013.02.012

Kowdley KV (2004) Iron, hemochromatosis, and hepatocellular carcinoma. Gastroenterology 127(5 Suppl 1):S79–S86

Laguna-Fernandez A, Carracedo M, Jeanson G, Nagy E, Eriksson P, Caligiuri G, Franco-Cereceda A, Bäck M (2016) Iron alters valvular interstitial cell function and is associated with calcification in aortic stenosis. Eur Heart J 37(47):3532–3535. https://doi.org/10.1093/eurheartj/ehw122

Lakhal-Littleton S, Wolna M, Carr CA, Miller JJ, Christian HC, Ball V, Santos A, Diaz R, Biggs D, Stillion R, Holdship P, Larner F, Tyler DJ, Clarke K, Davies B, Robbins PA (2015) Cardiac ferroportin regulates cellular iron homeostasis and is important for cardiac function. Proc Natl Acad Sci USA 112(10):3164–3169. https://doi.org/10.1073/pnas.1422373112

Lamas GA, Goertz C, Boineau R, Mark DB, Rozema T, Nahin RL, Lindblad L, Lewis EF, Drisko J, Lee KL, TACT Investigators ft (2013) Effect of disodium EDTA chelation regimen on cardiovascular events in patients with previous myocardial infarction: the TACT randomized trial. JAMA 309(12):1241–1250. https://doi.org/10.1001/jama.2013.2107

Lamas GA, Anstrom KJ, Navas-Acien A, Boineau R, Nemeth H, Huang Z, Wen J, Rosenberg Y, Stylianou M, Jones TLZ, Joubert BR, Yu Q, Santella RM, Mon AC, Ujueta F, Escolar E, Nathan DM, Fonseca VA, Aude YW, Ehrman JK, Elliott T, Prashad R, Lewis EF, Lopes RD, Farkouh ME, Elliott A-M, Newman JD, Mark DB, Investigators T (2024) Edetate disodium–based chelation for patients with a previous myocardial infarction and diabetes: TACT2 randomized clinical trial. JAMA 332(10):794–803. https://doi.org/10.1001/jama.2024.11463

Latunde-Dada GO, Simpson RJ, McKie AT (2008) Duodenal cytochrome B expression stimulates iron uptake by human intestinal epithelial cells. J Nutr 138(6):991–995. https://doi.org/10.1093/jn/138.6.991

Le Blanc S, Garrick MD, Arredondo M (2012) Heme carrier protein 1 transports heme and is involved in heme-Fe metabolism. Am J Phys Cell Phys 302(12):C1780–C1785. https://doi.org/10.1152/ajpcell.00080.2012

Leszek P, Sochanowicz B, Szperl M, Kolsut P, Brzóska K, Piotrowski W, Rywik TM, Danko B, Polkowska-Motrenko H, Różański JM, Kruszewski M (2012) Myocardial iron homeostasis in advanced chronic heart failure patients. Int J Cardiol 159(1):47–52. https://doi.org/10.1016/j.ijcard.2011.08.006

Lewis GD, Malhotra R, Hernandez AF, McNulty SE, Smith A, Felker GM, Tang WHW, LaRue SJ, Redfield MM, Semigran MJ, Givertz MM, Van Buren P, Whellan D, Anstrom KJ, Shah MR, Desvigne-Nickens P, Butler J, Braunwald E, Network ftNHFCR (2017) Effect of oral iron repletion on exercise capacity in patients with heart failure with reduced ejection fraction and iron deficiency: the IRONOUT HF randomized clinical trial. JAMA 317(19):1958–1966. https://doi.org/10.1001/jama.2017.5427

Lopez A, Cacoub P, Macdougall IC, Peyrin-Biroulet L (2016) Iron deficiency anaemia. Lancet 387(10021):907–916. https://doi.org/10.1016/s0140-6736(15)60865-0

Lupu M, Tudor DV, Filip GA (2019) Influence of mitochondrial and systemic iron levels in heart failure pathology. Heart Fail Rev 24(5):647–659. https://doi.org/10.1007/s10741-019-09788-z

Macdougall IC, Bhandari S, White C, Anker SD, Farrington K, Kalra PA, Mark PB, McMurray JJV, Reid C, Robertson M, Tomson CRV, Wheeler DC, Winearls CG, Ford I, Investigators obotP, Committees (2020) Intravenous iron dosing and infection risk in patients on hemodialysis: a prespecified secondary analysis of the PIVOTAL trial. J Am Soc

Nephrol 31(5):1118–1127. https://doi.org/10.1681/asn.2019090972
Madsen MC, Podieh F, Overboom MC, Thijs A, den Heijer M, Hordijk PL (2023) The effect of circulating iron on barrier integrity of primary human endothelial cells. Sci Rep 13(1):16857. https://doi.org/10.1038/s41598-023-44122-6
McDonagh TA, Metra M, Adamo M, Gardner RS, Baumbach A, Böhm M, Burri H, Butler J, Čelutkienė J, Chioncel O, Cleland JGF, Coats AJS, Crespo-Leiro MG, Farmakis D, Gilard M, Heymans S, Hoes AW, Jaarsma T, Jankowska EA, Lainscak M, Lam CSP, Lyon AR, McMurray JJV, Mebazaa A, Mindham R, Muneretto C, Francesco Piepoli M, Price S, Rosano GMC, Ruschitzka F, Kathrine Skibelund A (2021) 2021 ESC Guidelines for the diagnosis and treatment of acute and chronic heart failure. Eur Heart J 42(36):3599–3726. https://doi.org/10.1093/eurheartj/ehab368
McKie AT, Marciani P, Rolfs A, Brennan K, Wehr K, Barrow D, Miret S, Bomford A, Peters TJ, Farzaneh F, Hediger MA, Hentze MW, Simpson RJ (2000) A novel duodenal iron-regulated transporter, IREG1, implicated in the basolateral transfer of iron to the circulation. Mol Cell 5(2):299–309. https://doi.org/10.1016/s1097-2765(00)80425-6
McKie AT, Barrow D, Latunde-Dada GO, Rolfs A, Sager G, Mudaly E, Mudaly M, Richardson C, Barlow D, Bomford A, Peters TJ, Raja KB, Shirali S, Hediger MA, Farzaneh F, Simpson RJ (2001) An iron-regulated ferric reductase associated with the absorption of dietary iron. Science 291(5509):1755–1759. https://doi.org/10.1126/science.1057206
Mehta JL, Pothineni NV (2016) Inflammation in heart failure: the holy grail? Hypertension 68(1):27–29. https://doi.org/10.1161/hypertensionaha.116.07307
Melenovsky V, Petrak J, Mracek T, Benes J, Borlaug BA, Nuskova H, Pluhacek T, Spatenka J, Kovalcikova J, Drahota Z, Kautzner J, Pirk J, Houstek J (2017) Myocardial iron content and mitochondrial function in human heart failure: a direct tissue analysis. Eur J Heart Fail 19(4):522–530. https://doi.org/10.1002/ejhf.640
Members: ATF, McDonagh TA, Metra M, Adamo M, Gardner RS, Baumbach A, Böhm M, Burri H, Butler J, Čelutkienė J, Chioncel O, Cleland JGF, Coats AJS, Crespo-Leiro MG, Farmakis D, Gilard M, Heymans S, Hoes AW, Jaarsma T, Jankowska EA, Lainscak M, Lam CSP, Lyon AR, McMurray JJV, Mebazaa A, Mindham R, Muneretto C, Francesco Piepoli M, Price S, Rosano GMC, Ruschitzka F, Kathrine Skibelund A, Group ESD (2022) 2021 ESC Guidelines for the diagnosis and treatment of acute and chronic heart failure. Eur J Heart Fail 24(1):4–131. https://doi.org/10.1002/ejhf.2333
Mentz RJ, Garg J, Rockhold FW, Butler J, Pasquale CGD, Ezekowitz JA, Lewis GD, O'Meara E, Ponikowski P, Troughton RW, Wong YW, She L, Harrington J, Adamczyk R, Blackman N, Hernandez AF (2023) Ferric carboxymaltose in heart failure with iron deficiency. N Engl J Med 389(11):975–986. https://doi.org/10.1056/NEJMoa2304968
Miyazawa M, Bogdan AR, Hashimoto K, Tsuji Y (2018) Regulation of transferrin receptor-1 mRNA by the interplay between IRE-binding proteins and miR-7/miR-141 in the 3'-IRE stem-loops. RNA 24(4):468–479. https://doi.org/10.1261/rna.063941.117
Moretti D, Goede JS, Zeder C, Jiskra M, Chatzinakou V, Tjalsma H, Melse-Boonstra A, Brittenham G, Swinkels DW, Zimmermann MB (2015) Oral iron supplements increase hepcidin and decrease iron absorption from daily or twice-daily doses in iron-depleted young women. Blood 126(17):1981–1989. https://doi.org/10.1182/blood-2015-05-642223
Nemeth E, Ganz T (2021) Hepcidin-ferroportin interaction controls systemic iron homeostasis. Int J Mol Sci 22(12). https://doi.org/10.3390/ijms22126493
Nemeth E, Tuttle MS, Powelson J, Vaughn MB, Donovan A, Ward DM, Ganz T, Kaplan J (2004) Hepcidin regulates cellular iron efflux by binding to ferroportin and inducing its internalization. Science 306(5704):2090–2093. https://doi.org/10.1126/science.1104742
Neven E, De Schutter TM, Behets GJ, Gupta A, D'Haese PC (2011) Iron and vascular calcification. Is there a link? Nephrol Dial Transplant 26(4):1137–1145. https://doi.org/10.1093/ndt/gfq858
Ning S, Zeller MP (2019) Management of iron deficiency. Hematol Am Soc Hematol Educ Program 1:315–322. https://doi.org/10.1182/hematology.2019000034
Norrish G, Rance T, Montanes E, Field E, Brown E, Bhole V, Stuart G, Uzun O, McLeod KA, Ilina M, Adwani S, Daubeney P, Delle Donne G, Linter K, Jones CB, Bharucha T, Cervi E, Kaski JP (2022) Friedreich's ataxia-associated childhood hypertrophic cardiomyopathy: a national cohort study. Arch Dis Child 107(5):450–455. https://doi.org/10.1136/archdischild-2021-322455
Núñez J, Domínguez E, Ramón JM, Núñez E, Sanchis J, Santas E, Heredia R, González J, Miñana G, López L, Chorro FJ, Palau P (2016) Iron deficiency and functional capacity in patients with advanced heart failure with preserved ejection fraction. Int J Cardiol 207:365–367. https://doi.org/10.1016/j.ijcard.2016.01.187
Oudit GY, Trivieri MG, Khaper N, Liu PP, Backx PH (2006) Role of L-type Ca2+ channels in iron transport and iron-overload cardiomyopathy. J Mol Med 84(5):349–364. https://doi.org/10.1007/s00109-005-0029-x
Panjrath GS, Patel V, Valdiviezo CI, Narula N, Narula J, Jain D (2007) Potentiation of Doxorubicin cardiotoxicity by iron loading in a rodent model. J Am Coll Cardiol 49(25):2457–2464. https://doi.org/10.1016/j.jacc.2007.02.060
Pantopoulos K (2004) Iron metabolism and the IRE/IRP regulatory system: an update. Ann N Y Acad Sci 1012:1–13. https://doi.org/10.1196/annals.1306.001
Paterek A, Mackiewicz U, Mączewski M (2019) Iron and the heart: a paradigm shift from systemic

to cardiomyocyte abnormalities. J Cell Physiol 234(12):21613–21629. https://doi.org/10.1002/jcp.28820

Paterek A, Oknińska M, Chajduk E, Polkowska-Motrenko H, Mączewski M, Mackiewicz U (2021) Systemic iron deficiency does not affect the cardiac iron content and progression of heart failure. J Mol Cell Cardiol 159:16–27. https://doi.org/10.1016/j.yjmcc.2021.06.005

Perdomini M, Belbellaa B, Monassier L, Reutenauer L, Messaddeq N, Cartier N, Crystal RG, Aubourg P, Puccio H (2014) Prevention and reversal of severe mitochondrial cardiomyopathy by gene therapy in a mouse model of Friedreich's ataxia. Nat Med 20(5):542–547. https://doi.org/10.1038/nm.3510

Petit F, Drecourt A, Dussiot M, Zangarelli C, Hermine O, Munnich A, Rotig A (2021) Defective palmitoylation of transferrin receptor triggers iron overload in Friedreich ataxia fibroblasts. Blood 137(15):2090–2102. https://doi.org/10.1182/blood.2020006987

Piechnik SK, Polzella P, Shah A, Vera-Aviles M, Kabir SN, Desborough M, Ferreira VM, Lakhal-Littleton S (2024) Myocardial iron intake following intravenous iron therapy with ferric carboxymaltose is sustained at 1 year despite recurrence of iron deficiency. Br J Haematol. https://doi.org/10.1111/bjh.19915

Piperno A, Pelucchi S, Mariani R (2020) Inherited iron overload disorders. Transl Gastroenterol Hepatol 5:25. https://doi.org/10.21037/tgh.2019.11.15

Ponikowski P, van Veldhuisen DJ, Comin-Colet J, Ertl G, Komajda M, Mareev V, McDonagh T, Parkhomenko A, Tavazzi L, Levesque V, Mori C, Roubert B, Filippatos G, Ruschitzka F, Anker SD (2015) Beneficial effects of long-term intravenous iron therapy with ferric carboxymaltose in patients with symptomatic heart failure and iron deficiency†. Eur Heart J 36(11):657–668. https://doi.org/10.1093/eurheartj/ehu385

Ponikowski P, Kirwan BA, Anker SD, McDonagh T, Dorobantu M, Drozdz J, Fabien V, Filippatos G, Göhring UM, Keren A, Khintibidze I, Kragten H, Martinez FA, Metra M, Milicic D, Nicolau JC, Ohlsson M, Parkhomenko A, Pascual-Figal DA, Ruschitzka F, Sim D, Skouri H, van der Meer P, Lewis BS, Comin-Colet J, von Haehling S, Cohen-Solal A, Danchin N, Doehner W, Dargie HJ, Motro M, Butler J, Friede T, Jensen KH, Pocock S, Jankowska EA (2020) Ferric carboxymaltose for iron deficiency at discharge after acute heart failure: a multicentre, double-blind, randomised, controlled trial. Lancet 396(10266):1895–1904. https://doi.org/10.1016/s0140-6736(20)32339-4

Przybyszewska J, Zekanowska E (2014) The role of hepcidin, ferroportin, HCP1, and DMT1 protein in iron absorption in the human digestive tract. Prz Gastroenterol 9(4):208–213. https://doi.org/10.5114/pg.2014.45102

Radford-Smith DE, Powell EE, Powell LW (2018) Haemochromatosis: a clinical update for the practising physician. Intern Med J 48(5):509–516. https://doi.org/10.1111/imj.13784

Ramos-Polo R, Ras-Jiménez MDM, Francesch Manzano J, Jovells-Vaqué S, Morillas Climent H, Pons-Riverola A, Yun Viladomat S, Moliner Borja P, Diez-Lopez C, González-Costello J, Garcia-Romero E, Herrador L, de Frutos Seminario F, Enjuanes Grau C, Tajes Orduña M, Comin-Colet J (2024) Prognostic role of tissue iron deficiency measured by sTfR levels in heart failure patients without systemic iron deficiency or anemia. J Clin Med 13(16). https://doi.org/10.3390/jcm13164742

Rasel M, Hashmi MF, Mahboobi SK (2024) Transfusion iron overload. In: StatPearls. StatPearls Publishing. Copyright © 2024, StatPearls Publishing LLC., Treasure Island

Richardson KJ, McNamee AP, Simmonds MJ (2018) Haemochromatosis: pathophysiology and the red blood cell1. Clin Hemorheol Microcirc 69(1–2):295–304. https://doi.org/10.3233/ch-189128

Roetto A, Mezzanotte M, Pellegrino RM (2018) The functional versatility of transferrin receptor 2 and its therapeutic value. Pharmaceuticals (Basel) 11(4). https://doi.org/10.3390/ph11040115

Rooyakkers TM, Stroes ES, Kooistra MP, van Faassen EE, Hider RC, Rabelink TJ, Marx JJ (2002) Ferric saccharate induces oxygen radical stress and endothelial dysfunction in vivo. Eur J Clin Investig 32(Suppl 1):9–16. https://doi.org/10.1046/j.1365-2362.2002.0320s1009.x

Sato T, Chang HC, Bayeva M, Shapiro JS, Ramos-Alonso L, Kouzu H, Jiang X, Liu T, Yar S, Sawicki KT, Chen C, Martínez-Pastor MT, Stumpo DJ, Schumacker PT, Blackshear PJ, Ben-Sahra I, Puig S, Ardehali H (2018) mRNA-binding protein tristetraprolin is essential for cardiac response to iron deficiency by regulating mitochondrial function. Proc Natl Acad Sci USA 115(27):E6291–E6300. https://doi.org/10.1073/pnas.1804701115

Sawicki KT, Ardehali H (2021) Intravenous iron therapy in heart failure with reduced ejection fraction: tackling the deficiency. Circulation 144(4):253–255. https://doi.org/10.1161/CIRCULATIONAHA.121.054271

Sawicki KT, Chang HC, Ardehali H (2015) Role of heme in cardiovascular physiology and disease. J Am Heart Assoc 4(1):e001138. https://doi.org/10.1161/jaha.114.001138

Sawicki KT, Chang HC, Shapiro JS, Bayeva M, De Jesus A, Finck BN, Wertheim JA, Blackshear PJ, Ardehali H (2018) Hepatic tristetraprolin promotes insulin resistance through RNA destabilization of FGF21. JCI Insight 3(13). https://doi.org/10.1172/jci.insight.95948

Sawicki KT, De Jesus A, Ardehali H (2023) Iron metabolism in cardiovascular disease: physiology, mechanisms, and therapeutic targets. Circ Res 132(3):379–396. https://doi.org/10.1161/CIRCRESAHA.122.321667

Schaefer B, Tobiasch M, Wagner S, Glodny B, Tilg H, Wolf M, Zoller H (2022) Hypophosphatemia after intravenous iron therapy: comprehensive review of clinical findings and recommendations for management. Bone 154:116202. https://doi.org/10.1016/j.bone.2021.116202

Shah AA, Donovan K, Seeley C, Dickson EA, Palmer AJR, Doree C, Brunskill S, Reid J, Acheson AG, Sugavanam A, Litton E, Stanworth SJ (2021) Risk of infection associated with administration of intravenous iron: a systematic review and meta-analysis. JAMA Netw Open 4(11):e2133935. https://doi.org/10.1001/jamanetworkopen.2021.33935

Shapiro JS, Chang HC, Tatekoshi Y, Zhao Z, Waxali ZS, Hong BJ, Chen H, Geier JA, Bartom ET, De Jesus A, Nejad FK, Mahmoodzadeh A, Sato T, Ramos-Alonso L, Romero AM, Martinez-Pastor MT, Jiang SC, Sah-Teli SK, Li L, Bentrem D, Lopaschuk G, Ben-Sahra I, O'Halloran TV, Shilatifard A, Puig S, Bergelson J, Koivunen P, Ardehali H (2023) Iron drives anabolic metabolism through active histone demethylation and mTORC1. Nat Cell Biol 25(10):1478–1494. https://doi.org/10.1038/s41556-023-01225-6

Shen J, Fu H, Ding Y, Yuan Z, Xiang Z, Ding M, Huang M, Peng Y, Li T, Zha K, Ye Q (2024) The role of iron overload and ferroptosis in arrhythmia pathogenesis. IJC Heart Vasc 52:101414. https://doi.org/10.1016/j.ijcha.2024.101414

Shirazi LF, Bissett J, Romeo F, Mehta JL (2017) Role of inflammation in heart failure. Curr Atheroscler Rep 19(6):27. https://doi.org/10.1007/s11883-017-0660-3

Sierpinski R, Josiak K, Suchocki T, Wojtas-Polc K, Mazur G, Butrym A, Rozentryt P, van der Meer P, Comin-Colet J, von Haehling S, Kosmala W, Przewlocka-Kosmala M, Banasiak W, Nowak J, Voors AA, Anker SD, Cleland JGF, Ponikowski P, Jankowska EA (2021) High soluble transferrin receptor in patients with heart failure: a measure of iron deficiency and a strong predictor of mortality. Eur J Heart Fail 23(6):919–932. https://doi.org/10.1002/ejhf.2036

Siri-Angkul N, Xie L-H, Chattipakorn SC, Chattipakorn N (2018) Cellular electrophysiology of iron-overloaded cardiomyocytes. Front Physiol 9. https://doi.org/10.3389/fphys.2018.01615

Smith FM, Kosman DJ (2020) Molecular defects in Friedreich's Ataxia: convergence of oxidative stress and cytoskeletal abnormalities. Front Mol Biosci 7:569293. https://doi.org/10.3389/fmolb.2020.569293

Smith LA, Cornelius VR, Plummer CJ, Levitt G, Verrill M, Canney P, Jones A (2010) Cardiotoxicity of anthracycline agents for the treatment of cancer: systematic review and meta-analysis of randomised controlled trials. BMC Cancer 10:337. https://doi.org/10.1186/1471-2407-10-337

Song JX, Xu S, Chen Q, Gou Y, Zhao CB, Jia CL, Liu H, Zhang Z, Li BL, Gao Y, Zhao Y, Ji ES (2024) Cardiomyocyte-specific overexpression of FPN1 diminishes cardiac hypertrophy induced by chronic intermittent hypoxia. J Cell Mol Med 28(14):e18543. https://doi.org/10.1111/jcmm.18543

Stamenkovic A, Pierce GN, Ravandi A (2019) Phospholipid oxidation products in ferroptotic myocardial cell death. Am J Physiol Heart Circ Physiol 317(1):H156–h163. https://doi.org/10.1152/ajpheart.00076.2019

Sweet ME, Mestroni L, Taylor MR (2018) Genetic infiltrative cardiomyopathies. Heart Fail Clin 14(2):215–224

Taher AT, Musallam KM, Cappellini MD (2021) β-Thalassemias. N Engl J Med 384(8):727–743. https://doi.org/10.1056/NEJMra2021838

Tan C, Tasaka H, Yu K-P, Murphy ML, Karnofsky DA (1967) Daunomycin, an antitumor antibiotic, in the treatment of neoplastic disease. Clinical evaluation with special reference to childhood leukemia. Cancer 20(3):333–353. https://doi.org/10.1002/1097-0142(1967)20:3<333::AID-CNCR2820200302>3.0.CO;2-K

Tandy S, Williams M, Leggett A, Lopez-Jimenez M, Dedes M, Ramesh B, Srai SK, Sharp P (2000) Nramp2 expression is associated with pH-dependent iron uptake across the apical membrane of human intestinal Caco-2 cells. J Biol Chem 275(2):1023–1029. https://doi.org/10.1074/jbc.275.2.1023

Tsushima RG, Wickenden AD, Bouchard RA, Oudit GY, Liu PP, Backx PH (1999) Modulation of iron uptake in heart by L-type Ca2+ channel modifiers: possible implications in iron overload. Circ Res 84(11):1302–1309. https://doi.org/10.1161/01.res.84.11.1302

Udani K, Chris-Olaiya A, Ohadugha C, Malik A, Sansbury J, Paari D (2021) Cardiovascular manifestations in hospitalized patients with hemochromatosis in the United States. Int J Cardiol 342:117–124. https://doi.org/10.1016/j.ijcard.2021.07.060

Ursu ON, Sauter M, Ettischer N, Kandolf R, Klingel K (2014) Heme oxygenase-1 mediates oxidative stress and apoptosis in coxsackievirus B3-induced myocarditis. Cell Physiol Biochem 33(1):52–66. https://doi.org/10.1159/000356649

van Veldhuisen DJ, Ponikowski P, van der Meer P, Metra M, Böhm M, Doletsky A, Voors AA, Macdougall IC, Anker SD, Roubert B, Zakin L, Cohen-Solal A (2017) Effect of ferric carboxymaltose on exercise capacity in patients with chronic heart failure and iron deficiency. Circulation 136(15):1374–1383. https://doi.org/10.1161/CIRCULATIONAHA.117.027497

Vera-Aviles M, Kabir SN, Shah A, Polzella P, Lim DY, Buckley P, Ball C, Swinkels D, Matlung H, Blans C, Holdship P, Nugent J, Anderson E, Desborough M, Piechnik S, Ferreira V, Lakhal-Littleton S (2024) Intravenous iron therapy results in rapid and sustained rise in myocardial iron content through a novel pathway. Eur Heart J. https://doi.org/10.1093/eurheartj/ehae359

Vinchi F (2021) Non-transferrin-bound iron in the spotlight: novel mechanistic insights into the vasculotoxic and atherosclerotic effect of iron. Antioxid Redox Signal 35(6):387–414. https://doi.org/10.1089/ars.2020.8167

Wang W, Knovich MA, Coffman LG, Torti FM, Torti SV (2010) Serum ferritin: past, present and future. Biochim Biophys Acta 1800 8:760–769. https://doi.org/10.1016/j.bbagen.2010.03.011

Weinberg ED (1999) Iron therapy and cancer. Kidney Int Suppl 69:S131–S134. https://doi.org/10.1046/j.1523-1755.1999.055suppl.69131.x

Wolf M, Chertow GM, Macdougall IC, Kaper R, Krop J, Strauss W (2018) Randomized trial of intravenous iron-induced hypophosphatemia. JCI Insight 3(23). https://doi.org/10.1172/jci.insight.124486

Xu W, Barrientos T, Mao L, Rockman HA, Sauve AA, Andrews NC (2015) Lethal cardiomyopathy in mice lacking transferrin receptor in the heart. Cell Rep 13(3):533–545. https://doi.org/10.1016/j.celrep.2015.09.023

Xu R, Zhu D, Guo J, Wang C (2021) IL-18 promotes erythrophagocytosis and erythrocyte degradation by M1 macrophages in a calcific microenvironment. Can J Cardiol 37(9):1460–1471. https://doi.org/10.1016/j.cjca.2021.04.007

Zhao G, Arosio P, Chasteen ND (2006) Iron(II) and hydrogen peroxide detoxification by human H-chain ferritin. An EPR spin-trapping study. Biochemistry 45(10):3429–3436. https://doi.org/10.1021/bi052443r

Zheng H, Huang X, Zhang Q, Katz SD (2006) Iron sucrose augments homocysteine-induced endothelial dysfunction in normal subjects. Kidney Int 69(4):679–684. https://doi.org/10.1038/sj.ki.5000046

Iron and Liver Disease

16

Heinz Zoller

Abstract

The mutual relationship between iron and liver disease is highlighted by the fact that too much iron can cause liver disease, while hepatic iron overload can also result from acute or chronic liver diseases. As the principal storage site for iron, hepatic iron concentration is a reliable surrogate marker for total body iron in iron overload conditions. Assessment of hepatic iron is therefore important for the initiation and monitoring of treatment of iron overload by therapeutic phlebotomies or iron chelators. The present chapter reviews how the liver controls iron homeostasis, and discusses how liver iron can be assessed before primary and secondary liver diseases with iron overload.

Keywords

Hepatic iron overload · Hemochromatosis · Hepcidin · Liver fibrosis · Chronic liver disease

16.1 Introduction: Iron and Liver Diseases—Cause or Consequence?

The liver and iron are intimately connected because the liver is key to control iron homeostasis and excess iron can be stored in hepatocytes and Kupffer cells. Liver iron concentration is an excellent surrogate marker for total body iron, especially in patients with thalassemia (Angelucci et al. 2000). Elevated hepatic iron is a risk factor for progressive liver fibrosis, cirrhosis, and hepatocellular carcinoma, especially in patients with the genetic iron storage disease hemochromatosis. In addition, in patients with alcohol-related liver disease, metabolic dysfunction-associated steatotic liver disease (MASLD), or hepatitis C virus infection, hepatic iron has been associated with progressive fibrosis (Lupşor et al. 2008; Olynyk et al. 2005). Impaired liver function reduces the ability to control plasma iron homeostasis and thus causes hepatic iron overload (Sharma et al. 2024).

These observations highlight the mutual relationship between iron and liver diseases (Prieto et al. 1975). As liver disease progresses, the distinction between iron overload as cause or consequence gets more and more difficult. The majority of patients with cirrhosis presents with increased hepatic iron, regardless of the underlying etiology (Ludwig et al. 1997). Hepatic iron concentrations and serum iron parameters are therefore

H. Zoller (✉)
Department of Medicine I, Christian Doppler Laboratory for Iron and Phosphate Biology, Medical University of Innsbruck, Innsbruck, Austria
e-mail: heinz.zoller@i-med.ac.at; https://heplab.at

K. Pantopoulos (ed.), *Iron Metabolism in Human Health and Disease*, Advances in Experimental Medicine and Biology 1480, https://doi.org/10.1007/978-3-031-92033-2_16

highly relevant for the diagnosis, staging, and treatment of liver diseases.

16.2 The Liver Controls Iron Homeostasis by Hormonal Regulation and Removal of Excess Iron from the Circulation

Dietary iron is absorbed from the intestine, and iron from senescent red blood cells is released from recycling red-pulp macrophages of the spleen into the circulation. Both intestine and spleen deliver their venous blood to the liver via the portal circulation. Hence, hepatocytes are exposed to higher iron concentrations and higher transferrin saturation (TSAT) than any other organ (Vogel et al. 1989). Excess non-transferrin-bound iron (NTBI) (see Chap. 4) is subject to a high first-pass in the liver, where different NTBI transporters are expressed, for instance, Zip-14 (*SLC39A14*) and divalent metal transporter 1 (*SLC11A2*)—but also L- and T-type calcium channels (LTCC/TTCC)—Zip8 (*SLC39A8*) (Knutson 2019). Similar to cardiomyocytes and pancreatic cells, hepatocytes can become iron-overloaded when NTBI is present in the circulation. Studies have shown that redox-active forms of NTBI also known as labile plasma iron (LPI) occur when TSAT is >65% (Le Lan et al. 2005).

16.3 Control of Hepatic Expression of the Iron-Hormone Hepcidin

The liver responds to high circulating iron and inflammation with increased hepcidin expression, which mediates reduced iron egress from intestinal epithelial cells and macrophages (Chap. 3). Thereby, a negative feedback loop is closed which maintains plasma iron homeostasis (Nemeth and Ganz 2023). Although the molecular details on how the liver senses iron have not yet been fully elucidated, data from animal models suggest that the sinusoidal endothelial cells in the liver have a critical iron-sensing function that culminates in production and secretion of bone morphogenetic proteins BMP6 and BMP2, which in turn induce hepcidin expression in hepatocytes (Canali et al. 2017; Koch et al. 2017). Identification of heterozygous mutations in the propeptide region of BMP6 in families with increased hepatic iron supports a role of BMP6 also in humans (Daher et al. 2016). In mice, BMPs mainly induce hepcidin via heterodimerization of BMP type I and BMP type II receptors (Dogan et al. 2024). Genetic studies also support a role of BMP receptor type 1B in human iron overload (Schaefer et al. 2015). BMP signaling is also determined by the adequate expression of coreceptors of the BMP receptor complex. These coreceptors include the hemochromatosis protein *HFE*, *TFR2*, and *HJV* (Latour et al. 2016). The cell surface expression of hemojuvelin is in turn controlled by a cell surface serine protease matriptase-2—encoded by *TMPRSS6* (Du et al. 2008) (Fig. 16.1).

Hepatocellular hepcidin production is controlled by iron not only via bone morphogenetic proteins, but also via proinflammatory cytokines. Interleukin-6 (IL-6) signals via the JAK-STAT pathway (Nemeth et al. 2004).

16.4 Genetic Defects Resulting in Impaired Hepcidin Expression or Hepcidin Resistance Cause Hemochromatosis

Mutations in any of the known coreceptors of the BMP receptor complex (*HFE, TFR2, HJV*) cause reduced hepatic hepcidin expression, which can result in high TSAT and hepatocellular iron overload (Girelli et al. 2022). Severe and early-onset iron overload can also result from low or absent hepcidin due to recessive genetic defects in *HAMP* (the gene encoding hepcidin) (Roetto et al. 2003). In addition, certain mutations in *SLC40A1* that affect the hepcidin-binding site of the iron export protein ferroportin, such as p. Cys326Tyr or p.His504Arg, cause impaired hepcidin effector function (Drakesmith et al. 2005; Mayr et al. 2011). This so-called hepcidin resis-

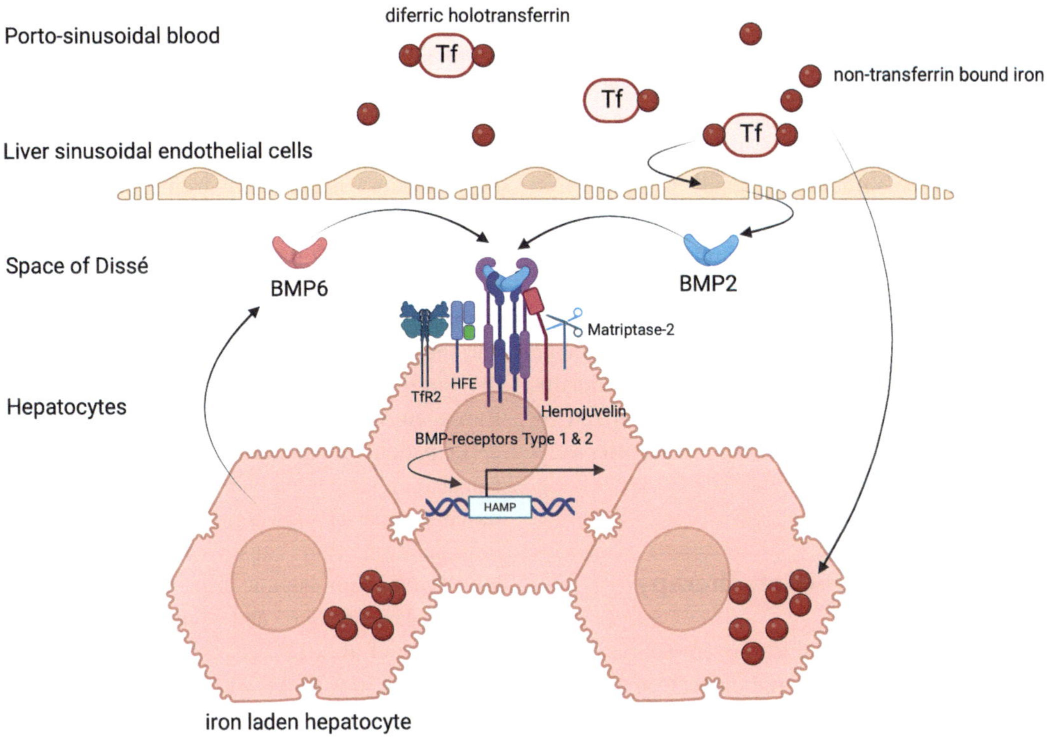

Fig. 16.1 Hemochromatosis can be caused by genetic defects in HFE, HJV, TFR2, hepcidin, or BMP6. Defects in any of these genes can cause inappropriately low hepcidin expression, which results in high TSAT and possibly non-transferrin-bound iron species, which are rapidly taken up by hepatocytes

tance causes increased iron uptake, increased macrophage iron release, and high TSAT, resulting in hepatocellular iron storage (Mayr et al. 2011; Drakesmith et al. 2005). In summary, genetic conditions affecting hepcidin, hepcidin production, or its effects on ferroportin can result in phenotypic hemochromatosis, which is characterized by inappropriate hepcidin response, high TSAT, and hepatocellular iron loading (see also Chaps. 9 and 10).

16.5 Acquired Causes of Reduced Hepatic Hepcidin Expression Also Cause Hepatocellular Iron Loading

Reduced hepcidin and consequent hepatocellular iron storage can result not only from impaired control of upstream factors regulating hepcidin gene expression, but also from reduced liver cell mass (Pietrangelo 2015). This explains why advanced acute or chronic liver disease is typically associated with increased iron absorption and hepatic iron loading (Ludwig et al. 1997). Hepcidin can also be suppressed by impaired signal transduction from the BMP receptor complex to the hepcidin promoter, which is mediated by SMAD-1/5/8 proteins (Corradini et al. 2011). Viral hepatitis and alcohol have been found to directly interfere with SMAD signaling and therefore promote hepatocellular iron overload (Foka et al. 2014; Gerjevic et al. 2012).

16.6 Hepatic Iron Overload in Iron-Loading Anemias

Signals known to reduce hepcidin expression include anemia and hypoxia, where iron absorption and iron release from intestinal epithelial cells are increased (Nicolas et al. 2002). This response can be physiological in patients with iron deficiency anemia, but in patients with anemia due to defective hemoglobin synthesis, hemolysis, or other disorders that impair red cell or their lifespan, and reduce hepcidin production, which can be sufficient to result in hepatic iron overload (Wang and Babitt 2019; see also Chap. 11). These so-called iron-loading anemias can be aggravated by transfusions. Each unit of red blood cells contains ca. 200–250 mg of iron. In contrast to low hepcidin states, where stainable iron is mainly localized to hepatocytes, transfusional iron overload is mainly characterized by iron overload in the spleen and in Kupffer cells (Shander et al. 2009).

16.7 Parenteral Iron Therapy

Similar to transfusional iron overload, parenteral iron therapy can also cause hepatic iron overload. Intravenous (IV) iron formulations are nanoparticles composed of complex carbohydrates and iron oxides (Schaefer et al. 2020; see also Chap. 24). After IV infusion, these nanoparticles are taken up by macrophages before being degraded and iron is slowly released into the circulation. In patients with chronic inflammation or chronic kidney disease, where hepcidin concentrations are high, this iron release from macrophages can be impaired (Weiss et al. 2009). Especially, dialysis patients are therefore at risk for the development of iatrogenic iron overload due to repeated IV iron transfusions that can also affect the liver (Ramanathan et al. 2017).

16.8 Diagnostic Approach to Patients with Iron-Related Liver Disease

The assessment of serum iron parameters (Chap. 25) is the first step for the diagnosis and staging of all patients with suspected liver iron overload, and serum iron parameters should be included when patients are being evaluated for liver disease (Newsome et al. 2018). High serum ferritin is suggestive of increased liver iron and has a reasonable sensitivity, but poor specificity for detecting hepatic iron overload. Alternative causes of high ferritin include inflammation, cell death, malignancy, and high alcohol intake. Serum iron parameters also include ferritin TSAT, which can be calculated from serum iron and transferrin. High TSAT is suggestive of hemochromatosis. Patients with high ferritin and high TSAT should be tested for hemochromatosis by *HFE* genotyping (Chap. 9) for the common disease-associated variant p.Cys282Tyr (European Association for the Study of the Liver. Electronic address and European Association for the Study of the 2022).

In patients with normal TSAT, high ferritin can also indicate hepatic iron overload. The commonest cause of high ferritin and low TSAT is inflammation. Rare causes of high ferritin and low TSAT with hepatic iron overload include aceruloplasminemia and ferroportin disease (Nashwan et al. 2023; also see Chap. 10). High ferritin and low TSAT without inflammation or hepatic iron overload should raise the suspicion of hyperferritinemia cataract syndrome, which can be confirmed by genetic testing for mutations in the iron-response element in L-ferritin (Aguilar-Martinez et al. 2005).

In patients with suspected hepatic iron overload, liver iron can be quantified noninvasively by magnetic resonance imaging (MRI) (Reeder et al. 2023). The preferred MRI sequences for iron quantification are T2* relaxation times. As T2* relaxation time is indirectly proportional to hepatic iron concentrations, R2* relaxation times are typically calculated, which is the reciprocal of T2* relaxation time expressed in seconds (R2* = 1/T2*). Hepatic R2* values are therefore expressed as s^{-1} (Hertz, Hz). R2* values are also indirectly proportional to the strength of the magnetic field of the MR scanner. Studies have proposed 70 s^{-1} as the upper limit of normal for 1.5 Tesla scanners (Viveiros et al. 2021). Hepatic iron can also be quantified on MRI scanners with the support of a vendor (FerriScan), where MR sequences are acquired locally, data are transmitted to the vendor, and a report is returned to the site (Sussman et al. 2024). Another noninvasive alternative method to quantify liver iron concentration is susceptometry, which requires a specially dedicated diagnostic device (superconducting paramagnetic quantum interference device—SQUID), that is only available in few sites worldwide (Brittenham et al. 1982).

Historically, the gold standards for the quantification of liver iron were invasive: liver biopsy with chemical iron quantification and iron staining, or quantitative phlebotomy. Different studies have shown a correlation of these methods (Phatak and Barton 2003; Reeder et al. 2023). For quantitative phlebotomy, 500 mL of blood is removed by weekly or fortnightly phlebotomy until ferritin decreases to <25 μg/L. The development of anemia should be avoided. Iron removed with each phlebotomy can be calculated as grams of iron using the following formula: [Hct/3 × weight of blood in grams × 0.0035]. Alternatively, the amount of iron removed can be estimated by multiplying the number of phlebotomies needed by 200 mg (Phatak and Barton 2003).

Atomic absorption spectrometry is used for liver iron quantification in tissue specimens. Upper limits of normal range are from 25 to 36 μmol iron per gram of dry liver tissue (Walker et al. 1971). As the iron concentration in the liver increases with age, dividing the hepatic iron concentration expressed as μmoles per gram of dry liver tissue by age results in a higher specificity to discriminate hemochromatosis from milder forms of iron overload caused by increased alcohol intake. This calculated ratio is known as a hepatic iron index (HII). Studies have shown that an HII >1.9 has a higher specificity for the diagnosis of hemochromatosis than hepatic iron concentration alone (Adams et al. 1997).

The cellular distribution of stainable iron within the liver on histological assessment by Perls' staining can provide etiological information. At early stages, iron overload caused by hemochromatosis is confined to hepatocytes with a portal to central gradient. Iron in Kupffer cells is typically found in patients with iron overload caused by increased alcohol intake, intravenous iron therapy, red blood cell transfusions, or ferroportin disease. At advanced stages of iron overload, a mixed type of increased stainable iron in both hepatocytes and Kupffer cells can be present (Batts 2007) (Fig. 16.2).

The noninvasive alternative to histological assessment of a liver biopsy to study the cell type predominantly affected by iron overload is T2*-based MR imaging of the abdomen, where R2* frequencies can be quantified in representative regions of interest in the liver, spleen, bone marrow, and pancreas. Iron confined to the liver is typical for hemochromatosis, non-transfusional iron-loading anemias, and aceruloplasminemia. Predominant iron in the spleen and bone marrow is typical for Kupffer cell and macrophage iron overload. This would be a typical finding in transfusional iron overload, iron overload caused by repeated iron infusions, or ferroportin disease (Fig. 16.3). Noninvasive iron quantification by MRI can also be used to quantify iron in the heart, pituitary gland, and brain (Ning et al. 2024).

16.9 Hemochromatosis

Hemochromatosis has been initially described as "bronze diabetes" in young patients with severe iron overload, dark gray-brown ("bronze") discoloration of the skin, diabetes mellitus, decompensated cirrhosis, heart failure, and high mortality (Trosseau 1865). Histological examination of affected tissues in patients with this condition showed a typical brown pigment, which was thought to be derived from the blood, and thus, the disease name—hemochromatosis—was coined in 1889 by Daniel Friedrich von Recklinghausen (1889). Later studies have shown that hemochromatosis follows autosomal recessive inheritance and the responsible variant is the c.845G>A variant in the HFE gene, which causes a cysteine-to-tyrosine substitution in position 282 (p.C282Y) of the predicted protein (Simon et al. 1976; Feder et al. 1997; see also Chap. 9). Homozygosity for p.C282Y has an allele frequency of up to 12.5% in Ireland and follows a north-west to south-east gradient in Europe (European Association for the Study of the Liver. Electronic address and European Association for the Study of the 2022). This variant is rare in non-Caucasians. Another variant in HFE is p.H63D, which has a global allele frequency of ca 10–15%

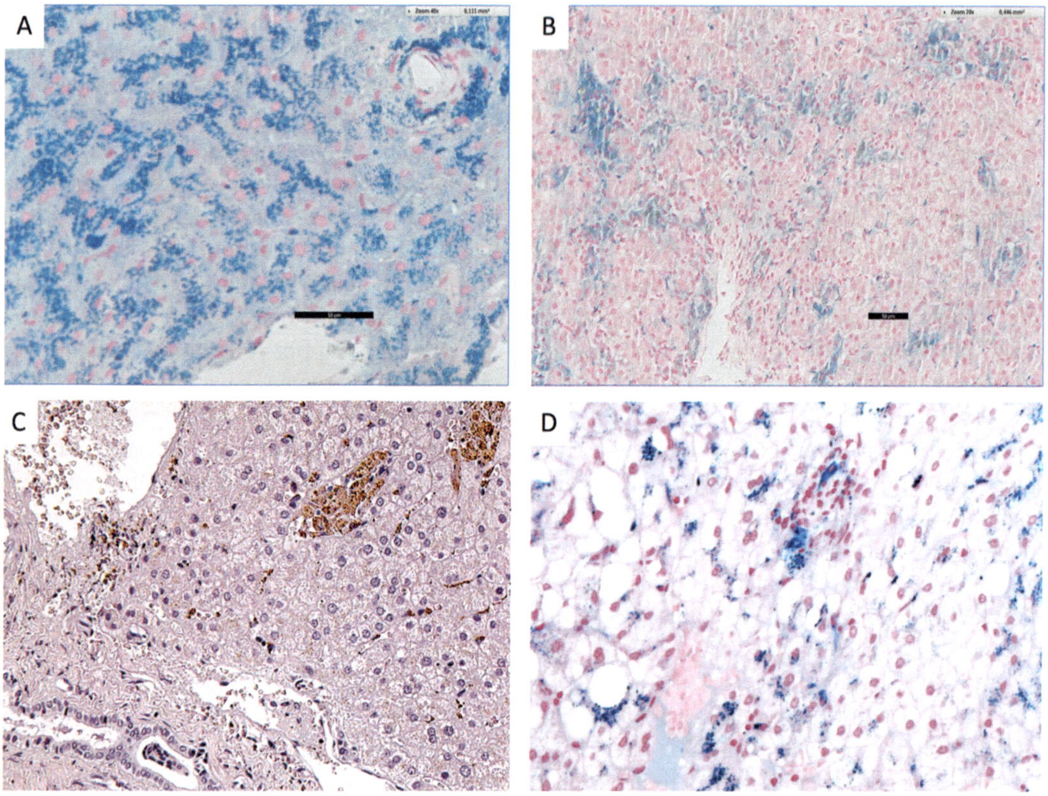

Fig. 16.2 Histological findings in patients with iron overload of various etiologies. (**a**) Severe hepatocellular iron overload in juvenile hemochromatosis caused by a homozygous pathogenic mutation in HAMP—Perls' stain. (**b**) Mixed patchy hepatocellular and Kupffer cell iron overload in a patient thalassemia intermedia—Perls' stain. (**c**) Kupffer cell iron overload with severe siderosis in a ferroportin disease patient—HE stain. (**d**) Predominant hepatocellular iron overload in a patient with aceruloplasminemia—Perls' stain

and is not associated with an increased hemochromatosis risk, unless in compound heterozygosity with p.C282Y. Even among individuals homozygous of p.C282Y, only ca 12–25% develop iron overload by the age of 65 years (Schaefer et al. 2024). Risk factors for the development of iron overload—other than age—include male sex and increased alcohol intake (Fletcher et al. 2002).

The diagnosis of hemochromatosis requires the presence of increased hepatic iron >36 μmol/g dry weight on liver biopsy, an R2* value >70 s^{-1} on hepatic MRI, or a mobilizable iron of >3–5 g. In clinical practice, therapeutic venesection can be started in patients homozygous for p.C282Y, high TSAT, and elevated ferritin. In such patients, disease staging should not only include a clinical examination for signs of liver disease, heart failure, diabetes, or arthropathy, but also include liver ultrasound and elastography of the liver for fibrosis staging (Hegde et al. 2024). Endocrine manifestations of hemochromatosis and heart failure can be present in hemochromatosis patients with severe iron overload. They include diabetes, hypopituitary hypogonadism, and adrenal insufficiency. Patients with early disease onset and severe disease manifestations should be assessed for such disease manifestations (European Association for the Study of the Liver. Electronic address and European Association for the Study of the 2022). Recessive pathogenic variants *HJV*, *HAMP*, or *TFR2* cause "juvenile hemochromatosis," which has also been reported as a digenic disease (Chap. 10). Juvenile hemochromatosis can be suspected if affected patients present before the age of 30 years (Girelli et al.

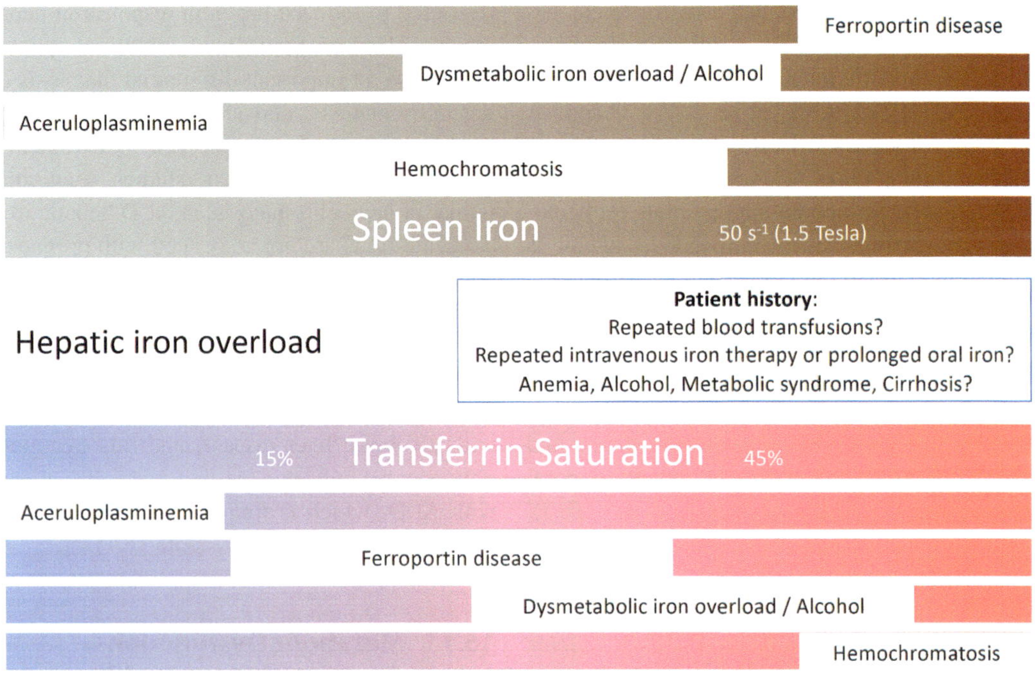

Fig. 16.3 Summary of phenotypic presentation in patients with diseases primarily presenting with hepatic iron overload. TSAT and spleen iron quantification by R2* on MRI can help distinguishing between iron overload caused by hemochromatosis, dysmetabolic iron overload, ferroportin disease, and aceruloplasminemia. The patient history can also provide important diagnostic information because blood transfusions, oral or intravenous iron therapy, excessive alcohol consumption, MASLD, and cirrhosis can also cause hepatic iron overload

2022). Alternative genetic causes of hemochromatosis include certain dominant variants in SLC40A1 and BMP6 (Daher et al. 2016; Viveiros et al. 2019).

The recommended treatment for hemochromatosis is therapeutic venesection, where 450–500 mL of blood is removed by weekly phlebotomy, while avoiding anemia, until the ferritin drops below 100 µg/L. After this induction phase, patients should undergo lifelong maintenance where serum ferritin should be kept <100 µg/L—usually by phlebotomy 2–6 times per year. Patients who underwent phlebotomy have been found to have better survival than untreated patients (Bomford and Williams 1976). Hemochromatosis is likely caused by a blunted hepcidin response to an increase in plasma iron and is therefore now considered an endocrine disease (Piperno et al. 2007). Even with appropriate removal of increased iron as indicated by a serum ferritin concentration <100 µg/L, TSAT can remain elevated (>45%). In such patients, guidelines recommend intensified treatment with a ferritin target <50 µg/L. This is supported by a retrospective study, where improved outcome was found when ferritin is <50 µg/L and a TSAT <50% was achieved during maintenance phlebotomy (Bardou-Jacquet et al. 2017). A possible explanation for this finding is that redox-active iron is present in the plasma of hemochromatosis patients with TSAT >65% (Le Lan et al. 2005).

The greatest challenges in hemochromatosis remain early diagnosis and timely initiation of preventative measures such as avoidance of alcohol consumption, reduction in dietary iron uptake, and prevention of infections with pathogens where virulence is increased in the presence of high iron such as vibrio vulnificus or yersinia pestis (Gerhard et al. 2001; Frank et al. 2011). Duodenal iron uptake can be reduced with proton-pump inhibitors, and this treatment has been shown to reduce the number of phlebotomies required to maintain normal serum iron parameters (Vanclooster et al. 2017). Diagnosis of hemochromatosis should also prompt family screening in siblings of diagnosed patients. The question whether population screening for hemochromatosis should be recommended, and how, is still being debated (Delatycki and Allen 2024).

In conclusion, hemochromatosis is an autosomal recessive disease characterized by iron overload in the liver and high TSAT. The principal clinical symptom in affected patients is fatigue, but patients can also have joint pain and require earlier joint replacement than the general populations (Elmberg et al. 2013). Hemochromatosis patients have an increased risk for the development of cirrhosis and hepatocellular carcinoma, which can be prevented with early phlebotomy (Schaefer et al. 2024). Patients with decompensated cirrhosis or early-stage hepatocellular carcinoma should be considered as candidates for liver transplantation, while mitigating their increased susceptibility to infections in the post-transplant period (Brandhagen et al. 2000). The best treatment for hemochromatosis is prevention of organ damage by early diagnosis and treatment.

16.10 Alcoholic Liver Disease

Increased alcohol consumption is also associated with an increased risk for the development of hepatic iron overload. Hepatic iron overload in alcoholic siderosis is less severe than in hemochromatosis, and in contrast to hemochromatosis, hepatic iron overload caused by chronic alcohol consumption primarily affects Kupffer cells (Chapman et al. 1982; Ali et al. 2023). The prevalence of increased hepatic iron concentrations defined as >25 μmol/g dry weight was 29% in one study. Noninvasive quantification of iron by R2*-MRI in patients with alcoholic liver disease (ALD) can show iron not restricted to the liver but also in the spleen (Zoller and Henninger 2016). The cellular mechanisms that cause hepatic iron loading in ALD include but are not restricted to reduced hepcidin response in hepatocytes (Harrison-Findik et al. 2009).

ALD is an important differential diagnosis of hemochromatosis, and alcohol is an important modifier of hemochromatosis penetrance and disease severity. More recent studies show that unlike in hemochromatosis, in ALD hepatic iron concentrations are not associated with prognosis (Jayaswal et al. 2020). In ALD, iron may rather be an indicator of advanced fibrosis and cirrhosis than a disease-modifying factor itself. Iron-directed therapies by phlebotomy or iron chelators are therefore not recommended in patients with alcoholic liver disease, and this treatment option is actually not feasible in many patients with ALD, which is frequently associated with anemia (Milic et al. 2016) (Table 16.1).

16.11 Metabolic Dysfunction-Associated Steatotic Liver Disease

Similar to iron in ALD, MASLD, which was formerly known as non-alcoholic fatty liver disease (NAFLD), can be associated with increased hepatic iron concentrations (Kanwal et al. 2024). Hepatic iron overload in patients with MASLD has been described as dysmetabolic iron overload syndrome (DIOS) (Moirand et al. 1997). The pathogenesis of DIOS has also been linked to reduced hepcidin expression in an animal model of insulin resistance, where reduced hepcidin response was found (Wang et al. 2014). The prevalence of increased hepatic iron concentrations among patients with MASLD is lower than the prevalence of elevated serum ferritin, which has been referred to as metabolic hyperferritinemia (MetHF) (Valenti et al. 2012; see also Chap. 17). High ferritin has been associated with a higher fibrosis risk, increased risk for the development of liver cancer, and higher overall mortality in a large prospective study (Armandi et al. 2024). A recent consensus statement suggested to use a grading system, based on ferritin, R2* MRI, hepatic iron stores, and the presence or absence of organ damage (Table 16.2) (Valenti et al. 2023).

Table 16.1 Methods and thresholds for the quantification of hepatic iron for hepatic iron overload

Serum iron parameters (European Association for the Study of the Liver. Electronic address and European Association for the Study of the 2022; Valent et al. 2018)	Suspected hemochromatosis Males >300 μg/L & TSAT >45% Females >200 μg/L & TSAT >45% For anemia with iron loading Ferritin >1000–2000 μg/L
Grading of Perls' iron staining according to Scheuer (Scheuer et al. 1962)	Grade 0: Granules absent or barely discernible at ×400 magnification Grade 1: Granules barely discernible at ×250 magnification and easily confirmed at ×100 Grade 2: Discrete granules resolved at ×100 magnification Threshold for the diagnosis of hemochromatosis Grade 3-Grade 4 Grade 3: Discrete granules resolved at ×25 magnification Grade 4: Masses visible at ×10 magnification or by the naked eye
Grading of Perls' iron staining according to Deugnier (Deugnier and Turlin 2011)	Hepatocytic iron score (HIS) 0, 3, 6, 9, or 12 according to granule size in each Rappaport area (sum score 0–36 points) Sinusoidal iron score (SIS) 0, 3, 6, 9, or 12 according to granule size in each Rappaport area (score 0–12) Portal iron score (PIS) 0, 1, 2, 3, or 4 according to the percentage of iron-overloaded macrophages, biliary cells, and vascular walls (score 0–12) Total iron score (TIS: 0–60)
Hepatic iron quantification by atomic absorption spectrometry (Walker et al. 1971)	Reference <25–36 μmol/g dry liver weight
Hepatic iron index (Adams et al. 1997)	>1.9 suggestive of hemochromatosis
Mobilizable iron on quantitative phlebotomy (Phatak and Barton 2003)	2.4 g females 3.5 g males
R2*-based quantification of hepatic iron (Viveiros et al. 2021)	Reference <70 s^{-1} (Hertz) on a 1.5 Tesla scanner

Although increased hepatic iron has been associated with advanced disease, controlled studies and meta-analyses investigating the effect of therapeutic venesection in patients with MetHF or DIOS have not demonstrated a beneficial effect of phlebotomy (Lainé et al. 2017; Murali et al. 2018). Experimental studies in animal models of MASLD even suggest that aberrant iron redistribution from hepatocytes to stellate cells via exosomes drive liver lipogenesis and fibrosis (Gao et al. 2022). In this model, hepatocellular iron deficiency is causing steatosis and fibrosis, which suggests that phlebotomies could even have negative effects on MASLD.

16.12 Liver Disease in Patients with Iron-Loading Anemias

Patients with iron-loading anemias such as thalassemia, sickle cell disease, aplastic anemia, or myelodysplastic syndrome are at increased risk for the development of cirrhosis—especially if they are transfusion-dependent. Historically, the majority of patients with repeated transfusions had chronic hepatitis C, which used to be a major risk factor for liver disease development. Therefore, there is no clear linear relationship between hepatic iron concentration and fibrosis (Berdoukas et al. 2005). In recent years, hepatitis C virus transmission by transfusions has become exceedingly rare after universal testing of blood

Table 16.2 Spectrum of iron metabolism disorders in individuals with metabolic dysfunction

	Physiological iron metabolism	Stage 1 (metabolic hyperferritinemia)	Stage 2 (dysmetabolic iron accumulation)	Stage 3 (dysmetabolic iron overload syndrome)
Ferritin (µg/L)	50 to upper limit of physiological iron metabolism	Upper limit of physiological iron metabolism to 550	550–1000	>1000
Liver iron stores ($R2^{*}s^{-1}$)	<70	<70	70–140	>140
Iron stores	Normal	Normal	Increased	Highly increased
Iron-related organ damage	Absent	Mostly absent	Rare	Present

Adapted from Valenti et al. (2023)

donors for HCV infection has been introduced, and the HCV prevalence decreased with effective direct antiviral treatments. Today, the development of fibrosis can be prevented by adequate chelation therapy in thalassemia patients free of hepatitis C virus infection (Di Marco et al. 2008). Guidelines therefore recommend monitoring for the development of hepatic iron overload by measuring liver serum ferritin every 3 months and conduct MRI when ferritin is >500–1000 µg/L. Liver iron concentration can be used to calculate total body iron using the following formula: Total body iron in µmol/kg = 10.6 × liver iron concentration in µmol/g dry tissue weight (Angelucci et al. 2000).

Chelation therapy should be initiated to prevent the development of iron overload after 10–12 transfusions for transfusion-dependent patients or when serum ferritin is >1000 µg/L. As chronic anemia suppresses hepcidin production, patients with anemia can develop iron overload even without or only a few transfusions. Such patients are known as non-transfusion-dependent and should be considered for iron chelation therapy if ferritin is above 800 µg/l or liver iron is >90 µmol/g dry weight. In patients with non-transfused rare inherited anemia (NTRIA), chelation therapy may even be considered at a ferritin concentration of >500 µg/L. Such rare anemias include Diamond-Blackfan anemia, congenital dyserythropoietic anemia, congenital sideroblastic anemia, and red cell enzyme defects (Shah et al. 2022).

High liver iron has also been associated with poor outcome in patients undergoing hematopoietic stem cell transplantation (HSCT). If present, iron overload should be treated with iron chelation prior to HSCT when ferritin is 1000 µg/L. After HSCT, iron chelation and/or phlebotomy should be considered when post-transplant liver iron concentration is >125 µmol/g dry weight (Rizzo et al. 2006). For low- and intermediate-risk patients with myelodysplastic syndrome, chelation therapy should be initiated after 20 units of blood have been transfused or when ferritin is >1000 µg/L (Killick et al. 2021).

16.13 Aceruloplasminemia

Aceruloplasminemia is caused by recessive pathogenic mutations in the ceruloplasmin gene (CP) (Harris et al. 2004). Disease manifestations include hepatic iron overload, diabetes, progressive neurodegeneration, and retinopathy (see also Chaps. 10 and 19). The disease is slowly progressive, and clinical manifestations before the fifth decade of life are rare (Loreal et al. 2002). Variable affection of Kupffer cells and the spleen has been reported, but iron overload confined to hepatocytes and low to normal spleen iron concentrations are typical for aceruloplasminemia. TSAT in patients with aceruloplasminemia is low. The diagnosis should be suspected in patients with high ferritin, low TSAT, and undetectable ceruloplasmin. Despite significant hepatic iron storage, progressive fibrosis or cirrhosis is exceedingly rare in patients with aceruloplasminemia. Deferasirox has been shown to effectively remove liver iron, but not brain iron (Finkenstedt et al. 2010). The primary treatment goal is prevention or delay in onset of neurological symptoms. Treatment with zinc and the oral iron chelator deferiprone has been tested (Vroegindeweij et al. 2020). Infusions of fresh-frozen plasma in order to replace the missing plasma protein have also been suggested, but these experimental treatment options have only been evaluated in single patients and small uncontrolled case series (Marchi et al. 2019).

16.14 Ferroportin Disease

High serum ferritin with normal TSAT is also a characteristic finding in patients with ferroportin disease, which is caused by dominant mutations in the gene encoding ferroportin *SLC40A1* (Njajou et al. 2001; see also Chap. 10). The characteristic histological finding in ferroportin disease is iron in macrophages with no or little stainable iron in hepatocytes. These histological findings are reminiscent of transfusional iron overload. On MRI, increased iron concentrations are typically present in the liver, spleen, and bone marrow (Zoller et al. 2005). The paradigm that mutations causing an impaired iron export function of ferroportin result in ferroportin disease with normal TSAT and predominant Kupffer cell iron, whereas mutations that affect hepcidin binding to ferroportin cause "SLC40A1-related hemochromatosis" with high TSAT and predominant hepatocellular iron accumulation has been challenged (Mayr et al. 2010). Newer studies suggest that certain mutations can affect both iron export function and hepcidin binding (Uguen et al. 2024; Praschberger et al. 2014).

The diagnosis of ferroportin disease is rarely based on clinical symptoms, because most patients are asymptomatic and the risk for the development of liver disease is low. Most patients are identified during evaluation for asymptomatic elevation in ferritin followed by MRI and genetic testing for "non-HFE hemochromatosis." The true prevalence of ferroportin disease is unknown, but a modeling study suggests a global prevalence of 0.04%—mainly among Africans (Wallace and Subramaniam 2016). Some authors suggest "non-aggressive" phlebotomy to reduce hepatic iron concentrations (Pietrangelo 2017). This management strategy is based on expert opinion and case reports. Data supporting this recommendation from controlled clinical trials are lacking.

16.15 Conclusions

The intricate relationship between iron and liver disease underscores the critical role of the liver in iron homeostasis. Hepatic iron overload can be both a cause and consequence of liver disorders, contributing to disease progression in conditions such as hemochromatosis, alcoholic liver disease, MASLD, and iron-loading anemias. Advances in noninvasive diagnostics, including MRI-based iron quantification, have improved the assessment and management of hepatic iron overload. While therapeutic interventions like phlebotomy and iron chelation are effective in select cases, their application must be carefully tailored to the underlying condition. Continued research into iron metabolism will enhance our understanding of its impact on liver disease and inform more precise therapeutic strategies.

References

Adams PC, Bradley C, Henderson AR (1997) Evaluation of the hepatic iron index as a diagnostic criterion for genetic hemochromatosis. J Lab Clin Med 130(5):509–514. https://doi.org/10.1016/s0022-2143(97)90127-x

Aguilar-Martinez P, Schved JF, Brissot P (2005) The evaluation of hyperferritinemia: an updated strategy based on advances in detecting genetic abnormalities. Am J Gastroenterol 100(5):1185–1194. https://doi.org/10.1111/j.1572-0241.2005.40998.x

Ali N, Ferrao K, Mehta KJ (2023) Liver iron loading in alcohol-associated liver disease. Am J Pathol 193(10):1427–1439. https://doi.org/10.1016/j.ajpath.2022.08.010

Angelucci E, Brittenham GM, McLaren CE, Ripalti M, Baronciani D, Giardini C, Galimberti M, Polchi P,

Lucarelli G (2000) Hepatic iron concentration and total body iron stores in thalassemia major. N Engl J Med 343(5):327–331
Armandi A, Sanavia T, Younes R, Caviglia GP, Rosso C, Govaere O, Liguori A, Francione P, Gallego-Duràn R, Ampuero J, Pennisi G, Aller R, Tiniakos D, Burt A, David E, Vecchio F, Maggioni M, Cabibi D, McLeod D, Pareja MJ, Zaki MYW, Grieco A, Stål P, Kechagias S, Fracanzani AL, Valenti L, Miele L, Fariselli P, Eslam M, Petta S, Hagström H, George J, Schattenberg JM, Romero-Gómez M, Anstee QM, Bugianesi E (2024) Serum ferritin levels can predict long-term outcomes in patients with metabolic dysfunction-associated steatotic liver disease. Gut 73(5):825–834. https://doi.org/10.1136/gutjnl-2023-330815
Bardou-Jacquet E, Laine F, Guggenbuhl P, Morcet J, Jezequel C, Guyader D, Moirand R, Deugnier Y (2017) Worse outcomes of patients with HFE hemochromatosis with persistent increases in transferrin saturation during maintenance therapy. Clin Gastroenterol Hepatol 15(10):1620–1627. https://doi.org/10.1016/j.cgh.2016.12.039
Batts KP (2007) Iron overload syndromes and the liver. Mod Pathol 20(Suppl 1):S31–S39. https://doi.org/10.1038/modpathol.3800715
Berdoukas V, Bohane T, Tobias V, De Silva K, Fraser I, Aessopos A, Lindeman R (2005) Liver iron concentration and fibrosis in a cohort of transfusion-dependent patients on long-term desferrioxamine therapy. Hematol J 5(7):572–578. https://doi.org/10.1038/sj.thj.6200569
Bomford A, Williams R (1976) Long term results of venesection therapy in idiopathic haemochromatosis. Q J Med 45(180):611–623
Brandhagen DJ, Alvarez W, Therneau TM, Kruckeberg KE, Thibodeau SN, Ludwig J, Porayko MK (2000) Iron overload in cirrhosis-HFE genotypes and outcome after liver transplantation. Hepatology 31(2):456–460. https://doi.org/10.1002/hep.510310227
Brittenham GM, Farrell DE, Harris JW, Feldman ES, Danish EH, Muir WA, Tripp JH, Bellon EM (1982) Magnetic-susceptibility measurement of human iron stores. N Engl J Med 307(27):1671–1675. https://doi.org/10.1056/nejm198212303072703
Canali S, Zumbrennen-Bullough KB, Core AB, Wang CY, Nairz M, Bouley R, Swirski FK, Babitt JL (2017) Endothelial cells produce bone morphogenetic protein 6 required for iron homeostasis in mice. Blood 129(4):405–414. https://doi.org/10.1182/blood-2016-06-721571
Chapman RW, Morgan MY, Laulicht M, Hoffbrand AV, Sherlock S (1982) Hepatic iron stores and markers of iron overload in alcoholics and patients with idiopathic hemochromatosis. Dig Dis Sci 27(10):909–916. https://doi.org/10.1007/bf01316575
Corradini E, Rozier M, Meynard D, Odhiambo A, Lin HY, Feng Q, Migas MC, Britton RS, Babitt JL, Fleming RE (2011) Iron regulation of hepcidin despite attenuated smad1,5,8 signaling in mice without transferrin receptor 2 or hfe. Gastroenterology 141(5):1907–1914. https://doi.org/10.1053/j.gastro.2011.06.077
Daher R, Kannengiesser C, Houamel D, Lefebvre T, Bardou-Jacquet E, Ducrot N, de Kerguenec C, Jouanolle AM, Robreau AM, Oudin C, Le Gac G, Moulouel B, Loustaud-Ratti V, Bedossa P, Valla D, Gouya L, Beaumont C, Brissot P, Puy H, Karim Z, Tchernitchko D (2016) Heterozygous mutations in BMP6 pro-peptide lead to inappropriate hepcidin synthesis and moderate iron overload in humans. Gastroenterology 150(3):672–683.e674. https://doi.org/10.1053/j.gastro.2015.10.049
Delatycki MB, Allen KJ (2024) Population screening for hereditary haemochromatosis-should it be carried out, and if so, how? Genes (Basel) 15(8). https://doi.org/10.3390/genes15080967
Deugnier Y, Turlin B (2011) Pathology of hepatic iron overload. Semin Liver Dis 31(3):260–271. https://doi.org/10.1055/s-0031-1286057
Di Marco V, Capra M, Gagliardotto F, Borsellino Z, Cabibi D, Barbaria F, Ferraro D, Cuccia L, Ruffo GB, Bronte F, Di Stefano R, Almasio PL, Craxì A (2008) Liver disease in chelated transfusion-dependent thalassemics: the role of iron overload and chronic hepatitis C. Haematologica 93(8):1243–1246. https://doi.org/10.3324/haematol.12554
Dogan DY, Urzica EI, Hornung I, Kastl P, Oguama D, Fette FM, Nguyen LH, Rosenbauer F, Zacharowski K, Klingmüller U, Gradhand E, von Knethen A, Popp R, Fleming I, Schrader L, Steinbicker AU (2024) Hemojuvelin-mediated hepcidin induction requires both bone morphogenetic protein type I receptors ALK2 and ALK3. Blood Adv 8(11):2870–2879. https://doi.org/10.1182/bloodadvances.2023012322
Drakesmith H, Schimanski LM, Ormerod E, Merryweather-Clarke AT, Viprakasit V, Edwards JP, Sweetland E, Bastin JM, Cowley D, Chinthammitr Y, Robson KJ, Townsend AR (2005) Resistance to hepcidin is conferred by hemochromatosis-associated mutations of ferroportin. Blood 106(3):1092–1097. https://doi.org/10.1182/blood-2005-02-0561
Du X, She E, Gelbart T, Truksa J, Lee P, Xia Y, Khovananth K, Mudd S, Mann N, Moresco EM, Beutler E, Beutler B (2008) The serine protease TMPRSS6 is required to sense iron deficiency. Science 320(5879):1088–1092. https://doi.org/10.1126/science.1157121
Elmberg M, Hultcrantz R, Simard JF, Carlsson Å, Askling J (2013) Increased risk of arthropathies and joint replacement surgery in patients with genetic hemochromatosis: a study of 3,531 patients and their 11,794 first-degree relatives. Arthritis Care Res 65(5):678–685. https://doi.org/10.1002/acr.21883
European Association for the Study of the Liver. Electronic address eee, European Association for the Study of the L (2022) EASL clinical practice guidelines on haemochromatosis. J Hepatol 77(2):479–502. https://doi.org/10.1016/j.jhep.2022.03.033
Feder JN, Tsuchihashi Z, Irrinki A, Lee VK, Mapa FA, Morikang E, Prass CE, Starnes SM, Wolff RK, Parkkila S, Sly WS, Schatzman RC (1997) The hemo-

chromatosis founder mutation in HLA-H disrupts beta2-microglobulin interaction and cell surface expression. J Biol Chem 272(22):14025–14028

Finkenstedt A, Wolf E, Hofner E, Gasser BI, Bosch S, Bakry R, Creus M, Kremser C, Schocke M, Theurl M, Moser P, Schranz M, Bonn G, Poewe W, Vogel W, Janecke AR, Zoller H (2010) Hepatic but not brain iron is rapidly chelated by deferasirox in aceruloplasminemia due to a novel gene mutation. J Hepatol 53(6):1101–1107. https://doi.org/10.1016/j.jhep.2010.04.039

Fletcher LM, Dixon JL, Purdie DM, Powell LW, Crawford DH (2002) Excess alcohol greatly increases the prevalence of cirrhosis in hereditary hemochromatosis. Gastroenterology 122(2):281–289

Foka P, Dimitriadis A, Kyratzopoulou E, Giannimaras DA, Sarno S, Simos G, Georgopoulou U, Mamalaki A (2014) A complex signaling network involving protein kinase CK2 is required for hepatitis C virus core protein-mediated modulation of the iron-regulatory hepcidin gene expression. Cell Mol Life Sci 71(21):4243–4258. https://doi.org/10.1007/s00018-014-1621-4

Frank KM, Schneewind O, Shieh WJ (2011) Investigation of a researcher's death due to septicemic plague. N Engl J Med 364(26):2563–2564. https://doi.org/10.1056/NEJMc1010939

Gao H, Jin Z, Bandyopadhyay G, Wang G, Zhang D, Rocha KCE, Liu X, Zhao H, Kisseleva T, Brenner DA, Karin M, Ying W (2022) Aberrant iron distribution via hepatocyte-stellate cell axis drives liver lipogenesis and fibrosis. Cell Metab 34(8):1201–1213.e1205. https://doi.org/10.1016/j.cmet.2022.07.006

Gerhard GS, Levin KA, Price Goldstein J, Wojnar MM, Chorney MJ, Belchis DA (2001) Vibrio vulnificus septicemia in a patient with the hemochromatosis HFE C282Y mutation. Arch Pathol Lab Med 125(8):1107–1109

Gerjevic LN, Liu N, Lu S, Harrison-Findik DD (2012) Alcohol activates TGF-beta but inhibits BMP receptor-mediated Smad signaling and Smad4 binding to hepcidin promoter in the liver. Int J Hepatol 2012:459278. https://doi.org/10.1155/2012/459278

Girelli D, Busti F, Brissot P, Cabantchik I, Muckenthaler MU, Porto G (2022) Hemochromatosis classification: update and recommendations by the BIOIRON Society. Blood 139(20):3018–3029. https://doi.org/10.1182/blood.2021011338

Harris ZL, Davis-Kaplan SR, Gitlin JD, Kaplan J (2004) A fungal multicopper oxidase restores iron homeostasis in aceruloplasminemia. Blood 103(12):4672–4673. https://doi.org/10.1182/blood-2003-11-4060

Harrison-Findik DD, Klein E, Evans J, Gollan J (2009) Regulation of liver hepcidin expression by alcohol in vivo does not involve Kupffer cell activation or TNF-alpha signaling. Am J Physiol Gastrointest Liver Physiol 296(1):G112–G118. https://doi.org/10.1152/ajpgi.90550.2008

Hegde S, Pierce TT, Heidari F, Ozturk A, Cheah E, Pope K, Blake MA, Shih A, Misdraji J, Samir AE (2024) Noninvasive assessment of liver fibrosis in patients with iron overload. Ultrasound Med Biol. https://doi.org/10.1016/j.ultrasmedbio.2024.11.017

Jayaswal ANA, Levick C, Selvaraj EA, Dennis A, Booth JC, Collier J, Cobbold J, Tunnicliffe EM, Kelly M, Barnes E, Neubauer S, Banerjee R, Pavlides M (2020) Prognostic value of multiparametric magnetic resonance imaging, transient elastography and blood-based fibrosis markers in patients with chronic liver disease. Liver Int 40(12):3071–3082. https://doi.org/10.1111/liv.14625

Kanwal F, Neuschwander-Tetri BA, Loomba R, Rinella ME (2024) Metabolic dysfunction-associated steatotic liver disease: update and impact of new nomenclature on the American Association for the Study of Liver Diseases practice guidance on nonalcoholic fatty liver disease. Hepatology 79(5):1212–1219. https://doi.org/10.1097/hep.0000000000000670

Killick SB, Ingram W, Culligan D, Enright H, Kell J, Payne EM, Krishnamurthy P, Kulasekararaj A, Raghavan M, Stanworth SJ, Green S, Mufti G, Quek L, Cargo C, Jones GL, Mills J, Sternberg A, Wiseman DH, Bowen D (2021) British Society for Haematology guidelines for the management of adult myelodysplastic syndromes. Br J Haematol 194(2):267–281. https://doi.org/10.1111/bjh.17612

Knutson MD (2019) Non-transferrin-bound iron transporters. Free Radic Biol Med 133:101–111. https://doi.org/10.1016/j.freeradbiomed.2018.10.413

Koch PS, Olsavszky V, Ulbrich F, Sticht C, Demory A, Leibing T, Henzler T, Meyer M, Zierow J, Schneider S, Breitkopf-Heinlein K, Gaitantzi H, Spencer-Dene B, Arnold B, Klapproth K, Schledzewski K, Goerdt S, Geraud C (2017) Angiocrine Bmp2 signaling in murine liver controls normal iron homeostasis. Blood 129(4):415–419. https://doi.org/10.1182/blood-2016-07-729822

Lainé F, Ruivard M, Loustaud-Ratti V, Bonnet F, Calès P, Bardou-Jacquet E, Sacher-Huvelin S, Causse X, Beusnel C, Renault A, Bellissant E, Deugnier Y (2017) Metabolic and hepatic effects of bloodletting in dysmetabolic iron overload syndrome: a randomized controlled study in 274 patients. Hepatology 65(2):465–474. https://doi.org/10.1002/hep.28856

Latour C, Besson-Fournier C, Meynard D, Silvestri L, Gourbeyre O, Aguilar-Martinez P, Schmidt PJ, Fleming MD, Roth MP, Coppin H (2016) Differing impact of the deletion of hemochromatosis-associated molecules HFE and transferrin receptor-2 on the iron phenotype of mice lacking bone morphogenetic protein 6 or hemojuvelin. Hepatology 63(1):126–137. https://doi.org/10.1002/hep.28254

Le Lan C, Loreal O, Cohen T, Ropert M, Glickstein H, Laine F, Pouchard M, Deugnier Y, Le Treut A, Breuer W, Cabantchik ZI, Brissot P (2005) Redox active plasma iron in C282Y/C282Y hemochromatosis. Blood 105(11):4527–4531

Loreal O, Turlin B, Pigeon C, Moisan A, Ropert M, Morice P, Gandon Y, Jouanolle AM, Verin M, Hider RC, Yoshida K, Brissot P (2002) Aceruloplasminemia:

new clinical, pathophysiological and therapeutic insights. J Hepatol 36(6):851–856

Ludwig J, Hashimoto E, Porayko MK, Moyer TP, Baldus WP (1997) Hemosiderosis in cirrhosis: a study of 447 native livers. Gastroenterology 112(3):882–888. https://doi.org/10.1053/gast.1997.v112.pm9041250

Lupşor M, Badea R, Stefănescu H, Grigorescu M, Sparchez Z, Serban A, Branda H, Iancu S, Maniu A (2008) Analysis of histopathological changes that influence liver stiffness in chronic hepatitis C. Results from a cohort of 324 patients. J Gastrointestin Liver Dis 17(2):155–163

Marchi G, Busti F, Lira Zidanes A, Castagna A, Girelli D (2019) Aceruloplasminemia: a severe neurodegenerative disorder deserving an early diagnosis. Front Neurosci 13:325. https://doi.org/10.3389/fnins.2019.00325

Mayr R, Janecke AR, Schranz M, Griffiths WJ, Vogel W, Pietrangelo A, Zoller H (2010) Ferroportin disease: a systematic meta-analysis of clinical and molecular findings. J Hepatol 53(5):941–949. https://doi.org/10.1016/j.jhep.2010.05.016

Mayr R, Griffiths WJ, Hermann M, McFarlane I, Halsall DJ, Finkenstedt A, Douds A, Davies SE, Janecke AR, Vogel W, Cox TM, Zoller H (2011) Identification of mutations in SLC40A1 that affect ferroportin function and phenotype of human ferroportin iron overload. Gastroenterology 140(7):2056–2063.e2051. https://doi.org/10.1053/j.gastro.2011.02.064

Milic S, Mikolasevic I, Orlic L, Devcic E, Starcevic-Cizmarevic N, Stimac D, Kapovic M, Ristic S (2016) The role of iron and iron overload in chronic liver disease. Med Sci Monit 22:2144–2151. https://doi.org/10.12659/msm.896494

Moirand R, Mortaji AM, Loreal O, Paillard F, Brissot P, Deugnier Y (1997) A new syndrome of liver iron overload with normal transferrin saturation. Lancet 349(9045):95–97

Murali AR, Gupta A, Brown K (2018) Systematic review and meta-analysis to determine the impact of iron depletion in dysmetabolic iron overload syndrome and non-alcoholic fatty liver disease. Hepatol Res 48(3):E30–E41. https://doi.org/10.1111/hepr.12921

Nashwan AJ, Yassin MA, Abd-Alrazaq A, Shuweihdi F, Othman M, Abdul Rahim HF, Shraim M (2023) Hepatic and cardiac iron overload quantified by magnetic resonance imaging in patients on hemodialysis: a systematic review and meta-analysis. Hemodial Int 27(1):3–11. https://doi.org/10.1111/hdi.13054

Nemeth E, Ganz T (2023) Hepcidin and iron in health and disease. Annu Rev Med 74:261–277. https://doi.org/10.1146/annurev-med-043021-032816

Nemeth E, Rivera S, Gabayan V, Keller C, Taudorf S, Pedersen BK, Ganz T (2004) IL-6 mediates hypoferremia of inflammation by inducing the synthesis of the iron regulatory hormone hepcidin. J Clin Invest 113(9):1271–1276. https://doi.org/10.1172/jci20945

Newsome PN, Cramb R, Davison SM, Dillon JF, Foulerton M, Godfrey EM, Hall R, Harrower U, Hudson M, Langford A, Mackie A, Mitchell-Thain R, Sennett K, Sheron NC, Verne J, Walmsley M, Yeoman A (2018) Guidelines on the management of abnormal liver blood tests. Gut 67(1):6–19. https://doi.org/10.1136/gutjnl-2017-314924

Nicolas G, Chauvet C, Viatte L, Danan JL, Bigard X, Devaux I, Beaumont C, Kahn A, Vaulont S (2002) The gene encoding the iron regulatory peptide hepcidin is regulated by anemia, hypoxia, and inflammation. J Clin Invest 110(7):1037–1044. https://doi.org/10.1172/JCI15686

Ning X, Peng F, Tan S, Tang C, Luo C, Xiao F, Zhao C, Peng P (2024) Quantitative MRI evaluation of iron deposition in patients with transfusion-dependent thalassemia: clinical management insights. Expert Rev Hematol 17(12):977–983. https://doi.org/10.1080/17474086.2024.2435353

Njajou OT, Vaessen N, Joosse M, Berghuis B, van Dongen JW, Breuning MH, Snijders PJ, Rutten WP, Sandkuijl LA, Oostra BA, van Duijn CM, Heutink P (2001) A mutation in SLC11A3 is associated with autosomal dominant hemochromatosis. Nat Genet 28(3):213–214

Olynyk JK, St Pierre TG, Britton RS, Brunt EM, Bacon BR (2005) Duration of hepatic iron exposure increases the risk of significant fibrosis in hereditary hemochromatosis: a new role for magnetic resonance imaging. Am J Gastroenterol 100(4):837–841. https://doi.org/10.1111/j.1572-0241.2005.41287.x

Phatak PD, Barton JC (2003) Phlebotomy-mobilized iron as a surrogate for liver iron content in hemochromatosis patients. Hematology 8(6):429–432. https://doi.org/10.1080/10245330310001621332

Pietrangelo A (2015) Genetics, genetic testing, and management of hemochromatosis: 15 years since hepcidin. Gastroenterology 149(5):1240–1251.e1244. https://doi.org/10.1053/j.gastro.2015.06.045

Pietrangelo A (2017) Ferroportin disease: pathogenesis, diagnosis and treatment. Haematologica 102(12):1972–1984. https://doi.org/10.3324/haematol.2017.170720

Piperno A, Girelli D, Nemeth E, Trombini P, Bozzini C, Poggiali E, Phung Y, Ganz T, Camaschella C (2007) Blunted hepcidin response to oral iron challenge in HFE-related hemochromatosis. Blood 110(12):4096–4100

Praschberger R, Schranz M, Griffiths WJ, Baumgartner N, Hermann M, Lomas DJ, Pietrangelo A, Cox TM, Vogel W, Zoller H (2014) Impact of D181V and A69T on the function of ferroportin as an iron export pump and hepcidin receptor. Biochim Biophys Acta. https://doi.org/10.1016/j.bbadis.2014.05.011

Prieto J, Barry M, Sherlock S (1975) Serum ferritin in patients with iron overload and with acute and chronic liver diseases. Gastroenterology 68(3):525–533

Ramanathan G, Olynyk JK, Ferrari P (2017) Diagnosing and preventing iron overload. Hemodial Int 21(Suppl 1):S58–s67. https://doi.org/10.1111/hdi.12555

Reeder SB, Yokoo T, França M, Hernando D, Alberich-Bayarri Á, Alústiza JM, Gandon Y, Henninger B, Hillenbrand C, Jhaveri K, Karçaaltıncaba M, Kühn JP, Mojtahed A, Serai SD, Ward R, Wood JC, Yamamura

J, Martí-Bonmatí L (2023) Quantification of liver iron overload with MRI: review and guidelines from the ESGAR and SAR. Radiology 307(1):e221856. https://doi.org/10.1148/radiol.221856

Rizzo JD, Wingard JR, Tichelli A, Lee SJ, Van Lint MT, Burns LJ, Davies SM, Ferrara JLM, Socié G (2006) Recommended screening and preventive practices for long-term survivors after hematopoietic cell transplantation: joint recommendations of the European Group for Blood and Marrow Transplantation, the Center for International Blood and Marrow Transplant Research, and the American Society of Blood and Marrow Transplantation. Biol Blood Marrow Transplant 12(2):138–151. https://doi.org/10.1016/j.bbmt.2005.09.012

Roetto A, Papanikolaou G, Politou M, Alberti F, Girelli D, Christakis J, Loukopoulos D, Camaschella C (2003) Mutant antimicrobial peptide hepcidin is associated with severe juvenile hemochromatosis. Nat Genet 33(1):21–22. https://doi.org/10.1038/ng1053

Schaefer B, Haschka D, Finkenstedt A, Petersen BS, Theurl I, Henninger B, Janecke AR, Wang CY, Lin HY, Veits L, Vogel W, Weiss G, Franke A, Zoller H (2015) Impaired hepcidin expression in alpha-1-antitrypsin deficiency associated with iron overload and progressive liver disease. Hum Mol Genet 24(21):6254–6263. https://doi.org/10.1093/hmg/ddv348

Schaefer B, Meindl E, Wagner S, Tilg H, Zoller H (2020) Intravenous iron supplementation therapy. Mol Asp Med 75:100862. https://doi.org/10.1016/j.mam.2020.100862

Schaefer B, Pammer LM, Pfeifer B, Neururer S, Troppmair MR, Panzer M, Wagner S, Pertler E, Gieger C, Kronenberg F, Lamina C, Tilg H, Zoller H (2024) Penetrance, cancer incidence and survival in HFE haemochromatosis—a population-based cohort study. Liver Int 44(3):838–847. https://doi.org/10.1111/liv.15797

Scheuer PJ, Williams R, Muir AR (1962) Hepatic pathology in relatives of patients with haemochromatosis. J Pathol Bacteriol 84:53–64

Shah FT, Porter JB, Sadasivam N, Kaya B, Moon JC, Velangi M, Ako E, Pancham S (2022) Guidelines for the monitoring and management of iron overload in patients with haemoglobinopathies and rare anaemias. Br J Haematol 196(2):336–350. https://doi.org/10.1111/bjh.17839

Shander A, Cappellini MD, Goodnough LT (2009) Iron overload and toxicity: the hidden risk of multiple blood transfusions. Vox Sang 97(3):185–197. https://doi.org/10.1111/j.1423-0410.2009.01207.x

Sharma R, Zhao W, Zafar Y, Murali AR, Brown KE (2024) Serum hepcidin levels in chronic liver disease: a systematic review and meta-analysis. Clin Chem Lab Med 62(3):373–384. https://doi.org/10.1515/cclm-2023-0540

Simon M, Bourel M, Fauchet R, Genetet B (1976) Association of HLA-A3 and HLA-B14 antigens with idiopathic haemochromatosis. Gut 17(5):332–334

Sussman MS, Kannengiesser SAR, Prasla S, Ward R, Jhaveri KS (2024) Comparison of R2* and FerriScan liver iron concentration (LIC) quantification in the clinical classification of iron overload states. Magn Reson Imaging 113:110216. https://doi.org/10.1016/j.mri.2024.110216

Trosseau A (1865) Gycosurie; diabète sucré, 2nd edn. Clinique Méd de l'Hotel-Dieu de Paris 2, pp 663–698

Uguen K, Le Tertre M, Tchernitchko D, Elbahnsi A, Maestri S, Gourlaouen I, Férec C, Ka C, Callebaut I, Le Gac G (2024) The dual loss and gain of function of the FPN1 iron exporter results in the ferroportin disease phenotype. HGG Adv 5(4):100335. https://doi.org/10.1016/j.xhgg.2024.100335

Valent P, Stauder R, Theurl I, Geissler K, Sliwa T, Sperr WR, Bettelheim P, Sill H, Pfeilstöcker M (2018) Diagnosis, management and response criteria of iron overload in myelodysplastic syndromes (MDS): updated recommendations of the Austrian MDS platform. Expert Rev Hematol 11(2):109–116. https://doi.org/10.1080/17474086.2018.1420473

Valenti L, Dongiovanni P, Fargion S (2012) Diagnostic and therapeutic implications of the association between ferritin level and severity of nonalcoholic fatty liver disease. World J Gastroenterol 18(29):3782–3786. https://doi.org/10.3748/wjg.v18.i29.3782

Valenti L, Corradini E, Adams LA, Aigner E, Alqahtani S, Arrese M, Bardou-Jacquet E, Bugianesi E, Fernandez-Real JM, Girelli D, Hagström H, Henninger B, Kowdley K, Ligabue G, McClain D, Lainé F, Miyanishi K, Muckenthaler MU, Pagani A, Pedrotti P, Pietrangelo A, Prati D, Ryan JD, Silvestri L, Spearman CW, Stål P, Tsochatzis EA, Vinchi F, Zheng MH, Zoller H (2023) Consensus statement on the definition and classification of metabolic hyperferritinaemia. Nat Rev Endocrinol 19(5):299–310. https://doi.org/10.1038/s41574-023-00807-6

Vanclooster A, van Deursen C, Jaspers R, Cassiman D, Koek G (2017) Proton pump inhibitors decrease phlebotomy need in HFE hemochromatosis: double-blind randomized placebo-controlled trial. Gastroenterology 153(3):678–680.e672. https://doi.org/10.1053/j.gastro.2017.06.006

Viveiros A, Schaefer B, Tilg H, Zoller H (2019) Iron matryoshka-haemochromatosis nested in ferroportin disease? Liver Int 39(6):1014–1015. https://doi.org/10.1111/liv.14061

Viveiros A, Schaefer B, Panzer M, Henninger B, Plaikner M, Kremser C, Franke A, Franzenburg S, Hoeppner MP, Stauder R, Janecke A, Tilg H, Zoller H (2021) MRI-based iron phenotyping and patient selection for next-generation sequencing of non-homeostatic iron regulator hemochromatosis genes. Hepatology 74(5):2424–2435. https://doi.org/10.1002/hep.31982

Vogel W, Herold M, Margreiter R, Bomford A (1989) Occupancy of the iron-binding sites of human transferrin in sera obtained from different anatomical sites. Klin Wochenschr 67(10):538–542. https://doi.org/10.1007/BF01719780

von Recklinghausen DF (1889) Über Hämochromatose. Tagblatt der Versammlung Deutscher Naturforscher und Ärzte in Heidelberg 62:324–325

Vroegindeweij LHP, Boon AJW, Wilson JHP, Langendonk JG (2020) Effects of iron chelation therapy on the clinical course of aceruloplasminemia: an analysis of aggregated case reports. Orphanet J Rare Dis 15(1):105. https://doi.org/10.1186/s13023-020-01385-w

Walker RJ, Miller JP, Dymock IW, Shilkin KB, Williams R (1971) Relationship of hepatic iron concentration to histochemical grading and to total chelatable body iron in conditions associated with iron overload. Gut 12(12):1011–1014. https://doi.org/10.1136/gut.12.12.1011

Wallace DF, Subramaniam VN (2016) The global prevalence of HFE and non-HFE hemochromatosis estimated from analysis of next-generation sequencing data. Genet Med 18(6):618–626. https://doi.org/10.1038/gim.2015.140

Wang CY, Babitt JL (2019) Liver iron sensing and body iron homeostasis. Blood 133(1):18–29. https://doi.org/10.1182/blood-2018-06-815894

Wang H, Li H, Jiang X, Shi W, Shen Z, Li M (2014) Hepcidin is directly regulated by insulin and plays an important role in iron overload in streptozotocin-induced diabetic rats. Diabetes 63(5):1506–1518. https://doi.org/10.2337/db13-1195

Weiss G, Theurl I, Eder S, Koppelstaetter C, Kurz K, Sonnweber T, Kobold U, Mayer G (2009) Serum hepcidin concentration in chronic haemodialysis patients: associations and effects of dialysis, iron and erythropoietin therapy. Eur J Clin Investig 39(10):883–890. https://doi.org/10.1111/j.1365-2362.2009.02182.x

Zoller H, Henninger B (2016) Pathogenesis, diagnosis and treatment of hemochromatosis. Dig Dis (Basel, Switzerland) 34(4):364–373. https://doi.org/10.1159/000444549

Zoller H, McFarlane I, Theurl I, Stadlmann S, Nemeth E, Oxley D, Ganz T, Halsall DJ, Cox TM, Vogel W (2005) Primary iron overload with inappropriate hepcidin expression in V162del ferroportin disease. Hepatology 42(2):466–472

17

The Impact of Iron Homeostasis in Insulin-Sensitive Tissues and Gut Microbiome on Obesity-Driven Metabolic Disorders

José María Moreno-Navarrete,
Isa Galvao Rodrigues,
and José Manuel Fernández-Real

Abstract

Iron is a crucial element for vital biological processes in both prokaryotic and eukaryotic cells, requiring precise regulation to maintain homeostasis. In humans and animal models, dysregulation of iron homeostasis is often linked to obesity-associated metabolic disturbances, which are characterized by elevated serum ferritin levels and excessive iron accumulation in insulin-dependent tissues like the liver, adipose tissue, and skeletal muscle. Prolonged iron overload in tissues induces oxidative stress, which impairs insulin sensitivity and promotes systemic insulin resistance and hyperglycemia. This creates a vicious cycle in which decreased serum hepcidin levels enhance intestinal iron absorption, further exacerbating iron accumulation. While the impact of iron on gut microbiota is well established, the role of gut microbiota in regulating body iron homeostasis is less studied. Recent studies have uncovered new mechanisms by which gut microbiota influence intestinal iron absorption and the regulation of body iron stores. In this chapter, we summarize recent findings on iron homeostasis in insulin-sensitive metabolic tissues and explore how gut microbiota can modulate body iron regulation in the context of obesity.

Keywords

Iron homeostasis · Obesity-related metabolic disorders · Gut microbiota · Insulin resistance · Hepcidin regulation

J. M. Moreno-Navarrete (✉)
Department of Diabetes, Endocrinology and Nutrition, Hospital of Girona "Dr Josep Trueta", Institut d'Investigació Biomèdica de Girona (IdIBGi), CIBEROBN (CB06/03/010) and Instituto de Salud Carlos III (ISCIII), Girona, Spain

Department of Medicine, Universitat de Girona, Girona, Spain

Nutrition, Eumetabolism and Health, Biomedical Research Institute of Girona "Dr Josep Trueta", Salt, Spain
e-mail: jmoreno@idibgi.org

I. G. Rodrigues
Nutrition Unit, Procape, Universidade de Pernambuco, Recife, Brazil

J. M. Fernández-Real (✉)
Department of Diabetes, Endocrinology and Nutrition, Hospital of Girona "Dr Josep Trueta", Institut d'Investigació Biomèdica de Girona (IdIBGi), CIBEROBN (CB06/03/010) and Instituto de Salud Carlos III (ISCIII), Girona, Spain

Department of Medicine, Universitat de Girona, Girona, Spain
e-mail: jmfreal@idibgi.org

K. Pantopoulos (ed.), *Iron Metabolism in Human Health and Disease*, Advances in Experimental Medicine and Biology 1480, https://doi.org/10.1007/978-3-031-92033-2_17

17.1 Iron Dysregulation in Obesity and Type 2 Diabetes

Iron is the most abundant metal in the human body and is essential in proper quantities for numerous critical biological functions, including oxygen transport, DNA synthesis regulation, cell growth and differentiation, electron transport, mitochondrial biogenesis, and other key metabolic processes (Fernández-Real et al. 2015, 2019); see also Chap. 1.

Disruptions in iron homeostasis have been associated with an elevated risk of type 2 diabetes (Fernández-Real et al. 2015). Extensive clinical and experimental research has documented this relationship, highlighting how increased body iron stores contribute to the progression of insulin resistance, reduced insulin secretion, and impaired glucose tolerance. These effects are well supported in several systematic reviews and meta-analyses (Zhao et al. 2012; Bao et al. 2012; Fernández-Real and Manco 2014; Fernández-Real et al. 2015; Simcox and McClain 2013).

An extensive body of clinical and experimental research has highlighted disrupted iron metabolism in obesity, particularly among women, where decreased iron absorption and reduced availability for erythropoiesis are common (Stoffel et al. 2020; Tussing-Humphreys et al. 2009). Notably, bariatric surgery-induced weight loss often restores these iron levels (Santos et al. 2014; Cepeda-Lopez et al. 2016; Tussing-Humphreys et al. 2010). Recent findings indicate that central obesity—especially the android fat-to-total body fat ratio—is more strongly associated with decreased iron absorption and availability, alongside elevated hepcidin levels (Stoffel et al. 2020).

Studies show that, in obese individuals, serum hepcidin correlates with markers of tissue iron accumulation, such as ferritin and hepatic iron, but not with iron availability (Moreno-Navarrete et al. 2017a). In patients with type 2 diabetes, genes related to non-heme iron intake in the small intestine increase, correlating positively with fasting glucose and the ferritin/hepcidin ratio; only ferroportin expression inversely correlates with serum hepcidin levels (Moreno-Navarrete et al. 2018). However, glucose tolerance did not significantly impact the expression of genes related to heme iron uptake (Moreno-Navarrete et al. 2018). The inhibitory effect of hepcidin on ferroportin mRNA and protein levels is well established in both mice and humans (Ganz 2013; Gulec et al. 2014; Bergamaschi et al. 2017).

Interestingly, while serum hepcidin generally increases with obesity (Moreno-Navarrete et al. 2017a; Stoffel et al. 2020; del Giudice et al. 2009), it significantly decreases in type 2 diabetes (Pechlaner et al. 2016; Aregbesola et al. 2015; Suárez-Ortegón et al. 2015). In diabetic rat models, administration of streptozotocin raises hepatic and serum iron content along with increased ferroportin expression and iron absorption but lowers serum hepcidin levels (Wang et al. 2014). These findings suggest that serum hepcidin may partially explain the contrasting patterns of iron excess and deficiency in obesity. In "healthy" obesity, elevated hepatic hepcidin and serum levels suppress iron absorption and availability but promote tissue iron accumulation, leading to increased serum ferritin and iron deposits in the liver, adipose tissue, and muscle (Fig. 17.1). Over time, this iron accumulation induces oxidative stress, impairing insulin action and contributing to systemic insulin resistance and hyperglycemia. This creates a feedback loop where hepcidin levels decline, enhancing iron absorption.

Research on male C57Bl/6 mice on a high-fat diet (HFD) has shown that insulin resistance precedes disruptions in iron homeostasis, with hepatic iron stores and hepcidin levels decreasing as iron builds up in epididymal (visceral) white adipose tissue over 24 weeks (Varghese et al. 2020). Hepatic non-heme iron, ferritin protein, and hepcidin gene expression were reduced in HFD-fed mice but increased with caloric restriction, which also raised markers of duodenal iron absorption (Park et al. 2017; Chung et al. 2016). Further, in HFD conditions, increasing iron availability through supplementation reduced body weight gain and hepatic lipid accumulation by improving mitochondrial function in the liver and skeletal muscle (Kitamura et al. 2021).

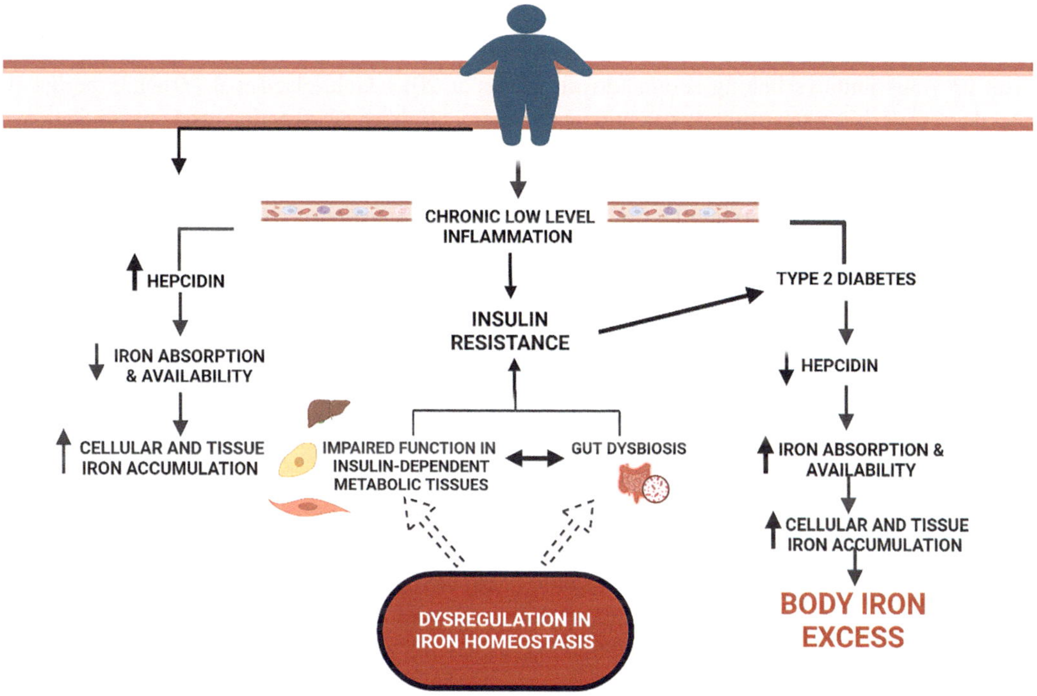

Fig. 17.1 Dysregulation in iron homeostasis in obesity and type 2 diabetes. Possible mechanism that underlies the dysregulation in iron homeostasis observed in subjects with obesity and type 2 diabetes. In obesity, elevated levels of liver-derived hepcidin in serum suppress iron absorption and availability but promote tissue iron accumulation, leading to increased serum ferritin and iron deposits in the liver, adipose tissue, and muscle. Over time, this iron accumulation induces oxidative stress, impairing insulin action and contributing to systemic insulin resistance and hyperglycemia. In the intestine, dysregulation in iron homeostasis leads to gut dysbiosis, which also promotes insulin resistance. This creates a feedback loop where hepcidin levels decline, enhancing iron absorption and resulting in increased body iron levels

Iron deficiency also has notable metabolic effects. In mice, perinatal iron deficiency predisposes offspring to increased fat accumulation, visceral fat depot weight, and higher arterial pressure when subjected to an HFD (Bourque et al. 2012). Mechanistically, iron is essential for the activity of brown adipose tissue and the browning of white adipose tissue, which are vital for energy regulation (Wang et al. 2017; Yook et al. 2021a, b; Li et al. 2020). Recent single-cell sequencing studies identified three distinct adipocyte subtypes, each with unique functions related to thermogenesis, lipid storage, and adipokine secretion. Notably, thermogenic adipocytes showed enriched iron accumulation and high oxidative stress protection, critical for optimal mitochondrial biogenesis and respiration (Min et al. 2019).

To better understand the links between iron metabolism and obesity-associated metabolic dysfunction, this chapter examines recent studies on iron homeostasis in insulin-dependent metabolic tissues and gut microbiota.

17.2 Iron Homeostasis Impacts on White Adipose Tissue Physiology

Several studies have shown that elevated body iron stores can contribute to adiposity and fat accumulation, negatively affecting adipogenesis and insulin signaling in adipose tissue. In 2005, Iwasaki et al. first observed a positive correlation between serum ferritin, a surrogate for body iron stores, and both subcutaneous and visceral fat areas, implying iron accumulation in adipose tissue (Iwasaki et al. 2005). This relationship aligns with previous findings linking serum ferritin to the degree of liver steatosis (Loguercio et al. 2004) and insulin resistance (Fernández-Real et al. 1998). In line with these findings, Dongiovanni et al. demonstrated that male mice on an iron-enriched diet developed hyperglycemia, hypertriglyceridemia, and insulin resistance, with increased iron accumulation in the liver and visceral adipose tissue as well as elevated serum hepcidin levels. Furthermore, insulin signaling was impaired in visceral adipose tissue in direct relation to iron excess (Dongiovanni et al. 2013). In models of obesity and diabetes, iron storage is associated with fat accumulation and type 2 diabetes, and administration of deferoxamine in these models reduced fat iron concentration, serum ferritin, adipocyte size, and markers of inflammation and oxidative stress, highlighting the benefits of iron chelation in adipose physiology (Tajima et al. 2012; Yan et al. 2018).

In recent studies on acute iron overload, visceral adipose tissue exhibited substantial iron accumulation, accompanied by reductions in adiponectin and leptin levels. Acute iron excess also led to dyslipidemia, which was partially reversed by adiponectin injection but not leptin (Tang et al. 2021). These findings suggest that iron overload may disrupt adiponectin expression, potentially contributing to dyslipidemia. Further supporting the link between adipose iron accumulation and type 2 diabetes, a recent study using inductively coupled plasma mass spectrometry identified Fe, Cr, and Cu as factors associated with increased diabetes risk (Rodríguez-Pérez et al. 2021).

Transcriptomic and proteomic studies from 2009 to 2012 showed that transferrin, typically expressed in adipose tissue under conditions of tissue expansion (e.g., overfeeding in lean subjects or rosiglitazone treatment in obese participants), was downregulated in visceral and subcutaneous adipose tissue in obese individuals once adipose expansion was maximized and exhausted (Shea et al. 2009; Ahmed et al. 2010; Insenser et al. 2012). Increased serum ferritin and iron correlated with markers of insulin resistance, such as elevated non-esterified fatty acids (NEFAs) and decreased adiponectin (Wlazlo et al. 2013; Gabrielsen et al. 2012). Experimental models have shown that iron accumulation in adipose tissue reduces adiponectin, while iron reduction increases adiponectin and improves insulin sensitivity (Gabrielsen et al. 2012). Consistently, phlebotomy in humans with altered glucose tolerance led to increased adiponectin and improved glucose tolerance (Gabrielsen et al. 2012). Elevated adipose tissue ferritin and ferroportin mRNA levels were linked to obesity and insulin resistance, reflecting adipose tissue dysfunction (Moreno-Navarrete et al. 2014a). Bariatric surgery-induced weight loss increased transferrin levels while reducing transferrin receptor, ferritin, and ferroportin mRNA in subcutaneous adipose tissue, accompanied by improved insulin action and adipogenesis (Moreno-Navarrete et al. 2014a).

Studies have further demonstrated the importance of iron balance in adipogenesis. During adipocyte differentiation, transferrin levels increase, supporting the development of mature adipocytes, while knockdown of the transferrin gene impairs this process (Moreno-Navarrete et al. 2014b). Iron chelation inhibits adipocyte differentiation, but this inhibition can be reversed by iron sources like transferrin or lactoferrin (Moreno-Navarrete et al. 2014b, c). Excess iron similarly impairs adipogenesis, underscoring the need for precisely regulated iron availability for adipocyte function and mitochondrial biogenesis (Moreno-Navarrete et al. 2014b). Palmitate treatment during adipocyte differentiation downregulates transferrin, leading to a cellular iron-deficient phenotype with inflammation markers (Moreno-Navarrete et al. 2014b). McClain et al. confirmed that transferrin mRNA levels correlate positively with insulin sensitivity, while markers of iron accumulation correlate negatively (McClain et al. 2018). This study also found that certain SNPs in the transferrin gene promoter region may influence adipose transferrin levels, suggesting a genetic component in its regulation.

Iron regulatory protein 1/cytosolic aconitase 1 (ACO1), essential for intracellular iron homeostasis, also plays a role in lipid biosynthesis via its aconitase activity. ACO1 expression increases during adipocyte differentiation and is downregulated by inflammatory conditions, highlighting its role in adipogenesis (Moreno et al. 2015). Knockdown of ACO1 reduces NADPH production, impairing adipogenesis and iron uptake-related gene expression. Both iron overload and iron deficiency decrease ACO1 expression, indicating the necessity of adequate intracellular iron levels for adipogenic functions (Moreno et al. 2015).

The importance of maintaining balanced iron levels in adipocytes is further supported by studies demonstrating that iron excess disrupts adipose function and contributes to carbohydrate metabolism disturbances. Excess iron in adipocytes, due to decreased expression of iron-exporting proteins like hephaestin and ceruloplasmin, is associated with reduced adiponectin and leptin levels and a prediabetic state (Zheng et al. 2018). Increased heme oxygenase 1 (HMOX1) activity, a potential source of intracellular iron, was found in obese individuals, correlating with higher fasting glucose, triglycerides, and markers of inflammation, but reduced adipogenesis and mitochondrial function (Moreno-Navarrete et al. 2017b). HMOX1 expression is also elevated in adipose tissue macrophages, where it contributes to iron accumulation, inflammation, and impaired glucose metabolism (Moreno-Navarrete et al. 2017b).

Reducing adipocyte iron stores can positively affect metabolic health. A study on mice with reduced adipocyte iron levels through transferrin receptor 1 deficiency or ferroportin 1 overexpression showed reduced body weight, fat mass, and liver steatosis, alongside improved insulin sensitivity and glucose tolerance, without altering food intake or energy expenditure (Zhang et al. 2021). Conversely, the deletion of adipocyte ferroportin 1 had minimal effects on iron storage and metabolic disturbances, suggesting additional mechanisms may regulate adipocyte iron homeostasis (Britton et al. 2018b). As stated above, iron reduction increases adiponectin and improves insulin sensitivity in animal models and in humans (Gabrielsen et al. 2012).

These findings underscore the role of adipose tissue iron balance in metabolic health, suggesting that targeting iron regulation may benefit obesity-related metabolic dysfunctions.

17.3 Modulation of Mitochondrial Iron Impacts on White Adipose Tissue

MitoNEET is a protein located on the outer mitochondrial membrane that regulates mitochondrial iron transport into the matrix, effectively limiting electron transport, beta-oxidation rates, and reducing mitochondrial membrane potential and reactive oxygen species (ROS) damage (Kusminski et al. 2012). In studies with adipose-specific mitoNEET overexpression in *ob/ob* mice, these mice exhibited increased body weight and fat mass but showed enhanced glucose tolerance and insulin sensitivity, along with decreased immune cell infiltration, smaller adipocyte size, reduced ROS-driven lipid peroxidation in subcutaneous white adipose tissue (WAT), and lower lipid and ceramide accumulation in the liver—resulting in a "healthy obesity" phenotype (Kusminski et al. 2012). This effect appears linked to an enhanced ability for fuel storage driven by adipocyte hyperplasia rather than hypertrophy. Mechanistically, mitoNEET reduces mitochondrial matrix iron levels, thereby decreasing ROS production from the electron transport chain and supporting mitochondrial biogenesis, both of which are essential for optimal adipose tissue function (Kusminski et al. 2012).

These beneficial effects of mitoNEET on adipose tissue and overall insulin sensitivity were also observed in HFD-fed mice (Kusminski et al. 2014). In humans, adipose tissue mitoNEET expression was found to decrease in obesity and positively correlate with insulin sensitivity and markers of adipose tissue health, including genes related to adipogenesis and mitochondrial biogenesis (Moreno-Navarrete et al. 2016b). Moreover, adipose tissue mitoNEET expression was inversely correlated with markers of intracellular iron accumulation and oxidative stress but positively correlated with factors involved in intracellular iron regulation, suggesting that mitoNEET may play a role in managing iron levels within human adipose tissue (Moreno-Navarrete et al. 2016b). In contrast, overexpression of mitochondrial ferritin in adipocytes, which increases mitochondrial iron, produced a lean yet insulin-resistant phenotype marked by low adiponectin and increased ROS (Kusminski et al. 2020).

Together, these findings indicate that limiting iron storage within adipocyte mitochondria could be a promising strategy to mitigate obesity-associated adipose tissue dysfunction and metabolic disturbances. However, it is also important to note that iron deficiency can negatively impact brown adipose tissue activity and the browning of WAT, highlighting the need for balanced iron levels to support healthy adipose function.

17.4 The Relevance of Iron Homeostasis in Brown Adipose Tissue Activity, Thermogenesis, and Energy Expenditure

Increased iron content has been observed in brown adipose tissue (BAT) upon acute cold exposure, hinting at a possible role for iron in thermogenic activity within this fat depot (Wang et al. 2017). Dietary iron deficiency, however, reduces iron levels in inguinal adipose tissue, impairs adaptive thermogenesis, and alters the expression of beige adipocyte markers in response to cold and β-adrenergic stimuli, leading to increased weight gain (Yook et al. 2021a, b). At a molecular level, iron uptake and regulation in adipose tissue are closely linked to thermogenesis and browning (Fig. 17.2). Specifically, iron regulatory proteins interact with iron-response elements to enhance iron uptake under β3-adrenoceptor stimulation, while thermoneutral conditions decrease these processes (Yook et al. 2021a, b).

During brown adipocyte differentiation, iron uptake is elevated to meet the high mitochondrial iron demand for optimal thermogenic function, whereas white adipocytes tend to favor iron storage in the cytosol (Yook et al. 2021a, b). Disrupting iron uptake in brown and beige adipocytes by depleting transferrin receptors results in mitochondrial dysfunction and impaired differentiation, severely limiting BAT activity (Li et al. 2020). Furthermore, a global deletion of ferritin heavy chain (Fth1), essential for cellular iron metabolism, in adult mice caused widespread iron imbalance. This included reduced circulating iron, increased tissue iron in organs such as the heart, liver, and kidneys, heightened oxidative stress and inflammation, and pronounced atrophy in both white and brown adipose tissues—ultimately compromising energy metabolism due to mitochondrial dysfunction (Blankenhaus et al. 2019).

Single-cell transcriptomic analysis of human adipose tissue has further revealed that "brite/beige" thermogenic adipocytes have unique mechanisms for iron handling and oxidative stress protection, necessary for mitochondrial biogenesis and respiration (Min et al. 2019). These studies highlight the differential roles of iron across adipose tissue types: while iron uptake is essential for white, beige, and brown adipocyte differentiation and sustaining high mitochondrial activity in thermogenic cells, reducing intracellular and mitochondrial iron levels optimizes function in mature white adipocytes. Local iron demands across adipose cell types might also be modulated by tissue-resident macrophages, which play a role in managing iron availability (Winn et al. 2020; Ameka and Hasty 2020).

17.5 Macrophages Modulate Tissue Iron Homeostasis

Macrophages play a crucial role in systemic iron homeostasis by either supplying iron for erythropoiesis or sequestering it to limit bacterial growth (Soares and Hamza 2016). Their involvement in tissue-specific iron regulation, particularly in adipose tissue, has been further elucidated by recent research from Hasty and colleagues (Winn et al. 2020).

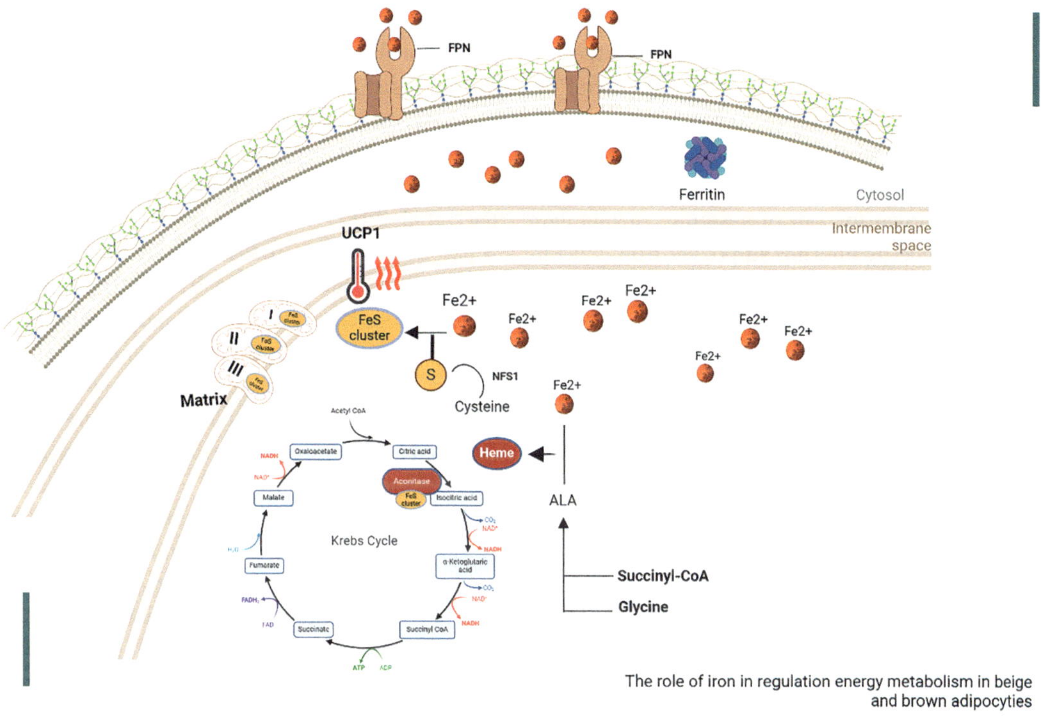

Fig. 17.2 The role of iron in regulating energy metabolism in beige and brown adipocytes. In the brown and beige adipocytes, iron participates in energy metabolism as a component of the iron-sulfur cluster and heme proteins. Uncoupling protein 1(UCP1) generates heat instead of ATP in the brown and beige adipose tissue. Iron deficiency, along with the reduced iron-sulfur cluster synthesis, results in impaired function of the ETC and impaired thermogenesis, leading to a decrease in energy expenditure

In adipose tissue, elevated iron levels drive a shift from low to high iron-recycling macrophages, helping to protect adipocytes from iron overload and reducing inflammation within adipose tissue (Orr et al. 2014; Hubler et al. 2018). Obesity, however, disrupts this balance by increasing low iron-recycling macrophages and reducing the expression of iron-handling genes across all macrophage populations. This shift is associated with a fourfold increase in adipocyte iron stores and a depletion of liver iron stores (Orr et al. 2014; Varghese et al. 2020).

Inhibition of iron export in macrophages through myeloid-specific deletion of ferroportin led to significant iron accumulation in bone marrow, liver, and adipose tissue. This iron buildup impaired mitochondrial respiratory function in liver and adipose tissue macrophages and was linked to reduced glucose tolerance in aged, but not young, lean, or obese mice (Winn et al. 2021). Similarly, selective disruption of macrophage ferroportin in muscle tissue caused iron to accumulate in muscle-infiltrating macrophages, resulting in impaired muscle regeneration with smaller myofibers and increased fat accumulation (Corna et al. 2016).

Conversely, depleting ferritin heavy chain (Fth1) in macrophages, which reduced intracellular iron levels and increased iron export, led to reduced body weight, inflammation, and oxidative stress in adipose tissue, with decreased iron accumulation in both adipose tissue and spleen. Under HFD conditions, macrophage-specific Fth1 deletion improved insulin sensitivity, enhanced energy expenditure, increased expression of thermogenic genes, and raised body temperature—suggesting that iron was redistributed to beige adipocytes and BAT, where it could support thermogenic activity (Ikeda et al. 2020).

These findings underscore the nuanced role of macrophages in iron management across tissues, with iron export capabilities that are crucial for balancing iron distribution and metabolic health in adipose and other tissues. Interestingly, macrophage mitochondrial iron levels have been recently shown to control systemic metabolism in male mice by altering adipocyte iron concentrations. This study showed that lowering macrophage mitochondrial matrix iron led to increased numbers of M2-like macrophages in adipose tissue, lowered iron levels in adipocytes, attenuated inflammation, and protected from high-fat-diet-induced metabolic deterioration. Very interestingly, these phenotypes were reproduced by transplantation of a small amount of fat from transgenic mice, in which mitochondrial iron levels were altered, to wild-type mice (Joffin et al. 2022).

17.6 Disruption in Iron Homeostasis Promotes Metabolic-Associated Fatty Liver Disease

Extensive experimental evidence highlights the negative effects of high-iron diets on liver iron accumulation, glucose metabolism, and insulin sensitivity (Dongiovanni et al. 2013; Ma et al. 2019; Meli et al. 2013; Tsuchiya et al. 2013). For example, genetically obese db/db mice fed a high-iron diet (1000 mg/kg chow) for 9 weeks showed increased iron accumulation in the liver and visceral adipose tissue, alongside elevated liver hepcidin, insulin resistance, upregulated liver gluconeogenesis genes, reduced adipose adiponectin mRNA, and decreased liver lipogenesis (Ma et al. 2019). In rats, high-fat diets also elevated hepatocyte iron, leading to greater liver inflammation and oxidative stress (Meli et al. 2013). Similarly, mice on high-fat and high-fructose diets for 16 weeks experienced early hepatic fat and insulin resistance, with iron accumulation and oxidative stress markers rising as early as 2 weeks, suggesting hepatic iron excess may contribute to the onset of liver steatosis and insulin resistance (Tsuchiya et al. 2013).

Conversely, other studies report an inverse association between obesity and hepatic iron levels (Kitamura et al. 2021; Varghese et al. 2020; Park et al. 2017). For instance, in HFD-fed mice, supplemental iron reduced body weight and hepatic lipid accumulation by improving mitochondrial function in the liver and skeletal muscle (Kitamura et al. 2021). In another study, insulin resistance onset occurred around week 12 of HFD feeding, while hepatic iron progressively declined after 16 weeks, correlating with reduced hepatic and serum hepcidin but increased iron in epididymal WAT, illustrating how obesity affects systemic iron distribution (Varghese et al. 2020). Additional research confirms reduced hepatic iron in obesity, with levels partially restored upon weight loss through caloric restriction (Park et al. 2017; Chung et al. 2016).

In mechanistic terms, hepatic mitoNEET overexpression was found to promote liver lipid accumulation while reducing mitochondrial iron and respiratory capacity; in contrast, mitoNEET knockdown decreased lipid accumulation and increased mitochondrial iron and respiration, enhancing electron transport chain (ETC) activity and beta-oxidation (Kusminski et al. 2012) (Fig. 17.3). Notably, in a methionine-choline-deficient model of non-alcoholic fatty liver disease (NAFLD), currently known as metabolic-associated steatotic liver disease (MASLD), iron supplementation raised hepatic iron concentration and lobular inflammation but reduced hepatic steatosis, triglycerides, and lipid peroxidation products (Kirsch et al. 2006).

The relationship between iron accumulation and MASLD in humans remains complex. Some studies have demonstrated an association between hepatic iron accumulation and MASLD (Kowdley et al. 2012; Tsuchiya et al. 2010), while others have found conflicting results (Britton et al. 2018a; Murali et al. 2018; Adams et al. 2015; Zhou et al. 2019); see also Chap. 16.

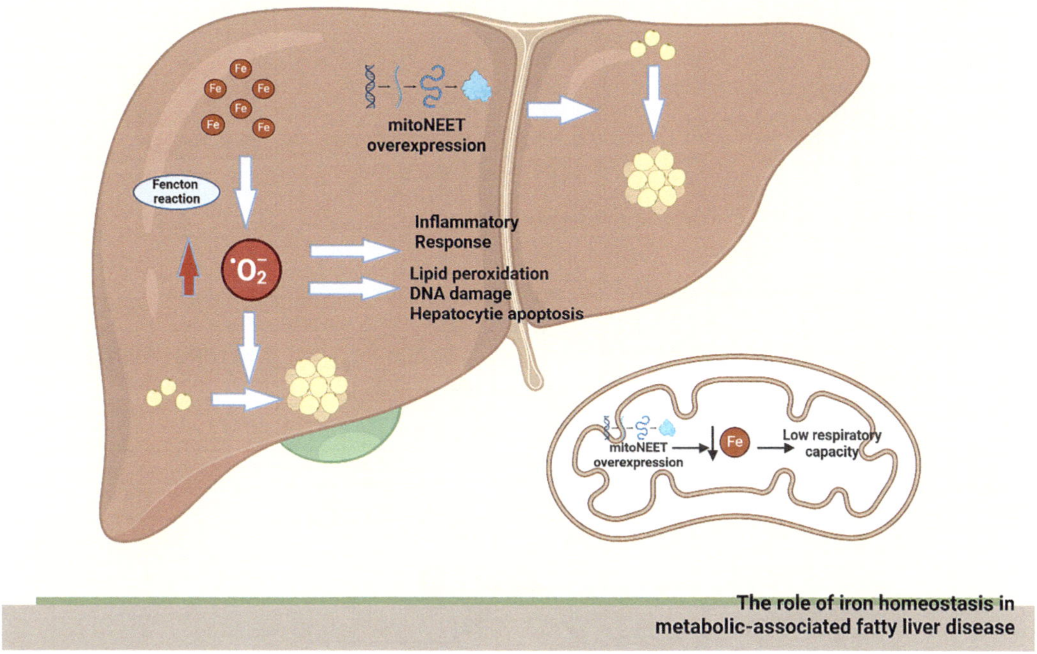

Fig. 17.3 The role of iron homeostasis in MASLD. Excessive iron accumulation in the liver is associated with increased production of ROS via the Fenton reaction. Excessive production of iron-induced reactive oxygen intermediates, such as peroxides and free radicals, can damage cellular proteins, lipids, and DNA, contributing to hepatic inflammation, hepatic steatosis, and MASLD. Conversely, hepatic overexpression of the mitochondrial Asn-Glu-Glu-Thr-containing protein (mitoNEET) promotes hepatic lipid accumulation while reducing mitochondrial iron and respiratory capacity

17.7 Iron Excess in Skeletal Muscle Leads to Muscle Insulin Resistance and Atrophy

Iron excess in skeletal muscle has been shown to negatively impact muscle insulin sensitivity and contribute to atrophy (Zhou et al. 2019; Dahyaleh et al. 2021; Cui et al. 2019; Jahng et al. 2019). For instance, in a model of endogenous iron overload using iron regulatory protein 2 knockout (IRP2 KO) mice, hyperglycemia and insulin resistance were observed alongside increased iron accumulation and oxidative stress in both liver and skeletal muscle (Zhou et al. 2019). Iron overload-induced oxidative stress in skeletal myocytes also leads to adiponectin resistance through inhibition of FOXO1 nuclear translocation, a key transcription factor for adiponectin receptor expression (Dahyaleh et al. 2021).

In further support, treatment with palmitate elevated intracellular iron levels and induced insulin resistance, whereas iron chelation via deferoxamine or transferrin receptor knockdown prevented these effects. Conversely, iron donors exacerbated palmitate-induced insulin resistance (Cui et al. 2019). Beyond oxidative stress, iron overload has been found to disrupt autophagy in skeletal muscle through mTORC1-UVRAG inhibition, further impairing insulin signaling (Jahng et al. 2019). Interestingly, iron deficiency appears protective, as it reduces high-fat diet-induced muscle insulin resistance and activates AMPK in rats, suggesting a role for iron reduction in

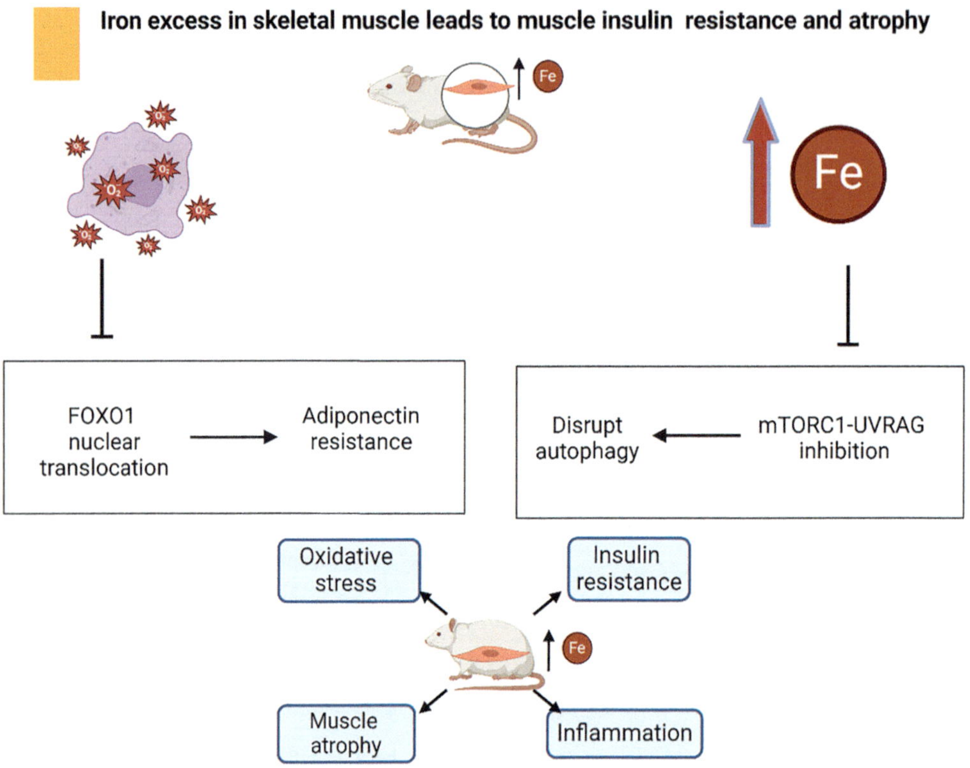

Fig. 17.4 Iron excess in skeletal muscle negatively impacts muscle insulin sensitivity and contributes to atrophy. Iron excess in skeletal muscle has been shown to negatively impact muscle insulin sensitivity and contribute to atrophy in a model of endogenous iron overload using iron regulatory protein 2 knockout (IRP2 KO) mice. Hyperglycemia and insulin resistance were observed alongside increased iron accumulation and oxidative stress in both liver and skeletal muscle. Iron overload-induced oxidative stress in skeletal myocytes also leads to adiponectin resistance through inhibition of FOXO1 nuclear translocation, a key transcription factor for adiponectin receptor expression. In addition to oxidative stress, iron overload disrupts autophagy in skeletal muscle through inhibition of mTORC1-UVRAG, further impairing insulin signaling. In genetically obese mice, increased muscle iron content was associated with inflammation and oxidative stress related to muscle atrophy, indicating a potential link between iron accumulation and obesity-related sarcopenia

enhancing insulin sensitivity in skeletal muscle (Han et al. 2011) (Fig. 17.4).

Notably, in genetically obese mice, increased muscle iron content and markers of iron accumulation (e.g., lipocalin 2 and ferritin) were associated with inflammation and oxidative stress related to muscle atrophy, indicating a potential link between iron accumulation and obesity-related sarcopenia (Choi et al. 2021). Similarly, in human studies, the paravertebral muscle R2* signal, indicative of iron content, and expression of iron excess-related genes in rectus abdominis muscle correlated positively with obesity metrics (body mass index, waist circumference, and fat mass), as well as markers of systemic iron stores (hepatic iron, serum ferritin, and hepcidin), even after adjusting for age and sex (Moreno-Navarrete et al. 2016a).

17.8 Iron Homeostasis Impacts on Gut Microbiome

The microbiota and host cells interact closely to regulate iron homeostasis within the body. Iron can influence the composition and activity of the gut microbiota (Tompkins et al. 2001; Werner et al. 2011; Dostal et al. 2014; Rusu et al. 2020); see also Chap. 22. For instance, excessive iron intake can promote the growth of pathogenic bacteria, leading to intestinal inflammation (Jaeggi et al. 2015; Paganini et al. 2016), while iron deficiency may reduce pathogen emergence (Deriu et al. 2013; Kortman et al. 2015). Iron supplementation, particularly in high doses, can disrupt gut microbiota composition by reducing beneficial lactic acid bacteria, such as bifidobacteria and lactobacilli, while increasing enteropathogenic Escherichia coli, which contributes to intestinal inflammation (Paganini and Zimmermann 2017). In Kenyan infants, iron supplementation increased pathogen abundance, including *Salmonella*, *Clostridium difficile*, *Clostridium perfringens*, and *Escherichia coli*, at the expense of beneficial *Bifidobacteriaceae*, promoting intestinal inflammation (Jaeggi et al. 2015). Notably, *Bifidobacteriaceae's* ability to bind iron and reduce iron-associated free radical formation may help lower colorectal cancer risk (Bielik and Kolisek 2021).

Iron replacement in patients with inflammatory bowel disease (IBD) has been linked to changes in gut microbiota, including decreased abundance of beneficial species like *Faecalibacterium prausnitzii*, *Ruminococcus bromii*, and *Dorea* sp., along with increased Bifidobacterium species (Lee et al. 2017). Moreover, gut microbiota also play a role in regulating host iron homeostasis. Short-chain fatty acids (especially propionate) produced by gut microbes enhance iron absorption in the proximal colon (Bouglé et al. 2002). Studies in germ-free mice have shown that microbial colonization alters iron handling in the small intestine, decreasing proteins related to iron uptake but increasing ferritin and ferroportin, proteins involved in iron storage and export (Deschemin et al. 2016). These findings highlight the role of the microbiota in modulating intestinal iron absorption.

Recent research by Das et al. (2020) identified specific microbial metabolites—1,3-diaminopropane and reuterin—that inhibit intestinal iron absorption. These metabolites are produced by indigenous gut microbiota and suppress HIF-2α activity, leading to reduced expression of iron transporters (Dmt1 and Dcytb) in the small intestine. Importantly, these effects were independent of hepcidin levels and were associated with reduced iron absorption under conditions of dietary iron deficiency. In contrast, antibiotic treatment in iron-deficient mice decreased Lactobacillus species and increased the expression of iron-absorptive genes, improving iron homeostasis and related blood parameters (Das et al. 2020). Additionally, gut microbiota was found to be essential for maintaining intestinal ferritin expression, which is crucial for regulating intestinal iron absorption (Vanoaica et al. 2010). In germ-free and antibiotic-treated mice, ferritin expression was decreased in the duodenum and peripheral tissues (liver and spleen), but this expression was restored after microbiome recolonization. Crude fecal metabolites from pathogen-free mice also restored ferritin expression in germ-free or antibiotic-treated mice, though the specific metabolites responsible remain unidentified. Bacterial lipopolysaccharide, known to induce ferritin expression (Siegert et al. 2015; Huang et al. 2019), could be a contributing factor.

These studies underscore the potential of microbiome-based therapies to treat iron-related disorders, highlighting the complex interplay between gut microbiota and iron metabolism.

17.9 Iron Excess Is Associated to Obesity-Induced Gut Dysbiosis

Patients with obesity-associated metabolic disturbances and insulin resistance often exhibit dysregulation in iron homeostasis (as discussed herein) and gut dysbiosis (Thingholm et al. 2019; Aron-Wisnewsky et al. 2019; Le Chatelier et al. 2013; Forslund et al. 2015; Karlsson et al. 2013). This suggests a potential link between altered iron homeostasis and gut microbiome changes in the progression of metabolic disease, a hypothesis supported by recent studies (Zhang et al. 2021; Mayneris-Perxachs et al. 2021). For example, reducing adipose tissue iron accumulation in a mouse model with adipocyte-specific deletion of the transferrin receptor resulted in decreased intestinal inflammation and oxidative stress under obesogenic conditions. This was accompanied by healthy compositional changes in the gut microbiota, such as a significant reduction in Proteobacteria (including pro-inflammatory groups like Erysipelotrichi, Clostridia, Deltaproteobacteria, Epsilonproteobacteria, and Gammaproteobacteria) and an increase in Coriobacteria and Bacilli (Zhang et al. 2021).

A recent study also demonstrated that serum ferritin, a marker of body iron stores, correlates with a specific microbial functional signature that might modulate liver gluconeogenesis and lipid accumulation, potentially impacting the progression of metabolic-associated fatty liver disease (Mayneris-Perxachs et al. 2021). In this study, microbial families associated with serum ferritin levels were linked to the expression of liver iron metabolism-related genes, such as transferrin receptor, hepcidin, mitoNEET, iron regulatory protein 1, and ferroportin. For instance, serum ferritin levels were negatively correlated with the abundance of Pasteurellaceae, Leuconostocaceae, and Micrococcaceae, while certain Veillonella, Bifidobacterium, Lactobacillus, Bacteroides, and Prevotella species were also negatively associated with serum ferritin levels (Mayneris-Perxachs et al. 2021). Interestingly, many of these species, including Veillonella, Bifidobacterium, Bacteroides, and Prevotella, have been previously linked to the progression of non-alcoholic steatohepatitis (NASH), currently known as metabolic-associated steatohepatitis (MASH) (Loomba et al. 2021; Zhu et al. 2013). Additionally, the abundance of Lactobacillus species is increased in diet-induced iron deficiency, and Lactobacillus reuteri can attenuate intestinal iron absorption through the production of reuterin (Das et al. 2020). These findings suggest that promoting Lactobacillus species in individuals with obesity-associated metabolic disturbances could mitigate the negative effects of iron excess on type 2 diabetes and MASH progression.

This study further revealed that microbiome functionality is also associated with serum ferritin levels. Specifically, the bacterial cytochrome b561, which is involved in iron absorption (Asard et al. 2013), was negatively correlated with serum ferritin levels, indicating that it may modulate intestinal iron uptake. In addition, bacterial iron transport and chelation functions, as well as enzymes involved in heme and siderophore biosynthesis, were reduced in individuals with high serum ferritin levels (Mayneris-Perxachs et al. 2021). Importantly, serum ferritin was also negatively associated with bacterial functions involved in glutathione biosynthesis, which is crucial in the context of metabolic disorders like MASLD. Specifically, serum ferritin levels were inversely correlated with the expression of the bacterial glycine transporter and glutamate-cysteine ligase, the first enzyme in the glutathione biosynthetic pathway. Plasma and liver glutathione levels are depleted in MASLD patients, and altered glutathione metabolism is a key feature of the disease (Mardinoglu et al. 2017).

In further support of these findings, fecal microbiome transplantation from "high ferritin donors" to recipient mice increased the expression of genes related to iron metabolism and fatty acid accumulation. Conversely, microbiota from "low ferritin donors" increased expression of Tfrc, a marker of low tissue iron levels (Mayneris-Perxachs et al. 2021).

Together, these studies suggest that, in addition to altered hepcidin levels, obesity-associated gut dysbiosis should be considered a significant factor in the dysregulation of body iron homeostasis, contributing to excessive iron accumulation in insulin-sensitive tissues such as adipose tissue, liver, and skeletal muscle.

17.10 Conclusions

In summary, iron homeostasis is critical for maintaining overall health. In humans and animal models, disruptions in iron regulation are frequently associated with obesity-related metabolic issues. These conditions are marked by increased serum ferritin levels and excessive iron accumulation in insulin-sensitive tissues such as the liver, adipose tissue, and skeletal muscle. Chronic iron overload induces oxidative stress, which impacts insulin sensitivity, contributing to systemic insulin resistance and hyperglycemia. This creates a feedback loop where reduced serum hepcidin levels drive more intestinal iron absorption, worsening iron accumulation.

While iron's effects on gut microbiota are well-known, the microbiota's role in regulating iron homeostasis is less understood. Recent studies have revealed mechanisms by which gut microbiota influence intestinal iron absorption and overall iron storage.

References

Adams LA, Crawford DH, Stuart K et al (2015) The impact of phlebotomy in nonalcoholic fatty liver disease: a prospective, randomized, controlled trial. Hepatology 61:1555–1564. https://doi.org/10.1002/hep.27662

Ahmed M, Neville MJ, Edelmann MJ et al (2010) Proteomic analysis of human adipose tissue after rosiglitazone treatment shows coordinated changes to promote glucose uptake. Obesity (Silver Spring) 18:27–34. https://doi.org/10.1038/oby.2009.208

Ameka MK, Hasty AH (2020) Fat and iron don't mix. Immunometabolism 2. https://doi.org/10.20900/immunometab20200034

Aregbesola A, Voutilainen S, Virtanen JK, Aregbesola A, Tuomainen TP (2015) Serum hepcidin concentrations and type 2 diabetes. World J Diabetes 6(7):978–982. https://doi.org/10.4239/wjd.v6.i7.978

Aron-Wisnewsky J, Prifti E, Belda E et al (2019) Major microbiota dysbiosis in severe obesity: fate after bariatric surgery. Gut 68:70–82. https://doi.org/10.1136/gutjnl-2018-316103

Asard H, Barbaro R, Trost P, Bérczi A (2013) Cytochromes b561: ascorbate-mediated trans-membrane electron transport. Antioxid Redox Signal 19:1026–1035. https://doi.org/10.1089/ars.2012.5065

Bao W, Rong Y, Rong S, Liu L (2012) Dietary iron intake, body iron stores, and the risk of type 2 diabetes: a systematic review and meta-analysis. BMC Med 10:119. https://doi.org/10.1186/1741-7015-10-119

Bergamaschi G, Di Sabatino A, Pasini A et al (2017) Intestinal expression of genes implicated in iron absorption and their regulation by hepcidin. Clin Nutr 36:1427–1433. https://doi.org/10.1016/j.clnu.2016.09.021

Bielik V, Kolisek M (2021) Bioaccessibility and bioavailability of minerals in relation to a healthy gut microbiome. Int J Mol Sci 22. https://doi.org/10.3390/ijms22136803

Blankenhaus B, Braza F, Martins R et al (2019) Ferritin regulates organismal energy balance and thermogenesis. Mol Metab 24:64–79. https://doi.org/10.1016/j.molmet.2019.03.008

Bouglé D, Vaghefi-Vaezzadeh N, Roland N et al (2002) Influence of short-chain fatty acids on iron absorption by proximal colon. Scand J Gastroenterol 37:1008–1011. https://doi.org/10.1080/003655202320378176

Bourque SL, Komolova M, McCabe K et al (2012) Perinatal iron deficiency combined with a high-fat diet causes obesity and cardiovascular dysregulation. Endocrinology 153:1174–1182. https://doi.org/10.1210/en.2011-1700

Britton L, Bridle K, Reiling J et al (2018a) Hepatic iron concentration correlates with insulin sensitivity in nonalcoholic fatty liver disease. Hepatol Commun 2:644–653. https://doi.org/10.1002/hep4.1190

Britton L, Jaskowski L-A, Bridle K et al (2018b) Ferroportin expression in adipocytes does not contribute to iron homeostasis or metabolic responses to a high calorie diet. Cell Mol Gastroenterol Hepatol 5:319–331. https://doi.org/10.1016/j.jcmgh.2018.01.005

Cepeda-Lopez AC, Allende-Labastida J, Melse-Boonstra A et al (2016) The effects of fat loss after bariatric surgery on inflammation, serum hepcidin, and iron absorption: a prospective 6-mo iron stable isotope study. Am J Clin Nutr 104:1030–1038. https://doi.org/10.3945/ajcn.115.115592

Choi EB, Jeong JH, Jang HM et al (2021) Skeletal lipocalin-2 is associated with iron-related oxidative stress in ob/ob mice with sarcopenia. Antioxidants (Basel) 10. https://doi.org/10.3390/antiox10050758

Chung H, Wu D, Smith D et al (2016) Lower hepatic iron storage associated with obesity in mice can be restored by decreasing body fat mass through feeding a low-fat diet. Nutr Res 36:955–963. https://doi.org/10.1016/j.nutres.2016.06.003

Corna G, Caserta I, Monno A et al (2016) The repair of skeletal muscle requires iron recycling through macrophage ferroportin. J Immunol 197:1914–1925. https://doi.org/10.4049/jimmunol.1501417

Cui R, Choi S-E, Kim TH et al (2019) Iron overload by transferrin receptor protein 1 regulation plays an important role in palmitate-induced insulin resistance in human skeletal muscle cells. FASEB J 33:1771–1786. https://doi.org/10.1096/fj.201800448R

Dahyaleh K, Sung HK, Prioriello M et al (2021) Iron overload reduces adiponectin receptor expression via a ROS/FOXO1-dependent mechanism leading to adiponectin resistance in skeletal muscle cells. J Cell Physiol 236:5339–5351. https://doi.org/10.1002/jcp.30240

Das NK, Schwartz AJ, Barthel G et al (2020) Microbial metabolite signaling is required for systemic iron homeostasis. Cell Metab 31:115–130.e6. https://doi.org/10.1016/j.cmet.2019.10.005

Del Giudice EM, Santoro N, Amato A et al (2009) Hepcidin in obese children as a potential mediator of the association between obesity and iron deficiency. J Clin Endocrinol Metab 94:5102–5107. https://doi.org/10.1210/jc.2009-1361

Deriu E, Liu JZ, Pezeshki M et al (2013) Probiotic bacteria reduce salmonella typhimurium intestinal colonization by competing for iron. Cell Host Microbe 14:26–37. https://doi.org/10.1016/j.chom.2013.06.007

Deschemin J-C, Noordine M-L, Remot A et al (2016) The microbiota shifts the iron sensing of intestinal cells. FASEB J 30:252–261. https://doi.org/10.1096/fj.15-276840

Dongiovanni P, Ruscica M, Rametta R et al (2013) Dietary iron overload induces visceral adipose tissue insulin resistance. Am J Pathol 182:2254–2263. https://doi.org/10.1016/j.ajpath.2013.02.019

Dostal A, Baumgartner J, Riesen N et al (2014) Effects of iron supplementation on dominant bacterial groups in the gut, faecal SCFA and gut inflammation: a randomised, placebo-controlled intervention trial in South African children. Br J Nutr 112:547–556. https://doi.org/10.1017/S0007114514001160

Fernández Real JM, Moreno-Navarrete JM, Manco M (2019) Iron influences on the gut-brain axis and development of type 2 diabetes. Crit Rev Food Sci Nutr 59:443–449. https://doi.org/10.1080/10408398.2017.1376616

Fernández-Real JM, Manco M (2014) Effects of iron overload on chronic metabolic diseases. Lancet Diabetes Endocrinol 2:513–526. https://doi.org/10.1016/S2213-8587(13)70174-8

Fernández-Real JM, Ricart-Engel W, Arroyo E et al (1998) Serum ferritin as a component of the insulin resistance syndrome. Diabetes Care 21:62–68. https://doi.org/10.2337/diacare.21.1.62

Fernández-Real JM, McClain D, Manco M (2015) Mechanisms linking glucose homeostasis and iron metabolism toward the onset and progression of type 2 diabetes. Diabetes Care 38:2169–2176. https://doi.org/10.2337/dc14-3082

Forslund K, Hildebrand F, Nielsen T et al (2015) Disentangling type 2 diabetes and metformin treatment signatures in the human gut microbiota. Nature 528:262–266. https://doi.org/10.1038/nature15766

Gabrielsen JS, Gao Y, Simcox JA et al (2012) Adipocyte iron regulates adiponectin and insulin sensitivity. J Clin Invest 122:3529–3540. https://doi.org/10.1172/JCI44421

Ganz T (2013) Systemic iron homeostasis. Physiol Rev 93:1721–1741. https://doi.org/10.1152/physrev.00008.2013

Gulec S, Anderson GJ, Collins JF (2014) Mechanistic and regulatory aspects of intestinal iron absorption. Am J Physiol Gastrointest Liver Physiol 307:G397–G409. https://doi.org/10.1152/ajpgi.00348.2013

Han D-H, Hancock CR, Jung SR et al (2011) Deficiency of the mitochondrial electron transport chain in muscle does not cause insulin resistance. PLoS One 6:e19739. https://doi.org/10.1371/journal.pone.0019739

Huang W, Zhan Y, Zheng Y et al (2019) Up-regulated ferritin in periodontitis promotes inflammatory cytokine expression in human periodontal ligament cells through transferrin receptor via ERK/P38 MAPK pathways. Clin Sci (Lond) 133:135–148. https://doi.org/10.1042/CS20180679

Hubler MJ, Erikson KM, Kennedy AJ, Hasty AH (2018) MFehi adipose tissue macrophages compensate for tissue iron perturbations in mice. Am J Phys Cell Phys 315:C319–C329. https://doi.org/10.1152/ajpcell.00103.2018

Ikeda Y, Watanabe H, Shiuchi T et al (2020) Deletion of H-ferritin in macrophages alleviates obesity and diabetes induced by high-fat diet in mice. Diabetologia 63:1588–1602. https://doi.org/10.1007/s00125-020-05153-0

Insenser M, Montes-Nieto R, Vilarrasa N et al (2012) A nontargeted proteomic approach to the study of visceral and subcutaneous adipose tissue in human obesity. Mol Cell Endocrinol 363:10–19. https://doi.org/10.1016/j.mce.2012.07.001

Iwasaki T, Nakajima A, Yoneda M et al (2005) Serum ferritin is associated with visceral fat area and subcutaneous fat area. Diabetes Care 28:2486–2491. https://doi.org/10.2337/diacare.28.10.2486

Jaeggi T, Kortman GAM, Moretti D et al (2015) Iron fortification adversely affects the gut microbiome, increases pathogen abundance and induces intestinal inflammation in Kenyan infants. Gut 64:731–742. https://doi.org/10.1136/gutjnl-2014-307720

Jahng JWS, Alsaadi RM, Palanivel R et al (2019) Iron overload inhibits late stage autophagic flux leading to insulin resistance. EMBO Rep 20:e47911. https://doi.org/10.15252/embr.201947911

Joffin N, Gliniak CM, Funcke JB et al (2022) Adipose tissue macrophages exert systemic metabolic control by manipulating local iron concentrations. Nat Metab 4(11):1474–1494. https://doi.org/10.1038/s42255-022-00664-z

Karlsson FH, Tremaroli V, Nookaew I et al (2013) Gut metagenome in European women with normal, impaired and diabetic glucose control. Nature 498:99–103. https://doi.org/10.1038/nature12198

Kirsch R, Sijtsema HP, Tlali M et al (2006) Effects of iron overload in a rat nutritional model of non-alcoholic fatty liver disease. Liver Int 26:1258–1267. https://doi.org/10.1111/j.1478-3231.2006.01329.x

Kitamura N, Yokoyama Y, Taoka H et al (2021) Iron supplementation regulates the progression of high fat diet induced obesity and hepatic steatosis via mitochondrial signaling pathways. Sci Rep 11:10753. https://doi.org/10.1038/s41598-021-89673-8

Kortman GAM, Mulder MLM, Richters TJW et al (2015) Low dietary iron intake restrains the intestinal inflammatory response and pathology of enteric infection by food-borne bacterial pathogens. Eur J Immunol 45:2553–2567. https://doi.org/10.1002/eji.201545642

Kowdley KV, Belt P, Wilson LA et al (2012) Serum ferritin is an independent predictor of histologic severity and advanced fibrosis in patients with nonalcoholic fatty liver disease. Hepatology 55:77–85. https://doi.org/10.1002/hep.24706

Kusminski CM, Holland WL, Sun K et al (2012) MitoNEET-driven alterations in adipocyte mitochondrial activity reveal a crucial adaptive process that preserves insulin sensitivity in obesity. Nat Med 18:1539–1549. https://doi.org/10.1038/nm.2899

Kusminski CM, Park J, Scherer PE (2014) MitoNEET-mediated effects on browning of white adipose tissue. Nat Commun 5:3962. https://doi.org/10.1038/ncomms4962

Kusminski CM, Ghaben AL, Morley TS et al (2020) A novel model of diabetic complications: adipocyte mitochondrial dysfunction triggers massive β-cell hyperplasia. Diabetes 69:313–330. https://doi.org/10.2337/db19-0327

Le Chatelier E, Nielsen T, Qin J et al (2013) Richness of human gut microbiome correlates with metabolic markers. Nature 500:541–546. https://doi.org/10.1038/nature12506

Lee T, Clavel T, Smirnov K et al (2017) Oral versus intravenous iron replacement therapy distinctly alters the gut microbiota and metabolome in patients with IBD. Gut 66:863–871. https://doi.org/10.1136/gutjnl-2015-309940

Li J, Pan X, Pan G et al (2020) Transferrin receptor 1 regulates thermogenic capacity and cell fate in brown/beige adipocytes. Adv Sci. https://doi.org/10.1002/advs.201903366

Loguercio C, De Simone T, D'Auria MV et al (2004) Non-alcoholic fatty liver disease: a multicentre clinical study by the Italian Association for the Study of the liver. Dig Liver Dis 36:398–405. https://doi.org/10.1016/j.dld.2004.01.022

Loomba R, Ling L, Dinh DM et al (2021) The commensal microbe veillonella as a marker for response to an FGF19 analog in NASH. Hepatology 73:126–143. https://doi.org/10.1002/hep.31523

Ma W, Feng Y, Jia L et al (2019) Dietary iron modulates glucose and lipid homeostasis in diabetic mice. Biol Trace Elem Res 189:194–200. https://doi.org/10.1007/s12011-018-1446-3

Mardinoglu A, Bjornson E, Zhang C et al (2017) Personal model-assisted identification of NAD+ and glutathione metabolism as intervention target in NAFLD. Mol Syst Biol 13:916. https://doi.org/10.15252/msb.20167422

Mayneris-Perxachs J, Cardellini M, Hoyles L et al (2021) Iron status influences non-alcoholic fatty liver disease in obesity through the gut microbiome. Microbiome 9:104. https://doi.org/10.1186/s40168-021-01052-7

McClain DA, Sharma NK, Jain S et al (2018) Adipose tissue transferrin and insulin resistance. J Clin Endocrinol Metab 103:4197–4208. https://doi.org/10.1210/jc.2018-00770

Meli R, Mattace Raso G, Irace C et al (2013) High fat diet induces liver steatosis and early dysregulation of iron metabolism in rats. PLoS One:e66570. https://doi.org/10.1371/journal.pone.0066570

Min SY, Desai A, Yang Z et al (2019) Diverse repertoire of human adipocyte subtypes develops from transcriptionally distinct mesenchymal progenitor cells. Proc Natl Acad Sci 116:17970–17979. https://doi.org/10.1073/pnas.1906512116

Moreno M, Ortega F, Xifra G et al (2015) Cytosolic aconitase activity sustains adipogenic capacity of adipose tissue connecting iron metabolism and adipogenesis. FASEB J 29:1529–1539. https://doi.org/10.1096/fj.14-258996

Moreno-Navarrete JM, Novelle MG, Catalán V et al (2014a) Insulin resistance modulates iron-related proteins in adipose tissue. Diabetes Care 37:1092–1100. https://doi.org/10.2337/dc13-1602

Moreno-Navarrete JM, Ortega F, Moreno M et al (2014b) Fine-tuned iron availability is essential to achieve optimal adipocyte differentiation and mitochondrial biogenesis. Diabetologia 57:1957–1967. https://doi.org/10.1007/s00125-014-3298-5

Moreno-Navarrete JM, Ortega F, Moreno M et al (2014c) Lactoferrin gene knockdown leads to similar effects to iron chelation in human adipocytes. J Cell Mol Med 18:391–395. https://doi.org/10.1111/jcmm.12234

Moreno-Navarrete JM, Blasco G, Xifra G et al (2016a) Obesity is associated with gene expression and imaging markers of iron accumulation in skeletal muscle. J Clin Endocrinol Metab 101:1282–1289. https://doi.org/10.1210/jc.2015-3303

Moreno-Navarrete JM, Moreno M, Ortega F et al (2016b) CISD1 in association with obesity-associated dysfunctional adipogenesis in human visceral adipose tissue. Obesity (Silver Spring) 24:139–147. https://doi.org/10.1002/oby.21334

Moreno-Navarrete JM, Moreno M, Puig J et al (2017a) Hepatic iron content is independently associated with serum hepcidin levels in subjects with obesity. Clin Nutr 36:1434–1439. https://doi.org/10.1016/j.clnu.2016.09.022

Moreno-Navarrete JM, Ortega F, Rodríguez A et al (2017b) HMOX1 as a marker of iron excess-induced adipose tissue dysfunction, affecting glucose uptake and respiratory capacity in human adipocytes. Diabetologia 60:915–926. https://doi.org/10.1007/s00125-017-4228-0

Moreno-Navarrete JM, Rodríguez A, Becerril S et al (2018) Increased small intestine expression of non-heme iron transporters in morbidly obese patients with newly diagnosed type 2 diabetes. Mol Nutr Food Res. https://doi.org/10.1002/mnfr.201700301

Murali AR, Gupta A, Brown K (2018) Systematic review and meta-analysis to determine the impact of iron depletion in dysmetabolic iron overload syndrome and non-alcoholic fatty liver disease. Hepatol Res:E30–E41. https://doi.org/10.1111/hepr.12921

Orr JS, Kennedy A, Anderson-Baucum EK et al (2014) Obesity alters adipose tissue macrophage iron content and tissue iron distribution. Diabetes 63:421–432. https://doi.org/10.2337/db13-0213

Paganini D, Zimmermann MB (2017) The effects of iron fortification and supplementation on the gut microbiome and diarrhea in infants and children: a review. Am J Clin Nutr 106:1688S–1693S. https://doi.org/10.3945/ajcn.117.156067

Paganini D, Uyoga MA, Zimmermann MB (2016) Iron fortification of foods for infants and children in low-income countries: effects on the gut microbiome, gut inflammation, and diarrhea. Nutrients 8. https://doi.org/10.3390/nu8080494

Park CY, Chung J, Koo K-O et al (2017) Hepatic iron storage is related to body adiposity and hepatic inflammation. Nutr Metab (Lond) 14:14. https://doi.org/10.1186/s12986-017-0169-3

Pechlaner R, Weiss G, Bansal S et al (2016) Inadequate hepcidin serum concentrations predict incident type 2 diabetes mellitus. Diabetes Metab Res Rev 32:187–192. https://doi.org/10.1002/dmrr.2711

Rodríguez-Pérez C, Gómez-Peña C, Pérez-Carrascosa FM et al (2021) Trace elements concentration in adipose tissue and the risk of incident type 2 diabetes in a prospective adult cohort. Environ Pollut 286:117496. https://doi.org/10.1016/j.envpol.2021.117496

Rusu IG, Suharoschi R, Vodnar DC et al (2020) Iron supplementation influence on the gut microbiota and probiotic intake effect in iron deficiency-a literature-based review. Nutrients 12. https://doi.org/10.3390/nu12071993

Santos J, Salgado P, Santos C et al (2014) Effect of bariatric surgery on weight loss, inflammation, iron metabolism, and lipid profile. Scand J Surg 103:21–25. https://doi.org/10.1177/1457496913490467

Shea J, French CR, Bishop J et al (2009) Changes in the transcriptome of abdominal subcutaneous adipose tissue in response to short-term overfeeding in lean and obese men. Am J Clin Nutr 89:407–415. https://doi.org/10.3945/ajcn.2008.25970

Siegert I, Schödel J, Nairz M et al (2015) Ferritin-mediated iron sequestration stabilizes hypoxia-inducible factor-1α upon LPS activation in the presence of ample oxygen. Cell Rep 13:2048–2055. https://doi.org/10.1016/j.celrep.2015.11.005

Simcox JA, McClain DA (2013) Iron and diabetes risk. Cell Metab 17:329–341. https://doi.org/10.1016/j.cmet.2013.02.007

Soares MP, Hamza I (2016) Macrophages and iron metabolism. Immunity 44:492–504. https://doi.org/10.1016/j.immuni.2016.02.016

Stoffel NU, El-Mallah C, Herter-Aeberli I et al (2020) The effect of central obesity on inflammation, hepcidin, and iron metabolism in young women. Int J Obes 44:1291–1300. https://doi.org/10.1038/s41366-020-0522-x

Suárez-Ortegón MF, Moreno M, Arbeláez A et al (2015) Circulating hepcidin in type 2 diabetes: a multivariate analysis and double blind evaluation of metformin effects. Mol Nutr Food Res 59:2460–2470. https://doi.org/10.1002/mnfr.201500310

Tajima S, Ikeda Y, Sawada K et al (2012) Iron reduction by deferoxamine leads to amelioration of adiposity via the regulation of oxidative stress and inflammation in obese and type 2 diabetes KKAy mice. Am J Physiol Endocrinol Metab:E77–E86. https://doi.org/10.1152/ajpendo.00033.2011

Tang Y, Wang D, Zhang H et al (2021) Rapid responses of adipocytes to iron overload increase serum TG level by decreasing adiponectin. J Cell Physiol 236:7544–7553. https://doi.org/10.1002/jcp.30391

Thingholm LB, Rühlemann MC, Koch M et al (2019) Obese individuals with and without type 2 diabetes show different gut microbial functional capacity and composition. Cell Host Microbe 26:252–264.e10. https://doi.org/10.1016/j.chom.2019.07.004

Tompkins GR, O'Dell NL, Bryson IT, Pennington CB (2001) The effects of dietary ferric iron and iron deprivation on the bacterial composition of the mouse intestine. Curr Microbiol 43:38–42. https://doi.org/10.1007/s002840010257

Tsuchiya H, Ashla AA, Hoshikawa Y, Matsumi Y, Kanki K, Enjoji M, Momosaki S, Nakamuta M, Taketomi A, Maehara Y, Shomori K, Kurimasa A, Hisatome I, Ito H, Shiota G (2010) Iron state in association with retinoid metabolism in non-alcoholic fatty liver disease. Hepatol Res 40(12):1227–1238. https://doi.org/10.1111/j.1872-034X.2010.00719.x

Tsuchiya H, Ebata Y, Sakabe T et al (2013) High-fat, high-fructose diet induces hepatic iron overload via a hepcidin-independent mechanism prior to the onset of liver steatosis and insulin resistance in mice. Metabolism 62:62–69. https://doi.org/10.1016/j.metabol.2012.06.008

Tussing-Humphreys LM, Liang H, Nemeth E, Freels S, Braunschweig CA (2009) Excess adiposity, inflammation, and iron-deficiency in female adolescents. J Am Diet Assoc 109(2):297–302. https://doi.org/10.1016/j.jada.2008.10.044

Tussing-Humphreys LM, Nemeth E, Fantuzzi G et al (2010) Decreased serum hepcidin and improved functional iron status 6 months after restrictive bariatric surgery. Obesity 18:2010–2016. https://doi.org/10.1038/oby.2009.490

Vanoaica L, Darshan D, Richman L et al (2010) Intestinal ferritin H is required for an accurate control of iron absorption. Cell Metab 12:273–282. https://doi.org/10.1016/j.cmet.2010.08.003

Varghese J, James JV, Anand R et al (2020) Development of insulin resistance preceded major changes in iron homeostasis in mice fed a high-fat diet. J Nutr Biochem 84:108441. https://doi.org/10.1016/j.jnutbio.2020.108441

Wang H, Li H, Jiang X et al (2014) Hepcidin is directly regulated by insulin and plays an important role in iron overload in streptozotocin-induced diabetic rats. Diabetes 63:1506–1518. https://doi.org/10.2337/db13-1195

Wang C, Liang X, Tao C et al (2017) Induction of copper and iron in acute cold-stimulated brown adipose tissues. Biochem Biophys Res Commun 488:496–500. https://doi.org/10.1016/j.bbrc.2017.05.073

Werner T, Wagner SJ, Martínez I et al (2011) Depletion of luminal iron alters the gut microbiota and prevents Crohn's disease-like ileitis. Gut 60:325–333. https://doi.org/10.1136/gut.2010.216929

Winn NC, Volk KM, Hasty AH (2020) Regulation of tissue iron homeostasis: the macrophage "ferrostat". JCI Insight 5. https://doi.org/10.1172/jci.insight.132964

Winn NC, Wolf EM, Cottam MA et al (2021) Myeloid-specific deletion of ferroportin impairs macrophage bioenergetics but is disconnected from systemic insulin action in adult mice. Am J Physiol Endocrinol Metab:E376–E391. https://doi.org/10.1152/ajpendo.00116.2021

Wlazlo N, van Greevenbroek MMJ, Ferreira I et al (2013) Iron metabolism is associated with adipocyte insulin resistance and plasma adiponectin: the Cohort on Diabetes and Atherosclerosis Maastricht (CODAM) study. Diabetes Care 36:309–315. https://doi.org/10.2337/dc12-0505

Yan H-F, Liu Z-Y, Guan Z-A, Guo C (2018) Deferoxamine ameliorates adipocyte dysfunction by modulating iron metabolism in ob/ob mice. Endocr Connect 7:604–616. https://doi.org/10.1530/EC-18-0054

Yook J-S, Thomas SS, Toney AM et al (2021a) Dietary iron deficiency modulates adipocyte iron homeostasis, adaptive thermogenesis, and obesity in C57BL/6 mice. J Nutr 151:2967–2975. https://doi.org/10.1093/jn/nxab222

Yook J-S, You M, Kim Y et al (2021b) The thermogenic characteristics of adipocytes are dependent on the regulation of iron homeostasis. J Biol Chem 296:100452. https://doi.org/10.1016/j.jbc.2021.100452

Zhang Z, Funcke J-B, Zi Z et al (2021) Adipocyte iron levels impinge on a fat-gut crosstalk to regulate intestinal lipid absorption and mediate protection from obesity. Cell Metab 33:1624–1639.e9. https://doi.org/10.1016/j.cmet.2021.06.001

Zhao Z, Li S, Liu G et al (2012) Body iron stores and heme-iron intake in relation to risk of type 2 diabetes: a systematic review and meta-analysis. PLoS One:e41641. https://doi.org/10.1371/journal.pone.0041641

Zheng J, Chen M, Liu G et al (2018) Ablation of hephaestin and ceruloplasmin results in iron accumulation in adipocytes and type 2 diabetes. FEBS Lett 592:394–401. https://doi.org/10.1002/1873-3468.12978

Zhou Y, Wu W, Xu Z et al (2019) Iron regulatory protein 2 deficiency may correlate with insulin resistance. Biochem Biophys Res Commun 510:191–197. https://doi.org/10.1016/j.bbrc.2019.01.022

Zhu L, Baker SS, Gill C et al (2013) Characterization of gut microbiomes in nonalcoholic steatohepatitis (NASH) patients: a connection between endogenous alcohol and NASH. Hepatology 57:601–609. https://doi.org/10.1002/hep.26093

18 Iron and Cancer

Suzy V. Torti, Lia Tesfay, and Frank M. Torti

Abstract

Iron is required for numerous essential processes, including DNA synthesis, DNA repair, and cellular metabolism. Cancer cells frequently demonstrate an enhanced demand for iron when compared to slowly cycling non-cancer cells due to their increased reliance on these processes. Manifestations of this demand include up-regulation of iron import, decrease in iron export, alterations in iron intracellular trafficking, as well as alterations in "iron gene" expression signatures that can predict prognosis. Cells of the tumor microenvironment, including T cells, tumor-associated macrophages, and cancer-associated fibroblasts, crosstalk with tumor cells to further modulate tumor iron status. Dietary iron, particularly heme iron, has been associated with increased cancer risk, although the influence of iron on immune cells of the microenvironment may modulate this risk. Tumor cell reliance on iron creates therapeutic opportunities. For example, the excess iron accumulated by cancer cells renders them susceptible to agents that induce ferroptosis, an iron-dependent form of cell death. In addition, significant progress has been made in the design of agents to target tumor cell iron dependence in other ways, including small molecule iron chelators and agents that target iron uptake, some of which are in current clinical trials. Recent discoveries, such as the key role of iron recycling in *KRAS*-mutated pancreatic cancer, are expected to further accelerate the transition of such agents to the clinic.

Keywords

Iron metabolism in cancer · Ferroptosis · Tumor microenvironment · Iron chelation therapy · Iron-dependent enzymes

This article is dedicated to Frank M. Torti MD, who passed away during its completion. His work unraveled many complexities in the relationship between iron and cancer and provided insights that continue to pave the way for research in this critical field.

S. V. Torti (✉) · L. Tesfay
Department of Molecular Biology and Biophysics, UConn Health, Farmington, CT, USA
e-mail: storti@uchc.edu

F. M. Torti
Department of Medicine, UConn Health, Farmington, CT, USA

18.1 Introduction

The relationship between iron and cancer has been the subject of exploration and debate for decades. In the 1940s and 1950s, exposure to environmental iron dust or the intramuscular injection of iron was shown to induce tumors in

K. Pantopoulos (ed.), *Iron Metabolism in Human Health and Disease*, Advances in Experimental Medicine and Biology 1480, https://doi.org/10.1007/978-3-031-92033-2_18

rodents (Campbell 1940; Richmond 1959), introducing the concept of iron as a tumor initiator. Work performed in ensuing decades showed that iron-rich diets promoted tumor growth in mice; conversely, iron-deficient diets reduced tumor growth (Hann et al. 1988, 1991, 1992; Radulescu et al. 2012), suggesting that iron can also serve as a tumor growth factor. Clinical observations that levels of iron in the blood are frequently reduced in patients with cancer inspired the idea that iron withholding is an inherent organismal strategy to combat malignancy (and infections) by depriving invaders of this essential nutrient, indirectly supporting the notion that iron is critical to malignant disease. These ideas were supported by many subsequent experiments that contributed to the broad concept (championed by ED Weinberg and others) that iron is an essential tumor nutrient that can be manipulated for therapeutic benefit (Weinberg 1983). An early case report (1987) in which DFO, an iron chelator, induced a partial response in a patient with neonatal leukemia further demonstrated the potential of this idea for translation to human cancer therapy (Estrov et al. 1987). The benefit of manipulation of iron in cancer patients has been supported by a substantial number of more modern studies (see below), although control of iron has yet to advance to a first-line therapy. Nevertheless, the core concept that iron plays a vital and potentially targetable role in cancer still stands.

18.2 Molecular Regulation of Iron in Cancer Cells

The ability of iron to cycle between reduced and oxidized states endows it with valuable properties in catalysis but also renders it capable of inducing oxidative stress (Chap. 1). Maintaining a balance between these features of iron—i.e., maintaining the bioavailability of iron for essential processes while preventing undesirable induction of oxidative stress—represents the major challenge in organismal and cellular handling of iron. Many of the fundamental proteins and processes used to meet this challenge have become well understood, primarily through the study of non-cancer cells and patients with iron overload disorders (see Fig. 18.1 and (Katsarou and Pantopoulos 2020; Galy et al. 2024) for excellent reviews of this topic). Here we focus on modulations of iron metabolism that have been described in cancer.

An early molecular insight into differences in iron handling between normal and neoplastic cells came from the study of the regulation of ferritin, a protein with a major role in the sequestration, storage, and recycling of intracellular iron. Ferritin is a ubiquitously expressed 24-subunit protein composed of two subunit types, termed H and L (reviewed in Arosio et al. 2024). The H and L subunits of this protein are subject to coordinate post-transcriptional regulation (Arosio et al. 2024). In a departure from this paradigm, early studies showed that the H chain of ferritin could be transcriptionally regulated by the adenoviral E1A oncogene (Tsuji et al. 1993) and *c-myc* (Wu et al. 1999), both of which were found to suppress ferritin H, reducing iron storage and enhancing its intracellular availability.

The discovery of the iron efflux pump ferroportin and its negative regulation by the liver peptide hormone hepcidin (Nemeth et al. 2004 and Chap. 3) provided a new dimension to our understanding of the role of iron in cancer. Following this seminal discovery in non-cancer cells, the study of ferroportin in breast cancer revealed that ferroportin downregulation was frequently observed in breast cancer cells and contributed to iron retention and tumor growth in mice (Pinnix et al. 2010). Directly tying this observation to human disease, transcriptional downregulation of ferroportin in tumors was associated with poor patient prognosis (Miller et al. 2011). Subsequently, it was observed that hepcidin was also synthesized by cancer cells and that its upregulation was linked to cancer cell growth, exemplifying another mechanism for enhanced tumor iron retention (Blanchette et al. 2016; Tesfay et al. 2015).

IRP1 and IRP2 are post-transcriptional regulators of several proteins of iron metabolism, including ferritin, ferroportin, and transferrin receptor (reviewed in Galy et al. (2024) and in Chap. 2), and have also been linked to cancer.

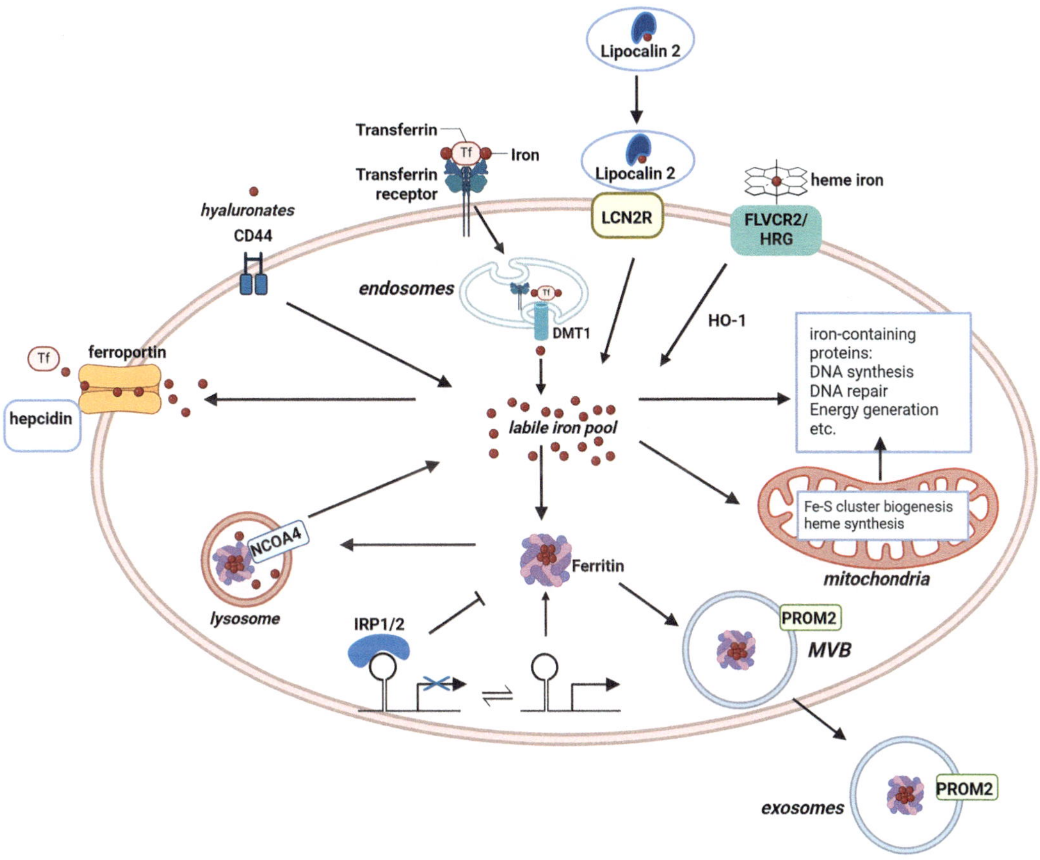

Fig. 18.1 Proteins and pathways are discussed in this review. Iron circulates in the blood bound to the glycoprotein transferrin. Diferric transferrin binds to transferrin receptor (TFRC or TfR1), which is ubiquitously expressed on cell surfaces. Following internalization of the TF-TFRC complex in the endosome, iron is released from Tf and transported out of the endosome by DMT1. Iron can also be taken up by alternative pathways, including CD44-mediated uptake of iron-bound hyaluronates and binding of lipocalin carrying an iron-loaded siderophore to its receptor LCN2R. Internalized iron is transferred to a metabolically labile low molecular weight iron pool (the "LIP") before being delivered to sites of utilization. Glutathione and proteins such as PCBP1 (Philpott and Jadhav 2019) (not depicted) are believed to chaperone iron to facilitate this process and prevent undesired redox cycling of iron. Iron not required for immediate metabolic needs is stored in ferritin, a 24-subunit protein consisting of H and L types of subunits that can assemble in variable ratios. As many as 4500 atoms of iron can be stored in a non-toxic but bioavailable form in the core of ferritin. Ferritin is regulated translationally through IRP1 and IRP2, which bind to an IRE in the 5' UTR of ferritin H and L mRNA to repress translation when iron levels are high. If required, iron-loaded ferritin can be degraded and its iron released intracellularly for utilization following binding to the cargo receptor NCOA4 in the autophagolysosome, a process termed ferritinophagy. When excess iron is present, it can be released from the cell by the iron efflux pump ferroportin, a protein whose mRNA is also stabilized by IRP1/IRP2 under conditions of high iron. Ferroportin is negatively regulated by the binding of hepcidin, a peptide hormone that circulates extracellularly, binds to ferroportin, and causes its degradation. Heme is taken up by the FLVCR2 receptor, and its iron is abstracted for cellular use by heme oxygenase (HO-1). Ferritin can be exported from the cell in endosomes, a process facilitated by prominin2 (PROM2). Abbreviations: *TFRC or TfR1* transferrin receptor 1, *DMT1* divalent metal transporter 1, *CD44* (***CD44*** Molecule (IN Blood Group)), *IRP1/2* iron-responsive element binding protein 1 or iron-responsive element binding protein 2, *IRE* iron-responsive element, *NCOA4* nuclear receptor coactivator 4, *FLVCR2* (Choline And Putative Heme Transporter 2)

IRP2, which represses ferritin translation (thus promoting iron storage) and positively regulates transferrin receptor mRNA stability (thus promoting iron uptake), was also associated with cancer cell growth in vitro and in animal models (Wang et al. 2014).

Cancer cells also deploy multiple mechanisms to amplify iron uptake relative to their non-cancer counterparts. For example, lipocalin (LCN2, (also termed 24p3)) is upregulated in breast and other cancers (Yang et al. 2009). Lipocalin binds iron-containing siderophores and delivers its iron payload following internalization and binding to its receptor (Bao et al. 2010; Devireddy et al. 2005), enabling metastatic cells to acquire iron in nutrient-limited microenvironments (Chi et al. 2020). Another example is CD44, a transmembrane glycoprotein that mediates endocytosis of hyaluronate-bound iron in breast tumors (Muller et al. 2020). Intriguingly, it was recently reported that ether lipids, prevalent in cancer stem cells and cells with a high metastatic capacity, confer a high membrane fluidity that is conducive to CD44-mediated iron uptake. This novel finding represents an entirely new mechanism of iron regulation in cancer cells that operates through lipid-mediated changes in iron uptake (Henry et al. 2024).

In addition to mechanisms affecting iron uptake and efflux, it has become apparent that modulation of intracellular pathways of iron trafficking also occurs in cancer cells. NCOA4 (nuclear receptor coactivator 4) is a cargo receptor that mediates delivery of ferritin to the lysosome for degradation and release of iron to the cytosol and mitochondria, a process termed ferritinophagy (Dowdle et al. 2014; Mancias et al. 2014). NCOA4 accelerates pancreatic tumorigenesis in mouse models (Santana-Codina et al. 2022). Further, an elevated ferritinophagy expression signature predicts poor prognosis in patients with pancreatic cancer (Santana-Codina et al. 2022). Pancreatic tumors that utilize the ferritinophagy pathway as a source of iron for mitochondrial iron-sulfur cluster protein synthesis become dependent upon this pathway following treatment with RAS-MAPK inhibitors (Ravichandran et al. 2022). Since disruption of ferritinophagy synergizes with KRAS-MAPK blockade to inhibit pancreatic tumor growth, ferritinophagy may represent a new targetable vulnerability in pancreatic cancer, a notoriously treatment-refractory disease (Ravichandran et al. 2022). Notably, targeting ferritinophagy also impaired quiescent cancer stem cells in in vivo models of acute myeloid leukemia (Larrue et al. 2024).

Iron balance can also be affected through the efflux of ferritin in extracellular vesicles via less well-studied pathways mediated by CD63 (Yanatori et al. 2021, 2023) and prominin2 (Brown et al. 2019). Ferritin released by these pathways may also impact tumorigenesis, since ferritin has the potential to interact with kininogen to affect angiogenesis (Coffman et al. 2009); in addition, the removal of iron-containing ferritin through this pathway may reduce cancer cell susceptibility to ferroptosis (Brown et al. 2019).

Overall, these observations are consistent with a picture in which cancer cells activate a complex variety of mechanisms, ranging from transcriptional regulation to membrane lipid content, that converge on remodeling pathways of iron uptake, efflux, and trafficking to optimize iron acquisition and retention while promoting cell survival.

18.3 Iron Signatures in Cancer

Bioinformatics has been a highly informative tool both in validating iron as a critical player in human malignancy and in identifying new genes that may play previously unappreciated roles in iron processing in tumors. Studying differences in patterns of gene expression between normal tissue and breast cancer tissue, our group uncovered a group of 16 genes encoding proteins involved in iron metabolism whose level of expression could provide prognostic guidance in breast cancer (Miller et al. 2011). This "iron regulatory gene signature" included genes not previously linked to cancer prognosis, such as *SLC25A37* (mitoferrin1) and *SFXN1*, as well as genes encoding proteins already known to be important in tumor iron metabolism, such as *TFRC*. An 8-gene iron regulatory gene signature composed of some of the same genes was also identified in infiltrating gliomas (Weston et al. 2016). As shown in Table 18.1, iron gene signatures have subsequently been constructed using tissue samples from multiple malignancies, including breast cancer, glioma, adrenocortical carcinoma, hepatocellular carcinoma, lung adenocarcinoma, sarcoma, endometriosis-related ovarian cancer, colon cancer, clear cell renal carcinoma, bladder cancer, papillary thyroid carcinoma, gastric cancer, and nasopharyngeal carcinoma (Zhu et al. 2019, 2024; Tang et al. 2020; Qin et al. 2021; Zhang et al. 2024a; Wang et al. 2023a; Yuan et al. 2021; Mou et al. 2020; Jin et al. 2023; Liu et al. 2023a; Su et al. 2024). Approximately 120 different genes are represented in these signatures, with the number of genes comprising the signature ranging from 2 to 21. Most genes (101) appear in only one signature; those that appear in multiple signatures are listed in Table 18.1. Since these studies used different criteria to define genes related to iron and evaluated different malignancies, variances in these prognostic signatures are not unexpected. Nevertheless, it is interesting that collections of genes related to iron metabolism are consistently associated with prognosis across different cancer types, with several genes related to aspects of iron metabolism, particularly iron transport, appearing with regularity in multiple different malignancies.

18.4 Ferroptosis

The role of iron in cancer has recently been extended to the study of ferroptosis, a term coined in 2012 (Dixon et al. 2012) to describe an iron-dependent form of cell death (reviewed in detail elsewhere (Berndt et al. 2024); see also Chap. 5). A myriad of studies have associated ferroptosis gene signatures with cancer patient prognosis in a variety of malignancies. It is worth noting that in ovarian and other cancers, "persister" cells and cancer cells with stem cell-like properties—treatment refractory cells thought to give rise to metastases—are particularly susceptible to ferroptosis due to their enhanced levels of iron (Rodriguez et al. 2022; Basuli et al. 2017). Observations such as these have led to the current concept that in some instances, iron repletion or provision of excess iron may push cancer cells into death pathways. This concept is being actively pursued in the design of novel therapeutics, as discussed below.

18.5 Iron-Dependent Enzymes and Processes Implicated in Malignancy

What is the fundamental metabolic need that drives cancer cells to acquire and retain excess iron? This simple question has been surprisingly complex to resolve, likely because there is more than one answer and because the requirement for iron may be both direct and indirect. Many iron-dependent proteins have been proposed as candidates for drivers of iron dependence. One of the earliest was ribonucleotide reductase, an iron-dependent enzyme that catalyzes the rate-limiting step in DNA synthesis and is required to sustain cancer cell proliferation. This remains a viable notion, and drugs that target this enzyme have been and remain in clinical use in the treatment of cancer patients (Huff et al. 2022). Other iron-containing proteins implicated in the malignant process include replicative DNA polymerases, helicases, primases, and DNA-modifying members of the Fe(II) 2-oxoglutarate-dependent enzyme family, such as histone demethylase JumonjiC domain-containing histone demethylases (Roatsch et al. 2019), and TET (ten-eleven translocation (TET) methylcytosine dioxygenases (Yang et al. 2024)). Iron has also been linked to the activation of a number of signaling pathways, including WNT, STAT3, EGFR, ERK 1/2, and Akt, as reviewed elsewhere (Torti and Torti 2020).

In addition to individual proteins and signaling pathways, metabolic processes linked to iron have been implicated in cancer. Mitochondria appear essential to cancer cell growth (Tan et al. 2015), and many of the numerous metabolic processes carried out by this organelle depend on iron, including iron-sulfur cluster biogenesis, heme synthesis, maintenance of redox balance, and oxidative phosphorylation (Fig. 18.2). Iron-sulfur clusters are present in numerous proteins, including enzymes involved in oxidative phosphorylation and DNA repair. Our group showed that SFXN4, an inner mitochondrial membrane protein upregulated in ovarian cancer cells and tissues, plays a role in iron-sulfur cluster biogenesis (Paul et al. 2019). Accordingly, knockdown

Table 18.1 Genes that appear in multiple prognostic "Iron gene signatures" derived by assessment of gene expression in human tissues from patients with a variety of malignancies

Gene	Number of signatures in which gene is present	Name	Function	Cancers in which reported	References
STEAP3	4	STEAP 3 metalloreductase	Iron and copper reductase involved in Tf-mediated iron uptake	Glioma; adrenocortical carcinoma	Weston et al. (2016), Zhang et al. (2024b), Zhu et al. (2019) and Miller et al. (2011)
FLVCR2	3	FLVCR choline and putative heme transporter 2	Heme transporter	Adrenocortical carcinoma, HCC, glioma	Zhu et al. (2019), Tang et al. (2020) and Xu et al. (2021)
LTF	3	Lactotransferrin	Iron binding protein involved in host defense	Breast, glioblastoma, thyroid carcinoma	Miller et al. (2011), Zhang et al. (2024a) and Jin et al. (2023)
NCOA4	3	Nuclear receptor coactivator 4	Receptor mediating autophagic ferritin degradation	Sarcoma, ovarian cancer, glioma	Xu et al. (2021), Wang et al. (2023a) and Li et al. (2020)
RRM2	3	Ribonucleotide reductase regulatory subunit M2	Iron-dependent reduction of NDP to dNDP	Ovarian cancer, glioma, HCC	Wang et al. (2023a), Zhu et al. (2024) and Xu et al. (2021)
SFXN1	3	Sideroflexin 1	Transport of alanine and serine into mitochondria; affects mitochondrial iron (Tifoun et al. 2021)	Breast, glioma, sarcoma	Miller et al. (2011), Weston et al. (2016) and Tang et al. (2020)
SLC48A1	3	Solute carrier family 48 member 1	Heme transmembrane transport	HCC sarcoma, colon	Tang et al. (2020), Yuan et al. (2021) and Li et al. (2020)
TFRC	3	Transferrin receptor	Transport of transferrin-bound iron	Breast, glioma, nasopharyngeal carcinoma	Miller et al. (2011), Weston et al. (2016) and Su et al. (2024)
ABCB6	2	ATP binding cassette subfamily B member 6	Porphyrin import	HCC	Tang et al. (2020) and Zhu et al. (2024)
CP	2	Ceruloplasmin	Oxidation of transferrin-bound iron	Adrenocortical carcinoma, thyroid carcinoma	Zhu et al. (2024) and Jin et al. (2023)
HFE	2	Hereditary hemochromatosis protein	Affects interaction of TfR with Tf	Breast, glioma	Miller et al. (2011) and Weston et al. (2016)

(continued)

Table 18.1 (continued)

Gene	Number of signatures in which gene is present	Name	Function	Cancers in which reported	References
HMOX1	2	Heme oxygenase 1	Heme catabolism	Glioblastoma, thyroid carcinoma	Zhang et al. (2024a) and Jin et al. (2023)
SCD5	2	Stearoyl-CoA desaturase 5	Formation of MUFAs from saturated FA; binds non-heme iron	Glioma, bladder	Xu et al. (2021) and Song et al. (2022)
SLC25A37	2	Solute carrier family 25 member 37	Mitochondrial iron transport (mitoferrin 1)	Breast, sarcoma	Miller et al. (2011) and Li et al. (2020)
SLC39A14	2	Solute carrier family 39 member 14	Transport of non-transferrin-bound iron (ZIP14)	Thyroid, nasopharyngeal	Jin et al. (2023) and Su et al. (2024)
SLC40A1	2	Solute carrier family 40 member 1	Cellular iron efflux (ferroportin)	Adrenocortical carcinoma, breast	Miller et al. (2011) and Zhu et al. (2019)
STEAP2	2	STEAP2 metalloreductase	Ferrireductase	Breast, thyroid	Miller et al. (2011) and Jin et al. (2023)
TMPRSS6	2	Transmembrane serine protease 6	Negative regulator of hepcidin	Glioma, thyroid	Weston et al. (2016) and Jin et al. (2023)
UROS	2	Uroporphyrinogen III synthase	Heme biosynthesis	Glioma	Weston et al. (2016) and Xu et al. (2021)

HCC hepatocellular carcinoma

of SFXN4 not only reduced respiration but also reduced levels of DNA repair and sensitized cells to chemotherapy in ovarian cancer models (Tesfay et al. 2022). Along similar lines, a mitochondrially targeted iron chelator inhibited iron-sulfur cluster as well as heme biogenesis in in vitro and in vivo models of breast cancer, leading to destabilization of proteins containing these prosthetic groups, inhibition of respiration, and inhibition of tumor growth and metastasis (Sandoval-Acuna et al. 2021). NEET proteins, including mitoNEET (CISD1), are a class of mitochondrial 2Fe-2S iron-sulfur proteins with a unique sequence motif and a labile and redox-active cluster. Although a unifying mechanism of action has not yet emerged (Mittler et al. 2019) NEET proteins affect iron/Fe-S homeostasis, protect OXPHOS proteins from oxidative stress, impact various signaling pathways, and have been implicated in tumor growth and metastasis in breast, cervical, and other cancers (Mittler et al. 2019). Ligands that stabilize the mitoNEET cluster under oxidative conditions reduce ovarian cancer cell proliferation, implying an important role for the lability of the cluster in mitoNEET function (Marjault et al. 2022).

In addition to examples of individual iron-sulfur cluster-containing proteins, genes encoding proteins involved in processes of iron-sulfur cluster biogenesis are frequently altered in cancer, leading to the suggestion that this may represent a critical target in cancer (Petronek et al. 2021). Intriguingly, a CRISPR-Cas9 screen revealed that genes required for the mitochondrial functions of iron-sulfur cluster biogenesis (e.g., NFU1) and oxidative phosphorylation (e.g., NDFUS1) facilitate survival in the acidic environment favored by tumors (Boedtkjer and Pedersen 2020), suggesting it may be possible to selectively target these processes in tumors despite their broad utilization in non-cancer cells (Michl et al. 2022).

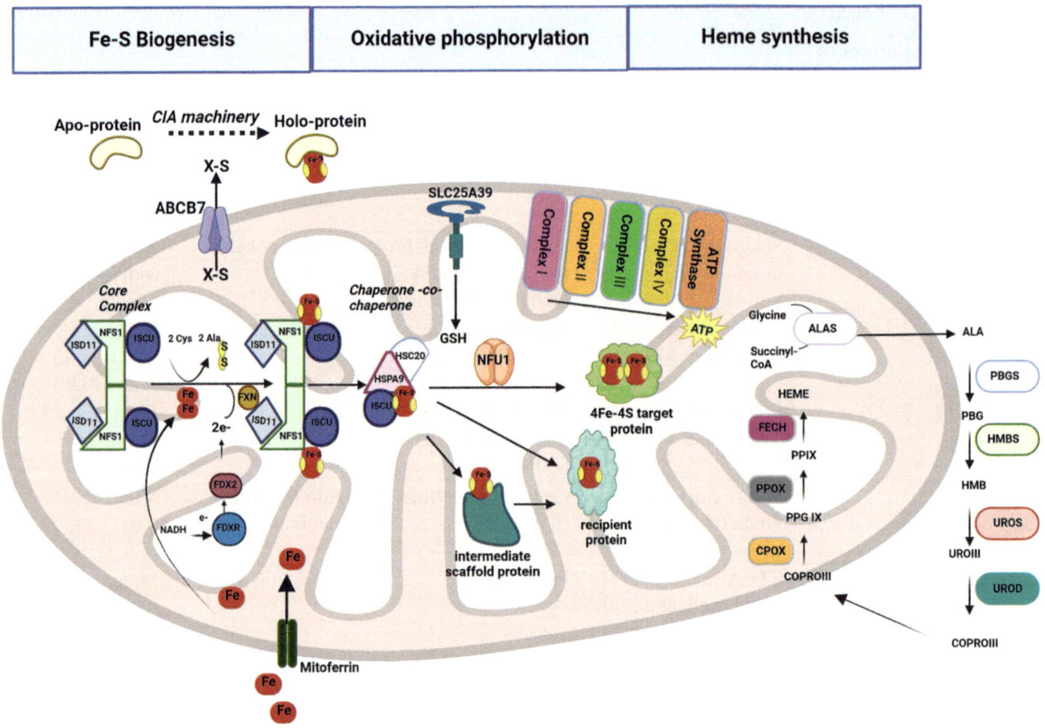

Fig. 18.2 Iron is required for multiple indispensable roles of mitochondria. *Iron-sulfur cluster biogenesis.* Mitochondria are required for the synthesis of iron-sulfur clusters, co-factors essential to the enzymatic activity of numerous proteins in the mitochondria, nucleus, and cytosol. Mitochondrial iron-sulfur cluster synthesis begins with a core complex consisting of cysteine desulfurase (NFS1), the scaffold protein ISCU, and regulatory proteins ISD11 and frataxin (FXN). NFS1 abstracts sulfur from cysteine for nascent Fe-S cluster assembly. Iron is transported into mitochondria by mitoferrin. Electrons provided by NADH, ferredoxin reductase (FDXR), and ferredoxin (FDX2), assisted by frataxin (FXN), facilitate assembly of the cluster. The nascent Fe-S cluster is then transferred to recipient proteins or intermediate carriers by a chaperone/co-chaperone (HSPA9/HSC20) system. Biogenesis of cytosolic Fe-S clusters involves the export of an uncharacterized sulfur-containing compound by ABCB7 (a step facilitated by glutathione (GSH)), transfer to the cytoplasmic Fe-S assembly machinery (CIA), and incorporation of clusters into target apoproteins. Mitochondria are essential to the synthesis of Fe-S clusters in all cellular compartments, including the cytosol and nucleus. *Oxidative phosphorylation.* OXPHOS is carried out in mitochondria to generate ATP; enzymatic complexes involved in this process (I-IV) require iron as Fe-S clusters or heme. *Heme synthesis.* Heme is a required cofactor for multiple enzymes. The first and last steps in heme synthesis occur in the mitochondria. Heme synthesis begins in the mitochondrial matrix with the condensation of glycine and succinyl CoA to form 5-aminolevulinic acid (ALA) by 5-aminolevulinate synthase (ALAS). ALA is exported to the cytosol, where it undergoes four steps of enzymatic modification to generate coproporphyrinogen III (COPROIII), which is then transported back to the cytosol to undergo the final three enzymatic reactions that produce heme. Abbreviations: *PBG* Porphobilinogen, *PBGS* Porphobilinogen synthase, *HMB* Hydroxymethylbilane, *HMBS* Hydroxymethylbilane synthase, *UROIII* Uroporphyrinogen III, *UROS* Uroporphyrinogen III synthase, *COPRO III* Coproporphyrinogen III, *UROD* Uroporphyrinogen decarboxylase, *PPG IX* Protoporphyrinogen IX, *CPOX* Coproporphyrinogen oxidase, *PP IX* Protoporphyrin IX, *PPOX* Protoporphyrinogen oxidase, *FECH* Ferrochelatase, *ABCB7* ATP Binding Cassette Subfamily B Member 7

Iron in iron-sulfur clusters can also play important regulatory roles in other metabolic processes. An interesting example is SLC25A39, a protein that was recently identified as a transporter of mitochondrial GSH required for proper mitochondrial respiration (Shi et al. 2022; Wang et al. 2021). The stability of this protein depends on an iron-sulfur cluster that facilitates its interaction with AFG3L2 (Shi et al. 2024; Liu et al. 2023b), a protease that degrades SLC25A39 (Shi et al. 2024; Liu et al. 2023b). Overexpression of mitochondrial iron transporters mitoferrin1 or mitoferrin2 stabilized SLC25A39 iron-sulfur cluster content, increased SLC25A39 degradation, and decreased mitochondrial GSH, thus linking iron metabolism and maintenance of mitochondrial GSH (Shi et al. 2024; Liu et al. 2023b). Upregulation of SLC25A39 promoted cell growth and metastasis in colorectal cancer models (Zhang et al. 2024b), suggesting that iron may be utilized by cancer cells to regulate mitochondrial GSH as well as respiration. These recent observations illustrate the indirect and complex way in which iron can affect cellular function.

18.6 Iron and the Tumor Microenvironment

In vivo, cancer cells inhabit a microenvironment composed of many different cell types, each of which can have positive or negative effects on tumor growth. A growing number of studies on the role of iron in cancer have begun to focus on cells that constitute the tumor microenvironment (Fig. 18.3). Such studies have examined not only the "iron phenotype" of the cellular microenvironment (i.e., expression of genes, proteins, and metabolites important to iron metabolism as well as iron content itself) but also how cells in the microenvironment interact with cancer cells to either provide or compete for iron. Although multiple cell types are present in the tumor microenvironment, including endothelial cells, fibroblasts, adipocytes, and immune cells, here we briefly examine tumor-associated fibroblasts (also termed cancer-associated fibroblasts) and cells of the immune system as examples.

Cancer-Associated Fibroblasts (CAFs) Our group found that cancer-associated fibroblasts stimulate hepcidin synthesis in breast cancer cells grown in 3D spheroid culture, reducing levels of the iron efflux pump ferroportin in cancer cells and potentially contributing to increased cancer cell iron retention (Blanchette-Farra et al. 2018). A complementary mechanism was suggested in pancreatic cancer cells, which depend on ferritinophagy to provide sufficient iron to sustain iron-sulfur cluster biogenesis and mitochondrial function (Mukhopadhyay et al. 2023; Santana-Codina et al. 2022). Co-culture experiments revealed that this requirement can be overridden by co-culture of pancreatic cancer cells with CAFS. In this scenario, elevated levels of ferroportin in CAFS enable them to efflux and supply iron to pancreatic cancer cells under conditions in which cancer cell ferritinophagy is blocked (Mukhopadhyay et al. 2023). Accordingly, restriction of dietary iron synergizes with autophagy inhibition in reducing the growth of pancreatic tumors (Mukhopadhyay et al. 2023).

Fibroblasts in the tumor microenvironment also contribute to the production of a collagenous matrix that stiffens the stroma and activates tumor cell invasion, a process facilitated by iron-dependent lysyl hydroxylase (Guo et al. 2018). The matrix may also contribute to cancer invasiveness by influencing the iron regulatory state of tumor cells. Thus, sequencing analysis of triple-negative breast cancer (TNBC) cells implanted in a dense collagen matrix mimicking the extracellular tumor microenvironment revealed an induction of heme oxygenase 1 and polarization of cancer cells into a non-invasive state, characterized by expression of genes associated with iron sequestration, or an invasive state, characterized by upregulation of genes involved in iron uptake such as *TFRC* and *SLC39A14* (Zip14). These alternative iron regulatory states were associated with patient survival outcomes (Leineweber et al. 2024).

Immune Cells Tumors are invested with multiple different immune cell types, which can exert either pro- or anti-tumor effects. These include innate immune cells such as macrophages, NK cells, neutrophils, and dendritic cells, as well as adaptive B and T cells. Many immune cells influence iron pathways in tumor cells (Zhang et al. 2024c). For example, alternatively activated mac-

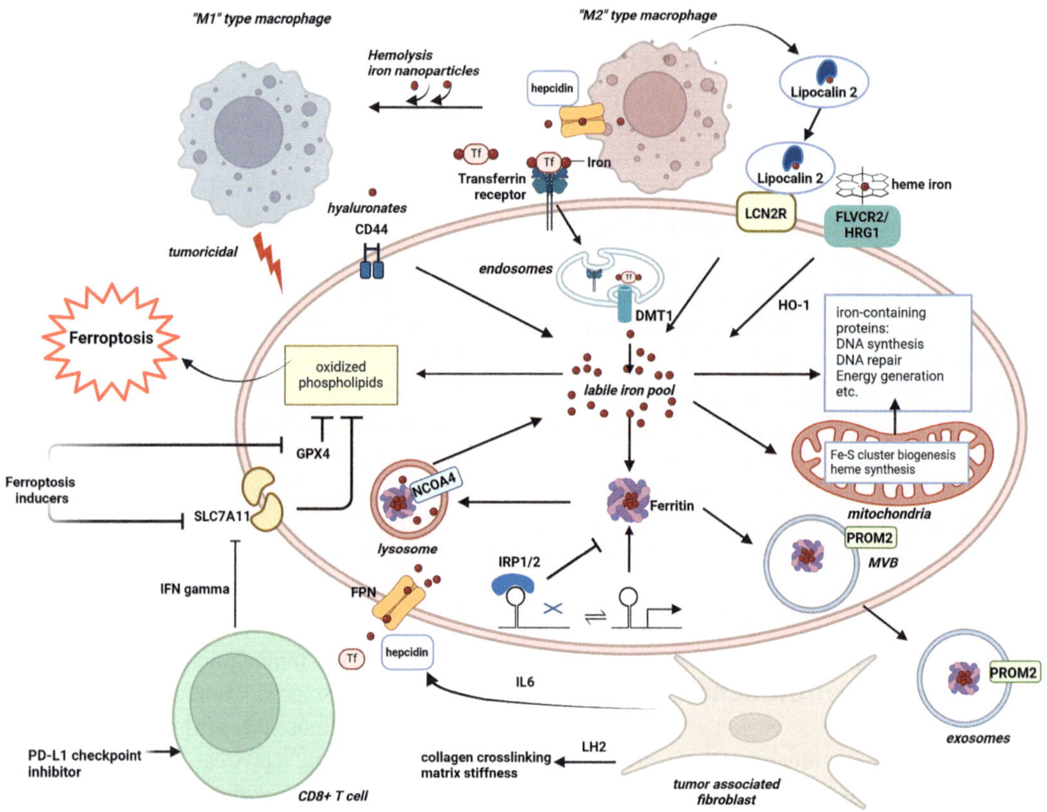

Fig. 18.3 Iron in the interaction between cancer cells and cells of the tumor microenvironment. Iron is required for the formation of oxidized lipids that characterize ferroptosis, an iron-dependent form of cell death. Ferroptosis can be induced by agents that inhibit GPX4 (glutathione peroxidase 4), an enzyme that reduces oxidized lipids, or by agents that inhibit SLC7A11, which transports cystine to produce intracellular glutathione, a cellular reductant and GPX4 co-factor. In response to treatment with checkpoint inhibitors, CD8+T cells in the tumor microenvironment secrete interferon gamma, which inhibits SLC7A11 and facilitates ferroptosis. Other cells in the microenvironment include tumor-associated fibroblasts, which produce LH2 (lysyl hydroxylase 2), an enzyme involved in collagen crosslinking and matrix stiffness, as well as IL6 (interleukin 6), a cytokine that induces hepcidin and thereby blocks ferroportin-mediated iron egress. Tumor-associated macrophages exhibit multiple phenotypes and plasticity among these phenotypes. In this simplified depiction, classically activated (depicted here as "M1" type macrophages) act to foster tumor cell cytolysis, whereas alternatively activated "M2" type macrophages support tumor cell growth by providing iron for tumor cell uptake. Excess iron provided by hemolysis or synthetic iron nanoparticles can foster the transition from M2 to M1-type macrophages, thus enhancing tumor cell killing

rophages ("M2-like") can deliver iron to tumor cells in vitro (Recalcati et al. 2010, 2019). Iron-laden tumor-associated macrophages were observed in breast tumors (Marques et al. 2016) and their depletion was associated with tumor reduction in animal models (Leftin et al. 2019). Both ferroportin (Recalcati et al. 2010, 2019) and lipocalin (Duan et al. 2018) have been implicated in mechanisms of iron delivery from macrophages to tumor cells. In sarcomas, an iron-rich subset of macrophages characterized by high iron content and a unique genetic signature was induced by heme and contributed to tumor growth and vascular density (Folkert et al. 2024). In non-small cell lung cancer, single-cell transcriptomics revealed a transcriptional reprogramming of macrophages in tumors that promoted iron efflux within the tumor microenvironment (De Zuani et al. 2024). In contrast, in hepatocellular carcinoma, reduced levels of iron in TAMs were observed and attributed to TFRC-mediated iron uptake by tumor cells (Sun et al. 2021). Iron provided by hemolysis or synthetic iron-containing nanoparticles can also directly polarize myeloid cells to a proinflammatory, tumor-suppressive phenotype (Ludwig et al. 2024), as well as modulate cell plasticity (Muller et al. 2024), lending further nuance to the conversation between tumor and myeloid cells in the tumor microenvironment. Cytokines released by T cells of the adaptive immune system, in particular CTLs activated by immune checkpoint blockade, can both promote and inhibit ferroptosis of cancer cells (Zhang et al. 2024c) (Fig. 18.3). Overall, immune cells play an active and significant role in tumor iron biology that is nevertheless complex, nuanced, and may differ among tumor types.

18.7 Dietary Iron and Cancer Risk

Another prediction of a link between iron and cancer is that levels of dietary iron in patient populations would affect cancer risk. There have been many attempts to address this prediction using epidemiological tools, but they have produced a stubbornly mixed picture, with some studies pointing to positive associations and others finding no correlation or even a protective effect of iron (reviewed in Zeidan et al. 2024; Charlebois and Pantopoulos 2023). The form of dietary iron (heme vs. non-heme iron) may be important, with a trend toward a more positive association seen between heme iron intake and cancer risk. For example, in a meta-analysis of 23 studies, high dietary heme iron significantly increased the risk of breast cancer (RR 1.12, 95% CI:1.04–1.22), whereas total iron intake did not (1.01, 95% CI: 0.89–1.15) (Chang et al. 2019). Similarly, in a meta-analysis of 20 studies of esophageal cancer, a 1 mg/day increase in heme iron intake was associated with a 21% increase in cancer risk (OR = 1.21, 95% CI: 1.02–1.45), whereas total iron intake reduced risk (OR = 0.81, 95% CI: 0.70–0.94) (Ma et al. 2018). Heme iron intake also increased the risk of lung cancer (Ward et al. 2019) in the European Prospective Investigation into Cancer and Nutrition (EPIC) study of 16,746 individuals from 10 countries (HR 1.16, 95% CI 1.02–1.32, highest versus lowest quintile), whereas non-heme iron intake did not. However, red and processed meats are the most prevalent dietary forms of heme, and it is difficult to dissociate the contribution of heme iron from other potentially cancer-promoting components of these foods. For example, in a study of meat consumption of over 175,000 US men, a heightened risk of prostate cancer was associated with increased heme iron intake (HR = 1.09 (95% CI: 1.02, 1.17) (Sinha et al. 2009)). However, prostate cancer risk was also positively associated with other aspects of dietary red meat, including cooking method (grilling/barbecuing), nitrite/nitrates, and benzo[a]pyrene. Other complexities associated with assessing the contribution of dietary iron to cancer risk include the influence of sex and variations among different malignancies (Zeidan et al. 2024). Disentangling these variables remains a challenge in assessing the role of dietary iron in cancer risk.

A related approach to assessing the relationship between body iron and cancer risk has been to evaluate cancer risk in patients with hemochromatosis, a disorder that results in iron overload, particularly in the liver (see Chap. 9). Although the increased risk of untreated hemochromatotic individuals for liver cancer has been known for decades (Niederau et al. 1985), whether hemochromatosis increases overall cancer risk is less clear. A study of approximately 28,000 individuals in the Melbourne Collaborative Cohort Study identified subjects with HFE C282Y homozygous mutations as having twice the risk of colorectal and breast cancer (Osborne et al. 2010) when compared to individuals without the mutation. However, a recent study that evaluated this question using a large community sample of more than 450,000 individuals of European ancestry in Great Britain (Atkins et al. 2022) provided suggestive evidence of increased prostate cancer incidence (14.4% of male p. C282Y homozygotes projected to develop prostate cancer versus 10.7% without mutations, 95% CI, 1.3–6.8), but no evidence of excess risk in other cancers.

The "flip side of the coin" (Aksan et al. 2021) has also been considered: i.e., whether iron *insufficiency* may contribute to cancer (Phipps et al. 2021; Aksan et al. 2021). Supporting this possibility, in a population-based study in Taiwan, an approximately twofold increased incidence of gastrointestinal and other cancers in patients with iron deficiency anemia was observed (Hung et al. 2015). In another population-based cohort study, the proportion of men and postmenopausal women developing GI malignancy was 31 times greater (95% confidence interval 9–107) in those with iron deficiency anemia compared with those with normal iron levels (Ioannou et al. 2002). These results are congruent with a study demonstrating a U-shaped association between iron intake and the development of adenomatous polyps (a precursor lesion to colorectal cancer), which demonstrated that consumption of either high or low amounts of iron led to an increased risk of developing polyps relative to iron-adequate diets (Bird et al. 1996). To explain the apparent ability of iron insufficiency to increase cancer incidence, it has been argued that iron is required for proper immunosurveillance, and that lack of iron may therefore create a more permissive environment for cancer initiation and progression (Phipps et al. 2021). Overall, these demonstrations of the potentially dual role of iron excess and deficiency in alternative facets of cancer risk are both a reminder of the need to consider the critical role of iron in the tumor microenvironment and a potential explanation for the mixed picture that sometimes emerges when evaluating iron and cancer risk in patient populations.

18.8 Targeting Iron in Cancer Therapy

The long-standing appreciation of links between iron and cancer has been accompanied by an equally long-standing effort to leverage iron in cancer therapy. Two opposing approaches have been taken: to inhibit tumor growth and metastases by depriving tumors of this essential nutrient or, conversely, to provide tumors with sufficient excess iron to drive them into iron-dependent cell death pathways.

Use of iron chelators has the longest history in the quest to manipulate iron to target cancer. Iron chelators are small molecules that selectively bind iron; several of these have an established safety record by virtue of their use in treating iron overload disorders in human patients (Chap. 23). Although chelator-mediated iron depletion has met with some efficacy in cancer patients using the iron chelator desferoxamine, responses were modest (Yamasaki et al. 2011), and problems of selective delivery to tumors without disruption of systemic iron homeostasis have yet to be overcome. An alternative strategy has been to design iron chelators that do not produce a chemically inert complex of iron but rather permit bound iron to redox cycle. Well-studied examples are the thiosemicarbazones such as Dp44mT and triapine (Jansson et al. 2010; Lovejoy et al. 2011), which also inhibit the essential iron-dependent enzyme ribonucleotide reductase. Dp44mT has the additional unanticipated property of inducing the metastasis suppressor NDRG1 (Wijesinghe et al. 2021). Newly designed derivatives of the Dp44mT chelator family continue to be actively investigated for their anti-cancer properties (Dharmasivam et al. 2023), and a Phase III trial of oral triapine with chemoradiation is currently underway for advanced stage cervical and vaginal cancers (NCT02466971) based on encouraging results obtained in Phase I and Phase II trials (Taylor et al. 2024; Kunos et al. 2019).

The discovery of ferroptosis and the susceptibility of cancer cells to this mode of cell death has prompted a vigorous effort to identify agents that increase tumor cell iron, foster ferroptosis-mediated cell death, and thus act as anti-tumor agents (Basuli et al. 2017). Iron-containing nanoparticles are a good example of the new candidate drugs being developed to exploit this vulnerability (Wang et al. 2023b). Along these lines, preclinical studies with ferumoxytol, a clinically approved iron supplement, showed substantial promise, significantly reducing disease burden and prolonging survival in a murine leukemia model (Su et al. 2020; Trujillo-Alonso et al. 2019). Similarly, carbon-nanoparticle-loaded iron (CNSI-Fe(II)) is currently being explored in a Phase I study of patients with solid tumors, particularly those with *KRAS* mutations (clinicaltrials.gov NCT06048367).

Approaches that target iron indirectly are also being explored. One is to use iron markers overexpressed in cancer cells, such as TFRC (TfR1), either as direct targets of anti-TfR1 antibodies (Daniels-Wells et al. 2020, 2024) or as targets for the preferential delivery of cytotoxic agents to tumor cells via conjugation to its ligand TF (Candelaria et al. 2021). However, this approach has been hindered by the presence of TF receptors on non-cancer cells. Gallium is an iron mimetic that can replace iron in key proteins such as iron-dependent mitochondrial enzymes and ribonucleotide reductase to generate nonfunctional catalytic centers (Chitambar 2018). Gallium maltolate, a contemporary version of gallium nitrate, the form of gallium originally used in this approach, has recently demonstrated promising anti-cancer activity in experimental models of glioblastoma (Al-Gizawiy et al. 2023). In addition to their potential use as direct tumoricidal agents, drugs that target iron metabolism may also be useful in combination chemotherapy. For example, as discussed above, iron-sulfur clusters are required for the function of DNA repair enzymes (Gari et al. 2012; Stehling et al. 2012). Since blocking DNA repair increases sensitivity to DNA-damaging drugs, impeding iron metabolism to block the formation of iron-sulfur clusters may increase the activity of DNA-damaging agents used in conventional chemotherapy (Tesfay et al. 2022).

18.9 Conclusions

Decades of work firmly support the core concept that iron plays a vital and potentially targetable role in cancer. We now understand at a much more granular (and hence more specifically targetable) level the complex interplay between iron and cancer cell metabolism. The continuing discovery of new players in cancer cell iron uptake and handling, such as CD44 and NCOA4, as well as the discovery of new roles for iron in controlling processes as diverse as ferroptotic cell death and mitochondrial glutathione, have further illuminated (and may be expected to continue to illuminate) the multifaceted role of iron in the tumor cell and its environment. These discoveries have already offered new opportunities for therapeutic intervention. However, with the discovery of each new iron-dependent protein or process comes additional complexity. Disentangling the contribution of each of these elements to patient outcome presents a challenge that will ultimately have to be overcome to optimize and tailor therapies to target iron in cancer.

Acknowledgements Supported in part by NCI grants R01 CA188025 and R01 CA233636. The authors declare no conflict of interest. AI was not used in generating this article.

References

Aksan A, Farrag K, Aksan S, Schroeder O, Stein J (2021) Flipside of the coin: iron deficiency and colorectal cancer. Front Immunol 12:635899. https://doi.org/10.3389/fimmu.2021.635899

Al-Gizawiy MM, Wujek RT, Alhajala HS, Cobb JM, Prah MA, Doan NB, Connelly JM, Chitambar CR, Schmainda KM (2023) Potent in vivo efficacy of oral gallium maltolate in treatment-resistant glioblastoma. Front Oncol 13:1278157. https://doi.org/10.3389/fonc.2023.1278157

Arosio P, Cairo G, Bou-Abdallah F (2024) A brief history of ferritin, an ancient and versatile protein. Int J Mol Sci 26(1). https://doi.org/10.3390/ijms26010206

Atkins JL, Pilling LC, Torti SV, Torti FM, Kuchel GA, Melzer D (2022) Hereditary hemochromatosis variant associations with incident nonliver malignancies: 11-year follow-up in UK biobank. Cancer Epidemiol Biomarkers Prev 31(9):1780–1787. https://doi.org/10.1158/1055-9965.EPI-22-0284

Bao G, Clifton M, Hoette TM, Mori K, Deng SX, Qiu A, Viltard M, Williams D, Paragas N, Leete T, Kulkarni R, Li X, Lee B, Kalandadze A, Ratner AJ, Pizarro JC, Schmidt-Ott KM, Landry DW, Raymond KN, Strong RK, Barasch J (2010) Iron traffics in circulation bound to a siderocalin (Ngal)-catechol complex. Nat Chem Biol 6(8):602–609. https://doi.org/10.1038/nchembio.402

Basuli D, Tesfay L, Deng Z, Paul B, Yamamoto Y, Ning G, Xian W, McKeon F, Lynch M, Crum CP, Hegde P, Brewer M, Wang X, Miller LD, Dyment N, Torti FM, Torti SV (2017) Iron addiction: a novel therapeutic target in ovarian cancer. Oncogene 36(29):4089–4099. https://doi.org/10.1038/onc.2017.11

Berndt C, Alborzinia H, Amen VS, Ayton S, Barayeu U, Bartelt A, Bayir H, Bebber CM, Birsoy K, Bottcher JP, Brabletz S, Brabletz T, Brown AR, Brune B, Bulli G, Bruneau A, Chen Q, DeNicola GM, Dick TP, Distefano A, Dixon SJ, Engler JB, Esser-von Bieren J, Fedorova M, Friedmann Angeli JP, Friese MA, Fuhrmann DC, Garcia-Saez AJ, Garbowicz K, Gotz M, Gu W, Hammerich L, Hassannia B, Jiang X, Jeridi A, Kang YP, Kagan VE, Konrad DB, Kotschi S, Lei P, Le Tertre M, Lev S, Liang D, Linkermann A, Lohr C, Lorenz S, Luedde T, Methner A, Michalke B, Milton AV, Min J, Mishima E, Muller S, Motohashi H, Muckenthaler MU, Murakami S, Olzmann JA, Pagnussat G, Pan Z, Papagiannakopoulos T, Pedrera Puentes L, Pratt DA, Proneth B, Ramsauer L, Rodriguez R, Saito Y, Schmidt F, Schmitt C, Schulze A, Schwab A, Schwantes A, Soula M, Spitzlberger B, Stockwell BR, Thewes L, Thorn-Seshold O, Toyokuni S, Tonnus W, Trumpp A, Vandenabeele P, Vanden Berghe T, Venkataramani V, Vogel FCE, von Karstedt S, Wang F, Westermann F, Wientjens C, Wilhelm C, Wolk M, Wu K, Yang X, Yu F, Zou Y, Conrad M (2024) Ferroptosis in health and disease. Redox Biol 75:103211. https://doi.org/10.1016/j.redox.2024.103211

Bird CL, Witte JS, Swendseid ME, Shikany JM, Hunt IF, Frankl HD, Lee ER, Longnecker MP, Haile RW (1996) Plasma ferritin, iron intake, and the risk of colorectal polyps. Am J Epidemiol 144(1):34–41. https://doi.org/10.1093/oxfordjournals.aje.a008852

Blanchette NL, Manz DH, Torti FM, Torti SV (2016) Modulation of hepcidin to treat iron deregulation: potential clinical applications. Expert Rev Hematol 9(2):169–186. https://doi.org/10.1586/17474086.2016.1124757

Blanchette-Farra N, Kita D, Konstorum A, Tesfay L, Lemler D, Hegde P, Claffey KP, Torti FM, Torti SV (2018) Contribution of three-dimensional architecture and tumor-associated fibroblasts to hepcidin regulation in breast cancer. Oncogene 37(29):4013–4032. https://doi.org/10.1038/s41388-018-0243-y

Boedtkjer E, Pedersen SF (2020) The acidic tumor microenvironment as a driver of cancer. Annu Rev Physiol 82:103–126. https://doi.org/10.1146/annurev-physiol-021119-034627

Brown CW, Amante JJ, Chhoy P, Elaimy AL, Liu H, Zhu LJ, Baer CE, Dixon SJ, Mercurio AM (2019) Prominin2 drives ferroptosis resistance by stimulating iron export. Dev Cell 51(5):575–586.e574. https://doi.org/10.1016/j.devcel.2019.10.007

Campbell JA (1940) Effects of precipitated silica and of iron oxide on the incidence of primary lung tumours in mice. Br Med J 2(4156):275–280. https://doi.org/10.1136/bmj.2.4156.275

Candelaria PV, Leoh LS, Penichet ML, Daniels-Wells TR (2021) Antibodies targeting the transferrin receptor 1 (TfR1) as direct anti-cancer agents. Front Immunol 12:607692. https://doi.org/10.3389/fimmu.2021.607692

Chang VC, Cotterchio M, Khoo E (2019) Iron intake, body iron status, and risk of breast cancer: a systematic review and meta-analysis. BMC Cancer 19(1):543. https://doi.org/10.1186/s12885-019-5642-0

Charlebois E, Pantopoulos K (2023) Nutritional aspects of iron in health and disease. Nutrients 15(11). https://doi.org/10.3390/nu15112441

Chi Y, Remsik J, Kiseliovas V, Derderian C, Sener U, Alghader M, Saadeh F, Nikishina K, Bale T, Iacobuzio-Donahue C, Thomas T, Pe'er D, Mazutis L, Boire A (2020) Cancer cells deploy lipocalin-2 to collect limiting iron in leptomeningeal metastasis. Science 369(6501):276–282. https://doi.org/10.1126/science.aaz2193

Chitambar CR (2018) Gallium complexes as anticancer drugs. Met Ions Life Sci 18. https://doi.org/10.1515/9783110470734-016

Coffman LG, Parsonage D, D'Agostino R Jr, Torti FM, Torti SV (2009) Regulatory effects of ferritin on angiogenesis. Proc Natl Acad Sci USA 106(2):570–575. https://doi.org/10.1073/pnas.0812010106

Daniels-Wells TR, Candelaria PV, Leoh LS, Nava M, Martinez-Maza O, Penichet ML (2020) An IgG1 version of the anti-transferrin receptor 1 antibody ch128.1 shows significant antitumor activity against different xenograft models of multiple myeloma: a brief communication. J Immunother 43(2):48–52. https://doi.org/10.1097/cji.0000000000000304

Daniels-Wells TR, Candelaria PV, Rodriguez AM, Poul MA, Penichet ML (2024) hu128.1, a humanized antibody targeting transferrin receptor 1, blocks erythroleukemia growth in xenograft mouse models. Haematologica. https://doi.org/10.3324/haematol.2024.286518

De Zuani M, Xue H, Park JS, Dentro SC, Seferbekova Z, Tessier J, Curras-Alonso S, Hadjipanayis A, Athanasiadis EI, Gerstung M, Bayraktar O, Cvejic A (2024) Single-cell and spatial transcriptomics analysis of non-small cell lung cancer. Nat Commun 15(1):4388. https://doi.org/10.1038/s41467-024-48700-8

Devireddy LR, Gazin C, Zhu X, Green MR (2005) A cell-surface receptor for lipocalin 24p3 selectively mediates apoptosis and iron uptake. Cell 123(7):1293–1305. https://doi.org/10.1016/j.cell.2005.10.027

Dharmasivam M, Kaya B, Wijesinghe T, Gholam Azad M, Gonzalvez MA, Hussaini M, Chekmarev J, Bernhardt PV, Richardson DR (2023) Designing tailored thiosemicarbazones with bespoke properties: the styrene moiety imparts potent activity, inhibits heme center oxidation, and results in a novel "stealth zinc(II) complex". J Med Chem 66(2):1426–1453. https://doi.org/10.1021/acs.jmedchem.2c01600

Dixon SJ, Lemberg KM, Lamprecht MR, Skouta R, Zaitsev EM, Gleason CE, Patel DN, Bauer AJ, Cantley AM, Yang WS, Morrison B 3rd, Stockwell BR (2012) Ferroptosis: an iron-dependent form of nonapoptotic cell death. Cell 149(5):1060–1072. https://doi.org/10.1016/j.cell.2012.03.042

Dowdle WE, Nyfeler B, Nagel J, Elling RA, Liu S, Triantafellow E, Menon S, Wang Z, Honda A, Pardee G, Cantwell J, Luu C, Cornella-Taracido I, Harrington E, Fekkes P, Lei H, Fang Q, Digan ME, Burdick D,

Powers AF, Helliwell SB, D'Aquin S, Bastien J, Wang H, Wiederschain D, Kuerth J, Bergman P, Schwalb D, Thomas J, Ugwonali S, Harbinski F, Tallarico J, Wilson CJ, Myer VE, Porter JA, Bussiere DE, Finan PM, Labow MA, Mao X, Hamann LG, Manning BD, Valdez RA, Nicholson T, Schirle M, Knapp MS, Keaney EP, Murphy LO (2014) Selective VPS34 inhibitor blocks autophagy and uncovers a role for NCOA4 in ferritin degradation and iron homeostasis in vivo. Nat Cell Biol 16(11):1069–1079. https://doi.org/10.1038/ncb3053

Duan X, He K, Li J, Cheng M, Song H, Liu J, Liu P (2018) Tumor associated macrophages deliver iron to tumor cells via Lcn2. Int J Physiol Pathophysiol Pharmacol 10(2):105–114

Estrov Z, Tawa A, Wang XH, Dube ID, Sulh H, Cohen A, Gelfand EW, Freedman MH (1987) In vitro and in vivo effects of deferoxamine in neonatal acute leukemia. Blood 69(3):757–761

Folkert IW, Molina Arocho WA, To TKJ, Devalaraja S, Molina IS, Shoush J, Mohei H, Zhai L, Akhtar MN, Kochat V, Arslan E, Lazar AJ, Wani K, Israel WP, Zhang Z, Chaluvadi VS, Norgard RJ, Liu Y, Fuller AM, Dang MT, Roses RE, Karakousis GC, Miura JT, Fraker DL, Eisinger-Mathason TSK, Simon MC, Weber K, Tan K, Fan Y, Rai K, Haldar M (2024) An iron-rich subset of macrophages promotes tumor growth through a Bach1-Ednrb axis. J Exp Med 221(10). https://doi.org/10.1084/jem.20230420

Galy B, Conrad M, Muckenthaler M (2024) Mechanisms controlling cellular and systemic iron homeostasis. Nat Rev Mol Cell Biol 25(2):133–155. https://doi.org/10.1038/s41580-023-00648-1

Gari K, Leon Ortiz AM, Borel V, Flynn H, Skehel JM, Boulton SJ (2012) MMS19 links cytoplasmic iron-sulfur cluster assembly to DNA metabolism. Science 337(6091):243–245. https://doi.org/10.1126/science.1219664

Guo HF, Tsai CL, Terajima M, Tan X, Banerjee P, Miller MD, Liu X, Yu J, Byemerwa J, Alvarado S, Kaoud TS, Dalby KN, Bota-Rabassedas N, Chen Y, Yamauchi M, Tainer JA, Phillips GN Jr, Kurie JM (2018) Pro-metastatic collagen lysyl hydroxylase dimer assemblies stabilized by Fe(2+)-binding. Nat Commun 9(1):512. https://doi.org/10.1038/s41467-018-02859-z

Hann HW, Stahlhut MW, Blumberg BS (1988) Iron nutrition and tumor growth: decreased tumor growth in iron-deficient mice. Cancer Res 48(15):4168–4170

Hann HW, Stahlhut MW, Menduke H (1991) Iron enhances tumor growth. Observation on spontaneous mammary tumors in mice. Cancer 68(11):2407–2410. https://doi.org/10.1002/1097-0142(19911201)68:11<2407::aid-cncr2820681113>3.0.co;2-n

Hann HW, Stahlhut MW, Rubin R, Maddrey WC (1992) Antitumor effect of deferoxamine on human hepatocellular carcinoma growing in athymic nude mice. Cancer 70(8):2051–2056. https://doi.org/10.1002/1097-0142(19921015)70:8<2051::aid-cncr2820700806>3.0.co;2-1

Henry WS, Müller S, Yang J-S, Innes-Gold S, Das S, Reinhardt F, Sigmund K, Phadnis VV, Wan Z, Eaton E, Sampaio JL, Bell GW, Viravalli A, Hammond PT, Kamm RD, Cohen AE, Boehnke N, Hsu VW, Levental KR, Rodriguez R, Weinberg RA (2024) Ether lipids influence cancer cell fate by modulating iron uptake. bioRxiv:2024.2003.2020.585922. https://doi.org/10.1101/2024.03.20.585922

Huff SE, Winter JM, Dealwis CG (2022) Inhibitors of the cancer target ribonucleotide reductase, past and present. Biomol Ther 12(6). https://doi.org/10.3390/biom12060815

Hung N, Shen CC, Hu YW, Hu LY, Yeh CM, Teng CJ, Kuan AS, Chen SC, Chen TJ, Liu CJ (2015) Risk of cancer in patients with iron deficiency anemia: a nationwide population-based study. PLoS One 10(3):e0119647. https://doi.org/10.1371/journal.pone.0119647

Ioannou GN, Rockey DC, Bryson CL, Weiss NS (2002) Iron deficiency and gastrointestinal malignancy: a population-based cohort study. Am J Med 113(4):276–280. https://doi.org/10.1016/s0002-9343(02)01214-7

Jansson PJ, Hawkins CL, Lovejoy DB, Richardson DR (2010) The iron complex of Dp44mT is redox-active and induces hydroxyl radical formation: an EPR study. J Inorg Biochem 104(11):1224–1228. https://doi.org/10.1016/j.jinorgbio.2010.07.012

Jin T, Ge L, Chen J, Wang W, Zhang L, Ge M (2023) Identification of iron metabolism-related genes as prognostic indicators for papillary thyroid carcinoma: a retrospective study. PeerJ 11:e15592. https://doi.org/10.7717/peerj.15592

Katsarou A, Pantopoulos K (2020) Basics and principles of cellular and systemic iron homeostasis. Mol Asp Med 75:100866. https://doi.org/10.1016/j.mam.2020.100866

Kunos CA, Andrews SJ, Moore KN, Chon HS, Ivy SP (2019) Randomized phase II trial of triapine-cisplatin-radiotherapy for locally advanced stage uterine cervix or vaginal cancers. Front Oncol 9:1067. https://doi.org/10.3389/fonc.2019.01067

Larrue C, Mouche S, Angelino P, Sajot M, Birsen R, Kosmider O, McKee T, Vergez F, Recher C, Mas VM, Gu Q, Xu J, Tsantoulis P, Sarry JE, Tamburini J (2024) Targeting ferritinophagy impairs quiescent cancer stem cells in acute myeloid leukemia in vitro and in vivo models. Sci Transl Med 16(757):eadk1731. https://doi.org/10.1126/scitranslmed.adk1731

Leftin A, Ben-Chetrit N, Joyce JA, Koutcher JA (2019) Imaging endogenous macrophage iron deposits reveals a metabolic biomarker of polarized tumor macrophage infiltration and response to CSF1R breast cancer immunotherapy. Sci Rep 9(1):857. https://doi.org/10.1038/s41598-018-37408-7

Leineweber WD, Rowell MZ, Ranamukhaarachchi SK, Walker A, Li Y, Villazon J, Mestre-Farrera A, Hu Z, Yang J, Shi L, Fraley SI (2024) Divergent iron regulatory states contribute to heterogeneity in breast cancer aggressiveness. iScience 27(9):110661. https://doi.org/10.1016/j.isci.2024.110661

Li J, Hu C, Du Y, Tang X, Shao C, Xu T, Zhao Z, Hu H, Sheng Y, Guo J, Xi Y (2020) Identification of iron metabolism-related gene signatures for predicting the prognosis of patients with sarcomas. Front Oncol 10:599816. https://doi.org/10.3389/fonc.2020.599816

Liu X, Ren J, Zhou R, Wen Z, Wen Z, Chen Z, He S, Zhang H (2023a) Construction of iron metabolism-related prognostic features of gastric cancer based on RNA sequencing and TCGA database. BMC Cancer 23(1):1106. https://doi.org/10.1186/s12885-023-11569-9

Liu Y, Liu S, Tomar A, Yen FS, Unlu G, Ropek N, Weber RA, Wang Y, Khan A, Gad M, Peng J, Terzi E, Alwaseem H, Pagano AE, Heissel S, Molina H, Allwein B, Kenny TC, Possemato RL, Zhao L, Hite RK, Vinogradova EV, Mansy SS, Birsoy K (2023b) Autoregulatory control of mitochondrial glutathione homeostasis. Science 382(6672):820–828. https://doi.org/10.1126/science.adf4154

Lovejoy DB, Jansson PJ, Brunk UT, Wong J, Ponka P, Richardson DR (2011) Antitumor activity of metal-chelating compound Dp44mT is mediated by formation of a redox-active copper complex that accumulates in lysosomes. Cancer Res 71(17):5871–5880. https://doi.org/10.1158/0008-5472.CAN-11-1218

Ludwig N, Cucinelli S, Hametner S, Muckenthaler MU, Schirmer L (2024) Iron scavenging and myeloid cell polarization. Trends Immunol 45(8):625–638. https://doi.org/10.1016/j.it.2024.06.006

Ma J, Li Q, Fang X, Chen L, Qiang Y, Wang J, Wang Q, Min J, Zhang S, Wang F (2018) Increased total iron and zinc intake and lower heme iron intake reduce the risk of esophageal cancer: a dose-response meta-analysis. Nutr Res 59:16–28. https://doi.org/10.1016/j.nutres.2018.07.007

Mancias JD, Wang X, Gygi SP, Harper JW, Kimmelman AC (2014) Quantitative proteomics identifies NCOA4 as the cargo receptor mediating ferritinophagy. Nature 509(7498):105–109. https://doi.org/10.1038/nature13148

Marjault HB, Yang-Sung S, Zuo K, Carloni P, Mittler R, Nechushtai R (2022) Structure-based screening reveals a ligand that stabilizes the [2Fe-2S] clusters of human mitoNEET and reduces ovarian cancer cell proliferation. J Phys Chem B 126(46):9559–9565. https://doi.org/10.1021/acs.jpcb.2c05728

Marques O, Porto G, Rema A, Faria F, Cruz Paula A, Gomez-Lazaro M, Silva P, Martins da Silva B, Lopes C (2016) Local iron homeostasis in the breast ductal carcinoma microenvironment. BMC Cancer 16:187. https://doi.org/10.1186/s12885-016-2228-y

Michl J, Wang Y, Monterisi S, Blaszczak W, Beveridge R, Bridges EM, Koth J, Bodmer WF, Swietach P (2022) CRISPR-Cas9 screen identifies oxidative phosphorylation as essential for cancer cell survival at low extracellular pH. Cell Rep 38(10):110493. https://doi.org/10.1016/j.celrep.2022.110493

Miller LD, Coffman LG, Chou JW, Black MA, Bergh J, D'Agostino R Jr, Torti SV, Torti FM (2011) An iron regulatory gene signature predicts outcome in breast cancer. Cancer Res 71(21):6728–6737. https://doi.org/10.1158/0008-5472.CAN-11-1870

Mittler R, Darash-Yahana M, Sohn YS, Bai F, Song L, Cabantchik IZ, Jennings PA, Onuchic JN, Nechushtai R (2019) NEET proteins: a new link between iron metabolism, reactive oxygen species, and cancer. Antioxid Redox Signal 30(8):1083–1095. https://doi.org/10.1089/ars.2018.7502

Mou Y, Zhang Y, Wu J, Hu B, Zhang C, Duan C, Li B (2020) The landscape of iron metabolism-related and methylated genes in the prognosis prediction of clear cell renal cell carcinoma. Front Oncol 10:788. https://doi.org/10.3389/fonc.2020.00788

Mukhopadhyay S, Encarnacion-Rosado J, Lin EY, Sohn ASW, Zhang H, Mancias JD, Kimmelman AC (2023) Autophagy supports mitochondrial metabolism through the regulation of iron homeostasis in pancreatic cancer. Sci Adv 9(16):eadf9284. https://doi.org/10.1126/sciadv.adf9284

Muller S, Sindikubwabo F, Caneque T, Lafon A, Versini A, Lombard B, Loew D, Wu TD, Ginestier C, Charafe-Jauffret E, Durand A, Vallot C, Baulande S, Servant N, Rodriguez R (2020) CD44 regulates epigenetic plasticity by mediating iron endocytosis. Nat Chem 12(10):929–938. https://doi.org/10.1038/s41557-020-0513-5

Muller S, Caneque T, Solier S, Rodriguez R (2024) Copper and iron orchestrate cell-state transitions in cancer and immunity. Trends Cell Biol. https://doi.org/10.1016/j.tcb.2024.07.005

Nemeth E, Tuttle MS, Powelson J, Vaughn MB, Donovan A, Ward DM, Ganz T, Kaplan J (2004) Hepcidin regulates cellular iron efflux by binding to ferroportin and inducing its internalization. Science 306(5704):2090–2093. https://doi.org/10.1126/science.1104742

Niederau C, Fischer R, Sonnenberg A, Stremmel W, Trampisch HJ, Strohmeyer G (1985) Survival and causes of death in cirrhotic and in noncirrhotic patients with primary hemochromatosis. N Engl J Med 313(20):1256–1262. https://doi.org/10.1056/NEJM198511143132004

Osborne NJ, Gurrin LC, Allen KJ, Constantine CC, Delatycki MB, McLaren CE, Gertig DM, Anderson GJ, Southey MC, Olynyk JK, Powell LW, Hopper JL, Giles GG, English DR (2010) HFE C282Y homozygotes are at increased risk of breast and colorectal cancer. Hepatology 51(4):1311–1318. https://doi.org/10.1002/hep.23448

Paul BT, Tesfay L, Winkler CR, Torti FM, Torti SV (2019) Sideroflexin 4 affects Fe-S cluster biogenesis, iron metabolism, mitochondrial respiration and heme biosynthetic enzymes. Sci Rep 9(1):19634. https://doi.org/10.1038/s41598-019-55907-z

Petronek MS, Spitz DR, Allen BG (2021) Iron-sulfur cluster biogenesis as a critical target in cancer. Antioxidants (Basel) 10(9). https://doi.org/10.3390/antiox10091458

Philpott CC, Jadhav S (2019) The ins and outs of iron: escorting iron through the mammalian cytosol.

Free Radic Biol Med 133:112–117. https://doi.org/10.1016/j.freeradbiomed.2018.10.411
Phipps O, Brookes MJ, Al-Hassi HO (2021) Iron deficiency, immunology, and colorectal cancer. Nutr Rev 79(1):88–97. https://doi.org/10.1093/nutrit/nuaa040
Pinnix ZK, Miller LD, Wang W, D'Agostino R Jr, Kute T, Willingham MC, Hatcher H, Tesfay L, Sui G, Di X, Torti SV, Torti FM (2010) Ferroportin and iron regulation in breast cancer progression and prognosis. Sci Transl Med 2(43):43ra56. https://doi.org/10.1126/scisignal.3001127
Qin J, Xu Z, Deng K, Qin F, Wei J, Yuan L, Sun Y, Zheng T, Li S (2021) Development of a gene signature associated with iron metabolism in lung adenocarcinoma. Bioengineered 12(1):4556–4568. https://doi.org/10.1080/21655979.2021.1954840
Radulescu S, Brookes MJ, Salgueiro P, Ridgway RA, McGhee E, Anderson K, Ford SJ, Stones DH, Iqbal TH, Tselepis C, Sansom OJ (2012) Luminal iron levels govern intestinal tumorigenesis after Apc loss in vivo. Cell Rep 2(2):270–282. https://doi.org/10.1016/j.celrep.2012.07.003
Ravichandran M, Hu J, Cai C, Ward NP, Venida A, Foakes C, Kuljanin M, Yang A, Hennessey CJ, Yang Y, Desousa BR, Rademaker G, Staes AAL, Cakir Z, Jain IH, Aguirre AJ, Mancias JD, Shen Y, DeNicola GM, Perera RM (2022) Coordinated transcriptional and catabolic programs support iron-dependent adaptation to RAS-MAPK pathway inhibition in pancreatic cancer. Cancer Discov 12(9):2198–2219. https://doi.org/10.1158/2159-8290.CD-22-0044
Recalcati S, Locati M, Marini A, Santambrogio P, Zaninotto F, De Pizzol M, Zammataro L, Girelli D, Cairo G (2010) Differential regulation of iron homeostasis during human macrophage polarized activation. Eur J Immunol 40(3):824–835. https://doi.org/10.1002/eji.200939889
Recalcati S, Gammella E, Cairo G (2019) Ironing out macrophage immunometabolism. Pharmaceuticals (Basel) 12(2). https://doi.org/10.3390/ph12020094
Richmond HG (1959) Induction of sarcoma in the rat by iron-dextran complex. Br Med J 1(5127):947–949. https://doi.org/10.1136/bmj.1.5127.947
Roatsch M, Hoffmann I, Abboud MI, Hancock RL, Tarhonskaya H, Hsu KF, Wilkins SE, Yeh TL, Lippl K, Serrer K, Moneke I, Ahrens TD, Robaa D, Wenzler S, Barthes NPF, Franz H, Sippl W, Lassmann S, Diederichs S, Schleicher E, Schofield CJ, Kawamura A, Schule R, Jung M (2019) The clinically used iron chelator deferasirox is an inhibitor of epigenetic JumonjiC domain-containing histone demethylases. ACS Chem Biol 14(8):1737–1750. https://doi.org/10.1021/acschembio.9b00289
Rodriguez R, Schreiber SL, Conrad M (2022) Persister cancer cells: iron addiction and vulnerability to ferroptosis. Mol Cell 82(4):728–740. https://doi.org/10.1016/j.molcel.2021.12.001
Sandoval-Acuna C, Torrealba N, Tomkova V, Jadhav SB, Blazkova K, Merta L, Lettlova S, Adamcova MK, Rosel D, Brabek J, Neuzil J, Stursa J, Werner L, Truksa J (2021) Targeting mitochondrial iron metabolism suppresses tumor growth and metastasis by inducing mitochondrial dysfunction and mitophagy. Cancer Res 81(9):2289–2303. https://doi.org/10.1158/0008-5472.CAN-20-1628
Santana-Codina N, Del Rey MQ, Kapner KS, Zhang H, Gikandi A, Malcolm C, Poupault C, Kuljanin M, John KM, Biancur DE, Chen B, Das NK, Lowder KE, Hennessey CJ, Huang W, Yang A, Shah YM, Nowak JA, Aguirre AJ, Mancias JD (2022) NCOA4-mediated ferritinophagy is a pancreatic cancer dependency via maintenance of iron bioavailability for iron-sulfur cluster proteins. Cancer Discov 12(9):2180–2197. https://doi.org/10.1158/2159-8290.CD-22-0043
Shi X, Reinstadler B, Shah H, To TL, Byrne K, Summer L, Calvo SE, Goldberger O, Doench JG, Mootha VK, Shen H (2022) Combinatorial GxGxE CRISPR screen identifies SLC25A39 in mitochondrial glutathione transport linking iron homeostasis to OXPHOS. Nat Commun 13(1):2483. https://doi.org/10.1038/s41467-022-30126-9
Shi X, DeCiucis M, Grabinska KA, Kanyo J, Liu A, Lam TT, Shen H (2024) Dual regulation of SLC25A39 by AFG3L2 and iron controls mitochondrial glutathione homeostasis. Mol Cell 84(4):802–810.e806. https://doi.org/10.1016/j.molcel.2023.12.008
Sinha R, Park Y, Graubard BI, Leitzmann MF, Hollenbeck A, Schatzkin A, Cross AJ (2009) Meat and meat-related compounds and risk of prostate cancer in a large prospective cohort study in the United States. Am J Epidemiol 170(9):1165–1177. https://doi.org/10.1093/aje/kwp280
Song X, Xin S, Zhang Y, Mao J, Duan C, Cui K, Chen L, Li F, Liu Z, Wang T, Liu J, Liu X, Song W (2022) Identification and quantification of iron metabolism landscape on therapy and prognosis in bladder cancer. Front Cell Dev Biol 10:810272. https://doi.org/10.3389/fcell.2022.810272
Stehling O, Vashisht AA, Mascarenhas J, Jonsson ZO, Sharma T, Netz DJ, Pierik AJ, Wohlschlegel JA, Lill R (2012) MMS19 assembles iron-sulfur proteins required for DNA metabolism and genomic integrity. Science 337(6091):195–199. https://doi.org/10.1126/science.1219723
Su Y, Zhao B, Zhou L, Zhang Z, Shen Y, Lv H, AlQudsy LHH, Shang P (2020) Ferroptosis, a novel pharmacological mechanism of anti-cancer drugs. Cancer Lett. https://doi.org/10.1016/j.canlet.2020.02.015
Su J, Zhong G, Qin W, Zhou L, Ye J, Ye Y, Chen C, Liang P, Zhao W, Xiao X, Wen W, Luo W, Zhou X, Zhang Z, Cai Y, Li C (2024) Integrating iron metabolism-related gene signature to evaluate prognosis and immune infiltration in nasopharyngeal carcinoma. Discov Oncol 15(1):112. https://doi.org/10.1007/s12672-024-00969-3
Sun JL, Zhang NP, Xu RC, Zhang GC, Liu ZY, Abuduwaili W, Wang F, Yu XN, Shi X, Song GQ, Wu H, Liu TT, Shen XZ, Deng B, Weng SQ, Dong L, Zhu JM (2021) Tumor cell-imposed iron restriction drives immunosuppressive polarization of tumor-associated

macrophages. J Transl Med 19(1):347. https://doi.org/10.1186/s12967-021-03034-7
Tan AS, Baty JW, Dong LF, Bezawork-Geleta A, Endaya B, Goodwin J, Bajzikova M, Kovarova J, Peterka M, Yan B, Pesdar EA, Sobol M, Filimonenko A, Stuart S, Vondrusova M, Kluckova K, Sachaphibulkij K, Rohlena J, Hozak P, Truksa J, Eccles D, Haupt LM, Griffiths LR, Neuzil J, Berridge MV (2015) Mitochondrial genome acquisition restores respiratory function and tumorigenic potential of cancer cells without mitochondrial DNA. Cell Metab 21(1):81–94. https://doi.org/10.1016/j.cmet.2014.12.003
Tang B, Zhu J, Li J, Fan K, Gao Y, Cheng S, Kong C, Zheng L, Wu F, Weng Q, Lu C, Ji J (2020) The ferroptosis and iron-metabolism signature robustly predicts clinical diagnosis, prognosis and immune microenvironment for hepatocellular carcinoma. Cell Commun Signal 18(1):174. https://doi.org/10.1186/s12964-020-00663-1
Taylor SE, Behr S, Cooper KL, Mahdi H, Fabian D, Gallion H, Ueland F, Vargo J, Orr B, Girda E, Courtney-Brooks M, Olawaiye AB, Randall LM, Richardson DL, Sullivan SA, Huang M, Christner SM, Beriwal S, Lin Y, Chauhan A, Chu E, Kohn EC, Kunos C, Ivy SP, Beumer JH (2024) Dose finding, bioavailability, and PK-PD of oral triapine with concurrent chemoradiation for locally advanced cervical cancer and vaginal cancer (ETCTN 9892). Cancer Chemother Pharmacol 95(1):4. https://doi.org/10.1007/s00280-024-04720-1
Tesfay L, Clausen KA, Kim JW, Hegde P, Wang X, Miller LD, Deng Z, Blanchette N, Arvedson T, Miranti CK, Babitt JL, Lin HY, Peehl DM, Torti FM, Torti SV (2015) Hepcidin regulation in prostate and its disruption in prostate cancer. Cancer Res 75(11):2254–2263. https://doi.org/10.1158/0008-5472.CAN-14-2465
Tesfay L, Paul BT, Hegde P, Brewer M, Habbani S, Jellison E, Moore T, Wu H, Torti SV, Torti FM (2022) Complementary anti-cancer pathways triggered by inhibition of sideroflexin 4 in ovarian cancer. Sci Rep 12(1):19936. https://doi.org/10.1038/s41598-022-24391-3
Tifoun N, De Las Heras JM, Guillaume A, Bouleau S, Mignotte B, Le Floch N (2021) Insights into the roles of the sideroflexins/SLC56 family in iron homeostasis and iron-sulfur biogenesis. Biomedicines 9(2). https://doi.org/10.3390/biomedicines9020103
Torti SV, Torti FM (2020) Iron and cancer: 2020 vision. Cancer Res 80(24):5435–5448. https://doi.org/10.1158/0008-5472.CAN-20-2017
Trujillo-Alonso V, Pratt EC, Zong H, Lara-Martinez A, Kaittanis C, Rabie MO, Longo V, Becker MW, Roboz GJ, Grimm J, Guzman ML (2019) FDA-approved ferumoxytol displays anti-leukaemia efficacy against cells with low ferroportin levels. Nat Nanotechnol 14(6):616–622. https://doi.org/10.1038/s41565-019-0406-1
Tsuji Y, Kwak E, Saika T, Torti SV, Torti FM (1993) Preferential repression of the H subunit of ferritin by adenovirus E1A in NIH-3T3 mouse fibroblasts. J Biol Chem 268(10):7270–7275
Wang W, Deng Z, Hatcher H, Miller LD, Di X, Tesfay L, Sui G, D'Agostino RB Jr, Torti FM, Torti SV (2014) IRP2 regulates breast tumor growth. Cancer Res 74(2):497–507. https://doi.org/10.1158/0008-5472.CAN-13-1224
Wang Y, Yen FS, Zhu XG, Timson RC, Weber R, Xing C, Liu Y, Allwein B, Luo H, Yeh HW, Heissel S, Unlu G, Gamazon ER, Kharas MG, Hite R, Birsoy K (2021) SLC25A39 is necessary for mitochondrial glutathione import in mammalian cells. Nature 599(7883):136–140. https://doi.org/10.1038/s41586-021-04025-w
Wang X, Zhou L, Dong Z, Wang G (2023a) Identification of iron metabolism-related predictive markers of endometriosis and endometriosis-relevant ovarian cancer. Medicine (Baltimore) 102(15):e33478. https://doi.org/10.1097/MD.0000000000033478
Wang Y, Wu X, Bao X, Mou X (2023b) Progress in the mechanism of the effect of Fe(3)O(4) nanomaterials on ferroptosis in tumor cells. Molecules 28(11). https://doi.org/10.3390/molecules28114562
Ward HA, Whitman J, Muller DC, Johansson M, Jakszyn P, Weiderpass E, Palli D, Fanidi A, Vermeulen R, Tjonneland A, Hansen L, Dahm CC, Overvad K, Severi G, Boutron-Ruault MC, Affret A, Kaaks R, Fortner R, Boeing H, Trichopoulou A, La Vecchia C, Kotanidou A, Berrino F, Krogh V, Tumino R, Ricceri F, Panico S, Bueno-de-Mesquita HB, Peeters PH, Nost TH, Sandanger TM, Quiros JR, Agudo A, Rodriguez-Barranco M, Larranaga N, Huerta JM, Ardanaz E, Drake I, Brunnstrom H, Johansson M, Grankvist K, Travis RC, Freisling H, Stepien M, Merritt MA, Riboli E, Cross AJ (2019) Haem iron intake and risk of lung cancer in the European Prospective Investigation into Cancer and Nutrition (EPIC) cohort. Eur J Clin Nutr 73(8):1122–1132. https://doi.org/10.1038/s41430-018-0271-2
Weinberg ED (1983) Iron in neoplastic disease. Nutr Cancer 4(3):223–233. https://doi.org/10.1080/01635588209513761
Weston C, Klobusicky J, Weston J, Connor J, Toms SA, Marko NF (2016) Aberrations in the iron regulatory gene signature are associated with decreased survival in diffuse infiltrating gliomas. PLoS One 11(11):e0166593. https://doi.org/10.1371/journal.pone.0166593
Wijesinghe TP, Dharmasivam M, Dai CC, Richardson DR (2021) Innovative therapies for neuroblastoma: the surprisingly potent role of iron chelation in up-regulating metastasis and tumor suppressors and down-regulating the key oncogene, N-myc. Pharmacol Res 173:105889. https://doi.org/10.1016/j.phrs.2021.105889
Wu KJ, Polack A, Dalla-Favera R (1999) Coordinated regulation of iron-controlling genes, H-ferritin and IRP2, by c-MYC. Science 283(5402):676–679. https://doi.org/10.1126/science.283.5402.676
Xu S, Wang Z, Ye J, Mei S, Zhang J (2021) Identification of iron metabolism-related genes as prognostic indicators for lower-grade glioma. Front Oncol 11:729103. https://doi.org/10.3389/fonc.2021.729103

Yamasaki T, Terai S, Sakaida I (2011) Deferoxamine for advanced hepatocellular carcinoma. N Engl J Med 365(6):576–578. https://doi.org/10.1056/NEJMc1105726

Yanatori I, Richardson DR, Dhekne HS, Toyokuni S, Kishi F (2021) CD63 is regulated by iron via the IRE-IRP system and is important for ferritin secretion by extracellular vesicles. Blood 138(16):1490–1503. https://doi.org/10.1182/blood.2021010995

Yanatori I, Kishi F, Toyokuni S (2023) New iron export pathways acting via holo-ferritin secretion. Arch Biochem Biophys 746:109737. https://doi.org/10.1016/j.abb.2023.109737

Yang J, Bielenberg DR, Rodig SJ, Doiron R, Clifton MC, Kung AL, Strong RK, Zurakowski D, Moses MA (2009) Lipocalin 2 promotes breast cancer progression. Proc Natl Acad Sci USA 106(10):3913–3918. https://doi.org/10.1073/pnas.0810617106

Yang S, Xing J, Liu D, Song Y, Yu H, Xu S, Zuo Y (2024) Review and new insights into the catalytic structural domains of the Fe(II) and 2-Oxoglutarate families. Int J Biol Macromol 278(Pt 4):134798. https://doi.org/10.1016/j.ijbiomac.2024.134798

Yuan J, Liu T, Zhang Y (2021) An iron metabolism-related gene signature for the prognosis of colon cancer. Front Cell Dev Biol 9:786684. https://doi.org/10.3389/fcell.2021.786684

Zeidan RS, Yoon HS, Yang JJ, Sobh A, Braithwaite D, Mankowski R, Leeuwenburgh C, Anton S (2024) Iron and cancer: overview of the evidence from population-based studies. Front Oncol 14:1393195. https://doi.org/10.3389/fonc.2024.1393195

Zhang J, Zhao L, Xuan S, Liu Z, Weng Z, Wang Y, Dai K, Gu A, Zhao P (2024a) Global analysis of iron metabolism-related genes identifies potential mechanisms of gliomagenesis and reveals novel targets. CNS Neurosci Ther 30(2):e14386. https://doi.org/10.1111/cns.14386

Zhang W, Ou Z, Tang T, Yang T, Li Y, Wu H, Li L, Liu M, Niu L, Zhu J (2024b) Up-regulated SLC25A39 promotes cell growth and metastasis via regulating ROS production in colorectal cancer. J Cancer 15(17):5841–5854. https://doi.org/10.7150/jca.98844

Zhang YY, Han Y, Li WN, Xu RH, Ju HQ (2024c) Tumor iron homeostasis and immune regulation. Trends Pharmacol Sci 45(2):145–156. https://doi.org/10.1016/j.tips.2023.12.003

Zhu B, Zhi Q, Xie Q, Wu X, Gao Y, Chen X, Shi L (2019) Reduced expression of ferroportin1 and ceruloplasmin predicts poor prognosis in adrenocortical carcinoma. J Trace Elem Med Biol 56:52–59. https://doi.org/10.1016/j.jtemb.2019.07.009

Zhu Z, Cao H, Yan H, Liu H, Hong Z, Sun A, Liu T, Mao F (2024) Prognostic iron-metabolism signature robustly stratifies single-cell characteristics of hepatocellular carcinoma. Comput Struct Biotechnol J 23:929–941. https://doi.org/10.1016/j.csbj.2024.01.022

19 Neurodegeneration with Brain Iron Accumulation

David A. Schipper and Hyman M. Schipper

Abstract

Iron participates in a wide array of cellular functions and is essential for normal neural development and physiology. However, if inappropriately managed, the transition metal is capable of generating neurotoxic reactive oxygen species. A number of hereditary conditions perturb body iron homeostasis and some, collectively referred to as neurodegeneration with brain iron accumulation (NBIA), promote pathological deposition of the metal predominantly or exclusively within the central nervous system (CNS). In this chapter, we discuss ten NBIA disorders with emphasis on the clinical syndromes and neuroimaging. The latter primarily entails magnetic resonance scanning using iron-sensitive sequences. The conditions considered are pantothenate kinase 2-associated neurodegeneration (PKAN), neuroferritinopathy, aceruloplasminemia, Kufor–Rakeb disease (KRD), PLA2G6-associated neurodegeneration (PLAN), FA2H-associated neurodegeneration (FAHN), Woodhouse–Sakati syndrome (WSS), beta-propeller protein-associated neurodegeneration (BPAN), mitochondrial membrane protein-associated neurodegeneration (MPAN), and coenzyme A synthase protein-associated neurodegeneration (CoPAN). An approach to differential diagnosis and the status of iron chelation therapy for several of these entities are presented.

Keywords

Ceruloplasmin · Ferritin · Movement disorder · Neuroimaging · Neurodegeneration with brain iron accumulation (NBIA)

19.1 Brain Iron

Iron is essential for normal neural development and function and contributes to a wide range of cellular activities including electron transport, cytokinesis, antioxidant enzyme activity, myelination, and biogenic amine metabolism (Schipper 2009). On the other hand, the redox (Fenton) chemistry of this transition metal may render it toxic to neural and other cellular substrates (see Chap. 1). Mechanisms subserving iron homeostasis in the mammalian CNS and various factors specifically predisposing the latter to iron-mediated injury have been reviewed elsewhere

D. A. Schipper
New York Medical College, Valhalla, NY, USA

H. M. Schipper (✉)
Departments of Neurology & Neurosurgery and Medicine, Lady Davis Institute, Jewish General Hospital, McGill University, Montreal, QC, Canada
e-mail: hyman.schipper@mcgill.ca

K. Pantopoulos (ed.), *Iron Metabolism in Human Health and Disease*, Advances in Experimental Medicine and Biology 1480, https://doi.org/10.1007/978-3-031-92033-2_19

Table 19.1 Neurodegeneration with brain iron accumulation (NBIA): diseases, genes, and modes of inheritance (MOI)

Disease	Gene	MOI
PKAN	*PANK2*	AR
Neuroferritinopathy	*FTL*	AD
Aceruloplasminemia	*CP*	AR
KRD	*ATP13A2*	AR
PLAN	*PLA2G6*	AR
FAHN	*FA2H*	AR
WSS	*DCAF17*	AR
BPAN	*WDR45*	XD
MPAN	*C19orf12*	AR (AD)
CoPAN	*COASY*	AR

AD autosomal dominant, *AR* autosomal recessive, *BPAN* beta-propeller protein-associated neurodegeneration, *CoPAN* coenzyme A synthase protein-associated neurodegeneration, *FAHN* fatty acid hydroxylase-associated neurodegeneration, *KRD* Kufor–Rakeb disease, *PKAN* pantothenate kinase-2-associated neurodegeneration, *MPAN* mitochondrial membrane protein-associated neurodegeneration, *PLAN* phospholipase A2G6-associated neurodegeneration, *WSS* Woodhouse–Sakati syndrome, *XD* X-linked dominant

(Schipper 2011, 2012). Iron-derived reactive oxygen species (ROS) may promote bioenergetic failure, proteasomal dysfunction, protein aggregation and inclusion biogenesis, pro-toxin bioactivation, aberrant cell signaling, electrophysiological derangements, and ferroptosis within the diseased CNS (Schipper 2004). In this chapter, we focus on the clinical and neuroimaging aspects of a group of heritable conditions referred to collectively as "neurodegeneration with brain iron accumulation (NBIA)" (Table 19.1).

Friedreich ataxia, an autosomal recessive neurodegenerative condition exhibiting pathological iron sequestration in cerebellar dentate nuclei (Schipper 2012), is not formally classified among the NBIA and will receive only brief mention here. Acquired conditions featuring secondary CNS iron overload, e.g., Parkinson disease (PD), Alzheimer disease (AD), amyotrophic lateral sclerosis (ALS), multiple sclerosis, corticobasal degeneration, progressive supranuclear palsy, and superficial siderosis, are reviewed elsewhere (Benarroch 2009; Connor 1997; Schipper 2004; Li et al. 2010; Stankiewicz et al. 2007).

19.2 Neurodegeneration with Brain Iron Accumulation

NBIA comprises an expanding group of genetic disorders encompassing a diverse array of molecular pathologies (Hayflick 2023). Prior to considering the NBIAs individually, several general principles germane to these conditions are noteworthy: (1) In contrast to Friedreich ataxia, the NBIAs are rare conditions which are infrequently encountered outside of a tertiary referral center. However, it is anticipated that established and novel NBIA syndromes will come to light more often as magnetic resonance imaging (MRI) using iron-sensitive sequences (T2*, susceptibility weighted imaging) and genetic testing become increasingly available. (2) In some conditions, the responsible mutations directly alter the structure and function of proteins with known roles in tissue iron homeostasis. Prime examples of this are mutant frataxin in Friedreich ataxia (although not a classical NBIA) and L-ferritin in neuroferritinopathy. In pantothenate-kinase 2-associated neurodegeneration (PKAN) and other NBIAs, the mechanism whereby the aberrant protein engenders excessive tissue iron sequestration remains unknown or conjectural. Iron dyshomeostasis in the latter may be due to an interplay between mitochondrial dysfunction, aberrant lipid metabolism, and altered autophagic activity (Agostini et al. 2024). (3) Certain NBIA syndromes implicate the CNS almost exclusively (e.g., PKAN), while others also target peripheral tissues (e.g., aceruloplasminemia). Interestingly, hereditary hemochromatosis (Chaps. 9 and 10), a systemic disease which may promote hepatic iron deposition sufficing to trigger airport metal detectors (Felitti and Beutler 1999), largely spares the CNS (Schipper and Ponka 2009). (4) Most cases of NBIA exhibit extrapyramidal and/or cerebellar signs (involuntary movements, incoordination) at some point in their natural history, and corresponding abnormalities in the basal ganglia and cerebellum on neuroimaging and histopathology. The predilection for involvement of these regions may relate to the relative susceptibility of the basal ganglia, cerebellar dentate nuclei, and red nuclei to the accumulation of iron during normal development and aging. (5) Many NBIA syndromes present in childhood or early adulthood (e.g., PKAN), while others may remain symptom-free until midlife or later (e.g., aceruloplasminemia). The latter suggests that, as in conditions such as familial AD, normal CNS aging and/or environmental factors may facilitate "unmasking" of latent genetic defects. (6) Clinical NBIA phenotypes may vary broadly in severity and tempo among affected individuals even within kindreds. In virtually all cases, these diseases unfortunately exhibit relentless progression and are uniformly fatal. (7) As a rule, although certain interventions (e.g., deep brain stimulation for tremor; baclofen for spasticity) may afford symptomatic relief in select patients, the NBIAs remain notoriously refractory to disease-modifying therapy. (8) Administration of clinically approved iron chelators (e.g., deferoxamine) have rarely yielded significant benefits and may, theoretically, expose unaffected neural tissues to the toxic effects of the metal. On the other hand, advances in the delineation of the molecular biology underpinning the neurodegenerative processes, and the development of orally bioavailable, brain-permeant antioxidants and iron chelators (e.g., deferiprone), raise hope that rational and effective treatments for NBIA may be soon be forthcoming (Marupudi and Xiong 2024).

19.2.1 Pantothenate Kinase-2-Associated Neurodegeneration

Formerly referred to as the Hallervorden–Spatz syndrome, pantothenate kinase-2-associated neurodegeneration (PKAN) is an autosomal recessive disorder resulting from loss-of-function mutations in the gene encoding brain-specific mitochondrial pantothenate kinase-2 (*PANK2*) at locus 20p13-p12.3 (Hayflick 2006; Johnson et al. 2004; Zhou et al. 2001). The prevalence of PKAN is approximately 1–3 per million (Rump et al. 2005). The majority of patients with this condition present as children or young adults with combinations of dysarthria, parkinsonism, dystonia, choreoathetosis, spasticity, seizures, optic atrophy, pigmentary retinopathy, mental retardation, and dementia (Hayflick 2003, 2006; Swaiman 2001). Classically, affected individuals at onset exhibit clumsiness and gait abnormalities around 3 years of age followed by asymmetric lower limb dystonia. A gradually progressive neuropsychiatric syndrome in addition to the movement disorder is observed in atypical cases (Hayflick 2006). Psychiatric morbidity includes depression, schizophrenia-like psychosis, and obsessive-compulsive disorder. Cognitive impairment, often implicating attention and executive functions, tends to be more severe in patients with earlier-onset disease and may precede motor impairment (Walterfang et al. 2011). Unusual cases of PKAN presenting in adulthood may resemble amyotrophic lateral sclerosis or Parkinson disease (Mak et al. 2011; Swaiman 2001; Vasconcelos et al. 2003; Seo et al. 2009). Rarely, individuals harboring *PANK2* mutations present with HARP (hypoprebetalipoproteinemia, acanthocytosis, retinitis pigmentosa, and pallidal degeneration) (Ching et al. 2002).

The majority of symptomatic (both classical and atypical) and many pre-clinical subjects with PKAN exhibit the "eye-of-the-tiger" sign on T2-weighted MRI (Fig. 19.1). The latter appears as a hyperintense signal surrounded by a ring of hypointensity in the globus pallidus attributed, respectively, to tissue vacuolization/gliosis and excessive iron sequestration (Savoiardo et al. 1993). At autopsy, the basal ganglia exhibit brown discoloration due to the pathological iron deposition. Granular iron concretions scattered throughout the neuropil and within astrocytes and microglial cells of the substantia nigra pars reticulata and globus pallidus (and occasionally in the subthalamic nucleus and within cerebral white and grey matter) are noted on histochemical staining with Perls' method. Peripheral blood mononuclear cells may exhibit vacuoles, granules, fingerprint bodies, and multilamellar inclusions in the absence of systemic symptoms (Swaiman et al. 1983; Zupanc et al. 1990).

PANK2 catalyzes the phosphorylation of pantothenate (vitamin B_5) in the coenzyme A biosynthetic pathway. The product of the PANK2

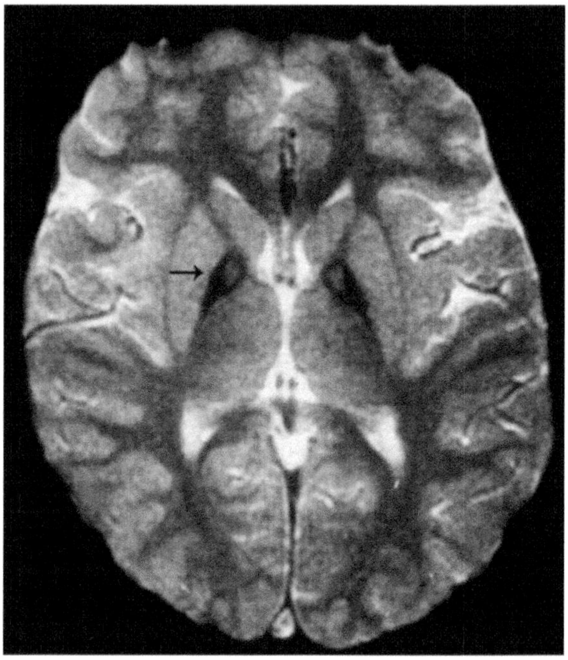

Fig. 19.1 Brain MRI of patient with pantothenate kinase-2 associated neurodegeneration (PKAN). T2*-weighted echo-planar sequence demonstrates characteristic "eye-of-the-tiger" sign in the globus pallidus (arrow). (Modified from Castelnau et al. (2001), with permission)

reaction, phosphopantothenate, normally condenses with cysteine in the next step of the pathway. Cysteine accumulates in the globus pallidus of patients with PANK2 deficiency (Perry et al. 1985). The iron-chelating properties of cysteine initially suggested that the excessive accumulation of this amino acid may mediate the pathological storage of redox-active iron in the basal ganglia of patients with PANK2 deficiency. However, the identification of an NBIA syndrome (phospholipase A2G6-associated neurodegeneration (PLAN)—see below) phenotypically indistinguishable from PKAN, but lacking anticipated perturbations in brain cysteine metabolism, argues against the "cysteine hypothesis" of iron trapping in PKAN.

There is currently no specific, disease-modifying treatment with proven efficacy for PKAN sufferers, although there are hopeful signs on the horizon. Initial studies investigating deferoxamine chelation therapy in PKAN were disappointing (Schneider et al. 2009), but the ability of deferoxamine to cross the blood–brain barrier is limited. More recently, evidence has accrued suggesting that the orally available, brain-permeant iron chelator, deferiprone may attenuate iron trapping in the basal ganglia and possibly slow disease progression (Klopstock et al. 2019). To date, administration of vitamin B_5 or fosmetpantotenate (RE-024), a substrate designed to bypass the enzymatic block in PKAN, yielded no salutary effects (Schneider et al. 2009; Iankova et al. 2021). 4′-Phosphopantetheine, on the other hand, showed significant benefits in an animal model of PKAN and in human PKAN cell cultures; the compound is currently being tested in the human condition (Iankova et al. 2021). Oral or intrathecal baclofen, dopaminergic medications, stereotactic pallidotomy, and deep brain stimulation may provide symptomatic relief but confer no known disease-modifying benefits in this condition (Justesen et al. 1999; Zimmerman and Schenck 2007; Zhai et al. 2024).

19.2.2 Neuroferritinopathy

Mammalian ferritin, a protein responsible for intracellular sequestration of metabolically inert (Fe^{3+}) iron, can store up to 4500 iron atoms in its internal cavity. The apoprotein shell consists of a 19 kDa light subunit (L-subunit) and a 21 kDa heavy subunit (H-subunit) (Arosio and Levi 2002; Ponka et al. 1998). Ferrous iron inserted into the ferritin shell is oxidized to Fe^{3+} by the ferroxidase activity of the H-subunit; the L-subunit subserves iron-core nucleation. A dominantly inherited neurodegenerative disease featuring extrapyramidal signs and excessive iron deposition in the basal ganglia, now termed neuroferritinopathy, was first reported by Curtis et al. (2001). The initial cohort, originating from the Cumbrian region of Northern England, presented at 40–55 years of age with rigidity, dystonia, choreoathetosis, and spasticity, but no cerebellar or cognitive deficits (Burn and Chinnery 2006; Chinnery et al. 2007). Mixed-signal abnormalities are typically observed in the basal ganglia, thalami, and red nuclei of these patients which may correlate with disease progression (Fig. 19.2) (Curtis et al. 2001; Wills et al. 2002; Chinnery et al. 2007).

The genetic anomaly in the Cumbrian cases was mapped to locus 19q13.3 encompassing the gene for ferritin L-chain (Curtis et al. 2001). The same mutation appears to have arisen independently in France yielding a somewhat different neurophenotype characterized by blepharospasm, dystonic dysphonia, and cerebellar signs (Chinnery et al. 2003; Mancuso et al. 2005).

On neuropathological examination, there is cystic necrosis of the globus pallidus with variable involvement of surrounding regions. Prominent deposition of ferritin protein and stainable iron, often appearing as discrete concretions, is observed in pallidal neurons and within oligodendrocytes and microglia of the forebrain and cerebellum. The Cumbrian patients notably exhibited no evidence of systemic iron overload or diabetes, and their circulating ferritin levels were unusually low (hypoferritinemia). Although liver function tests in these subjects were unremarkable (Curtis et al. 2001), unique hepatocytic inclusions have been observed on biopsy (Professor John Burn, personal communication cited in Schipper and Ponka (2009)). In one family, abundant intracytoplasmic and intranuclear bodies containing ferritin and iron were documented in the cytoplasm and nuclei of skin fibroblasts, renal tubular epithelial cells, and muscle capillary endothelial cells (Vidal et al. 2004) potentially providing peripheral diagnostic mark-

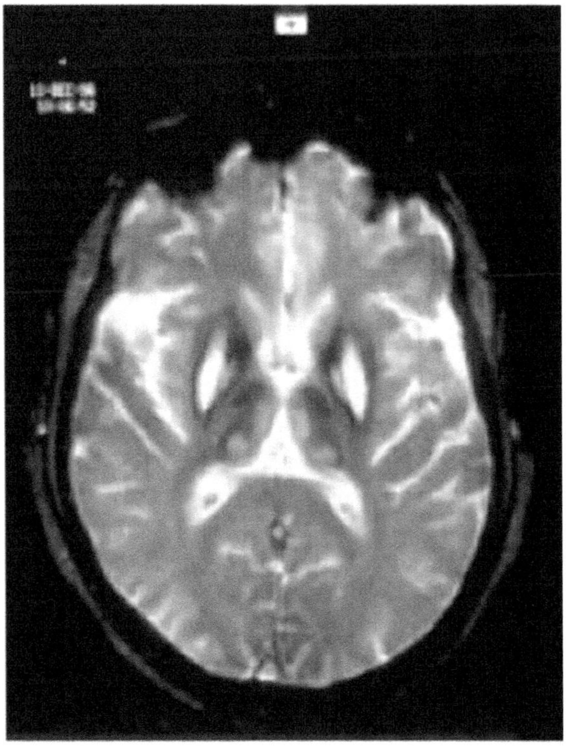

Fig. 19.2 A case of neuroferritinopathy. T2-weighted MRI of the brain shows mixed hyper- and hypointensities affecting the putamen, globus pallidus, and thalamus symmetrically. (From Wills et al. (2002), with permission)

ers of the disease (Schroder 2005). Regarding pathophysiology, it is has been posited that compromise of the C-terminus of L-ferritin and holoprotein stability in neuroferritinopathy facilitates oxidative tissue damage by "unshielded" cytosolic iron (Levi et al. 2005). Attempts at iron chelation in neuroferritinopathy have not yielded any clinical benefits to date (Burn and Chinnery 2006). Chelation therapy may need to be implemented in preclinical stages of neuroferritinopathy because 7T MRI revealed increased magnetic susceptibility indicative of augmented cerebral iron in a *pre-symptomatic* carrier of the FTL mutation (Murley et al. 2024). In 2022, Marchand et al. reported significant clinical improvement in four persons with neuroferritinopathy following deferiprone administration, especially when chelation therapy was initiated early in the course of the disease. In one patient, disease progression was stabilized for more than 11 years; in another case, neurological symptoms were reversed within several months of treatment (Marchand et al. 2022).

19.2.3 Aceruloplasminemia

Ceruloplasmin is a plasma α_2-glycoprotein that is synthesized primarily, but not exclusively, in the liver. The holoprotein contains six copper atoms, three of which form a trinuclear cluster that activates oxygen required for catalytic activity. The ferroxidase activity of ceruloplasmin oxidizes ferrous (Fe^{2+}) to ferric (Fe^{3+}) iron, thereby facilitating iron efflux from cells and its delivery to transferrin in the extracellular *milieu* (Hellman and Gitlin 2002; Nittis and Gitlin 2002). A glycosylphosphatidylinositol (GPI)-anchored ceruloplasmin bound to astroglial cell membranes is a major isoform of this protein in the CNS. GPI-anchored ceruloplasmin is an important player in CNS iron homeostasis as evidenced by the fact that egress of iron is significantly compromised in astrocytes isolated from ceruloplasmin-deficient mice (Patel and David 1997).

In 1987, Miyajima and associates (1987) reported a 52-year-old Japanese woman with extrapyramidal symptoms and MRI signal hypointensities of the basal ganglia and cerebellum, indicative of excessive iron deposition (Fig. 19.3). Additionally, the patient had diabetes mellitus, retinal degeneration, anemia, low plasma iron levels, elevated plasma ferritin concentrations, severe hepatic iron overload, and undetectable serum ceruloplasmin. Nucleotide sequence analysis revealed a loss-of-function mutation in the ceruloplasmin gene (chromosome 3q25) (Harris et al. 1995). Cases harboring other point mutations and splice-variants in the ceruloplasmin gene were subsequently disclosed, with the majority causing defective assembly of the trinuclear copper cluster (Daimon et al. 1995; Okamoto et al. 1996; Yoshida et al. 1995). Homozygosity for this defect engenders serious parenchymal iron overload involving the liver, pancreas, and CNS. CNS implications are commonly diagnosed at midlife, are progressive, and preferentially target the basal ganglia (dysarthria, dystonia, choreathetosis, dementia), the cerebellum (ataxia), and retina (pigmentary degeneration, visual loss). In the homozygous condition, circulating ceruloplasmin concentrations and attendant ferroxidase activity are nil or negligible. Individuals heterozygous for mutant ceruloplasmin manifest ~50% reduction in circulating protein levels ("hypoceruloplasminemia") and exhibit subtle signs of cerebellar impairment on careful testing (Miyajima et al. 2001).

Initial trials to purge brain and body iron with deferoxamine in aceruloplasminemic patients were largely futile, possibly because deferoxamine predominately chelates Fe^{3+} iron and, absent normal ferroxidase activity, the iron burden in these individuals favors the Fe^{2+} state (Yonekawa et al. 1999). In a wise attempt to circumvent this conundrum, Yonekawa et al. (1999) first administered fresh frozen plasma for 6 weeks to replenish blood ceruloplasmin levels and restore ferroxidase activity. They then administered deferoxamine for 6 weeks to deplete *ferric* iron stores. Although MR images of the brain in this test case remained unchanged following treatment, there was unprecedented improvement in choreoathetosis and ataxia and resolution of abnormal high-voltage sharp activity on electroencephalography. Amelioration of neurological

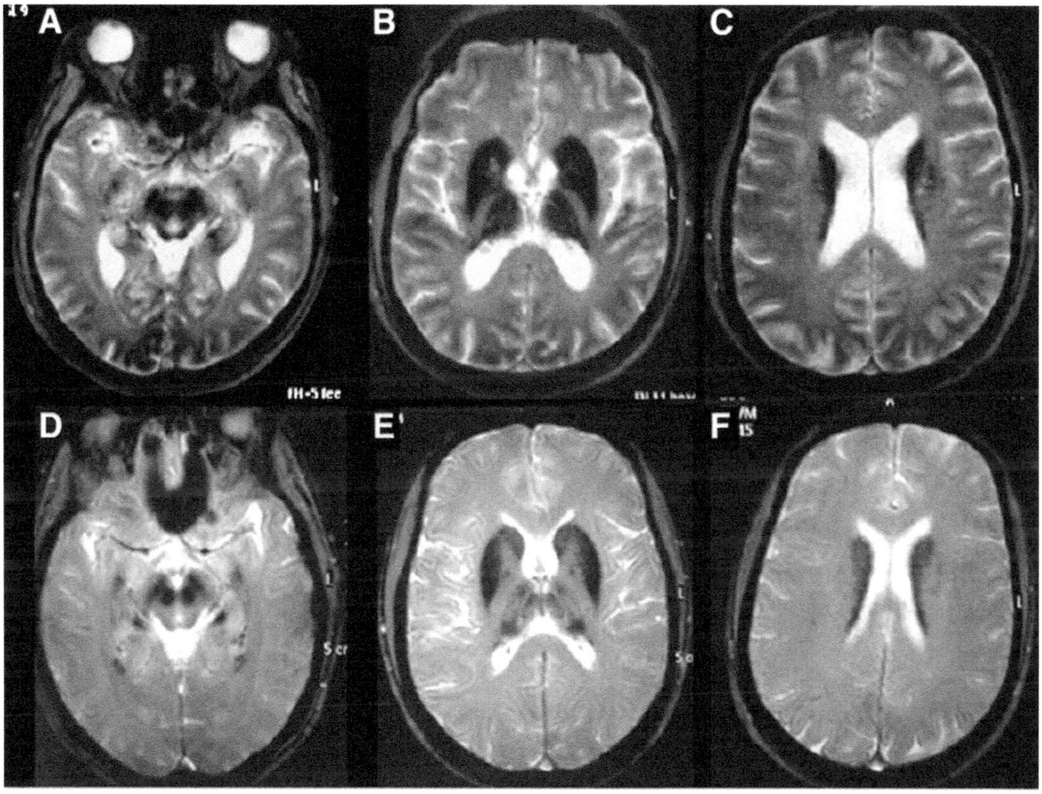

Fig. 19.3 Aceruloplasminemia. Gradient echo (GE) T2*-weighted images reveal profound hypointensities (iron deposition) in the caudate nucleus, putamen, globus pallidus, thalamus, and midbrain with ventriculomegaly in a 56 year-old male proband (**a–c**). Similar, but less florid, lesions of the basal ganglia and thalamus are demonstrable on neuroimaging of the proband's sister (**d–f**). (From Fasano et al. (2008), with permission)

deficits in this condition has also been reported after administration of oral zinc sulfate and/or the iron chelator, deferasirox (McNeill et al. 2008; Miyake et al. 2020). In 2001, substantial improvement in hematological parameters without significant adverse effects was documented in a 19-year-old aceruloplasminemic man who received a combination of fresh frozen plasma and deferasirox for 3 consecutive years (Tridimas et al. 2021).

19.2.4 Kufor–Rakeb Disease

Kufor–Rakeb disease (KRD) is an autosomal recessive NBIA disorder caused by mutations in the *ATP13A2* gene on chromosome 1p36 (Ramirez et al. 2006). The gene codes for a putative lysosomal P-type transmembrane cation-transporting ATPase of uncertain function (Gitler et al. 2009; Schneider et al. 2010). The condition often presents in the second or third decade of life with insomnia, mania, delusions, and other neuropsychiatric symptoms. Parkinsonism (which may transiently improve with L-dopa treatment), supranuclear gaze palsies and pyramidal tract deficits (spastic weakness) are typical clinical features of the disease. Cognitive decline and dementia are variable. The disease is usually slowly progressive, but may evolve in subacute fashion rendering victims bed-bound within the first few years of onset (Najim al-Din et al. 1994; Schneider et al. 2010). Brain MRI in KRD shows hypointensities of the caudate and putamen on T2* sequences in keeping with enhanced iron storage and generalized cortical and subcortical atrophy. Dopamine transporter (DAT) positron emission tomography scans show diminished

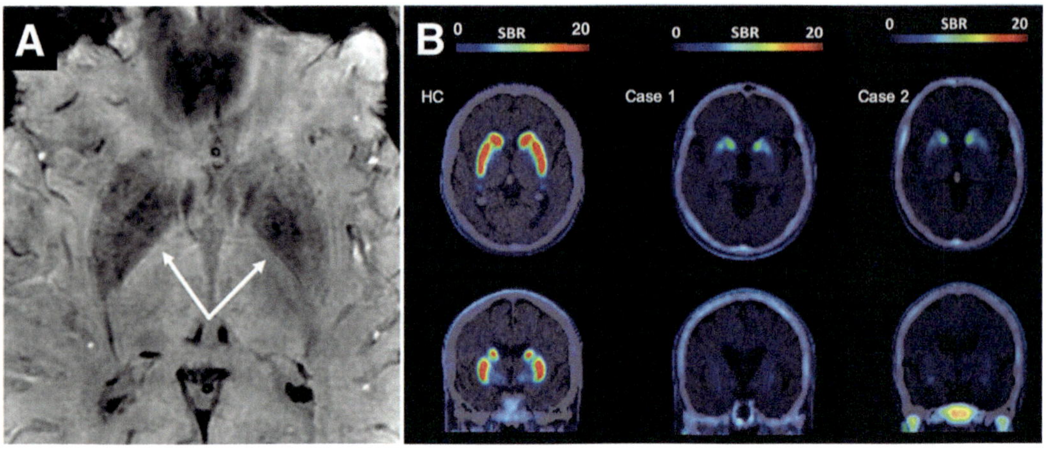

Fig. 19.4 Kufor–Rakeb disease (KRD). (**a**) T2*-weighted MRI demonstrating bilateral caudate and putaminal hypointensities (arrow) consistent with augmented iron deposition. (**b**) DAT PET in two cases of KRD shows marked bilateral reduction of dopamine transporters in the nigrostriatum. *DAT* dopamine transporter, *HC* healthy control, *PET* positron emission tomography. (From Satolli et al. (2023) and Saffie et al. (2023), with permission)

radiotracer uptake in the basal ganglia indicative of depleted dopamine stores (Fig. 19.4) (Saffie et al. 2023). Children heterozygous for mutant *ATP13A2* may manifest subtle signs of parkinsonism (mild postural tremor, limb rigidity, and reduced arm swing) and normal basal ganglia on T2* imaging (Bruggemann et al. 2010).

19.2.5 Phospholipase A2G6-Associated Neurodegeneration

Mutations in the gene encoding a calcium-independent group VI phospholipase A_2 (*PLA2G6*) have been recognized in a number of children with autosomal recessive NBIA initially diagnosed as PKAN (Morgan et al. 2006; Hayflick 2006). Individuals with phospholipase A2G6-associated neurodegeneration (PLAN; a.k.a. NBIA-2) may present with early psychomotor retardation, spasticity, cerebellar signs, and visual abnormalities; others exhibit parkinsonism-dystonia and dementia with frontotemporal atrophy (Yoshino et al. 2010). In a recent study, levodopa responsiveness was noted in 12 of 14 patients with parkinsonism/dystonia (Holla et al. 2024). Many, but not all, patients display T2 signal attenuation indicative of iron deposition in the striatum and substantia nigra (Gregory et al. 2008). Unlike PKAN, the "eye-of-the-tiger" sign is not characteristic of PLAN (Fig. 19.5).

The PLA2G6 enzyme plays a role in phospholipid remodeling, arachidonic acid release, prostaglandin and leukotriene synthesis, and apoptosis (Balsinde and Balboa 2005). Prior to the identification of PLAN as a unique NBIA, mutations of the *PLA2G6* gene were reportedly associated with subtypes of infantile neuroaxonal dystrophy (Aicardi and Castelein 1979) and the Karak syndrome (spastic paraparesis, intellectual impairment, seizures, and dysmorphic features) (Mubaidin et al. 2003). Whether pathological brain iron accumulation in patients with PLAN results from aberrant lipid neurochemistry or from some undiscovered (gain-of-function?) action of the mutant enzyme remains unknown.

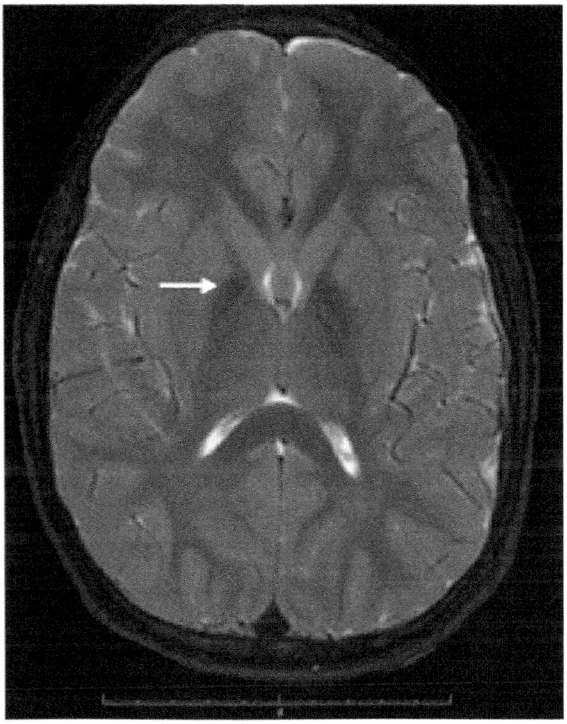

Fig. 19.5 T2-weighted MRI of a 9-year-old patient with phospholipase A2G6-associated neurodegeneration (PLAN). There is symmetrical signal attenuation in the globus pallidus compatible with excessive iron deposition (arrow). (From Gregory et al. (2008), with permission)

19.2.6 Fatty Acid Hydroxylase-Associated Neurodegeneration

Mutations in the fatty acid hydroxylase gene, *FA2H*, are another autosomal recessive etiology of NBIA (Kruer et al. 2010). Mutant *FA2H* has previously been associated with forms of leukodystrophy and hereditary spastic paraplegia (Edvardson et al. 2008). Persons with FA2H-associated neurodegeneration (FAHN) display childhood-onset gait disturbances, dystonia, spastic paraparesis, and profound ataxia, but no pigmentary retinopathy, optic atrophy, or cognitive dysfunction. MRI of FAHN patients is notable for bilateral T2 hypointensities of the globus pallidus commensurate with abundant iron, marked pontocerebellar degeneration, mild diffuse cortical atrophy, narrowing of the corpus callosum, and confluent T2 hyperintensities in periventricular white matter (Dick et al. 2010; Edvardson et al. 2008; Kruer et al. 2010) (Fig. 19.6). FA2H catalyzes the formation of 2-hydroxylated fatty acid precursors of ceramide, galactosylceramide, and sulfatide constituents of normal CNS myelin (Hama 2010).

19.2.7 Woodhouse–Sakati Syndrome

Woodhouse–Sakati syndrome (WSS) is a rare childhood-onset NBIA initially reported in two Saudi families in 1983. Affected individuals exhibit triangular facies, hypertelorism, alopecia, diabetes mellitus, hypogonadism, hypothyroidism, low serum IGF-1 levels, and neurological symptoms (Woodhouse and Sakati 1983; Messina 2024). WSS is an autosomal recessive condition resulting from mutation of the DDB1- and CUL4-associated factor 17 (DCAF17; a.k.a. C2orf37) gene located on chromosome 2q31.1 (Bohlega et al. 2019). Abnormal expression of the DCAF17 protein disrupts the integrity of the nuclear envelope, renders nucleoli hypersensitive to transcriptional blockade, and may impact cell cycle kinetics, mRNA transport, and apoptosis (Alazami et al. 2008; Messina 2024).

Neurological manifestations of WSS include dysarthria, dystonia (oromandibular, craniocervical or limb), dysphagia, sensorineural hearing loss, intellectual deficits, and, less commonly, seizures and sensorimotor polyneuropathy (Messina 2024). MRI of the head

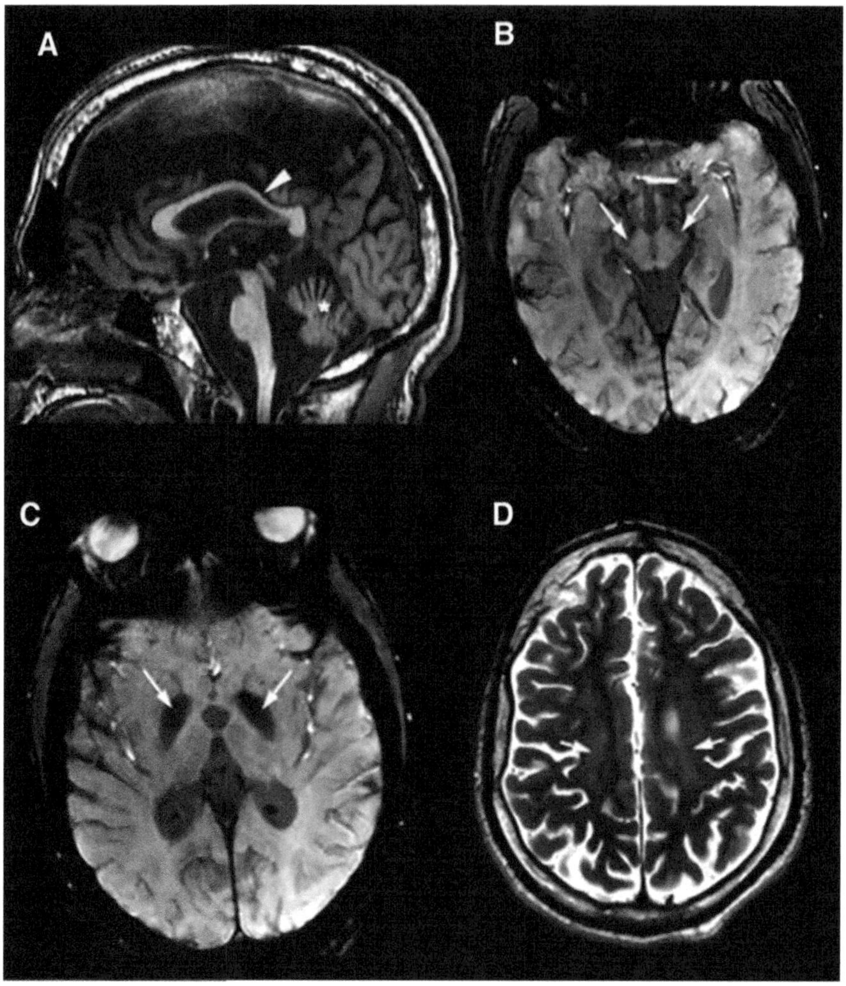

Fig. 19.6 A 29-year-old man with FAHN. (**a**) Sagittal T1-weighted image showing cerebellar (asterisk) and brainstem atrophy and thinning of the corpus callosum (arrow). (**b**, **c**) Axial SWI images showing hypointense signal indicative of iron deposition (arrows) in the substantia nigra (**b**) and the globus pallidus (**c**). (**d**) Axial T2-weighted image showing hyperintense signal in the subcortical white matter (double arrows). (Adapted from Lehericy et al. (2020), with permission)

may show iron deposition in the globus pallidus, substantia nigra, and red nucleus; bilateral white matter T2 hyperintensities in frontotemporal regions and cerebellum; and pituitary hypoplasia (Lehericy et al. 2020) (Fig. 19.7). Prolonged latencies on visual, somatosensory, and auditory evoked potentials may occur (Messina 2024).

As with other forms of NBIA, there is currently no specific pharmacological or gene therapy for WSS. Available interventions mainly target clinical neurologic (e.g., anticholinergic medications, botulinum toxin injections, or deep brain stimulation for dystonia) and endocrinologic (antidiabetic agents, sex steroid and thyroid hormone replacement) manifestations.

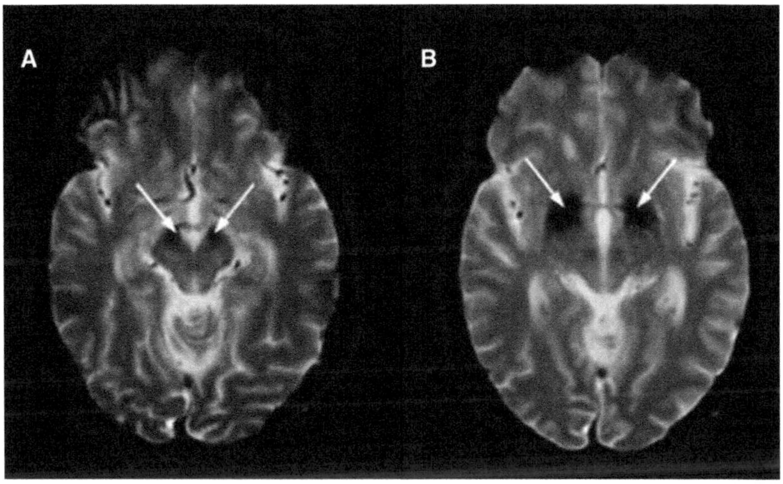

Fig. 19.7 Patient with Woodhouse–Sakati syndrome. T2-weighted images showing hypointense signals indicating iron deposition (arrows) in the substantia nigra (**a**) and the globus pallidus (**b**). (Adapted from Lehericy et al. (2020), with permission)

19.2.8 Beta-Propeller Protein-Associated Neurodegeneration

Beta-propeller protein-associated neurodegeneration (BPAN) is an X-linked dominant disorder resulting from heterozygous or hemizygous germline mutations of the WDR45 gene. The gene codes for WIPI4 (WD40 repeat protein interacting with phosphoinositides 4) which is important for autolysosome biogenesis and autophagy (Hor and Tang 2019). The clinical course of the disease is biphasic: Infancy/childhood is characterized by focal and generalized seizures, intellectual delay, language impairment, and autism. These symptoms are followed in adolescence and early adulthood by progressive dystonia-parkinsonism and cognitive decline. MRI of the brain with T2* or quantitative susceptibility mapping reveals iron deposition in the globus pallidus and substantia nigra accompanied by cerebral cortical and hippocampal atrophy (Fig. 19.8) (Lehericy et al. 2020). There is no specific treatment for BPAN. Levodopa can be considered for management of dystonia-parkinsonism, antiepileptic drugs are administered to control seizures, muscle relaxants or botulinum toxin injections may be used for spasticity, and behavioral therapy and medications (e.g., antidepressants, antipsychotics) may be considered for alleviation of anxiety, irritability, and other psychiatric symptoms (Hoffjan et al. 2016).

19.2.9 Mitochondrial Membrane Protein-Associated Neurodegeneration

Mitochondrial membrane protein-associated neurodegeneration (MPAN) is generally described as an autosomal recessive disease arising from homozygous loss-of-function mutations of the *C19orf12* gene. However, evidence of autosomal dominant inheritance resulting from several C19orf12 pathogenic protein variants with putative dominant negative properties has also been adduced in rare cases (Chini 2019; Gregory et al. 1993a). While the function of the C19orf12 mitochondrial membrane protein remains unknown, a role in lipid metabolism has been proposed (Hartig et al. 2013). Although MPAN typically presents in childhood or adolescence, certain alleles may manifest in young or even middle-aged adults (Gregory et al. 2019).

Clinical manifestations of MPAN are characterized by gait changes early in the disease course, followed by progressive dystonia, spastic lower limb or quadriparesis, parkinsonism, sensorineural hearing loss, optic atrophy, cognitive decline, and various neuropsychiatric sequelae (Gregory et al. 1993a, 2019).

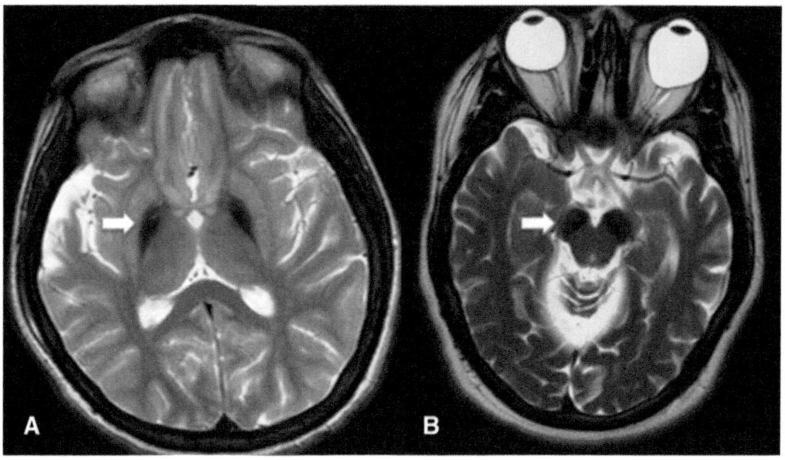

Fig. 19.8 Beta-propeller protein-associated neurodegeneration (BPAN). T_2-weighted sequences demonstrating hypointensity in the globus pallidus (**a**) and substantia nigra (**b**). The findings depicted are not typically present in early childhood when imaging may be normal or show only hypomyelination, a thin corpus callosum, or cerebral/cerebellar volume loss. (Adapted from Gregory et al. (1993b), with permission)

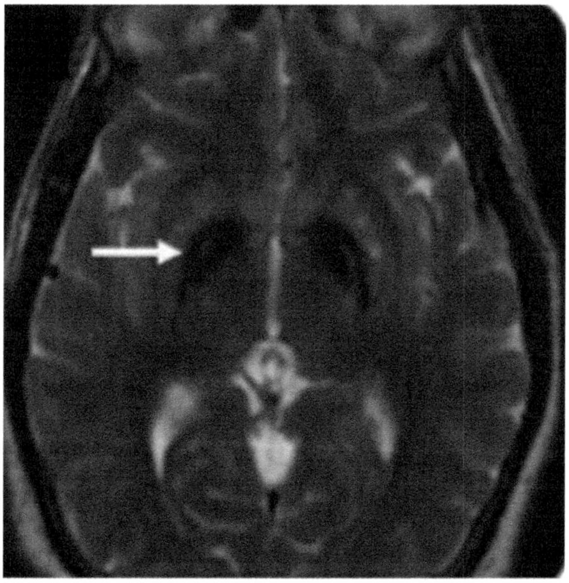

Fig. 19.9 T_2-weighted imaging in mitochondrial membrane protein-associated neurodegeneration (MPAN). Isointense streaking of the medial medullary lamina between the hypointense signal regions in globus pallidus externa and interna, observed in most persons with MPAN, may be mistaken for an "eye-of-the-tiger" sign. (Adapted from Gregory et al. (1993a), with permission)

MRI with iron-sensitive sequences in patients with MPAN reveals deposition of the metal in the globus pallidus, striatum, and substantia nigra. With prolonged disease duration, cerebral cortical or cerebellar atrophy may also be apparent (Fig. 19.9) (Lehericy et al. 2020).

Clinical management of MPAN is primarily symptomatic. In addition to pharmacologic therapy commonly used for parkinsonism, spasticity, dystonia, and neuropsychiatric manifestations, patients are often referred for speech, physical, and occupational therapy (Gregory et al. 1993a). There is little evidence regarding the efficacy of iron chelation in patients with MPAN.

19.2.10 Coenzyme A Synthase Protein-Associated Neurodegeneration

Coenzyme A synthase protein-associated neurodegeneration (CoPAN) is a very rare autosomal

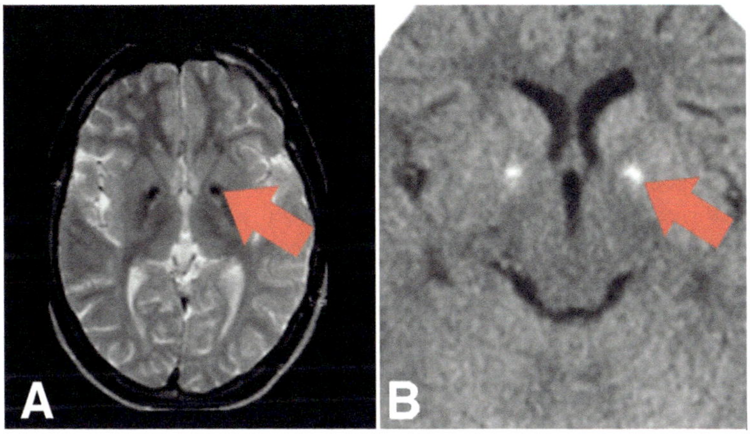

Fig. 19.10 Coenzyme A synthase protein-associated neurodegeneration (CoPAN). (**a**) T2-weighted MRI of the head shows "eye-of-of-the-tiger"-like appearance of the globus pallidus (arrow) consistent with pathological iron deposition. (**b**) CT of the head shows bilateral globus pallidus calcifications (arrow) which may contribute to the hyperintense component of the "eye-of-of-the-tiger" sign in this condition. (Adapted from Dusi et al. (2014), with permission)

recessive NBIA disorder caused by mutations in the *COASY* gene located on chromosome 17. The *COASY* gene encodes the enzyme, coenzyme A synthase which catalyzes CoA biosynthesis. CoA plays a critical role in energy metabolism and fatty acid synthesis. CoPAN typically presents in childhood or early adolescence with a combination of motor and cognitive symptoms. Hallmark clinical features include early-onset parkinsonism, dysarthria, oromandibular dystonia, spasticity, and cognitive decline. Seizures, optic atrophy, and axonal neuropathy are less commonly observed (Hogarth 2015).

Brain MRI in young children with CoPAN demonstrates striatal and thalamic "swelling" and hyperintensity on T2 weighted images which regress over time (Fig. 19.10). After age 7, augmented iron stores are demonstrable in the globus pallidus and substantia nigra. Pallidal calcifications may occur and simulate the "eye-of-the-tiger" sign commonly observed in PKAN. Disease progression is associated with generalized brain atrophy, particularly in cerebral cortex and basal ganglia (Lehericy et al. 2020). Treatment is limited to antiparkinsonian drugs, anticonvulsants, physical therapy, and supportive care.

19.3 Differential Diagnosis

Prominent extrapyramidal dysfunction and iron deposition in basal ganglia are characteristics of PKAN, PLAN, FAHN, KRD, and neuroferritinopathy. The absence of peripheral iron overload in these conditions, and possible hypoferritinemia in neuroferritinopathy, distinguishes these forms of NBIA from aceruloplasminemia (Curtis et al. 2001; Nittis and Gitlin 2002). Although abnormal iron-sensitive MRI sequences of the caudate, putamen, and/or globus pallidus are typical of most NBIA disorders, the "eye-of-the-tiger" sign most frequently denotes PKAN and should prompt molecular testing for *PANK2* mutations. Rarely, this MRI sign may be seen in NBIA cases harboring normal *PANK2* genes such as CoPAN (due to pallidal calcifications) and may occur in non-NBIA conditions including certain acquired dystonias, organic acidurias, Leigh disease, and Wilson disease (Savoiardo et al. 1993; van der Knaap et al. 1991; Rutledge et al. 1988; Kumar et al. 2024). Genetic testing for PLAN or other NBIA may be warranted in the absence of the "eye-of-the-tiger" sign when homogeneous T2 signal attenuation in basal ganglia is encountered in the appropriate clinical setting (Hayflick 2006).

Mixed hyper- and hypointense T2 signals in the thalamus and basal ganglia in the context of autosomal-dominant inheritance favor a diagnosis of neuroferritinopathy. Brain iron overload may not be evident in some individuals with PLAN; the absence of MRI hypointensities in basal ganglia does not rule out this diagnosis in cases presenting with infantile neuroaxonal dystrophy (Hayflick 2006; Yoshino et al. 2010).

Systemic abnormalities of iron metabolism may be appreciated prior to the onset of neurological symptoms in persons with aceruloplasminemia. Such cases may be misdiagnosed as hereditary hemochromatosis. Elevation of plasma transferrin saturation in hemochromatosis and its diminution in aceruloplasminemia assist in the differentiation of these conditions (Nittis and Gitlin 2002). Patients with Wilson disease may exhibit symptoms referable to the basal ganglia and suppressed plasma ceruloplasmin concentrations. However, individuals with aceruloplasminemia do not display increased hepatic and urinary copper levels or corneal copper deposition (Kaiser-Fleischer rings) on slit-lamp examination, in contradistinction to Wilson disease (Nittis and Gitlin 2002). Patients with either hypoceruloplasminemia or Friedreich ataxia may exhibit cerebellar atrophy/dysfunction and T2 signal loss (iron deposition) in the dentate nuclei. Both disorders exhibit autosomal recessive inheritance but can be readily differentiated by diminished circulating ceruloplasmin in the former and dorsal column and cardiac involvement in the latter. Patients with acquired copper deficiency may also manifest low blood ceruloplasmin levels (Henri-Bhargava et al. 2008), but the myelopathy which often complicates this condition is not characteristic of heritable hypoceruloplasminemia. The facial dysmorphism, alopecia, and multiple endocrinopathies help distinguish WSS from other NBIA syndromes (Messina 2024).

19.4 Conclusions

Genetic disruptions of brain and body iron homeostasis and pathological deposition of this redox-active metal in the CNS have been documented in a number of pediatric and adult neurodegenerative conditions. In this chapter, we reviewed the clinical, pathological, and neuroimaging features of ten NBIA syndromes: PKAN, neuroferritinopathy, aceruloplasminemia, KRD, PLAN, FAHN, WSS, BPAN, MPAN, and CoPAN. Although heterozygosity for common gene (*HFE*) mutations responsible for hereditary hemochromatosis have been proposed as risk factors for "acquired" human neurological disorders (PD, AD, ALS, stroke), epidemiological data regarding this matter are highly conflicting, and conjecture regarding the potential impact of the defective *HFE* alleles on brain iron regulation remains unsubstantiated (Schipper and Ponka 2009). Given the rapid pace of discovery of novel proteins governing diverse aspects of iron metabolism, ongoing refinements in MRI and other neuroimaging techniques responsive to tissue iron, and advances in molecular genetic testing, it is almost certain that additional, inheritable NBIA conditions will be identified. Persons with genetic NBIA syndromes may be discovered after pathological brain iron deposition and neurological deficits are first described in animal models bearing specific mutations in genes controlling iron metabolism. For example, after demonstrating that targeted deletion of *Ireb2* engenders iron-deficiency anemia, erythropoietic protoporphyria, and a movement disorder in mice (Cooperman et al. 2005), Rouault and co-workers reported commensurate hematological abnormalities and refractory choreoathetosis in a 16-year-old man with a loss-of-function variant in *IREB2* and absence of iron regulatory protein 2 (IRP2) (Costain et al. 2019). The advent of effective metal chelation therapies, substrate replacement strategies, gene-editing technologies, and other disease-modifying approaches will represent critical milestones in the management of these devastating hereditary disorders of brain iron homeostasis (Marupudi and Xiong 2024; Romano et al. 2022; Spaull et al. 2021).

Acknowledgements The HMS laboratory is supported by the Canadian Institutes of Health Research.

Disclosures HMS served as consultant to Osta Biotechnologies, TEVA Neurosciences, and Caprion Pharmaceuticals. He served as an officer and held stock options in Molecular Biometrics Inc. and HemOx Biotechnologies. HMS has no financial interests related to the material contained in this book chapter. DAS has no disclosures to declare.

References

Agostini F, Sgalletta B, Bisaglia M (2024) Iron dyshomeostasis in neurodegeneration with brain iron accumulation (NBIA): is it the cause or the effect? Cells 13(16). https://doi.org/10.3390/cells13161376

Aicardi J, Castelein P (1979) Infantile neuroaxonal dystrophy. Brain 102(4):727–748

Alazami AM, Al-Saif A, Al-Semari A, Bohlega S, Zlitni S, Alzahrani F, Bavi P, Kaya N, Colak D, Khalak H, Baltus A, Peterlin B, Danda S, Bhatia KP, Schneider SA, Sakati N, Walsh CA, Al-Mohanna F, Meyer B, Alkuraya FS (2008) Mutations in C2orf37, encoding a nucleolar protein, cause hypogonadism, alopecia, diabetes mellitus, mental retardation, and extrapyramidal syndrome. Am J Hum Genet 83(6):684–691. https://doi.org/10.1016/j.ajhg.2008.10.018

Arosio P, Levi S (2002) Ferritin, iron homeostasis, and oxidative damage. Free Radic Biol Med 33(4):457–463

Balsinde J, Balboa MA (2005) Cellular regulation and proposed biological functions of group VIA calcium-independent phospholipase A2 in activated cells. Cell Signal 17(9):1052–1062

Benarroch EE (2009) Brain iron homeostasis and neurodegenerative disease. Neurology 72(16):1436–1440. https://doi.org/10.1212/WNL.0b013e3181a26b30

Bohlega S, Abusrair AH, Al-Ajlan FS, Alharbi N, Al-Semari A, Bohlega B, Abualsaud D, Alkuraya F (2019) Patterns of neurological manifestations in Woodhouse-Sakati syndrome. Parkinsonism Relat Disord 69:99–103. https://doi.org/10.1016/j.parkreldis.2019.10.007

Bruggemann N, Hagenah J, Reetz K, Schmidt A, Kasten M, Buchmann I, Eckerle S, Bahre M, Munchau A, Djarmati A, van der Vegt J, Siebner H, Binkofski F, Ramirez A, Behrens MI, Klein C (2010) Recessively inherited parkinsonism: effect of ATP13A2 mutations on the clinical and neuroimaging phenotype. Arch Neurol 67(11):1357–1363. https://doi.org/10.1001/archneurol.2010.281

Burn J, Chinnery PF (2006) Neuroferritinopathy. Semin Pediatr Neurol 13(3):176–181. https://doi.org/10.1016/j.spen.2006.08.006

Castelnau P, Zilbovicius M, Ribeiro MJ, Hertz-Pannier L, Ogier H, Evrard P (2001) Striatal and pontocerebellar hypoperfusion in Hallervorden-Spatz syndrome. Pediatr Neurol 25(2):170–174

Ching KH, Westaway SK, Gitschier J, Higgins JJ, Hayflick SJ (2002) HARP syndrome is allelic with pantothenate kinase-associated neurodegeneration. Neurology 58(11):1673–1674

Chini B (2019) Expanding neuropeptide signalling by multiplying receptor functional states and sub-cellular locations. Cell Tissue Res 375(1):49–56. https://doi.org/10.1007/s00441-018-2923-x

Chinnery PF, Curtis AR, Fey C, Coulthard A, Crompton D, Curtis A, Lombes A, Burn J (2003) Neuroferritinopathy in a French family with late onset dominant dystonia. J Med Genet 40(5):e69

Chinnery PF, Crompton DE, Birchall D, Jackson MJ, Coulthard A, Lombes A, Quinn N, Wills A, Fletcher N, Mottershead JP, Cooper P, Kellett M, Bates D, Burn J (2007) Clinical features and natural history of neuroferritinopathy caused by the FTL1 460InsA mutation. Brain 130(Pt 1):110–119

Connor JR (1997) Metals and oxidative damage in neurological disorders. Plenum Press, New York

Cooperman SS, Meyron-Holtz EG, Olivierre-Wilson H, Ghosh MC, McConnell JP, Rouault TA (2005) Microcytic anemia, erythropoietic protoporphyria, and neurodegeneration in mice with targeted deletion of iron-regulatory protein 2. Blood 106(3):1084–1091

Costain G, Ghosh MC, Maio N, Carnevale A, Si YC, Rouault TA, Yoon G (2019) Absence of iron-responsive element-binding protein 2 causes a novel neurodegenerative syndrome. Brain 142(5):1195–1202. https://doi.org/10.1093/brain/awz072

Curtis AR, Fey C, Morris CM, Bindoff LA, Ince PG, Chinnery PF, Coulthard A, Jackson MJ, Jackson AP, McHale DP, Hay D, Barker WA, Markham AF, Bates D, Curtis A, Burn J (2001) Mutation in the gene encoding ferritin light polypeptide causes dominant adult-onset basal ganglia disease. Nat Genet 28(4):350–354

Daimon M, Kato T, Kawanami T, Tominaga M, Igarashi M, Yamatani K, Sasaki H (1995) A nonsense mutation of the ceruloplasmin gene in hereditary ceruloplasmin deficiency with diabetes mellitus. Biochem Biophys Res Commun 217(1):89–95

Dick KJ, Eckhardt M, Paisan-Ruiz C, Alshehhi AA, Proukakis C, Sibtain NA, Maier H, Sharifi R, Patton MA, Bashir W, Koul R, Raeburn S, Gieselmann V, Houlden H, Crosby AH (2010) Mutation of FA2H underlies a complicated form of hereditary spastic paraplegia (SPG35). Hum Mutat 31(4):E1251–E1260. https://doi.org/10.1002/humu.21205

Dusi S, Valletta L, Haack TB, Tsuchiya Y, Venco P, Pasqualato S, Goffrini P, Tigano M, Demchenko N, Wieland T, Schwarzmayr T, Strom TM, Invernizzi F, Garavaglia B, Gregory A, Sanford L, Hamada J, Bettencourt C, Houlden H, Chiapparini L, Zorzi G, Kurian MA, Nardocci N, Prokisch H, Hayflick S, Gout I, Tiranti V (2014) Exome sequence reveals mutations in CoA synthase as a cause of neurodegeneration with brain iron accumulation. Am J Hum Genet 94(1):11–22. https://doi.org/10.1016/j.ajhg.2013.11.008

Edvardson S, Hama H, Shaag A, Gomori JM, Berger I, Soffer D, Korman SH, Taustein I, Saada A, Elpeleg O (2008) Mutations in the fatty acid 2-hydroxylase gene are associated with leukodystrophy with spastic paraparesis and dystonia. Am J Hum Genet 83(5):643–648. https://doi.org/10.1016/j.ajhg.2008.10.010

Fasano A, Colosimo C, Miyajima H, Tonali PA, Re TJ, Bentivoglio AR (2008) Aceruloplasminemia: a novel mutation in a family with marked phenotypic variability. Mov Disord 23(5):751–755. https://doi.org/10.1002/mds.21938

Felitti VJ, Beutler E (1999) New developments in hereditary hemochromatosis. Am J Med Sci 318(4):257–268

Gitler AD, Chesi A, Geddie ML, Strathearn KE, Hamamichi S, Hill KJ, Caldwell KA, Caldwell GA, Cooper AA, Rochet JC, Lindquist S (2009) Alpha-synuclein is part of a diverse and highly conserved interaction network that includes PARK9 and manganese toxicity. Nat Genet 41(3):308–315. https://doi.org/10.1038/ng.300
Gregory A, Klopstock T, Kmiec T, Hogarth P, Hayflick SJ (1993a) Mitochondrial membrane protein-associated neurodegeneration. In: Adam MP, Feldman J, Mirzaa GM et al (eds) GeneReviews®. University of Washington, Seattle
Gregory A, Kurian MA, Haack T, Hayflick SJ, Hogarth P (1993b) Beta-propeller protein-associated neurodegeneration. In: Adam MP, Feldman J, Mirzaa GM et al (eds) GeneReviews®. University of Washington, Seattle
Gregory A, Westaway SK, Holm IE, Kotzbauer PT, Hogarth P, Sonek S, Coryell JC, Nguyen TM, Nardocci N, Zorzi G, Rodriguez D, Desguerre I, Bertini E, Simonati A, Levinson B, Dias C, Barbot C, Carrilho I, Santos M, Malik I, Gitschier J, Hayflick SJ (2008) Neurodegeneration associated with genetic defects in phospholipase A(2). Neurology 71(18):1402–1409. https://doi.org/10.1212/01.wnl.0000327094.67726.28
Gregory A, Lotia M, Jeong SY, Fox R, Zhen D, Sanford L, Hamada J, Jahic A, Beetz C, Freed A, Kurian MA, Cullup T, van der Weijden MCM, Nguyen V, Setthavongsack N, Garcia D, Krajbich V, Pham T, Woltjer R, George BP, Minks KQ, Paciorkowski AR, Hogarth P, Jankovic J, Hayflick SJ (2019) Autosomal dominant mitochondrial membrane protein-associated neurodegeneration (MPAN). Mol Genet Genomic Med 7(7):e00736. https://doi.org/10.1002/mgg3.736
Hama H (2010) Fatty acid 2-hydroxylation in mammalian sphingolipid biology. Biochim Biophys Acta 1801(4):405–414. https://doi.org/10.1016/j.bbalip.2009.12.004
Harris ZL, Takahashi Y, Miyajima H, Serizawa M, MacGillivray RT, Gitlin JD (1995) Aceruloplasminemia: molecular characterization of this disorder of iron metabolism. Proc Natl Acad Sci USA 92(7):2539–2543
Hartig M, Prokisch H, Meitinger T, Klopstock T (2013) Mitochondrial membrane protein-associated neurodegeneration (MPAN). Int Rev Neurobiol 110:73–84. https://doi.org/10.1016/B978-0-12-410502-7.00004-1
Hayflick SJ (2003) Unraveling the Hallervorden-Spatz syndrome: pantothenate kinase-associated neurodegeneration is the name. Curr Opin Pediatr 15(6):572–577
Hayflick SJ (2006) Neurodegeneration with brain iron accumulation: from genes to pathogenesis. Semin Pediatr Neurol 13(3):182–185
Hayflick SJ (2023) A brief history of NBIA gene discovery. J Mov Disord 16(2):133–137. https://doi.org/10.14802/jmd.23014
Hellman NE, Gitlin JD (2002) Ceruloplasmin metabolism and function. Annu Rev Nutr 22:439–458
Henri-Bhargava A, Melmed C, Glikstein R, Schipper HM (2008) Neurologic impairment due to vitamin E and copper deficiencies in celiac disease. Neurology 71(11):860–861. https://doi.org/10.1212/01.wnl.0000325473.13088.18
Hoffjan S, Ibisler A, Tschentscher A, Dekomien G, Bidinost C, Rosa AL (2016) WDR45 mutations in Rett (-like) syndrome and developmental delay: case report and an appraisal of the literature. Mol Cell Probes 30(1):44–49. https://doi.org/10.1016/j.mcp.2016.01.003
Hogarth P (2015) Neurodegeneration with brain iron accumulation: diagnosis and management. J Mov Disord 8(1):1–13. https://doi.org/10.14802/jmd.14034
Holla VV, Samim MM, Kumari R, Dhar D, Phulpagar P, Sriram N, Prasad S, Saini J, Kamble N, Yadav R, Muthusamy B, Pal PK (2024) The clinical, radiological and genetic spectrum of PLA2G6-associated neurodegeneration: an experience from a tertiary center. Tremor Other Hyperkinet Mov (N Y) 14:41. https://doi.org/10.5334/tohm.897
Hor CHH, Tang BL (2019) Beta-propeller protein-associated neurodegeneration (BPAN) as a genetically simple model of multifaceted neuropathology resulting from defects in autophagy. Rev Neurosci 30(3):261–277. https://doi.org/10.1515/revneuro-2018-0045
Iankova V, Karin I, Klopstock T, Schneider SA (2021) Emerging disease-modifying therapies in neurodegeneration with brain iron accumulation (NBIA) disorders. Front Neurol 12:629414. https://doi.org/10.3389/fneur.2021.629414
Johnson MA, Kuo YM, Westaway SK, Parker SM, Ching KH, Gitschier J, Hayflick SJ (2004) Mitochondrial localization of human PANK2 and hypotheses of secondary iron accumulation in pantothenate kinase-associated neurodegeneration. Ann N Y Acad Sci 1012:282–298
Justesen CR, Penn RD, Kroin JS, Egel RT (1999) Stereotactic pallidotomy in a child with Hallervorden-Spatz disease. Case report. J Neurosurg 90(3):551–554
Klopstock T, Tricta F, Neumayr L, Karin I, Zorzi G, Fradette C, Kmiec T, Buchner B, Steele HE, Horvath R, Chinnery PF, Basu A, Kupper C, Neuhofer C, Kalman B, Dusek P, Yapici Z, Wilson I, Zhao F, Zibordi F, Nardocci N, Aguilar C, Hayflick SJ, Spino M, Blamire AM, Hogarth P, Vichinsky E (2019) Safety and efficacy of deferiprone for pantothenate kinase-associated neurodegeneration: a randomised, double-blind, controlled trial and an open-label extension study. Lancet Neurol 18(7):631–642. https://doi.org/10.1016/S1474-4422(19)30142-5
Kruer MC, Paisan-Ruiz C, Boddaert N, Yoon MY, Hama H, Gregory A, Malandrini A, Woltjer RL, Munnich A, Gobin S, Polster BJ, Palmeri S, Edvardson S, Hardy J, Houlden H, Hayflick SJ (2010) Defective FA2H leads to a novel form of neurodegeneration with brain iron accumulation (NBIA). Ann Neurol 68(5):611–618. https://doi.org/10.1002/ana.22122

Kumar S, Pandey S, Malhotra HS, Garg RK (2024) Is the "eye of tiger" really emblematic of pantothenate kinase-associated neurodegeneration type 1? An uncommon MR image in Wilson's disease. Neurol India 72(1):221–222. https://doi.org/10.4103/neurol--india.Neurol-India-D-23-00696
Lehericy S, Roze E, Goizet C, Mochel F (2020) MRI of neurodegeneration with brain iron accumulation. Curr Opin Neurol 33(4):462–473. https://doi.org/10.1097/WCO.0000000000000844
Levi S, Cozzi A, Arosio P (2005) Neuroferritinopathy: a neurodegenerative disorder associated with L-ferritin mutation. Best Pract Res Clin Haematol 18(2):265–276. https://doi.org/10.1016/j.beha.2004.08.021
Li X, Jankovic J, Le W (2010) Iron chelation and neuroprotection in neurodegenerative diseases. J Neural Transm. https://doi.org/10.1007/s00702-010-0518-0
Mak CM, Sheng B, Lee HH, Lau KK, Chan WT, Lam CW, Chan YW (2011) Young-onset parkinsonism in a Hong Kong Chinese man with adult-onset Hallervorden-Spatz syndrome. Int J Neurosci. https://doi.org/10.3109/00207454.2010.542843
Mancuso M, Davidzon G, Kurlan RM, Tawil R, Bonilla E, Di Mauro S, Powers JM (2005) Hereditary ferritinopathy: a novel mutation, its cellular pathology, and pathogenetic insights. J Neuropathol Exp Neurol 64(4):280–294
Marchand F, Moreau C, Kuchcinski G, Huin V, Defebvre L, Devos D (2022) Conservative iron chelation for neuroferritinopathy. Mov Disord 37(9):1948–1952. https://doi.org/10.1002/mds.29145
Marupudi N, Xiong MP (2024) Genetic targets and applications of iron chelators for neurodegeneration with brain iron accumulation. ACS Bio Med Chem Au 4(3):119–130. https://doi.org/10.1021/acsbiomedchemau.3c00066
McNeill A, Pandolfo M, Kuhn J, Shang H, Miyajima H (2008) The neurological presentation of ceruloplasmin gene mutations. Eur Neurol 60(4):200–205. https://doi.org/10.1159/000148691
Messina C (2024) Woodhouse-Sakati syndrome: a review. Rev Neurol (Paris). https://doi.org/10.1016/j.neurol.2023.11.008
Miyajima H, Nishimura Y, Mizoguchi K, Sakamoto M, Shimizu T, Honda N (1987) Familial apoceruloplasmin deficiency associated with blepharospasm and retinal degeneration. Neurology 37(5):761–767
Miyajima H, Kono S, Takahashi Y, Sugimoto M, Sakamoto M, Sakai N (2001) Cerebellar ataxia associated with heteroallelic ceruloplasmin gene mutation. Neurology 57(12):2205–2210
Miyake Z, Nakamagoe K, Yoshida K, Kondo T, Tamaoka A (2020) Deferasirox might be effective for microcytic anemia and neurological symptoms associated with aceruloplasminemia: a case report and review of the literature. Intern Med 59(14):1755–1761. https://doi.org/10.2169/internalmedicine.4178-19
Morgan NV, Westaway SK, Morton JE, Gregory A, Gissen P, Sonek S, Cangul H, Coryell J, Canham N, Nardocci N, Zorzi G, Pasha S, Rodriguez D, Desguerre I, Mubaidin A, Bertini E, Trembath RC, Simonati A, Schanen C, Johnson CA, Levinson B, Woods CG, Wilmot B, Kramer P, Gitschier J, Maher ER, Hayflick SJ (2006) PLA2G6, encoding a phospholipase A2, is mutated in neurodegenerative disorders with high brain iron. Nat Genet 38(7):752–754
Mubaidin A, Roberts E, Hampshire D, Dehyyat M, Shurbaji A, Mubaidien M, Jamil A, Al-Din A, Kurdi A, Woods CG (2003) Karak syndrome: a novel degenerative disorder of the basal ganglia and cerebellum. J Med Genet 40(7):543–546
Murley AG, Rua C, Biggs H, Rodgers CT, Matys T, van den Ameele J, Horvath R, Chinnery PF (2024) 7T MRI detects widespread brain iron deposition in neuroferritinopathy. Ann Clin Transl Neurol 11(5):1359–1364. https://doi.org/10.1002/acn3.52053
Najim al-Din AS, Wriekat A, Mubaidin A, Dasouki M, Hiari M (1994) Pallido-pyramidal degeneration, supranuclear upgaze paresis and dementia: Kufor-Rakeb syndrome. Acta Neurol Scand 89(5):347–352
Nittis T, Gitlin JD (2002) The copper-iron connection: hereditary aceruloplasminemia. Semin Hematol 39(4):282–289
Okamoto N, Wada S, Oga T, Kawabata Y, Baba Y, Habu D, Takeda Z, Wada Y (1996) Hereditary ceruloplasmin deficiency with hemosiderosis. Hum Genet 97(6):755–758
Patel BN, David S (1997) A novel glycosylphosphatidylinositol-anchored form of ceruloplasmin is expressed by mammalian astrocytes. J Biol Chem 272(32):20185–20190
Perry TL, Norman MG, Yong VW, Whiting S, Crichton JU, Hansen S, Kish SJ (1985) Hallervorden-Spatz disease: cysteine accumulation and cysteine dioxygenase deficiency in the globus pallidus. Ann Neurol 18(4):482–489
Ponka P, Beaumont C, Richardson DR (1998) Function and regulation of transferrin and ferritin. Semin Hematol 35(1):35–54
Ramirez A, Heimbach A, Grundemann J, Stiller B, Hampshire D, Cid LP, Goebel I, Mubaidin AF, Wriekat AL, Roeper J, Al-Din A, Hillmer AM, Karsak M, Liss B, Woods CG, Behrens MI, Kubisch C (2006) Hereditary parkinsonism with dementia is caused by mutations in ATP13A2, encoding a lysosomal type 5 P-type ATPase. Nat Genet 38(10):1184–1191. https://doi.org/10.1038/ng1884
Romano N, Baiardi G, Pinto VM, Quintino S, Gianesin B, Sasso R, Diociasi A, Mattioli F, Marchese R, Abbruzzese G, Castaldi A, Forni GL (2022) Long-term neuroradiological and clinical evaluation of NBIA patients treated with a deferiprone based iron-chelation therapy. J Clin Med 11(15). https://doi.org/10.3390/jcm11154524
Rump P, Lemmink HH, Verschuuren-Bemelmans CC, Grootscholten PM, Fock JM, Hayflick SJ, Westaway SK, Vos YJ, van Essen AJ (2005) A novel 3-bp deletion in the PANK2 gene of Dutch patients with pantothenate kinase-associated neurodegeneration: evidence

for a founder effect. Neurogenetics 6(4):201–207. https://doi.org/10.1007/s10048-005-0018-9
Rutledge JN, Hilal SK, Silver AJ, Defendini R, Fahn S (1988) Magnetic resonance imaging of dystonic states. Adv Neurol 50:265–275
Saffie P, Fernandez E, Mariacca S, Salles P, Chana P (2023) Kufor-Rakef syndrome, report of two unrelated cases from Chile studied with DAT-PET. Mov Disord 2023:38
Satolli S, Di Fonzo A, Zanobio M, Pezzullo G, De Micco R (2023) Kufor Rakeb syndrome without gaze palsy and pyramidal signs due to novel ATP13A2 mutations. Neurol Sci 44(10):3723–3725. https://doi.org/10.1007/s10072-023-06899-2
Savoiardo M, Halliday WC, Nardocci N, Strada L, D'Incerti L, Angelini L, Rumi V, Tesoro-Tess JD (1993) Hallervorden-Spatz disease: MR and pathologic findings. AJNR Am J Neuroradiol 14(1):155–162
Schipper HM (2004) Brain iron deposition and the free radical-mitochondrial theory of ageing. Ageing Res Rev 3:265–301
Schipper HM (2009) Chapter 123. Metal accumulation during aging. In: Squire LR (ed) Encyclopedia of neuroscience. Academic, Oxford, pp 811–818
Schipper HM (2011) Redox neurology. In: Pantopoulos K, Schipper HM (eds) Principles of free radical biomedicine. Nova Science Publishers, Hauppauge
Schipper HM (2012) Neurodegeneration with brain iron accumulation—clinical syndromes and neuroimaging. Biochim Biophys Acta 1822(3):350–360. https://doi.org/10.1016/j.bbadis.2011.06.016
Schipper HM, Ponka P (2009) Inherited disorders of brain iron homeostasis. In: Yehudah S, Mostofsky DI (eds) Iron deficiency and overload: from basic biology to clinical medicine. Human Press Inc., Totowa, pp 251–276
Schneider SA, Bhatia KP, Hardy J (2009) Complicated recessive dystonia parkinsonism syndromes. Mov Disord 24(4):490–499. https://doi.org/10.1002/mds.22314
Schneider SA, Paisan-Ruiz C, Quinn NP, Lees AJ, Houlden H, Hardy J, Bhatia KP (2010) ATP13A2 mutations (PARK9) cause neurodegeneration with brain iron accumulation. Mov Disord 25(8):979–984. https://doi.org/10.1002/mds.22947
Schroder JM (2005) Ferritinopathy: diagnosis by muscle or nerve biopsy, with a note on other nuclear inclusion body diseases. Acta Neuropathol 109(1):109–114
Seo JH, Song SK, Lee PH (2009) A novel PANK2 mutation in a patient with atypical pantothenate-kinase-associated neurodegeneration presenting with adult-onset parkinsonism. J Clin Neurol 5(4):192–194. https://doi.org/10.3988/jcn.2009.5.4.192
Spaull RVV, Soo AKS, Hogarth P, Hayflick SJ, Kurian MA (2021) Towards precision therapies for inherited disorders of neurodegeneration with brain iron accumulation. Tremor Other Hyperkinet Mov (N Y) 11:51. https://doi.org/10.5334/tohm.661
Stankiewicz J, Panter SS, Neema M, Arora A, Batt CE, Bakshi R (2007) Iron in chronic brain disorders: imaging and neurotherapeutic implications. Neurotherapeutics 4(3):371–386. https://doi.org/10.1016/j.nurt.2007.05.006
Swaiman KF (2001) Hallervorden-Spatz syndrome. Pediatr Neurol 25(2):102–108
Swaiman KF, Smith SA, Trock GL, Siddiqui AR (1983) Sea-blue histiocytes, lymphocytic cytosomes, movement disorder and 59Fe-uptake in basal ganglia: Hallervorden-Spatz disease or ceroid storage disease with abnormal isotope scan? Neurology 33(3):301–305
Tridimas A, Gillett GT, Pollard S, Sadasivam N, Williams A, Mellor K, Catchpole A, Stepien KM (2021) Three-year follow up of using combination therapy with fresh-frozen plasma and iron chelation in a patient with acaeruloplasminemia. JIMD Rep 57(1):23–28. https://doi.org/10.1002/jmd2.12176
van der Knaap MS, Valk J, de Neeling N, Nauta JJ (1991) Pattern recognition in magnetic resonance imaging of white matter disorders in children and young adults. Neuroradiology 33(6):478–493
Vasconcelos OM, Harter DH, Duffy C, McDonough B, Seidman JG, Seidman CE, Campbell WW (2003) Adult Hallervorden-Spatz syndrome simulating amyotrophic lateral sclerosis. Muscle Nerve 28(1):118–122. https://doi.org/10.1002/mus.10389
Vidal R, Ghetti B, Takao M, Brefel-Courbon C, Uro-Coste E, Glazier BS, Siani V, Benson MD, Calvas P, Miravalle L, Rascol O, Delisle MB (2004) Intracellular ferritin accumulation in neural and extraneural tissue characterizes a neurodegenerative disease associated with a mutation in the ferritin light polypeptide gene. J Neuropathol Exp Neurol 63(4):363–380
Walterfang M, Evans A, Looi J, Jung HH, Danek A, Walker R, Velakoulis D (2011) The neuropsychiatry of neuroacanthocytosis syndromes. Neurosci Biobehav Rev. https://doi.org/10.1016/j.neubiorev.2011.01.001
Wills AJ, Sawle GV, Guilbert PR, Curtis AR (2002) Palatal tremor and cognitive decline in neuroferritinopathy. J Neurol Neurosurg Psychiatry 73(1):91–92
Woodhouse NJ, Sakati NA (1983) A syndrome of hypogonadism, alopecia, diabetes mellitus, mental retardation, deafness, and ECG abnormalities. J Med Genet 20(3):216–219. https://doi.org/10.1136/jmg.20.3.216
Yonekawa M, Okabe T, Asamoto Y, Ohta M (1999) A case of hereditary ceruloplasmin deficiency with iron deposition in the brain associated with chorea, dementia, diabetes mellitus and retinal pigmentation: administration of fresh-frozen human plasma. Eur Neurol 42(3):157–162
Yoshida K, Furihata K, Takeda S, Nakamura A, Yamamoto K, Morita H, Hiyamuta S, Ikeda S, Shimizu N, Yanagisawa N (1995) A mutation in the ceruloplasmin gene is associated with systemic hemosiderosis in humans. Nat Genet 9(3):267–272
Yoshino H, Tomiyama H, Tachibana N, Ogaki K, Li Y, Funayama M, Hashimoto T, Takashima S, Hattori N (2010) Phenotypic spectrum of patients with PLA2G6 mutation and PARK14-linked parkinson-

ism. Neurology 75(15):1356–1361. https://doi.org/10.1212/WNL.0b013e3181f73649
Zhai Z, Sun K, Liu T, Liang S, Ding C, Ren S, Wei S, Zhai F, Zhang G (2024) Deep brain stimulation for pediatric pantothenate kinase-associated neurodegeneration with status dystonicus: a case report and literature review. Clin Neurol Neurosurg 241:108306. https://doi.org/10.1016/j.clineuro.2024.108306
Zhou B, Westaway SK, Levinson B, Johnson MA, Gitschier J, Hayflick SJ (2001) A novel pantothenate kinase gene (PANK2) is defective in Hallervorden-Spatz syndrome. Nat Genet 28(4):345–349. https://doi.org/10.1038/ng572ng572
Zimmerman EA, Schenck JF (2007) Neurodegeneration with brain iron accumulation. MedLink Neurology. http://www.medlink.com/cip.asp?UID=mlt002h8&src=Search&ref=19850753
Zupanc ML, Chun RW, Gilbert-Barness EF (1990) Osmiophilic deposits in cytosomes in Hallervorden-Spatz syndrome. Pediatr Neurol 6(5):349–352

20 Iron and Bone Pathophysiology

Martina Rauner

Abstract

Clinical and preclinical studies indicate that unbalanced systemic iron concentrations negatively affect bone health. Despite distinct underlying mechanisms, both iron deficiency and iron overload result in decreased bone mineral density and an increased fracture risk, the latter being associated with higher morbidity and mortality rates. In iron deficiency, both osteoclasts and osteoblasts suffer from the lack of iron to efficiently execute metabolic processes and produce enough energy for bone resorption and formation, respectively. In contrast, iron overload inhibits osteoblast differentiation and function by increasing the production of reactive oxygen species and lipid peroxidation, while it stimulates osteoclastogenesis and bone resorption. Besides the direct effects of iron on bone, regulators of systemic iron homeostasis have also been shown to impact bone homeostasis. This chapter summarizes current knowledge on the regulation of bone cell function and bone homeostasis by iron. Furthermore, it discusses new avenues for the treatment of iron-related bone loss.

M. Rauner (✉)
Department of Medicine III & Center for Healthy Aging, Technische Universität Dresden, Dresden, Germany
e-mail: Martina.Rauner@ukdd.de

Keywords

Iron · Hemochromatosis proteins · Osteoblast · Osteoclast · Osteoporosis

20.1 Introduction

Dysregulation of iron metabolism is associated with several negative outcomes on health, the most prominent being anemia due to iron deficiency (Chap. 12) or liver disease due to iron overload conditions (Chap. 16). However, besides the liver, red blood cell compartment, and endocrine organs, such as the pancreas, gonads, and pituitary gland, also bone is an organ that suffers from dysbalanced iron levels (Ledesma-Colunga et al. 2021; Baschant et al. 2022). Both iron deficiency and iron overload have been associated with loss of bone mineral density (BMD) and an increased risk of fractures. While iron overload is frequently associated with increased bone resorption and decreased bone formation, in iron deficiency, bone resorption and formation are both impaired, leading to low bone mass and increased fracture rates. In addition, recent evidence suggests that intravenous administration of certain forms of iron, in particular ferric carboxymaltose, which is used to treat severe or refractory forms of iron-deficiency anemia, can be associated with transient fibroblast growth factor (FGF)-23-induced hypophosphatemia (Wolf

K. Pantopoulos (ed.), *Iron Metabolism in Human Health and Disease*, Advances in Experimental Medicine and Biology 1480, https://doi.org/10.1007/978-3-031-92033-2_20

et al. 2020); see also Chap. 24. Case reports indicate that this can lead to phosphate wasting and hypophosphatemic osteomalacia (Klein et al. 2018; Arora et al. 2023; Bartko et al. 2018). Thus, considering these manifold clinical associations of iron and bone, it is pertinent to better understand how iron and its imbalance affect bone biology in order to prevent iron-associated bone fragility or identify treatment schemes to ameliorate iron-induced bone loss.

In this chapter, bone homeostasis will be briefly reviewed, followed by a presentation of the clinical evidence linking iron dysregulation to bone disease. Potential underlying mechanisms of iron deficiency and iron overload-induced bone loss based on preclinical and cell culture-based models will be discussed.

20.2 Bone Homeostasis

Bone health is an integral part of our overall well-being as it ensures mobility and protection of internal organs and is involved in mineral homeostasis, endocrine functions, hematopoiesis, and immune functions. Bone is a highly dynamic tissue that undergoes constant remodeling to adapt to the changing mechanical loads or metabolic demands throughout life. About 10% of the bone surfaces are being remodeled at a given time point, ensuring that microcracks and other small damages are instantly repaired and matrix quality maintained (Parfitt 2002).

Bone remodeling is carried out by the spatio-temporally coupled action of two main cell types: osteoclasts, which resorb bone, and osteoblasts, which form a new bone matrix (Fig. 20.1). Osteoclasts are derived from hematopoietic stem cells, more specifically, the monocyte/macrophage lineage (Feng and Teitelbaum 2013; Jacome-Galarza et al. 2019). The exposure of these mononuclear precursors to the receptor activator of NF-κB ligand (RANKL) induces signaling via its receptor RANK, which initiates the differentiation and fusion of several precursors into large multinucleated syncytia, which, once attached to the bone surface, differentiate into fully mature, bone-resorbing osteoclasts. Osteoclasts resorb the bone matrix by acidifying the resorption lacunae to dissolve the inorganic hydroxyapatite mineral and by secreting enzymes such as cathepsin K, which work best in an acidic environment and break down the type I collagen matrix. Once bone resorption is completed, a reversal phase is initiated that allows osteoblasts to take over and refill the resorption lacunae with newly formed bone. Osteoblasts derive from the mesenchymal skeletal stem and progenitor cell under the influence of various osteogenic signaling pathways, such as the Wnt and bone morphogenetic protein (BMP) pathways that activate critical transcription factors for osteogenic differentiation (e.g., Runx2/Cbfa1, Osterix/Sp7) (Chan et al. 2015, 2018). Osteoblasts are responsible for the production of the organic matrix composed mainly of type I collagen (also referred to as osteoid) and the induction of its mineralization. In addition, they are critical regulators of osteoclast differentiation and function by secreting the osteoclast-promoting factor RANKL as well as its decoy receptor osteoprotegerin (OPG) and are implicated in maintaining hematopoietic stem cell niches (Calvi et al. 2003; Ponzetti and Rucci 2021) (Fig. 20.1). Once they become embedded by the extracellular matrix they secreted, osteoblasts turn into osteocytes. With their stellate appearance and processes spanning throughout the bone tissue, osteocytes are regarded as the mechanosensors and detectors of microdamages in bone (Bonewald 2011). They initiate the bone remodeling process and coordinate the actions of osteoclasts and osteoblasts via secreting factors such as RANKL and Wnt inhibitors, respectively. Besides their role in bone remodeling, osteocytes are the main producers of FGF-23, which regulates phosphate levels in the blood.

Bone remodeling and consequently bone homeostasis are under the strict control of various external (e.g., mechanical loading, diet), systemic (e.g., vitamin D, parathyroid hormone, sex hormones, etc.), and local factors (cytokines, growth factors) (Fig. 20.1) (Siddiqui and Partridge 2016). In addition, age and sex, as well as diseases and medications, can affect bone remodeling (Fig. 20.1). Estrogen deficiency after menopause is a main driver of bone loss in post-

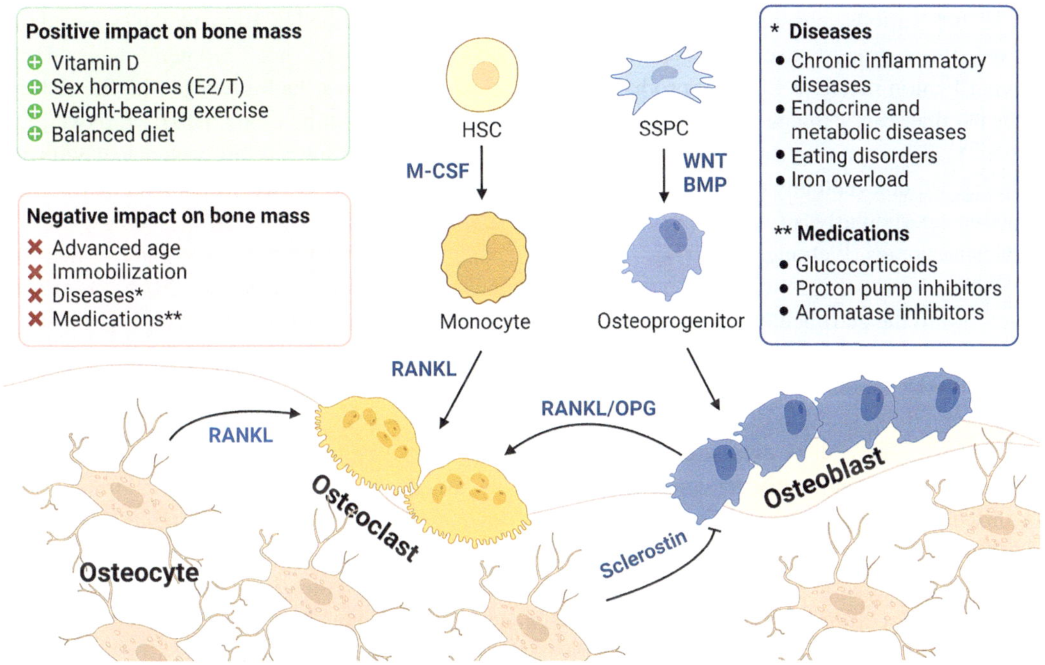

Fig. 20.1 Bone remodeling and its control. Bone remodeling is carried out by osteoclasts, derived from the hematopoietic stem cell (HSC) under the influence of macrophage-colony stimulating factors (M-CSF) and receptor activator of NF-κB ligand (RANKL), and osteoblasts, stemming from the mesenchymal skeletal stem and progenitor cell (SSPC). Wnt and the BMP pathway drive the differentiation of SSPC into osteoblasts, which produce the new bone matrix. Once embedded by the mineralized matrix, osteoblasts turn into osteocytes, which, with their long dendrites, act as mechanosensors in bone. As they can detect microcracks and produce factors that stimulate osteoclasts (RANKL) and inhibit osteoblasts (sclerostin), they are the main initiators of bone remodeling. Various hormones, including vitamin D and sex hormones (estrogens (E2) and testosterone (T)), as well as weight-bearing exercise and a balanced diet, contribute to healthy bone (green box). Advanced age, a sedentary life style, poor diet, disease, and medications can negatively impact bone (red box). A non-exhaustive list of diseases and medications that negatively affect bone is shown in the blue box. (Created with Biorender)

menopausal osteoporosis, the most frequent form of osteoporosis. However, chronic inflammatory diseases (e.g., rheumatoid arthritis), metabolic diseases (e.g., diabetes mellitus, thyroid diseases), or malignant diseases as well as medications including glucocorticoids, aromatase inhibitors, and proton pump inhibitors, can decrease bone mass and increase fracture risk (Rachner et al. 2011) (Fig. 20.1). In recent years, several studies have shown that patients with iron overload and iron deficiency also have a higher risk for fractures (Jorgensen et al. 2010; Looker 2014; Merriman et al. 2010; Valderrabano et al. 2017; Kim et al. 2013; Lipovetzki et al. 2017). As osteoporotic fractures impose not only a substantial burden on affected individuals, being associated with pain, immobility, and higher mortality, but are also associated with high healthcare costs due to hospitalization, rehabilitation, and loss of working force (Abrahamsen et al. 2009), the prevention of osteoporotic fractures is of utmost importance. This includes obtaining a better understanding of the cellular and molecular mechanisms leading to impaired bone quality and increased fracture risk.

20.3 Iron Overload-Induced Bone Pathologies

20.3.1 Clinical Evidence

Even though the capacity of iron to easily exchange electrons makes it an indispensable metal for various physiological cellular processes, these properties can be harmful when driving Fenton reactions, thereby producing free radicals that can damage cellular components (Chap. 1). Thus, iron overload can result in tissue damage, which eventually can lead to organ dysfunction, particularly of the liver, heart, and endocrine organs (Pietrangelo 2007).

Studies from patients with iron overload also point toward the detrimental role of iron overload on bone. Excessive iron accumulation can be a result of interruptions in systemic iron regulation, enhanced red blood cell lysis occurring in certain types of anemias, or liver diseases, in which iron regulation and hepcidin production are disturbed. In this chapter, the main focus will be on iron overload due to dysregulation in systemic iron regulators and hemolytic anemias, even though patients with chronic liver disease frequently present with iron overload and osteoporosis (Guanabens and Pares 2018; VanWagner and Rinella 2016).

Mutations in various systemic iron regulators such as hemochromatosis protein (HFE), hemojuvelin (HJV), transferrin receptor 2 (TFR2), hepcidin (HAMP), and loss-of-function mutations in ferroportin (FPN) cause various degrees of iron overload (Galy et al. 2024; Muckenthaler et al. 2017). With HFE-hemochromatosis being the most common form of hereditary hemochromatosis (Chap. 9), this is also the most investigated type of hemochromatosis with regard to bone. Osteoporosis was reported in about 25% of patients with HFE-hemochromatosis, and over 50% of patients presented with low BMD (Guggenbuhl et al. 2005; Valenti et al. 2009; Jandl et al. 2020), with the lumbar spine being more affected than the femoral neck. Also, the risk of vertebral and wrist fractures is increased (Diamond et al. 1989; Richette et al. 2010). A small study using high-resolution imaging showed that cortical bone is more severely affected than trabecular bone at the distal radius and distal tibia (Jandl et al. 2020). The increased incidence of osteoporosis was confirmed by larger studies including data from the UK Biobank and a cross-sectional study with over 300 patients (Pilling et al. 2019; Richette et al. 2010). Of note, the UK Biobank data showed that only men, but not women with HFE-hemochromatosis, had a twofold higher risk for osteoporosis (Pilling et al. 2019). Moreover, low BMD and fracture risk are associated with the severity of iron overload, as patients with ferritin levels above 1000 μg/L had the highest rates of bone loss and fractures (Guggenbuhl et al. 2005; Richette et al. 2010; Kim et al. 2012, 2013). This was further confirmed in a large population-based matched cohort study with UK data, showing a twofold higher risk for fractures in patients with ferritin levels above 1000 μg/L. The highest risk was identified for vertebral fractures. Treatment with phlebotomy ameliorated bone loss at the lumbar spine in patients with HFE-hemochromatosis (Diamond et al. 1989; Valenti et al. 2009). To date, the contribution of iron toxicity *per se* versus the endocrinopathies related to iron overload is not fully understood and disentangled. Studies indicate higher rates of bone loss in HFE-hemochromatosis patients with hypogonadism than without (Guggenbuhl et al. 2005; Valenti et al. 2009; Diamond et al. 1989). However, even HFE-hemochromatosis patients without hypogonadism present with bone loss (Valenti et al. 2009), indicating that iron overload-induced endocrinopathies likely exaggerate the direct negative effects of iron on bone.

Hemoglobinopathies are a group of hereditary diseases affecting hemoglobin synthesis. Thalassemia and sickle cell disease are the most common types of hemoglobinopathies and present with various degrees of anemia (Chap. 11). In case of severe or persistent anemia, patients obtain blood transfusions to manage their anemia, which, however, is associated with iron overload (Weatherall 1997). To avoid iron overload, patients receive iron chelation therapy (Bayanzay and Alzoebie 2016); see also Chap. 23. However, in most patients, iron overload is inevitable, leading to end-organ damage. Osteoporosis is one of the most frequent co-morbidities in these patients, with up to 70% of patients showing low bone mass and an increased risk for fracture (De Sanctis et al. 2018; Almeida and Roberts 2005; Sarrai et al. 2007; Dede et al. 2016). Of note, also other forms of hemolytic anemias, such as hemolytic anemia caused by pyruvate kinase deficiency, are associated with iron overload and bone loss in up to 20% of the patients (Glenthoj et al. 2024). Bone disease in these patients is likely multifactorial, with the anemia *per se* potentially adding to the increased fracture risk, as well as the transfusion-induced iron overload with its associated endocrinopathies including hypogonadism (Vogiatzi et al. 2009). Moreover, thalassemia is associated with high serum levels of erythropoietin as well as FGF-23, which both have negative effects on bone. While persistently high levels of erythropoietin suppress bone formation and stimulate bone resorption, FGF-23 induces poor bone mineralization by suppressing osteoblast functions (Aprile et al. 2023; Hiram-Bab et al. 2015; Rauner et al. 2016). In addition, FGF-23 also exerts negative effects on the erythroid compartment, thus being an interesting target to simultaneously treat bone and erythroid effects in diseases with ineffective erythropoiesis (Aprile et al. 2023; Weidner et al. 2020). Finally, insulin-like growth factor (IGF-1), which is produced in the liver and has important bone-anabolic effects, is reduced in the serum of patients with thalassemia and negatively correlates with vertebral fracture risk (Costanzo et al. 2024).

20.3.2 Preclinical Evidence and Underlying Mechanisms

At a mechanistic level, iron overload can cause oxidative stress due to an excessive amount of circulating non-transferrin-bound iron (NTBI), which facilitates the production of reactive oxygen species (ROS) and lipid peroxidation (Winterbourn 1995). The important role of increased ROS levels and lipid peroxidation, which is a hallmark of ferroptosis (i.e., an iron-dependent form of cell death), in iron overload-induced bone loss has been verified in multiple preclinical and cell culture-based models (Chap. 5). By treating iron-overloaded mice or rats with anti-oxidants (e.g., niacin or icariin) or blocking important ROS pathways including the Nrf2 pathway or the Nox redox-regulatory system, bone loss is mitigated (Isomura et al. 2004; Jia et al. 2012; Jing et al. 2019; Ma et al. 2022; Tian et al. 2016; Tsay et al. 2010; Wang et al. 2018b; Tao et al. 2024; Dong et al. 2024). Importantly, it should be noted that while increased levels of ROS are detrimental for osteoblasts, increased ROS production in osteoclasts enhances their differentiation potential, resorptive capacity, and lifespan (Ha et al. 2004; Lee et al. 2005) (Fig. 20.2). In fact, RANKL stimulation of osteoclast precursors rapidly induces ROS formation, complex I activity, and mitochondrial respiration, as well as the biogenesis of new mitochondria (Ishii et al. 2009). Deletion of factors involved in ROS generation or treatment of anti-oxidants reduces osteoclast numbers and bone resorption, thus implying that blocking ROS may be an effective strategy to prevent bone loss (Marques-Carvalho et al. 2023). Besides targeting ROS, also direct inhibition of lipid peroxidation and ferroptosis using ferrostatin-1 has been proven effective in inhibiting iron-induced bone loss (Xiang et al. 2024).

Exogenously induced iron overload in rodent and zebrafish models leads to enhanced osteoclast formation, reduced osteoblast differentiation and function, and increased osteoblast and osteocyte apoptosis (Jia et al. 2012; Ma et al. 2022; Zarjou et al. 2010; Zhang et al. 2018). Moreover, mouse models of genetically induced iron overload (e.g., $Hfe^{-/-}$, $Hamp^{-/-}$) confirm suppressed bone formation as a main mechanism of iron-induced bone loss, while the increase in osteoclast numbers is not consistent in all models (Li et al. 2020; Shen et al. 2014; Doyard et al. 2016; Ledesma-Colunga et al. 2020). As mentioned above, osteoclasts rely heavily on ROS production and mitochondrial function to fulfill their bone-degrading function. Thus, transferrin-bound iron uptake mediated via transferrin receptor 1 (TFR1) is crucial to maintain osteoclast differentiation and function by controlling mitochondrial metabolism and cytoskeletal reorganization (Das et al. 2022; Ishii et al. 2009). In line with this, increasing intracellular levels of iron via genetic deletion of FPN in osteoclast precursors or treatment with hepcidin increases mitochondrial metabolism and osteoclast numbers

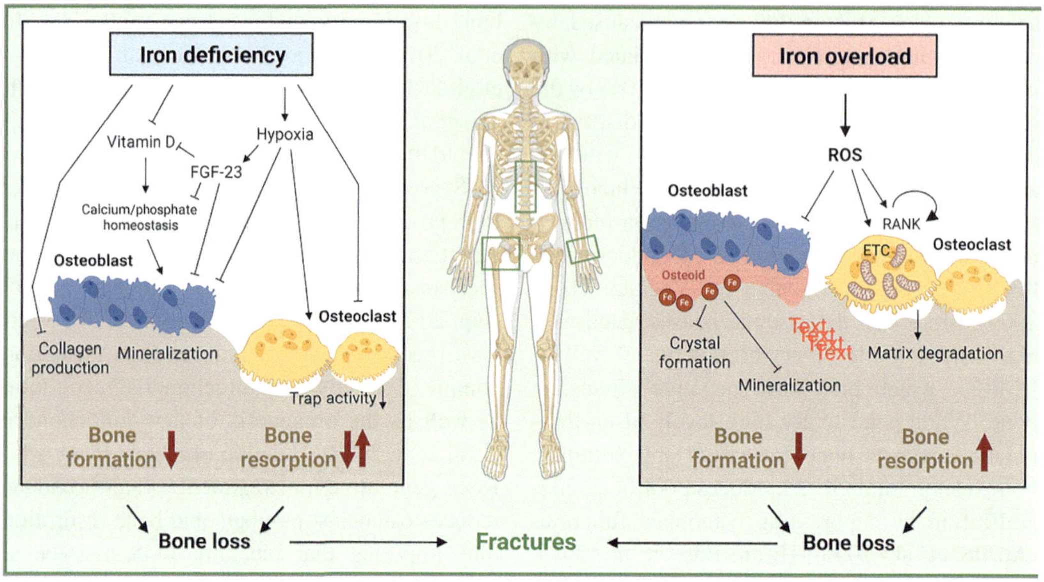

Fig. 20.2 Mechanisms of bone loss due to iron overload and iron deficiency. Iron overload and iron deficiency both lead to bone loss and increased fracture risk, in particular at the lumbar spine, wrist, and hip. In iron deficiency, iron-dependent enzymes such as TRAP are inhibited, which limits bone resorption. However, iron deficiency also leads to the induction of hypoxia, which stimulates bone resorption. Iron deficiency also reduces the production of vitamin D and collagen, and, via the induction of hypoxia, induces fibroblast growth factor-23 (FGF-23), which further inhibits vitamin D. This leads to deranged mineral metabolism and limited bone mineralization. Iron overload causes an increased production of ROS, which stimulates osteoclastogenesis by increasing mitochondrial biogenesis and the production of energy via the electron transport chain (ETC) and amplifying RANK signaling. At the same time, ROS inhibits osteoblast function. Excessive iron that is incorporated at the mineralization fronts inhibits hydroxyapatite crystal formation and mineralization. Collectively, both iron deficiency and iron overload lead to bone loss and an increased risk for fractures. (Created with Biorender)

and reduces bone mass (Wang et al. 2018a, b; Guo et al. 2021; Zhao et al. 2015).

Osteoblast function and mineralization are consistently and profoundly inhibited by iron overload (Fig. 20.2). Gene expression levels of type I collagen, alkaline phosphatase, and osteocalcin are reduced in osteoblast cultures treated with iron and in bone tissue from iron-treated mice (Zarjou et al. 2009; Baschant et al. 2016; Jing et al. 2019). One of the underlying mechanisms of inhibited bone formation is impaired Wnt signaling, which in part is also related to the iron-induced increase of oxidative stress (Luo et al. 2022; Li et al. 2020; Baschant et al. 2016). Both osteoblasts and osteocytes have been shown to undergo apoptosis during iron overload conditions, partially increasing the RANKL/OPG ratio to stimulate osteoclasts and bone resorption (Ma et al. 2022; Ke et al. 2017; Liu et al. 2017).

In addition, iron has been shown to accumulate at the mineralization fronts of newly formed bone, where it may interfere with proper calcium incorporation into bone (Hiratsuka et al. 1997; Pierides and Myli 1984). In fact, *in vitro* studies showed that iron treatment led to decreased levels of calcium and phosphate in hydroxyapatite crystals and that it inhibited crystal growth (Malecki et al. 2000; Guggenbuhl et al. 2008; Borriello et al. 2016) (Fig. 20.2).

Taken together, there is strong evidence that iron overload reduces BMD and increases fracture risk, which is mechanistically explained by suppressed bone formation and increased bone resorption. Besides inhibited endocrine functions due to iron overload, which may contribute to iron-induced bone loss, iron directly impacts bone cells via increasing oxidative stress, thereby inhibiting osteoblasts and promoting osteoclasts.

20.4 Iron Deficiency-Induced Bone Pathology

20.4.1 Clinical Evidence

Iron deficiency is common in the general population, particularly in children, premenopausal or pregnant women, and individuals aged over 65 years. Iron deficiency, typically attributed to inadequate dietary iron intake or blood loss, is the most common cause of anemia due to a lack of iron for heme biosynthesis. Considering that iron is vital for several physiological and biochemical processes, it is not surprising that iron-deficiency anemia (IDA) affects various organ systems including the brain, the immune system, and muscles. Also, bone relies on iron for various cellular processes including collagen production and vitamin D synthesis. Thus, a deficiency of iron is likely to have detrimental effects on bone.

Studies investigating the relationship between dietary iron intake and bone in healthy, postmenopausal women revealed that increasing levels of iron intake were associated with a greater BMD and slower postmenopausal bone loss at the lumbar spine (Abraham et al. 2006; Harris et al. 2003). Moreover, data from the National Health and Nutrition Examination Survey with over 11,600 participants showed that in women, but not men, the risk of osteopenia or osteoporosis declined with a moderate increase in dietary iron consumption (Liu and An 2023). In young women with iron deficiency, increased bone resorption markers were found (Toxqui et al. 2014). Thus, these data suggest that adequate levels of dietary iron may be protective for preventing bone loss later in life.

Clinically, anemia is defined by low hemoglobin levels. Thus, several studies have investigated the relationship between BMD and hemoglobin as a surrogate marker. In several studies, hemoglobin levels were shown to positively correlate with BMD (Cesari et al. 2005; Laudisio et al. 2009; Korkmaz et al. 2012; Pan et al. 2017). Moreover, a large population-based study showed that patients with IDA had a twofold higher risk for osteoporosis compared to patients without IDA. Levels of hemoglobin have also been shown to be associated with fractures. In men aged over 65 years, the lowest and highest deciles of hemoglobin concentrations were found to be associated with an increased fracture risk (Looker 2014). In support of that finding, three large studies have identified that the risk of sustaining a hip fracture is up to twofold higher in anemic subjects compared to controls (Jorgensen et al. 2010; Merriman et al. 2010; Valderrabano et al. 2017; Lee et al. 2019) (Fig. 20.2).

IDA is further a common complication in patients with chronic kidney disease (CKD), occurring in about 20% of the patients (Awan et al. 2021). Iron deficiency in CKD includes both absolute and functional iron deficiency. While absolute iron deficiency may stem from increased iron losses during hemodialysis or various causes of blood loss, functional iron deficiency is related to increased hepcidin levels due to chronic inflammation, decreased renal clearance, and reduced erythropoietin levels (Chap. 13). Importantly, osteopenia and osteoporosis occur in up to 57% and 30% of patients, respectively (Abdalbary et al. 2022; Najar et al. 2017). Fracture risk has been reported to be increased twofold in pre-dialysis patients and even eightfold in patients with end-stage renal disease (CKD) (Ensrud et al. 2007; Jadoul et al. 2006). The complex pathophysiology of bone disease in CKD is described elsewhere, but briefly, it is multifactorial and likely due to the abnormal uremic milieu, abnormalities in vitamin D metabolism, parathyroid hormone, calcium, and phosphorus balance, as well as the use of drugs that can cause bone loss, such as glucocorticoids, diuretics, and proton pump inhibitors (Pazianas and Miller 2021).

20.4.2 Preclinical Evidence and Underlying Mechanisms

Iron is a critical co-factor for several cellular processes linked to bone formation and bone resorption. As such, a deficit of iron is likely to affect specific functions of osteoblasts and osteoclasts. Besides being involved in processes that affect every cell type, such as DNA synthesis, replication, and repair, as well as epigenetic control and mitochondrial respiration (Teh et al. 2024), iron is an essential co-factor for enzymes involved in type I collagen production, vitamin D metabolism, and tartrate-resistant acid phosphatase (TRAP) activity (Fig. 20.2). Prolyl- and lysyl-hydroxylases are key enzymes in collagen biosynthesis that require iron, ascorbate, oxygen, and 2-oxoglutarate for proper function (Hausmann 1967; Hutton et al. 1966). In addition, iron is critically involved in cholesterol biosynthesis, which is an important precursor for several steroid hormones including sex hormones and vitamin D. Vitamin D production in particular requires sequential hydroxylation steps that require the heme-dependent reactions of CYP2R1 in the liver (25-hydroxylation) and CYP27B1 (1α-hydroxylation) in the kidney (Machalz et al. 2021). Vitamin D deficiency has been repeatedly reported in the context of IDA (Blanco-Rojo et al. 2013; Jin et al. 2013; Qiu et al. 2022). Vitamin D is vital for calcium homeostasis and thus for the induction of bone mineralization. In line, treatment of rat osteoblasts with the iron chelator deferoxamine suppressed the osteoblastic phenotype indicated by a low expression of osteoblast markers and reduced mineralization (Messer et al. 2009). *In vivo*, iron deficiency in rats reduced bone mineral content due to impaired osteoblast activity (Katsumata et al. 2009; Medeiros et al. 2002). The loss of bone mass and strength as well as poor bone mineralization in rats was further exacerbated with a combined iron- and calcium-deficient diet (Medeiros et al. 2002). Further, in a zebrafish model of severe iron deficiency and hypochromic anemia due to a mutation in the FPN gene, a decreased number of mineralized vertebrae was observed, along with reduced expression of osteoblastic markers (Bo et al. 2016), indicating that iron deficiency suppresses bone formation activity by osteoblasts.

Also osteoclasts produce enzymes that are iron-dependent. TRAP is a metalloenzyme containing a binuclear iron (Fe^{3+}/Fe^{2+}) center that facilitates the hydrolysis of phosphate esters and phosphoproteins (e.g., osteopontin, bone sialoprotein) and the generation of ROS (Boorsma et al. 2017) (Fig. 20.2). As outlined above, ROS are critical for osteoclast generation. Accordingly, treatment of osteoclast precursors with deferoxamine inhibits osteoclastogenesis *in vitro* (Ishii et al. 2009; Zhang et al. 2019; Guo et al. 2015). Interestingly, studies in rodents have shown that IDA increases levels of bone resorption markers (similar to humans) (Diaz-Castro et al. 2012; Toxqui et al. 2014). However, to date, there are only very few studies investigating bone turnover markers in individuals with iron deficiency, and thus, more research is needed to understand the discrepancy between *in vitro* and *in vivo* data (Fig. 20.2).

One important aspect in the context of iron deficiency is that it induces hypoxia, due to the critical role of iron as a co-factor of the iron-sensing prolyl-4-hydroxylases (PHDs). Under normal conditions, PHDs hydroxylate hypoxia-inducible factors (HIFs) and mark them for proteasomal degradation. Thus, reduced PHD activity results in HIF-1α and HIF-2α accumulation and induction of the hypoxic response. Prolonged HIF activation enhances bone resorption and suppresses bone formation (Arnett et al. 2003; Hulley et al. 2017; Knowles et al. 2010; Rauner et al. 2016); thus, potentially contributing to the bone loss observed in individuals with IDA (Fig. 20.2).

Finally, hypoxia also induces the transcription of FGF-23, a hormone that is produced by osteocytes and acts on the kidney to regulate phosphate and vitamin D metabolism (Ratsma et al. 2021). FGF-23 inhibits 1α-hydroxylase expression in the kidney, which is required to produce the active form of vitamin D (Shimada et al. 2004). Thus, besides inhibiting enzyme activities for vitamin D metabolism, iron deficiency further reduces vitamin D levels by increasing FGF-23 expression and thereby impacts bone mineralization. In addition, FGF-23 can directly inhibit osteoblast mineralization (Lyu et al. 2023) (Fig. 20.2). The tight connection between iron and FGF-23 levels was demonstrated in multiple studies, showing that low serum iron is associated with high intact FGF-23 (iFGF-23) levels in healthy elderly men (Lewerin et al. 2017). Interestingly, in states of iron deficiency, transcription of FGF-23 is increased, yet levels of iFGF-23 are normal. This is because iron deficiency simultaneously stimulates cleavage of FGF-23, resulting in the accumulation of C-terminal FGF-23 (cFGF-23) fragments (David et al. 2016; Farrow et al. 2011; Clinkenbeard et al. 2014; Imel et al. 2011; Wolf et al. 2013). Recent evidence suggests that cFGF-23 can antagonize BMP-induced hepcidin production, thereby increasing iron availability and preventing IDA during inflammation (Courbon et al. 2023).

Paradoxically, besides iron deficiency, certain forms of intravenous iron used to treat anemia have been shown to induce iFGF-23 levels, resulting in hypophosphatemia. Even though this phenomenon has been reported with multiple iron formulations, a particularly high incidence of hypophosphatemia was reported in patients treated with ferric carboxymaltose (FCM) (~50%) (Wolf et al. 2020; Okada et al. 1982). Pathophysiologically, FCM appears to inhibit the cleavage of FGF-23, resulting in high iFGF-23, which is associated with its typical downstream manifestations including reduction of 1,25-dihydroxy vitamin D, secondary hyperparathyroidism, and hypophosphatemia. Due to the suppression of vitamin D activity, blood calcium levels also are decreased. Thus, abnormal phosphate and calcium levels may result in defects in bone mineralization, which have been reported in case studies (Okada et al. 1982; Bartko et al. 2018). Of note, while some patient groups are relatively resistant to iron-induced hypophosphatemia, such as patients with CKD due to their reduced ability to excrete phosphate, patients with other underlying diseases, such as inflammatory bowel disease, appear to be particularly susceptible (Schaefer et al. 2022). Thus, more research is required to fully understand the mechanisms of action and determine if a single treatment or continuous treatment with intravenous iron can lead to low bone mass and an increase in fracture rates.

Taken together, iron deficiency renders individuals vulnerable to fractures. Although the underlying mechanisms have not been studied in great detail in patients, preclinical studies to date suggest that low bone mass may be related to direct and indirect effects on osteoclasts and osteoblasts. Important mechanisms include the reduction of collagen production, vitamin D and mineral metabolism, and induction of hypoxia. Importantly, also certain forms of intravenous iron may alter bone biology by inducing FGF-23 levels.

20.5 Conclusions

Bone metabolism is critically dependent on balanced iron levels. Both iron excess and iron deficiency negatively impact bone. Thus, it is imperative to maintain iron homeostasis to prevent bone loss in the first place. However, once imbalances occur, which may be the result of diseases or specific treatment regimens, a detailed understanding of how iron and bone metabolism are interwoven is critical to mitigate bone loss and potentially intervene with both systems. One such example is FGF-23. Targeting excess FGF-23 in conditions of ineffective erythropoiesis and anemia in preclinical models does not only improve erythroid outcomes but also bone health. Thus, identifying additional shared pathways may give rise to novel strategies to treat iron-related bone pathologies. Future research will need to further disentangle the contributions of iron overload-induced endocrinopathies to bone loss and explore the utility of targeting oxidative stress to prevent iron overload-induced bone loss in humans. Finally, clinicians should be aware of the close interactions of iron and bone and assess bone health in patients with iron disorders or patients requiring continuous treatment with intravenous iron.

Acknowledgements I would like to thank Vanessa Passin for her help with creating the figures and proofreading.

FundingThis work was supported by funds from the German Research Foundation—DFG—FOR5146.

References

Abdalbary M, Sobh M, Elnagar S, Elhadedy MA, Elshabrawy N, Abdelsalam M, Asadipooya K, Sabry A, Halawa A, El-Husseini A (2022) Management of osteoporosis in patients with chronic kidney disease. Osteoporos Int 33(11):2259–2274. https://doi.org/10.1007/s00198-022-06462-3

Abraham R, Walton J, Russell L, Wolman R, Wardley-Smith B, Green JR, Mitchell A, Reeve J (2006) Dietary determinants of post-menopausal bone loss at the lumbar spine: a possible beneficial effect of iron. Osteoporos Int 17(8):1165–1173. https://doi.org/10.1007/s00198-005-0033-6

Abrahamsen B, van Staa T, Ariely R, Olson M, Cooper C (2009) Excess mortality following hip fracture: a systematic epidemiological review. Osteoporos Int 20(10):1633–1650. https://doi.org/10.1007/s00198-009-0920-3

Almeida A, Roberts I (2005) Bone involvement in sickle cell disease. Br J Haematol 129(4):482–490. https://doi.org/10.1111/j.1365-2141.2005.05476.x

Aprile A, Raggi L, Bolamperti S, Villa I, Storto M, Morello G, Marktel S, Tripodo C, Cappellini MD, Motta I, Rubinacci A, Ferrari G (2023) Inhibition of FGF23 is a therapeutic strategy to target hematopoietic stem cell niche defects in beta-thalassemia. Sci Transl Med 15(698):eabq3679. https://doi.org/10.1126/scitranslmed.abq3679

Arnett TR, Gibbons DC, Utting JC, Orriss IR, Hoebertz A, Rosendaal M, Meghji S (2003) Hypoxia is a major stimulator of osteoclast formation and bone resorption. J Cell Physiol 196(1):2–8. https://doi.org/10.1002/jcp.10321

Arora I, Kaprove A, Perrone R, Ceglia L (2023) An unusually prolonged case of FGF23-mediated hypophosphatemia secondary to ferric carboxymaltose use. JCEM Case Rep 1(5):luad117. https://doi.org/10.1210/jcemcr/luad117

Awan AA, Walther CP, Richardson PA, Shah M, Winkelmayer WC, Navaneethan SD (2021) Prevalence, correlates and outcomes of absolute and functional iron deficiency anemia in nondialysis-dependent chronic kidney disease. Nephrol Dial Transplant 36(1):129–136. https://doi.org/10.1093/ndt/gfz192

Bartko J, Roschger P, Zandieh S, Brehm A, Zwerina J, Klaushofer K (2018) Hypophosphatemia, severe bone pain, gait disturbance, and fatigue fractures after iron substitution in inflammatory bowel disease: a case report. J Bone Miner Res 33(3):534–539. https://doi.org/10.1002/jbmr.3319

Baschant U, Rauner M, Balaian E, Weidner H, Roetto A, Platzbecker U, Hofbauer LC (2016) Wnt5a is a key target for the pro-osteogenic effects of iron chelation on osteoblast progenitors. Haematologica 101(12):1499–1507. https://doi.org/10.3324/haematol.2016.144808

Baschant U, Altamura S, Steele-Perkins P, Muckenthaler MU, Spasic MV, Hofbauer LC, Steinbicker AU, Rauner M (2022) Iron effects versus metabolic alterations in hereditary hemochromatosis driven bone loss. Trends Endocrinol Metab 33(9):652–663. https://doi.org/10.1016/j.tem.2022.06.004

Bayanzay K, Alzoebie L (2016) Reducing the iron burden and improving survival in transfusion-dependent thalassemia patients: current perspectives. J Blood Med 7:159–169. https://doi.org/10.2147/JBM.S61540

Blanco-Rojo R, Perez-Granados AM, Toxqui L, Zazo P, de la Piedra C, Vaquero MP (2013) Relationship between vitamin D deficiency, bone remodelling and iron status in iron-deficient young women consuming an iron-fortified food. Eur J Nutr 52(2):695–703. https://doi.org/10.1007/s00394-012-0375-8

Bo L, Liu Z, Zhong Y, Huang J, Chen B, Wang H, Xu Y (2016) Iron deficiency anemia's effect on bone formation in zebrafish mutant. Biochem Biophys Res Commun 475(3):271–276. https://doi.org/10.1016/j.bbrc.2016.05.069

Bonewald LF (2011) The amazing osteocyte. J Bone Miner Res 26(2):229–238. https://doi.org/10.1002/jbmr.320

Boorsma CE, van der Veen TA, Putri KSS, de Almeida A, Draijer C, Mauad T, Fejer G, Brandsma CA, van den Berge M, Bosse Y, Sin D, Hao K, Reithmeier A, Andersson G, Olinga P, Timens W, Casini A, Melgert BN (2017) A potent tartrate resistant acid phosphatase inhibitor to study the function of TRAP in alveolar macrophages. Sci Rep 7(1):12570. https://doi.org/10.1038/s41598-017-12623-w

Borriello A, Caldarelli I, Speranza MC, Scianguetta S, Tramontano A, Bencivenga D, Stampone E, Negri A, Nobili B, Locatelli F, Perrotta S, Oliva A, Della Ragione F (2016) Iron overload enhances human mesenchymal stromal cell growth and hampers matrix calcification. Biochim Biophys Acta 1860 6:1211–1223. https://doi.org/10.1016/j.bbagen.2016.01.025

Calvi LM, Adams GB, Weibrecht KW, Weber JM, Olson DP, Knight MC, Martin RP, Schipani E, Divieti P, Bringhurst FR, Milner LA, Kronenberg HM, Scadden DT (2003) Osteoblastic cells regulate the haematopoietic stem cell niche. Nature 425(6960):841–846. https://doi.org/10.1038/nature02040

Cesari M, Pahor M, Lauretani F, Penninx BW, Bartali B, Russo R, Cherubini A, Woodman R, Bandinelli S, Guralnik JM, Ferrucci L (2005) Bone density and hemoglobin levels in older persons: results from the InCHIANTI study. Osteoporos Int 16(6):691–699. https://doi.org/10.1007/s00198-004-1739-6

Chan CK, Seo EY, Chen JY, Lo D, McArdle A, Sinha R, Tevlin R, Seita J, Vincent-Tompkins J, Wearda T, Lu WJ, Senarath-Yapa K, Chung MT, Marecic O, Tran M, Yan KS, Upton R, Walmsley GG, Lee AS, Sahoo D, Kuo CJ, Weissman IL, Longaker MT (2015) Identification and specification of the mouse skeletal stem cell. Cell 160(1–2):285–298. https://doi.org/10.1016/j.cell.2014.12.002

Chan CKF, Gulati GS, Sinha R, Tompkins JV, Lopez M, Carter AC, Ransom RC, Reinisch A, Wearda T, Murphy M, Brewer RE, Koepke LS, Marecic O, Manjunath A, Seo EY, Leavitt T, Lu WJ, Nguyen A, Conley SD, Salhotra A, Ambrosi TH, Borrelli MR, Siebel T, Chan K, Schallmoser K, Seita J, Sahoo D, Goodnough H, Bishop J, Gardner M, Majeti R, Wan DC, Goodman S, Weissman IL, Chang HY, Longaker MT (2018) Identification of the human skeletal stem cell. Cell 175(1):43–56.e21. https://doi.org/10.1016/j.cell.2018.07.029

Clinkenbeard EL, Farrow EG, Summers LJ, Cass TA, Roberts JL, Bayt CA, Lahm T, Albrecht M, Allen MR, Peacock M, White KE (2014) Neonatal iron deficiency causes abnormal phosphate metabolism by elevating FGF23 in normal and ADHR mice. J Bone Miner Res 29(2):361–369. https://doi.org/10.1002/jbmr.2049

Costanzo G, Naselli A, Arpi ML, Piticchio T, Le Moli R, Belfiore A, Frasca F (2024) Very low serum IGF-1 levels are associated with vertebral fractures in adult males with beta-thalassemia major. J Endocrinol Investig 47(7):1691–1700. https://doi.org/10.1007/s40618-023-02270-6

Courbon G, Thomas JJ, Martinez-Calle M, Wang X, Spindler J, Von Drasek J, Hunt-Tobey B, Mehta R, Isakova T, Chang W, Creemers JWM, Ji P, Martin A, David V (2023) Bone-derived C-terminal FGF23 cleaved peptides increase iron availability in acute inflammation. Blood 142(1):106–118. https://doi.org/10.1182/blood.2022018475

Das BK, Wang L, Fujiwara T, Zhou J, Aykin-Burns N, Krager KJ, Lan R, Mackintosh SG, Edmondson R, Jennings ML, Wang X, Feng JQ, Barrientos T, Gogoi J, Kannan A, Gao L, Xing W, Mohan S, Zhao H (2022) Transferrin receptor 1-mediated iron uptake regulates bone mass in mice via osteoclast mitochondria and cytoskeleton. elife 11. https://doi.org/10.7554/eLife.73539

David V, Martin A, Isakova T, Spaulding C, Qi L, Ramirez V, Zumbrennen-Bullough KB, Sun CC, Lin HY, Babitt JL, Wolf M (2016) Inflammation and functional iron deficiency regulate fibroblast growth factor 23 production. Kidney Int 89(1):135–146. https://doi.org/10.1038/ki.2015.290

De Sanctis V, Soliman AT, Elsefdy H, Soliman N, Bedair E, Fiscina B, Kattamis C (2018) Bone disease in beta thalassemia patients: past, present and future perspectives. Metabolism 80:66–79. https://doi.org/10.1016/j.metabol.2017.09.012

Dede AD, Trovas G, Chronopoulos E, Triantafyllopoulos IK, Dontas I, Papaioannou N, Tournis S (2016) Thalassemia-associated osteoporosis: a systematic review on treatment and brief overview of the disease. Osteoporos Int 27(12):3409–3425. https://doi.org/10.1007/s00198-016-3719-z

Diamond T, Stiel D, Posen S (1989) Osteoporosis in hemochromatosis: iron excess, gonadal deficiency, or other factors? Ann Intern Med 110(6):430–436

Diaz-Castro J, Lopez-Frias MR, Campos MS, Lopez-Frias M, Alferez MJ, Nestares T, Ojeda ML, Lopez-Aliaga I (2012) Severe nutritional iron-deficiency anaemia has a negative effect on some bone turnover biomarkers in rats. Eur J Nutr 51(2):241–247. https://doi.org/10.1007/s00394-011-0212-5

Dong Y, Kang H, Peng R, Liu Z, Liao F, Hu SA, Ding W, Wang P, Yang P, Zhu M, Wang S, Wu M, Ye D, Gan X, Li F, Song K (2024) A clinical-stage Nrf2 activator suppresses osteoclast differentiation via the iron-ornithine axis. Cell Metab 36(8):1679–1695. e1676. https://doi.org/10.1016/j.cmet.2024.03.005

Doyard M, Chappard D, Leroyer P, Roth MP, Loreal O, Guggenbuhl P (2016) Decreased bone formation explains osteoporosis in a genetic mouse model of hemochromatosiss. PLoS One 11(2):e0148292. https://doi.org/10.1371/journal.pone.0148292

Ensrud KE, Lui LY, Taylor BC, Ishani A, Shlipak MG, Stone KL, Cauley JA, Jamal SA, Antoniucci DM, Cummings SR, Osteoporotic Fractures Research G (2007) Renal function and risk of hip and vertebral fractures in older women. Arch Intern Med 167(2):133–139. https://doi.org/10.1001/archinte.167.2.133

Farrow EG, Yu X, Summers LJ, Davis SI, Fleet JC, Allen MR, Robling AG, Stayrook KR, Jideonwo V, Magers MJ, Garringer HJ, Vidal R, Chan RJ, Goodwin CB, Hui SL, Peacock M, White KE (2011) Iron deficiency drives an autosomal dominant hypophosphatemic rickets (ADHR) phenotype in fibroblast growth factor-23 (Fgf23) knock-in mice. Proc Natl Acad Sci USA 108(46):E1146–E1155. https://doi.org/10.1073/pnas.1110905108

Feng X, Teitelbaum SL (2013) Osteoclasts: new insights. Bone Res 1(1):11–26. https://doi.org/10.4248/BR201301003

Galy B, Conrad M, Muckenthaler M (2024) Mechanisms controlling cellular and systemic iron homeostasis. Nat Rev Mol Cell Biol 25(2):133–155. https://doi.org/10.1038/s41580-023-00648-1

Glenthoj A, Grace RF, Lander C, van Beers EJ, Glader B, Kuo KHM, Yan Y, McGee B, Boscoe AN, Li J, Bianchi P (2024) Comorbidities and complications in adult and paediatric patients with pyruvate kinase deficiency: analysis from the Peak Registry. Br J Haematol 205(2):613–623. https://doi.org/10.1111/bjh.19601

Guanabens N, Pares A (2018) Osteoporosis in chronic liver disease. Liver Int 38(5):776–785. https://doi.org/10.1111/liv.13730

Guggenbuhl P, Deugnier Y, Boisdet JF, Rolland Y, Perdriger A, Pawlotsky Y, Chales G (2005) Bone mineral density in men with genetic hemochromatosis and HFE gene mutation. Osteoporos Int 16(12):1809–1814

Guggenbuhl P, Filmon R, Mabilleau G, Basle MF, Chappard D (2008) Iron inhibits hydroxyapatite crystal growth in vitro. Metabolism 57(7):903–910. https://doi.org/10.1016/j.metabol.2008.02.004

Guo JP, Pan JX, Xiong L, Xia WF, Cui S, Xiong WC (2015) Iron chelation inhibits osteoclastic differentiation in vitro and in Tg2576 mouse model of Alzheimer's disease. PLoS One 10(11):e0139395. https://doi.org/10.1371/journal.pone.0139395

Guo HH, Xiong L, Pan JX, Lee D, Liu K, Ren X, Wang B, Yang X, Cui S, Mei L, Xiong WC (2021) Hepcidin contributes to Swedish mutant APP-induced osteoclastogenesis and trabecular bone loss. Bone Res 9(1):31. https://doi.org/10.1038/s41413-021-00146-0

Ha H, Kwak HB, Lee SW, Jin HM, Kim HM, Kim HH, Lee ZH (2004) Reactive oxygen species mediate RANK signaling in osteoclasts. Exp Cell Res 301(2):119–127. https://doi.org/10.1016/j.yexcr.2004.07.035

Harris MM, Houtkooper LB, Stanford VA, Parkhill C, Weber JL, Flint-Wagner H, Weiss L, Going SB, Lohman TG (2003) Dietary iron is associated with bone mineral density in healthy postmenopausal women. J Nutr 133(11):3598–3602. https://doi.org/10.1093/jn/133.11.3598

Hausmann E (1967) Cofactor requirements for the enzymatic hydroxylation of lysine in a polypeptide precursor of collagen. Biochim Biophys Acta 133(3):591–593. https://doi.org/10.1016/0005-2795(67)90566-1

Hiram-Bab S, Liron T, Deshet-Unger N, Mittelman M, Gassmann M, Rauner M, Franke K, Wielockx B, Neumann D, Gabet Y (2015) Erythropoietin directly stimulates osteoclast precursors and induces bone loss. FASEB J 29(5):1890–1900. https://doi.org/10.1096/fj.14-259085

Hiratsuka H, Katsuta O, Toyota N, Tsuchitani M, Akiba T, Marumo F, Umemura T (1997) Iron deposition at mineralization fronts and osteoid formation following chronic cadmium exposure in ovariectomized rats. Toxicol Appl Pharmacol 143(2):348–356. https://doi.org/10.1006/taap.1996.8101

Hulley PA, Bishop T, Vernet A, Schneider JE, Edwards JR, Athanasou NA, Knowles HJ (2017) Hypoxia-inducible factor 1-alpha does not regulate osteoclastogenesis but enhances bone resorption activity via prolyl-4-hydroxylase 2. J Pathol 242(3):322–333. https://doi.org/10.1002/path.4906

Hutton JJ Jr, Trappel AL, Udenfriend S (1966) Requirements for alpha-ketoglutarate, ferrous ion and ascorbate by collagen proline hydroxylase. Biochem Biophys Res Commun 24(2):179–184. https://doi.org/10.1016/0006-291x(66)90716-9

Imel EA, Peacock M, Gray AK, Padgett LR, Hui SL, Econs MJ (2011) Iron modifies plasma FGF23 differently in autosomal dominant hypophosphatemic rickets and healthy humans. J Clin Endocrinol Metab 96(11):3541–3549. https://doi.org/10.1210/jc.2011-1239

Ishii KA, Fumoto T, Iwai K, Takeshita S, Ito M, Shimohata N, Aburatani H, Taketani S, Lelliott CJ, Vidal-Puig A, Ikeda K (2009) Coordination of PGC-1beta and iron uptake in mitochondrial biogenesis and osteoclast activation. Nat Med 15(3):259–266. https://doi.org/10.1038/nm.1910

Isomura H, Fujie K, Shibata K, Inoue N, Iizuka T, Takebe G, Takahashi K, Nishihira J, Izumi H, Sakamoto W (2004) Bone metabolism and oxidative stress in postmenopausal rats with iron overload. Toxicology 197(2):93–100

Jacome-Galarza CE, Percin GI, Muller JT, Mass E, Lazarov T, Eitler J, Rauner M, Yadav VK, Crozet L, Bohm M, Loyher PL, Karsenty G, Waskow C, Geissmann F (2019) Developmental origin, functional maintenance and genetic rescue of osteoclasts. Nature 568(7753):541–545. https://doi.org/10.1038/s41586-019-1105-7

Jadoul M, Albert JM, Akiba T, Akizawa T, Arab L, Bragg-Gresham JL, Mason N, Prutz KG, Young EW, Pisoni RL (2006) Incidence and risk factors for hip or other bone fractures among hemodialysis patients in the Dialysis Outcomes and Practice Patterns Study. Kidney Int 70(7):1358–1366. https://doi.org/10.1038/sj.ki.5001754

Jandl NM, Rolvien T, Schmidt T, Mussawy H, Nielsen P, Oheim R, Amling M, Barvencik F (2020) Impaired bone microarchitecture in patients with hereditary hemochromatosis and skeletal complications. Calcif Tissue Int 106(5):465–475. https://doi.org/10.1007/s00223-020-00658-7

Jia P, Xu YJ, Zhang ZL, Li K, Li B, Zhang W, Yang H (2012) Ferric ion could facilitate osteoclast differentiation and bone resorption through the production of reactive oxygen species. J Orthop Res 30(11):1843–1852. https://doi.org/10.1002/jor.22133

Jin HJ, Lee JH, Kim MK (2013) The prevalence of vitamin D deficiency in iron-deficient and normal children under the age of 24 months. Blood Res 48(1):40–45. https://doi.org/10.5045/br.2013.48.1.40

Jing X, Du T, Chen K, Guo J, Xiang W, Yao X, Sun K, Ye Y, Guo F (2019) Icariin protects against iron overload-induced bone loss via suppressing oxidative stress. J Cell Physiol 234(7):10123–10137. https://doi.org/10.1002/jcp.27678

Jorgensen L, Skjelbakken T, Lochen ML, Ahmed L, Bjornerem A, Joakimsen R, Jacobsen BK (2010) Anemia and the risk of non-vertebral fractures: the Tromso Study. Osteoporos Int 21(10):1761–1768. https://doi.org/10.1007/s00198-009-1131-7

Katsumata S, Katsumata-Tsuboi R, Uehara M, Suzuki K (2009) Severe iron deficiency decreases both bone formation and bone resorption in rats. J Nutr 139(2):238–243. https://doi.org/10.3945/jn.108.093757

Ke JY, Cen WJ, Zhou XZ, Li YR, Kong WD, Jiang JW (2017) Iron overload induces apoptosis of murine preosteoblast cells via ROS and inhibition of AKT pathway. Oral Dis 23(6):784–794. https://doi.org/10.1111/odi.12662

Kim BJ, Ahn SH, Bae SJ, Kim EH, Lee SH, Kim HK, Choe JW, Koh JM, Kim GS (2012) Iron overload accelerates bone loss in healthy postmenopausal women and middle-aged men: a 3-year retrospective longitudinal study. J Bone Miner Res 27(11):2279–2290. https://doi.org/10.1002/jbmr.1692

Kim BJ, Lee SH, Koh JM, Kim GS (2013) The association between higher serum ferritin level and lower bone mineral density is prominent in women >/=45 years of age (KNHANES 2008-2010). Osteoporos

Int 24(10):2627–2637. https://doi.org/10.1007/s00198-013-2363-0

Klein K, Asaad S, Econs M, Rubin JE (2018) Severe FGF23-based hypophosphataemic osteomalacia due to ferric carboxymaltose administration. BMJ Case Rep 2018. https://doi.org/10.1136/bcr-2017-222851

Knowles HJ, Cleton-Jansen AM, Korsching E, Athanasou NA (2010) Hypoxia-inducible factor regulates osteoclast-mediated bone resorption: role of angiopoietin-like 4. FASEB J 24(12):4648–4659. https://doi.org/10.1096/fj.10-162230

Korkmaz U, Korkmaz N, Yazici S, Erkan M, Baki AE, Yazici M, Ozhan H, Ataoglu S (2012) Anemia as a risk factor for low bone mineral density in postmenopausal Turkish women. Eur J Intern Med 23(2):154–158. https://doi.org/10.1016/j.ejim.2011.11.009

Laudisio A, Marzetti E, Pagano F, Bernabei R, Zuccala G (2009) Haemoglobin levels are associated with bone mineral density in the elderly: a population-based study. Clin Rheumatol 28(2):145–151. https://doi.org/10.1007/s10067-008-0998-6

Ledesma-Colunga MG, Baschant U, Fiedler IAK, Busse B, Hofbauer LC, Muckenthaler MU, Altamura S, Rauner M (2020) Disruption of the hepcidin/ferroportin regulatory circuitry causes low axial bone mass in mice. Bone 137:115400. https://doi.org/10.1016/j.bone.2020.115400

Ledesma-Colunga MG, Weidner H, Vujic Spasic M, Hofbauer LC, Baschant U, Rauner M (2021) Shaping the bone through iron and iron-related proteins. Semin Hematol 58(3):188–200. https://doi.org/10.1053/j.seminhematol.2021.06.002

Lee NK, Choi YG, Baik JY, Han SY, Jeong DW, Bae YS, Kim N, Lee SY (2005) A crucial role for reactive oxygen species in RANKL-induced osteoclast differentiation. Blood 106(3):852–859. https://doi.org/10.1182/blood-2004-09-3662

Lee EA, Shin DW, Yoo JH, Ko HY, Jeong SM (2019) Anemia and risk of fractures in older korean adults: a nationwide population-based study. J Bone Miner Res 34(6):1049–1057. https://doi.org/10.1002/jbmr.3675

Lewerin C, Ljunggren O, Nilsson-Ehle H, Karlsson MK, Herlitz H, Lorentzon M, Ohlsson C, Mellstrom D (2017) Low serum iron is associated with high serum intact FGF23 in elderly men: the Swedish MrOS study. Bone 98:1–8. https://doi.org/10.1016/j.bone.2017.02.005

Li G, Zhang H, Wu J, Wang A, Yang F, Chen B, Gao Y, Ma X, Xu Y (2020) Hepcidin deficiency causes bone loss through interfering with the canonical Wnt/beta-catenin pathway via Forkhead box O3a. J Orthop Translat 23:67–76. https://doi.org/10.1016/j.jot.2020.03.012

Lipovetzki Y, Zandman-Goddard G, Feldbrin Z, Shargorodsky M (2017) Elevated ferritin and circulating osteoprotegerin levels as independent predictors of hip fracture in postmenopausal women admitted for fragility fracture: time for new screening strategies? Immunol Res 65(1):423–427. https://doi.org/10.1007/s12026-016-8849-z

Liu X, An J (2023) Dietary iron intake and its impact on osteopenia/osteoporosis. BMC Endocr Disord 23(1):154. https://doi.org/10.1186/s12902-023-01389-0

Liu F, Zhang WL, Meng HZ, Cai ZY, Yang MW (2017) Regulation of DMT1 on autophagy and apoptosis in osteoblast. Int J Med Sci 14(3):275–283. https://doi.org/10.7150/ijms.17860

Looker AC (2014) Hemoglobin and hip fracture risk in older non-Hispanic white adults. Osteoporos Int 25(10):2389–2398. https://doi.org/10.1007/s00198-014-2769-3

Luo C, Xu W, Tang X, Liu X, Cheng Y, Wu Y, Xie Z, Wu X, He X, Wang Q, Xiao Y, Qiu X, Tang Z, Shao G, Tu X (2022) Canonical Wnt signaling works downstream of iron overload to prevent ferroptosis from damaging osteoblast differentiation. Free Radic Biol Med 188:337–350. https://doi.org/10.1016/j.freeradbiomed.2022.06.236

Lyu Z, Li H, Li X, Wang H, Jiao H, Wang X, Zhao J, Lin H (2023) Fibroblast growth factor 23 inhibits osteogenic differentiation and mineralization of chicken bone marrow mesenchymal stem cells. Poult Sci 102(1):102287. https://doi.org/10.1016/j.psj.2022.102287

Ma J, Wang A, Zhang H, Liu B, Geng Y, Xu Y, Zuo G, Jia P (2022) Iron overload induced osteocytes apoptosis and led to bone loss in Hepcidin(−/−) mice through increasing sclerostin and RANKL/OPG. Bone 164:116511. https://doi.org/10.1016/j.bone.2022.116511

Machalz D, Pach S, Bermudez M, Bureik M, Wolber G (2021) Structural insights into understudied human cytochrome P450 enzymes. Drug Discov Today 26(10):2456–2464. https://doi.org/10.1016/j.drudis.2021.06.006

Malecki EA, Buhl KM, Beard JL, Jacobs CR, Connor JR, Donahue HJ (2000) Bone structural and mechanical properties are affected by hypotransferrinemia but not by iron deficiency in mice. J Bone Miner Res 15(2):271–277. https://doi.org/10.1359/jbmr.2000.15.2.271

Marques-Carvalho A, Kim HN, Almeida M (2023) The role of reactive oxygen species in bone cell physiology and pathophysiology. Bone Rep 19:101664. https://doi.org/10.1016/j.bonr.2023.101664

Medeiros DM, Plattner A, Jennings D, Stoecker B (2002) Bone morphology, strength and density are compromised in iron-deficient rats and exacerbated by calcium restriction. J Nutr 132(10):3135–3141. https://doi.org/10.1093/jn/131.10.3135

Merriman NA, Putt ME, Metz DC, Yang YX (2010) Hip fracture risk in patients with a diagnosis of pernicious anemia. Gastroenterology 138(4):1330–1337. https://doi.org/10.1053/j.gastro.2009.12.007

Messer JG, Kilbarger AK, Erikson KM, Kipp DE (2009) Iron overload alters iron-regulatory genes and proteins, down-regulates osteoblastic phenotype, and is associated with apoptosis in fetal rat calvaria cul-

tures. Bone 45(5):972–979. https://doi.org/10.1016/j.bone.2009.07.073
Muckenthaler MU, Rivella S, Hentze MW, Galy B (2017) A red carpet for iron metabolism. Cell 168(3):344–361. https://doi.org/10.1016/j.cell.2016.12.034
Najar MS, Mir MM, Muzamil M (2017) Prevalence of osteoporosis in patients with chronic kidney disease (stages 3-5) in comparison with age- and sex-matched controls: a study from Kashmir Valley Tertiary Care Center. Saudi J Kidney Dis Transpl 28(3):538–544. https://doi.org/10.4103/1319-2442.206439
Okada M, Imamura K, Fuchigami T, Omae T, Iida M, Nanishi F, Murakami M, Ohgushi H, Yao T, Fujita K, Ogawa K (1982) 2 cases of nonspecific multiple ulcers of the small intestine associated with osteomalacia caused by long-term intravenous administration of saccharated ferric oxide. Nippon Naika Gakkai Zasshi 71(11):1566–1572
Pan ML, Chen LR, Tsao HM, Chen KH (2017) Iron deficiency anemia as a risk factor for osteoporosis in Taiwan: a nationwide population-based study. Nutrients 9(6). https://doi.org/10.3390/nu9060616
Parfitt AM (2002) Misconceptions (2): turnover is always higher in cancellous than in cortical bone. Bone 30(6):807–809. https://doi.org/10.1016/s8756-3282(02)00735-4
Pazianas M, Miller PD (2021) Osteoporosis and chronic kidney disease-mineral and bone disorder (CKD-MBD): back to basics. Am J Kidney Dis 78(4):582–589. https://doi.org/10.1053/j.ajkd.2020.12.024
Pierides AM, Myli MP (1984) Iron and aluminum osteomalacia in hemodialysis patients. N Engl J Med 310(5):323. https://doi.org/10.1056/NEJM198402023100518
Pietrangelo A (2007) Hemochromatosis: an endocrine liver disease. Hepatology 46(4):1291–1301. https://doi.org/10.1002/hep.21886
Pilling LC, Tamosauskaite J, Jones G, Wood AR, Jones L, Kuo CL, Kuchel GA, Ferrucci L, Melzer D (2019) Common conditions associated with hereditary haemochromatosis genetic variants: cohort study in UK Biobank. BMJ 364:k5222. https://doi.org/10.1136/bmj.k5222
Ponzetti M, Rucci N (2021) Osteoblast differentiation and signaling: established concepts and emerging topics. Int J Mol Sci 22(13). https://doi.org/10.3390/ijms22136651
Qiu F, Li R, Gu S, Zhao Y, Yang L (2022) The effect of iron dextran on vitamin D(3) metabolism in SD rats. Nutr Metab (Lond) 19(1):47. https://doi.org/10.1186/s12986-022-00681-5
Rachner TD, Khosla S, Hofbauer LC (2011) Osteoporosis: now and the future. Lancet 377(9773):1276–1287. https://doi.org/10.1016/S0140-6736(10)62349-5
Ratsma DMA, Zillikens MC, van der Eerden BCJ (2021) Upstream regulators of fibroblast growth factor 23. Front Endocrinol (Lausanne) 12:588096. https://doi.org/10.3389/fendo.2021.588096
Rauner M, Franke K, Murray M, Singh RP, Hiram-Bab S, Platzbecker U, Gassmann M, Socolovsky M, Neumann D, Gabet Y, Chavakis T, Hofbauer LC, Wielockx B (2016) Increased EPO levels are associated with bone loss in mice lacking PHD2 in EPO-producing cells. J Bone Miner Res 31(10):1877–1887. https://doi.org/10.1002/jbmr.2857
Richette P, Ottaviani S, Vicaut E, Bardin T (2010) Musculoskeletal complications of hereditary hemochromatosis: a case-control study. J Rheumatol 37(10):2145–2150. https://doi.org/10.3899/jrheum.100234
Sarrai M, Duroseau H, D'Augustine J, Moktan S, Bellevue R (2007) Bone mass density in adults with sickle cell disease. Br J Haematol 136(4):666–672. https://doi.org/10.1111/j.1365-2141.2006.06487.x
Schaefer B, Tobiasch M, Wagner S, Glodny B, Tilg H, Wolf M, Zoller H (2022) Hypophosphatemia after intravenous iron therapy: comprehensive review of clinical findings and recommendations for management. Bone 154:116202. https://doi.org/10.1016/j.bone.2021.116202
Shen GS, Yang Q, Jian JL, Zhao GY, Liu LL, Wang X, Zhang W, Huang X, Xu YJ (2014) Hepcidin1 knockout mice display defects in bone microarchitecture and changes of bone formation markers. Calcif Tissue Int 94(6):632–639. https://doi.org/10.1007/s00223-014-9845-8
Shimada T, Hasegawa H, Yamazaki Y, Muto T, Hino R, Takeuchi Y, Fujita T, Nakahara K, Fukumoto S, Yamashita T (2004) FGF-23 is a potent regulator of vitamin D metabolism and phosphate homeostasis. J Bone Miner Res 19(3):429–435. https://doi.org/10.1359/JBMR.0301264
Siddiqui JA, Partridge NC (2016) Physiological bone remodeling: systemic regulation and growth factor involvement. Physiology (Bethesda) 31(3):233–245. https://doi.org/10.1152/physiol.00061.2014
Tao Z, Tao M, Zhou M, Wu XJ (2024) Niacin treatment prevents bone loss in iron overload osteoporotic rats via activation of SIRT1 signaling pathway. Chem Biol Interact 388:110827. https://doi.org/10.1016/j.cbi.2023.110827
Teh MR, Armitage AE, Drakesmith H (2024) Why cells need iron: a compendium of iron utilisation. Trends Endocrinol Metab. https://doi.org/10.1016/j.tem.2024.04.015
Tian Q, Wu S, Dai Z, Yang J, Zheng J, Zheng Q, Liu Y (2016) Iron overload induced death of osteoblasts in vitro: involvement of the mitochondrial apoptotic pathway. PeerJ 4:e2611. https://doi.org/10.7717/peerj.2611
Toxqui L, Perez-Granados AM, Blanco-Rojo R, Wright I, de la Piedra C, Vaquero MP (2014) Low iron status as a factor of increased bone resorption and effects of an iron and vitamin D-fortified skimmed milk on bone remodelling in young Spanish women. Eur J Nutr 53(2):441–448. https://doi.org/10.1007/s00394-013-0544-4
Tsay J, Yang Z, Ross FP, Cunningham-Rundles S, Lin H, Coleman R, Mayer-Kuckuk P, Doty SB, Grady RW, Giardina PJ, Boskey AL, Vogiatzi MG (2010) Bone

loss caused by iron overload in a murine model: importance of oxidative stress. Blood 116(14):2582–2589. https://doi.org/10.1182/blood-2009-12-260083

Valderrabano RJ, Lee J, Lui LY, Hoffman AR, Cummings SR, Orwoll ES, Wu JY, Osteoporotic Fractures in Men Study Research G (2017) Older men with anemia have increased fracture risk independent of bone mineral density. J Clin Endocrinol Metab 102(7):2199–2206. https://doi.org/10.1210/jc.2017-00266

Valenti L, Varenna M, Fracanzani AL, Rossi V, Fargion S, Sinigaglia L (2009) Association between iron overload and osteoporosis in patients with hereditary hemochromatosis. Osteoporos Int 20(4):549–555. https://doi.org/10.1007/s00198-008-0701-4

VanWagner LB, Rinella ME (2016) Extrahepatic manifestations of nonalcoholic fatty liver disease. Curr Hepatol Rep 15(2):75–85. https://doi.org/10.1007/s11901-016-0295-9

Vogiatzi MG, Macklin EA, Fung EB, Cheung AM, Vichinsky E, Olivieri N, Kirby M, Kwiatkowski JL, Cunningham M, Holm IA, Lane J, Schneider R, Fleisher M, Grady RW, Peterson CC, Giardina PJ, Thalassemia Clinical Research N (2009) Bone disease in thalassemia: a frequent and still unresolved problem. J Bone Miner Res 24(3):543–557. https://doi.org/10.1359/jbmr.080505

Wang L, Fang B, Fujiwara T, Krager K, Gorantla A, Li C, Feng JQ, Jennings ML, Zhou J, Aykin-Burns N, Zhao H (2018a) Deletion of ferroportin in murine myeloid cells increases iron accumulation and stimulates osteoclastogenesis in vitro and in vivo. J Biol Chem 293(24):9248–9264. https://doi.org/10.1074/jbc.RA117.000834

Wang X, Chen B, Sun J, Jiang Y, Zhang H, Zhang P, Fei B, Xu Y (2018b) Iron-induced oxidative stress stimulates osteoclast differentiation via NF-kappaB signaling pathway in mouse model. Metabolism 83:167–176. https://doi.org/10.1016/j.metabol.2018.01.005

Weatherall DJ (1997) The thalassaemias. BMJ 314(7095):1675–1678. https://doi.org/10.1136/bmj.314.7095.1675

Weidner H, Baschant U, Lademann F, Ledesma Colunga MG, Balaian E, Hofbauer C, Misof BM, Roschger P, Blouin S, Richards WG, Platzbecker U, Hofbauer LC, Rauner M (2020) Increased FGF-23 levels are linked to ineffective erythropoiesis and impaired bone mineralization in myelodysplastic syndromes. JCI Insight 5(15). https://doi.org/10.1172/jci.insight.137062

Winterbourn CC (1995) Toxicity of iron and hydrogen peroxide: the Fenton reaction. Toxicol Lett 82-83:969–974. https://doi.org/10.1016/0378-4274(95)03532-x

Wolf M, Koch TA, Bregman DB (2013) Effects of iron deficiency anemia and its treatment on fibroblast growth factor 23 and phosphate homeostasis in women. J Bone Miner Res 28(8):1793–1803. https://doi.org/10.1002/jbmr.1923

Wolf M, Rubin J, Achebe M, Econs MJ, Peacock M, Imel EA, Thomsen LL, Carpenter TO, Weber T, Brandenburg V, Zoller H (2020) Effects of iron isomaltoside vs ferric carboxymaltose on hypophosphatemia in iron-deficiency anemia: two randomized clinical trials. JAMA 323(5):432–443. https://doi.org/10.1001/jama.2019.22450

Xiang S, Zhao L, Tang C, Ling L, Xie C, Shi Y, Liu W, Li X, Cao Y (2024) Icariin inhibits osteoblast ferroptosis via Nrf2/HO-1 signaling and enhances healing of osteoporotic fractures. Eur J Pharmacol 965:176244. https://doi.org/10.1016/j.ejphar.2023.176244

Zarjou A, Jeney V, Arosio P, Poli M, Antal-Szalmas P, Agarwal A, Balla G, Balla J (2009) Ferritin prevents calcification and osteoblastic differentiation of vascular smooth muscle cells. J Am Soc Nephrol 20(6):1254–1263. https://doi.org/10.1681/ASN.2008070788

Zarjou A, Jeney V, Arosio P, Poli M, Zavaczki E, Balla G, Balla J (2010) Ferritin ferroxidase activity: a potent inhibitor of osteogenesis. J Bone Miner Res 25(1):164–172. https://doi.org/10.1359/jbmr.091002

Zhang W, Xu J, Qiu J, Xing C, Li X, Leng B, Su Y, Lin J, Lin J, Mei X, Huang Y, Pan Y, Xue Y (2018) Novel and rapid osteoporosis model established in zebrafish using high iron stress. Biochem Biophys Res Commun 496(2):654–660. https://doi.org/10.1016/j.bbrc.2017.12.172

Zhang J, Hu W, Ding C, Yao G, Zhao H, Wu S (2019) Deferoxamine inhibits iron-uptake stimulated osteoclast differentiation by suppressing electron transport chain and MAPKs signaling. Toxicol Lett 313:50–59. https://doi.org/10.1016/j.toxlet.2019.06.007

Zhao GY, Di DH, Wang B, Huang X, Xu YJ (2015) Effects of mouse hepcidin 1 treatment on osteoclast differentiation and intracellular iron concentration. Inflammation 38(2):718–727. https://doi.org/10.1007/s10753-014-9982-2

Iron and Pregnancy

21

Veena Sangkhae and Elizabeta Nemeth

Abstract

Iron is essential for nearly all living organisms, but iron requirements vary throughout one's lifetime. Pregnancy is a period notable for vastly increased iron requirements. Iron is needed to support maternal adaptations throughout pregnancy and to enable the growth and development of both the placenta and fetus. Insufficient iron has long been linked to adverse pregnancy outcomes; thus, universal iron supplementation is common practice before and during pregnancy. In high-income countries, however, where foods are often iron-fortified and red meat consumption is common, too much iron supplementation during pregnancy has become a concern, as iron excess has also been linked to adverse pregnancy outcomes. Advances in clinical management have enabled more women with iron-related disorders to conceive; thus, it is important to understand iron physiology and how iron disorders affect pregnant women and their offspring to help inform clinical practice and optimize outcomes. This chapter will address the physiologic iron homeostasis of pregnancy and discuss what is currently known about iron-related disease and their consequences on human pregnancy.

Keywords

Iron · Pregnancy · Hepcidin · Placenta

21.1 Iron Requirements for Pregnancy

Pregnancy requires a large influx of iron. On average, just over 1 gram of additional iron is required to sustain a healthy pregnancy. For comparison, an adult human body contains around 3–4 grams of iron (Aggett 2012), so over 9 months of gestation, a pregnant person must acquire an additional 30% of their total iron content. Many women of reproductive age have insufficient body iron stores to meet the pregnancy iron demands. On average, 270 mg of iron is required for fetal development, 46 mg for placenta development, and 450 mg for expansion of maternal red blood cell mass. An additional 224 mg of iron is needed to compensate for basal iron losses and 150 mg for blood loss during delivery (Sangkhae et al. 2023). Although 450 mg of iron for red cell production is required during pregnancy, some of this iron is recovered postpartum when the red cell mass decreases. Table 21.1 summarizes iron balances in menstru-

V. Sangkhae · E. Nemeth (✉)
Center for Iron Disorders, Department of Medicine, David Geffen School of Medicine, University of California, Los Angeles, Los Angeles, CA, USA
e-mail: ENemeth@mednet.ucla.edu

K. Pantopoulos (ed.), *Iron Metabolism in Human Health and Disease*, Advances in Experimental Medicine and Biology 1480, https://doi.org/10.1007/978-3-031-92033-2_21

Table 21.1 Physiological iron balance in menstruating and pregnant females by gestation

Iron fate (amount, mg)	Menstruating females (28 days)	First trimester (0–13 weeks, 13 weeks)	Second trimester (14–26 weeks, 13 weeks)	Third trimester (27–40 weeks, 14 weeks)	Pregnancy sum (0–40 weeks, 40 weeks)
Basal iron losses	23 mg (0.8 mg/d)[a] (Hallberg and Rossander-Hulten 1991)	73 mg (0.8 mg/d)[a]	73 mg (0.8 mg/d)[a]	78 mg (0.8 mg/d)[a]	**224 mg**
Menstruation	13 mg[a]				
Placental iron			16 mg (0.2 mg/d)[e,f]	30 mg (0.3 mg/d)[b,e,g]	**46 mg**
Fetal iron			60 mg (0.7 mg/d)[b,c]	210 mg (2.1 mg/d)[b,c]	**270 mg**
Expansion of maternal RBC mass			112 mg (1.2 mg/d)[h,i]	338 mg (3.5 mg/d)[h,i]	**450 mg**[h]
Delivery blood loss					**150mg**[h]
Total iron needs	**36 mg (1.3 mg/d)**	**73 mg (0.8 mg/d)**	**261 mg (2.9 mg/d)**	**656 (6.7 mg/d)**	**1140 mg**
RBC-mass contraction after delivery					**−450mg**[h]

Based on data reported by [a](Hallberg and Rossander-Hulten 1991), [b](Widdowson and Spray 1951), [c](Alexander et al. 1996), [d](Salomon et al. 2007), [e](Barad et al. 2022), [f](Hecht et al. 2007), [g](Hayward et al. 2016), [h](Bothwell 2000), [i](Hytten 1985), assumed 55-kg woman

Menstruation calculation (Hallberg and Rossander-Hulten 1991): Hb 135 g/L blood, Fe 3.34 mg/g Hb

Fetal iron: Iron content/g fetus (Widdowson and Spray 1951); Avg fetal weights 2nd trimester (Alexander et al. 1996; Salomon et al. 2007), 3rd trimester (Alexander et al. 1996)

Placental iron: Iron content/g placenta (71.1 μg/g) (Barad et al. 2022); Avg placenta weight 2nd trimester (Hecht et al. 2007), 3rd trimester (Widdowson and Spray 1951; Barad et al. 2022; Hayward et al. 2016)

Maternal RBC expansion: Total iron required (Bothwell 2000); 2nd and 3rd trimesters area under curve calculation assuming a linear increase in requirement (Hytten 1985)

ating females and during pregnancy, with estimates based on a 120 lb (54 kg) woman. Pregnancy iron requirements are not constant across gestation; daily requirements increase during the period of rapid fetal development in the 2nd and 3rd trimesters, changing from 1 mg/day pre-pregnancy to 3 mg/day in the 2nd and 7 mg/day in the 3rd trimester (Sangkhae et al. 2023) (Fig. 21.1).

21.2 Pregnancy Iron Homeostasis

Hepcidin is a key regulator of systemic iron homeostasis (Chap. 3); this regulation is maintained during pregnancy. Maternal hepcidin concentrations control circulating iron levels and determine how much iron is delivered to the placenta and the fetus. During pregnancy, hepcidin levels initially increase due to cessation of menstruation but progressively decrease in the second and third trimesters to nearly undetectable levels (van Santen et al. 2013). This decrease in hepcidin promotes increased intestinal iron absorption and release of iron from liver stores to maximize circulating iron levels (Sangkhae and Nemeth 2017; Sangkhae et al. 2020b). Increased iron mobilization allows for expanded maternal hematopoiesis and iron uptake by the placenta and transfer to the developing fetus (Fig. 21.2). Several studies in mouse models highlight the importance of maternal hepcidin regulation during pregnancy. Absence of maternal hepcidin results in viable but iron-overloaded offspring, while elevated maternal hepcidin results in embryonal iron deficiency anemia (IDA), smaller embryos, and even embryonal death (Sangkhae et al. 2020a). These data suggest that suppression of maternal hepcidin is essential for a healthy pregnancy.

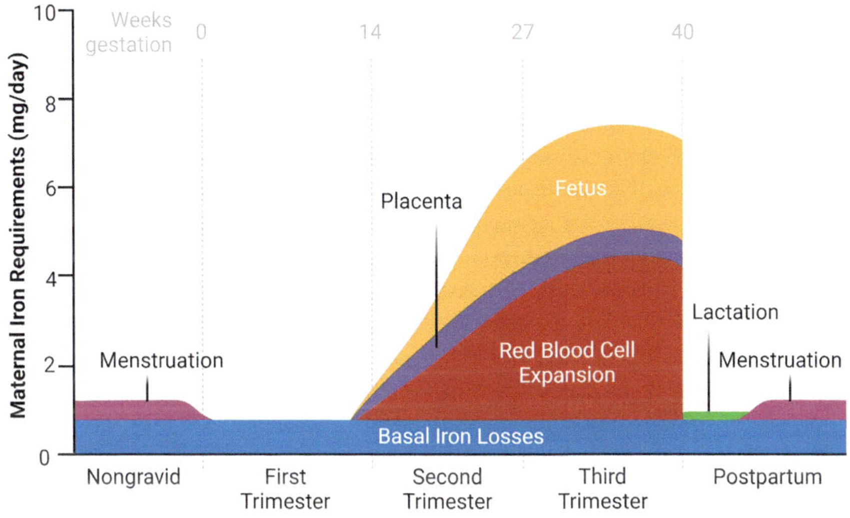

Fig. 21.1 Pregnancy iron requirements

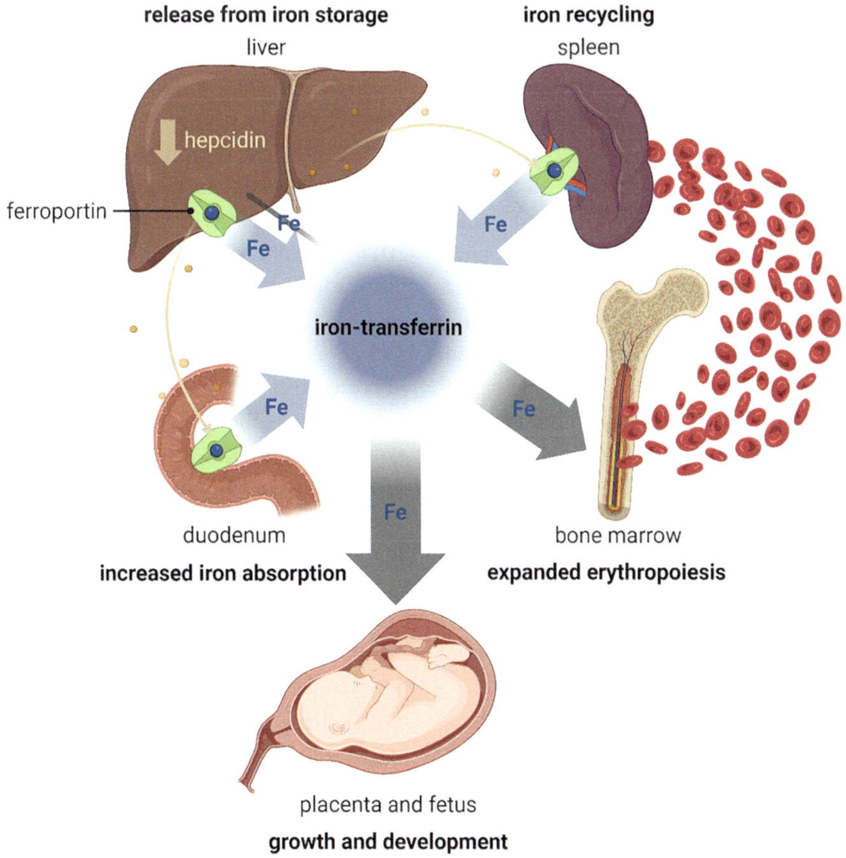

Fig. 21.2 Pregnancy iron homeostasis

21.2.1 Regulation of Maternal Hepcidin

Despite its importance, the mechanism(s) governing maternal hepcidin suppression in pregnancy are still being investigated. Nevertheless, it is established that known regulatory mechanisms are also involved in modulating hepcidin levels.

Hepcidin regulation by iron status is preserved during pregnancy. Maternal hepcidin positively correlates with maternal serum iron concentrations and ferritin (Finkenstedt et al. 2012) and is increased by iron supplementation (Flores-Quijano et al. 2016; Guo et al. 2019). Conversely, serum hepcidin levels are lower in pregnant women with IDA compared to non-anemic or non-IDA pregnant women (Zaman et al. 2019).

Erythroferrone (Chap. 3), an erythroid regulator of iron homeostasis, suppresses hepcidin by sequestering bone morphogenetic protein 2 and 6 (BMP2, 6) (Kautz et al. 2014; Arezes et al. 2018). ERFE is increased in anemic pregnancies (Delaney et al. 2021) but not in the absence of anemia (Wei et al. 2021), suggesting that maternal ERFE only plays a role in maternal hepcidin suppression in the presence of maternal anemia. Deletion of ERFE in mouse pregnancy had only a minor effect, and mostly under iron-deficient conditions (Sangkhae et al. 2022). Soluble hemojuvelin (sHJV), a cleavage product of membrane-bound HJV, is thought to suppress hepcidin by antagonizing the BMP-SMAD signaling pathway (Babitt et al. 2006). Maternal sHJV concentrations increase significantly in the third trimester of pregnancy and correlate with hepcidin (Finkenstedt et al. 2012); however, additional studies are necessary to determine if maternal sHJV suppresses hepcidin during pregnancy.

Pregnancy is characterized by large changes in hormone levels, including estrogen, progesterone, and prolactin, all of which have been implicated in hepcidin regulation (Sangkhae et al. 2023). However, deletion of prolactin or estrogen receptors in the maternal liver in mouse studies did not affect maternal hepcidin suppression or maternal and embryo iron or hematological status at term, suggesting that these hormones are not essential for pregnancy-related hepcidin suppression (Zhang et al. 2024).

21.3 Iron Transport Across the Placenta

The placenta is a temporary organ whose primary function is to supply the fetus with oxygen and nutrients and remove waste, without allowing for maternal and fetal blood to mix. Nutrient and waste exchange occurs through the syncytiotrophoblast, a single-cell layer that interfaces the maternal and fetal circulation. Iron is transported unidirectionally from the maternal to fetal circulation across the syncytiotrophoblast. Over the course of an average human pregnancy, the placenta transports approximately 270 mg of iron to the fetus and retains around 45 mg of iron (Bothwell 2000; Widdowson and Spray 1951; Fisher and Nemeth 2017; Barad et al. 2022).

The transport of non-heme iron across the placenta is depicted in Fig. 21.3 and reviewed in greater detail elsewhere (Sangkhae and Nemeth 2018). Transferrin-bound ferric iron (Tf-Fe^{3+}) is the predominant form of iron in the maternal circulation (also referred to as holo-transferrin). Tf-Fe^{3+} is imported into the placenta by the iron importer transferrin receptor 1 (TFRC or TfR) expressed on the maternal-facing membrane of syncytiotrophoblast. Once Tf-Fe^{3+} binds to TFRC, the complex is internalized via clathrin-mediated endocytosis (see also Chap. 2). In the acidic endosome, Fe^{3+} dissociates from Tf, and Fe^{3+} is reduced to Fe^{2+} by a ferrireductase. Ferrous iron (Fe^{2+}) in the endosome is subsequently exported across the endosomal membrane into the cytoplasm. The apoTf-TFRC complex is then recycled back to the syncytiotrophoblast membrane, apoTf is released from TFRC, and TFRC can undergo a subsequent round of iron import. Within the syncytiotrophoblast cytoplasm, iron can be utilized by the placenta, stored in ferritin, or exported to the fetal circulation. Iron is transported out of the syncytiotrophoblast by the iron exporter ferroportin (FPN, SLC40A1) (Bastin

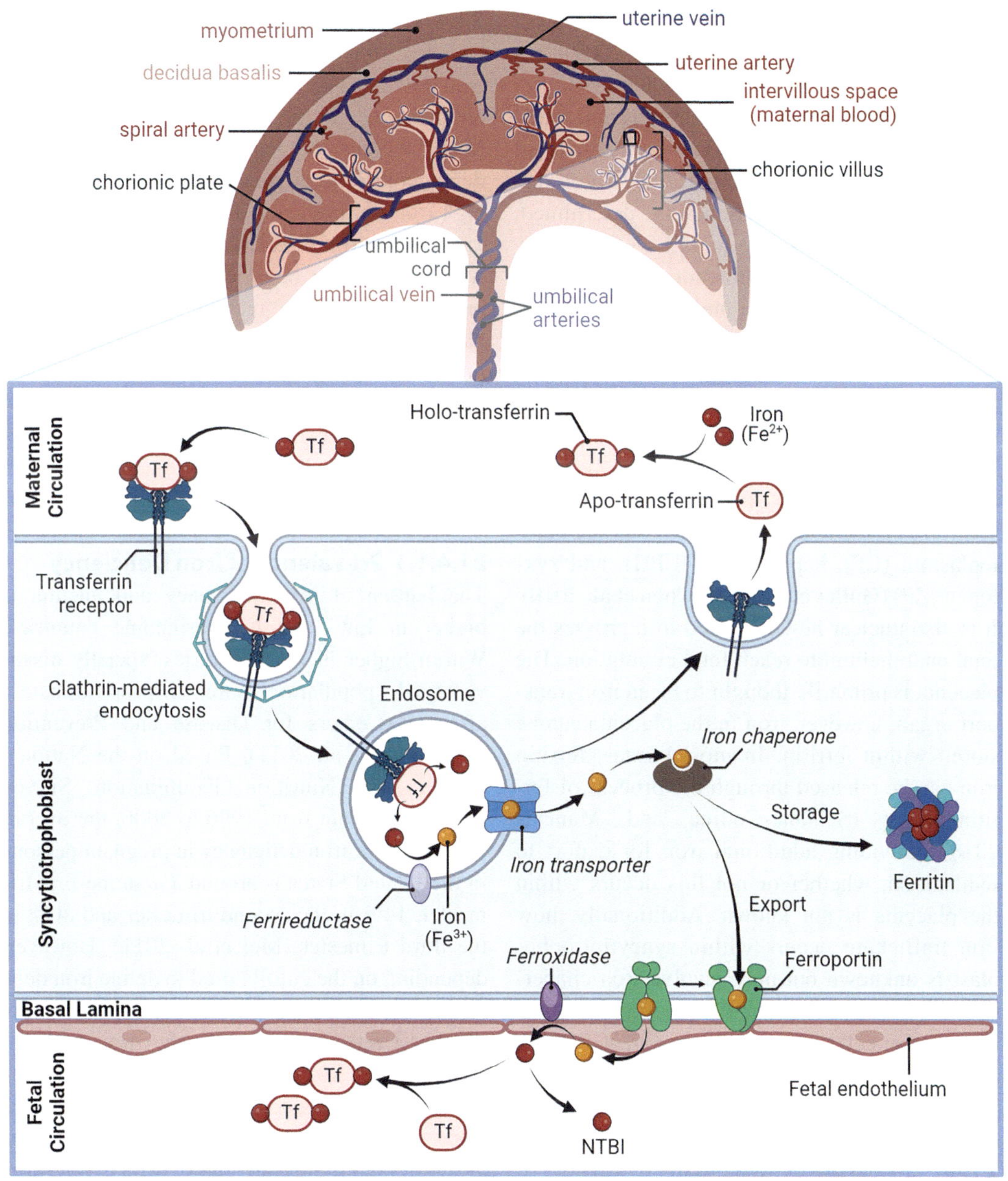

Fig. 21.3 Placental iron trafficking

et al. 2006), expressed on the fetal facing membrane (Sangkhae et al. 2020b). After iron is exported out of the syncytiotrophoblast, it undergoes oxidation to the ferric form (Fe^{3+}) by a ferroxidase. Once oxidized, ferric iron can be loaded onto fetal Tf, where it can be utilized by tissues or stored in the liver.

Our understanding of iron transport across the placenta and iron handling within the placenta is still incomplete. Circulating holoTf (Tf-Fe^{3+}) is thought to be the main species of iron taken up by the placenta; however, transporters for other iron species such as non-transferrin-bound iron (NTBI), heme, and ferritin are expressed in the placenta (Cao et al. 2014; Keel et al. 2008; Lyall et al. 2000; Li et al. 2009). It is unknown whether any of these molecules represent a significant iron source for placental transport or fetal development. Additionally, many of the key players involved in placental iron transport have not been definitively determined, although a number of proteins have been proposed: the endosomal ferrireductase could be STEAP3 (six-transmembrane epithelial of prostate) or STEAP4 (Ohgami et al. 2006), the endosomal iron exporter could be divalent metal transporter 1 (DMT1, SLC11A2) (Georgieff et al. 2000; Gunshin et al. 2005; Li et al. 2012), ZRT/IRT-like protein (ZIP)8 (SLC39A8) (Wang et al. 2012) and ZIP14 (SLC39A14) (Hojyo et al. 2011), the fetal ferroxidase could be ceruloplasmin (CP), hephaestin (HEPH), and zyklopen (ZP) (Guller et al. 2008; Chen et al. 2010). It is also unclear how exported iron crosses the fetal endothelium to reach fetal circulation. The placenta is primarily thought to be an iron transport organ; however, iron in the placenta can be stored within ferritin. In most tissues, ferritin iron can be released through the process of ferritinophagy (Santana-Codina and Mancias 2018), providing additional iron for export to fetal blood; whether or not this occurs within the placenta is not known. Additionally, how iron trafficking occurs within syncytiotrophoblast is unknown but may involve iron chaperones PCBP1 and 2 (poly(rC)-binding proteins), like in other cell types (Yanatori et al. 2016; Ryu et al. 2017).

21.4 Pregnancy and Iron Disorders

Pregnancy increases the metabolic demands of both mother and fetus. Thus, many conditions can be exacerbated by the demands of pregnancy. Pregnancy complications associated with iron and iron-related diseases are summarized in Fig. 21.4.

21.4.1 Iron Deficiency

Iron deficiency is the most common micronutrient deficiency in the world (Organization 2012). Many women of reproductive age are iron-deficient prior to pregnancy. Inability to meet the high iron requirements during pregnancy results in maternal iron deficiency and IDA (Chap. 12), which have long been associated with adverse pregnancy outcomes, including increased maternal morbidity and mortality, preterm birth, low birth weight, cognitive defects in newborns, and impaired immune function (Arija et al. 2019; Cogswell et al. 2003; Corwin et al. 2003; Georgieff 2011; McArdle et al. 2014; Scanlon et al. 2000; Scholl and Reilly 2000; Allen 2000).

21.4.1.1 Prevalence of Iron Deficiency

The burden of iron deficiency and anemia is higher in low- and middle-income countries. Within higher income countries, socially disadvantaged populations are disproportionately affected (Centers for Disease and Prevention 2002; Mei et al. 2011). Based on the National Health and Nutrition Examination Survey (NHANES) data from 1996 to 2006, the overall prevalence of iron deficiency in pregnant persons in the United States is around 7% in the first trimester, 14% in the second trimester and 30% in the third trimester (Mei et al. 2011). However, depending on the cutoffs used to define iron deficiency, up to 40% of first-trimester pregnancies in the United States may be iron-deficient (Benson et al. 2022). An additional study reporting on 2039 pregnant females, reported prevalence of iron deficiency varied from 5.44% for the Hemochromatosis and Iron Overload Screening Study (HEIRS), 18.05% for the World Health Organization (WHO) to 36.10% for iron deficiency erythropoiesis (IDE) defined cutoffs (HEIRS tsat <10% + sFerritin <15 ng/mL; WHO sFerritin <15 ng/mL; IDE, sFerritin <25 ng/mL) (Barton et al. 2024). Iron deficiency with anemia is estimated to affect 5% of pregnant persons in the United States, although the prevalence also

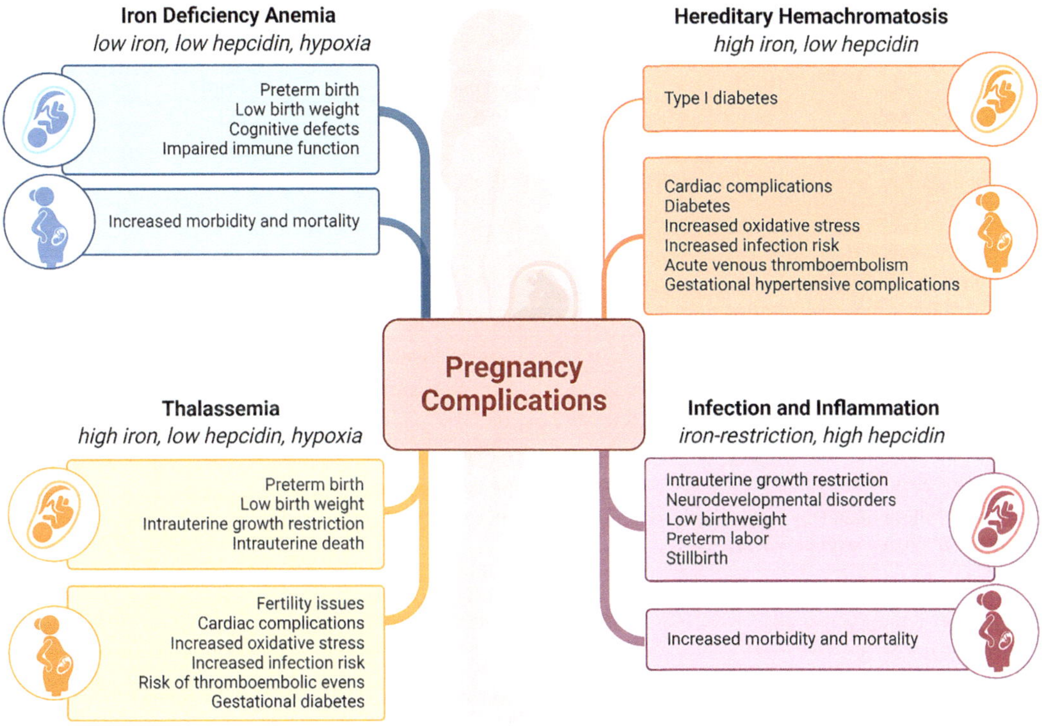

Fig. 21.4 Iron-associated pregnancy complication

varies across trimesters (Mei et al. 2011). Globally, the prevalence of pregnancy iron deficiency is less clear, but the prevalence of anemia is well characterized. An estimated 36% of all pregnant women have anemia (Stevens et al. 2022), with half of these cases attributed to iron deficiency.

21.4.1.2 Screening for Pregnancy Iron Deficiency

Given the importance of iron for pregnancy and the known adverse effects of pregnancy iron deficiency and IDA, the American College of Obstetricians and Gynecologists (ACOG) (American College of and Gynecologists' Committee on Practice 2021), American Academy of Family Physicians (AAFP) (Short and Domagalski 2013), U.S. Preventive Services Task Force (Cantor et al. 2024), and Centers for Disease Control and Prevention (CDC) (Recommendations to prevent and control iron deficiency in the United States. Centers for Disease Control and Prevention 1998) all recommend routine screening of pregnant women for anemia. ACOG recommends a complete blood count in the first trimester and again at 24 0/7–28 6/7 weeks of gestation (American College of and Gynecologists' Committee on Practice 2021). Hemoglobin (Hb) values are commonly used to define anemia. WHO defines pregnancy anemia as Hb concentrations less than 11 g/dL in the first trimester, less than 10.5 g/dL in the second trimester, and less than 11 g/dL in the third trimester (Table 21.2). For pregnant females >15 years of age, WHO anemia severity is defined as: mild (10–10.9 g/dL), moderate (7–9.9 g/d), and severe (<7 g/dL). ACOG considers Hb concentration of less than 6 g/dL as severe anemia (American College of and Gynecologists' Committee on Practice 2021).

When it comes to screening for iron deficiency, there are no specific recommendations aside from additional workup following anemia diagnosis (Table 21.3). If anemia is diagnosed, ACOG recommends follow-up to determine the cause if hematocrit is less than 33% in the first or

Table 21.2 Anemia cutoffs during pregnancy

	Trimester		
Test	First	Second	Third
Hemoglobin (g/dL)	11.0	10.5	11.0
Hematocrit (%)	33.0	32.0	33.0

Recommendations to prevent and control iron deficiency in the United States. (Centers for Disease Control and Prevention 1998; WHO 2024)

Table 21.3 Iron reference ranges for pregnancy

Test	Normal value
Plasma iron level	40–175 μg/dL
Plasma total iron-binding capacity	216–400 μg/dL
Transferrin saturation	16–60%
Serum ferritin level	>30 μg/L[a]
Free erythrocyte protoporphyrin level	<3 μg/g

American College of and Gynecologists' Committee on Practice (2021)
[a]No upper reference level provided

Table 21.4 Anemia follow-up

	Results indicating		
Test	Iron deficiency anemia	Anemia of chronic disease	Thalassemia
Serum iron	Decreased	Decreased	Normal-to-increased
Total iron-binding capacity	Increased	Decreased	Normal-to-decreased
Serum ferritin	Decreased	Increased	Normal-to-increased
Transferrin saturation	Decreased	Decreased	Normal-to-increased

Modified from American College of and Gynecologists' Committee on Practice (2021) and Vlachodimitropoulou et al. (2024)

third trimester or less than 32% in the second trimester and provide guidelines to distinguish IDA from thalassemia and anemia of chronic disease (Table 21.4) (American College of and Gynecologists' Committee on Practice 2021). The AAFP recommends serum ferritin as the most accurate test to diagnose IDA (Short and Domagalski 2013). However, the efficacy of these individual markers for evaluating maternal iron status is not entirely clear (Bah et al. 2017; Ssewanyana et al. 2023).

Analysis of iron biomarkers in pregnant and postpartum women in India (Nair et al. 2023) found only weak or no relationship between hemoglobin and iron biomarkers: ferritin, soluble transferrin receptor (sTfR), transferrin saturation (TSAT), and hepcidin. This indicated that hemoglobin concentration alone is not an effective marker of pregnancy iron deficiency. Low serum ferritin was more common in anemic than non-anemic women, but high ferritin was also more common in women with severe anemia. In addition to being an iron storage marker, ferritin is also an acute phase reactant and could thus misrepresent iron status in the presence of inflammation (Madu and Ughasoro 2017). Furthermore, hemoglobin measurements can be confounded by the presence of hemoglobin variants (HbE, beta-thalassemia trait, sickle cell trait) (Nair et al. 2023). In a study conducted in The Gambia, hepcidin was a superior indicator of maternal iron deficiency compared to hemoglobin and sTfR (Bah et al. 2017). Work is still needed to define and implement a biomarker or a panel of biomarkers to accurately assess iron deficiency during pregnancy, including in a resource-constrained setting.

21.4.1.3 Pregnancy Iron Supplementation

Routine prenatal iron supplementation reduces the incidence of iron deficiency and IDA during pregnancy (Cantor et al. 2024). Maternal iron supplementation is shown to improve anemia and ferritin levels in pregnant women but had little effect on cord blood ferritin at delivery. However, infants of mothers who were iron-supplemented during pregnancy had detectable improvement in serum ferritin and anemia several months after delivery (Milman et al. 1991; Preziosi et al. 1997). Thus, ACOG and the CDC recommend universal low-dose iron supplementation for pregnant women beginning at the first prenatal visit (Benson et al. 2022). Often, it is recommended that women supplement for several months prior to pregnancy. Iron in prenatal supplements is typically in the form of iron salts (non-heme iron) providing 26–34 mg of elemental iron per dose (range: 4.5–106 mg/dose) (Saldanha et al. 2019). In the US, more than 75% of pregnant women use prenatal supplements, with use increasing as pregnancy progresses (Branum et al. 2013). Although oral iron is typically the first line of treatment for maternal iron deficiency and IDA, for some patients, the uptake of iron is suboptimal, and for others, gastrointestinal side effects lead to nonadherence (Chap. 24).

Intravenous iron is an alternative to oral iron and is effective in increasing maternal hemoglobin and maternal iron stores (Govindappagari and Burwick 2019; Lewkowitz et al. 2019). Compared to oral iron, intravenous iron was associated with increased neonatal hemoglobin and ferritin (Lewkowitz et al. 2019). ACOG recommends considering parenteral iron treatment for patients with severe anemia (Hb less than 6 g/dL) or for those who cannot tolerate or do not respond to oral iron. Multiple formulations of parenteral iron are currently used in the clinic (Chap. 24).

However, it might be important to consider maternal iron status prior to recommending supplementation to avoid unnecessary or ineffective treatment and associated side effects. In higher income countries where foods are often iron-fortified and red meat consumption is common, the majority of women have sufficient iron stores prior to pregnancy. More recently, too much iron supplementation during pregnancy has become a concern as elevated maternal serum ferritin has also been linked to adverse pregnancy outcomes including preterm birth, low birth weight, gestational diabetes, and neurodevelopmental deficiencies in children (Bowers et al. 2016; Goldenberg et al. 1996; Guo et al. 2019; Khambalia et al. 2015; Lao et al. 2000; Scholl 1998). However, the attribution of increased serum ferritin to iron overload requires great caution because increased ferritin can also result from inflammation. Consequently, the U.S. Preventive Services Task Force concluded that there was inadequate evidence to recommend for or against routine iron supplementation in pregnancy to improve maternal or neonatal outcomes (Cantor et al. 2024).

21.4.2 Infectious and Inflammatory Diseases

Maternal infection during pregnancy is associated with increased maternal morbidity and mortality, low birth weight, stillbirth, miscarriage, and preterm labor (Kumar et al. 2022). Even maternal systemic inflammation in the absence of infection is associated with offspring neurodevelopmental disorders including autism, schizophrenia, attention-deficit hyperactivity disorder, and depression (Estes and McAllister 2016; Jones et al. 2017; Allswede et al. 2020). Interestingly, many of these adverse pregnancy outcomes are also associated with anemia during pregnancy (Sangkhae et al. 2023).

Iron homeostasis and inflammation are intimately linked; hepcidin is potently induced by inflammation (discussed in greater detail in Chaps. 13 and 14) (Ganz and Nemeth 2024). During infection, iron sequestration is an innate host defense strategy. Hepcidin transcription is rapidly induced by interleukin-6 (IL-6) and inhibits the release of iron into the plasma resulting in hypoferremia (Ganz and Nemeth 2024). Long-term activation of this pathway results in anemia of inflammation. Although the mechanisms are well described in the non-pregnant state, the effects of inflammation on iron homeostasis during pregnancy are only recently being examined. Hepcidin and inflammatory markers (CRP or IL-6) correlate in pregnancies complicated by infection or inflammation (Gonzalez-Fernandez et al. 2021) but do not correlate in healthy pregnancies (Schulze et al. 2008; Simavli et al. 2014; van Santen et al. 2013), suggesting that inflammation-mediated regulation of hepcidin is preserved during pregnancy.

It is possible that some of the consequences of infection or inflammation during pregnancy are attributable to elevated hepcidin. In fact, in a cohort of pregnancies complicated by inflammation and nutrient deficiencies, higher maternal hepcidin was identified as a main determinant of intrauterine growth restriction (IUGR) (Gonzalez-Fernandez et al. 2021), suggesting that hepcidin-mediated iron restriction, particularly when chronic, contributes to adverse pregnancy outcome. Additionally, the induction of fetal hepcidin during intraamniotic infection or inflammation (IAI) resulted in fetal hypoferremia, suggesting that iron sequestration may already be important for fetal host defense (Fisher et al. 2020).

21.4.3 Hemochromatosis

Hemochromatosis (formerly referred to as hereditary hemochromatosis) is a genetic disorder characterized by systemic iron overload primarily due to dysregulated hepcidin production (discussed in greater detail in Chap. 9). The most common cause of hemochromatosis is mutation of the homeostatic iron regulator (*HFE*) gene, accounting for 90% of cases in people of Northern European descent. Other genes involved in the pathophysiology of hemochromatosis are hepcidin (*HAMP*), hemojuvelin (*HJV*), ferroportin (*FPN*), and transferrin receptor 2 (*TFR2*) (Girelli et al. 2022). Clinically, women are less affected than men (1:1.8 to 1:3 ratio) and are diagnosed later in life (52.6 v 47.7 years) (Scotet et al. 2018), potentially due to iron losses through menstruation.

The prevalence of recorded hemochromatosis in pregnant persons is increasing (from 1.91 per 100,000 hospitalizations for delivery in 2010 to 8.65 cases per 100,000 in 2016 and ~7.0 cases per 100,000 in 2019) (Niu et al. 2023). However, few studies have reported on the consequences of hemochromatosis on pregnancy. Because hepcidin is already inappropriately low in hemochromatosis and would be further suppressed by pregnancy, this could lead to excessive iron absorption and mobilization from maternal stores, resulting in high circulating iron levels and iron loading of the placenta and fetus. Indeed, based on mouse studies, maternal iron excess does lead to placental and fetal iron loading (Sangkhae et al. 2020a); however, this has not been systematically studied in human pregnancies.

21.4.3.1 Hemochromatosis Effects on Maternal Health

In the handful of data on hemochromatosis pregnancy, some pregnancies were reportedly healthy and uneventful. The first description of hemochromatosis pregnancy was a 1991 case report (Baynes et al. 1991). Prior to pregnancy, a 33-year-old patient presented with a history of irregular menses and had high TSAT (96%) and serum ferritin (2311 ng/mL). Therapeutic phlebotomies removed 10.5 g of iron; the patient subsequently became pregnant, and phlebotomy was discontinued during pregnancy. Serum ferritin increased from 8 to 91.6 ng/mL during pregnancy, with increased TSAT (66%) and hemoglobin (13.9 g/dL) at delivery. The pregnancy was reported as uneventful; she delivered a normal infant, and cord blood iron parameters were within the reference range with moderately elevated ferritin (250 ng/ml) and TSAT (88%).

In 2009, a case report characterized a pregnancy from a 36-year-old female homozygous for *HFE* C282Y (Alisa Sokoloff and Cooper 2009). The diagnosis was made 5 months prior to pregnancy, and at the time, the patient's TSAT was 75% and ferritin was 320 ng/mL. During pregnancy, the patient did not receive iron supplementation or phlebotomy treatment and did not develop anemia. Ferritin decreased over gestation, from 258 ng/mL at 7 weeks to 145 ng/mL at 37 weeks; however, TSAT was continually increased, from 77% at 7 weeks to 93% at 37 weeks, as would be expected from profoundly lowered hepcidin. The child was delivered at term via normal vaginal delivery and considered healthy; however, an iron panel on the child was not obtained.

Although these cases reported healthy newborns, no long-term follow-up studies were published. Transferrin saturations above 70% lead to the generation of highly reactive NTBI, resulting in increased oxidative stress with negative impacts on organs and an increased risk of certain infections (Ganz and Nemeth 2024). Additional studies are necessary to understand the impact of hemochromatosis on maternal and fetal health.

In other case reports, adverse maternal outcomes associated with hemochromatosis were observed including cardiac complications (Durand et al. 2003; Hoefnagel and Wissler 2012) and severe diabetes insipidus (Krysiak et al. 2010). In a 2003 case report, a pregnant 26-year-old woman was admitted for congestive heart failure and rapid atrial fibrillation. She had major cardiac iron overload and was subsequently diagnosed with hemochromatosis. She was negative for HFE mutations (C282Y and H63D), but familial data suggested that the patient suffered from juvenile hemochromatosis (Durand et al. 2003). The fetus was not characterized in the study. It is likely that the increased cardiac stress of pregnancy exacerbated underlying cardiac defects from hemochromatosis.

In a 2010 case report, a 35-year-old homozygous *HFE* C282Y patient was pregnant with her second child (Krysiak et al. 2010). Her first pregnancy at 29 years old was uneventful and ended at term via spontaneous vaginal delivery. During her second pregnancy, at 28 weeks, hemoglobin was high (15.8 g/dL), TSAT >100%, and ferritin was 937 μg/L. Her second pregnancy was complicated by severe pregnancy-induced diabetes insipidus (DI). She responded to treatment with desmopressin acetate and delivered a week before the calculated term; delivery was spontaneous, and the baby was fully developed. Following delivery, she was officially diagnosed with hemochromatosis and underwent deferoxamine (DFO) treatment. For her third pregnancy, DFO was discontinued and she received phlebotomy treatment. Her third pregnancy was still complicated by pregnancy-induced DI, although milder, likely due to iron-chelation therapy prior to the third pregnancy. Her third baby was delivered at 38 weeks and was reportedly healthy. Newborns' iron parameters were not measured.

A large 2023 retrospective study investigating the impact of hemochromatosis on maternal and perinatal outcomes found that during delivery, hemochromatosis patients had increased prevalence of acute venous thromboembolism (aOR 20.35), gestational hypertensive complications (aOR 1.50), and gestational diabetes mellitus (aOR 1.61) (Niu et al. 2023). Fetal outcomes such as preterm birth, fetal growth restriction, large for gestational age, and fetal death were not different between control and hemochromatosis pregnancies, although newborn iron status and longer term follow-up were not reported.

21.4.3.2 Maternal Hemochromatosis, Effects on Offspring

Very few studies have reported specifically on the effects of maternal hemochromatosis or maternal iron overload on offspring.

In 2018, a study out of Norway found the incidence of type I diabetes to be higher in children from *HFE* (C282Y and H63D) mothers (aOR 1.69), and it was unaffected by offspring HFE genotype (aOR 0.73) (Stordal et al. 2018). Maternal iron supplementation alone was a risk factor for offspring type I diabetes (aOR 1.33). A 2016 study in Taiwan analyzing the incidence and risks of congenital anomalies of kidney and urinary tract (CAKUT) in newborns found that a maternal risk factor associated with CAKUT included thalassemia or hemochromatosis (OR 5.69). In both conditions, maternal iron overload is likely (Tain et al. 2016). Additionally, in a 2022 Dutch study, high maternal ferritin of unspecified etiology during early pregnancy (<18 weeks gestation), even adjusted for high CRP, was associated with 2.54 points lower child intelligence quotient and 16.02 cm^2 smaller brain volume (Sammallahti et al. 2022).

Overall, iron overload and its effects on pregnancy outcomes are alarmingly understudied. Of the few published studies, data suggest that maternal hemochromatosis could have long-term adverse consequences that might not be overt at birth. These data also raise questions regarding universal iron supplementation, especially in regions where most women are iron-sufficient.

21.4.4 Thalassemia

Thalassemias (discussed in Chap. 11) are a genetically and clinically heterogeneous group of hereditary blood disorders that are characterized by ineffective erythropoiesis. β-thalassemia affects 1.5% of the global population, although regional prevalence differs considerably. Although the prevalence of β-thalassemia pregnancies is not well documented, a recent population study of uncomplicated pregnancies in India found that 4% of pregnant women had homozygous HbE variant and 9% heterozygous (Nair et al. 2023). Apart from anemia, another hallmark of thalassemia is iron overload, resulting from chronic transfusions and/or from increased intestinal iron absorption due to hepcidin suppression in response to ineffective erythropoiesis (Kautz et al. 2015).

Thalassemia can have negative impacts on fertility. Hypogonadotropic hypogonadism, the most frequent endocrinopathy in transfused thalassemia patients, predisposes patients to pubertal failure, sexual dysfunction, short stature, and infertility (Petrakos et al. 2016). Oxidative stress from tissue iron loading can result in impaired gonadotropin and growth hormone secretion from the anterior pituitary gland or may interfere with multiple physiological processes within the ovary including oocyte maturation (Petrakos et al. 2016). Advances in clinical management have allowed more women with β-thalassemia to have children, but their pregnancies are considered high-risk.

Transfusion requirements increase during pregnancy, but iron chelation is discontinued due to potential teratogenicity, and these together lead to exacerbation of iron loading. Indeed, a recent systematic review and meta-analysis revealed that compared to pre-pregnancy levels, in β-thalassemia major pregnancies at term, plasma ferritin increased by ~1000 ng/ml, liver iron concentration tripled, and myocardial T2* MRI signal decreased, consistent with heart iron loading (Vlachodimitropoulou et al. 2024). Pregnant women with β-thalassemia are known to have an increased risk of cardiac complications, increased risk of infections, increased risk of thromboembolic events, and complications associated with iron loading of the liver and endocrine glands (Petrakos et al. 2016). There are also increased risks for the developing fetus, including IUGR, low birth weight, prematurity, abortion, and intrauterine fetal death (Fozza et al. 2017; Tuck et al. 1998).

It is not clear to which extent maternal iron overload, anemia, and/or pre-existing disease-related conditions contribute to adverse pregnancy outcomes; thus, a more thorough understanding of the pathophysiology in pregnancies affected by β-thalassemia is essential.

21.5 Conclusions

Although iron is essential for healthy pregnancy, large gaps exist in both our basic and clinical understanding of iron homeostasis and disorders in pregnancy. In basic research, important questions include understanding regulatory mechanisms that ensure iron availability during pregnancy, iron transport mechanisms across the placenta, iron forms contributing to placental and fetal development, and cellular consequences of iron deficiency or excess during fetal development. Clinically, reference ranges for iron parameters and regulators at different stages of pregnancy are not well documented, screening for iron deficiency is not routinely performed except in the context of anemia, and cutoffs are not defined based on functional outcomes. Abnormal maternal iron status at either extreme, either deficiency or excess of iron, is associated with adverse pregnancy outcomes for both the mother and the baby. However, the underlying mechanisms causing adverse pregnancy outcomes are understudied, as are the consequences of iron dysregulation on postnatal development or the risk of disease later in life. Particularly understudied is the effect of iron excess, which is concerning considering that universal iron supplementation is recommended in many countries even without screening for iron deficiency and that the rate of pregnancies in conditions associated with high iron is increasing (thalassemia, hemochromatosis).

References

Aggett PJ (2012) Iron. In: Present knowledge in nutrition. pp 506–520. https://doi.org/10.1002/9781119946045.ch33

Alexander GR, Himes JH, Kaufman RB, Mor J, Kogan M (1996) A United States national reference for fetal growth. Obstet Gynecol 87(2):163–168. https://doi.org/10.1016/0029-7844(95)00386-X

Alisa Sokoloff SB, Cooper M (2009) Iron studies in hemochromatosis during pregnancy. Blood 114(22):5099. https://doi.org/10.1182/blood.V114.22.5099.5099

Allen LH (2000) Anemia and iron deficiency: effects on pregnancy outcome. Am J Clin Nutr 71(5 Suppl):1280S–1284S

Allswede DM, Yolken RH, Buka SL, Cannon TD (2020) Cytokine concentrations throughout pregnancy and risk for psychosis in adult offspring: a longitudinal case-control study. Lancet Psychiatry 7(3):254–261. https://doi.org/10.1016/S2215-0366(20)30006-7

American College of O, Gynecologists' Committee on Practice B-O (2021) Anemia in pregnancy: ACOG practice bulletin, number 233. Obstet Gynecol 138(2):e55–e64. https://doi.org/10.1097/AOG.0000000000004477

Arezes J, Foy N, McHugh K, Sawant A, Quinkert D, Terraube V, Brinth A, Tam M, LaVallie ER, Taylor S, Armitage AE, Pasricha SR, Cunningham O, Lambert M, Draper SJ, Jasuja R, Drakesmith H (2018) Erythroferrone inhibits the induction of hepcidin by BMP6. Blood 132(14):1473–1477. https://doi.org/10.1182/blood-2018-06-857995

Arija V, Hernandez-Martinez C, Tous M, Canals J, Guxens M, Fernandez-Barres S, Ibarluzea J, Babarro I, Soler-Blasco R, Llop S, Vioque J, Sunyer J, Julvez J (2019) Association of iron status and intake during pregnancy with neuropsychological outcomes in children aged 7 years: the prospective birth cohort infancia y medio ambiente (INMA) study. Nutrients 11(12):2999. https://doi.org/10.3390/nu11122999

Babitt JL, Huang FW, Wrighting DM, Xia Y, Sidis Y, Samad TA, Campagna JA, Chung RT, Schneyer AL, Woolf CJ, Andrews NC, Lin HY (2006) Bone morphogenetic protein signaling by hemojuvelin regulates hepcidin expression. Nat Genet 38(5):531–539. https://doi.org/10.1038/ng1777

Bah A, Pasricha SR, Jallow MW, Sise EA, Wegmuller R, Armitage AE, Drakesmith H, Moore SE, Prentice AM (2017) Serum hepcidin concentrations decline during pregnancy and may identify iron deficiency: analysis of a longitudinal pregnancy cohort in the Gambia. J Nutr 147(6):1131–1137. https://doi.org/10.3945/jn.116.245373

Barad A, Guillet R, Pressman EK, Katzman PJ, Miller RK, Darrah TH, O'Brien KO (2022) Placental iron content is lower than previously estimated and is associated with maternal iron status in women at greater risk of gestational iron deficiency and anemia. J Nutr 152(3):737–746. https://doi.org/10.1093/jn/nxab416

Barton JC, Wiener HW, Barton JC, Acton RT (2024) Prevalence of iron deficiency using 3 definitions among women in the US and Canada. JAMA Netw Open 7(6):e2413967. https://doi.org/10.1001/jamanetworkopen.2024.13967

Bastin J, Drakesmith H, Rees M, Sargent I, Townsend A (2006) Localisation of proteins of iron metabolism in the human placenta and liver. Br J Haematol 134(5):532–543. https://doi.org/10.1111/j.1365-2141.2006.06216.x

Baynes RD, Meyer TE, Bothwell TH, Lamparelli RD (1991) Maternal and fetal iron measurements in a hemochromatotic pregnancy. Am J Hematol 36(1):48–49. https://doi.org/10.1002/ajh.2830360110

Benson AE, Shatzel JJ, Ryan KS, Hedges MA, Martens K, Aslan JE, Lo JO (2022) The incidence, complications, and treatment of iron deficiency in pregnancy. Eur J Haematol. https://doi.org/10.1111/ejh.13870

Bothwell TH (2000) Iron requirements in pregnancy and strategies to meet them. Am J Clin Nutr 72(1 Suppl):257S–264S. https://doi.org/10.1093/ajcn/72.1.257S

Bowers KA, Olsen SF, Bao W, Halldorsson TI, Strom M, Zhang C (2016) Plasma concentrations of ferritin in early pregnancy are associated with risk of gestational diabetes mellitus in women in the Danish national birth cohort. J Nutr 146(9):1756–1761. https://doi.org/10.3945/jn.115.227793

Branum AM, Bailey R, Singer BJ (2013) Dietary supplement use and folate status during pregnancy in the United States. J Nutr 143(4):486–492. https://doi.org/10.3945/jn.112.169987

Cantor AG, Holmes R, Bougatsos C, Atchison C, DeLoughery T, Chou R (2024) Screening and supplementation for iron deficiency and iron deficiency anemia during pregnancy: updated evidence report and systematic review for the US Preventive Services Task Force. JAMA 332(11):914–928. https://doi.org/10.1001/jama.2024.13546

Cao C, Pressman EK, Cooper EM, Guillet R, Westerman M, O'Brien KO (2014) Placental heme receptor LRP1 correlates with the heme exporter FLVCR1 and neonatal iron status. Reproduction 148(3):295–302. https://doi.org/10.1530/REP-14-0053

Centers for Disease C, Prevention (2002) Iron deficiency—United States, 1999-2000. MMWR Morb Mortal Wkly Rep 51(40):897–899

Chen H, Attieh ZK, Syed BA, Kuo YM, Stevens V, Fuqua BK, Andersen HS, Naylor CE, Evans RW, Gambling L, Danzeisen R, Bacouri-Haidar M, Usta J, Vulpe CD, McArdle HJ (2010) Identification of zyklopen, a new member of the vertebrate multicopper ferroxidase family, and characterization in rodents and human cells. J Nutr 140(10):1728–1735. https://doi.org/10.3945/jn.109.117531

Cogswell ME, Parvanta I, Ickes L, Yip R, Brittenham GM (2003) Iron supplementation during pregnancy, anemia, and birth weight: a randomized controlled trial. Am J Clin Nutr 78(4):773–781. https://doi.org/10.1093/ajcn/78.4.773

Corwin EJ, Murray-Kolb LE, Beard JL (2003) Low hemoglobin level is a risk factor for postpartum depression. J Nutr 133(12):4139–4142. https://doi.org/10.1093/jn/133.12.4139

Delaney KM, Guillet R, Pressman EK, Ganz T, Nemeth E, O'Brien KO (2021) Serum erythroferrone during pregnancy is related to erythropoietin but does not predict the risk of anemia. J Nutr 151(7):1824–1833. https://doi.org/10.1093/jn/nxab093

Durand E, Grinda JM, Bruneval P (2003) Cardiogenic shock in a young woman revealing juvenile haemochromatosis. Heart 89(8):870. https://doi.org/10.1136/heart.89.8.870

Estes ML, McAllister AK (2016) Maternal immune activation: implications for neuropsychiatric disorders. Science 353(6301):772–777. https://doi.org/10.1126/science.aag3194

Finkenstedt A, Widschwendter A, Brasse-Lagnel CG, Theurl I, Hubalek M, Dieplinger H, Tselepis C, Ward DG, Vogel W, Zoller H (2012) Hepcidin is correlated to soluble hemojuvelin but not to increased GDF15 during pregnancy. Blood Cells Mol Dis 48(4):233–237. https://doi.org/10.1016/j.bcmd.2012.02.001

Fisher AL, Nemeth E (2017) Iron homeostasis during pregnancy. Am J Clin Nutr 106(Suppl 6):1567S–1574S. https://doi.org/10.3945/ajcn.117.155812

Fisher AL, Sangkhae V, Presicce P, Chougnet CA, Jobe AH, Kallapur SG, Tabbah S, Buhimschi CS, Buhimschi IA, Ganz T, Nemeth E (2020) Fetal and amniotic fluid iron homeostasis in healthy and complicated murine, macaque, and human pregnancy. JCI Insight 5(4):e135321. https://doi.org/10.1172/jci.insight.135321

Flores-Quijano ME, Montalvo-Velarde I, Vital-Reyes VS, Rodriguez-Cruz M, Rendon-Macias ME, Lopez-Alarcon M (2016) Longitudinal analysis of the interaction between obesity and pregnancy on iron homeostasis: role of hepcidin. Arch Med Res 47(7):550–556. https://doi.org/10.1016/j.arcmed.2016.11.011

Fozza C, Asara MA, Vacca N, Caggiari S, Monti A, Zaccheddu F, Capobianco G, Dessole S, Dore F, Antonucci R (2017) Pregnancy outcome among women with beta-thalassemia major in north sardinia. Acta Haematol 138(3):166–167. https://doi.org/10.1159/000480450

Ganz T, Nemeth E (2024) Hypoferremia of inflammation: innate host defense against infections. Blood Cells Mol Dis 104:102777. https://doi.org/10.1016/j.bcmd.2023.102777

Georgieff MK (2011) Long-term brain and behavioral consequences of early iron deficiency. Nutr Rev 69(Suppl 1):S43–S48. https://doi.org/10.1111/j.1753-4887.2011.00432.x

Georgieff MK, Wobken JK, Welle J, Burdo JR, Connor JR (2000) Identification and localization of divalent metal transporter-1 (DMT-1) in term human placenta. Placenta 21(8):799–804. https://doi.org/10.1053/plac.2000.0566

Girelli D, Busti F, Brissot P, Cabantchik I, Muckenthaler MU, Porto G (2022) Hemochromatosis classification: update and recommendations by the BIOIRON Society. Blood 139(20):3018–3029. https://doi.org/10.1182/blood.2021011338

Goldenberg RL, Tamura T, DuBard M, Johnston KE, Copper RL, Neggers Y (1996) Plasma ferritin and pregnancy outcome. Am J Obstet Gynecol 175(5):1356–1359

Gonzalez-Fernandez D, Nemeth E, Pons EDC, Rueda D, Sinisterra OT, Murillo E, Sangkhae V, Starr LM, Scott ME, Koski KG (2021) INTERGROWTH-21 identifies high prevalence of low symphysis-fundal height in indigenous pregnant women experiencing multiple infections, nutrient deficiencies, and inflammation: the maternal infections, nutrient deficiencies, and inflammation (MINDI) cohort. Curr Dev Nutr 5(4):nzab012. https://doi.org/10.1093/cdn/nzab012

Govindappagari S, Burwick RM (2019) Treatment of iron deficiency anemia in pregnancy with intravenous versus oral iron: systematic review and meta-analysis. Am J Perinatol 36(4):366–376. https://doi.org/10.1055/s-0038-1668555

Guller S, Buhimschi CS, Ma YY, Huang ST, Yang L, Kuczynski E, Zambrano E, Lockwood CJ, Buhimschi IA (2008) Placental expression of ceruloplasmin in pregnancies complicated by severe preeclampsia. Lab Investig 88(10):1057–1067. https://doi.org/10.1038/labinvest.2008.74

Gunshin H, Fujiwara Y, Custodio AO, Direnzo C, Robine S, Andrews NC (2005) Slc11a2 is required for intestinal iron absorption and erythropoiesis but dispensable in placenta and liver. J Clin Invest 115(5):1258–1266. https://doi.org/10.1172/JCI24356

Guo Y, Zhang N, Zhang D, Ren Q, Ganz T, Liu S, Nemeth E (2019) Iron homeostasis in pregnancy and spontaneous abortion. Am J Hematol 94(2):184–188. https://doi.org/10.1002/ajh.25341

Hallberg L, Rossander-Hulten L (1991) Iron requirements in menstruating women. Am J Clin Nutr 54(6):1047–1058. https://doi.org/10.1093/ajcn/54.6.1047

Hayward CE, Lean S, Sibley CP, Jones RL, Wareing M, Greenwood SL, Dilworth MR (2016) Placental adaptation: what can we learn from birthweight: placental weight ratio? Front Physiol 7:28. https://doi.org/10.3389/fphys.2016.00028

Hecht JL, Kliman HJ, Allred EN, Pflueger SM, Chang CH, Doss BJ, Roberts D, Livasy CA, Bhan I, Zambrano E, Ross DW, Senagore P, Husain AN, Leviton A (2007) Reference weights for placentas delivered before the 28th week of gestation. Placenta 28(10):987–990. https://doi.org/10.1016/j.placenta.2007.04.009

Hoefnagel AL, Wissler R (2012) Anesthetic management of vaginal delivery in a parturient with hemochromatosis induced end-organ failure. Int J Obstet Anesth 21(1):83–85. https://doi.org/10.1016/j.ijoa.2011.09.001

Hojyo S, Fukada T, Shimoda S, Ohashi W, Bin BH, Koseki H, Hirano T (2011) The zinc transporter SLC39A14/ZIP14 controls G-protein coupled receptor-mediated signaling required for systemic growth. PLoS One 6(3):e18059. https://doi.org/10.1371/journal.pone.0018059

Hytten F (1985) Blood volume changes in normal pregnancy. Clin Haematol 14(3):601–612

Jones KL, Croen LA, Yoshida CK, Heuer L, Hansen R, Zerbo O, DeLorenze GN, Kharrazi M, Yolken R, Ashwood P, Van de Water J (2017) Autism with intellectual disability is associated with increased levels of maternal cytokines and chemokines during gestation. Mol Psychiatry 22(2):273–279. https://doi.org/10.1038/mp.2016.77

Kautz L, Jung G, Valore EV, Rivella S, Nemeth E, Ganz T (2014) Identification of erythroferrone as an erythroid regulator of iron metabolism. Nat Genet 46(7):678–684. https://doi.org/10.1038/ng.2996

Kautz L, Jung G, Du X, Gabayan V, Chapman J, Nasoff M, Nemeth E, Ganz T (2015) Erythroferrone contributes to hepcidin suppression and iron overload in a mouse model of beta-thalassemia. Blood 126(17):2031–2037. https://doi.org/10.1182/blood-2015-07-658419

Keel SB, Doty RT, Yang Z, Quigley JG, Chen J, Knoblaugh S, Kingsley PD, De Domenico I, Vaughn MB, Kaplan J, Palis J, Abkowitz JL (2008) A heme export protein is required for red blood cell differentiation and iron homeostasis. Science 319(5864):825–828. https://doi.org/10.1126/science.1151133

Khambalia AZ, Collins CE, Roberts CL, Morris JM, Powell KL, Tasevski V, Nassar N (2015) High maternal serum ferritin in early pregnancy and risk of spontaneous preterm birth. Br J Nutr 114(3):455–461. https://doi.org/10.1017/S0007114515001932

Krysiak R, Kobielusz-Gembala I, Okopien B (2010) Recurrent pregnancy-induced diabetes insipidus in a woman with hemochromatosis. Endocr J 57(12):1023–1028. https://doi.org/10.1507/endocrj.k10e-125

Kumar M, Saadaoui M, Al Khodor S (2022) Infections and pregnancy: effects on maternal and child health. Front Cell Infect Microbiol 12:873253. https://doi.org/10.3389/fcimb.2022.873253

Lao TT, Tam KF, Chan LY (2000) Third trimester iron status and pregnancy outcome in non-anaemic women; pregnancy unfavourably affected by maternal iron excess. Hum Reprod 15(8):1843–1848

Lewkowitz AK, Gupta A, Simon L, Sabol BA, Stoll C, Cooke E, Rampersad RA, Tuuli MG (2019) Intravenous compared with oral iron for the treatment of iron-deficiency anemia in pregnancy: a systematic review and meta-analysis. J Perinatol 39(4):519–532. https://doi.org/10.1038/s41372-019-0320-2

Li JY, Paragas N, Ned RM, Qiu AD, Viltard M, Leete T, Drexler IR, Chen X, Sanna-Cherchi S, Mohammed F, Williams D, Lin CS, Schmidt-Ott KM, Andrews NC, Barasch J (2009) Scara5 Is a ferritin receptor mediating non-transferrin iron delivery. Dev Cell 16(1):35–46. https://doi.org/10.1016/j.devcel.2008.12.002

Li YQ, Bai B, Cao XX, Zhang YH, Yan H, Zheng QQ, Zhuang GH (2012) Divalent metal transporter 1 expression and regulation in human placenta. Biol Trace Elem Res 146(1):6–12. https://doi.org/10.1007/s12011-011-9214-7

Lyall F, Barber A, Myatt L, Bulmer JN, Robson SC (2000) Hemeoxygenase expression in human placenta and placental bed implies a role in regulation of trophoblast invasion and placental function. FASEB J 14(1):208–219. https://doi.org/10.1096/fasebj.14.1.208

Madu AJ, Ughasoro MD (2017) Anaemia of chronic disease: an in-depth review. Med Princ Pract 26(1):1–9. https://doi.org/10.1159/000452104

McArdle HJ, Gambling L, Kennedy C (2014) Iron deficiency during pregnancy: the consequences for placental function and fetal outcome. Proc Nutr Soc 73(1):9–15. https://doi.org/10.1017/S0029665113003637

Mei Z, Cogswell ME, Looker AC, Pfeiffer CM, Cusick SE, Lacher DA, Grummer-Strawn LM (2011) Assessment of iron status in US pregnant women from the National Health and Nutrition Examination Survey (NHANES), 1999-2006. Am J Clin Nutr 93(6):1312–1320. https://doi.org/10.3945/ajcn.110.007195

Milman N, Agger AO, Nielsen OJ (1991) Iron supplementation during pregnancy. Effect on iron status markers, serum erythropoietin and human placental lactogen. A placebo controlled study in 207 Danish women. Dan Med Bull 38(6):471–476

Nair M, Choudhury SS, Rani A, Solomi CT, Kakoty SD, Medhi R, Rao S, Mahanta P, Zahir F, Roy I, Chhabra S, Deka G, Minz B, Deka R, Opondo C, Churchill D, Lakhal-Littleton S, Nemeth E, Maat HRIC (2023) The complex relationship between iron status and anemia in pregnant and postpartum women in India: analysis of two Indian study cohorts of uncomplicated pregnancies. Am J Hematol 98(11):1721–1731. https://doi.org/10.1002/ajh.27059

Niu C, Zhang J, Goldenberg I, Gill S, Saeed H, Iyer C, Dunnigan K (2023) Maternal and prenatal outcomes of hemochromatosis in pregnancy: a population-based study. Clin Res Hepatol Gastroenterol 47(9):102221. https://doi.org/10.1016/j.clinre.2023.102221

Ohgami RS, Campagna DR, McDonald A, Fleming MD (2006) The Steap proteins are metalloreductases. Blood 108(4):1388–1394. https://doi.org/10.1182/blood-2006-02-003681

Organization WH (2012) Guideline: daily iron and folic acid supplementation in pregnant women. WHO Guidelines Approved by the Guidelines Review Committee, Geneva

Petrakos G, Andriopoulos P, Tsironi M (2016) Pregnancy in women with thalassemia: challenges and solutions. Int J Women's Health 8:441–451. https://doi.org/10.2147/IJWH.S89308

Preziosi P, Prual A, Galan P, Daouda H, Boureima H, Hercberg S (1997) Effect of iron supplementation on the iron status of pregnant women: consequences for newborns. Am J Clin Nutr 66(5):1178–1182

Recommendations to prevent and control iron deficiency in the United States. Centers for Disease Control and Prevention (1998) MMWR Recommendations and reports: morbidity and mortality weekly report Recommendations and reports 47 (RR-3):1–29

Ryu MS, Zhang D, Protchenko O, Shakoury-Elizeh M, Philpott CC (2017) PCBP1 and NCOA4 regulate erythroid iron storage and heme biosynthesis. J Clin Invest 127(5):1786–1797. https://doi.org/10.1172/JCI90519

Saldanha LG, Dwyer JT, Andrews KW, Brown LL (2019) The chemical forms of iron in commercial prenatal supplements are not always the same as those tested in clinical trials. J Nutr 149(6):890–893. https://doi.org/10.1093/jn/nxz042

Salomon LJ, Bernard JP, Ville Y (2007) Estimation of fetal weight: reference range at 20-36 weeks' gestation and comparison with actual birth-weight reference range. Ultrasound Obstet Gynecol 29(5):550–555. https://doi.org/10.1002/uog.4019

Sammallahti S, Tiemeier H, Reiss IKM, Muckenthaler MU, El Marroun H, Vermeulen M (2022) Maternal early-pregnancy ferritin and offspring neurodevelopment: a prospective cohort study from gestation to school age. Paediatr Perinat Epidemiol 36(3):425–434. https://doi.org/10.1111/ppe.12854

Sangkhae V, Nemeth E (2017) Regulation of the iron homeostatic hormone hepcidin. Adv Nutr (Bethesda, Md) 8(1):126–136. https://doi.org/10.3945/an.116.013961

Sangkhae V, Nemeth E (2018) Placental iron transport: the mechanism and regulatory circuits. Free Radic Biol Med. https://doi.org/10.1016/j.freeradbiomed.2018.07.001

Sangkhae V, Fisher AL, Chua KJ, Ruchala P, Ganz T, Nemeth E (2020a) Maternal hepcidin determines embryo iron homeostasis in mice. Blood 136(19):2206–2216. https://doi.org/10.1182/blood.2020005745

Sangkhae V, Fisher AL, Wong S, Koenig MD, Tussing-Humphreys L, Chu A, Lelic M, Ganz T, Nemeth E (2020b) Effects of maternal iron status on placental and fetal iron homeostasis. J Clin Invest 130(2):625–640. https://doi.org/10.1172/JCI127341

Sangkhae V, Yu V, Coffey R, O'Brien KO, Ganz T, Nemeth E (2022) Erythroferrone contributes to iron mobilization for embryo erythropoiesis in iron-deficient mouse pregnancies. Am J Hematol 97(10):1348–1358. https://doi.org/10.1002/ajh.26680

Sangkhae V, Fisher AL, Ganz T, Nemeth E (2023) Iron homeostasis during pregnancy: maternal, placental, and fetal regulatory mechanisms. Annu Rev Nutr 43:279–300. https://doi.org/10.1146/annurev-nutr-061021-030404

Santana-Codina N, Mancias JD (2018) The role of NCOA4-mediated ferritinophagy in health and disease. Pharmaceuticals (Basel) 11(4). https://doi.org/10.3390/ph11040114

Scanlon KS, Yip R, Schieve LA, Cogswell ME (2000) High and low hemoglobin levels during pregnancy: differential risks for preterm birth and small for gestational age. Obstet Gynecol 96(5 Pt 1):741–748. https://doi.org/10.1016/s0029-7844(00)00982-0

Scholl TO (1998) High third-trimester ferritin concentration: associations with very preterm delivery, infection, and maternal nutritional status. Obstet Gynecol 92(2):161–166

Scholl TO, Reilly T (2000) Anemia, iron and pregnancy outcome. J Nutr 130(2S Suppl):443S–447S. https://doi.org/10.1093/jn/130.2.443S

Schulze KJ, Christian P, Ruczinski I, Ray AL, Nath A, Wu LS, Semba RD (2008) Hepcidin and iron status among pregnant women in Bangladesh. Asia Pac J Clin Nutr 17(3):451–456

Scotet V, Saliou P, Uguen M, L'Hostis C, Merour MC, Triponey C, Chanu B, Nousbaum JB, Le Gac G, Ferec C (2018) Do pregnancies reduce iron overload in HFE hemochromatosis women? Results from an observational prospective study. BMC Pregnancy Childbirth 18(1):53. https://doi.org/10.1186/s12884-018-1684-6

Short MW, Domagalski JE (2013) Iron deficiency anemia: evaluation and management. Am Fam Physician 87(2):98–104

Simavli S, Derbent AU, Uysal S, Turhan NO (2014) Hepcidin, iron status, and inflammation variables among healthy pregnant women in the Turkish population. J Matern Fetal Neonatal Med 27(1):75–79. https://doi.org/10.3109/14767058.2013.804054

Ssewanyana D, Borque SL, Lye SJ, Matthews SG (2023) Hepcidin across pregnancy and its correlation with maternal markers of iron and inflammation, maternal body weight outcomes, and offspring neurodevelopmental outcomes: a systematic review and meta-analysis. AJOG Glob Rep 3(3):100222. https://doi.org/10.1016/j.xagr.2023.100222

Stevens GA, Paciorek CJ, Flores-Urrutia MC, Borghi E, Namaste S, Wirth JP, Suchdev PS, Ezzati M, Rohner F, Flaxman SR, Rogers LM (2022) National, regional, and global estimates of anaemia by severity in women and children for 2000-19: a pooled analysis of population-representative data. Lancet Glob Health 10(5):e627–e639. https://doi.org/10.1016/S2214-109X(22)00084-5

Stordal K, McArdle HJ, Hayes H, Tapia G, Viken MK, Lund-Blix NA, Haugen M, Joner G, Skrivarhaug T, Marild K, Njolstad PR, Eggesbo M, Mandal S, Page CM, London SJ, Lie BA, Stene LC (2018) Prenatal iron exposure and childhood type 1 diabetes. Sci Rep 8(1):9067. https://doi.org/10.1038/s41598-018-27391-4

Tain YL, Luh H, Lin CY, Hsu CN (2016) Incidence and risks of congenital anomalies of kidney and urinary tract in newborns: a population-based case-control study in Taiwan. Medicine (Baltimore) 95(5):e2659. https://doi.org/10.1097/MD.0000000000002659

Tuck SM, Jensen CE, Wonke B, Yardumian A (1998) Pregnancy management and outcomes in women with thalassaemia major. J Pediatr Endocrinol Metab 11(Suppl 3):923–928

van Santen S, Kroot JJ, Zijderveld G, Wiegerinck ET, Spaanderman ME, Swinkels DW (2013) The iron regulatory hormone hepcidin is decreased in pregnancy: a prospective longitudinal study. Clin Chem Lab Med 51(7):1395–1401. https://doi.org/10.1515/cclm-2012-0576

Vlachodimitropoulou E, Mogharbel H, Kuo KHM, Hwang M, Ward R, Shehata N, Malinowski AK (2024) Pregnancy outcomes and iron status in beta-thalassemia major and intermedia: a systematic review and meta-analysis. Blood Adv 8(3):746–757. https://doi.org/10.1182/bloodadvances.2023011636

Wang CY, Jenkitkasemwong S, Duarte S, Sparkman BK, Shawki A, Mackenzie B, Knutson MD (2012) ZIP8 is an iron and zinc transporter whose cell-surface expression is up-regulated by cellular iron loading. J Biol Chem 287(41):34032–34043. https://doi.org/10.1074/jbc.M112.367284

Wei S, Liu W, Qi Y, Guo Y, Zhang S, Wang L, Zhuang T, Zhang N, Liu S (2021) Disordered serum erythroferrone and hepcidin levels as indicators of the spontaneous abortion occurrence during early pregnancy in humans. Br J Haematol 192(3):643–651. https://doi.org/10.1111/bjh.17049

WHO (2024) Guideline on haemoglobin cutoffs to define anaemia in individuals and populations. World Health Organization, Geneva. Licence: CC BY-NC-SA 3.0 IGO

Widdowson EM, Spray CM (1951) Chemical development in utero. Arch Dis Child 26(127):205–214

Yanatori I, Richardson DR, Imada K, Kishi F (2016) Iron export through the transporter ferroportin 1 is modulated by the iron chaperone PCBP2. J Biol Chem 291(33):17303–17318. https://doi.org/10.1074/jbc.M116.721936

Zaman B, Rasool S, Jasim S, Abdulah D (2019) Hepcidin as a diagnostic biomarker of iron deficiency anemia during pregnancy. J Matern Fetal Neonatal Med:1–9. https://doi.org/10.1080/14767058.2019.1635112

Zhang V, Fisher AL, Hewitt MS, Ganz T, Nemeth E, Sangkhae V (2024) Maternal prolactin or estrogen signaling in hepatocytes does not regulate iron homeostasis during pregnancy. Haematologica. https://doi.org/10.3324/haematol.2024.285456

Iron and the Intestinal Microbiome 22

Huihui Li

Abstract

Iron in the gut is like a double-edged sword. On one side, it is essential for numerous physiological processes, including oxygen transport and energy production, and is crucial for addressing conditions like anemia. However, on the other side, excess or unabsorbed iron can disrupt gut homeostasis, fuel harmful pathogens, contribute to dysbiosis, and promote gut inflammation. The challenge lies in finding a balance that ensures adequate iron absorption while minimizing its adverse effects on the gut microbiota. Host and microbiota in the gut have evolved sophisticated strategies to maintain iron homeostasis within the bacterial community and balance iron needs between host and gut microbiota. Disruption of this balance by excess iron could lead to serious consequences by promoting intestinal inflammation and disease progression. Current research points to promising therapeutical approaches that enhance iron absorption while suppressing iron-induced adverse effects, such as oxidative stress and dysbiosis, in the gut. These strategies offer the potential to reduce intestinal inflammation and improve gut health, paving the way for more effective therapies to mitigate disease complications.

Keywords

Iron · Heme · Gut microbiota · Dysbiosis · Inflammation · Siderophore · Inflammatory bowel disease · Colorectal cancer · Sickle cell disease · Obesity

H. Li (✉)
Icahn School of Medicine at Mount Sinai, New York, NY, USA
e-mail: Huihui.Li@mountsinai.org

Gut microbiota is a complex and dynamic population of microorganisms that harbors in the gastrointestinal (GI) tract. It is composed of trillions of microorganisms, including bacteria, viruses, fungi, and parasites, with bacteria comprising the majority of the microbial community. These microorganisms help digest food, produce short-chain fatty acids (SCFAs) that are vital for intestinal health, and synthesize essential vitamins. Given that bacteria constitute the largest portion of this microbial ecosystem, this chapter will focus on exploring the bacterial components of the gut microbiota, which significantly influence the host's health.

There is increasing evidence suggesting that gut microbiota influences immune homeostasis. Microbiota could regulate disease progression

K. Pantopoulos (ed.), *Iron Metabolism in Human Health and Disease*, Advances in Experimental Medicine and Biology 1480, https://doi.org/10.1007/978-3-031-92033-2_22

in disorders such as inflammatory bowel disease, colorectal cancer, sickle cell disease, and obesity (Lee and Chang 2021; Tilg and Moschen 2014; Manfredo Vieira et al. 2018; Meisel et al. 2018; Li et al. 2023; Liu et al. 2023). When gut microbiota significantly changes its composition and function, this is a state known as dysbiosis. Dysbiosis can lead to a compromised intestinal barrier, excessive immune activation, and chronic inflammation, all of which contribute to disease pathogenesis. In parallel, the beneficial microbiota, also known as probiotics, are known to prevent pathogenic infections, protect the intestinal barrier, and modulate immune host function (Raheem et al. 2021; Mazziotta et al. 2023).

The human GI tract hosts approximately 10^{14} bacteria, making it one of the most densely populated bacterial regions in the body (Sender et al. 2016). These bacteria are highly dynamic, capable of rapid expansion, and adaptation to changes in the gut environment (Eckburg et al. 2005). Iron is one of the vital elements that support bacterial survival and expansion by participating in essential biological processes such as the electron transport chain, DNA biosynthesis, and SCFAs production (Dostal et al. 2015). It is estimated that 5–20% of dietary iron is absorbed in the duodenum, and the remaining >80% of iron stays in the gut (McCance and Widdowson 1938) (see also Chap. 6). Given that free iron is very toxic and reactive (Chap. 1), the host secretes iron-binding proteins such as transferrin, ferritin, or hemoproteins to lower free iron bioavailability, reduce the production of reactive oxygen species (ROS), and, consequently, reduce damage to cells and tissue (Raymond et al. 2003). The concentration of free iron in the gut ($\sim 10^{-24}$ M) is significantly lower than the amount optimal for bacterial growth ($\sim 10^{-6}$ M) (Raymond et al. 2003). In these iron-limiting conditions, there is intense competition for iron among bacterial species, as well as between microorganisms and the host. Iron levels in the gut can significantly impact microbiota composition, the host's iron homeostasis, and disease progression.

22.1 Bacterial Iron Homeostasis

Bacterial survival and virulence are closely tied to their ability to acquire and regulate iron in the gut. To maintain iron homeostasis, bacteria have evolved a range of specialized mechanisms for iron uptake, utilization, and storage in response to iron scarcity in the intestinal environment (Fig. 22.1).

1. Bacterial iron uptake

Free iron in the intestine is largely in the form of ferric iron (Fe^{3+}), which is insoluble due to the intestine's non-acidic pH, making it difficult for bacteria to directly take up. To obtain iron from the gut environment, bacteria have evolved multiple mechanisms to acquire and utilize this limited but essential nutrient.

A. Siderophores:

The siderophores are synthesized and secreted by bacteria in response to iron restriction (see also Chap. 1). They are characterized by their high specificity and affinity toward ferric iron. Hundreds of siderophores have been characterized, with the most well-known being enterobactin (Andrews et al. 2003). Fe^{3+}-siderophore complexes need to be internalized through Fe^{3+}-siderophore receptors. Using gram-negative bacteria as an example, the receptor on the outer membrane acquires energy from the TonB-ExbB-ExbD protein complex and translocates the Fe^{3+}-siderophore to periplasmic space, where it binds with periplasmic binding protein (PBP). Next, the PBP-Fe^{3+}-siderophore complex passes to the bacterial cytoplasm through the cytoplasmic membrane receptor (ABC permease complexes) and gets utilized by the bacteria (Andrews et al. 2003). Once internalized, ferric iron from Fe^{3+}-siderophore complex gets reduced to ferrous iron (Fe^{2+}) and released into the cytoplasm (Heidinger et al. 1983; Venuti et al. 1979).

Interestingly, due to their dynamic diversity, siderophores secreted by one microorganism may selectively deliver iron to species

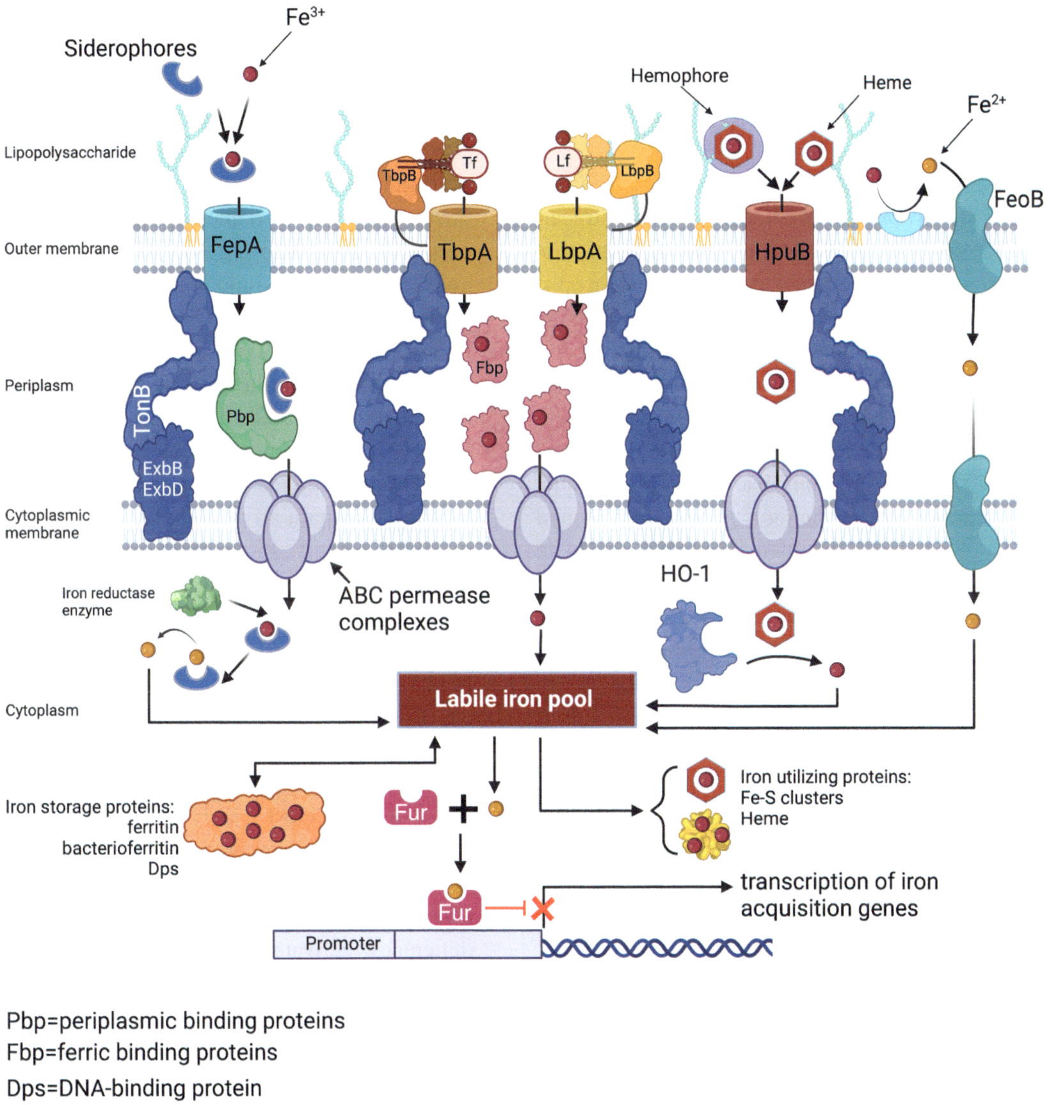

Fig. 22.1 Iron homeostasis in a gram-negative bacterium: mechanisms for iron uptake, iron utilization, iron storage, and iron acquisition genes regulation

with matching receptors while causing iron starvation in other microorganisms that lack the appropriate receptors (Kramer et al. 2020).

B. Transferrin and lactoferrin receptors and heme uptake receptor

Once released, free iron and heme are rapidly bound by host-chelating proteins such as transferrin (Tf), lactoferrin (Lf), and hemopexin. During iron starvation, bacteria, primarily pathogens, upregulate independent transferrin receptors (TfR) and lactoferrin receptors (LfR) to acquire iron directly from host iron-binding proteins (Tf and Lf) (Cornelissen and Sparling 1994). Additionally, bacteria can also take up heme, hemoglobin, or the hemopexin-heme complex through receptors on the bacterial outer membrane. Some bacteria can also secrete "hemophores" that bind with hemoglobin and hemopexin to facilitate the translocation of heme (Genco and

Dixon 2001). Once in the cytoplasm, heme is degraded by heme oxygenase (HO-1), which releases the iron molecule (Genco and Dixon 2001).

C. Feo system

Some bacteria, such as *Escherichia coli* and *Helicobacter pylori*, can convert extracellular ferric iron (Fe^{3+}) to ferrous iron (Fe^{2+}) on their surface. The ferrous iron is then transported into the bacteria through ferrous iron transporters, specifically FeoA or FeoB (Kammler et al. 1993; Velayudhan et al. 2000). This Feo-mediated iron transport is essential for the survival of certain bacteria. For example, mutations in the *feoB* gene in *E. coli* and *Salmonella* result in a reduced ability to colonize the mouse intestine, likely due to their impaired capacity to take up ferrous iron under the gut anaerobic environment (Stojiljkovic et al. 1993; Tsolis et al. 1996). However, the mutation in the *feoB* gene in *Salmonella* doesn't influence its virulence, suggesting that iron-uptake mechanisms are not dependent on Feo system during infection (Tsolis et al. 1996). These data imply that under different circumstances, bacteria can switch Feo iron-uptake mechanisms on and off.

2. Bacterial iron utilization and storage

Once inside the cytoplasm, ferrous iron can be utilized for incorporation into proteins. Iron is needed for heme biosynthesis, which is crucial for bacteria to sustain key metabolic and respiratory processes, enabling them to efficiently adapt and switch between aerobic and anaerobic environments (Dailey et al. 2017). Additionally, iron is required for the synthesis of iron-sulfur (Fe-S) clusters. These Fe-S clusters play crucial roles as essential metal cofactors in vital biological processes, including respiration, nitrogen fixation, and DNA repair (Przybyla-Toscano et al. 2018; Rouault 2019). Inhibition of Fe-S cluster synthesis leads to extensive impairments in carbohydrate metabolism under anaerobic conditions and failure of bacteria colonization in the gut (Xu et al. 2024).

It is imperative for bacteria to acquire sufficient iron for growth while also safely storing excess iron in a non-toxic form. There are three types of iron storage proteins in bacteria: the archetypal ferritin, which is found in both eukaryotes and prokaryotes (Theil 1987); bacterioferritin, a heme-containing protein common in eubacteria (Romao et al. 2000); and a starvation-inducible DNA-binding protein-Dps, which is unique to prokaryotes (Almiron et al. 1992). These iron storage reservoirs can store iron (in ferric form) within the bacterial cytoplasm and supply iron to support bacterial growth when external iron supplies are depleted. A bacterium may simultaneously contain all three types of iron storage proteins and can express multiple ferritin or bacterioferritin genes (Andrews et al. 2003).

3. Iron-responsive regulator for bacterial iron homeostasis

Bacteria regulate their iron metabolism in response to iron availability. This regulation is mediated by the ferric-uptake regulator protein (Fur), the most widely conserved master transcriptional regulator controlling iron homeostasis in bacteria (Bradley et al. 2020; Caza and Kronstad 2013). Specifically, iron deficiency activates iron acquisition genes, including those involved in the biosynthesis of siderophores, iron-uptake receptors, PBP, iron storage protein, and Fe-proteins (Andrews et al. 2003). Consequently, bacteria enhance their ability to acquire more iron during this period of scarcity. Once the bacterial cytoplasm labile iron pool reaches certain levels, free ferrous iron binds with Fur, forming a Fe^{2+}-Fur complex known as a "Fur box." Next, the Fur box binds to the iron acquisition genes, repressing their expression to reduce iron uptake and maintain iron homeostasis (Andrews et al. 2003; Abdul-Tehrani et al. 1999; Masse and Gottesman 2002). Although Fur functions as the major iron-responsive regulator in many bacteria, high-guanine-cytosine-content gram-positive bacteria use the diphtheria toxin regulator (DtxR) to control the iron acquisition genes. DtxR has little or no sequence similarity to Fur but still acts in a ferrous iron-dependent manner to repress iron acquisition genes (Schmitt et al. 1997; Andrews et al. 2003).

22.2 Maintaining Iron Balance in the Gut: Strategies to Restrict Bacterial Virulence

Iron balance is carefully regulated in the microbiota community and between the host and pathogens (Fig. 22.2). Iron is essential to support the enteropathogens to replicate, sustain their virulence (e.g., adhesion, invasion, and induction of virulence factor), and further promote infection and inflammation (Kortman et al. 2014; Andrews et al. 2003). For instance, *Salmonella*, *Shigella*, pathogenic *Escherichia coli,* or *Clostridium difficile* contain multiple ferric, ferrous, and siderophore-bound iron-uptake systems that are Fur-regulated and highly responsive to changes in iron supply (Andrews et al. 2003; Kortman et al. 2014; Ho and Ellermeier 2015; Berges et al. 2018). The fact that approximately 7% of the *Salmonella typhimurium* genome is directly or indirectly regulated by iron highlights how integral iron is to the organism's physiology (Bjarnason et al. 2003).

To restrict the expansion of pathogenic bacteria, many probiotic bacteria develop iron-related mechanisms to compete for iron resources. Many bacteria in the *Bifidobacteriaceae* and *Lactobacillaceae* families have significantly different iron requirements compared to pathogens, typically requiring very little or no iron for their growth and maintenance (Vazquez-Gutierrez et al. 2015, 2016). They also often possess mechanisms to highly sequester iron, reducing iron availability for potentially harmful pathogens and providing a protective barrier against pathogen colonization (Vazquez-Gutierrez et al.

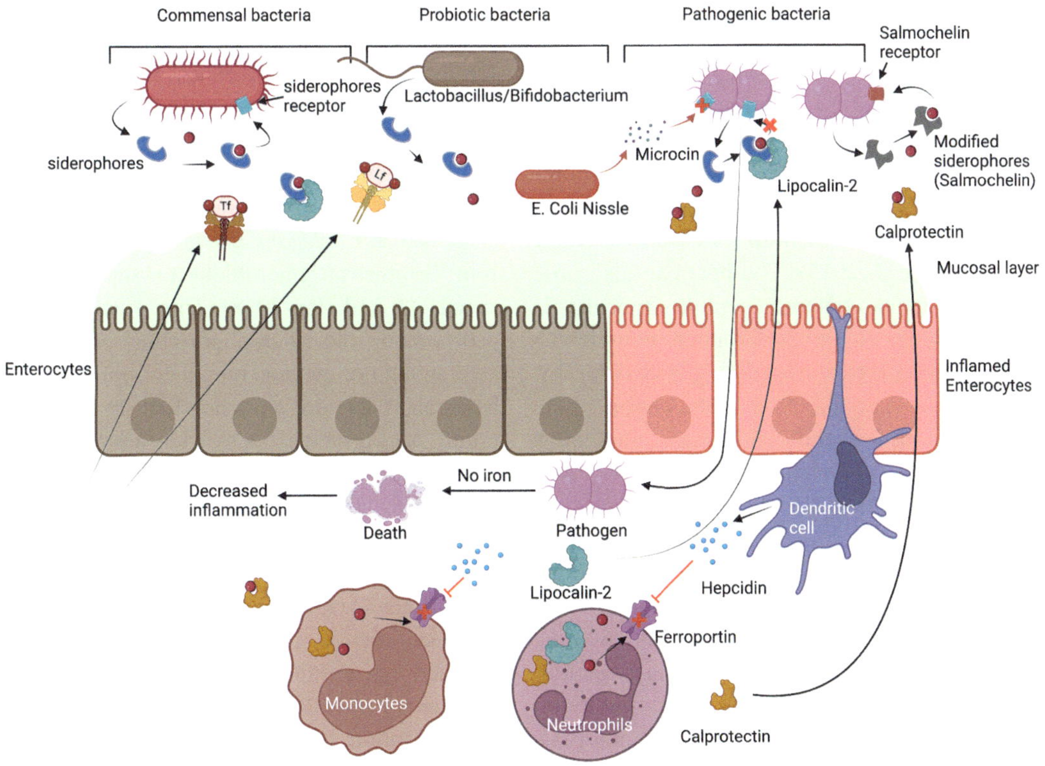

Fig. 22.2 Maintaining iron balance between microbiota and the host

2015). Additionally, probiotic commensal bacteria, *E. coli Nissle*, can secrete various types of siderophores to compete for iron, inhibiting the growth of other competing pathogens such as *Salmonella* (Deriu et al. 2013; Valdebenito et al. 2006). *E. coli Nissle* also suppresses iron uptake of pathogens by secreting a small protein-microcin, which damages the siderophore receptors of its competing pathogens and restricts their growth (Sassone-Corsi et al. 2016).

Mammalian hosts have also evolved several strategies to reduce iron availability during infections and inflammation, a process known as nutritional immunity. During gut inflammation, several host proteins are upregulated to sequester iron and deprive pathogens of this essential nutrient. Hepcidin is the master iron regulator that systemically controls iron levels in our body (see Chaps. 3, 7, and 14). In the GI tract, local dendritic cell-derived hepcidin restricts iron released from the local myeloid cells, limiting the increase in free iron during gut inflammation (Bessman et al. 2020). Failure of local hepcidin secretion results in dysbiosis, increased microbial translocation, and more severe intestinal inflammation (Bessman et al. 2020). Calprotectin, as another example, is identified as an iron-chelating antimicrobial protein secreted from host immune cells during gut inflammation (Konikoff and Denson 2006; Nakashige et al. 2015). The host can also produce the innate defense peptide lipocalin-2, which binds to the Fe^{3+}-siderophore complex to prevent the complex from being accessible by pathogens (Flo et al. 2004). In the ongoing battle for iron, several pathogens, including *Salmonella typhimurium, Klebsiella,* and *Shigella,* can evade this host defense mechanism by producing modified siderophores (e.g., salmochelin and aerobactin) that resist binding by host lipocalin-2 (Kortman et al. 2014). Despite these sophisticated strategies to limit iron availability, pathogens continuously evolve mechanisms to circumvent these defenses, highlighting the dynamic and ongoing interplay between host and microbial iron acquisition systems.

22.3 Microbiota Is Required for Host Iron Homeostasis

Although iron concentration influences microbiota composition, the microbiota itself also plays a role in regulating host iron homeostasis. Depending on iron availability, the microbiota can either cooperate with or compete against the host in iron regulation (Fig. 22.3). In iron-abundant conditions, microbiota facilitates iron storage in the duodenum and supports iron delivery to the host (Noordine et al. 2024; Qi and Han 2018). When iron is limited, the microbiota competes against the host for iron by suppressing the host intestine iron transporter via the pathway activated by hypoxia-inducible factor (HIF)-2α (Das et al. 2020; Noordine et al. 2024).

In the cooperative condition, with iron-enriched chow, intestinal iron storage protein-ferritin levels are significantly higher in regular mice compared to germ-free mice, implying that the microbiota plays a role in facilitating intestinal iron storage (Noordine et al. 2024; Das et al. 2020). Germ-free rats and mice are more likely to develop anemia when fed an iron-deficient diet (Reddy et al. 1972; Deschemin et al. 2016; Noordine et al. 2024), suggesting the absence of microbiota results in reduced iron intestinal storage, leaving the animals poorly resistant to the iron-deficient diet (Noordine et al. 2024). Some hosts, such as *C. elegans*, can utilize siderophores from commensal microbiota to support their growth (Qi and Han 2018). This is achieved by incorporating the siderophore enterobactin into the α-subunit of the host mitochondrial ATP synthases and elevating mitochondrial iron concentration to promote host development; the mechanism is also conserved in mammalian cells (Qi and Han 2018).

Host intestinal iron uptake is regulated by HIF-2α, a protein that responds to cellular iron and oxygen levels. When activated, HIF-2α upregulates key iron regulatory proteins, such as duodenal cytochrome b (DcytB), divalent metal transporter 1 (DMT1), and ferroportin, to enhance iron absorption into the host (Taylor et al. 2011; Shah et al. 2009). In competitive conditions, in an iron-scarce environment, bacteria, such as

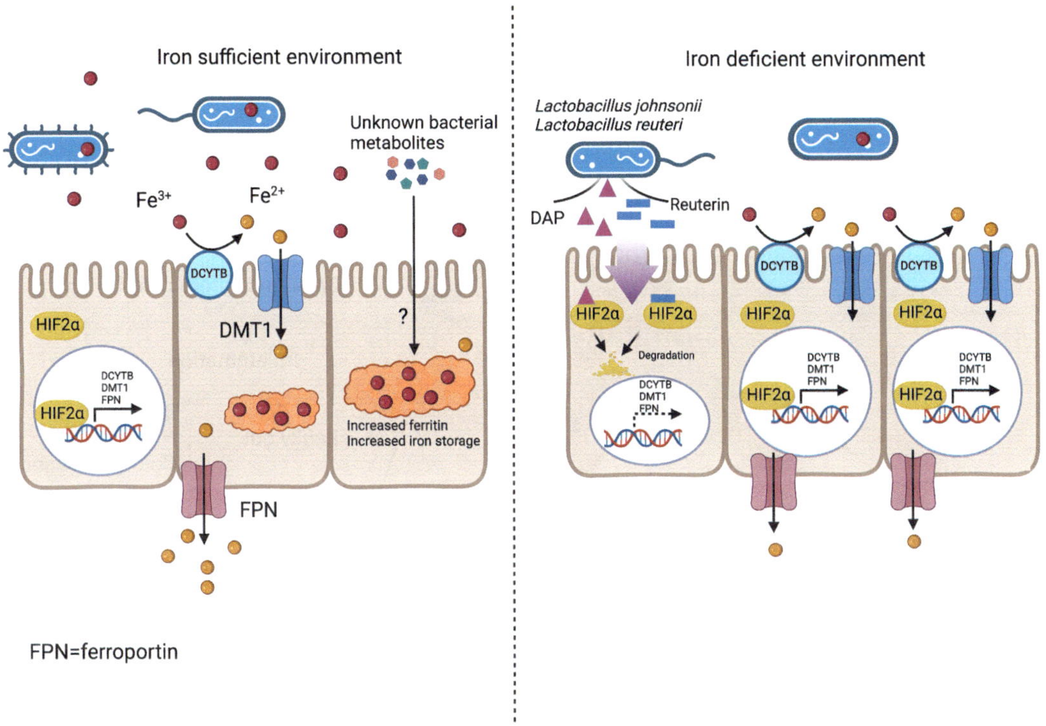

Fig. 22.3 Microbiota and the host cooperate in an iron-sufficient environment and compete in an iron-deficient environment

Lactobacillus species, sense iron levels and attenuate host iron absorption by secreting bacterial metabolites, such as reuterin and 1,3-diaminopropane (DAP) (Das et al. 2020). These metabolites degrade host intestine HIF-2α and consequently downregulate the expression of iron absorption proteins (Das et al. 2020; Noordine et al. 2024). When microbiota is depleted in germ-free mice or by antibiotic treatment, DcytB, DMT1, and their transcriptional regulator HIF-2α are further upregulated, allowing more iron uptake (Das et al. 2020).

22.4 Iron-Enriched Diet Promotes Dysbiosis and Disease Progression

Regular whole foods and supplements are the primary sources of iron. Heme iron is derived from animal-based foods such as meats, poultry, and seafood, while non-heme iron is found in plant-based foods like beans, nuts, dark chocolate, legumes, and spinach (Moustarah and Daley 2024). Iron supplements (Chap. 24) are recommended for individuals whose dietary intake does not meet nutritional needs and for populations at risk of iron deficiency. These groups include infants, breastfeeding mothers, vegetarians, and those with limited access to diverse diets (Moustarah and Daley 2024; Friel et al. 2018). Additionally, individuals with anemia, blood loss, or medical conditions with impaired iron absorption (celiac disease, Crohn's disease, malabsorptive disorder, or a history of gastric bypass surgery (Moustarah and Daley 2024)) also rely on iron supplements to support their iron needs. While iron is an essential nutrient for many body functions, its supplementation needs careful consideration due to potential impacts on health beyond nutrient deficiency. Excessive iron intake can disrupt the balance of the gut microbiota, leading to dysbiosis and a damaged gut barrier, promoting disease progression (Fig. 22.4).

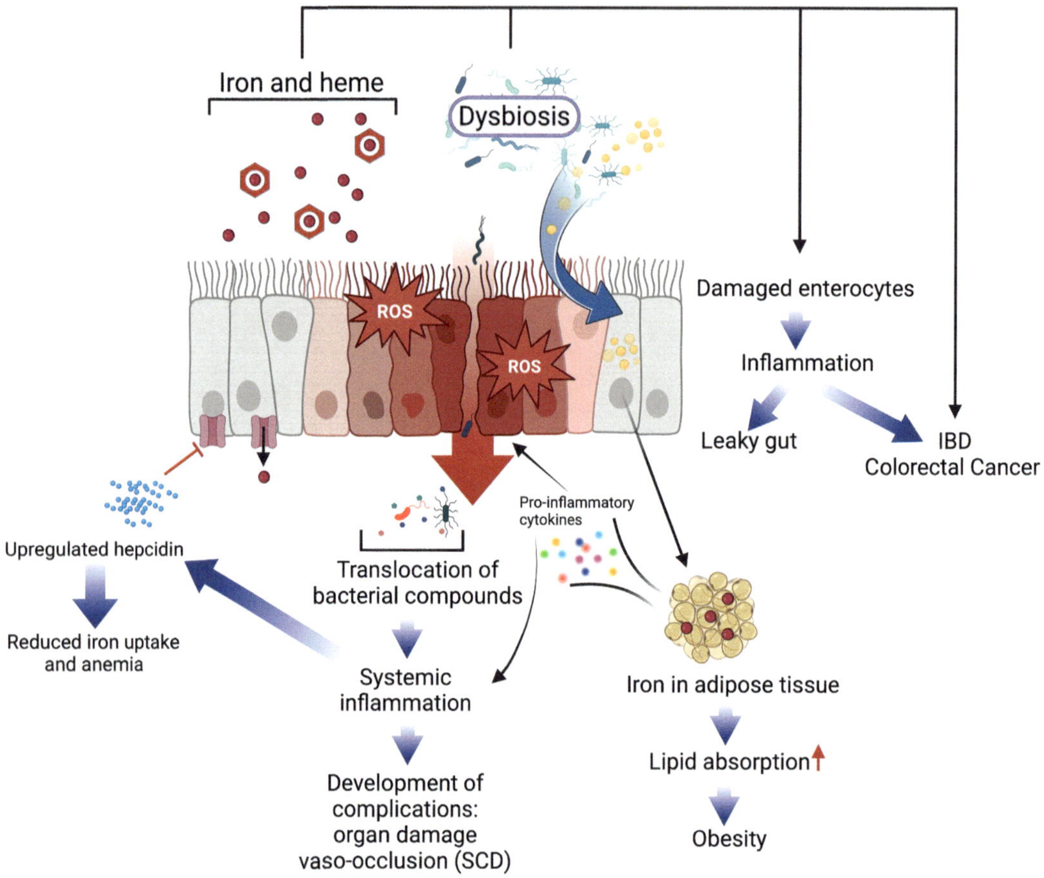

Fig. 22.4 Excess iron and heme promote dysbiosis and disease progression

1. Iron supplement and dysbiosis

Iron-fortified food is a type of iron supplement. The WHO recommends iron fortification with iron-containing micronutrient powders (MPNs) for infants aged 6–23 months. As African babies absorb iron at a low rate (Tondeur et al. 2004), they are usually given a high dose of iron (12.5 mg) in their formulation. Studies reported that, in 6-month-old infants from Kenya, there was an overall high prevalence of potential enteropathogens at baseline. The treatment group with daily 12.5 mg iron contained MPN (Fe-MPN) further promotes the abundance of enterobacteria (Jaeggi et al. 2015; Tang et al. 2017). Specifically, the abundance of *Escherichia/Shigella* increased in the Fe-MPN group compared to the MPN group with no additional iron. Additionally, the abundance of probiotic *Bifidobacterium* was decreased in the Fe-MPN group (Jaeggi et al. 2015; Tang et al. 2017). One study also reported that fecal calprotectin was elevated in Fe-MPN, implying elevated intestinal inflammation (Jaeggi et al. 2015; Tang et al. 2017). Compared to these infants from Kenya, two studies from high-income countries indicated that daily 6–6.6 mg iron altered infants' microbiota composition but had no effect on the abundance of pathogenic bacteria or fecal calprotectin (Simonyte Sjodin et al. 2019; Tang et al. 2017). Notably, baseline fecal calprotectin in Kenyan infants was 4–6 times higher than that of U.S. infants (200–300ug/g vs. 50ug/g) (Tang et al. 2017; Jaeggi et al. 2015; Paganini et al. 2017). These data suggest that the adverse effects of iron may be particularly noticeable in locations with poor hygiene and a high burden of infection and inflammation, such as in rural parts of Africa.

A study in Australia reported that, compared to a low iron supplement (≤10 mg/d), a high iron supplement (≥60 mg/d) does not significantly alter the overall composition of the gut microbiota in women during early pregnancy (Dekker Nitert et al. 2018). However, women in the low iron supplement group show a slightly higher abundance of SCFAs, which may benefit gut barrier integrity (Dekker Nitert et al. 2018). The lactating women living in rural Kenya were highly colonized by enteropathogens (Paganini et al. 2019), highlighting the need for further investigation into the impact of iron supplementation in this population.

Given that iron supplements promote dysbiosis in infants from areas with poor hygiene, the risk of diarrhea is also increased. In Kenyan infants, 27% of infants on 12.5 mg Fe-MPN required treatment for diarrhea compared to 8% in the control group (Jaeggi et al. 2015). Additionally, in areas with malaria, three studies reported that a combination of iron-zinc supplements increases the risk of diarrhea by 15%, one trial showed more days with fever among iron-treated participants, and one trial reported more infectious episodes in the iron-treatment arm (Neuberger et al. 2016). These findings emphasize the need for tailored strategies when implementing iron supplementation programs.

2. Iron, microbiota, and inflammatory bowel disease/colorectal cancer

Inflammatory bowel disease (IBD) is a group of conditions, including ulcerative colitis and Crohn's disease, characterized by chronic, relapsing inflammation in the GI tract, while colorectal cancer (CRC), another disease originating in the intestine, is the third most commonly diagnosed cancer globally and the second major cause of mortality (Nielsen et al. 2018; Mahadea et al. 2021; Phipps et al. 2021). Iron-deficient anemia (IDA) is commonly found in both IBD and CRC patients (see Chaps. 12 and 13) and is a consequence of chronic intestinal hemorrhages as well as impaired iron absorption due to inflammation or medication (Nielsen et al. 2018). Iron supplements are given to these patients as a therapeutic intervention for anemia (Chap. 24).

Oral iron administration may lead to side effects due to a large amount of non-absorbed iron remaining in the intestines, including gastroduodenitis, nausea, bloating, vomiting, dyspepsia, constipation, diarrhea, abdominal pain, or darkening of the stools (Qi et al. 2020). In addition to its impact on promoting dysbiosis, iron may directly damage intestinal cells through lipid peroxidation of cellular membranes and disrupt energy processes in the cell due to mitochondrial damage, as well as by promoting endoplasmic reticulum dysfunction (Li et al. 2016). In an IBD animal model, dietary iron administration exacerbated colitis by promoting oxidative stress, which increased neutrophil infiltration, activated the NF-κB pathway, and upregulated pro-inflammatory cytokines such as interferon-γ (IFN-γ), tumor necrosis factor α (TNF-α), and inducible nitric oxide synthase (iNOS) (Carrier et al. 2006). In a CRC mouse model, dietary iron promoted colorectal tumorigenesis via crosstalk with microbiota (Liu et al. 2023). When microbiota was depleted by antibiotic treatment, dietary iron alone had a less pronounced effect on CRC development, highlighting the synergistic role of both iron and microbiota in promoting CRC (Liu et al. 2023). Mechanistically, microbiota modulated by dietary iron impaired the gut barrier and upregulated expression of secretory leukocyte protease inhibitor (SLPI), which activated MAPK signaling pathways in the intestinal epithelial cells, exacerbating tumorigenesis (Liu et al. 2023).

In addition, growing evidence suggests that heme iron contributes to the weakening of the gut mucosal barrier and damaging the colon epithelium. Dietary heme, a component normally found in red meat, could worsen colitis and potentiate the development of adenomas in mice (Constante et al. 2017). A heme-enriched diet is also correlated with reduced microbiota diversity and onsite dysbiosis, accompanied by chronic gut inflammation and hyperproliferation of the intestinal epithelium (Seiwert et al. 2021). In 2015, the International Agency for Research on Cancer classified the consumption of processed meat as carcinogenic and red meat as probably carcinogenic in humans (Malesza et al. 2022).

Although studies in animal models suggest dietary iron alters microbiota composition, damages the gut barrier, and promotes disease progression in IBD and CRC, the evidence from human studies remains limited. Clinically, intravenous iron injection is recommended instead of oral iron supplements for IBD patients with IDA, because it causes fewer alterations in bacterial diversity and lower toxicity (Niepel et al. 2018; Stein et al. 2018). However, other studies comparing oral and intravenous iron therapy in IBD patients have shown that neither approach alters intestinal inflammation or disease activities (Lee et al. 2017). Additionally, unlike the oral iron supplement-ferrous sulfate, oral ferric maltol does not adversely affect the intestinal microbiome (Mahalhal et al. 2021). These variations may be attributed to differences in the bioavailability of specific iron supplements and treatment duration, highlighting the need for further research.

3. Iron, microbiota, and other chronic inflammatory diseases

The role of iron in promoting chronic inflammation and disease complications through alterations in microbiota is also evident in other diseases, such as sickle cell disease (SCD). SCD is the most commonly inherited hemoglobinopathy and results from a mutation in the β-globin gene, leading to red blood cell (RBC) sickling, vaso-occlusive episodes (VOE), and organ damage (Piel et al. 2013). Recently, damaged gut barrier and dysbiosis have been found in both SCD patients and a mouse model, and gut microbiota emerged as an important factor triggering disease complications (Zhang et al. 2015; Dutta et al. 2019; Xu et al. 2020). Iron from the diet promotes dysbiosis and weakens the gut barrier, which consequently allows increased translocation of bacterial compounds and exacerbates disease complications (Li et al. 2023). When treating SCD mice with an iron-restricted diet (IRD), dysbiosis, evidenced by reduced firmicutes/bacterioidetes ratio, was normalized to levels similar to healthy mice (Li et al. 2023). Additionally, gut barrier integrity was restored, inflammation cytokines were reduced, and disease-related complications, such as VOE and organ damage, were all relieved in SCD animals treated with IRD (Li et al. 2023). Iron overload is common in SCD patients undergoing repeated transfusions and in SCD mouse models; blocking iron absorption and lowering systemic iron are areas of intense research (Das et al. 2015; Nyffenegger et al. 2022; Parrow et al. 2021). However, the impact of iron on the gut microbiota and gut barrier integrity in SCD patients remains less explored and requires in-depth investigation.

Obesity is another condition characterized by low-grade systemic, chronic inflammation, driven by pathological adipose tissue accumulation in the body. Adipocytes secrete proinflammatory cytokines involved in triggering insulin resistance (Coelho et al. 2014; Monteiro et al. 2010). The gut microbiota in obese individuals differs significantly compared to that of a metabolically healthy group (Kim et al. 2020), and this dysbiosis is one of the key factors contributing to the development of obesity. Importantly, the microbiota from obese animals shows an increased capacity to harvest energy from the diet (Turnbaugh et al. 2006). In germ-free mice, animals that received the "obese microbiota" exhibited considerable weight gain compared to animals that received the "lean microbiota" (Turnbaugh et al. 2006). Iron can influence insulin resistance in white adipose tissue. Administration of dietary iron led to increased tissue iron levels in white adipose tissue and insulin resistance (Dongiovanni et al. 2013). Increased iron deposition is also found in the adipose tissue of individuals with insulin resistance (Fernandez-Real et al. 2016) (see also Chap. 17). Furthermore, when iron uptake was restricted in adipocytes by adipocyte-specific knockout of transferrin receptor 1, the animals were protected from high-fat-diet-induced metabolic disorders; this was accompanied by reduced inflammation and oxidative stress in the small intestine (Zhang et al. 2021). Interestingly, the improvement in intestinal health was associated with partial correction of gut dysbiosis in animals with low iron in adipose tissue. Specifically, the proportion of *Coriobacteriia* and *Bacilli*, which are decreased in obese individuals (Kim et al. 2020), was expanded in animals with low cellular iron in adipose tissue (Zhang et al. 2021). Mechanistically, reduced cellular iron in adipose tissue decreased vesicular transport in the intestinal enterocytes, limiting lipid absorption in response to high-fat diet challenges (Zhang et al. 2021). Consistent with findings from this animal model, clinical studies demonstrate that systemic reduction of iron levels through iron chelation or phlebotomy can improve the diabetic phenotype (Fernandez-Real et al. 2002; Gamberini et al. 2008).

22.5 Strategies to Balance the Need for Iron and Minimize Adverse Effects

The essential role of iron in supporting daily life must be carefully balanced with the risks of its potential toxicity. One way to mitigate the adverse effects of iron on the gut microbiota is by reducing the amount of unabsorbed iron entering the colon. In the Kenyan infant studies, MPNs containing 5 mg of iron are sufficient to improve hemoglobin concentration and decrease anemia prevalence (Paganini et al. 2017). This dose has a less pronounced adverse impact on the infant gut microbiota and doesn't elevate intestinal inflammatory markers, such as fecal calprotectin, compared to 12.5 mg Fe-MPNs (Jaeggi et al. 2015; Paganini et al. 2017). Additionally, iron formulations, like sodium iron ethylenediaminetetraacetate (NaFeEDTA) or ferric maltol, are more bioavailable to the host while having less impact on the gut microbiota compared to traditional supplements like ferrous fumarate or ferrous sulfate (Mahalhal et al. 2021; Paganini et al. 2017). Moreover, intravenous iron injection is another method of delivering iron with less impact on the microbiota (McCance and Widdowson 1938; Stein et al. 2018). Inflammation upregulates hepcidin, a master iron-regulating hormone that reduces iron absorption (Chap. 3), making intravenous iron delivery potentially more effective in treating inflammation-induced anemia. However, oxidative stress associated with intravenous iron administration should be carefully monitored.

Administration of probiotics, prebiotics, and vitamin E also shows promising effects in enhancing host iron uptake while reducing iron-related toxicity in the gut. Probiotics, particularly many bacteria from *Bifidobacteriaceae* and *Lactobacillaceae* families, don't rely on iron for growth and can secrete siderophores to chelate iron, reducing iron availability to other pathogens (Vazquez-Gutierrez et al. 2015, 2016). The analysis of randomized trials shows that using probiotics, especially the *Bifidobacterium* and *Lactobacilli* species, reduces postoperative complications and accelerates recovery in CRC patients (Pitsillides et al. 2021). Prebiotics, non-digestible compounds that promote the growth or activity of beneficial bacteria in the gut, also play a role in regulating intestinal iron homeostasis. Studies demonstrate that prebiotics, such as galactooligosaccharides (GOS) and fructooligosaccharides (FOS), enhance iron absorption and reduce the adverse effects of iron on the gut microbiome and inflammation (Mikulic et al. 2024; Paganini et al. 2017; Chen et al. 2022). Lastly, while iron supplements can alleviate anemia in animal models of colitis, oxidative stress, and intestinal inflammation triggered by iron supplements are reduced by administering vitamin E (Carrier et al. 2006). Vitamin E can also prevent the development of dysbiosis in anemic infants and toddlers receiving iron supplementation (Tang et al. 2016).

22.6 Conclusions

Iron in the gut is indeed like a double-edged sword. On one side, it is essential for numerous physiological processes, including oxygen transport and energy production, and is crucial for preventing anemia. However, on the other side, excess or unabsorbed iron can disrupt gut homeostasis, fuel harmful pathogens, contribute to dysbiosis, and promote gut inflammation. Novel therapeutic strategies hold promise for iron supplementation without interrupting gut microbiota homeostasis. These strategies may also offer further benefits to patients. Clinical trials and rigorous planning will be required to ensure the safety and efficiency of these approaches to deliver the best outcome.

References

Abdul-Tehrani H, Hudson AJ, Chang YS, Timms AR, Hawkins C, Williams JM, Harrison PM, Guest JR, Andrews SC (1999) Ferritin mutants of Escherichia coli are iron deficient and growth impaired, and fur mutants are iron deficient. J Bacteriol 181(5):1415–1428. https://doi.org/10.1128/JB.181.5.1415-1428.1999

Almiron M, Link AJ, Furlong D, Kolter R (1992) A novel DNA-binding protein with regulatory and protective roles in starved Escherichia coli. Genes Dev 6(12B):2646–2654. https://doi.org/10.1101/gad.6.12b.2646

Andrews SC, Robinson AK, Rodriguez-Quinones F (2003) Bacterial iron homeostasis. FEMS Microbiol Rev 27(2–3):215–237. https://doi.org/10.1016/S0168-6445(03)00055-X

Berges M, Michel AM, Lassek C, Nuss AM, Beckstette M, Dersch P, Riedel K, Sievers S, Becher D, Otto A, Maass S, Rohde M, Eckweiler D, Borrero-de Acuna JM, Jahn M, Neumann-Schaal M, Jahn D (2018) Iron regulation in Clostridioides difficile. Front Microbiol 9:3183. https://doi.org/10.3389/fmicb.2018.03183

Bessman NJ, Mathieu JRR, Renassia C, Zhou L, Fung TC, Fernandez KC, Austin C, Moeller JB, Zumerle S, Louis S, Vaulont S, Ajami NJ, Sokol H, Putzel GG, Arvedson T, Sockolow RE, Lakhal-Littleton S, Cloonan SM, Arora M, Peyssonnaux C, Sonnenberg GF (2020) Dendritic cell-derived hepcidin sequesters iron from the microbiota to promote mucosal healing. Science 368(6487):186–189. https://doi.org/10.1126/science.aau6481

Bjarnason J, Southward CM, Surette MG (2003) Genomic profiling of iron-responsive genes in Salmonella enterica serovar typhimurium by high-throughput screening of a random promoter library. J Bacteriol 185(16):4973–4982. https://doi.org/10.1128/JB.185.16.4973-4982.2003

Bradley JM, Svistunenko DA, Wilson MT, Hemmings AM, Moore GR, Le Brun NE (2020) Bacterial iron detoxification at the molecular level. J Biol Chem 295(51):17602–17623. https://doi.org/10.1074/jbc.REV120.007746

Carrier JC, Aghdassi E, Jeejeebhoy K, Allard JP (2006) Exacerbation of dextran sulfate sodium-induced colitis by dietary iron supplementation: role of NF-kappaB. Int J Color Dis 21(4):381–387. https://doi.org/10.1007/s00384-005-0011-7

Caza M, Kronstad JW (2013) Shared and distinct mechanisms of iron acquisition by bacterial and fungal pathogens of humans. Front Cell Infect Microbiol 3:80. https://doi.org/10.3389/fcimb.2013.00080

Chen JH, Wang Y, Pan J, Lu LW, Yu J, Liu B, Chen F, Deng H (2022) Prebiotic oligosaccharides enhance iron absorption via modulation of protein expression and gut microbiota in a dose-response manner in iron-deficient growing rats. Mol Nutr Food Res 66(10):e2101064. https://doi.org/10.1002/mnfr.202101064

Coelho R, Viola TW, Walss-Bass C, Brietzke E, Grassi-Oliveira R (2014) Childhood maltreatment and inflammatory markers: a systematic review. Acta Psychiatr Scand 129(3):180–192. https://doi.org/10.1111/acps.12217

Constante M, Fragoso G, Calve A, Samba-Mondonga M, Santos MM (2017) Dietary heme induces gut dysbiosis, aggravates colitis, and potentiates the development of adenomas in mice. Front Microbiol 8:1809. https://doi.org/10.3389/fmicb.2017.01809

Cornelissen CN, Sparling PF (1994) Iron piracy: acquisition of transferrin-bound iron by bacterial pathogens. Mol Microbiol 14(5):843–850. https://doi.org/10.1111/j.1365-2958.1994.tb01320.x

Dailey HA, Dailey TA, Gerdes S, Jahn D, Jahn M, O'Brian MR, Warren MJ (2017) Prokaryotic heme biosynthesis: multiple pathways to a common essential product. Microbiol Mol Biol Rev 81(1). https://doi.org/10.1128/MMBR.00048-16

Das N, Xie L, Ramakrishnan SK, Campbell A, Rivella S, Shah YM (2015) Intestine-specific disruption of hypoxia-inducible factor (HIF)-2alpha improves anemia in sickle cell disease. J Biol Chem 290(39):23523–23527. https://doi.org/10.1074/jbc.C115.681643

Das NK, Schwartz AJ, Barthel G, Inohara N, Liu Q, Sankar A, Hill DR, Ma X, Lamberg O, Schnizlein MK, Arques JL, Spence JR, Nunez G, Patterson AD, Sun D, Young VB, Shah YM (2020) Microbial metabolite signaling is required for systemic iron homeostasis. Cell Metab 31(1):115–130.e116. https://doi.org/10.1016/j.cmet.2019.10.005

Dekker Nitert M, Gomez-Arango LF, Barrett HL, McIntyre HD, Anderson GJ, Frazer DM, Callaway LK (2018) Iron supplementation has minor effects on gut microbiota composition in overweight and obese

women in early pregnancy. Br J Nutr 120(3):283–289. https://doi.org/10.1017/S0007114518001149
Deriu E, Liu JZ, Pezeshki M, Edwards RA, Ochoa RJ, Contreras H, Libby SJ, Fang FC, Raffatellu M (2013) Probiotic bacteria reduce salmonella typhimurium intestinal colonization by competing for iron. Cell Host Microbe 14(1):26–37. https://doi.org/10.1016/j.chom.2013.06.007
Deschemin JC, Noordine ML, Remot A, Willemetz A, Afif C, Canonne-Hergaux F, Langella P, Karim Z, Vaulont S, Thomas M, Nicolas G (2016) The microbiota shifts the iron sensing of intestinal cells. FASEB J 30(1):252–261. https://doi.org/10.1096/fj.15-276840
Dongiovanni P, Ruscica M, Rametta R, Recalcati S, Steffani L, Gatti S, Girelli D, Cairo G, Magni P, Fargion S, Valenti L (2013) Dietary iron overload induces visceral adipose tissue insulin resistance. Am J Pathol 182(6):2254–2263. https://doi.org/10.1016/j.ajpath.2013.02.019
Dostal A, Lacroix C, Bircher L, Pham VT, Follador R, Zimmermann MB, Chassard C (2015) Iron modulates butyrate production by a child gut microbiota in vitro. MBio 6(6):e01453–e01415. https://doi.org/10.1128/mBio.01453-15
Dutta D, Methe B, Amar S, Morris A, Lim SH (2019) Intestinal injury and gut permeability in sickle cell disease. J Transl Med 17(1):183. https://doi.org/10.1186/s12967-019-1938-8
Eckburg PB, Bik EM, Bernstein CN, Purdom E, Dethlefsen L, Sargent M, Gill SR, Nelson KE, Relman DA (2005) Diversity of the human intestinal microbial flora. Science 308(5728):1635–1638. https://doi.org/10.1126/science.1110591
Fernandez-Real JM, Lopez-Bermejo A, Ricart W (2002) Cross-talk between iron metabolism and diabetes. Diabetes 51(8):2348–2354. https://doi.org/10.2337/diabetes.51.8.2348
Fernandez-Real JM, Blasco G, Puig J, Moreno M, Xifra G, Sanchez-Gonzalez J, Maria Alustiza J, Pedraza S, Ricart W, Maria Moreno-Navarrete J (2016) Adipose tissue R2* signal is increased in subjects with obesity: a preliminary MRI study. Obesity (Silver Spring) 24(2):352–358. https://doi.org/10.1002/oby.21347
Flo TH, Smith KD, Sato S, Rodriguez DJ, Holmes MA, Strong RK, Akira S, Aderem A (2004) Lipocalin 2 mediates an innate immune response to bacterial infection by sequestrating iron. Nature 432(7019):917–921. https://doi.org/10.1038/nature03104
Friel J, Qasem W, Cai C (2018) Iron and the breast-fed infant. Antioxidants (Basel) 7(4). https://doi.org/10.3390/antiox7040054
Gamberini MR, De Sanctis V, Gilli G (2008) Hypogonadism, diabetes mellitus, hypothyroidism, hypoparathyroidism: incidence and prevalence related to iron overload and chelation therapy in patients with thalassaemia major followed from 1980 to 2007 in the Ferrara Centre. Pediatr Endocrinol Rev 6(Suppl 1):158–169
Genco CA, Dixon DW (2001) Emerging strategies in microbial haem capture. Mol Microbiol 39(1):1–11. https://doi.org/10.1046/j.1365-2958.2001.02231.x
Heidinger S, Braun V, Pecoraro VL, Raymond KN (1983) Iron supply to Escherichia coli by synthetic analogs of enterochelin. J Bacteriol 153(1):109–115. https://doi.org/10.1128/jb.153.1.109-115.1983
Ho TD, Ellermeier CD (2015) Ferric uptake regulator fur control of putative iron acquisition systems in Clostridium difficile. J Bacteriol 197(18):2930–2940. https://doi.org/10.1128/JB.00098-15
Jaeggi T, Kortman GA, Moretti D, Chassard C, Holding P, Dostal A, Boekhorst J, Timmerman HM, Swinkels DW, Tjalsma H, Njenga J, Mwangi A, Kvalsvig J, Lacroix C, Zimmermann MB (2015) Iron fortification adversely affects the gut microbiome, increases pathogen abundance and induces intestinal inflammation in Kenyan infants. Gut 64(5):731–742. https://doi.org/10.1136/gutjnl-2014-307720
Kammler M, Schon C, Hantke K (1993) Characterization of the ferrous iron uptake system of Escherichia coli. J Bacteriol 175(19):6212–6219. https://doi.org/10.1128/jb.175.19.6212-6219.1993
Kim MH, Yun KE, Kim J, Park E, Chang Y, Ryu S, Kim HL, Kim HN (2020) Gut microbiota and metabolic health among overweight and obese individuals. Sci Rep 10(1):19417. https://doi.org/10.1038/s41598-020-76474-8
Konikoff MR, Denson LA (2006) Role of fecal calprotectin as a biomarker of intestinal inflammation in inflammatory bowel disease. Inflamm Bowel Dis 12(6):524–534. https://doi.org/10.1097/00054725-200606000-00013
Kortman GA, Raffatellu M, Swinkels DW, Tjalsma H (2014) Nutritional iron turned inside out: intestinal stress from a gut microbial perspective. FEMS Microbiol Rev 38(6):1202–1234. https://doi.org/10.1111/1574-6976.12086
Kramer J, Ozkaya O, Kummerli R (2020) Bacterial siderophores in community and host interactions. Nat Rev Microbiol 18(3):152–163. https://doi.org/10.1038/s41579-019-0284-4
Lee M, Chang EB (2021) Inflammatory bowel diseases (IBD) and the microbiome-searching the crime scene for clues. Gastroenterology 160(2):524–537. https://doi.org/10.1053/j.gastro.2020.09.056
Lee T, Clavel T, Smirnov K, Schmidt A, Lagkouvardos I, Walker A, Lucio M, Michalke B, Schmitt-Kopplin P, Fedorak R, Haller D (2017) Oral versus intravenous iron replacement therapy distinctly alters the gut microbiota and metabolome in patients with IBD. Gut 66(5):863–871. https://doi.org/10.1136/gutjnl-2015-309940
Li Y, Hansen SL, Borst LB, Spears JW, Moeser AJ (2016) Dietary iron deficiency and oversupplementation increase intestinal permeability, ion transport, and inflammation in pigs. J Nutr 146(8):1499–1505. https://doi.org/10.3945/jn.116.231621
Li H, Kazmi JS, Lee S, Zhang D, Gao X, Maryanovich M, Torres L, Verma D, Kelly L, Ginzburg YZ, Frenette

PS, Manwani D (2023) Dietary iron restriction protects against vaso-occlusion and organ damage in murine sickle cell disease. Blood 141(2):194–199. https://doi.org/10.1182/blood.2022016218

Liu C, Gong J, Zhang Q, Chen G, Yin S, Luo Z, Zeng W, Yu J, Lan P, He Z (2023) Dietary iron modulates gut microbiota and induces SLPI secretion to promote colorectal tumorigenesis. Gut Microbes 15(1):2221978. https://doi.org/10.1080/19490976.2023.2221978

Mahadea D, Adamczewska E, Ratajczak AE, Rychter AM, Zawada A, Eder P, Dobrowolska A, Krela-Kazmierczak I (2021) Iron deficiency anemia in inflammatory bowel diseases-a narrative review. Nutrients 13(11). https://doi.org/10.3390/nu13114008

Mahalhal A, Frau A, Burkitt MD, Ijaz UZ, Lamb CA, Mansfield JC, Lewis S, Pritchard DM, Probert CS (2021) Oral ferric maltol does not adversely affect the intestinal microbiome of patients or mice, but ferrous sulphate does. Nutrients 13(7). https://doi.org/10.3390/nu13072269

Malesza IJ, Bartkowiak-Wieczorek J, Winkler-Galicki J, Nowicka A, Dzieciolowska D, Blaszczyk M, Gajniak P, Slowinska K, Niepolski L, Walkowiak J, Madry E (2022) The dark side of iron: the relationship between iron, inflammation and gut microbiota in selected diseases associated with iron deficiency anaemia-a narrative review. Nutrients 14(17). https://doi.org/10.3390/nu14173478

Manfredo Vieira S, Hiltensperger M, Kumar V, Zegarra-Ruiz D, Dehner C, Khan N, Costa FRC, Tiniakou E, Greiling T, Ruff W, Barbieri A, Kriegel C, Mehta SS, Knight JR, Jain D, Goodman AL, Kriegel MA (2018) Translocation of a gut pathobiont drives autoimmunity in mice and humans. Science 359(6380):1156–1161. https://doi.org/10.1126/science.aar7201

Masse E, Gottesman S (2002) A small RNA regulates the expression of genes involved in iron metabolism in Escherichia coli. Proc Natl Acad Sci USA 99(7):4620–4625. https://doi.org/10.1073/pnas.032066599

Mazziotta C, Tognon M, Martini F, Torreggiani E, Rotondo JC (2023) Probiotics mechanism of action on immune cells and beneficial effects on human health. Cells 12(1). https://doi.org/10.3390/cells12010184

McCance RA, Widdowson EM (1938) The absorption and excretion of iron following oral and intravenous administration. J Physiol 94(1):148–154. https://doi.org/10.1113/jphysiol.1938.sp003669

Meisel M, Hinterleitner R, Pacis A, Chen L, Earley ZM, Mayassi T, Pierre JF, Ernest JD, Galipeau HJ, Thuille N, Bouziat R, Buscarlet M, Ringus DL, Wang Y, Li Y, Dinh V, Kim SM, McDonald BD, Zurenski MA, Musch MW, Furtado GC, Lira SA, Baier G, Chang EB, Eren AM, Weber CR, Busque L, Godley LA, Verdu EF, Barreiro LB, Jabri B (2018) Microbial signals drive pre-leukaemic myeloproliferation in a Tet2-deficient host. Nature 557(7706):580–584. https://doi.org/10.1038/s41586-018-0125-z

Mikulic N, Uyoga MA, Stoffel NU, Derrien M, Nyilima S, Kostopoulos I, Roeselers G, Chenoll E, Mwasi E, Pironaci G, Karanja S, Bourdet-Sicard R, Zimmermann MB (2024) Prebiotics increase iron absorption and reduce the adverse effects of iron on the gut microbiome and inflammation: a randomized controlled trial using iron stable isotopes in Kenyan infants. Am J Clin Nutr 119(2):456–469. https://doi.org/10.1016/j.ajcnut.2023.11.018

Monteiro CA, Levy RB, Claro RM, Castro IR, Cannon G (2010) A new classification of foods based on the extent and purpose of their processing. Cad Saude Publica 26(11):2039–2049. https://doi.org/10.1590/s0102-311x2010001100005

Moustarah F, Daley SF (2024) Dietary iron. In: Disclosure: Sharon Daley declares no relevant financial relationships with ineligible companies. StatPearls, Treasure Island ineligible companies

Nakashige TG, Zhang B, Krebs C, Nolan EM (2015) Human calprotectin is an iron-sequestering host-defense protein. Nat Chem Biol 11(10):765–771. https://doi.org/10.1038/nchembio.1891

Neuberger A, Okebe J, Yahav D, Paul M (2016) Oral iron supplements for children in malaria-endemic areas. Cochrane Database Syst Rev 2(2):CD006589. https://doi.org/10.1002/14651858.CD006589.pub4

Nielsen OH, Soendergaard C, Vikner ME, Weiss G (2018) Rational management of iron-deficiency anaemia in inflammatory bowel disease. Nutrients 10(1). https://doi.org/10.3390/nu10010082

Niepel D, Klag T, Malek NP, Wehkamp J (2018) Practical guidance for the management of iron deficiency in patients with inflammatory bowel disease. Ther Adv Gastroenterol 11:1756284818769074. https://doi.org/10.1177/1756284818769074

Noordine ML, Seyoum Y, Bruneau A, Baye K, Lefebvre T, Cherbuy C, Canonne-Hergaux F, Nicolas G, Humblot C, Thomas M (2024) The microbiota and the host organism switch between cooperation and competition based on dietary iron levels. Gut Microbes 16(1):2361660. https://doi.org/10.1080/19490976.2024.2361660

Nyffenegger N, Zennadi R, Kalleda N, Flace A, Ingoglia G, Buzzi RM, Doucerain C, Buehler PW, Schaer DJ, Durrenberger F, Manolova V (2022) The oral ferroportin inhibitor vamifeport improves hemodynamics in a mouse model of sickle cell disease. Blood 140(7):769–781. https://doi.org/10.1182/blood.2021014716

Paganini D, Uyoga MA, Kortman GAM, Cercamondi CI, Moretti D, Barth-Jaeggi T, Schwab C, Boekhorst J, Timmerman HM, Lacroix C, Karanja S, Zimmermann MB (2017) Prebiotic galacto-oligosaccharides mitigate the adverse effects of iron fortification on the gut microbiome: a randomised controlled study in Kenyan infants. Gut 66(11):1956–1967. https://doi.org/10.1136/gutjnl-2017-314418

Paganini D, Uyoga MA, Kortman GAM, Boekhorst J, Schneeberger S, Karanja S, Hennet T, Zimmermann MB (2019) Maternal human milk oligosaccharide profile modulates the impact of an intervention with iron and galacto-oligosaccharides in Kenyan infants. Nutrients 11(11). https://doi.org/10.3390/nu11112596

Parrow NL, Violet PC, George NA, Ali F, Bhanvadia S, Wong R, Tisdale JF, Fitzhugh C, Levine M, Thein SL, Fleming RE (2021) Dietary iron restriction improves markers of disease severity in murine sickle cell anemia. Blood 137(11):1553–1555. https://doi.org/10.1182/blood.2020006919

Phipps O, Brookes MJ, Al-Hassi HO (2021) Iron deficiency, immunology, and colorectal cancer. Nutr Rev 79(1):88–97. https://doi.org/10.1093/nutrit/nuaa040

Piel FB, Hay SI, Gupta S, Weatherall DJ, Williams TN (2013) Global burden of sickle cell anaemia in children under five, 2010-2050: modelling based on demographics, excess mortality, and interventions. PLoS Med 10(7):e1001484. https://doi.org/10.1371/journal.pmed.1001484

Pitsillides L, Pellino G, Tekkis P, Kontovounisios C (2021) The effect of perioperative administration of probiotics on colorectal cancer surgery outcomes. Nutrients 13(5). https://doi.org/10.3390/nu13051451

Przybyla-Toscano J, Roland M, Gaymard F, Couturier J, Rouhier N (2018) Roles and maturation of iron-sulfur proteins in plastids. J Biol Inorg Chem 23(4):545–566. https://doi.org/10.1007/s00775-018-1532-1

Qi B, Han M (2018) Microbial siderophore enterobactin promotes mitochondrial iron uptake and development of the host via interaction with ATP synthase. Cell 175(2):571–582.e511. https://doi.org/10.1016/j.cell.2018.07.032

Qi X, Zhang Y, Guo H, Hai Y, Luo Y, Yue T (2020) Mechanism and intervention measures of iron side effects on the intestine. Crit Rev Food Sci Nutr 60(12):2113–2125. https://doi.org/10.1080/10408398.2019.1630599

Raheem A, Liang L, Zhang G, Cui S (2021) Modulatory effects of probiotics during pathogenic infections with emphasis on immune regulation. Front Immunol 12:616713. https://doi.org/10.3389/fimmu.2021.616713

Raymond KN, Dertz EA, Kim SS (2003) Enterobactin: an archetype for microbial iron transport. Proc Natl Acad Sci USA 100(7):3584–3588. https://doi.org/10.1073/pnas.0630018100

Reddy BS, Pleasants JR, Wostmann BS (1972) Effect of intestinal microflora on iron and zinc metabolism, and on activities of metalloenzymes in rats. J Nutr 102(1):101–107. https://doi.org/10.1093/jn/102.1.101

Romao CV, Louro R, Timkovich R, Lubben M, Liu MY, LeGall J, Xavier AV, Teixeira M (2000) Iron-coproporphyrin III is a natural cofactor in bacterioferritin from the anaerobic bacterium Desulfovibrio desulfuricans. FEBS Lett 480(2–3):213–216. https://doi.org/10.1016/s0014-5793(00)01939-6

Rouault TA (2019) The indispensable role of mammalian iron sulfur proteins in function and regulation of multiple diverse metabolic pathways. Biometals 32(3):343–353. https://doi.org/10.1007/s10534-019-00191-7

Sassone-Corsi M, Nuccio SP, Liu H, Hernandez D, Vu CT, Takahashi AA, Edwards RA, Raffatellu M (2016) Microcins mediate competition among Enterobacteriaceae in the inflamed gut. Nature 540(7632):280–283. https://doi.org/10.1038/nature20557

Schmitt MP, Talley BG, Holmes RK (1997) Characterization of lipoprotein IRP1 from Corynebacterium diphtheriae, which is regulated by the diphtheria toxin repressor (DtxR) and iron. Infect Immun 65(12):5364–5367. https://doi.org/10.1128/iai.65.12.5364-5367.1997

Seiwert N, Adam J, Steinberg P, Wirtz S, Schwerdtle T, Adams-Quack P, Hovelmeyer N, Kaina B, Foersch S, Fahrer J (2021) Chronic intestinal inflammation drives colorectal tumor formation triggered by dietary heme iron in vivo. Arch Toxicol 95(7):2507–2522. https://doi.org/10.1007/s00204-021-03064-6

Sender R, Fuchs S, Milo R (2016) Revised estimates for the number of human and bacteria cells in the body. PLoS Biol 14(8):e1002533. https://doi.org/10.1371/journal.pbio.1002533

Shah YM, Matsubara T, Ito S, Yim SH, Gonzalez FJ (2009) Intestinal hypoxia-inducible transcription factors are essential for iron absorption following iron deficiency. Cell Metab 9(2):152–164. https://doi.org/10.1016/j.cmet.2008.12.012

Simonyte Sjodin K, Domellof M, Lagerqvist C, Hernell O, Lonnerdal B, Szymlek-Gay EA, Sjodin A, West CE, Lind T (2019) Administration of ferrous sulfate drops has significant effects on the gut microbiota of iron-sufficient infants: a randomised controlled study. Gut 68(11):2095–2097. https://doi.org/10.1136/gutjnl-2018-316988

Stein J, Walper A, Klemm W, Farrag K, Aksan A, Dignass A (2018) Safety and efficacy of intravenous iron isomaltoside for correction of anaemia in patients with inflammatory bowel disease in everyday clinical practice. Scand J Gastroenterol 53(9):1059–1065. https://doi.org/10.1080/00365521.2018.1498914

Stojiljkovic I, Cobeljic M, Hantke K (1993) Escherichia coli K-12 ferrous iron uptake mutants are impaired in their ability to colonize the mouse intestine. FEMS Microbiol Lett 108(1):111–115. https://doi.org/10.1111/j.1574-6968.1993.tb06082.x

Tang M, Frank DN, Sherlock L, Ir D, Robertson CE, Krebs NF (2016) Effect of vitamin E with therapeutic iron supplementation on iron repletion and gut microbiome in US iron deficient infants and toddlers. J Pediatr Gastroenterol Nutr 63(3):379–385. https://doi.org/10.1097/MPG.0000000000001154

Tang M, Frank DN, Hendricks AE, Ir D, Esamai F, Liechty E, Hambidge KM, Krebs NF (2017) Iron in micronutrient powder promotes an unfavorable gut microbiota in Kenyan infants. Nutrients 9(7). https://doi.org/10.3390/nu9070776

Taylor M, Qu A, Anderson ER, Matsubara T, Martin A, Gonzalez FJ, Shah YM (2011) Hypoxia-inducible factor-2alpha mediates the adaptive increase of intestinal ferroportin during iron deficiency in mice.

Gastroenterology 140(7):2044–2055. https://doi.org/10.1053/j.gastro.2011.03.007

Theil EC (1987) Ferritin: structure, gene regulation, and cellular function in animals, plants, and microorganisms. Annu Rev Biochem 56:289–315. https://doi.org/10.1146/annurev.bi.56.070187.001445

Tilg H, Moschen AR (2014) Microbiota and diabetes: an evolving relationship. Gut 63(9):1513–1521. https://doi.org/10.1136/gutjnl-2014-306928

Tondeur MC, Schauer CS, Christofides AL, Asante KP, Newton S, Serfass RE, Zlotkin SH (2004) Determination of iron absorption from intrinsically labeled microencapsulated ferrous fumarate (sprinkles) in infants with different iron and hematologic status by using a dual-stable-isotope method. Am J Clin Nutr 80(5):1436–1444. https://doi.org/10.1093/ajcn/80.5.1436

Tsolis RM, Baumler AJ, Heffron F, Stojiljkovic I (1996) Contribution of TonB- and Feo-mediated iron uptake to growth of Salmonella typhimurium in the mouse. Infect Immun 64(11):4549–4556. https://doi.org/10.1128/iai.64.11.4549-4556.1996

Turnbaugh PJ, Ley RE, Mahowald MA, Magrini V, Mardis ER, Gordon JI (2006) An obesity-associated gut microbiome with increased capacity for energy harvest. Nature 444(7122):1027–1031. https://doi.org/10.1038/nature05414

Valdebenito M, Crumbliss AL, Winkelmann G, Hantke K (2006) Environmental factors influence the production of enterobactin, salmochelin, aerobactin, and yersiniabactin in Escherichia coli strain Nissle 1917. Int J Med Microbiol 296(8):513–520. https://doi.org/10.1016/j.ijmm.2006.06.003

Vazquez-Gutierrez P, Lacroix C, Jaeggi T, Zeder C, Zimmerman MB, Chassard C (2015) Bifidobacteria strains isolated from stools of iron deficient infants can efficiently sequester iron. BMC Microbiol 15(1):3. https://doi.org/10.1186/s12866-014-0334-z

Vazquez-Gutierrez P, de Wouters T, Werder J, Chassard C, Lacroix C (2016) High iron-sequestrating bifidobacteria inhibit enteropathogen growth and adhesion to intestinal epithelial cells in vitro. Front Microbiol 7:1480. https://doi.org/10.3389/fmicb.2016.01480

Velayudhan J, Hughes NJ, McColm AA, Bagshaw J, Clayton CL, Andrews SC, Kelly DJ (2000) Iron acquisition and virulence in Helicobacter pylori: a major role for FeoB, a high-affinity ferrous iron transporter. Mol Microbiol 37(2):274–286. https://doi.org/10.1046/j.1365-2958.2000.01987.x

Venuti MC, Rastetter WH, Neilands JB (1979) 1,3,5-Tris(N,N',N''-2,3-dihydroxybenzoyl)aminomethylbenzene, a synthetic iron chelator related to enterobactin. J Med Chem 22(2):123–124. https://doi.org/10.1021/jm00188a002

Xu C, Lee SK, Zhang D, Frenette PS (2020) The gut microbiome regulates psychological-stress-induced inflammation. Immunity 53(2):417–428.e414. https://doi.org/10.1016/j.immuni.2020.06.025

Xu L, Wu Y, Yang X, Pang X, Wu Y, Li X, Liu X, Zhao Y, Yu L, Wang P, Ye B, Jiang S, Ma J, Zhang X (2024) The Fe-S cluster biosynthesis in Enterococcus faecium is essential for anaerobic growth and gastrointestinal colonization. Gut Microbes 16(1):2359665. https://doi.org/10.1080/19490976.2024.2359665

Zhang D, Chen G, Manwani D, Mortha A, Xu C, Faith JJ, Burk RD, Kunisaki Y, Jang JE, Scheiermann C, Merad M, Frenette PS (2015) Neutrophil ageing is regulated by the microbiome. Nature 525(7570):528–532. https://doi.org/10.1038/nature15367

Zhang Z, Funcke JB, Zi Z, Zhao S, Straub LG, Zhu Y, Zhu Q, Crewe C, An YA, Chen S, Li N, Wang MY, Ghaben AL, Lee C, Gautron L, Engelking LJ, Raj P, Deng Y, Gordillo R, Kusminski CM, Scherer PE (2021) Adipocyte iron levels impinge on a fat-gut crosstalk to regulate intestinal lipid absorption and mediate protection from obesity. Cell Metab 33(8):1624–1639.e1629. https://doi.org/10.1016/j.cmet.2021.06.001

23 Iron Chelation Therapy

Maria Domenica Cappellini, Natalia Scaramellini, Simona Leoni, and Irene Motta

Abstract

Iron overload is a serious complication associated with transfusion-dependent anemias and anemias characterized by ineffective erythropoiesis. Iron accumulation is toxic to many tissues, causing heart failure, cirrhosis, liver cancer, growth retardation, and multiple endocrine abnormalities.

Chelation therapy aims to balance the rate of iron accumulation from blood transfusions by increasing iron excretion in urine and/or feces with chelators. Three iron chelators are currently available to treat patients: one administered parenterally called deferoxamine (or desferrioxamine) and two oral chelators, deferiprone and deferasirox. Careful dose adjustment is necessary to avoid excess chelation as iron levels fall.

A key factor for successful iron chelation is to achieve regular adherence to treatment regimens throughout life: while the convenience and tolerability of individual chelators are important in achieving this goal, other factors such as psychological well-being and family and institutional support also impact adherence and outcomes.

Keywords

Iron overload · Transfusion-dependent anemias · Deferoxamine · Deferiprone · Deferasirox

23.1 Introduction

Red cell transfusion is a life-saving treatment for patients affected by chronic anemias, such as beta-thalassemia, sickle cell anemia, myelodysplastic syndrome, and dyserythropoietic anemias.

Blood transfusions compensate for chronic anemia, prevent bone deformities, facilitate normal growth and activity levels, and allow patients a good quality of life.

Since a single unit of red blood cell contains 200–250 mg of iron and the human body has no mechanism to excrete excess iron, cumulative iron overload is an inevitable consequence of chronic transfusion therapy (Porter 2001). Moreover, inef-

M. D. Cappellini (✉)
SC Medicina ad Indirizzo Metabolico, Fondazione IRCCS Ca' Granda Ospedale Maggiore Policlinico, Milan, Italy
e-mail: maria.cappellini@unimi.it

N. Scaramellini · S. Leoni · I. Motta
SC Medicina ad Indirizzo Metabolico, Fondazione IRCCS Ca' Granda Ospedale Maggiore Policlinico, Milan, Italy

Dipartimento di Scienze Cliniche e di Comunità, Dipartimento di Eccellenza 2023–2027, University of Milan, Milan, Italy

K. Pantopoulos (ed.), *Iron Metabolism in Human Health and Disease*, Advances in Experimental Medicine and Biology 1480, https://doi.org/10.1007/978-3-031-92033-2_23

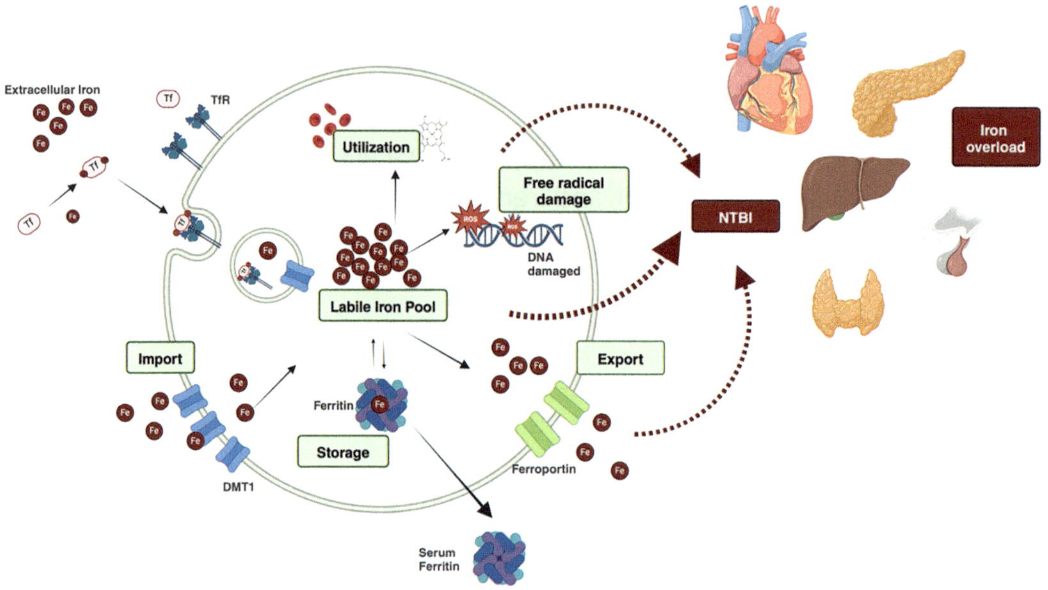

Fig. 23.1 Intracellular Iron Homeostasis: *DMT1* divalent metal ion transporter 1, *Tf* Transferrin, *TfR* Transferrin Receptor

fective erythropoiesis reduces hepcidin synthesis in chronic anemia patients, increasing iron absorption (Chap. 11). Iron toxicity is mainly mediated by the generation of free hydroxyl radicals, which propagate oxygen-related damage (Chap. 1). Hydroxyl radicals induce lipid peroxidation of cellular organelles, including mitochondria, lysosomes, and sarcoplasmic membranes (Halliwell and Gutteridge 1984).

In homeostatic conditions, iron is transported bound to transferrin, but in iron overload conditions, when transferrin bounding capacity exceeds, iron is "free" and forms non-transferrin bound iron (NTBI), a highly toxic form that correlates with the appearance of oxidation products and reduced plasma antioxidant capacity (Cighetti et al. 2002); see also Chaps. 9, 10 and 11. Labile iron pool, a form of NTBI, can easily be taken up by the liver, heart, and endocrine glands and causes serious organ damage (Taher 2018) (Fig. 23.1).

23.2 Iron Overload Monitoring

Serum ferritin is the most used diagnostic tool for assessing body iron storage; its normal level varies based on sex and age (Camaschella 2013). However, serum ferritin is a reactive protein, and it is increased in inflammatory conditions; thus, it does not always represent the iron stores (Camaschella 2013); see also Chap. 25.

Moreover, in non-transfusion-dependent thalassemia patients in which iron overload is due to increased absorption and ineffective erythropoiesis, the linear correlation between tissue iron and serum ferritin is lost (Taher and Cappellini 2014).

Nevertheless, ferritin is easy to measure and non-expensive; thus, regular monitoring of serum ferritin (e.g., every 2–3 months) (Farmakis et al. 2022) can provide helpful information to assess iron burden in thalassemia patients. A more accurate method to estimate iron overload is evaluating its presence in target organs, such as the liver and the heart. Liver iron concentration (LIC) closely correlates with total body iron storage. LIC value is expressed in mg/g/dry weight, and the normal range is 0.2–2.0 mg/g/dry weight in males and 0.2–1.6 mg/g/dry weight in females. LIC value >7 mg/g/dry weight is associated with an increased risk of clinical complications caused by hemosiderosis, and a LIC >15 mg/g/dry weight is considered the risk threshold for cardiac disease and premature death, due to cardiac arrhythmias, heart failure, or cirrhosis (Musallam et al. 2024). LIC was initially evaluated by liver biopsy, which remains an invasive method and at present should be limited to patients with persistently high levels of serum ferritin and/or suspected co-existing hepatitis. Magnetic resonance imaging (MRI) is the most widely used method to measure iron overload in different organs. Different MRI techniques have been validated based on measuring tissue's relaxation time after applying a radio-frequency magnetic field pulse. For the liver, values may be expressed directly by mentioning the relaxation time or by using the relaxation rates, the R1 rate (the same as 1/T1) and the R2 rate (the same as 1/T2). The T2* technique was the first widely adopted method (Anderson et al. 2001), utilizing liver biopsy for calibration. Another method is the R2-based FerriScan® technique, which appears to have acceptable linearity and reproducibility up to LIC values of about 30 mg/g dry weight (Garbowski et al. 2014). MRI techniques are also widely used to measure myocardial iron overload. The physical principles of iron measurement for the heart by MRI are the same as for the liver, with the additional challenge of measuring the moving, contracting myocardial tissue. The myocardial T2* (mT2*) MRI values were originally established based on shortened T2* values <20 ms in patients with decreased left ventricular ejection fraction (LVEF) (Anderson et al. 2001). The relationship between biochemically measured myocardial iron concentration and myocardial T2* was confirmed on post-mortem myocardial material (Carpenter et al. 2011). Heart failure and cirrhosis were the most common causes of death from iron overload before the introduction of iron chelation therapy. Monitoring cardiac function as well as the function of endocrine organs should also be considered an indicator of iron overload.

23.3 Iron Chelation

With the advent of iron chelation, the overall survival of thalassemia patients improved (Taher and Cappellini 2014; Farmakis et al. 2020). Iron chelation therapy aims to prevent iron deposition or to remove iron overload. The iron removal rate should be equal to the transfusion rate when there is no iron accumulation or higher than the transfusion rate when there is already iron overload. An emergency treatment should be considered if clinical heart failure develops.

Moreover, iron chelation aims to provide constant, 24-h protection from the harmful effects of toxic iron (i.e., NTBI). Iron chelation therapy must be effective in binding non-transferrin-bound iron over sustained periods, decreasing tissue uptake and iron-catalyzed toxic reactions (Olivieri and Brittenham 1997).

According to the guidelines, iron chelation therapy should be started when serum ferritin is higher than 1000 μg/L, or the patient has undergone at least 10 red blood cell transfusions, or LIC is above 3 mg/g dry weight (Farmakis et al. 2022). To be efficacious, iron chelation must be taken regularly; thus, good adherence is mandatory for treatment success.

Three iron chelators are currently available and patented in Europe and the USA for pediatric and adult thalassemia populations. Deferoxamine (DFO), deferiprone (DFP), and deferasirox (DFX) differ in iron-binding properties, routes of absorption, elimination, and metabolism ((EMA) EMA. Deferasirox n.d.; (EMA) EMa. Deferiprone n.d.) (Table 23.1).

23.3.1 Deferoxamine

Deferoxamine (DFO) was the first iron chelator, licensed for the treatment of chronic transfusion iron overload worldwide for patients over 2 years old. Patients born in the 1980s had access to DFO from an early age: these patients are the first cohort in which the impact of iron chelation in reducing the disease burden is seen (Olivieri and Brittenham 1997; Borgna-Pignatti et al. 2006; Borgna-Pignatti and Gamberini 2011; Brittenham et al. 1993; Bronspiegel-Weintrob et al. 1990).

DFO is a hexadentate molecule binding iron 1:1, which results in a very stable chelator-iron complex. DFO was initially developed and licensed for intramuscular injection; however, studies conducted in the 1970s demonstrated that subcutaneous injection of DFO was more effi-

Table 23.1 Summary of the characteristics of the three available iron chelators

	Deferoxamine	Deferiprone	Deferasirox
Administration method	Subcutaneously or intravenously	Oral	Oral
Administration frequency	8–15 h, 5–7 days per week	3 times daily	Once daily
Half-time of iron-free drug	20–30 min	3–4 h	12–16 h
Metabolism	Intrahepatic	>90% eliminated in feces, 60% unmetabolized.	Intrahepatic
Recommended dose	30–60 mg/kg per day	75–100 mg/kg per day	14–28 mg/kg per day
TIF guidelines indications TDT	>2 years: First line	>6 years, if other chelators are not tolerated	2–6 years: First-line (USA) second line (EU); >6 years: First line
TIF guidelines indications NTDT	No sufficient data	No sufficient data	>10 years: First line
Main adverse events	Ocular and auditory symptoms, bone-growth retardation; local reactions, allergy	Gastrointestinal symptoms; arthralgia; agranulocytosis or neutropenia	Gastrointestinal symptoms; increased creatinine; increase hepatic enzymes
Pregnancy	Contraindicated (but has been used in third trimester)	Contraindicated	Contraindicated
Precautions	Monitor ferritin, if it falls to <100 ng/mL, reduce dose, monitor audiometry regularly Monitor eyes regularly	Measure neutrophil count before starting and monitor ANC weekly	Monitor creatinine trend for 1st 4 weeks after starting or after dose escalation, then monthly If a rapid fall in serum ferritin to >1000 ng/mL dose is reduced. If ferritin <500 ng/mL consider very low doses Monitor urine protein regularly Monitor liver functions regularly

TDT transfusion-dependent thalassemia, *NTDT* non-transfusion-dependent thalassemia

cient in excreting iron via urine or feces in patients regularly transfused (Graziano 1978). It is a large, hydrophilic molecule and cannot be absorbed from the gut. It also has a very short plasma half-life (20–30 min) and must be administered parenterally, either intravenously or subcutaneously, with a long infusion time over at least 8 h a day at least 5 days a week in regularly transfused patients to secure the effect. Treatment is costly and inconvenient, and adherence to therapy has been the main limiting factor to successful outcomes. Constant subcutaneous deferoxamine administration is an effective and practical way of eliminating large quantities of iron in siderosis (Propper et al. 1977), and, when DFO is stopped, NTBI levels increase rapidly in the blood. Its rate of reappearance is rapid, reaching pretreatment levels within 6 h of cessation of DFO, supporting that continuous chelation therapy is the most beneficial approach to reduce iron burden (Porter 1997).

The recommended dosage is 30–60 mg/kg/day 5–7 days per week, and the dosage of 50 mg/kg/day 5 times per week has been demonstrated to reduce the LIC by 5 μg in 1 year (Musallam et al. 2024). Finetuned management of DFO dosage is required if serum ferritin is below 500 μg/L: dose reductions could be guided using a therapeutic index calculated as mean dose (mg/Kg)/serum ferritin (μg/L), which must be kept below 0.0023. DFO-related toxicities must be monitored. DFO is overall a safe drug and could even be administered during the third trimester of pregnancy or in patients with renal failure.

Unwanted effects, other than allergic reactions, are usually related to excessive chelation. Organs involved are the hearing system, the eye, and the skeleton, especially in children, in which growth retardation may occur. Yearly screening with audiometry could help to diagnose high-frequency sensorineural loss, tinnitus, and deafness. The eye involvement could present as night blindness, impaired color vision, impaired visual fields, and reduced visual acuity, with the most severe case presenting as retinitis pigmentosa. An excessive dosage could lead to skeletal changes, presented as vertebral demineralization, the flatness of vertebral bodies, or, in the worst cases, rickets-like body lesions (Porter 1997).

In case of cardiac iron overload, higher dosages (50–60 mg/kg/day) are considered sufficient to reduce cardiac iron burden, while for very severe T2* double chelation therapy is recommended.

In high-risk cases with decreased LVEF, continuous infusion is potentially more beneficial than periodic infusions because it reduces exposure to toxic free iron (NTBI) (Anderson et al. 2001). Intensification through continuous, 24-h intravenous DFO via an implanted intravenous delivery system (e.g., Port-a-cath) (Anderson et al. 2001) or subcutaneously has been shown to normalize heart function, reverse heart failure, and improve mT2*. At least 50 mg/kg/day (not exceeding 60 mg/kg/day) is recommended as a 24-h infusion (Farmakis et al. 2022; Ceci et al. 2002).

The administration of DFO via subcutaneous route by pump has been one of the primary obstacles to patient compliance; the infusion can be challenging, time-consuming, and painful. Furthermore, adolescents represent a particularly difficult patient group due to their desire to assert individuality and independence. During childhood, drug administration and management are typically the responsibility of parents and guardians. When patients reach adulthood, there is a shift in the responsibility for managing treatment, which can be problematic and may lead to non-adherence to therapy. For some adolescent patients, a directive from a physician to comply with a treatment regimen becomes just another occasion to express their independence, often resulting in non-compliance (Cappellini 2007). However, adolescents are also the patient group most likely to benefit from effective chelation therapy.

Published data suggest that compliance for DFO is between 60% and 80%. Some solutions could improve adherence: variations in the injection site could decrease the likelihood of local skin reactions, butterfly needles can be used, the infusate should be kept to a 10% solution, and patients should be properly trained to ensure that subcutaneous, rather than intradermal, infusions are achieved, and ready-made balloon infusers can be provided (Porter 2001). Studies demonstrated that patients who comply with chelation therapy have significantly better survival rates than those who do not (Brittenham et al. 1993; Gabutti and Piga 1996). Poor compliance leads to gaps in chelation coverage, during which NTBI levels can increase and cause further tissue damage.

23.3.2 Deferiprone Monotherapy

Deferiprone (DFP) was licensed in several countries and the European Union in the 1990s and in the USA in October 2011 ((EMA) EMa. Deferiprone n.d.). The accepted indications for treatment differ slightly in different countries. It is suitable for oral absorption but has lower efficiency in iron binding (DFP 3:1). DFP has a plasma half-life of 3–4 h, requiring it to be administered three times per day, with a recommended and effective dosage of 75–100 mg/kg/day. Various analyses suggested no difference in outcomes between DFP and DFO therapy at 1 year (Olivieri and Brittenham 1997; Pennell et al. 2006; Cohen et al. 2003; El-Beshlawy et al. 2014). Interestingly, it was demonstrated that DFP at higher doses has a higher effect in increasing myocardial T2* and improving left ventricle ejection fractions (Pennell et al. 2006; Maggio et al. 2002). DFP could be administered in pediatric patients, and it showed efficacy in inducing control of iron overload during 12 months of treatment (Maggio et al. 2020). Retrospective studies suggest that DFP monotherapy gives superior cardiac protection compared to DFO (Borgna-Pignatti et al. 2006). The main limitation of DFP is related to the risk of agranulocytosis and neutropenia: it is contraindicated during pregnancy and with patients with a history of neutropenia or with an underlying risk of cytopenia. Measuring the neutrophil count before starting and weekly during treatment is advisable. It is important to know that the risk of agranulocytosis is not dose- or time-related (Olivieri et al. 1995).

The registration studies by the US Food and Drug Administration (FDA) in 2011 concluded that with DFP monotherapy, a minimum 20% reduction was met in only 50% of patients.

Nonetheless, deferiprone reduced body iron concentrations in all the patients with a history of ineffective chelation therapy with deferoxamine. Also, studies that used serum ferritin concentration as an indirect estimate of the body's iron load suggest that the drug is more effective in patients with serum ferritin >2500 μg/L (Taher et al. 2001). Overall negative iron balance (decrease in LIC) at standard transfusion rates is achieved in only about a third of patients receiving 75 mg/kg/day.

In a randomized study, a higher dose of DFP (92 mg/kg/day) was compared with DFO at 43 mg/kg 5–7 days/week (mean daily dose 35 mg/kg/day). The increase in mT2* was greater than in the DFO group, although relatively low DFO doses were used as the comparator. LVEF showed a better improvement compared to patients treated with DFO (Musallam et al. 2024). In another trial, however, the results were similar to the two approaches, while 3 years of DFP monotherapy was associated with a significant increase in LVEF in patients with normal LVEF at baseline. A retrospective analysis suggested that DFP monotherapy offered superior cardiac protection to DFO (Farmakis et al. 2022).

All patients previously treated with DFO, when switched to DFP, stated a clear preference for DFP compared to subcutaneous infusion of DFO, improving the compliance.

23.3.3 Deferasirox Monotherapy

Deferasirox is the most recently developed iron chelator, and it is now licensed as first-line monotherapy for transfusion-dependent thalassemia in over 100 countries worldwide. It is orally absorbed and can be administered once daily. The original formulation, a tablet dispersible in water, has been replaced by a film-coated tablet (FCT), a formulation that is better absorbed and better tolerated, with less gastrointestinal disturbance (Taher 2018), ameliorating patient compliance to therapy. Since the absorption of the sprinkle is enhanced, the formulation needs to be decreased to 0.7 times those previously recommended with the classical formulation. Moreover, with sprinkles therapy, absorption and clearance kinetics are not the same in all patients. Studies on patients recorded FCT DFX plasma levels of 60% and even 70% of individuals 12 h after a single dose. This would result in inadequate NTBI removal, equivalent to not chelating iron a third of the time. Splitting the dose twice a day may significantly increase the area-under-the-curve exposure and improve chelation (Cappellini 2007).

Studies comparing DFX to DFO demonstrated that patient-reported satisfaction and convenience were significantly higher for the once-daily oral FCT deferasirox than for DFO infusions, impacting patients' daily lives (Cappellini 2007).

Deferasirox is licensed for patients affected by transfusion-dependent thalassemia above 6 years old, or above 2 years old if deferoxamine cannot be used or is inadequate. Moreover, it is licensed for patients from 2 years of age who suffer from other types of anemia and who receive blood transfusions, when deferoxamine cannot be used or is inadequate. It has also been approved for patients from 10 years of age with non-transfusion-dependent thalassemia syndromes when deferoxamine cannot be used or is inadequate ((EMA) EMA. Deferasirox n.d.).

The registration trial was the largest ever done for an iron chelating agent: patients were included from 2 years of age with a diagnosis of thalassemia and chronic iron overload from blood transfusions as indicated by a LIC value of 2 mg Fe/g dry weight or higher. Patients needed to be receiving at least 8 blood transfusions per year and could be enrolled irrespective of prior chelation therapy. Patients were excluded from this trial if they had one of the following conditions: an alanine aminotransferase (ALT) level greater than 250 U/L during the year before enrollment, chronic hepatitis B infection, active hepatitis C infection, a history of a positive HIV test, serum creatinine above the upper limit of normal (ULN), a urinary protein–creatinine ratio of greater than 0.5 mg/mg, nephrotic syndrome, uncontrolled systemic hypertension, a prolonged corrected QT interval, or systemic infection within the 10 days before entry. The primary objective of the study was non-inferiority to deferoxamine: a dosing algorithm was constructed according to baseline LIC values to attempt to appropriately treat patients across the spectrum of iron loading (Piga et al. 2006). A short-term iron balance study indicated that 20 mg/kg deferasirox daily maintained a stable iron balance and could reduce LIC comparably to 40 mg/kg DFO given 5 days a week. Moreover, changes in serum ferritin, LIC, and net body iron balance over a 1-year interval were consistent. In addition, a relationship between the dose of the drug administered and the response was clear for deferasirox (Cappellini et al. 2011). Subsequent studies focused on patient-reported satisfaction and convenience of DFX administration; it was clear that they were significantly higher for the once-daily, oral FCT deferasirox than for DFO infusions. Among patients who had received DFO before the study, the majority indicated a preference for deferasirox over DFO.

The ESCALATOR study was conducted primarily on heavily iron-overloaded patients, and it confirmed previous data, highlighting the importance of timely deferasirox dose adjustments based on serum ferritin levels and transfusion iron intake to ensure that patients achieve their therapeutic goal of maintenance or reduction in iron burden (Taher et al. 2009).

Several studies demonstrated that DFX has a dose-dependent effect on serum ferritin: a randomized study found that 20 mg/kg of the original formulation daily stabilized serum ferritin close to 2000 μg/L and at 30 mg/kg, serum ferritin was reduced with an average fall of 1249 μg/L over 1 year (Cappellini et al. 2011; Porter et al. 2016), while data from clinical observation suggest that, in patients without iron overload, 14 mg/kg of the new formulation can counterbalance an iron intake of 0.3 mg/kg day (Scaramellini et al. 2020).

DFX is effective in reducing liver iron content, with changes in LIC over 1 year (Cappellini 2007) and myocardial iron content. Prospective data demonstrated the efficacy of DFX in improving myocardial T2* over 1 year, with a baseline myocardial T2* range of 5–20 ms (Pennell et al. 2011), while a steady, significant myocardial T2* improvement, from a mean of 12.0 ms at baseline to 17.1 ms, was observed after 3 years of treatment, despite a high mean baseline LIC (mean 28 mg/g dry wt) (Pennell et al. 2006).

DFX is contraindicated during pregnancy and in patients with renal dysfunction, and careful dosage adjustment is recommended when there is an increase in serum creatinine. Clinical records reported cases of acquired Fanconi syndrome in patients, mostly children treated with deferasirox (Cappellini 2007; Yacobovich et al. 2010; Wong et al. 2016). As a precaution, it is recommended to monitor not only renal function but also urine protein regularly. Caution is recommended for patients with advanced liver disease and hepatic decompensation. It is important to enlighten that the elevation of alanine transferase and/or aminotransferase is frequently linked to damage due to iron overload.

As mentioned, currently DFX is the first-line treatment for transfusion-dependent thalassemia (TDT) patients, but mostly choosing between the three depends on patients' conditions, comorbidities, and organ involvement.

23.3.4 Combination Therapy

In case of severe iron overload or when rapid reduction of serum ferritin is needed, iron chelators could be administered in combination. The combination of DFP and DFO was the most commonly used when monotherapy regimens were not followed consistently and/or failed to control iron overload in the liver or heart. When there is severe heart iron overload, DFP at maximum dosage with 24-h DFO infusion is suggested as a rescue treatment. This combined therapy was effective in reducing iron burden, as assessed by serum ferritin levels (Balveer et al. 2000; Wonke et al. 1998). Recent evidence suggests that deferiprone may serve as a shuttle, entering cells and bringing iron from tissue compartments to the bloodstream, where it is then transferred to DFO, which is given simultaneously, and excreted bound to DFO (Taher et al. 2001; Breuer and Cabantchik 2001). Prospective studies have shown rapid removal of liver iron and improvement in cardiac iron in patients receiving DFX-DFO combinations.

Most recently, it has been shown that the sequential DFP-DFX therapy (Voskaridou et al. 2011) is tolerable, can control levels, and decreases cardiac and liver iron overload in TDT patients at a cost that is between that of DFX and DFP monotherapy (Vitrano et al. 2019). A combination of DFX and DFO has been demonstrated to be effective in fast iron removal from either the liver or the heart (Lal et al. 2013; Tanner et al. 2006).

23.4 Conclusions

The introduction of iron chelation therapy dramatically changed the natural history of transfusion-dependent anemias. Strict monitoring of serum ferritin and LIC is needed to tune iron chelation therapy dosage to prevent iron overload in the organs and therapy-related complications.

References

Anderson LJ, Holden S, Davis B, Prescott E, Charrier CC, Bunce NH et al (2001) Cardiovascular T2-star (T2*) magnetic resonance for the early diagnosis of myocardial iron overload. Eur Heart J 22(23):2171–2179

Balveer K, Pyar K, Wonke B (2000) Combined oral and parenteral iron chelation in beta thalassaemia major. Med J Malaysia 55(4):493–497

Borgna-Pignatti C, Gamberini MR (2011) Complications of thalassemia major and their treatment. Expert Rev Hematol 4(3):353–366

Borgna-Pignatti C, Cappellini MD, De Stefano P, Del Vecchio GC, Forni GL, Gamberini MR et al (2006) Cardiac morbidity and mortality in deferoxamine- or deferiprone-treated patients with thalassemia major. Blood 107(9):3733–3737

Breuer W, Cabantchik ZI (2001) A fluorescence-based one-step assay for serum non-transferrin-bound iron. Anal Biochem 299(2):194–202

Brittenham GM, Cohen AR, McLaren CE, Martin MB, Griffith PM, Nienhuis AW et al (1993) Hepatic iron stores and plasma ferritin concentration in patients with sickle cell anemia and thalassemia major. Am J Hematol 42(1):81–85

Bronspiegel-Weintrob N, Olivieri NF, Tyler B, Andrews DF, Freedman MH, Holland FJ (1990) Effect of age at the start of iron chelation therapy on gonadal function in beta-thalassemia major. N Engl J Med 323(11):713–719

Camaschella C (2013) Treating iron overload. N Engl J Med 368(24):2325–2327

Cappellini MD (2007) Exjade(R) (deferasirox, ICL670) in the treatment of chronic iron overload associated with blood transfusion. Ther Clin Risk Manag 3(2):291–299

Cappellini MD, Bejaoui M, Agaoglu L, Canatan D, Capra M, Cohen A et al (2011) Iron chelation with deferasirox in adult and pediatric patients with thalassemia major: efficacy and safety during 5 years' follow-up. Blood 118(4):884–893

Carpenter JP, He T, Kirk P, Roughton M, Anderson LJ, de Noronha SV et al (2011) On T2* magnetic resonance and cardiac iron. Circulation 123(14):1519–1528

Ceci A, Baiardi P, Felisi M, Cappellini MD, Carnelli V, De Sanctis V et al (2002) The safety and effectiveness of deferiprone in a large-scale, 3-year study in Italian patients. Br J Haematol 118(1):330–336

Cighetti G, Duca L, Bortone L, Sala S, Nava I, Fiorelli G et al (2002) Oxidative status and malondialdehyde in beta-thalassaemia patients. Eur J Clin Investig 32(Suppl 1):55–60

Cohen AR, Galanello R, Piga A, De Sanctis V, Tricta F (2003) Safety and effectiveness of long-term therapy with the oral iron chelator deferiprone. Blood 102(5):1583–1587

El-Beshlawy AM, El-Alfy MS, Sari TT, Chan LL, Tricta F (2014) Continuation of deferiprone therapy in patients with mild neutropenia may not lead to a more severe drop in neutrophil count. Eur J Haematol 92(4):337–340

(EMA) EMA. Deferasirox (n.d.) https://www.ema.europa.eu/en/medicines/human/EPAR/exjade

(EMA) EMa. Deferiprone (n.d.) https://www.ema.europa.eu/en/medicines/human/EPAR/exjade

Farmakis D, Giakoumis A, Angastiniotis M, Eleftheriou A (2020) The changing epidemiology of the ageing thalassaemia populations: a position statement of the Thalassaemia international federation. Eur J Haematol 105(1):16–23

Farmakis D, Porter J, Taher A, Domenica Cappellini M, Angastiniotis M, Eleftheriou A (2022) 2021 Thalassaemia International Federation Guidelines for the management of transfusion-dependent thalassemia. Hemasphere 6(8):e732

Gabutti V, Piga A (1996) Results of long-term iron-chelating therapy. Acta Haematol 95(1):26–36

Garbowski MW, Carpenter JP, Smith G, Roughton M, Alam MH, He T et al (2014) Biopsy-based calibration of T2* magnetic resonance for estimation of liver iron concentration and comparison with R2 Ferriscan. J Cardiovasc Magn Reson 16(1):40

Graziano JH (1978) Iron metabolism and chelation therapy in hemosiderosis. Curr Top Hematol 1:127–150

Halliwell B, Gutteridge JM (1984) Oxygen toxicity, oxygen radicals, transition metals and disease. Biochem J 219(1):1–14

Lal A, Porter J, Sweeters N, Ng V, Evans P, Neumayr L et al (2013) Combined chelation therapy with deferasirox and deferoxamine in thalassemia. Blood Cells Mol Dis 50(2):99–104

Maggio A, D'Amico G, Morabito A, Capra M, Ciaccio C, Cianciulli P et al (2002) Deferiprone versus deferoxamine in patients with thalassemia major: a randomized clinical trial. Blood Cells Mol Dis 28(2):196–208

Maggio A, Kattamis A, Felisi M, Reggiardo G, El-Beshlawy A, Bejaoui M et al (2020) Evaluation of the efficacy and safety of deferiprone compared with deferasirox in paediatric patients with transfusion-dependent haemoglobinopathies (DEEP-2): a multicentre, randomised, open-label, non-inferiority, phase 3 trial. Lancet Haematol 7(6):e469–ee78

Musallam KM, Barella S, Origa R, Ferrero GB, Lisi R, Pasanisi A et al (2024) Differential effects of iron chelators on iron burden and long-term morbidity and mortality outcomes in a large cohort of transfusion-dependent beta-thalassemia patients who remained on the same monotherapy over 10 years. Blood Cells Mol Dis 107:102859

Olivieri NF, Brittenham GM (1997) Iron-chelating therapy and the treatment of thalassemia. Blood 89(3):739–761

Olivieri NF, Brittenham GM, Matsui D, Berkovitch M, Blendis LM, Cameron RG et al (1995) Iron-chelation therapy with oral deferiprone in patients with thalassemia major. N Engl J Med 332(14):918–922

Organization Wh (n.d.) Serum ferritin concentrations for the assessment of iron status in individuals and populations: technical brief

Pennell DJ, Berdoukas V, Karagiorga M, Ladis V, Piga A, Aessopos A et al (2006) Randomized controlled trial of deferiprone or deferoxamine in beta-thalassemia major patients with asymptomatic myocardial siderosis. Blood 107(9):3738–3744

Pennell DJ, Porter JB, Cappellini MD, Chan LL, El-Beshlawy A, Aydinok Y et al (2011) Continued improvement in myocardial T2* over two years of deferasirox therapy in beta-thalassemia major patients with cardiac iron overload. Haematologica 96(1):48–54

Piga A, Galanello R, Forni GL, Cappellini MD, Origa R, Zappu A et al (2006) Randomized phase II trial of deferasirox (Exjade, ICL670), a once-daily, orally-administered iron chelator, in comparison to deferoxamine in thalassemia patients with transfusional iron overload. Haematologica 91(7):873–880

Porter JB (1997) A risk-benefit assessment of iron-chelation therapy. Drug Saf 17(6):407–421

Porter JB (2001) Practical management of iron overload. Br J Haematol 115(2):239–252

Porter JB, de Witte T, Cappellini MD, Gattermann N (2016) New insights into transfusion-related iron toxicity: implications for the oncologist. Crit Rev Oncol Hematol 99:261–271

Propper RD, Cooper B, Rufo RR, Nienhuis AW, Anderson WF, Bunn HF et al (1977) Continuous subcutaneous administration of deferoxamine in patients with iron overload. N Engl J Med 297(8):418–423

Scaramellini N, Consonni D, Cassinerio E, Arighi C, Marcon A, Graziadei G et al (2020) A holistic approach to iron chelation therapy in transfusion-dependent thalassemia patients with serum ferritin below 500 mug/L. Am J Hematol 95(9):E230–E2E2

Taher AT (2018) Thalassemia. Hematol Oncol Clin North Am 32(2):xv–xvi

Taher AT, Cappellini MD (2014) Management of non-transfusion-dependent thalassemia: a practical guide. Drugs 74(15):1719–1729

Taher A, Sheikh-Taha M, Koussa S, Inati A, Neeman R, Mourad F (2001) Comparison between deferoxamine and deferiprone (L1) in iron-loaded thalassemia patients. Eur J Haematol 67(1):30–34

Taher A, El-Beshlawy A, Elalfy MS, Al Zir K, Daar S, Habr D et al (2009) Efficacy and safety of deferasirox, an oral iron chelator, in heavily iron-overloaded patients with beta-thalassaemia: the ESCALATOR study. Eur J Haematol 82(6):458–465

Tanner MA, He T, Westwood MA, Firmin DN, Pennell DJ, Thalassemia International Federation Heart TI (2006) Multi-center validation of the transferability of the magnetic resonance T2* technique for the quantification of tissue iron. Haematologica 91(10):1388–1391

Vitrano A, Ruffo GB, Pepe A, D'Ascola DG, Caruso V, Filosa A et al (2019) Long-term sequential deferiprone and deferasirox therapy in transfusion-dependent thalassaemia patients: a prospective clinical trial. Br J Haematol 186(6):e209–ee11

Voskaridou E, Christoulas D, Terpos E (2011) Successful chelation therapy with the combination of deferasirox and deferiprone in a patient with thalassaemia major and persisting severe iron overload after single-agent chelation therapies. Br J Haematol 154(5):654–656

Wong P, Fuller PJ, Gillespie MT, Milat F (2016) Bone disease in thalassemia: a molecular and clinical overview. Endocr Rev 37(4):320–346

Wonke B, Wright C, Hoffbrand AV (1998) Combined therapy with deferiprone and desferrioxamine. Br J Haematol 103(2):361–364

Yacobovich J, Stark P, Barzilai-Birenbaum S, Krause I, Pazgal I, Yaniv I et al (2010) Acquired proximal renal tubular dysfunction in beta-thalassemia patients treated with deferasirox. J Pediatr Hematol Oncol 32(7):564–567

Oral and Intravenous Iron Therapy

24

Andrea U. Steinbicker and Kostas Pantopoulos

Abstract

Pre-anemic iron deficiency and iron deficiency anemia are the most common health conditions. They are mostly caused by blood loss and/or inadequate dietary iron absorption and are managed with iron replacement therapy. The goal of iron replacement therapy is to correct anemia and replenish iron stores. In most patients, oral iron is typically the first choice for treatment due to its ease of use, over-the-counter availability, and low cost of most common formulations. Ferrous sulfate remains the gold standard for oral iron therapy, but alternative formulations have also been developed. Effective iron repletion requires the daily intake of up to 200 mg elemental iron for 3–12 weeks. However, only a small fraction of it gets absorbed. Thus, accumulation of unabsorbed iron in the intestinal lumen may cause constipation and other gastrointestinal side effects. Intravenous iron is the preferred treatment for patients who are intolerant or refractory to oral iron or require rapid correction of anemia and repletion of iron stores, for example prior to surgery. The newest formulations such as ferric carboxymaltose and ferric derisomaltose have excellent safety profiles and can be provided at high doses of 500–1000 mg in a single infusion. Intravenous iron should always be administered in medical facilities by healthcare providers trained to manage potential but rare hypersensitivity reactions. Moreover, phosphate levels should be monitored in patients because intravenous iron formulations, especially ferric carboxymaltose, have been associated with increased risk for hypophosphatemia.

Keywords

Iron deficiency anemia (IDA) · Iron replacement therapy · Oral iron supplements · Intravenous (IV) iron therapy · Hypophosphatemia

A. U. Steinbicker
Department of Anesthesiology and Intensive Care Medicine, University of Cologne, Cologne University Hospital, Cologne, Germany
e-mail: andrea.steinbicker@uk-koeln.de

K. Pantopoulos (✉)
Lady Davis Institute for Medical Research, Jewish General Hospital and Department of Medicine, McGill University, Montreal, QC, Canada
e-mail: kostas.pantopoulos@mcgill.ca

24.1 Iron Deficiency

Iron deficiency (ID) occurs when body iron stores are depleted, and iron becomes insufficient to meet physiologic needs (Pasricha et al. 2021; Camaschella 2019). It is the most common nutri-

K. Pantopoulos (ed.), *Iron Metabolism in Human Health and Disease*, Advances in Experimental Medicine and Biology 1480, https://doi.org/10.1007/978-3-031-92033-2_24

tional deficiency and primarily affects menstruating and pregnant women, but also children and vulnerable populations in developing countries. While ID may be asymptomatic in early stages, it usually manifests with mild fatigue, weakness, and concentration difficulty. ID has also been linked to reduced libido and sexual dysfunction in women (Serati et al. 2023; Nikzad et al. 2018). When iron levels drop too low and cannot sustain physiological hemoglobin concentration, ID progresses to *iron deficiency anemia* (IDA), a major contributor to global burden of disease (Camaschella 2015; Collaborators 2023); see also Chap. 12. Symptoms of IDA include pale skin, shortness of breath, cold hands and feet, palpitations, koilonychia, and hair loss. Severe ID and IDA have also been associated with developmental defects and impaired cognitive and immune functions.

In most cases, ID and IDA are caused by blood loss, accompanied by inadequate dietary iron absorption, which results in mobilization and depletion of tissue iron stores (Fig. 24.1, red panel, IDA). In pregnant women, iron depletion occurs in response to hemodilution and to increased iron utilization by the growing fetus (Chap. 21). The second most common form of anemia occurs in patients with acute or chronic inflammation, cancer or infections, and is therefore named anemia of inflammation (AI, Fig. 24.1, blue panel, AI) or anemia of chronic disease (ACD); see also Chap. 13. A major contributor to AI is inflammatory induction of the hepatic hormone hepcidin (Chaps. 3 and 14), which promotes degradation of the iron exporter ferroportin in target tissues and thereby inhibits iron entry into plasma. This can cause a state of functional ID, where total body iron levels are physiological, but a significant fraction of iron is trapped within tissues and cannot be effectively utilized for erythropoiesis (Marques et al. 2022; Cappellini et al. 2017). In patients with inflammatory bowel disease (IBD) or other chronic inflammatory conditions, functional ID may be combined with true ID, or not. All other types of anemia, mainly caused by vitamin deficiencies or genetic mutations (Fig. 24.1, bottom panel), are rare and cannot be cured with iron supplementation.

For long time, ID and IDA have been viewed as a secondary condition of relatively minor importance that mostly affects women. However, research and clinical practice in recent years have shown that anemia is an independent predictor of a worse course of an underlying disease and a poorer patient outcome in both, men and women. For instance, anemia in heart failure is associated with more severe symptoms, impaired functional capacity, and increased mortality (Goel et al. 2021). In IBD, anemia may exacerbate the underlying condition leading to frequent hospitaliza-

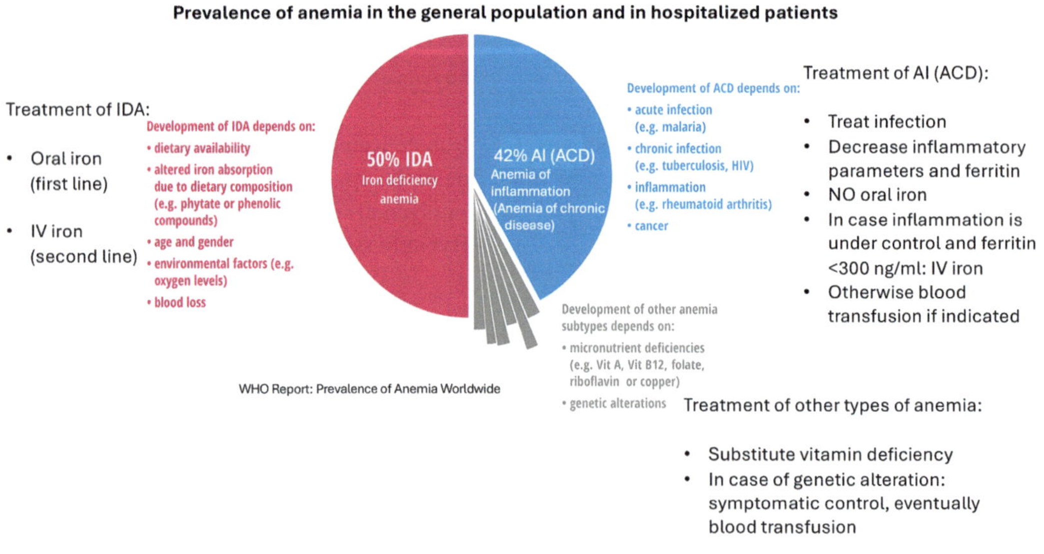

Fig. 24.1 Prevalence, causes, and treatment of various types of anemia

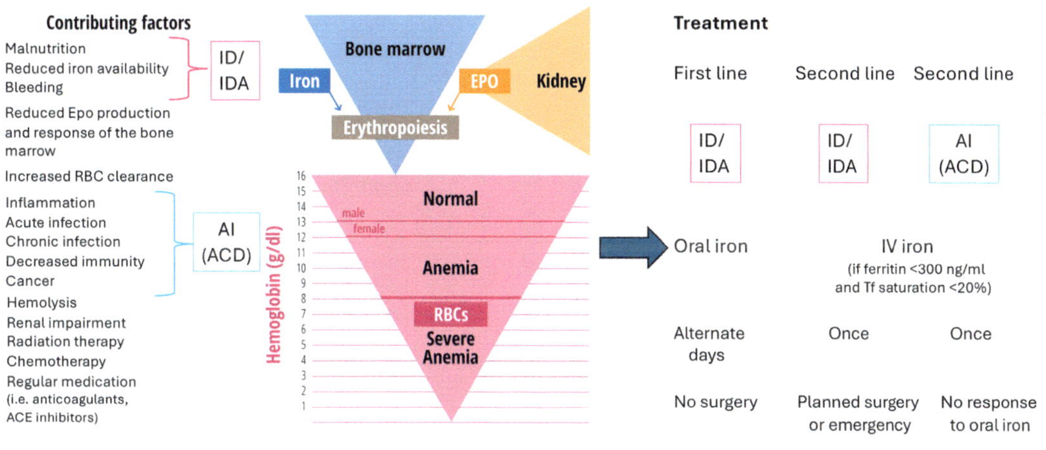

Fig. 24.2 Absolute or relative deficiency of iron causes ID/IDA; inflammation or cancer provokes functional iron deficiency and AI (ACD). A combination of both types is frequent. Other types of anemia (such as that induced by chemotherapy or radiation) are caused by suppression of bone marrow production and cannot be cured by IV/oral iron

tions (Maas et al. 2023). Likewise, in chronic kidney disease (CKD), anemia is a predictor of higher mortality and cardiovascular events (Badura et al. 2024). Thus, there is a need to correct ID and IDA with iron replacement therapy, although studies on long-term outcomes are lacking. The primary objective is to treat anemia and restore iron levels. Iron can be administered either orally or intravenously.

There are a variety of causes that provoke anemia (Fig. 24.2). Only in case of ID/IDA or AI with ferritin levels below 300 ng/ml and a transferrin saturation below 20%, treatment with iron is indicated. In hemolysis, or in scenarios with impaired bone marrow production, the body contains too much iron, as red blood cells production is decreased (Fig. 24.2); under these conditions, iron supplementation in any form is even contraindicated.

24.2 Oral Iron Therapy

In patients with modest ID or IDA, oral iron is typically the first choice due to its ease of use, over-the-counter availability, and lower cost (Pantopoulos 2024). Oral iron supplements fall into two main categories: ferrous (Fe^{2+}) salts and ferric (Fe^{3+}) complexes, along with other types like carbonyl iron and heme iron polypeptide (Table 24.1).

Ferrous salts are the most widely used. They include ferrous sulfate, fumarate, and gluconate and are available as tablets or syrups. To prevent oxidation of Fe^{2+} to insoluble Fe^{3+} in liquid forms, excipients like sodium bisulfite are added.

Ferric complexes include ferric polysaccharide complexes and ferric protein succinylate. Formulations developed for improved bioavailability and absorption, such as ferric maltol (Kontoghiorghes et al. 2021), sucrosomial® iron (Gomez-Ramirez et al. 2023) and liposomal iron (Tan et al. 2023), are substantially more expensive. Ferric citrate is mainly used for CKD patients to reduce phosphate levels and correct anemia (Ganz et al. 2019).

Carbonyl iron, a form of elemental iron, and heme iron polypeptide, derived from bovine hemoglobin, offer additional options with varying absorption rates.

Table 24.1 Oral iron supplements used for iron replacement therapy

	Type of supplements	Typical amount of elemental iron[a] (mg per tablet)	Advantages	Disadvantages
Ferrous salts	Ferrous sulfate	65	Low cost, high efficacy, wide availability	Poor tolerability
	Ferrous fumarate	100		
	Ferrous gluconate	35		
	Ferrous bisglycinate	50		
	Ferrous ascorbate	65		
Ferric complexes	Ferric polysaccharide	150	Possibly fair tolerability	Relatively low efficacy
	Ferric polymaltose	100		
	Ferric polydextrose	150		
	Iron protein succinylate	40[b]	High efficacy, fair tolerability	High cost
	Ferric citrate	210	Efficacy in CKD patients	
	Ferric maltol	30	Efficacy and tolerability in IBD patients	High cost, risk for IBD flare
	Sucrosomial® iron	30		High cost, no iron stores rebuild
	Liposomal iron	30		Scarce data, no iron stores rebuild
Other	Carbonyl iron	18	Low cost, wide availability	Scarce data, no iron stores rebuild
	Heme iron polypeptide	12	Efficacy in CKD patients	

[a]Elemental iron content may vary in different commercial products
[b]In liquid form (elemental iron content per 15 ml)

24.3 Efficacy and Tolerability of Oral Iron

Iron supplementation efficacy is typically measured by hemoglobin increase and iron absorption (Wienk et al. 1999). An optimal response to oral iron therapy is an increase in hemoglobin by 2 g/dl within 3–4 weeks, though 1 g/dl within 4 weeks is considered as reasonable. Ferrous salts, especially ferrous sulfate, are the mostly prescribed supplements and recommended for treatment of ID and IDA (Pantopoulos 2024). Slow-release forms are also available; however, they are less effective due to delayed iron release. Ferric complexes, while slower to absorb, can also increase hemoglobin and iron stores, although ferrous salts tend to perform better.

Ferric maltol has shown efficacy in anemic patients with IBD or CKD (Schmidt et al. 2021; Howaldt et al. 2022). Sucrosomial® iron and liposomal iron can improve hemoglobinization; however, without consistently replenishing iron stores (Abbati et al. 2019; de Alvarenga Antunes et al. 2020). Ferric citrate has improved hemoglobinization and iron parameters in studies with CKD patients (Fishbane et al. 2017; Womack et al. 2020). Carbonyl iron and heme iron polypeptide are less studied but have shown similar effectiveness to ferrous sulfate in some cases (Gordeuk et al. 1987; Nagaraju et al. 2013). These iron formulations have been evaluated in many studies with patients who are intolerant or refractory to oral iron; however, none of them has achieved non-inferiority versus intravenous (IV) iron.

Oral iron supplements often cause gastrointestinal side effects like constipation, diarrhea, or nausea (Pantopoulos 2024). This is because only a fraction of ingested iron (10–20%) is efficiently absorbed, leading to excess iron accumulation in the digestive system. Additionally, the unabsorbed surplus iron may disrupt gut microbiota, favoring growth of harmful bacteria (Jaeggi et al. 2015; Lee et al. 2017). Gut dysbiosis has been linked to increased infection risk, intestinal inflammation, and metabolic syndrome pathologies (Fan and Pedersen 2021); see also Chap. 22.

Ferrous sulfate, the gold standard among ferrous salts, though effective, can have limited tolerability, independently of the ingested dose (Tolkien et al. 2015). This often reduces adherence to therapy, especially if longer treatment is needed. Ferric complexes are generally thought to be better tolerated but have similar side effects (Powers et al. 2017; Berber et al. 2014). Syrups with high doses of ferrous or ferric iron formulations also tend to cause tooth staining (Pani et al. 2015). Ferric maltol and sucrosomial® iron appear to show good tolerability, especially in patients who are intolerant to ferrous salts (Gomez-Ramirez et al. 2023; Pergola and Kopyt 2021); nevertheless, ferric maltol may increase the risk for IBD flare and iron overload (Pantopoulos 2024). Carbonyl iron and heme iron polypeptide have not shown significant tolerability benefits compared to other iron supplements (Gordeuk et al. 1987; Nagaraju et al. 2013).

24.4 Optimal Dosage and Frequence for Oral Iron Supplements

The amount of elemental iron (the actual absorbable iron) varies between iron supplements (Table 24.1), so dosages must be adjusted accordingly. Oral iron therapy often starts with ferrous sulfate or similar salts, which provide about 60–70 mg of elemental iron. Daily absorption of 10–20 mg of iron is enough to raise hemoglobin levels by 2 g/dl within 3–4 weeks.

Several studies suggested that multiple daily doses of oral iron supplements are not beneficial. Iron intake triggers induction of hepcidin (Chap. 3), which blocks further intestinal iron absorption for up to 24 h. A pilot study with non-anemic, iron-deficient women found that a 60 mg dose of ferrous sulfate raised hepcidin levels, reducing absorption from subsequent doses (Moretti et al. 2015). Hepcidin returned to baseline after 48 h, suggesting that taking iron every other day might be more effective than daily dosing by bypassing the hepcidin effect.

Further research from randomized controlled trials (RCTs) supports that intermittent iron supplementation may be as effective as daily intake on anemic patients, with fewer side effects, especially in women of reproductive age (Stoffel et al. 2017, 2020; Kaundal et al. 2020; Kaynar et al. 2022; Pasupathy et al. 2023). The guidelines of the British Society of Gastroenterology now recommend starting with one tablet of ferrous salts (such as ferrous sulfate) daily or every other day if not tolerated (Snook et al. 2021). The American Gastroenterological Association clinical practice update on the management of IDA likewise indicates that every-other-day iron dosing of ferrous sulfate "may be better tolerated for some patients with similar or equal rates of iron absorption as daily dosing" (DeLoughery et al. 2024). The effects of intermittent dosing with other iron supplements, such as ferric polysaccharides, ferric protein succinylate, ferric maltol, ferric citrate, or sucrosomial® iron, have not yet been investigated.

24.5 Intravenous Iron Therapy

Compared to oral, IV iron bypasses the limiting step of gastrointestinal absorption and therefore raises hemoglobin levels and replenishes iron stores faster, if intact iron regulation and a productive bone marrow are secured (Girelli et al. 2018). Therefore, IV iron is the preferred option in several clinical settings.

24.5.1 IV Iron Therapy in Pre- or Post-operative Patients

The prevalence of preoperative anemia prior to elective surgery is high (up to 50–60%, depending on population and type of surgery) and has a negative impact on clinical and functional outcomes, as well as on post-operative survival (Shander et al. 2004; Taylor et al. 2022; Burton et al. 2019; Sicat et al. 2022; Liu et al. 2019; Baron et al. 2014). It affects both men and women. Historically, the primary approach to managing severe IDA has been allogeneic blood transfusion. However, this procedure poses an immunological challenge (Muller et al. 2015) and is associated with 1:21,000 risk for major morbidity (Bolton-Maggs and Cohen 2013). Therefore, healthcare systems worldwide have adopted Patient Blood Management (PBM) programs aiming to optimize patient outcomes by minimizing the need for transfusions (Mueller et al. 2019). In Germany, implementation of PBM programs has been associated with a reduced number of blood transfusions without higher morbidity or mortality rates (Meybohm et al. 2016). Data from over one million patients have shown that PBM measures are safe and improve patients` health and outcomes (Meybohm et al. 2023). Transfusion rates are also decreasing in pediatric and adolescent patients (Piekarski et al. 2020), although RCT data on the indications for transfusions in children are still sparse (Wittenmeier et al. 2024).

IV iron therapy can be applied in the context of PBM for rapid restoration of iron stores (Elhenawy et al. 2021). Data from RCTs suggest that preoperative correction of anemia with IV iron and erythropoietin can reduce the number of packed red blood cell transfusions (Spahn et al. 2019; Kong et al. 2022). This is in line with studies showing that IV iron can increase hemoglobin preoperatively, regardless of the underlying cause, in adult as well as in pediatric patients (Otsuka et al. 2022; Richards et al. 2020). IV iron supplementation 22–28 days prior to surgery was shown to be most effective in gynecological and orthopedic preoperative patients (Ellermann et al. 2018). In an observational pilot study, postoperative IV iron supplementation improved hemoglobin levels in cardiac surgery patients but the effect was considered minor (Peters et al. 2018). The patient cohort and the average blood loss during surgery are important markers to predict an increase in hemoglobin and a reduction in red blood cell transfusion. A meta-analysis suggested that IV iron more effectively reduces the need for blood transfusion in preoperative anemia patients when combined with erythropoiesis-stimulating agents (Van Remoortel et al. 2021).

24.5.2 IV Iron Therapy in Patients with Poor Response to Oral Iron

Low tolerance or inappropriate response to oral iron is often observed in anemic patients with IBD (Dignass et al. 2015), CKD (Use of iron to treat anemia in CKD 2012), cancer-associated anemia (Lebrun et al. 2017), genetic iron-refractory iron deficiency anemia (IRIDA) (Heeney and Finberg 2014), or refractory iron deficiency anemia due to autoimmune atrophic gastritis, celiac disease, or other rare immune-mediated gastrointestinal disorders (Bergamaschi et al. 2018). In IBD, anemia is caused by true iron deficiency due to chronic intestinal bleeding, and/or by functional iron deficiency due to chronic intestinal inflammation (Nielsen et al. 2018). Oral iron supplements may be ineffective or poorly tolerated because they can exacerbate gastrointestinal symptoms. Moreover, they can trigger unfavorable alterations in the gut microbiota (Lee et al. 2017). In CKD, anemia is mainly caused by reduced erythropoietin production and functional iron deficiency due to inflammation and impaired clearance of hepcidin, which inhibits intestinal iron absorption (Badura et al. 2024). Cancer patients often exhibit anemia due to functional iron deficiency that can be corrected with IV iron with or without erythropoiesis-stimulating agents (Aapro et al. 2018). IRIDA is caused by hepcidin overexpression due to mutations in the *TMPRSS6* gene, encoding the hepcidin suppressor matriptase (see Chaps. 3 and 12), while in autoimmune atrophic gastritis and immune-mediated gastrointestinal disorders, anemia and ineffectiveness of oral iron therapy are linked to inflammation (Bergamaschi et al. 2018).

24.5.3 IV Iron Therapy in Cardiology

Iron deficiency is common in patients with heart failure and is associated with poor outcomes and quality of life, as well as with increased hospitalization and mortality (Dhaliwal and Kalogeropoulos 2023). The cause of iron deficiency is inadequate iron intake, malabsorption due to medications such as proton pump inhibitors, chronic blood loss, or inflammation (Alnuwaysir et al. 2021). The diagnostic criteria for assessing iron deficiency in patients with heart failure are not clearly defined (Packer et al. 2024), but there is evidence from pilot studies and RCTs that IV iron administration is beneficial (Gale et al. 2024). Nevertheless, several RCTs have not shown improved clinical outcomes following IV iron therapy (Cheema et al. 2024), and there is an ongoing debate on how to manage iron deficiency in cardiological patients.

24.5.4 IV Iron in Gynecology and Obstetrics

Candidates for IV iron therapy are women with heavy menstrual or uterine bleeding, gynecologic bleeding caused by diseases such as adenomyosis, uterine fibroids, or endometrial hyperplasia, or women in late pregnancy (Breymann and Auerbach 2017; Kirthan and Somannavar 2024).

24.6 Intravenous Iron Formulations

IV iron formulations are colloidal nanoparticles consisting of a polynuclear Fe^{3+} oxyhydroxide core within a carbohydrate shell. They mainly differ in the type and strength of the bonds between the iron core and the carbohydrate shell (Girelli et al. 2018; Funk et al. 2022). This chemistry affects the stability of the iron complex, which in turn determines how much iron can be safely administered in a single infusion.

Early IV iron formulations, such as iron saccharide and high molecular weight (HMW) iron dextran, were introduced for the treatment of IDA around the mid-twentieth century. However, despite their effectiveness, iron saccharide and HMW iron dextran drugs were withdrawn from the market because their use has been associated with significant adverse effects. This spurred the development of newer formulations with improved safety profiles (Auerbach and Ballard 2010) that are appropriate for clinical applications. An overview on currently available IV iron pharmaceuticals is given in Table 24.2. Iron gluconate (Ferlixit®) and iron sucrose (Venofer®) are second-generation drugs. While still in use, these complexes have relatively low stability and tend to release excessive free labile iron at doses above 200–250 mg, raising safety concerns. As a result, to safely reach a total iron dose of 1000 mg, which is typically required for adequate iron replenishment, multiple infusions over several sessions are needed but not recommend any more in the clinics.

Contemporary formulations of IV iron exhibit high stability (Van Doren et al. 2024; Girelli et al. 2018). They include low molecular weight (LMW) iron dextran (INFeD® in the US, CosmoFer® in Europe and generic equivalents), ferric carboxymaltose (Injectafer® in the US and Ferinject® in Europe), ferric derisomaltose (Monofer®), and ferumoxytol (Feraheme® and generic equivalents). Ferric carboxymaltose (FCM) and ferric derisomaltose (FDI) are the newest IV iron formulations and can be administered in a single dose within 15–30 min (Van Doren et al. 2024). FCM consists of a macromolecular ferric hydroxide carbohydrate complex and was initially used for the treatment of iron deficiency in heart failure (Anker et al. 2009). FDI consists of ferric iron bound to derisomaltose, a carbohydrate complex, which stabilizes the iron and allows it to be released gradually in the bloodstream (Kassianides et al. 2021).

Ferumoxytol is an IV iron formulation with unique properties that distinguish it from other IV iron drugs. It is not ready to be used in all countries (e.g., Germany has no approved use in clinical settings). It is composed by a superparamagnetic iron oxide core enclosed within a

Table 24.2 IV iron formulations currently used for iron replacement therapy

	Iron sucrose	Ferric carboxymaltose	Iron isomaltoside	Iron dextran	Sodium ferric gluconate	Ferumoxytol	CosmoFer	Ferapplic	Feryxa
Manufacturer	Vifor Pharma	Vifor Pharma	Pharmacosmos	Pharmacosmos	Sanofi-Aventis	AMAG Pharmaceuticals	Pharmacosmos	Hexal	Viforpharma
Trade name	Venofer Fermed	Ferinject (Europe)/ Injectafer (USA)	Monofer	Sanofi-Aventis	Ferrlecit	Feraheme	CosmoFer (Europe)/ INFeD® (USA)	FerApplic	FERYXA
Carbohydrate	Sucrose	Carboxymaltose	Isomaltoside	Low molecular weight Dextran polysaccharides	Gluconate + sucrose	Polyglucose sorbitol carboxymethylether	Hydroxid-Dextran-complex	Carboxymaltose	Carboxymaltose
Molecular weight (kDa)	34–60 (FDA)	150 (FDA)	150 (12)	165 (13)	289–444 (FDA)	750 (FDA)	73	155–187	120–190
Iron content (elemental iron, mg/mL)	20	50	100	50	12.5	30	50	50	50
Half-life/ ferritin peak	6 h/24 h after last infusion	16 h/24 days	N/A	4–5 days/ 7–9 days	1 h/24 h after last infusion	8 days/ 28–35 days	7–9 days	11–18 h	11–18 h
Max. dose approved (mg)	200	1000 mg or 20/kg	20/kg	100 (daily)	125	510	20 mg/kg	15 mg/kg injection, 20 mg/kg infusion or 1000 mg	15 mg/kg injection, 20 mg/kg infusion or 1000 mg/week
Interval between 2 applications	200 mg/session; cumulative 1000 mg/14 days	1 week	200–1000 mg once/week	Daily, until calculated dose reached	Up to 8 injections in 14 days	Second injection after 3–8 days	100–200 mg 2–3 times/week	if needed after 7 days	if needed after 7 days
Allowance of total dose infusion (TDI)	No	Yes	Yes	Yes	No	Yes	Yes		
Requierement of a test dose	No	No	No	Yes	No	No	Yes	No	No
Preservative	None	None	None	None	Benzyl alcohol	None	None	None	None

carbohydrate shell and was originally developed for use as a magnetic resonance imaging (MRI) contrast agent (Long et al. 2024). Details on the structures and physicochemical properties of older and new IV iron formulations can be found elsewhere (Funk et al. 2022; Neiser et al. 2015; Blumenstein et al. 2021) and in Table 24.2.

24.7 Mechanism for IV Iron Utilization

After IV iron enters the bloodstream, the iron carbohydrate complexes are ingested by hepatic and splenic macrophages via phagocytosis (Funk et al. 2022). Macrophages degrade the carbohydrate shell in acidified phagolysosomes liberating free iron, which is released to plasma via ferroportin. Iron is then captured by transferrin and delivered to the bone marrow for erythropoiesis. Thus, IV iron eventually follows the same fate as physiologically recycled iron from senescent red blood cells by erythrophagocytic macrophages (see Chap. 3).

Nevertheless, infusion of even stable IV iron drugs results in a transient increase in labile non-transferrin bound iron (NTBI) in plasma (Garbowski et al. 2021). A recent study with patients and mouse models showed that NTBI released during IV administration of FCM is directly taken up by cardiomyocytes in the myocardium; moreover, excess iron is not efficiently stored within cardiac ferritin, but remains in a labile form (Vera-Aviles et al. 2024). Contrary to transferrin-bound iron (TBI) that is delivered to cells via iron-regulated transferrin receptor 1 (see Chap. 2), NTBI is internalized via NTBI transporters that are not subjected to iron regulation (Knutson 2019). Thus, the data reported in (Vera-Aviles et al. 2024) suggest that, beyond the intended use of replenishing iron stores, IV iron formulations like FCM could potentially result in cardiac iron overload. When given prior to cardiac surgery, almost all FCM was incorporated into the iron stores within 48 h (Olivier et al. 2023).

24.8 Potential Adverse Effects of IV Iron Therapy

The early use of HMW iron dextran has often been associated with severe hypersensitivity reactions, with an estimated 11.3 cases per million rate of life-threatening adverse drug events (ADEs) (Chertow et al. 2006). However, the rate of serious ADEs was drastically reduced to 3.3 cases per million with the replacement of HMW with LMW iron dextran, and further to 0.9 and 0.6 with the use of iron sucrose and ferric gluconate complexes, all second-generation IV iron formulations (Chertow et al. 2006). Contemporary IV iron formulations exhibit further improved safety profiles (Blumenstein et al. 2021). A meta-analysis of 21 prospective trials with 8599 participants concluded that the risk of serious or severe hypersensitivity reactions was lower with FDI relative to FCM and iron sucrose (Pollock and Biggar 2020). The use of ferumoxytol for the treatment of IDA has triggered in some occasions severe allergic reactions (Trumbo et al. 2021); therefore, safety alerts have been issued for this drug in several countries, including the US, Canada, UK, and Germany.

Most hypersensitivity reactions to IV iron are not true allergic (IgE-mediated) but rather caused by complement activation-related pseudo-allergy (CARPA), also known as Fishbane reaction (Macdougall and Vernon 2017; Szebeni et al. 2015). This may be triggered by labile iron released from the iron-carbohydrate nanoparticle, or by the nanoparticle itself. CARPA typically manifests at the start of the infusion as flushing, muscle or joint pain, back pain, or chest pressure. The symptoms depend on the type, dosage, and infusion rate of IV iron formulation and usually resolve without treatment.

On the other hand, IgE-mediated anaphylactic reactions may become life-threatening with severe symptoms such as airway compromise or systemic hypotension. These allergic reactions require prior sensitization with an allergen and have been described in patients treated with HMW iron dextran (Auerbach and Ballard 2010). The capacity of HMW dextran to act as allergen and trigger the production of antidextran antibodies has led to its discontinuation. The risk of anaphylaxis is extremely low with currently used IV iron formulations; nevertheless, infusions should always be performed by trained healthcare personnel in a well-equipped medical facility. Moreover, patients should be closely observed for any allergic reactions during, and in general 30 min after the iron infusion.

Hypophosphatemia is another potential adverse effect of IV iron therapy and has been documented in patients treated with some contemporary IV iron formulations and, particularly, with FCM (Schaefer et al. 2022). Mechanistically, FCM treatment induces expression of intact fibroblast growth factor 23 (iFGF23). This suppresses calcium and 1,25-dihydroxyvitamin D levels and induces parathyroid hormone, which in turn triggers phosphate excretion in urine (phosphaturia), eventually resulting in hypophosphatemia.

In a double-blind randomized controlled trial with anemic IBD patients comparing the effects of FCM and FDI, both IV iron formulations were effective in the treatment of anemia. However, the use of FCM was associated with a significantly higher rate of hypophosphatemia (Zoller et al. 2023). In another two randomized controlled trials with IDA patients who were unresponsive or intolerant to oral iron, FCM likewise resulted in significantly higher incidence of hypophosphatemia compared to FDI (Wolf et al. 2020). A systematic review and meta-analysis validated the higher propensity of FCM vs FDI to cause hypophosphatemia (Schaefer et al. 2021).

Specific complications of prolonged and severe FCM-induced hypophosphatemia include osteomalacia, bone fractures, and muscle weakness, which are of particular concern in patients undergoing orthopedic surgery (Schaefer et al. 2022). In the perioperative setting, FCM-induced hypophosphatemia may result in prolonged weaning from ventilation and increased length of stay in an intensive care unit (Wozniak et al. 2022). Further research is needed for a comprehensive characterization of clinical consequences of hypophosphatemia induced by FCM or other IV iron formulations.

24.9 Iron Therapy and the Risk for Infection

In developing countries, treating anemia with oral iron supplementation can increase the risk of infections in children (Mwangi et al. 2021). This is because many pathogens rely on iron for growth, and their ability to cause disease is linked to their access to iron. As part of the body's defense mechanism during infections, iron is often sequestered to limit its availability to pathogens, a strategy known as "nutritional immunity" (Murdoch and Skaar 2022). Therefore, iron supplementation may heighten the risk for individuals with ongoing infections. Along these lines, a systematic review and meta-analysis with a total of 154 RCTs and 32,920 participants identified an increased risk for infection in patients receiving IV vs oral iron therapy, or IV vs no iron (RR, 1.17; 95% CI 1.04–1.31; $I2 = 37\%$ moderate certainty of evidence), but there was no effect on mortality or hospital length of stay (Shah et al. 2021). However, iron is also vital for immune function, including the proliferation of T lymphocytes and neutrophils, which are crucial for adaptive immunity (Frost et al. 2021, 2022); see also Chap. 14. These contrasting effects highlight the importance of maintaining a careful balance of iron levels to support immune function while avoiding infection risk.

24.10 Conclusions

Iron deficiency with or without anemia are common medical conditions that negatively affect patients' quality of life and disease outcomes. They can be effectively managed with iron replacement therapy. Oral iron supplementation is usually the first line of treatment but may be associated with mild gastrointestinal side effects that could reduce compliance. When IDA is severe or refractory to oral iron supplements, or when a quick repletion of iron stores is needed, IV iron infusion is the preferred alternative. The newest IV iron formulations have excellent safety profiles and should be administered in a clinical setting with medical personnel trained to manage the low risk for hypersensitivity reactions. Phosphate levels should be monitored in patients receiving IV iron therapy, especially with FCM, which has been associated with increased risk for hypophosphatemia.

References

Aapro M, Beguin Y, Bokemeyer C, Dicato M, Gascon P, Glaspy J, Hofmann A, Link H, Littlewood T, Ludwig H, Osterborg A, Pronzato P, Santini V, Schrijvers D, Stauder R, Jordan K, Herrstedt J, Committee EG (2018) Management of anaemia and iron deficiency in patients with cancer: ESMO Clinical Practice Guidelines. Ann Oncol 29(Suppl 4):iv96–iv110. https://doi.org/10.1093/annonc/mdx758

Abbati G, Incerti F, Boarini C, Pileri F, Bocchi D, Ventura P, Buzzetti E, Pietrangelo A (2019) Safety and efficacy of sucrosomial iron in inflammatory bowel disease patients with iron deficiency anemia. Intern Emerg Med 14(3):423–431. https://doi.org/10.1007/s11739-018-1993-9

Alnuwaysir RIS, Hoes MF, van Veldhuisen DJ, van der Meer P, Grote Beverborg N (2021) Iron deficiency in heart failure: mechanisms and pathophysiology. J Clin Med 11(1). https://doi.org/10.3390/jcm11010125

Anker SD, Comin Colet J, Filippatos G, Willenheimer R, Dickstein K, Drexler H, Luscher TF, Bart B, Banasiak W, Niegowska J, Kirwan BA, Mori C, von Eisenhart Rothe B, Pocock SJ, Poole-Wilson PA, Ponikowski P, Investigators F-HT (2009) Ferric carboxymaltose in patients with heart failure and iron deficiency. N Engl J Med 361(25):2436–2448. https://doi.org/10.1056/NEJMoa0908355

Auerbach M, Ballard H (2010) Clinical use of intravenous iron: administration, efficacy, and safety. Hematology Am Soc Hematol Educ Program 2010:338–347. https://doi.org/10.1182/asheducation-2010.1.338

Badura K, Janc J, Wasik J, Gnitecki S, Skwira S, Mlynarska E, Rysz J, Franczyk B (2024) Anemia of chronic kidney disease—a narrative review of its pathophysiology, diagnosis, and management. Biomedicine 12(6). https://doi.org/10.3390/biomedicines12061191

Baron DM, Hochrieser H, Posch M, Metnitz B, Rhodes A, Moreno RP, Pearse RM, Metnitz P, European Surgical Outcomes Study group for Trials Groups of European Society of Intensive Care M, European Society of A (2014) Preoperative anaemia is associated with poor clinical outcome in non-cardiac surgery patients. Br J Anaesth 113(3):416–423. https://doi.org/10.1093/bja/aeu098

Berber I, Diri H, Erkurt MA, Aydogdu I, Kaya E, Kuku I (2014) Evaluation of ferric and ferrous iron therapies in women with iron deficiency anaemia. Adv Hematol 2014:297057. https://doi.org/10.1155/2014/297057

Bergamaschi G, Di Sabatino A, Corazza GR (2018) Pathogenesis, diagnosis and treatment of anaemia in immune-mediated gastrointestinal disorders. Br J Haematol 182(3):319–329. https://doi.org/10.1111/bjh.15254

Blumenstein I, Shanbhag S, Langguth P, Kalra PA, Zoller H, Lim W (2021) Newer formulations of intravenous iron: a review of their chemistry and key safety aspects—hypersensitivity, hypophosphatemia, and cardiovascular safety. Expert Opin Drug Saf 20(7):757–769. https://doi.org/10.1080/14740338.2021.1912010

Bolton-Maggs PH, Cohen H (2013) Serious Hazards of Transfusion (SHOT) haemovigilance and progress is improving transfusion safety. Br J Haematol 163(3):303–314. https://doi.org/10.1111/bjh.12547

Breymann C, Auerbach M (2017) Iron deficiency in gynecology and obstetrics: clinical implications and management. Hematol Am Soc Hematol Educ Program 1:152–159. https://doi.org/10.1182/asheducation-2017.1.152

Burton BN, Okwuegbuna O, Jafari A, Califano J, Brumund KT, Gabriel RA (2019) Association of preoperative anemia with 30-day morbidity and mortality among patients with thyroid cancer who undergo thyroidectomy. JAMA Otolaryngol Head Neck Surg 145(2):124–131. https://doi.org/10.1001/jamaoto.2018.3099

Camaschella C (2015) Iron-deficiency anemia. N Engl J Med 372(19):1832–1843. https://doi.org/10.1056/NEJMra1401038

Camaschella C (2019) Iron deficiency. Blood 133(1):30–39. https://doi.org/10.1182/blood-2018-05-815944

Cappellini MD, Comin-Colet J, de Francisco A, Dignass A, Doehner W, Lam CS, Macdougall IC, Rogler G, Camaschella C, Kadir R, Kassebaum NJ, Spahn DR, Taher AT, Musallam KM, Group IC (2017) Iron deficiency across chronic inflammatory conditions: International expert opinion on definition, diagnosis, and management. Am J Hematol 92(10):1068–1078. https://doi.org/10.1002/ajh.24820

Cheema B, Chokshi A, Orimoloye O, Ardehali H (2024) Intravenous iron repletion for patients with heart failure and iron deficiency: JACC state-of-the-art review. J Am Coll Cardiol 83(25):2674–2689. https://doi.org/10.1016/j.jacc.2024.03.431

Chertow GM, Mason PD, Vaage-Nilsen O, Ahlmen J (2006) Update on adverse drug events associated with parenteral iron. Nephrol Dial Transplant 21(2):378–382. https://doi.org/10.1093/ndt/gfi253

Collaborators GBDA (2023) Prevalence, years lived with disability, and trends in anaemia burden by severity and cause, 1990-2021: findings from the Global Burden of Disease Study 2021. Lancet Haematol 10(9):e713–e734. https://doi.org/10.1016/S2352-3026(23)00160-6

de Alvarenga Antunes CV, de Alvarenga Nascimento CR, da Rocha C, Ribeiro T, de Alvarenga Antunes P, de Andrade Chebli L, Martins Goncalves Fava L,

Malaguti C, Maria Fonseca Chebli J (2020) Treatment of iron deficiency anemia with liposomal iron in inflammatory bowel disease: efficacy and impact on quality of life. Int J Clin Pharm 42(3):895–902. https://doi.org/10.1007/s11096-020-01044-x

DeLoughery TG, Jackson CS, Ko CW, Rockey DC (2024) AGA clinical practice update on management of iron deficiency anemia: expert review. Clin Gastroenterol Hepatol 22(8):1575–1583. https://doi.org/10.1016/j.cgh.2024.03.046

Dhaliwal S, Kalogeropoulos AP (2023) Markers of iron metabolism and outcomes in patients with heart failure: a systematic review. Int J Mol Sci 24(6). https://doi.org/10.3390/ijms24065645

Dignass AU, Gasche C, Bettenworth D, Birgegard G, Danese S, Gisbert JP, Gomollon F, Iqbal T, Katsanos K, Koutroubakis I, Magro F, Savoye G, Stein J, Vavricka S, European Cs, Colitis O (2015) European consensus on the diagnosis and management of iron deficiency and anaemia in inflammatory bowel diseases. J Crohns Colitis 9(3):211–222. https://doi.org/10.1093/ecco-jcc/jju009

Elhenawy AM, Meyer SR, Bagshaw SM, MacArthur RG, Carroll LJ (2021) Role of preoperative intravenous iron therapy to correct anemia before major surgery: a systematic review and meta-analysis. Syst Rev 10(1):36. https://doi.org/10.1186/s13643-021-01579-8

Ellermann I, Bueckmann A, Eveslage M, Buddendick H, Latal T, Niehoff D, Geissler RG, Hempel G, Kerkhoff A, Berdel WE, Roeder N, Van Aken HK, Zarbock A, Steinbicker AU (2018) Treating anemia in the preanesthesia assessment clinic: results of a retrospective evaluation. Anesth Analg 127(5):1202–1210. https://doi.org/10.1213/ANE.0000000000003583

Fan Y, Pedersen O (2021) Gut microbiota in human metabolic health and disease. Nat Rev Microbiol 19(1):55–71. https://doi.org/10.1038/s41579-020-0433-9

Fishbane S, Block GA, Loram L, Neylan J, Pergola PE, Uhlig K, Chertow GM (2017) Effects of ferric citrate in patients with nondialysis-dependent CKD and iron deficiency anemia. J Am Soc Nephrol 28(6):1851–1858. https://doi.org/10.1681/ASN.2016101053

Frost JN, Tan TK, Abbas M, Wideman SK, Bonadonna M, Stoffel NU, Wray K, Kronsteiner B, Smits G, Campagna DR, Duarte TL, Lopes JM, Shah A, Armitage AE, Arezes J, Lim PJ, Preston AE, Ahern D, Teh M, Naylor C, Salio M, Gileadi U, Andrews SC, Dunachie SJ, Zimmermann MB, van der Klis FRM, Cerundolo V, Bannard O, Draper SJ, Townsend ARM, Galy B, Fleming MD, Lewis MC, Drakesmith H (2021) Hepcidin-mediated hypoferremia disrupts immune responses to vaccination and infection. Med (N Y) 2(2):164–179. e112. https://doi.org/10.1016/j.medj.2020.10.004

Frost JN, Wideman SK, Preston AE, Teh MR, Ai Z, Wang L, Cross A, White N, Yazicioglu Y, Bonadonna M, Clarke AJ, Armitage AE, Galy B, Udalova IA, Drakesmith H (2022) Plasma iron controls neutrophil production and function. Sci Adv 8(40):eabq5384. https://doi.org/10.1126/sciadv.abq5384

Funk F, Fluhmann B, Barton AE (2022) Criticality of surface characteristics of intravenous iron-carbohydrate nanoparticle complexes: implications for pharmacokinetics and pharmacodynamics. Int J Mol Sci 23(4). https://doi.org/10.3390/ijms23042140

Gale SE, Willeford A, Sandquist K, Watson K (2024) Intravenous iron in patients with iron deficiency and heart failure: a review of modern evidence. Curr Opin Cardiol 39(3):178–187. https://doi.org/10.1097/HCO.0000000000001121

Ganz T, Bino A, Salusky IB (2019) Mechanism of action and clinical attributes of Auryxia((R)) (ferric citrate). Drugs 79(9):957–968. https://doi.org/10.1007/s40265-019-01125-w

Garbowski MW, Bansal S, Porter JB, Mori C, Burckhardt S, Hider RC (2021) Intravenous iron preparations transiently generate non-transferrin-bound iron from two proposed pathways. Haematologica 106(11):2885–2896. https://doi.org/10.3324/haematol.2020.250803

Girelli D, Ugolini S, Busti F, Marchi G, Castagna A (2018) Modern iron replacement therapy: clinical and pathophysiological insights. Int J Hematol 107(1):16–30. https://doi.org/10.1007/s12185-017-2373-3

Goel H, Hirsch JR, Deswal A, Hassan SA (2021) Anemia in cardiovascular disease: marker of disease severity or disease-modifying therapeutic target? Curr Atheroscler Rep 23(10):61. https://doi.org/10.1007/s11883-021-00960-1

Gomez-Ramirez S, Brilli E, Tarantino G, Girelli D, Munoz M (2023) Sucrosomial® iron: an updated review of its clinical efficacy for the treatment of iron deficiency. Pharmaceuticals (Basel) 16(6). https://doi.org/10.3390/ph16060847

Gordeuk VR, Brittenham GM, Hughes M, Keating LJ, Opplt JJ (1987) High-dose carbonyl iron for iron deficiency anemia: a randomized double-blind trial. Am J Clin Nutr 46(6):1029–1034. https://doi.org/10.1093/ajcn/46.6.1029

Heeney MM, Finberg KE (2014) Iron-refractory iron deficiency anemia (IRIDA). Hematol Oncol Clin North Am 28(4):637–652. https://doi.org/10.1016/j.hoc.2014.04.009

Howaldt S, Domenech E, Martinez N, Schmidt C, Bokemeyer B (2022) Long-term effectiveness of oral ferric maltol vs intravenous ferric carboxymaltose for the treatment of iron-deficiency anemia in patients with inflammatory bowel disease: a randomized controlled noninferiority trial. Inflamm Bowel Dis 28(3):373–384. https://doi.org/10.1093/ibd/izab073

Jaeggi T, Kortman GA, Moretti D, Chassard C, Holding P, Dostal A, Boekhorst J, Timmerman HM, Swinkels DW, Tjalsma H, Njenga J, Mwangi A, Kvalsvig J, Lacroix C, Zimmermann MB (2015) Iron fortification adversely affects the gut microbiome, increases pathogen abundance and induces intestinal inflammation in Kenyan infants. Gut 64(5):731–742. https://doi.org/10.1136/gutjnl-2014-307720

Kassianides X, Bodington R, Bhandari S (2021) An evaluation of ferric derisomaltose as a treatment for anemia. Expert Rev Hematol 14(1):7–29. https://doi.org/10.1080/17474086.2021.1858406

Kaundal R, Bhatia P, Jain A, Jain A, Nampoothiri RV, Mishra K, Jandial A, Goni D, Sandal R, Jindal N, Meshram A, Sharma R, Khaire N, Singh C, Khadwal A, Prakash G, Das R, Varma N, Varma S, Malhotra P, Lad DP (2020) Randomized controlled trial of twice-daily versus alternate-day oral iron therapy in the treatment of iron-deficiency anemia. Ann Hematol 99(1):57–63. https://doi.org/10.1007/s00277-019-03871-z

Kaynar LA, Gokcen S, Can F, Yegin ZA, Ozkurt ZN, Yagci M (2022) Comparison of daily oral iron replacement therapy with every other day treatment in female reproductive age patients with iron-deficiency anemia. Ann Hematol 101(7):1459–1464. https://doi.org/10.1007/s00277-022-04835-6

Kirthan JPA, Somannavar MS (2024) Pathophysiology and management of iron deficiency anaemia in pregnancy: a review. Ann Hematol 103(8):2637–2646. https://doi.org/10.1007/s00277-023-05481-2

Knutson MD (2019) Non-transferrin-bound iron transporters. Free Radic Biol Med 133:101–111. https://doi.org/10.1016/j.freeradbiomed.2018.10.413

Kong R, Hutchinson N, Hill A, Ingoldby F, Skipper N, Jones C, Bremner S, Bruce C, Wright J, Lewis M, Newman S, Chevassut T, Hildick-Smith D (2022) Randomised open-label trial comparing intravenous iron and an erythropoiesis-stimulating agent versus oral iron to treat preoperative anaemia in cardiac surgery (INITIATE trial). Br J Anaesth 128(5):796–805. https://doi.org/10.1016/j.bja.2022.01.034

Kontoghiorghes GJ, Kolnagou A, Demetriou T, Neocleous M, Kontoghiorghe CN (2021) New era in the treatment of iron deficiency anaemia using trimaltol iron and other lipophilic iron chelator complexes: historical perspectives of discovery and future applications. Int J Mol Sci 22(11). https://doi.org/10.3390/ijms22115546

Lebrun F, Klastersky J, Levacq D, Wissam Y, Paesmans M (2017) Intravenous iron therapy for anemic cancer patients: a review of recently published clinical studies. Support Care Cancer 25(7):2313–2319. https://doi.org/10.1007/s00520-017-3672-1

Lee T, Clavel T, Smirnov K, Schmidt A, Lagkouvardos I, Walker A, Lucio M, Michalke B, Schmitt-Kopplin P, Fedorak R, Haller D (2017) Oral versus intravenous iron replacement therapy distinctly alters the gut microbiota and metabolome in patients with IBD. Gut 66(5):863–871. https://doi.org/10.1136/gutjnl-2015-309940

Liu Y, Bai Y-P, Zhou Z-F, Jiang C-R, Xu Z, Fan X-X (2019) Preoperative anemia as a prognostic factor in patients with lung cancer: a systematic review and meta-analysis of epidemiological studies. J Cancer 10(9):2047–2056. https://doi.org/10.7150/jca.29410

Long M, Li Y, He H, Gu N (2024) The story of ferumoxytol: synthesis production, current clinical applications, and therapeutic potential. Adv Healthc Mater 13(6):e2302773. https://doi.org/10.1002/adhm.202302773

Maas LA, Krishna M, Parian AM (2023) Ironing it all out: a comprehensive review of iron deficiency anemia in inflammatory bowel disease patients. Dig Dis Sci 68(2):357–369. https://doi.org/10.1007/s10620-022-07599-1

Macdougall IC, Vernon K (2017) Complement activation-related pseudo-allergy: a fresh look at hypersensitivity reactions to intravenous iron. Am J Nephrol 45(1):60–62. https://doi.org/10.1159/000451069

Marques O, Weiss G, Muckenthaler MU (2022) The role of iron in chronic inflammatory diseases: from mechanisms to treatment options in anemia of inflammation. Blood 140(19):2011–2023. https://doi.org/10.1182/blood.2021013472

Meybohm P, Herrmann E, Steinbicker AU, Wittmann M, Gruenewald M, Fischer D, Baumgarten G, Renner J, Van Aken HK, Weber CF, Mueller MM, Geisen C, Rey J, Bon D, Hintereder G, Choorapoikayil S, Oldenburg J, Brockmann C, Geissler RG, Seifried E, Zacharowski K, Collaborators PB-s (2016) Patient blood management is associated with a substantial reduction of red blood cell utilization and safe for patient's outcome: a prospective, multicenter cohort study with a noninferiority design. Ann Surg 264(2):203–211. https://doi.org/10.1097/SLA.0000000000001747

Meybohm P, Schmitt E, Choorapoikayil S, Hof L, Old O, Muller MM, Geisen C, Seifried E, Baumhove O, de Leeuw van Weenen S, Bayer A, Friederich P, Brautigam B, Friedrich J, Gruenewald M, Elke G, Molter GP, Narita D, Raadts A, Haas C, Schwendner K, Steinbicker AU, Jenke DJ, Thoma J, Weber V, Velten M, Wittmann M, Weigt H, Lange B, Herrmann E, Zacharowski K, German Patient Blood Management Network C (2023) German patient blood management network: effectiveness and safety analysis in 1.2 million patients. Br J Anaesth 131(3):472–481. https://doi.org/10.1016/j.bja.2023.05.006

Moretti D, Goede JS, Zeder C, Jiskra M, Chatzinakou V, Tjalsma H, Melse-Boonstra A, Brittenham G, Swinkels DW, Zimmermann MB (2015) Oral iron supplements increase hepcidin and decrease iron absorption from daily or twice-daily doses in iron-depleted young women. Blood 126(17):1981–1989. https://doi.org/10.1182/blood-2015-05-642223

Mueller MM, Van Remoortel H, Meybohm P, Aranko K, Aubron C, Burger R, Carson JL, Cichutek K, De Buck E, Devine D, Fergusson D, Follea G, French C, Frey KP, Gammon R, Levy JH, Murphy MF, Ozier Y, Pavenski K, So-Osman C, Tiberghien P, Volmink J, Waters JH, Wood EM, Seifried E, Group IPF (2019) Patient blood management: recommendations from the 2018 frankfurt consensus conference. JAMA 321(10):983–997. https://doi.org/10.1001/jama.2019.0554

Muller MM, Geisen C, Zacharowski K, Tonn T, Seifried E (2015) Transfusion of packed red cells: indications, triggers and adverse events. Dtsch Arztebl Int 112(29–30):507–517.; quiz 518. https://doi.org/10.3238/arztebl.2015.0507

Murdoch CC, Skaar EP (2022) Nutritional immunity: the battle for nutrient metals at the host-pathogen interface. Nat Rev Microbiol 20(11):657–670. https://doi.org/10.1038/s41579-022-00745-6

Mwangi MN, Mzembe G, Moya E, Verhoef H (2021) Iron deficiency anaemia in sub-Saharan Africa: a review of current evidence and primary care recommendations for high-risk groups. Lancet Haematol 8(10):e732–e743. https://doi.org/10.1016/S2352-3026(21)00193-9

Nagaraju SP, Cohn A, Akbari A, Davis JL, Zimmerman DL (2013) Heme iron polypeptide for the treatment of iron deficiency anemia in non-dialysis chronic kidney disease patients: a randomized controlled trial. BMC Nephrol 14:64. https://doi.org/10.1186/1471-2369-14-64

Neiser S, Rentsch D, Dippon U, Kappler A, Weidler PG, Gottlicher J, Steininger R, Wilhelm M, Braitsch M, Funk F, Philipp E, Burckhardt S (2015) Physico-chemical properties of the new generation IV iron preparations ferumoxytol, iron isomaltoside 1000 and ferric carboxymaltose. Biometals 28(4):615–635. https://doi.org/10.1007/s10534-015-9845-9

Nielsen OH, Soendergaard C, Vikner ME, Weiss G (2018) Rational management of iron-deficiency anaemia in inflammatory bowel disease. Nutrients 10(1). https://doi.org/10.3390/nu10010082

Nikzad Z, Iravani M, Abedi P, Shahbazian N, Saki A (2018) The relationship between iron deficiency anemia and sexual function and satisfaction among reproductive-aged Iranian women. PLoS One 13(12):e0208485. https://doi.org/10.1371/journal.pone.0208485

Olivier RMR, Macke M, Muller JC, Schrader L, Eveslage M, Rauer M, Wempe C, Martens S, Zarbock A, Wagner NM, Karst U, Dogan DY, Steinbicker AU (2023) Perioperative tracking of intravenous iron in patients undergoing on-pump cardiac surgery: a prospective, single-center pilot trial. Anesth Analg 136(3):578–587. https://doi.org/10.1213/ANE.0000000000006372

Otsuka Y, Naraine N, Switzer T, Faraoni D (2022) Preoperative iron supplementation in pediatric cardiac surgical patients: a preliminary single-center experience. J Cardiothorac Vasc Anesth 36(6):1565–1570. https://doi.org/10.1053/j.jvca.2021.12.022

Packer M, Anker SD, Butler J, Cleland JGF, Kalra PR, Mentz RJ, Ponikowski P, Talha KM (2024) Redefining iron deficiency in patients with chronic heart failure. Circulation 150(2):151–161. https://doi.org/10.1161/CIRCULATIONAHA.124.068883

Pani SC, Alenazi FM, Alotain AM, Alanazi HD, Alasmari AS (2015) Extrinsic tooth staining potential of high dose and sustained release iron syrups on primary teeth. BMC Oral Health 15:90. https://doi.org/10.1186/s12903-015-0072-0

Pantopoulos K (2024) Oral iron supplementation: new formulations, old questions. Haematologica 109(9):2790–2801. https://doi.org/10.3324/haematol.2024.284967

Pasricha SR, Tye-Din J, Muckenthaler MU, Swinkels DW (2021) Iron deficiency. Lancet 397(10270):233–248. https://doi.org/10.1016/S0140-6736(20)32594-0

Pasupathy E, Kandasamy R, Thomas K, Basheer A (2023) Alternate day versus daily oral iron for treatment of iron deficiency anemia: a randomized controlled trial. Sci Rep 13(1):1818. https://doi.org/10.1038/s41598-023-29034-9

Pergola PE, Kopyt NP (2021) Oral ferric maltol for the treatment of iron-deficiency anemia in patients with CKD: a randomized trial and open-label extension. Am J Kidney Dis 78(6):846–856.e841. https://doi.org/10.1053/j.ajkd.2021.03.020

Peters F, Eveslage M, Gallitz I, Wempe C, Meybohm P, Van Aken HK, Steinbicker AU (2018) Post-operative iron carboxymaltose may have an effect on haemoglobin levels in cardiothoracic surgical patients on the ICU—an observational pilot study about anaemia treatment with intravenous iron. Transfus Med Hemother 45(1):42–46. https://doi.org/10.1159/000481143

Piekarski F, Steinbicker A, Zacharowski K, Meybohm P (2020) Changes in transfusion practice in children and adolescents over time. Transfus Med Hemother 47(5):379–384. https://doi.org/10.1159/000511231

Pollock RF, Biggar P (2020) Indirect methods of comparison of the safety of ferric derisomaltose, iron sucrose and ferric carboxymaltose in the treatment of iron deficiency anemia. Expert Rev Hematol 13(2):187–195. https://doi.org/10.1080/17474086.2020.1709437

Powers JM, Buchanan GR, Adix L, Zhang S, Gao A, McCavit TL (2017) Effect of low-dose ferrous sulfate vs iron polysaccharide complex on hemoglobin concentration in young children with nutritional iron-deficiency anemia: a randomized clinical trial. JAMA 317(22):2297–2304. https://doi.org/10.1001/jama.2017.6846

Richards T, Baikady RR, Clevenger B, Butcher A, Abeysiri S, Chau M, Macdougall IC, Murphy G, Swinson R, Collier T, Van Dyck L, Browne J, Bradbury A, Dodd M, Evans R, Brealey D, Anker SD, Klein A (2020) Preoperative intravenous iron to treat anaemia before major abdominal surgery (PREVENTT): a randomised, double-blind, controlled trial. Lancet 396(10259):1353–1361. https://doi.org/10.1016/S0140-6736(20)31539-7

Schaefer B, Tobiasch M, Viveiros A, Tilg H, Kennedy NA, Wolf M, Zoller H (2021) Hypophosphataemia after treatment of iron deficiency with intravenous ferric carboxymaltose or iron isomaltoside—a systematic review and meta-analysis. Br J Clin Pharmacol 87(5):2256–2273. https://doi.org/10.1111/bcp.14643

Schaefer B, Tobiasch M, Wagner S, Glodny B, Tilg H, Wolf M, Zoller H (2022) Hypophosphatemia after intravenous iron therapy: comprehensive review of clinical findings and recommendations for management. Bone 154:116202. https://doi.org/10.1016/j.bone.2021.116202

Schmidt C, Allen S, Kopyt N, Pergola P (2021) Iron replacement therapy with oral ferric maltol: review of the evidence and expert opinion. J Clin Med 10(19). https://doi.org/10.3390/jcm10194448

Serati M, Espuna-Pons M, Mouton-Puglisi A, Padoa A (2023) Iron deficiency and sexual dysfunction in women. Sex Med Rev 11(4):342–348. https://doi.org/10.1093/sxmrev/qead028

Shah AA, Donovan K, Seeley C, Dickson EA, Palmer AJR, Doree C, Brunskill S, Reid J, Acheson AG, Sugavanam A, Litton E, Stanworth SJ (2021) Risk of infection associated with administration of intravenous iron: a systematic review and meta-analysis. JAMA Netw Open 4(11):e2133935. https://doi.org/10.1001/jamanetworkopen.2021.33935

Shander A, Knight K, Thurer R, Adamson J, Spence R (2004) Prevalence and outcomes of anemia in surgery: a systematic review of the literature. Am J Med 116(Suppl 7A):58S–69S. https://doi.org/10.1016/j.amjmed.2003.12.013

Sicat CS, Muthusamy N, Singh V, Davidovitch RI, Slover JD, Schwarzkopf R (2022) Impact of preoperative anemia severity on primary total hip arthroplasty outcomes. J Arthroplast 37(4):721–726. https://doi.org/10.1016/j.arth.2021.12.042

Snook J, Bhala N, Beales ILP, Cannings D, Kightley C, Logan RP, Pritchard DM, Sidhu R, Surgenor S, Thomas W, Verma AM, Goddard AF (2021) British Society of Gastroenterology guidelines for the management of iron deficiency anaemia in adults. Gut 70(11):2030–2051. https://doi.org/10.1136/gutjnl-2021-325210

Spahn DR, Schoenrath F, Spahn GH, Seifert B, Stein P, Theusinger OM, Kaserer A, Hegemann I, Hofmann A, Maisano F, Falk V (2019) Effect of ultra-short-term treatment of patients with iron deficiency or anaemia undergoing cardiac surgery: a prospective randomised trial. Lancet 393(10187):2201–2212. https://doi.org/10.1016/S0140-6736(18)32555-8

Stoffel NU, Cercamondi CI, Brittenham G, Zeder C, Geurts-Moespot AJ, Swinkels DW, Moretti D, Zimmermann MB (2017) Iron absorption from oral iron supplements given on consecutive versus alternate days and as single morning doses versus twice-daily split dosing in iron-depleted women: two open-label, randomised controlled trials. Lancet Haematol 4(11):e524–e533. https://doi.org/10.1016/S2352-3026(17)30182-5

Stoffel NU, Zeder C, Brittenham GM, Moretti D, Zimmermann MB (2020) Iron absorption from supplements is greater with alternate day than with consecutive day dosing in iron-deficient anemic women. Haematologica 105(5):1232–1239. https://doi.org/10.3324/haematol.2019.220830

Szebeni J, Fishbane S, Hedenus M, Howaldt S, Locatelli F, Patni S, Rampton D, Weiss G, Folkersen J (2015) Hypersensitivity to intravenous iron: classification, terminology, mechanisms and management. Br J Pharmacol 172(21):5025–5036. https://doi.org/10.1111/bph.13268

Tan C, Karaca AC, Assadpour E, Jafari SM (2023) Influence of different nano/micro-carriers on the bioavailability of iron: focus on in vitro-in vivo studies. Adv Colloid Interf Sci 318:102949. https://doi.org/10.1016/j.cis.2023.102949

Taylor M, Abah U, Hayes T, Eadington T, Smith M, Shackcloth M, Granato F, Shah R, Booton R, Grant SW, North West Thoracic Surgery C (2022) Preoperative anemia is associated with worse long-term survival after lung cancer resection: a multicenter cohort study of 5,029 patients. J Cardiothorac Vasc Anesth 36(5):1373–1379. https://doi.org/10.1053/j.jvca.2021.08.029

Tolkien Z, Stecher L, Mander AP, Pereira DI, Powell JJ (2015) Ferrous sulfate supplementation causes significant gastrointestinal side-effects in adults: a systematic review and meta-analysis. PLoS One 10(2):e0117383. https://doi.org/10.1371/journal.pone.0117383

Trumbo H, Kaluza K, Numan S, Goodnough LT (2021) Frequency and associated costs of anaphylaxis- and hypersensitivity-related adverse events for intravenous iron products in the USA: an analysis using the US Food and Drug Administration adverse event reporting system. Drug Saf 44(1):107–119. https://doi.org/10.1007/s40264-020-01022-2

Use of iron to treat anemia in CKD (2012) Kidney. Int Suppl (2011) 2(4):292–298. https://doi.org/10.1038/kisup.2012.34

Van Doren L, Steinheiser M, Boykin K, Taylor KJ, Menendez M, Auerbach M (2024) Expert consensus guidelines: intravenous iron uses, formulations, administration, and management of reactions. Am J Hematol 99(7):1338–1348. https://doi.org/10.1002/ajh.27220

Van Remoortel H, Laermans J, Avau B, Bekkering G, Georgsen J, Manzini PM, Meybohm P, Ozier Y, De Buck E, Compernolle V, Vandekerckhove P (2021) Effectiveness of iron supplementation with or without erythropoiesis-stimulating agents on red blood cell utilization in patients with preoperative anaemia undergoing elective surgery: a systematic review and meta-analysis. Transfus Med Rev 35(2):103–124. https://doi.org/10.1016/j.tmrv.2021.03.004

Vera-Aviles M, Kabir SN, Shah A, Polzella P, Lim DY, Buckley P, Ball C, Swinkels D, Matlung H, Blans C, Holdship P, Nugent J, Anderson E, Desborough M, Piechnik S, Ferreira V, Lakhal-Littleton S (2024) Intravenous iron therapy results in rapid and sustained rise in myocardial iron content through a novel pathway. Eur Heart J. https://doi.org/10.1093/eurheartj/ehae359

Wienk KJ, Marx JJ, Beynen AC (1999) The concept of iron bioavailability and its assessment. Eur J Nutr 38(2):51–75. https://doi.org/10.1007/s003940050046

Wittenmeier E, Piekarski F, Steinbicker AU (2024) Blood product transfusions for children in the perioperative period and for critically ill children. Dtsch Arztebl

Int 121(2):58–65. https://doi.org/10.3238/arztebl.m2023.0243
Wolf M, Rubin J, Achebe M, Econs MJ, Peacock M, Imel EA, Thomsen LL, Carpenter TO, Weber T, Brandenburg V, Zoller H (2020) Effects of iron isomaltoside vs ferric carboxymaltose on hypophosphatemia in iron-deficiency anemia: two randomized clinical trials. JAMA 323(5):432–443. https://doi.org/10.1001/jama.2019.22450
Womack R, Berru F, Panwar B, Gutierrez OM (2020) Effect of ferric citrate versus ferrous sulfate on iron and phosphate parameters in patients with iron deficiency and CKD: a randomized trial. Clin J Am Soc Nephrol 15(9):1251–1258. https://doi.org/10.2215/CJN.15291219
Wozniak H, Dos Santos Rocha A, Beckmann TS, Larpin C, Buetti N, Quintard H, Pugin J, Heidegger CP (2022) Hypophosphatemia on ICU admission is associated with an increased length of stay in the ICU and time under mechanical ventilation. J Clin Med 11(3). https://doi.org/10.3390/jcm11030581
Zoller H, Wolf M, Blumenstein I, Primas C, Lindgren S, Thomsen LL, Reinisch W, Iqbal T (2023) Hypophosphataemia following ferric derisomaltose and ferric carboxymaltose in patients with iron deficiency anaemia due to inflammatory bowel disease (PHOSPHARE-IBD): a randomised clinical trial. Gut 72(4):644–653. https://doi.org/10.1136/gutjnl-2022-327897

Diagnostics: Markers of Body Iron Status

25

Domenico Girelli, Giacomo Marchi, Fabiana Busti, Fabio Chesini, and Annalisa Castagna

Abstract

Iron is a micronutrient essential to nearly all living organisms. It plays a pleiotropic role in many vital metabolic processes beyond the classical function in hemoglobin synthesis. Both iron deficiency and iron overload are detrimental in humans, as iron excess can favor the generation of toxic oxygen radicals and cell death by ferroptosis. Therefore, an accurate assessment of iron status is paramount in many clinical settings. The classical biochemical parameters represented by serum ferritin and transferrin saturation (TSAT) are sufficiently informative in simple cases but do not work well whenever overt or subclinical inflammation is present, including acute diseases and many common chronic conditions (e.g., heart failure, dysmetabolic disorders, and chronic kidney disease). In these cases, they are neither sensitive nor specific enough to capture both iron deficiency and excess with sufficient accuracy. A typical controversy is represented by the ferritin thresholds to diagnose iron deficiency in different inflammatory disorders, which vary consistently in guidelines for specific diseases. To obtain meaningful information, clinicians can rely on the integration of additional indicators, including erythrocyte indices from automated analyzers, soluble transferrin receptor (sTfR), and hepcidin. The latter, however, are not universally available and often lack robust standardization. A further problem is represented by the difficulties in measuring non-transferrin-bound iron (NTBI), which plays a pivotal pathophysiological role in iron overload disorders. Herein, we provide a concise narrative overview of the available laboratory tools for the assessment of iron status, highlighting the current challenges and exploring the possible future solutions.

Keywords

Iron · Anemia · Iron deficiency anemia · Ferritin · Hepcidin

25.1 Background

Iron is known to be essential to nearly all living organisms, as it is required for many crucial metabolic functions due to its unique oxidoreductive properties and the high propensity to exchange electrons ($Fe^{2+} \longleftrightarrow Fe^{3+}$); see Chap. 1. Indeed, 100 years ago, the Nobel Prize winner

D. Girelli (✉) · G. Marchi · F. Busti · F. Chesini
A. Castagna
Department of Medicine, University of Verona and EuroBloodNet Referral Center for Iron Disorders, Azienda Ospedaliera Universitaria Integrata Verona, Verona, Italy
e-mail: domenico.girelli@univr.it

K. Pantopoulos (ed.), *Iron Metabolism in Human Health and Disease*, Advances in Experimental Medicine and Biology 1480, https://doi.org/10.1007/978-3-031-92033-2_25

Otto Warburg stated that "*life without iron is impossible*" (Warburg 1925). The importance of iron extends far beyond its fundamental role in oxygen transport through hemoglobin, as an array of ferroproteins and enzymes requires the metal for vital biochemical functions, including cell respiration, DNA synthesis, and many others (for a recent comprehensive review, see Teh et al. (2024)). This explains the multiple systemic consequences of iron deficiency (ID) that can appear independently and before classical microcytic and hypochromic anemia (Chap. 12). As a matter of fact, mounting evidence indicates that ID is a disease by itself, long before the late appearance of anemia (Pasricha et al. 2021). In humans, iron excess can also be detrimental because of the potential of iron to generate toxic oxygen radicals through the classical Fenton's reaction. For these reasons, body iron content needs strict regulation (Galy et al. 2024; Nemeth and Ganz 2023), and an accurate assessment of iron status is of paramount importance in clinical medicine (Ginzburg 2019). The classical iron parameters represented by serum ferritin and transferrin saturation generally work well but are subjected to numerous influences and potential confounders that can make a proper evaluation challenging. They include erythropoietic activity (Ginzburg et al. 2023), hypoxia (Lakhal-Littleton and Robbins 2017), liver function, and, in particular, inflammation, which markedly affects body iron homeostasis (Fig. 25.1) (Ganz 2019). From an evolutionary perspective, the iron-inflammatory connection reflects a conserved mechanism to subtract iron from invading pathogens during infections as a part of innate host defense (Ganz and Nemeth 2024); see also Chap. 14. This is orchestrated by hepcidin, the master regulator of iron metabolism that is chemically similar to defensins, a family of antimicrobial peptides of innate immunity (Ganz 2003). For example, during acute infection, hypoferremia and hyperferritinemia do not reflect ID or iron overload, respectively, but merely denote iron trapping into macrophages, irrespective of true body iron stores. This phenomenon is replicated in non-infectious hyperinflammatory acute disorders like adult-onset Still's disease and other macrophage activation syndromes (Gansner and Berliner 2018) and, at an attenuated level, in several conditions characterized by subclinical inflammation (Camaschella and Girelli 2020). The latter include common disorders, especially in the elderly population, like chronic heart failure, chronic kidney disease, and many others. Moreover, aging per se is accompanied by a low-grade sterile chronic inflammation designed as *inflammaging* (Franceschi et al. 2000), which is thought to reflect heterogeneous mechanisms related to the aging of the immune system (Villaume and Savona 2024), and also, more recently, to clonal hematopoiesis of indeterminate potential (CHIP) (Girelli and Busti 2019; Walsh 2024).

In light of global epidemiological shifts, characterized by aging populations with increasing multimorbidity (Camaschella and Girelli 2020; Girelli et al. 2018), there is an urgent need to reassess the diagnostic thresholds of classical iron parameters and to incorporate novel indicators for a more accurate evaluation of individuals' iron status.

This chapter reviews classical and other potentially useful tools to assess iron status from a clinical perspective, with a focus on circulating iron markers. However, it has to be considered that iron status in the clinic is increasingly assessed by using magnetic resonance (MR)-based methods, which are grounded on the disruptions in the magnetic field induced by ferromagnetic iron (Feng et al. 2024). They are particularly useful when an iron overload condition is suspected (Girelli et al. 2022) and promise to be further improved by implementing artificial intelligence algorithms (Nashwan et al. 2023). A detailed review of MR-based assessment of body iron is beyond the scope of this review and can be found elsewhere (Feng et al. 2024; Guglielmo et al. 2023).

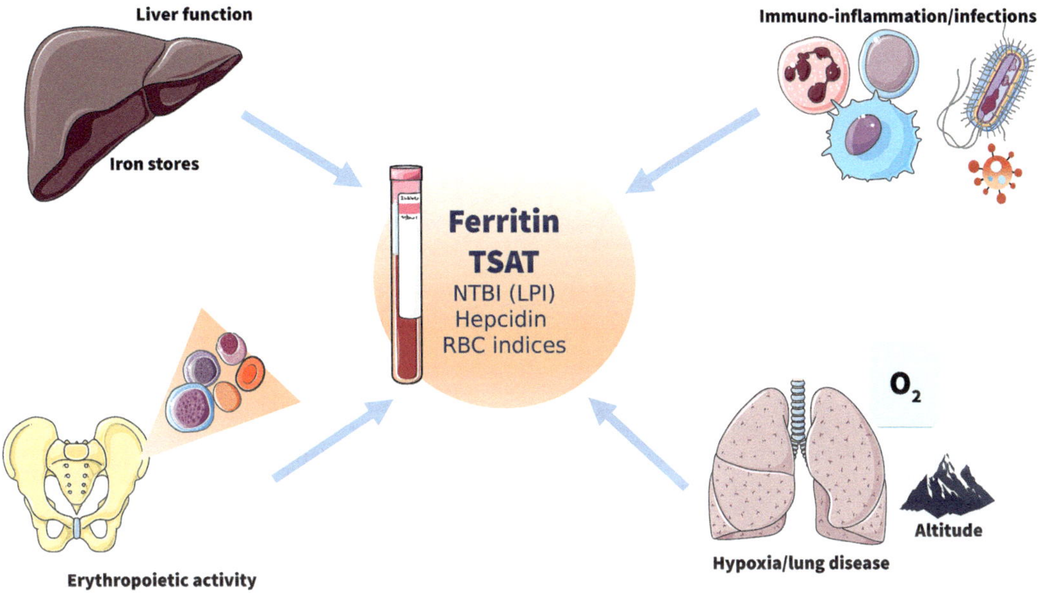

Fig. 25.1 The figure illustrates factors influencing the levels of iron biomarkers, such as ferritin, transferrin saturation (TSAT), non-transferrin-bound iron (NTBI) and labile plasma iron (LPI), hepcidin, and red blood cell (RBC) indices. These factors include liver function and iron stores, immune-inflammation and infections, erythropoietic activity in the bone marrow, and hypoxia

25.2 The Classical Iron Parameters, Ferritin and Transferrin Saturation

25.2.1 Ferritin

Ferritin is a ubiquitously expressed intracellular protein that is also present in small amounts (µg/L) in the circulation. It is a 24-subunit heteropolymer composed of H (heavy) and L (light) chains assembled to constitute a shell where iron is stored in an insoluble form (Chap. 2). Within the cells, ferritin protects against the potentially harmful effects of free iron through the generation of reactive oxygen species (ROS) (Arosio et al. 2017). The H-subunits possess ferroxidase activity to convert Fe^{2+} to Fe^{3+} and promote Fe^{3+} incorporation within the shell, while the L-subunits contain acidic residues crucial for the nucleation of Fe^{3+} within the core. The H:L ratio varies according to tissues and pathophysiological conditions (Chiou and Connor 2018). For example, liver ferritin contains more L-subunits than cardiac ferritin, and the proportion of L-subunits is further increased in iron overload (IO). Circulating ferritin is typically iron-poor, predominantly constituted by L-subunits, and is nearly 50% glycosylated (Cragg et al. 1981). An increase in the H-subunit proportion of circulating ferritin has been proposed to occur during inflammatory disorders (Kernan and Carcillo 2017). Indeed, ferritin is a classical acute-phase reactant, being positively regulated by several proinflammatory cytokines, particularly interleukin-6 (IL-6), IL-1α, and TNF-α (Arosio et al. 2017). In the absence of inflammation, serum ferritin is a good marker of body iron stores. Soon after its introduction as a clinical test in the 1970s (Addison et al. 1972), it was estimated that 1 µg/L of ferritin corresponds to nearly 8 mg of stored iron (Walters et al. 1973). Thus, a serum ferritin value of 100 µg/L should correspond to approximately 800 mg of iron stored in the liver, which reflects the normal condition in adult men (estimated to have ~1000 mg of iron stores, while in adult women, iron stores are ~300 mg) (Lakhal-Littleton and Cleland 2024). The origin of serum ferritin is still not completely under-

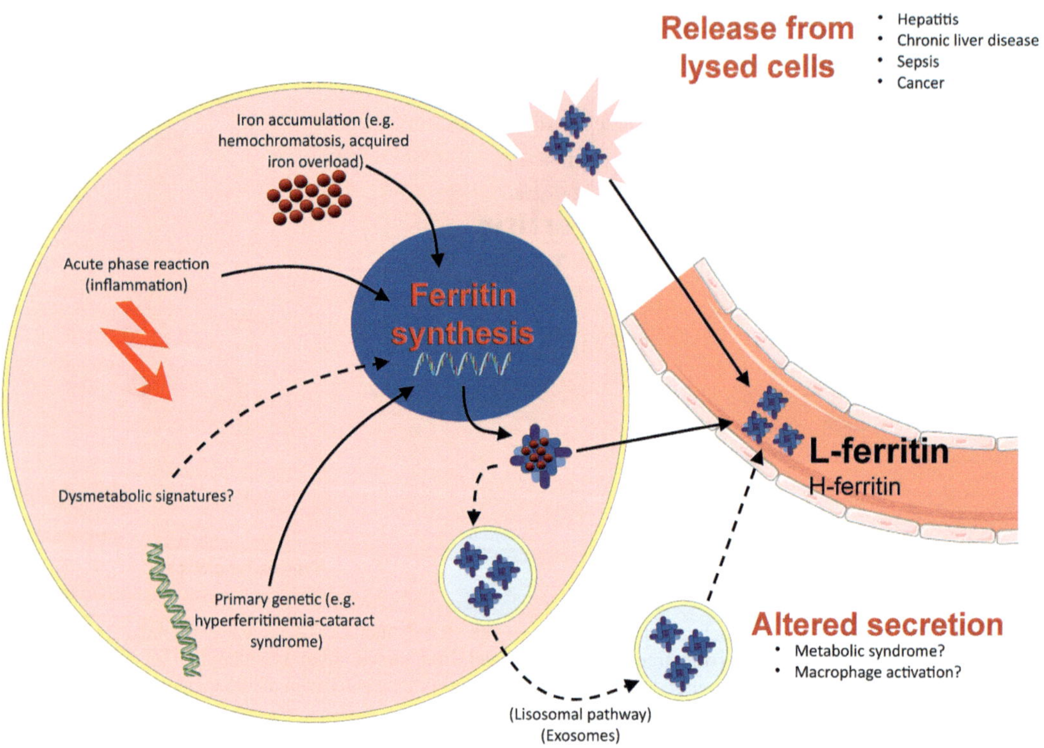

Fig. 25.2 The figure illustrates different mechanisms that can increase serum ferritin levels by inducing its synthesis or release from cells

stood, but current data suggest that it primarily derives from secretion by macrophages through the lysosomal secretory pathway (Cohen et al. 2010) or via the multivesicular body-exosome route (Truman-Rosentsvit et al. 2018; Gehrer et al. 2023). Nevertheless, several mechanisms beyond inflammation can cause an increase in serum ferritin levels, including leakage from damaged hepatocytes in acute and chronic liver diseases (Fig. 25.2). The biological meaning of serum ferritin is largely unknown, although, in recent years, some evidence suggests that it may serve as a cellular mediator, especially in inflammatory conditions where it could exert immunomodulatory properties (Chiou and Connor 2018; Ruscitti et al. 2020). The mechanisms of ferritin uptake by cells are also still controversial. While some studies have demonstrated clathrin-dependent endocytosis of ferritin, there is not yet a consensus on the specific receptor mediating the uptake. H- and L-subunits have been proposed to bind to different receptors, such as Scara5 (L subunit) or Tim-1, CXCR4, and soluble transferrin receptor 1 (H-subunit) (Galy et al. 2024).

Serum ferritin is a commonly requested laboratory test in primary care settings (Jager et al. 2024). The measurement of serum ferritin is considered the mainstay of the diagnosis of ID, although the threshold(s) vary consistently according to different guidelines and clinical conditions. Ideally, any threshold should be based on the classical "gold standard" for ID, that is, the absence of stainable iron in the bone marrow biopsy/aspirate. However, this is an unacceptably invasive procedure no longer used for comparative studies. In the absence of inflammation, a recent World Health Organization (WHO) statement suggests a threshold of 15 μg/L, which is increased to 70 μg/L when an inflammatory condition coexists (WHO 2023). The WHO expert panel has also proposed the use of equations based on inflammatory biomarkers like C-reactive protein (CRP) and/or α-1 acid glycoprotein

(AGP) levels as correction factors to adjust the ferritin cutoff for ID during inflammation. From a practical standpoint, there is no doubt that when ferritin is <15 μg/L, ID is present at a level in which anemia is incipient, if not already apparent. However, anemia is known to be a late consequence of ID, and a too-stringent ferritin cutoff, such as 15 μg/L, does not fully capture the disease spectrum of non-anemic iron deficiency (Pratt and Khan 2016; Al-Naseem et al. 2021). Based on a few studies that used bone marrow iron staining, a ferritin value ≤30 μg/L appears to combine a sufficiently high sensitivity (92%) with a still optimal specificity (98%) for ID (Guyatt et al. 1992; Ning and Zeller 2019). This cutoff is suggested in the recently published guidelines by the European Hematology Association (Iolascon et al. 2024). An elegant study by Galetti and co-workers established that the inflection point at which hepcidin levels begin to be suppressed to feedback upregulate iron absorption corresponds to a ferritin level of nearly 50 μg/L (Galetti et al. 2021). Based on the recent advance in the knowledge of the pathophysiology of iron homeostasis (Nemeth and Ganz 2023), this ferritin level should represent the earliest stage of an ID status. In line with this view, both the British Society of Gastroenterology and the American Gastroenterology Association guidelines set the ferritin level diagnostic for ID at 45 μg/L (Snook et al. 2021; Ko et al. 2020).

Another criticism of using too-stringent criteria for defining iron deficiency is that ferritin reference intervals have traditionally been based on population studies that included ID-prone populations, such as young women and/or blood donors (Spencer et al. 2019; Weyand et al. 2023). This lowered the "normal" threshold, especially in young women, potentially creating a gender bias that may prevent a substantial fraction of women from being properly treated (Weyand et al. 2022).

Defining a ferritin cutoff for ID during inflammation, especially in common chronic disorders, is even more difficult. A systematic meta-analysis of 38 studies including >1000 participants reporting data on serum ferritin levels and bone marrow iron staining in individuals with different disorders found a ferritin mean value of 82.4 μg/L (range 34–159 μg/L) in patients with documented iron depletion (Garcia-Casal et al. 2018).

Actually, most guidelines relying on expert consensus set up a ferritin cutoff of 100 μg/L for detecting ID in heart failure (McDonagh et al. 2021), inflammatory bowel diseases (Gordon et al. 2024), and cancer (Aapro et al. 2018a). A peculiar condition is represented by chronic kidney disease (CKD), where the ferritin threshold for iron therapy is up to <500 μg/L (Hain et al. 2023).

Table 25.1 summarizes the ferritin thresholds for defining ID and/or the indication of iron supplementation according to different diseases proposed by several International Scientific Societies.

Regarding iron overload, the upper normal limit of ferritin in adults is commonly considered 200 μg/L in women and 300 μg/L in men (Girelli et al. 2022), although some authors have proposed higher values in both genders, especially in elderly people (McKinnon et al. 2014). It's important to recognize that serum ferritin can increase because of multiple mechanisms unrelated to iron status (Fig. 25.2). Indeed, less than 10% of patients referred for hyperferritinemia have an IO disorder, such as hemochromatosis (Adams and Barton 2011). Metabolic hyperferritinemia (Valenti et al. 2023) is by far the most common cause in daily practice, which, at variance with hemochromatosis, is characterized by normal TSAT. A practical approach to patients referring for hyperferritinemia is discussed in detail elsewhere (Girelli et al. 2024).

25.2.2 Transferrin Saturation

Transferrin is a bilobal glycoprotein containing two binding sites with a high affinity for ferric ions (Fe^{3+}), which serve to transport iron in circulation. At variance with ferritin, it is a negative acute-phase reactant, being downregulated during inflammation (Mantovani and Garlanda 2023). The measurement of serum iron shows substantial diurnal fluctuations, which prevents its feasibility as a single test. On the other hand,

Table 25.1 Ferritin thresholds for the definition of iron deficiency (ID) or the indication of iron treatment according to international guidelines and recommendations

International Society	Ferritin threshold (μg/L)	Patients' category
World Health Organization (https://www.who.int/publications/i/item/9789240000124)	<15	Adults
	<70	Adults with infection or inflammation
European Hematology Association (Iolascon et al. 2024)	<30	Adults
	<100	Elderly and/or inflammatory disorders
British Society for Haematology (Fletcher et al. 2022)	<15–30	Adults
British Society of Gastroenterology (Snook et al. 2021)	<45	Adults
American Gastroenterology Association (Ko et al. 2020)	<45	Adults
British Columbia (2023) (https://www2.gov.bc.ca/gov/content/health/practitioner-professional-resources/bc-guidelines/iron-deficiency)	<15–30	Adults
European Society of Cardiology (McDonagh et al. 2021)	<100 or 100–299 with TSAT <20%	CHF
American Heart Association (Heidenreich et al. 2022)	<100 or 100–299 with TSAT <20%	CHF
European Society of Medical Oncology (Aapro et al. 2018b)	<100 or 500–800 with TSAT <40%	Cancer ESA-unresponsive patients receiving ChT
National Comprehensive Cancer Network (https://www.nccn.org/guidelines/guidelines-detail?category=1&id=1422)	<500-800 with TSAT <50%	Cancer patients receiving ChT
European Crohn's and Colitis Organization (ECCO) (Gordon et al. 2024)	<100	Active IBD
Kidney Disease: Improving Global Outcomes (KDIGO) (2012 and 2025 draft) (https://kdigo.org/wp-content/uploads/2016/10/KDIGO-2012-Anemia-Guideline-English.pdf; https://kdigo.org/wp-content/uploads/2024/11/KDIGO-2025-Anemia-in-CKD-Guideline_Public-Review-Draft_Nov42024.pdf)	≤500 with TSAT ≤30%	CKD treated with hemodialysis
	<100 with TSAT <40% or 100–299 with TSAT <25%	CKD not treated with hemodialysis

CHF chronic heart failure, *ChT* chemotherapy, *CKD* chronic kidney disease, *IBD* Inflammatory bowel disorders, *TSAT* transferrin saturation, *ESA* erythropoiesis-stimulating agents

its incorporation within the transferrin saturation (TSAT) calculation makes this parameter very useful for clinical purposes. TSAT reflects the fraction of all available transferrin binding sites occupied by iron ions. It can be calculated through the ratio of serum iron concentration (mmol/L) to total iron binding capacity (TIBC; mmol/L). Alternatively, if the concentration of the transferrin protein is measured, TSAT results from the ratio between serum iron and transferrin multiplied by the correction factor 1.42 (Girelli et al. 2022). Increased TSAT (>45%) is considered the hallmark of hemochromatosis, reflecting the expansion of the circulating iron pool due to hyperabsorption related to hepcidin insufficiency (Girelli et al. 2022). On the other hand, TSAT measures iron available for erythropoiesis, which becomes iron-restricted for values <15–20%. The role of TSAT in the diagnosis of ID has been extensively reviewed elsewhere (Cacoub et al. 2019). The currently accepted threshold of TSAT is less than 20%, regardless of the presence of inflammation (Cappellini et al. 2017).

25.3 Iron Deficiency in Heart Failure as an Example of the Current Limitations in the Laboratory Diagnosis of ID in Chronic Inflammatory Disorders

Chronic heart failure (CHF) can be considered a paradigmatic common condition characterized by subclinical inflammation (Braunwald 2013) and a high risk of developing ID, which negatively impacts disease outcomes (Lakhal-Littleton and Cleland 2024); see also Chap. 15. The European Society of Cardiology (ESC) guidelines define ID as a ferritin level below 100 μg/L or a ferritin level between 100 and 299 μg/L combined with TSAT below 20% (McDonagh et al. 2021). Such a definition is mainly based on initial consensus among experts and subsequent use as inclusion criteria in early intervention studies, demonstrating the efficacy of iron repletion strategies in improving symptoms and functional capacity in patients with CHF (Cheema et al. 2024). However, this definition lacks a robust validation against the gold standard of bone marrow iron staining. Noteworthy, recently published large trials have not confirmed the consistent benefit of iron supplementation based on the guideline definition, particularly when hard cardiovascular outcomes were considered. An individual patient data meta-analysis has revealed that greater benefits in terms of reduced hospitalizations and deaths are observed in patients with TSAT below 20% before iron supplementation (Ponikowski et al. 2023). This analysis also indicated that serum ferritin levels above or below 100 μg/L were not helpful in identifying either the response to treatment or potential harm. For this reason, several experts have recently suggested reconsidering the guideline definition of ID in CHF. Accordingly, TSAT <20% now looks like the most reliable single biomarker for diagnosing clinically relevant ID in CHF (Graham et al. 2024).

25.4 Additional Iron Biomarkers

25.4.1 NTBI and LPI

In certain IO conditions like hemochromatosis, iron-loading anemias, and myelodysplastic syndromes, non-transferrin-bound iron (NTBI) becomes detectable in the circulation and is believed to play a pivotal role in driving tissue damage (Chap. 4). NTBI is promptly taken up by hepatocytes, cardiomyocytes, and endocrine cells, which are rich in potential NTBI transporters like DMT1, zinc transporters 8 and 14, and L-type calcium channels (Knutson 2019). Cells are eventually damaged through oxidant stress and/or ferroptosis (Stockwell 2022). NTBI can also induce the proinflammatory activation of the vascular endothelium, thus contributing to atherosclerosis (Vinchi et al. 2020; Vinchi 2021). Only a small proportion of cellular iron, typically <5%, constitutes the so-called "labile iron pool" (LIP), which refers to NTBI species that can readily engage in redox cycling. NTBI and LPI reflect extracellular and intracellular labile iron species, respectively. Due to their pathophysiological role, reliable measurements of these iron species would be highly useful for clinicians and researchers. Although the precise chemical nature of heterogenous NTBI forms is not yet fully understood, the majority is likely represented by citrate-bound Fe^{3+} or albumin-bound Fe^{2+} or Fe^{3+} (Cabantchik and Hershko 2022). NTBI is generally detected in the 1–10 μM range when TSAT reaches values greater than 50%, at least in hemochromatosis (Ryan et al. 2022). Measuring NTBI is complex and can be done through a variety of analytical approaches that are difficult to standardize (Jacobs et al. 2005). The most used assay involves mobilizing NTBI with a chelating agent, typically nitrilotriacetic acid (NTA), under conditions that minimize transferrin-bound iron (TBI) mobilization (Silva and Rangel 2022). The resulting Fe-NTA complex is then separated from TBI using ultrafiltration, and iron is quantified in the ultrafiltrate. Alternative methods, such as the bleomycin assay and the LPI assay, exploit the reactive properties of iron and rely on NTBI reduction (Silva and Rangel 2022). Two international round-robin studies have been conducted in which samples were distributed among multiple laboratories for the measurement of NTBI (Jacobs et al. 2005; de Swart et al. 2016). They found significant variability across different assays and even between laboratories using the same method. Further studies are needed to establish NTBI as a valuable biochemical biomarker for the clinical management of IO and related conditions.

25.4.2 Soluble Transferrin Receptor

The transferrin receptor is highly expressed in rapidly dividing cells, particularly in erythroid precursors (Beguin 2003; Ponka and Lok 1999). In ID, such cells further overexpress and secrete a measurable soluble form (sTfR). Being scarcely influenced by inflammation, the sTFR ratio to log-transformed ferritin has been proposed as a valuable tool to detect ID in the context of inflammation (Weiss 2015). Moreover, a pilot study has suggested that the sTfR to hepcidin ratio may represent an early marker of ID since it begins to correlate significantly with ferritin at a threshold level below 50 μg/L (Tarancon-Diez et al. 2022) and can predict the response to IV iron (Tarancon-Diez et al. 2023). However, the lack of sufficient standardization of different sTfR assays still limits its widespread application in clinical practice (Bohn et al. 2024).

25.4.3 Hepcidin

Hepcidin is a major regulator of iron homeostasis and plays a pathophysiological role in many human disorders (Nemeth and Ganz 2023); see also Chap. 3. It is a small peptide (25 amino acids) that finely regulates circulating iron levels by controlling cellular iron export into the plasma through ferroportin; it is highly expressed by duodenal enterocytes and splenic reticuloendothelial macrophages, hence modulating dietary iron absorption and erythrocyte iron recycling, respectively (Galy et al. 2024; Nemeth and Ganz 2023). Circulating hepcidin is primarily produced by the liver hepatocytes, while many other cell types, including adipocytes (Chirumbolo et al. 2015), cardiomyocytes (Lakhal-Littleton and Cleland 2024; Lakhal-Littleton et al. 2016), and skin keratinocytes (Abboud et al. 2024), proportionally control cellular iron levels through local hepcidin production (Galy et al. 2024). Hepcidin is a hormone that influences iron levels in the circulation in a way similar to the effects of insulin on glycemia (Pietrangelo 2007). As such, and like every other hormone, serum hepcidin levels are influenced by a variety of physiological and pathological signals beyond iron status, including, among others, erythropoietic activity, inflammation, alcohol intake, liver diseases, and sexual hormones (Girelli et al. 2016). Mass spectrometry-based or immunological assays can measure serum hepcidin with a consistent degree of accuracy, although full harmonization is still lacking (van der Vorm et al. 2016).

Serum hepcidin levels are typically suppressed in ID to maximize iron absorption and mobilization to support erythropoiesis. Paradoxically, increased serum hepcidin levels can be virtually diagnostic in the rare iron-refractory iron deficiency anemia (IRIDA), due to mutations in the *TMPRSS6* gene encoding an upstream hepcidin suppressor (De Falco et al. 2010). Another rare condition in which measuring hepcidin could be particularly useful is represented by dominant hemochromatosis due to the gain-of-function mutations in the *SLC40A1* gene encoding ferroportin (Girelli et al. 2016, 2022). Nevertheless, the most promising application of hepcidin measurement is the diagnosis of absolute ID during inflammation (Girelli et al. 2016; Han and Wang 2021). While hepcidin synthesis is stimulated by inflammation, both in vitro and in vivo studies have demonstrated that the opposing suppression by ID prevails (Sangkhae and Nemeth 2019; Stoffel et al. 2019). This has been repeatedly demonstrated in inflammatory bowel diseases, where low hepcidin levels can distinguish absolute ID in the context of systemic inflammation

(Bergamaschi et al. 2013; Martinelli et al. 2016; Loveikyte et al. 2023; Ferrari et al. 2024). Similar results have been found in cancer-related anemia, where low hepcidin may be useful in predicting the response to intravenous iron (Bozzini et al. 2024; Steensma et al. 2015). In a study focusing on ID in critically ill patients who recovered after a prolonged stay in the Intensive Care Unit (ICU), hepcidin-driven treatment was associated with improved 1-year survival (Lasocki et al. 2021). While hepcidin assays are currently available primarily in specialized centers, their wider adoption in the future could greatly enhance the laboratory diagnosis of ID, at least in selected cases.

25.4.4 CHr and %HYPO

Modern automatic flow cytometry cell analyzers can assess some cheap but frequently overlooked parameters of the complete blood count, like reticulocyte hemoglobin content (CHr) and percentage of hypochromic red cells (%HYPO). Reticulocytes have a normal life span of 1–2 days in circulation. Therefore, CHr is an early indicator of iron-restricted erythropoiesis, less influenced by inflammation compared to other markers of iron status. Precisely, low CHr (<28 pg) provides an indirect measure of the functional iron available for new red blood cell production over the previous 3–4 days (Ogawa et al. 2020). It is also useful as an early measure of the response to iron therapy, helping to monitor intravenous iron supplementation (Ogawa et al. 2020). %HYPO is a red blood cell parameter and a strong predictor of ID. Increased %HYPO (>5%) accurately detects insufficient bone marrow iron supply during the previous 2–3 months (Amir et al. 2019), and it is particularly used to detect ID in patients with CKD undergoing dialysis and treatment with erythropoiesis-stimulating agents (Tessitore et al. 2001). However, its measured values can be influenced by preanalytical bias, i.e., the time between sample collection and analysis. Indeed, erythrocytes in tubes stored at room temperature tend to swell, increasing progressively HYPO%.

25.5 Conclusions

With the notable exception of ferritin levels <15–30 μg/L, which undoubtedly identify absolute ID, the laboratory diagnosis of disorders of iron metabolism is often complex and requires an integrated approach using different parameters. In particular, the interference from inflammation makes the diagnosis of ID in patients with chronic disorders and/or multimorbidity particularly challenging. This represents an important unmet clinical need, with uncertainties in the identification of the patients who may significantly benefit from iron supplementation. A thoughtful use of biomarkers in addition to classical ferritin and TSAT can be helpful, but more research is needed. Whether upcoming artificial intelligence-based approaches (Nashwan et al. 2023) may further aid in the future remains to be determined.

References

Aapro M, Beguin Y, Bokemeyer C, Dicato M, Gascon P et al (2018a) Management of anaemia and iron deficiency in patients with cancer: ESMO Clinical Practice Guidelines. Ann Oncol 29:iv96–iv110

Aapro M, Beguin Y, Bokemeyer C, Dicato M, Gascon P et al (2018b) Management of anaemia and iron deficiency in patients with cancer: ESMO Clinical Practice Guidelines. Ann Oncol 29:iv271

Abboud E, Chrayteh D, Boussetta N, Dalle H, Malerba M et al (2024) Skin hepcidin initiates psoriasiform skin inflammation via Fe-driven hyperproliferation and neutrophil recruitment. Nat Commun 15:6718

Adams PC, Barton JC (2011) A diagnostic approach to hyperferritinemia with a non-elevated transferrin saturation. J Hepatol 55:453–458

Addison GM, Beamish MR, Hales CN, Hodgkins M, Jacobs A, Llewellin P (1972) An immunoradiometric assay for ferritin in the serum of normal subjects and patients with iron deficiency and iron overload. J Clin Pathol 25:326–329

Al-Naseem A, Sallam A, Choudhury S, Thachil J (2021) Iron deficiency without anaemia: a diagnosis that matters. Clin Med (Lond) 21:107–113

Amir N, Md Noor S, Subbiah I, Osman M, Seman Z (2019) Percentage of hypochromic red cells as a potential screening test to evaluate iron status in blood donors. Int J Lab Hematol 41:418–423

Arosio P, Elia L, Poli M (2017) Ferritin, cellular iron storage and regulation. IUBMB Life 69:414–422

Beguin Y (2003) Soluble transferrin receptor for the evaluation of erythropoiesis and iron status. Clin Chim Acta 329:9–22
Bergamaschi G, Di Sabatino A, Albertini R, Costanzo F, Guerci M et al (2013) Serum hepcidin in inflammatory bowel diseases: biological and clinical significance. Inflamm Bowel Dis 19:2166–2172
Bohn MK, Berman M, Ali S, Cheng PL, Wang XY et al (2024) Evaluation of a new soluble transferrin receptor assay and comparison to three measurement procedures. Clin Biochem 135:110862
Bozzini C, Busti F, Marchi G, Vianello A, Cerchione C et al (2024) Anemia in patients receiving anticancer treatments: focus on novel therapeutic approaches. Front Oncol 14:1380358
Braunwald E (2013) Heart failure. JACC Heart Fail 1:1–20
Cabantchik IZ, Hershko C (2022) Plasma nontransferrin bound iron-nontransferrin bound iron revisited: implications for systemic iron overload and in iv iron supplementation. Am J Hematol 97:7–9
Cacoub P, Vandewalle C, Peoc'h K (2019) Using transferrin saturation as a diagnostic criterion for iron deficiency: a systematic review. Crit Rev Clin Lab Sci 56:526–532
Camaschella C, Girelli D (2020) The changing landscape of iron deficiency. Mol Asp Med 75:100861
Cappellini MD, Comin-Colet J, de Francisco A, Dignass A, Doehner W et al (2017) Iron deficiency across chronic inflammatory conditions: international expert opinion on definition, diagnosis, and management. Am J Hematol 92:1068–1078
Cheema B, Chokshi A, Orimoloye O, Ardehali H (2024) Intravenous iron repletion for patients with heart failure and iron deficiency: JACC state-of-the-art review. J Am Coll Cardiol 83:2674–2689
Chiou B, Connor JR (2018) Emerging and dynamic biomedical uses of ferritin. Pharmaceuticals (Basel) 11(4):124
Chirumbolo S, Rossi AP, Rizzatti V, Zoico E, Franceschetti G et al (2015) Iron primes 3T3-L1 adipocytes to a TLR4-mediated inflammatory response. Nutrition 31:1266–1274
Cohen LA, Gutierrez L, Weiss A, Leichtmann-Bardoogo Y, Zhang DL et al (2010) Serum ferritin is derived primarily from macrophages through a nonclassical secretory pathway. Blood 116:1574–1584
Cragg SJ, Wagstaff M, Worwood M (1981) Detection of a glycosylated subunit in human serum ferritin. Biochem J 199:565–571
De Falco L, Totaro F, Nai A, Pagani A, Girelli D et al (2010) Novel TMPRSS6 mutations associated with iron-refractory iron deficiency anemia (IRIDA). Hum Mutat 31:E1390–E1405
de Swart L, Hendriks JC, van der Vorm LN, Cabantchik ZI, Evans PJ et al (2016) Second international round robin for the quantification of serum non-transferrin-bound iron and labile plasma iron in patients with iron-overload disorders. Haematologica 101:38–45
Feng Q, Yi J, Li T, Liang B, Xu F, Peng P (2024) Narrative review of magnetic resonance imaging in quantifying liver iron load. Front Med (Lausanne) 11:1321513
Ferrari F, Carini M, Zanella I, Treglia G, Luglio G et al (2024) Potential diagnostic role of hepcidin in anemic patients affected by inflammatory bowel disease: a systematic review. Diagnostics (Basel) 14(4):375
Fletcher A, Forbes A, Svenson N, Wayne Thomas D (2022) Guideline for the laboratory diagnosis of iron deficiency in adults (excluding pregnancy) and children. Br J Haematol 196:523–529
Franceschi C, Bonafè M, Valensin S, Olivieri F, De Luca M et al (2000) Inflamm-aging. An evolutionary perspective on immunosenescence. Ann N Y Acad Sci 908:244–254
Galetti V, Stoffel NU, Sieber C, Zeder C, Moretti D, Zimmermann MB (2021) Threshold ferritin and hepcidin concentrations indicating early iron deficiency in young women based on upregulation of iron absorption. EClinicalMedicine 39:101052
Galy B, Conrad M, Muckenthaler M (2024) Mechanisms controlling cellular and systemic iron homeostasis. Nat Rev Mol Cell Biol 25:133–155
Gansner JM, Berliner N (2018) The rheumatology/hematology interface: CAPS and MAS diagnosis and management. Hematology 2018:313–317
Ganz T (2003) Defensins: antimicrobial peptides of innate immunity. Nat Rev Immunol 3:710–720
Ganz T (2019) Anemia of inflammation. N Engl J Med 381:1148–1157
Ganz T, Nemeth E (2024) Hypoferremia of inflammation: innate host defense against infections. Blood Cells Mol Dis 104:102777
Garcia-Casal MN, Pasricha SR, Martinez RX, Lopez-Perez L, Pena-Rosas JP (2018) Are current serum and plasma ferritin cut-offs for iron deficiency and overload accurate and reflecting iron status? A systematic review. Arch Med Res 49:405–417
Gehrer CM, Mitterstiller AM, Grubwieser P, Meyron-Holtz EG, Weiss G, Nairz M (2023) Advances in ferritin physiology and possible implications in bacterial infection. Int J Mol Sci 24(5):4659
Ginzburg YZ (2019) New diagnostic tools for delineating iron status. Hematology 2019:327–336
Ginzburg Y, An X, Rivella S, Goldfarb A (2023) Normal and dysregulated crosstalk between iron metabolism and erythropoiesis. elife 12:e90189
Girelli D, Busti F (2019) Anemia and adverse outcomes in the elderly: a detrimental inflammatory loop? Haematologica 104:417–419
Girelli D, Nemeth E, Swinkels DW (2016) Hepcidin in the diagnosis of iron disorders. Blood 127:2809–2813
Girelli D, Marchi G, Camaschella C (2018) Anemia in the elderly. Hemasphere 2:e40
Girelli D, Busti F, Brissot P, Cabantchik I, Muckenthaler MU, Porto G (2022) Hemochromatosis classification: update and recommendations by the BIOIRON Society. Blood 139:3018–3029
Girelli D, Marchi G, Busti F (2024) Diagnosis and management of hereditary hemochromatosis: life-

style modification, phlebotomy, and blood donation. Hematology 2024:434–442
Gordon H, Burisch J, Ellul P, Karmiris K, Katsanos K et al (2024) ECCO guidelines on extraintestinal manifestations in inflammatory bowel disease. J Crohns Colitis 18:1–37
Graham FJ, Guha K, Cleland JG, Kalra PR (2024) Treating iron deficiency in patients with heart failure: what, why, when, how, where and who. Heart 110:1201–1207
Guglielmo FF, Barr RG, Yokoo T, Ferraioli G, Lee JT et al (2023) Liver fibrosis, fat, and iron evaluation with MRI and fibrosis and fat evaluation with US: a practical guide for radiologists. Radiographics 43:e220181
Guyatt GH, Oxman AD, Ali M, Willan A, McIlroy W, Patterson C (1992) Laboratory diagnosis of iron-deficiency anemia: an overview. J Gen Intern Med 7:145–153
Hain D, Bednarski D, Cahill M, Dix A, Foote B et al (2023) Iron-deficiency anemia in CKD: a narrative review for the kidney care team. Kidney Med 5:100677
Han J, Wang K (2021) Clinical significance of serum hepcidin in the diagnosis and treatment of patients with anemia of chronic disease: a meta-analysis. Biomarkers 26:296–301
Heidenreich PA, Bozkurt B, Aguilar D, Allen LA, Byun JJ et al (2022) 2022 AHA/ACC/HFSA Guideline for the Management of Heart Failure: executive summary: a report of the American College of Cardiology/American Heart Association Joint Committee on Clinical Practice Guidelines. Circulation 145:e876–ee94
Iolascon A, Andolfo I, Russo R, Sanchez M, Busti F et al (2024) Recommendations for diagnosis, treatment, and prevention of iron deficiency and iron deficiency anemia. Hemasphere 8:e108
Jacobs EM, Hendriks JC, van Tits BL, Evans PJ, Breuer W et al (2005) Results of an international round robin for the quantification of serum non-transferrin-bound iron: need for defining standardization and a clinically relevant isoform. Anal Biochem 341:241–250
Jager L, Rachamin Y, Senn O, Burgstaller JM, Rosemann T, Markun S (2024) Ferritin cutoffs and diagnosis of iron deficiency in primary care. JAMA Netw Open 7:e2425692
Kernan KF, Carcillo JA (2017) Hyperferritinemia and inflammation. Int Immunol 29:401–409
Knutson MD (2019) Non-transferrin-bound iron transporters. Free Radic Biol Med 133:101–111
Ko CW, Siddique SM, Patel A, Harris A, Sultan S et al (2020) AGA clinical practice guidelines on the gastrointestinal evaluation of iron deficiency anemia. Gastroenterology 159:1085–1094
Lakhal-Littleton S, Cleland JGF (2024) Iron deficiency and supplementation in heart failure. Nat Rev Cardiol 21:463–486
Lakhal-Littleton S, Robbins PA (2017) The interplay between iron and oxygen homeostasis with a particular focus on the heart. J Appl Physiol (1985) 123:967–973
Lakhal-Littleton S, Wolna M, Chung YJ, Christian HC, Heather LC et al (2016) An essential cell-autonomous role for hepcidin in cardiac iron homeostasis. elife 5:e19804
Lasocki S, Asfar P, Jaber S, Ferrandiere M, Kerforne T et al (2021) Impact of treating iron deficiency, diagnosed according to hepcidin quantification, on outcomes after a prolonged ICU stay compared to standard care: a multicenter, randomized, single-blinded trial. Crit Care 25:62
Loveikyte R, Bourgonje AR, van der Reijden JJ, Bulthuis MLC, Hawinkels L et al (2023) Hepcidin and iron status in patients with inflammatory bowel disease undergoing induction therapy with vedolizumab or infliximab. Inflamm Bowel Dis 29:1272–1284
Mantovani A, Garlanda C (2023) Humoral innate immunity and acute-phase proteins. N Engl J Med 388:439–452
Martinelli M, Strisciuglio C, Alessandrella A, Rossi F, Auricchio R et al (2016) Serum hepcidin and iron absorption in paediatric inflammatory bowel disease. J Crohns Colitis 10(5):566–574
McDonagh TA, Metra M, Adamo M, Gardner RS, Baumbach A et al (2021) 2021 ESC Guidelines for the diagnosis and treatment of acute and chronic heart failure. Eur Heart J 42:3599–3726
McKinnon EJ, Rossi E, Beilby JP, Trinder D, Olynyk JK (2014) Factors that affect serum levels of ferritin in Australian adults and implications for follow-up. Clin Gastroenterol Hepatol 12:101–8.e4
Nashwan AJ, Alkhawaldeh IM, Shaheen N, Albalkhi I, Serag I et al (2023) Using artificial intelligence to improve body iron quantification: a scoping review. Blood Rev 62:101133
Nemeth E, Ganz T (2023) Hepcidin and iron in health and disease. Annu Rev Med 74:261–277
Ning S, Zeller MP (2019) Management of iron deficiency. Hematology 2019:315–322
Ogawa C, Tsuchiya K, Maeda K (2020) Reticulocyte hemoglobin content. Clin Chim Acta 504:138–145
Pasricha SR, Tye-Din J, Muckenthaler MU, Swinkels DW (2021) Iron deficiency. Lancet 397:233–248
Pietrangelo A (2007) Hemochromatosis: an endocrine liver disease. Hepatology 46:1291–1301
Ponikowski P, Mentz RJ, Hernandez AF, Butler J, Khan MS et al (2023) Efficacy of ferric carboxymaltose in heart failure with iron deficiency: an individual patient data meta-analysis. Eur Heart J 44:5077–5091
Ponka P, Lok CN (1999) The transferrin receptor: role in health and disease. Int J Biochem Cell Biol 31:1111–1137
Pratt JJ, Khan KS (2016) Non-anaemic iron deficiency—a disease looking for recognition of diagnosis: a systematic review. Eur J Haematol 96:618–628
Ruscitti P, Di Benedetto P, Berardicurti O, Panzera N, Grazia N et al (2020) Pro-inflammatory properties of H-ferritin on human macrophages, ex vivo and in vitro observations. Sci Rep 10:12232
Ryan E, Mulready K, Wiegerinck E, Russell J, Swinkels DW, Stewart S (2022) NTBI levels in C282Y homo-

zygotes after therapeutic phlebotomy. EJHaem 3:644–652
Sangkhae V, Nemeth E (2019) To induce or not to induce: the fight over hepcidin regulation. Haematologica 104:1093–1095
Silva AMN, Rangel M (2022) The (bio)chemistry of non-transferrin-bound iron. Molecules 27(6):1784
Snook J, Bhala N, Beales ILP, Cannings D, Kightley C et al (2021) British Society of Gastroenterology guidelines for the management of iron deficiency anaemia in adults. Gut 70:2030–2051
Spencer BR, Bialkowski W, Creel DV, Cable RG, Kiss JE et al (2019) Elevated risk for iron depletion in high-school age blood donors. Transfusion 59:1706–1716
Steensma DP, Sasu BJ, Sloan JA, Tomita DK, Loprinzi CL (2015) Serum hepcidin levels predict response to intravenous iron and darbepoetin in chemotherapy-associated anemia. Blood 125:3669–3671
Stockwell BR (2022) Ferroptosis turns 10: emerging mechanisms, physiological functions, and therapeutic applications. Cell 185:2401–2421
Stoffel NU, Lazrak M, Bellitir S, Mir NE, Hamdouchi AE et al (2019) The opposing effects of acute inflammation and iron deficiency anemia on serum hepcidin and iron absorption in young women. Haematologica 104:1143–1149
Tarancon-Diez L, Genebat M, Roman-Enry M, Vázquez-Alejo E, Espinar-Buitrago MS et al (2022) Threshold ferritin concentrations reflecting early iron deficiency based on hepcidin and soluble transferrin receptor serum levels in patients with absolute iron deficiency. Nutrients 14(22):4739
Tarancon-Diez L, Iriarte-Gahete M, Sanchez-Mingo P, Perez-Cabeza G, Romero-Candau F et al (2023) Real-world experience of intravenous iron sucrose supplementation and dynamics of soluble transferrin receptor and hepcidin in a Spanish cohort of absolute iron deficient patients. Biomed Pharmacother 167:115510
Teh MR, Armitage AE, Drakesmith H (2024) Why cells need iron: a compendium of iron utilisation. Trends Endocrinol Metab 35(12):1026–1049
Tessitore N, Solero GP, Lippi G, Bassi A, Faccini GB et al (2001) The role of iron status markers in predicting response to intravenous iron in haemodialysis patients on maintenance erythropoietin. Nephrol Dial Transplant 16:1416–1423
Truman-Rosentsvit M, Berenbaum D, Spektor L, Cohen LA, Belizowsky-Moshe S et al (2018) Ferritin is secreted via 2 distinct nonclassical vesicular pathways. Blood 131:342–352
Valenti L, Corradini E, Adams LA, Aigner E, Alqahtani S et al (2023) Consensus statement on the definition and classification of metabolic hyperferritinaemia. Nat Rev Endocrinol 19:299–310
van der Vorm LN, Hendriks JCM, Laarakkers CM, Klaver S, Armitage AE et al (2016) Toward worldwide hepcidin assay harmonization: identification of a commutable secondary reference material. Clin Chem 62:993–1001
Villaume MT, Savona MR (2024) Pathogenesis and inflammaging in myelodysplastic syndrome. Haematologica 110(2):283–299
Vinchi F (2021) Non-transferrin-bound iron in the spotlight: novel mechanistic insights into the vasculotoxic and atherosclerotic effect of iron. Antioxid Redox Signal 35:387–414
Vinchi F, Porto G, Simmelbauer A, Altamura S, Passos ST et al (2020) Atherosclerosis is aggravated by iron overload and ameliorated by dietary and pharmacological iron restriction. Eur Heart J 41:2681–2695
Walsh K (2024) The emergence of clonal hematopoiesis as a disease determinant. J Clin Invest 134(19):e180063
Walters GO, Miller FM, Worwood M (1973) Serum ferritin concentration and iron stores in normal subjects. J Clin Pathol 26:770–772
Warburg O (1925) Iron, the oxygen-carrier of respiration-ferment. Science 61:575–582
Weiss G (2015) Anemia of chronic disorders: new diagnostic tools and new treatment strategies. Semin Hematol 52:313–320
Weyand AC, McGann PT, Sholzberg M (2022) Sex specific definitions of anaemia contribute to health inequity and sociomedical injustice. Lancet Haematol 9:e6–e8
Weyand AC, Chaitoff A, Freed GL, Sholzberg M, Choi SW, McGann PT (2023) Prevalence of iron deficiency and iron-deficiency anemia in US females aged 12-21 years, 2003-2020. JAMA 329:2191–2193
WHO (2023) Use of ferritin concentrations to assess iron status in individuals and populations. https://www.who.int/tools/elena/interventions/ferritin-concentrations

Current Landscape of Hepcidin Therapeutics

26

Antonella Nai, Laura Silvestri, Michela Asperti, and Francesca Vinchi

Abstract

Hepcidin, the master regulator of systemic iron homeostasis, modulates the absorption of dietary iron as well as its release from stores by blocking and inducing the degradation of ferroportin, the only known cellular iron exporter. Thus, the interaction between hepcidin and ferroportin negatively regulates iron transport into plasma and reduces systemic iron availability. Several genetic and acquired iron-related disorders are characterized by dysregulated hepcidin expression. While hereditary hemochromatosis and β-thalassemia exhibit hepcidin suppression, causing iron overload, iron-refractory iron-deficiency anemia and anemia of chronic diseases show hepcidin induction, promoting iron deficiency. In the last decade, experimental evidence demonstrated that targeting the hepcidin-ferroportin axis and restoring normal hepcidin levels offer potential therapeutic benefits to patients with iron-related disorders. Due to these reasons, the development of hepcidin therapeutics has been exponentially growing in recent years, with the aim to design hepcidin modulators that either mimic or inhibit hepcidin action, leading to reduced or increased iron availability, respectively. This chapter summarizes the current landscape of hepcidin therapeutics that reached clinical development and the pathologies that could benefit from these pharmacological approaches. Moreover, it provides novel experimental insights into new potential therapies combining hepcidin modulators with existing or novel approaches such as luspatercept or TfR2 targeting as a way to achieve further improvement of anemia in pathologic conditions.

Authors "Michela Asperti" and "Francesca Vinchi" have equally contributed to this chapter.

A. Nai · L. Silvestri
Regulation of Iron Metabolism Unit, Division of Genetics and Cell Biology, IRCCS Ospedale San Raffaele, Milan, Italy

Vita-Salute San Raffaele University, Milan, Italy

M. Asperti
Department of Molecular and Translational Medicine, University of Brescia, Brescia, Italy

Iron Research Laboratory, Lindsley F. Kimball Research Institute, New York Blood Center, New York, NY, USA

F. Vinchi (✉)
Iron Research Laboratory, Lindsley F. Kimball Research Institute, New York Blood Center, New York, NY, USA

Department of Pathology and Laboratory Medicine, Weill Cornell Medicine, New York, NY, USA
e-mail: FVinchi@nybc.org

K. Pantopoulos (ed.), *Iron Metabolism in Human Health and Disease*, Advances in Experimental Medicine and Biology 1480, https://doi.org/10.1007/978-3-031-92033-2_26

Keywords

Hepcidin-ferroportin axis · Iron homeostasis · Iron-related disorders · Hepcidin agonists and antagonists · Therapeutic strategies for anemia and iron overload

26.1 Introduction

Biological organisms closely regulate intracellular and extracellular iron concentrations in order to provide adequate iron supply for critical cellular functions but, at the same time, prevent toxicities due to iron excess (Chap. 1). Systemic iron homeostasis (Chap. 3) is maintained by regulating intestinal iron absorption, the concentration of iron in blood plasma and extracellular fluid, the distribution of iron among tissues, and the amount of stored iron. Overall, iron balance is primarily affected by iron utilization by the erythron in the bone marrow, iron storage in hepatocytes and reticulo-endothelial macrophages, iron transport in the blood, and iron absorption by enterocytes. Importantly, the primary mechanism controlling the absorption and tissue distribution of iron is the hepcidin-ferroportin axis. The hepatic hormone hepcidin controls iron influx into plasma by binding to, occluding, and inducing the internalization and degradation of ferroportin, the sole known cellular exporter of elemental iron responsible for iron release from enterocytes and macrophages. The interaction between hepcidin and ferroportin prevents dietary iron absorption and the release of iron from macrophages, negatively regulating iron transport into plasma. Therefore, the modulation of hepcidin production is essential to control intestinal iron absorption and mobilization from stores to meet body iron needs. Stimuli including increased liver iron stores, circulating iron in the form of diferric transferrin, and inflammation induce hepcidin production, thus decreasing systemic iron levels. On the other hand, hypoxia and increased erythropoietic demand reduce hepcidin levels to increase iron availability for the production of red blood cells. Abnormal hepcidin regulation is responsible for iron disturbances in paradigmatic human disorders. Thus, in hereditary hemochromatosis and β-thalassemia (Chaps. 9, 10 and 11), hepcidin suppression promotes iron overload due to ferroportin stabilization, whereas in genetic diseases characterized by increased hepcidin synthesis, such as iron-refractory iron-deficiency anemia (IRIDA), or acquired conditions, such as anemia of chronic diseases or anemia of inflammation, hepcidin excess causes iron deficiency and consequent anemia (Chaps. 12 and 13). Experiments with preclinical animal models provided evidence that restoration of appropriate hepcidin levels can be leveraged for the treatment of these conditions. Targeting the hepcidin-ferroportin axis could offer therapeutic benefits to patients with iron-related disorders due to aberrant hepcidin expression. This notion has rapidly fueled the production of hepcidin therapeutics, including several hepcidin agonists and antagonists as well as inducers and inhibitors with various therapeutic indications. This chapter provides an update on the current landscape of hepcidin therapeutics, with a focus on molecules of recent development, drugs that have reached advanced clinical trial stage, and combinatorial approaches.

26.2 Therapeutic Indications for Hepcidin Mimetics and Inducers

Beyond hereditary hemochromatosis, several diseases could benefit from increased hepcidin levels and iron restriction (Fig. 26.1). Of particular relevance are iron-loading anemias, such as thalassemias and sickle cell disease (SCD), whereby ineffective erythropoiesis attenuates hepcidin expression through the production of its inhibitor, the erythroid hormone erythroferrone (Kautz et al. 2014, 2015; Arezes et al. 2018; Wang et al. 2020). This leads to the development of secondary iron overload, a life-threatening complication of the disease that further worsens the hematologic phenotype and overall disease pathophysiology (Ganz and Nemeth 2023). Experimental increase of hepcidin in mouse models of β-thalassemia proved not only to reduce organ

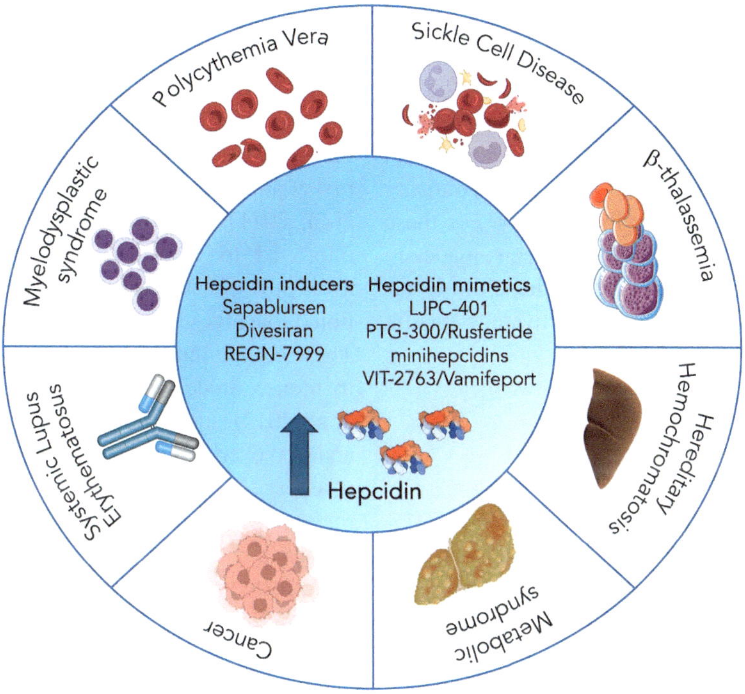

Fig. 26.1 Therapeutic indications for hepcidin mimetics and inducers. Diseases which may benefit from iron restriction and augmented hepcidin levels include SCD, PV, β-thalassemia, hereditary hemochromatosis, metabolic syndrome, cancer, systemic lupus erythematosus, and myelodysplastic syndromes. Both hepcidin inducers (Sapablursen, Divesiran, REGN-7999) and hepcidin mimetics (LJPC-401, PTG-300/Rusfertide, minihepcidins, VIT-2763/Vamifeport) reduce iron availability by increasing or mimicking hepcidin. (Figures were created using BioRender.com)

damage but also to improve ineffective erythropoiesis (Gardenghi et al. 2010). A recently identified target for this kind of therapeutic approach is polycythemia vera (PV), characterized by excessive red blood cell (RBC) production due to constitutive activation of the erythropoietin (EPO) signaling pathway. In this disorder, an increase of hepcidin may limit the availability of iron for erythropoiesis, thus correcting erythrocytosis (Patel et al. 2024).

Recent findings suggest that hepcidin modulation is likely therapeutic for non-erythroid diseases as well (Fig. 26.1). Iron overload is a common complication of myelodysplastic syndromes (MDS), hematopoietic neoplasms characterized by impaired hematopoiesis, dysplastic erythropoiesis, and propensity to acute myeloid leukemia evolution. Iron chelation has been shown to significantly prolong patients' overall survival (Rose et al. 2010; Neukirchen et al. 2012), suggesting that strategies aimed at increasing hepcidin levels may achieve similar benefits. Systemic lupus erythematosus is characterized by lupus nephritis, the end organ manifestation of the disease. Treatment with hepcidin decreased renal iron overload, inflammation, and immune infiltration in a mouse model of spontaneous lupus nephritis, without worsening anemia of inflammation (Scindia et al. 2020). In addition, diet-induced obesity and hepatic steatosis were reduced in mice characterized by high hepcidin levels because of deficiency of the inhibitor *Tmprss6* (Folgueras et al. 2018), raising the possibility that targeting the hepcidin-FPN axis might ameliorate metabolic disorders. Finally, hepcidin expression has been reported to be drastically reduced in samples of cancerous liver tissue compared to non-cancerous tissue, due to promoter hypermethylation (Kijima et al. 2008; Udali et al. 2018). Since low hepcidin levels are associated with greater proliferation, tumor growth, and metastasis of hepatocellular carcinoma (Shen et al. 2019), increasing hepcidin production/levels might represent a promising therapeutic approach for liver cancer. For these reasons, several efforts have been made to develop therapeutics that restore hepcidin levels (hepcidin mimetics) and/or activate hepcidin transcription (hepcidin inducers).

26.2.1 Hepcidin Mimetics

Two hepcidin mimetics, LJPC-401 and PTG-300/Rusfertide, have been developed in recent years and proved to be effective in controlling iron levels in preclinical models of hemochromatosis, β-thalassemia, and PV (Taranath et al. 2019, 2022), as well as in healthy volunteers, without severe side effects (Porter et al. 2018; Nicholls et al. 2018) (Fig. 26.2). However, LJPC, which mimics full-length hepcidin and whose administration causes frequent injection site adverse events, was discontinued after failing to meet both the primary and secondary endpoints of a Phase 2 clinical trial in β-thalassemia patients (NCT03381833). On the contrary, PTG-300/Rusfertide, which mimics only the N-terminal portion of hepcidin, effectively prevented further iron overload in hemochromatosis patients and caused iron redistribution to macrophages. The data suggest that PTG-300/Rusfertide may be useful during the initial phases of hemochromatosis treatment, as a supplement to the current standard of care therapy, phlebotomy (Kowdley et al. 2023). Although results of the TRANSCEND Phase 2 clinical trial in thalassemic patients are not yet available (NCT03802201), the drug reduced hematocrit levels and frequency of phlebotomies in PV patients, as reported in the Phase 2 PACIFIC and REVIVE trials (Kremyanskaya et al. 2024; Hoffman et al. 2022), demonstrating that hepcidin mimetics may represent a successful therapeutic for the control of erythrocytosis. The Phase 3 VERIFY trial is currently ongoing to evaluate the long-term efficacy of Rusfertide in PV (NCT05210790).

Other promising hepcidin mimetics are the so-called mini-hepcidins, short peptides designed on the bioactive hepcidin epitope encompassing the first 9 amino acids of the mature protein (Preza et al. 2011), whose structure has been recently improved to increase oral bioavailability and avoid side effects from subcutaneous administration (Goncalves Monteiro et al. 2021) (Fig. 26.2). These compounds effectively lowered iron levels in mouse models of hemochromatosis (Ramos et al. 2012), improved ineffective erythropoiesis and iron overload in β-thalassemia and SCD mice (Casu et al. 2016b, 2020a), and corrected erythrocytosis in PV models (Stetka et al. 2023). However, no clinical trial has been planned so far to evaluate their therapeutic potential in patients.

Another oral mimetic of hepcidin is VIT-2763/Vamifeport, a small molecule that competes with hepcidin for the binding to ferroportin (Lehmann et al. 2023) (Fig. 26.2). This drug had similar efficacy as mini-hepcidins in improving disease phenotype in mouse models of hemoglobinopathies (Manolova et al. 2019; Nyffenegger et al. 2021, 2022; Kalleda et al. 2023), hemochromatosis (Nyffenegger et al. 2024), and PV (Stetka et al. 2023). Recently, Vamifeport has been shown to not only reduce iron overload and ameliorate ineffective erythropoiesis and anemia but also to improve myeloid expansion and delay MDS-to-AML evolution in a mouse model of MDS (Antypiuk et al. 2025; Vinchi et al. 2020). In a Phase 1 clinical trial, the drug was shown to effectively lower serum iron levels and transferrin saturation in healthy volunteers, without severe adverse events. This was accompanied by a transient increase in serum hepcidin levels, which could help to reduce circulating iron (Richard et al. 2020). Unfortunately, the trial evaluating the efficacy and safety of the drug in β-thalassemia patients was terminated due to lack of efficacy in improving anemia (NCT04938635) (Pilo and Angelucci 2024), while the results of the Phase 2 ViSionSerenity trial in SCD patients are not yet available (NCT04817670).

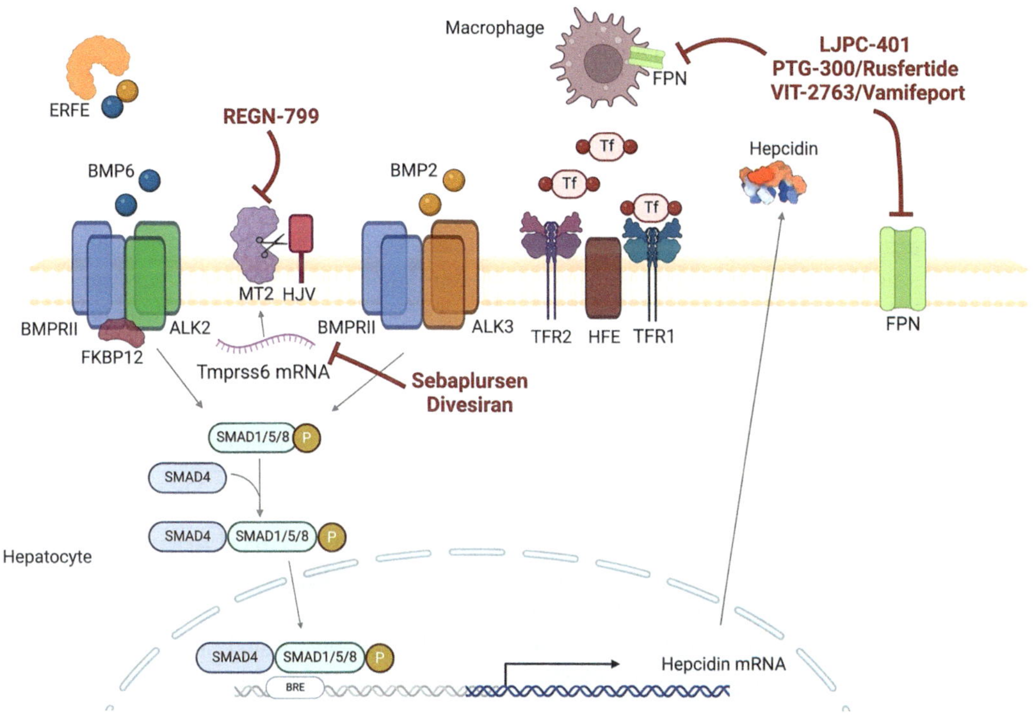

Fig. 26.2 Hepcidin mimetics and inducers in Phase 1/2 clinical development. Schematic representation of the pathways regulating the hepcidin/ferroportin (FPN) axis showing hepcidin mimetics or inducers with their respective targets. Among the hepcidin mimetics, LJPC-401, PTG-300/Rusfertide, and mini-hepcidins act by binding/occluding FPN and inducing its internalization and degradation; Vamifeport is an FPN inhibitor that blocks FPN. Among the hepcidin inducers, REGN-799 is an anti-TMPRSS6 (MT2) antibody that blocks TMPRSS6 activity, thus inducing hepcidin production; Sebaplursen, a TMPRSS6-ASO, and Divesiran, a TMPRSS6 siRNA, degrade Tmprss6 mRNA with consequent hepcidin induction. (Figures were created using BioRender.com)

26.2.2 Hepcidin Inducers

An alternative to the direct administration of hepcidin or its mimetic peptides is the induction of endogenous hepcidin production (Fig. 26.2). Due to the pleiotropic functions of the major drivers of hepcidin synthesis, the BMP ligands, efforts have focused on interfering with hepcidin inhibitors and negative regulators rather than directly with BMPs. Of particular relevance is the hepato-specific serine protease TMPRSS6/Matriptase-2, which inhibits hepcidin expression by cleaving the BMP co-receptor hemojuvelin (Du et al. 2008; Silvestri et al. 2008). Genetic inactivation of the protease was able to ameliorate anemia, ineffective erythropoiesis, and iron overload in a mouse model of β-thalassemia by increasing hepcidin levels and triggering iron restriction (Nai et al. 2012). Given the easy accessibility of the liver as the target organ (Sehgal et al. 2013), several molecules have been developed to interfere with TMPRSS6 function, including antisense oligonucleotides (ASOs, Sapablursen), small-interfering RNAs (siRNAs, Divesiran), and blocking antibodies (REGN-7999) (Fig. 26.2). All these compounds effectively lowered iron levels in hemochromatosis mice (Guo et al. 2013; Altamura et al. 2019; Chen et al. 2021; Lob et al. 2022) and healthy volunteers (McCaleb et al. 2018; Porter et al. 2023; Rottey et al. 2023) and improved anemia, ineffective erythropoiesis, and iron overload in thalassemic mice, both alone or in combination with other strategies (Guo et al. 2013; Casu et al. 2016a, 2020b; Schmidt et al. 2015, 2018; Vadolas et al. 2021; Guerra et al. 2024; Wake et al. 2024). However, the Phase 2 clinical trial of *Tmprss6*-ASOs in β-thalassemia patients was discontinued due to lack of efficacy (NCT04059406), while, based on promising results from preclinical studies (Casu et al. 2021) and the PTG-300/Rusfertide trials (Kremyanskaya et al. 2024; Hoffman et al. 2022), two different studies are currently investigating the ability of anti-*TMPRSS6* ASOs (NCT05143957) and siRNAs (NCT05499013) to improve PV pathophysiology. Interestingly, the U.S. FDA has granted Divesiran Fast Track and orphan drug designation for PV (https://silence-therapeutics.com/our-pipeline/sln124/default.aspx), pointing to favorable expectations. Based on the therapeutic potential of anti-TMPRSS6 agents, the search for novel compounds is still active and has led to the identification of promising chemicals (Béliveau et al. 2019) and selective small molecules (Desgagné et al. 2024) that inhibit TMPRSS6 activity *in vitro*, but whose efficacy *in vivo* remains unexplored. The immunophilin FKBP12 was recently identified as a novel hepcidin inhibitor, which lowers iron-dependent hepcidin induction by attenuating the activation of the BMP-SMAD-signaling pathway via BMP receptor ALK2 binding (Colucci et al. 2017). Pharmacologic displacement of FKBP12 from ALK2 through the administration of FK506/tacrolimus increased hepcidin production in hemochromatosis mice (Pettinato et al. 2023). However, this approach cannot be translated into the clinics due to the immunomodulatory properties of the drug. In an attempt to provide an FKBP12 modulator without immunosuppressive action, specific *Fkbp12*-ASOs have been developed, which unluckily failed in increasing hepcidin expression in wild-type mice due to a compensatory upregulation of *Tmprss6* expression. Interestingly, administration of *Fkbp12*-ASOs enhanced the hepcidin modulatory activity of *Tmprss6*-ASOs (Pettinato et al. 2024), suggesting that, although FKBP12 *per se* is not a therapeutic target, the combined downregulation of the two inhibitors may represent a promising tool for finely tuning the activation of the BMP-SMAD-hepcidin pathway *in vivo*.

A compelling possibility to increase hepcidin transcription in disorders characterized by ineffective erythropoiesis is targeting the hepcidin inhibitor ERFE. Although genetic ablation of *Erfe* in thalassemic mice resulted in only a transient increase in hepcidin levels and modest correction of iron overload (Kautz et al. 2015), antibodies against the N-terminal domain of the protein significantly improved anemia, ineffective erythropoiesis, and iron overload in the same model (Arezes et al. 2020). These promising results and the absence of signs of toxicity in germline *Erfe* deficiency (Tang et al. 2010) argue in favor of a clinical translation of this approach.

26.3 Combinatorial Approaches with Hepcidin Agonists

Several strategies aimed at increasing hepcidin levels or mimicking its function have shown promising results in preclinical models for the treatment of various hematologic diseases. Currently, available data from clinical trials with hepcidin agonists are summarized in Table 26.1. While Phase 1 clinical trials were generally successful in reducing circulating iron levels in healthy volunteers, the Phase 2 trials in β-thalassemia patients were mostly terminated due to lack of efficacy. However, the lack of available data from these trials prevents understanding the reasons for this failure. Among potential explanations, humans might exhibit increased sensitivity to erythroid iron restriction than mice, thus requiring a 'fine tuning' in the use of these therapeutics in β-thalassemia patients to provide enough iron to erythropoiesis. Inducing temporary rather than continuous iron restriction through 'pulsed' drug administration could ensure sufficient macrophage iron recycling to support erythron activity. Additional differences between human and murine erythropoiesis, including reliance on splenic erythropoiesis, might account for the limited anemia improvement in β-thalassemia patients treated with iron-restrictive compounds. Although the clinical development of hepcidin inducers/mimet-

Table 26.1 Hepcidin agonists and relative clinical trials

Agent	Mechanism of action	Phase 1 clinical trial	Phase 2 clinical trial	
			Disease/s	Successful?
Hepcidin mimetics				
LJPC-401	Hepcidin mimetic/Binds to, occludes, and induces internalization of ferroportin	Yes, successful	Hemochromatosis	Not performed
			Polycythemia Vera	Not performed
			Sickle cell disease	Not performed
			β-thalassemia	No
PTG300/Rusfertide	Hepcidin mimetic/Binds to, occludes, and induces internalization of ferroportin	Yes, successful	Sickle cell disease	Not performed
			Hemochromatosis	Yes
			β-thalassemia	Results not available
			Polycythemia Vera	Yes, Phase 3 trial ongoing
Minihepcidin	Hepcidin mimetic/Binds to, occludes, and induces internalization of ferroportin	Not performed	/	/
Vamifeport	Ferroportin inhibitor/Binds to and blocks ferroportin	Yes, successful	Hemochromatosis/	Not performed
			Polycythemia Vera	Not performed
			β-thalassemia	No
			Sickle cell disease	Ongoing
Hepcidin inducers				
Sebaplursen (*TMPRSS6*-ASO)	Activates hepcidin expression inducing the degradation of *TMPRSS6* mRNA	Yes, successful	Hemochromatosis	Not performed
			Sickle cell disease	Not performed
			β-thalassemia	No
			Polycythemia Vera	Ongoing
Divesiran (*TMPRSS6* siRNA)	Activates hepcidin production inducing the degradation of *TMPRSS6* mRNA	Yes, successful	Hemochromatosis	Not performed
			Sickle cell disease	Not performed
			β-thalassemia	Results not available
			Polycythemia Vera	Ongoing
REGN-7999 (TMPRSS6 antibody)	Induces hepcidin production blocking TMPRSS6 activity	Yes, successful	Hemochromatosis	Not performed
			Sickle cell disease	Not performed
			Polycythemia Vera	Not performed
			β-thalassemia	Ongoing
FKBP12-ASO	Activates hepcidin production inducing the degradation of *FKBP12* mRNA	Not performed	/	/
ERFE antibody	Induces hepcidin production blocking ERFE activity	Not performed	/	/

ics for single use in β-thalassemia was terminated, their use in combination with drugs against other targets may offer an alternative therapeutic option. Recently, the combinatorial administration of *Tmprss6*-ASOs with a luspatercept-like TGF-β ligand trap, an erythroid maturation agent (Guerra et al. 2024), or with the erythroid deletion of the iron sensor transferrin receptor 2 (*Tfr2*) proved to be superior in the correction of anemia and improvement of iron overload compared to the single approaches in preclinical models of β-thalassemia (Tanzi et al. 2024; Guerra et al. 2024). These studies paved the way for the clinical development of more robust approaches for the treatment of β-thalassemia based on the combinatorial application of iron-restrictive compounds.

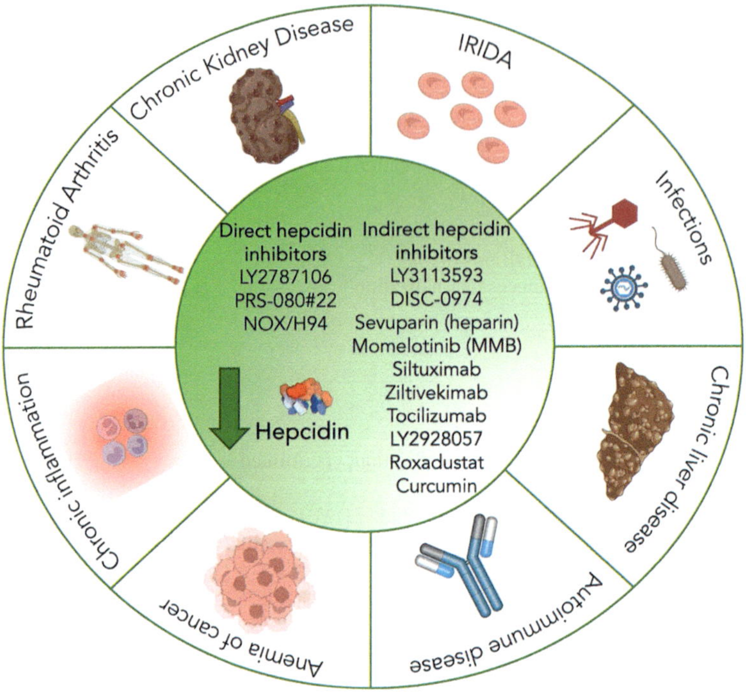

Fig. 26.3 Therapeutic indications for hepcidin inhibitors. Diseases which may benefit from increased iron availability through hepcidin inhibition and improved iron absorption and release from storage include IRIDA and anemia of chronic diseases associated with inflammation (infections, CKD, autoimmune diseases, cancer, and rheumatoid arthritis). Both direct (LY2787106, PRS-080#22, NOX/H94) and indirect (LY3113593, DISC-0974, Sevuparin/heparins, Momelotinib (MMB), Siltuximab, Ziltivekimab, Tocilizumab, LY2928057, Roxadustat, Curcumin) hepcidin inhibitors increase iron availability by reducing hepcidin levels. (Figures were created using BioRender.com)

26.4 Therapeutic Indications for Hepcidin Inhibitors

The development of hepcidin antagonists has clinical relevance for several forms of anemias, genetic and acquired, characterized by excess circulating hepcidin and functional iron deficiency (Fig. 26.3). Currently, available data from clinical trials with hepcidin antagonists are summarized in Table 26.2. The inherited iron-refractory iron deficiency anemia (IRIDA), due to mutations in TMPRSS6, is characterized by abnormally high hepcidin levels and severe systemic iron deficiency with anemia (Finberg et al. 2008; Du et al. 2008; Folgueras et al. 2008). Elevated hepcidin prevents dietary iron absorption in these patients, making them unresponsive to oral iron administration. The most common form of acquired anemia is the anemia of inflammation (AI), or anemia of chronic disease (ACD); see Chap. 13. This is a complication usually associated with infections and chronic diseases such as cancer, kidney diseases, chronic liver diseases, rheumatoid arthritis, autoimmune diseases, chronic obstructive pulmonary disease, pulmonary arterial hypertension, obesity, and advanced atherosclerosis (Weiss et al. 2019). In these inflammatory conditions, the production of interleukin-6 (IL-6) stimulates hepatic hepcidin synthesis (Verga

Table 26.2 Hepcidin antagonists and relative clinical trials

Agent	Mechanism of action	Phase 1 clinical trial	Phase 2 clinical trial	
			Disease/s	Successful?
Direct hepcidin inhibitors				
LY2787106	Anti-hepcidin antibody	Yes, successful	/	/
PRS-080#22	Anticalin protein that binds and sequester hepcidin	Yes, successful	Chronic Kidney Disease	Results not available yet
NOX/H94 (Lexaptepid)	Spiegelmer that binds and neutralizes hepcidin	Yes, successful	Multiple myeloma Chronic Kidney Disease	Partially. Yes, successful
Guanosine 5′-diphosphate (GDP)	Hepcidin-binding agent	Not performed	/	/
Indirect hepcidin inhibitors				
LY3113593	Anti-BMP6 antibody. Prevents the binding of BMP6 to BMPR	Yes, successful. Phase 1b in CKD patient: ongoing	Chronic Kidney Disease	Not performed
DISC-0974	Anti-HJV antibody	Yes, successful. Phase 1b in CKD patient: ongoing	Chronic Kidney Disease	Not performed
Sevuparin (heparin)	Inhibitors of BMPs/SMAD pathway	Yes, successful	/	/
Momelotinib (MMB)	Jak1/2 and ALK2 inhibitor	Yes, successful	/	/
LJ000328	ALK2/3 kinase inhibitor	Not performed		
Siltuximab	Anti-IL-6 antibody	Yes, successful	FDA approved for Castleman's disease	
Ziltivekimab	Anti-IL-6 antibody	Yes, successful	Chronic Kidney Disease	Yes, successful
			Rheumatoid arthritis	Not performed
Tocilizumab	Anti-IL-6 receptor antibody	Yes, successful	Rheumatoid arthritis	FDA approved. Phase 4
Curcumin	STAT3 inhibitor	Yes, successful	Chronic Kidney Disease	Yes, successful
AG490, PpYLKTK, Aspirin	STAT3 inhibitors	Not performed	/	/
LY2928057	Anti-FPN antibody. It binds FPN and blocks its interaction with hepcidin.	Yes, successful. Phase 1b in CKD patient: ongoing	Chronic Kidney Disease	Not performed
Roxadustat (FG-4592)	Prolyl-hydroxylase inhibitor. Modulate the hypoxia-responsive pathway	Yes, successful	Chronic Kidney Disease	Yes, successful. Phase 3. Adverse events

Falzacappa et al. 2007) and reduces systemic iron influx. This, together with the reduced production and/or reduced biological activity of erythropoietin (EPO) upon inflammatory conditions (Akchurin et al. 2019), causes anemia. All the aforementioned pathologies are often difficult to treat, and many patients do not respond to erythropoiesis-stimulating agents (ESAs) or iron supplementation commonly used in clinics. For these reasons, the use of hepcidin antagonists, targeting different steps implicated in hepcidin production, is of potential benefit for the treatment of anemias with high circulating hepcidin. Hepcidin antagonists can be divided into molecules that target hepcidin directly and indirectly, inhibiting the BMP/SMAD or IL-6/STAT3 signaling pathways or interfering with hepcidin-FPN interaction.

26.4.1 Direct Hepcidin Inhibitors

Among the direct approaches to inhibit hepcidin, four strategies have been reported, involving anti-hepcidin antibodies (Ab129m and LY2787106), the anticalin protein PRS080#022, the spiegelmer NOX/H94 (Lexaptepid), or guanosine 5′ diphosphate (GDP) (Fig. 26.4).

Anti-hepcidin antibodies have been generated by different research groups (Vadhan-Raj et al. 2017; Cooke et al. 2013). The Ab12B9m antibody showed a beneficial effect in increasing serum iron and consequently hemoglobin levels in the modified heat-killed *Brucella abortus* mouse model of AI, both alone and in combination with low doses of ESA. Also, repeated treatments with Ab12B9m antibody in non-human primates were effective in increasing serum iron (Cooke et al. 2013). The efficacy of another anti-hepcidin antibody, LY2787106, has been demonstrated in a Phase 1 clinical trial (NCT01340976) in cancer patients with anemia. LY2787106 was well tolerated and led to hepcidin neutralization with consequent iron mobilization (Vadhan-Raj et al. 2017).

PRS080#022 is a PEGylated (polyethylene glycol-bound) anticalin protein that targets and antagonizes hepcidin with picomolar affinity. Of interest, monkeys treated with PRS-080#022 showed effective and prolonged iron mobilization (Hohlbaum et al. 2018). In two randomized double-blind Phase 1 trials, a single IV infusion of PRS-080#22 was safe and well tolerated when administered to 48 healthy volunteers (Phase 1a NCT02340572) and to 24 patients with end-stage chronic kidney disease (CKD) on hemodialysis (Phase 1b NCT02754167) at different ascending doses. The drug exhibited linear pharmacokinetics and a longer half-life in CKD patients in comparison to healthy volunteers as well as expected pharmacodynamic effects, which hold promise for further clinical development (Renders et al. 2019). A Phase 2a study aimed at investigating the safety, pharmacokinetics, and pharmacodynamics of repeated administrations of PRS-080#022 in anemic patients with stage 5 CKD undergoing hemodialysis has already been completed (NCT03325621). However, no results have been published yet.

The spiegelmer NOX/H94 (Lexaptepid) is a synthetic PEGylated L-stereoisomer RNA aptamer, nuclease resistant and immunologically inactive, that binds and neutralizes human hepcidin with high affinity, thus blocking its biological function, acting similarly to antibodies or other scaffold proteins. In non-human primates, NOX-H94-mediated hepcidin neutralization prevented the onset of anemia and reversed hypoferremia induced by repeated IL-6 administration (Schwoebel et al. 2013). A Phase 1 clinical trial (NCT01372137) demonstrated the safety and tolerability of single and repeated IV infusion of NOX-H94 in healthy human volunteers. The treatment inhibited hepcidin production with a concomitant dose-dependent increase of serum iron and transferrin saturation (Boyce et al. 2016). Similar results were observed in another Phase 1 trial (NCT01522794) under conditions of endotoxemia (van Eijk et al. 2014), where NOX-H94 was effective in restoring the decreased serum iron levels in individuals with LPS-induced systemic inflammation. In a Phase 2 clinical trial (NCT01691040), the effect of repeated administrations of NOX-H94 was tested in patients with multiple myeloma and low-grade non-Hodgkin or Hodgkin lymphoma characterized by ACD. Five of 12 patients showed significant response, with a 1 g/dL increase in Hb, reticulocyte Hb, and RBCs (Georgiev et al. 2014). In a Phase 2 clinical trial (NCT02079896) in ESA-resistant CKD patients, the administration of a single IV dose of NOX/H94 at the end of a routine dialysis session achieved a significant and rapid increase of serum iron, consistent with its anti-hepcidin action, with no evident adverse events (Macdougall et al. 2015).

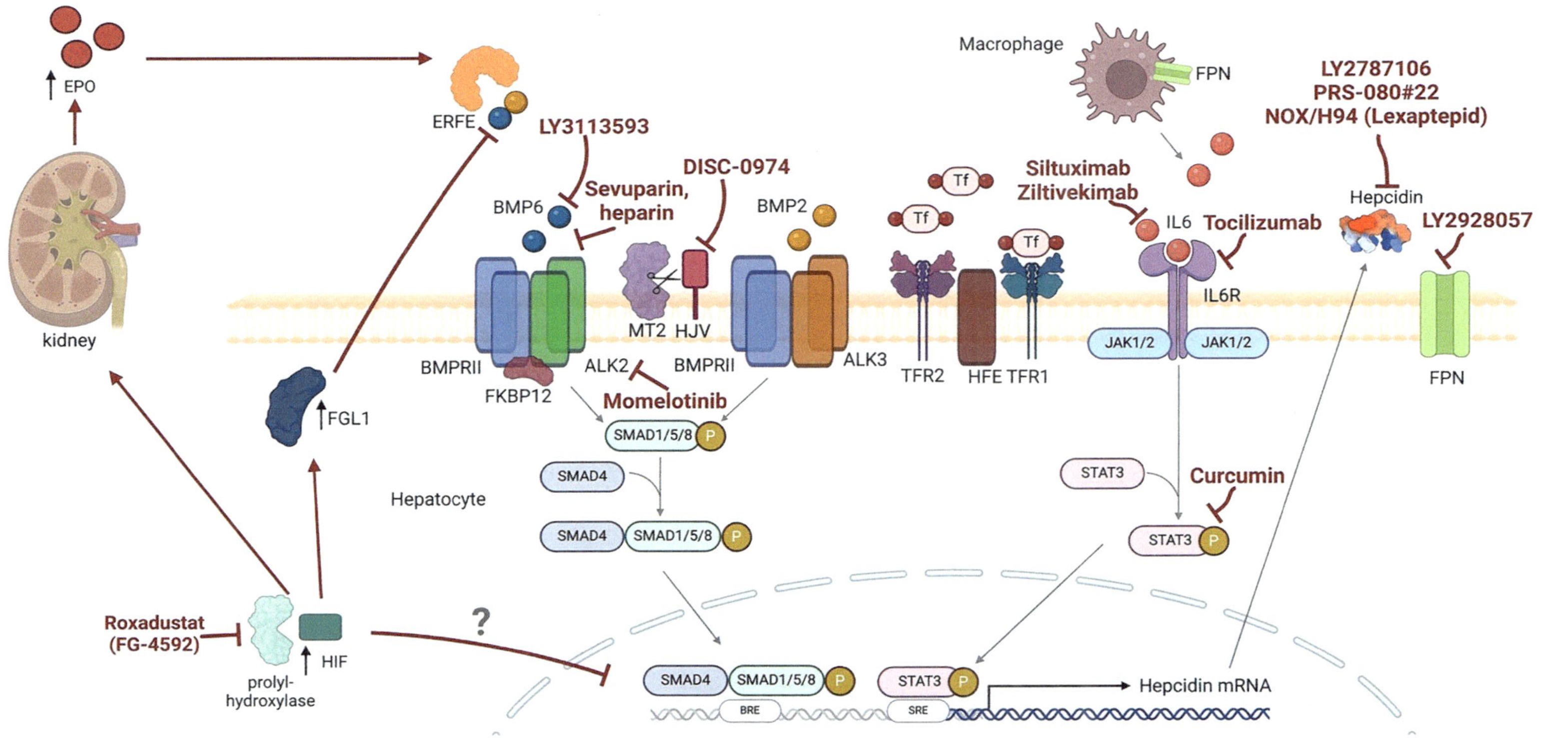

Fig. 26.4 Hepcidin inhibitors in Phase 1/2 clinical development. Schematic representation of the pathways regulating the hepcidin/ferroportin (FPN) axis showing direct and indirect hepcidin inhibitors with their respective targets. Among the direct hepcidin inhibitors, LY2787106 is an anti-hepcidin antibody, the anticalin PRS-080#22 binds and sequester hepcidin, and the spiegelmer NOX/H94 binds and neutralizes hepcidin. The indirect hepcidin inhibitors act by targeting the BMPs/HJV/BMPR or IL-6/IL-6R/STAT3 signaling pathways that control hepcidin expression or hepcidin/FPN interaction. Among the inhibitors of BMPs/HJV/BMPR, LY3113593 is an anti-BMP6 antibody preventing BMP6 binding to its receptor complex (BMPR); DISC-0974 is an anti-hemojuvelin (HJV) antibody that binds HJV, thus preventing the activation of the pSMAD1/5/8 pathway; Sevuparin is a heparin compound that inhibits the BMPs/SMAD pathway, most likely by binding and sequestering BMP6; Momelotinib (MMB) is a Jak1/2 and ALK2 inhibitor that prevents the activation of the SMAD pathway. Among the IL-6/IL-6R/STAT3 inhibitors, Siltuximab and Ziltivekimab are anti-IL-6 antibodies, Tocilizumab is an anti-IL-6 receptor antibody, and Curcumin is a STAT3 inhibitor. Among other indirect inhibitors, LY2928057 is an anti-FPN antibody; Roxadustat (FG-4592) is a prolyl-hydroxylase inhibitor that suppresses hepcidin by increasing kidney EPO production with consequent ERFE and/or FGL-1 elevation. (Figures were created using BioRender.com)

Finally, a recent integrated approach using *in silico*, *in vitro*, and *in vivo* studies demonstrated that the small natural molecule guanosine 5′-diphosphate (GDP) could be a promising hepcidin-binding agent. Treatment with GDP improved anemia in a turpentine-induced AI mouse model (Angmo et al. 2017), and the use of liposome-encapsulated GDP was even more effective (Angmo et al. 2018). GDP has been proposed to act by binding and inhibiting hepcidin function but also attenuating the secretion of the inflammatory cytokine IL-6 from macrophages that in turn decrease inflammation-driven hepcidin induction (Angmo et al. 2018). Further studies are ongoing to elucidate the underlying molecular mechanism and to understand if GDP treatment could have an impact on the GTP/GDP ratio in the organism.

26.4.2 Indirect Hepcidin Inhibitors

Alternative approaches to direct hepcidin inhibitors involve the use of molecules targeting the BMPs/HJV/BMPR or IL-6/IL-6R/STAT3 signaling pathways that control hepcidin expression as well as the development of molecules disrupting hepcidin/FPN interaction (Fig. 26.4).

The first effective attempt to use anti-BMP6 antibodies to prevent BMP6/BMPR interaction was reported in 2010 in HFE transgenic mice with iron-restricted anemia due to high hepcidin (Corradini et al. 2010). Recent studies showed the effectiveness of an anti-BMP6 antibody (KY1070) in improving anemia in murine models of ACD (Petzer et al. 2020). By specifically blocking BMP6, the anti-BMP6 antibody LY3113593 increased iron and transferrin saturation and reduced hepcidin in non-human primates. In a Phase 1 study (NCT02144285), the safety and tolerability profile of this drug was shown in healthy human volunteers and hemodialyzed patients with CKD. In addition, a follow-up Phase 1 clinical trial (NCT02604160) using repeated LY3113593 infusions resulted in a significant improvement of anemia in CKD patients (Sheetz et al. 2019).

Another target of the BMP/BMPR complex which can be leveraged to reduce hepcidin levels is the essential co-receptor hemojuvelin (HJV) (Fig. 26.4). Treatment with anti-HJV antibodies is effective *in vitro* and *in vivo* (healthy mice and preclinical disease models) in blocking SMAD1/5/8 activation, reducing hepcidin expression, and increasing serum iron (Babitt et al. 2007). Chronic treatment improved anemia in rodent models of ACD (Theurl et al. 2011), CKD (Wu et al. 2022), and IRIDA (Kovac et al. 2016), with non-toxic effects. In a recent Phase 1 clinical trial in healthy human volunteers (NCT04999527), DISC-0974, a first-in-class monoclonal antibody that blocks HJV, showed a dose-dependent reduction in serum hepcidin, an increase in serum iron, and a trend toward higher Hb and RBCs, with a safe and tolerable profile (Novikov et al. 2024). A Phase 1b clinical trial (NCT05745883) enrolling non-dialysis-dependent anemic patients with CKD is currently ongoing (Arora et al. 2023).

Commercial heparins, highly sulfated glycosaminoglycans used in clinics for their anticoagulant properties, and modified heparins with low or absent anticoagulant activity (RO heparins or oversulfated heparins) have potent anti-hepcidin activity both *in vitro* and *in vivo* (Poli et al. 2011, 2014a, b) (Fig. 26.4). Heparin acts by binding and sequestering BMP6 ligands and other components of the BMPs/BMPRs complex, thus preventing SMAD1/5/8 pathway activation that is essential for hepcidin expression (Denardo et al. 2021). Sevuparin, a novel clinical stage, low-molecular-weight heparinoid compound with reduced anticoagulant activity, effectively suppressed basal and LPS-induced hepcidin expression in mice, improved inflammatory markers, and improved anemia and kidney damage in a CKD mouse model (Asperti et al. 2023). In a Phase 1 clinical trial (NCT03853421), a single dose of sevuparin administered subcutaneously caused a significant reduction of plasma hepcidin in healthy volunteers in a safe and tolerable manner (Asperti et al. 2024).

The Jak1/2 inhibitor Momelotinib (MMB), recently approved for the treatment of myelofibrosis (Tefferi and Pardanani 2024), inhibits the kinase activity of the BMP Activin A receptor, type I (ACVR1, also known as activin receptor-like kinase 2, ALK2) (Fig. 26.4). In a rat model of ACD, MMB normalized hemoglobin and RBC count, likely by inhibiting the ACVR1-mediated hepcidin expression, thus increasing iron mobilization for erythropoiesis (Asshoff et al. 2017). The ALK2/3 kinase inhibitor LJ000328 suppressed hepcidin both in healthy mice and in a mouse model of IRIDA, improving anemia and increasing serum iron and transferrin saturation (Belot et al. 2019).

Targeting the IL-6/IL-6R/STAT3 signaling pathway represents an alternative approach to improve anemia upon inflammatory conditions (Fig. 26.4). Promising strategies include the use of anti-IL-6 antibodies (siltuximab and ziltivekimab), anti-IL-6 receptor antibodies (tocilizumab), and STAT3 inhibitors.

The anti-IL-6 monoclonal antibody Siltuximab was effective in inhibiting hepcidin expression in Castleman's disease (CD) patients, ameliorating their anemic status (van Rhee et al. 2010; Kurzrock et al. 2013) (Fig. 26.4). Ziltivekimab, another human antibody directed against IL-6, significantly decreased plasma hepcidin, improved the inflammatory status, and reduced ESA requirements in a Phase 1/2 clinical trial conducted in patients on hemodialysis with inflammation and resistant to ESA (NCT02868229) (Pergola et al. 2021). The efficacy of the drug in reducing inflammation has also been validated in recent Phase 1 clinical trials involving patients with rheumatoid arthritis (NCT01559103) or CKD (NCT03126318) (Nowak et al. 2021). Long-term treatment with the anti-IL-6R antibody tocilizumab, already approved for the treatment of rheumatoid arthritis (Hashizume et al. 2010) and giant cell arteritis, leads to hepcidin suppression and anemia improvement in chronic disease patients (Song et al. 2010). However, further studies are required to evaluate the possible adverse effect of the prolonged inhibition of the IL-6 pathway, due to its key roles in physiologic functions including cardio-protection, metabolic regulation, and tissue regeneration (Schett 2018).

STAT3 inhibitors include chemical compounds such as AG490, which suppresses STAT3 phosphorylation (Caceres-Cortes 2008), and PpYLKTK, which blocks pSTAT3 dimerization and DNA binding (Fig. 26.4). These drugs decreased IL-6-dependent hepcidin production in mouse hepatocytes (Fatih et al. 2010) and *in vivo* in healthy mice (Zhang et al. 2011). Curcumin (Lainé et al. 2017) and aspirin (Huang et al. 2018) have also been reported to have a STAT3 inhibitory effect. Curcumin likewise decreased hepcidin levels in healthy volunteers (NCT01489592) (Lainé et al. 2017) and inflammation in CKD patients (NCT03475017) (Alvarenga et al. 2020). Among the therapeutic strategies proposed to counteract high hepcidin levels, FPN stabilizers or antibodies blocking the hepcidin-binding site of FPN are promising tools. LY2928057 is a humanized antibody that binds FPN and blocks its interaction with hepcidin, with no effect on the channel efflux. Its administration in non-human primates significantly increased serum iron levels (Witcher et al. 2013). A Phase 1 clinical trial in healthy human volunteers (NCT01330953) and in CKD patients (NCT01991483) showed that LY2928057 is safe and well tolerated. Of note, LY2928057 counteracted hemoglobin decline and ferritin reduction in CKD patients by increasing serum iron (Sheetz et al. 2019).

Another approach involves the use of prolyl-hydroxylase inhibitors or hypoxia-inducible factor (HIF) stabilizers to modulate the hypoxia-responsive pathway and increase HIF2α (Fig. 26.4). These drugs likely suppress hepcidin through a dual action, the elevation of EPO production by the kidney with consequent ERFE increase, and/or the induction of FGL-1 synthesis in the liver (Sardo et al. 2024). For example, Roxadustat (FG-4592) is a potent first-in-class oral inhibitor of HIF prolyl-hydroxylase. In the Phase 3 clinical trial (NCT02652819), Roxadustat treatment in non-dialyzed CKD patients resulted in hepcidin reduction and anemia correction. However, patients reported adverse events such as higher frequencies of hyperkalemia and metabolic acidosis than in the placebo group (Chen et al. 2019; Li et al. 2020).

26.5 Combinatorial Approaches with Hepcidin Inhibitors

Due to the high prevalence of inflammatory conditions, AI is estimated to be the second most prevalent anemia worldwide, after IDA, with a negative impact on the quality of life of patients and a poor prognosis (Chap. 13). Current therapies based on ESAs or IV iron administration (Chap. 24) are often associated with ineffective anemia improvement and/or the development of adverse events. Thus, in recent years major research efforts have been dedicated to the preclinical and clinical development of therapeutics able to counteract AI clinical manifestation, including anemia. Among these, anti-hepcidin, anti-calin, anti-BMP6, anti-FPN, anti-HJV, anti-IL-6, and IL-6-R antibodies, and Roxadustat have reached more advanced stages of clinical development. Strategies combining the human anti-BMP6 antibody KY1070 with EPO have been reported to improve anemia and systemic iron availability in rodent models of ACD. Importantly, the administration of KY1070 lowered EPO requirements in this model, suggesting a superior benefit of combinatorial strategies and providing an approach to limit high-dose EPO-mediated adverse events (Petzer et al. 2020).

26.6 Future Perspectives

Ultimately, new key players involved in the regulation of the hepcidin-ferroportin axis are continuously discovered. Among them, the hepatic BMP inhibitor FGL1 (Sardo et al. 2024), the regulator of micronutrient metabolism and hepcidin activator FOXO1A (Xu et al. 2024), the ferroportin regulator PGCα (Banerjee et al. 2024), and the hepatic heparan sulfates (Asperti et al. 2016; Poli et al. 2019). The design of novel drugs directed to such new targets will likely contribute in the near future to the identification of additional safe and tolerable approaches for diseases characterized by the deregulation of hepcidin and iron homeostasis.

References

Akchurin O, Patino E, Dalal V, Meza K, Bhatia D, Brovender S, Zhu YS, Cunningham-Rundles S, Perelstein E, Kumar J, Rivella S, Choi ME (2019) Interleukin-6 contributes to the development of anemia in Juvenile CKD. Kidney Int Rep 4(3):470–483. https://doi.org/10.1016/j.ekir.2018.12.006

Altamura S, Schaeper U, Dames S, Löffler K, Eisermann M, Frauendorf C, Müdder K, Neves J, Muckenthaler MU (2019) SLN124, a GalNAc-siRNA conjugate targeting TMPRSS6, efficiently prevents iron overload in hereditary haemochromatosis type 1. Hemasphere 3(6):e301. https://doi.org/10.1097/hs9.0000000000000301

Alvarenga L, Salarolli R, Cardozo LFMF, Santos RS, de Brito JS, Kemp JA, Reis D, de Paiva BR, Stenvinkel P, Lindholm B, Fouque D, Mafra D (2020) Impact of curcumin supplementation on expression of inflammatory transcription factors in hemodialysis patients: a pilot randomized, double-blind, controlled study. Clin Nutr 39(12):3594–3600. https://doi.org/10.1016/j.clnu.2020.03.007

Angmo S, Tripathi N, Abbat S, Sharma S, Singh SS, Halder A, Yadav K, Shukla G, Sandhir R, Rishi V, Bharatam PV, Yadav H, Singhal NK (2017) Identification of guanosine 5′-diphosphate as potential iron mobilizer: preventing the hepcidin-ferroportin interaction and modulating the interleukin-6/Stat-3 pathway. Sci Rep 7:40097. https://doi.org/10.1038/srep40097

Angmo S, Rana S, Yadav K, Sandhir R, Singhal NK (2018) Novel liposome eencapsulated guanosine Di phosphate based therapeutic target against anemia of inflammation. Sci Rep 8(1):17684. https://doi.org/10.1038/s41598-018-35992-2

Antypiuk A, Vance S, Sharma R, Passos ST, Asperti M, Navaneethabalakrishnan S, Durrenberger F, Manolova V, Vinchi F (2025) Genetic iron overload aggravates and pharmacological iron restriction improves MDS pathophysiology in a preclinical study. Blood 145(2):155–169. https://doi.org/10.1182/blood.2024026135

Arezes J, Foy N, McHugh K, Sawant A, Quinkert D, Terraube V, Brinth A, Tam M, LaVallie ER, Taylor S, Armitage AE, Pasricha SR, Cunningham O, Lambert M, Draper SJ, Jasuja R, Drakesmith H (2018) Erythroferrone inhibits the induction of hepcidin by BMP6. Blood 132(14):1473–1477. https://doi.org/10.1182/blood-2018-06-857995

Arezes J, Foy N, McHugh K, Quinkert D, Benard S, Sawant A, Frost JN, Armitage AE, Pasricha SR, Lim PJ, Tam MS, Lavallie E, Pittman DD, Cunningham O, Lambert M, Murphy JE, Draper SJ, Jasuja R, Drakesmith H (2020) Antibodies against the erythroferrone N-terminal domain prevent hepcidin suppression and ameliorate murine thalassemia. Blood 135(8):547–557. https://doi.org/10.1182/blood.2019003140

Arora S, Monroy J, Pergola P, Silva A, Buch A, Novikov N, Pelletier O, Savage W (2023) A phase 1b double-blind, placebo-controlled study of DISC-0974, an anti-hemojuvelin antibody, in patients with non-dialysis dependent chronic kidney disease and anemia. Blood 142(Suppl 1):5236–5236. https://doi.org/10.1182/blood-2023-178033

Asperti M, Stuemler T, Poli M, Gryzik M, Lifshitz L, Meyron-Holtz EG, Vlodavsky I, Arosio P (2016) Heparanase overexpression reduces hepcidin expression, affects iron homeostasis and alters the response to inflammation. PLoS One 11(10):e0164183. https://doi.org/10.1371/journal.pone.0164183

Asperti M, Gryzik M, Cominelli M, Pagani F, Poliani PL, Girelli D, Westerberg G, Öhd J, Poli M (2023) The heparinoid sevuparin improves anemia and kidney status in a mouse model of chronic kidney disease. Blood 142(Suppl 1):2463–2463. https://doi.org/10.1182/blood-2023-181587

Asperti M, Denardo A, Gryzik M, Persson KEM, Westerberg G, Öhd J, Poli M (2024) Sevuparin strongly reduces hepcidin expression in cells, mice, and healthy human volunteers. Hemasphere 8(12):e70035. https://doi.org/10.1002/hem3.70035

Asshoff M, Petzer V, Warr MR, Haschka D, Tymoszuk P, Demetz E, Seifert M, Posch W, Nairz M, Maciejewski P, Fowles P, Burns CJ, Smith G, Wagner KU, Weiss G, Whitney JA, Theurl I (2017) Momelotinib inhibits ACVR1/ALK2, decreases hepcidin production, and ameliorates anemia of chronic disease in rodents. Blood 129(13):1823–1830. https://doi.org/10.1182/blood-2016-09-740092

Babitt JL, Huang FW, Xia Y, Sidis Y, Andrews NC, Lin HY (2007) Modulation of bone morphogenetic protein signaling in vivo regulates systemic iron balance. J Clin Invest 117(7):1933–1939. https://doi.org/10.1172/jci31342

Banerjee S, Lu S, Jain A, Wang I, Tao H, Srinivasan S, Nemeth E, He P (2024) Targeting PKCα alleviates iron overload in diabetes and hemochromatosis through the inhibition of ferroportin. Blood 144(13):1433–1444. https://doi.org/10.1182/blood.2024023829

Béliveau F, Tarkar A, Dion SP, Désilets A, Ghinet MG, Boudreault PL, St-Georges C, Marsault É, Paone D, Collins J, Macphee CH, Campobasso N, Groy A, Cottom J, Ouellette M, Pope AJ, Leduc R (2019) Discovery and development of TMPRSS6 inhibitors modulating hepcidin levels in human hepatocytes. Cell Chem Biol 26(11):1559–1572.e1559. https://doi.org/10.1016/j.chembiol.2019.09.004

Belot A, Gourbeyre O, Fay A, Palin A, Besson C, Latour C, Hopkins CR, Tidmarsh GF, Coppin H, Roth MP, Ritter MR, Hong CC, Meynard D (2019) LJ000328, a novel ALK2/3 kinase inhibitor, represses hepcidin and significantly improves the phenotype of IRIDA. Haematologica. https://doi.org/10.3324/haematol.2019.236133

Boyce M, Warrington S, Cortezi B, Zöllner S, Vauléon S, Swinkels DW, Summo L, Schwoebel F, Riecke K (2016) Safety, pharmacokinetics and pharmacodynamics of the anti-hepcidin Spiegelmer lexaptepid pegol in healthy subjects. Br J Pharmacol 173(10):1580–1588. https://doi.org/10.1111/bph.13433

Caceres-Cortes JR (2008) A potent anti-carcinoma and anti-acute myeloblastic leukemia agent, AG490. Anti Cancer Agents Med Chem 8(7):717–722

Casu C, Aghajan M, Oikonomidou PR, Guo S, Monia BP, Rivella S (2016a) Combination of Tmprss6- ASO and the iron chelator deferiprone improves erythropoiesis and reduces iron overload in a mouse model of beta-thalassemia intermedia. Haematologica 101(1):e8–e11. https://doi.org/10.3324/haematol.2015.133348

Casu C, Oikonomidou PR, Chen H, Nandi V, Ginzburg Y, Prasad P, Fleming RE, Shah YM, Valore EV, Nemeth E, Ganz T, MacDonald B, Rivella S (2016b) Minihepcidin peptides as disease modifiers in mice affected by β-thalassemia and polycythemia vera. Blood 128(2):265–276. https://doi.org/10.1182/blood-2015-10-676742

Casu C, Chessa R, Liu A, Gupta R, Drakesmith H, Fleming R, Ginzburg YZ, MacDonald B, Rivella S (2020a) Minihepcidins improve ineffective erythropoiesis and splenomegaly in a new mouse model of adult β-thalassemia major. Haematologica 105(7):1835–1844. https://doi.org/10.3324/haematol.2018.212589

Casu C, Pettinato M, Liu A, Aghajan M, Lo Presti V, Lidonnici MR, Munoz KA, O'Hara E, Olivari V, Di Modica SM, Booten S, Guo S, Neil G, Miari R, Shapir N, Zafir-Lavie I, Domev H, Ferrari G, Sitara D, Nai A, Rivella S (2020b) Correcting β-thalassemia by combined therapies that restrict iron and modulate erythropoietin activity. Blood 136(17):1968–1979. https://doi.org/10.1182/blood.2019004719

Casu C, Liu A, De Rosa G, Low A, Suzuki A, Sinha S, Ginzburg YZ, Abrams C, Aghajan M, Guo S, Rivella S (2021) Tmprss6-ASO as a tool for the treatment of Polycythemia Vera mice. PLoS One 16(12):e0251995. https://doi.org/10.1371/journal.pone.0251995

Chen N, Hao C, Peng X, Lin H, Yin A, Hao L, Tao Y, Liang X, Liu Z, Xing C, Chen J, Luo L, Zuo L, Liao Y, Liu BC, Leong R, Wang C, Liu C, Neff T, Szczech L, Yu KP (2019) Roxadustat for anemia in patients with kidney disease not receiving dialysis. N Engl J Med 381(11):1001–1010. https://doi.org/10.1056/NEJMoa1813599

Chen B, Wang J, Zheng B, Huang L, Mao Y, Ouyang Z, Du X (2021) Treating iron overload disorders with a novel therapeutic antibody targeting TMPRSS6. Blood 138(Suppl 1):941–941. https://doi.org/10.1182/blood-2021-151418

Colucci S, Pagani A, Pettinato M, Artuso I, Nai A, Camaschella C, Silvestri L (2017) The immunophilin FKBP12 inhibits hepcidin expression by binding the BMP type I receptor ALK2 in hepatocytes. Blood 130(19):2111–2120. https://doi.org/10.1182/blood-2017-04-780692

Cooke KS, Hinkle B, Salimi-Moosavi H, Foltz I, King C, Rathanaswami P, Winters A, Steavenson S,

Begley CG, Molineux G, Sasu BJ (2013) A fully human anti-hepcidin antibody modulates iron metabolism in both mice and nonhuman primates. Blood 122(17):3054–3061. https://doi.org/10.1182/blood-2013-06-505792

Corradini E, Schmidt PJ, Meynard D, Garuti C, Montosi G, Chen S, Vukicevic S, Pietrangelo A, Lin HY, Babitt JL (2010) BMP6 treatment compensates for the molecular defect and ameliorates hemochromatosis in Hfe knockout mice. Gastroenterology 139(5):1721–1729. https://doi.org/10.1053/j.gastro.2010.07.044

Denardo A, Elli S, Federici S, Asperti M, Gryzik M, Ruzzenenti P, Carmona F, Bergese P, Naggi A, Arosio P, Poli M (2021) BMP6 binding to heparin and heparan sulfate is mediated by N-terminal and C-terminal clustered basic residues. Biochim Biophys Acta Gen Subj 1865 2:129799. https://doi.org/10.1016/j.bbagen.2020.129799

Desgagné M, Désilets A, Ferková S, Lepage M, Perreault O, Joushomme A, Lemieux G, Guerrab W, Froehlich U, Comeau C, Sarret P, Leduc R, Boudreault PL (2024) Rational in silico design of selective TMPRSS6 peptidomimetic inhibitors via exploitation of the S2 subpocket. J Med Chem 67(15):12969–12983. https://doi.org/10.1021/acs.jmedchem.4c00922

Du X, She E, Gelbart T, Truksa J, Lee P, Xia Y, Khovananth K, Mudd S, Mann N, Moresco EM, Beutler E, Beutler B (2008) The serine protease TMPRSS6 is required to sense iron deficiency. Science 320(5879):1088–1092. https://doi.org/10.1126/science.1157121

Fatih N, Camberlein E, Island ML, Corlu A, Abgueguen E, Détivaud L, Leroyer P, Brissot P, Loréal O (2010) Natural and synthetic STAT3 inhibitors reduce hepcidin expression in differentiated mouse hepatocytes expressing the active phosphorylated STAT3 form. J Mol Med (Berl) 88(5):477–486. https://doi.org/10.1007/s00109-009-0588-3

Finberg KE, Heeney MM, Campagna DR, Aydinok Y, Pearson HA, Hartman KR, Mayo MM, Samuel SM, Strouse JJ, Markianos K, Andrews NC, Fleming MD (2008) Mutations in TMPRSS6 cause iron-refractory iron deficiency anemia (IRIDA). Nat Genet 40(5):569–571. https://doi.org/10.1038/ng.130

Folgueras AR, de Lara FM, Pendás AM, Garabaya C, Rodríguez F, Astudillo A, Bernal T, Cabanillas R, López-Otín C, Velasco G (2008) Membrane-bound serine protease matriptase-2 (Tmprss6) is an essential regulator of iron homeostasis. Blood 112(6):2539–2545. https://doi.org/10.1182/blood-2008-04-149773

Folgueras AR, Freitas-Rodríguez S, Ramsay AJ, Garabaya C, Rodríguez F, Velasco G, López-Otín C (2018) Matriptase-2 deficiency protects from obesity by modulating iron homeostasis. Nat Commun 9(1):1350. https://doi.org/10.1038/s41467-018-03853-1

Ganz T, Nemeth E (2023) Pathogenic mechanisms in thalassemia II: iron overload. Hematol Oncol Clin North Am 37(2):353–363. https://doi.org/10.1016/j.hoc.2022.12.006

Gardenghi S, Ramos P, Marongiu MF, Melchiori L, Breda L, Guy E, Muirhead K, Rao N, Roy CN, Andrews NC, Nemeth E, Follenzi A, An X, Mohandas N, Ginzburg Y, Rachmilewitz EA, Giardina PJ, Grady RW, Rivella S (2010) Hepcidin as a therapeutic tool to limit iron overload and improve anemia in β-thalassemic mice. J Clin Invest 120(12):4466–4477. https://doi.org/10.1172/jci41717

Georgiev P, Lazaroiu M, Ocroteala L, Grudeva-Popova J, Gheorghita E, Vasilica M, Popescu SM, Cucuianu A, Summo L, Schwoebel F, Riecke K, Ludwig H (2014) Abstract 3847: the anti-hepcidin Spiegelmer® Lexaptepid Pegol (NOX-H94) as treatment of anemia of chronic disease in patients with multiple myeloma, low grade lymphoma, and CLL: a phase II pilot study. Cancer Res 74(19_Suppl):3847–3847. https://doi.org/10.1158/1538-7445.am2014-3847

Goncalves Monteiro D, van Dijk JWA, Aliyanto R, Fung E, Nemeth E, Ganz T, Rosengren J, Clark RJ (2021) Pursuing orally bioavailable hepcidin analogues via cyclic. Biomedicine 9(2). https://doi.org/10.3390/biomedicines9020164

Guerra A, Hamilton N, Rivera A, Demsko P, Guo S, Rivella S (2024) Combination of a TGF-β ligand trap (RAP-GRL) and TMPRSS6-ASO is superior for correcting β-thalassemia. Am J Hematol 99(7):1300–1312. https://doi.org/10.1002/ajh.27332

Guo S, Casu C, Gardenghi S, Booten S, Aghajan M, Peralta R, Watt A, Freier S, Monia BP, Rivella S (2013) Reducing TMPRSS6 ameliorates hemochromatosis and β-thalassemia in mice. J Clin Invest 123(4):1531–1541. https://doi.org/10.1172/jci66969

Hashizume M, Uchiyama Y, Horai N, Tomosugi N, Mihara M (2010) Tocilizumab, a humanized anti-interleukin-6 receptor antibody, improved anemia in monkey arthritis by suppressing IL-6-induced hepcidin production. Rheumatol Int 30(7):917–923. https://doi.org/10.1007/s00296-009-1075-4

Hoffman R, Ginzburg Y, Kremyanskaya M, Khanna S, Modi N, Valone FH, O'Connor PG, Gupta S, Saks SR (2022) Rusfertide (PTG-300) treatment in phlebotomy-dependent polycythemia vera patients. J Clin Oncol 40(16_Suppl):7003. https://doi.org/10.1200/JCO.2022.40.16_suppl.7003

Hohlbaum AM, Gille H, Trentmann S, Kolodziejczyk M, Rattenstetter B, Laarakkers CM, Katzmann G, Christian HJ, Andersen N, Allersdorfer A, Olwill SA, Meibohm B, Audoly LP, Swinkels DW, van Swelm RPL (2018) Sustained plasma hepcidin suppression and iron elevation by Anticalin-derived hepcidin antagonist in cynomolgus monkey. Br J Pharmacol 175(7):1054–1065. https://doi.org/10.1111/bph.14143

Huang SN, Ruan HZ, Chen MY, Zhou G, Qian ZM (2018) Aspirin increases ferroportin 1 expression by inhibiting hepcidin via the JAK/STAT3 pathway in interleukin 6-treated PC-12 cells. Neurosci Lett 662:1–5. https://doi.org/10.1016/j.neulet.2017.10.001

Kalleda N, Flace A, Altermatt P, Ingoglia G, Doucerain C, Nyffenegger N, Dürrenberger F, Manolova V (2023) Ferroportin inhibitor vamifeport ameliorates ineffective erythropoiesis in a mouse model of β-thalassemia with blood transfusions. Haematologica 108(10):2703–2714. https://doi.org/10.3324/haematol.2022.282328

Kautz L, Jung G, Valore EV, Rivella S, Nemeth E, Ganz T (2014) Identification of erythroferrone as an erythroid regulator of iron metabolism. Nat Genet. https://doi.org/10.1038/ng.2996

Kautz L, Jung G, Du X, Gabayan V, Chapman J, Nasoff M, Nemeth E, Ganz T (2015) Erythroferrone contributes to hepcidin suppression and iron overload in a mouse model of β-thalassemia. Blood 126(17):2031–2037. https://doi.org/10.1182/blood-2015-07-658419

Kijima H, Sawada T, Tomosugi N, Kubota K (2008) Expression of hepcidin mRNA is uniformly suppressed in hepatocellular carcinoma. BMC Cancer 8:167. https://doi.org/10.1186/1471-2407-8-167

Kovac S, Böser P, Cui Y, Ferring-Appel D, Casarrubea D, Huang L, Fung E, Popp A, Mueller BK, Hentze MW (2016) Anti-hemojuvelin antibody corrects anemia caused by inappropriately high hepcidin levels. Haematologica 101(5):e173–e176. https://doi.org/10.3324/haematol.2015.140772

Kowdley KV, Modi NB, Peltekian K, Vierling JM, Ferris C, Valone FH, Gupta S (2023) Rusfertide for the treatment of iron overload in HFE-related haemochromatosis: an open-label, multicentre, proof-of-concept phase 2 trial. Lancet Gastroenterol Hepatol 8(12):1118–1128. https://doi.org/10.1016/S2468-1253(23)00250-9

Kremyanskaya M, Kuykendall AT, Pemmaraju N, Ritchie EK, Gotlib J, Gerds A, Palmer J, Pettit K, Nath UK, Yacoub A, Molina A, Saks SR, Modi NB, Valone FH, Khanna S, Gupta S, Verstovsek S, Ginzburg YZ, Hoffman R, Investigators RT (2024) Rusfertide, a hepcidin mimetic, for control of erythrocytosis in polycythemia vera. N Engl J Med 390(8):723–735. https://doi.org/10.1056/NEJMoa2308809

Kurzrock R, Voorhees PM, Casper C, Furman RR, Fayad L, Lonial S, Borghaei H, Jagannath S, Sokol L, Usmani SZ, van de Velde H, Qin X, Puchalski TA, Hall B, Reddy M, Qi M, van Rhee F (2013) A phase I, open-label study of siltuximab, an anti-IL-6 monoclonal antibody, in patients with B-cell non-Hodgkin lymphoma, multiple myeloma, or Castleman disease. Clin Cancer Res 19(13):3659–3670. https://doi.org/10.1158/1078-0432.ccr-12-3349

Lainé F, Laviolle B, Bardou-Jacquet E, Fatih N, Jezequel C, Collet N, Ropert M, Morcet J, Hamon C, Reymann JM, Loréal O (2017) Curcuma decreases serum hepcidin levels in healthy volunteers: a placebo-controlled, randomized, double-blind, cross-over study. Fundam Clin Pharmacol 31(5):567–573. https://doi.org/10.1111/fcp.12288

Lehmann EF, Liziczai M, Drożdżyk K, Altermatt P, Langini C, Manolova V, Sundstrom H, Dürrenberger F, Dutzler R, Manatschal C (2023) Structures of ferroportin in complex with its specific inhibitor vamifeport. elife 12. https://doi.org/10.7554/eLife.83053

Li ZL, Tu Y, Liu BC (2020) Treatment of renal anemia with roxadustat: advantages and achievement. Kidney Dis (Basel) 6(2):65–73. https://doi.org/10.1159/000504850

Lob HE, Ivanova L, Crowell B, Kim H, Idone V, Economides AN, Hatsell SJ (2022) TMPRSS6 inhibition with a monoclonal antibody improves red blood cell health and reduces hepatic iron loading in mouse models of iron overload diseases. Blood 140(Suppl 1):8178–8178. https://doi.org/10.1182/blood-2022-162702

Macdougall IC, Rumjon A, Cinco J, Goldstein L, Summo L, Vauleon S, Riecke K (2015) FP660 pharmacokinetics and pharmacodynamics of lexaptepid, a novel anti-hepcidin molecule, in ESA-resistant haemodialysis patients. Nephrol Dial Transplant 30(Suppl_3):iii294–iii295. https://doi.org/10.1093/ndt/gfv182.08

Manolova V, Nyffenegger N, Flace A, Altermatt P, Varol A, Doucerain C, Sundstrom H, Dürrenberger F (2019) Oral ferroportin inhibitor ameliorates ineffective erythropoiesis in a model of β-thalassemia. J Clin Invest 130(1):491–506. https://doi.org/10.1172/JCI129382

McCaleb M, Lickliter J, Dibble A, Schneider E, Aghajan M, Guo S, Hughes S, Geary RS, Monia BP (2018) Transmembrane protease, serine 6 (TMPRSS6) antisense oligonucleotide (IONIS-TMPRSS6-LRX) reduces plasma iron levels of healthy volunteers in a phase 1 clinical study. Blood 132(Suppl 1):3634–3634. https://doi.org/10.1182/blood-2018-99-115339

Nai A, Pagani A, Mandelli G, Lidonnici MR, Silvestri L, Ferrari G, Camaschella C (2012) Deletion of TMPRSS6 attenuates the phenotype in a mouse model of β-thalassemia. Blood 119(21):5021–5029. https://doi.org/10.1182/blood-2012-01-401885

Neukirchen J, Fox F, Kündgen A, Nachtkamp K, Strupp C, Haas R, Germing U, Gattermann N (2012) Improved survival in MDS patients receiving iron chelation therapy—a matched pair analysis of 188 patients from the Düsseldorf MDS registry. Leuk Res 36(8):1067–1070. https://doi.org/10.1016/j.leukres.2012.04.006

Nicholls A, Lickliter J, Tozzi L, Liu D, Shames R (2018) Hepcidin mimetic ptg-300 induces dose-related and sustained reductions in serum iron and transferrin saturation in healthy subjects. In: EHA Library (https://library.ehaweb.org/) of the 23rd EHA Congress

Novikov N, Buch A, Yang H, Andruk M, Liu G, Wu M, Howell H, MacDonald B, Savage W (2024) First-in-human phase 1 study evaluating the safety, pharmacokinetics, and pharmacodynamics of DISC-0974, an anti-hemojuvelin antibody, in healthy participants. J Clin Pharmacol 64(8):953–962. https://doi.org/10.1002/jcph.2432

Nowak KL, Kakkar R, Devalaraja M, Lo L, Park W, Gobburu J, Kling D, Davidson M, Chonchol M (2021) A phase 1 randomized dose-escalation study of a human monoclonal antibody to IL-6 in CKD. Kidney 360 2(2):224–235. https://doi.org/10.34067/KID.0005862020

Nyffenegger N, Flace A, Doucerain C, Dürrenberger F, Manolova V (2021) The oral ferroportin inhibitor VIT-2763 improves erythropoiesis without interfering with iron chelation therapy in a mouse model of β-thalassemia. Int J Mol Sci 22(2). https://doi.org/10.3390/ijms22020873

Nyffenegger N, Zennadi R, Kalleda N, Flace A, Ingoglia G, Buzzi RM, Doucerain C, Buehler PW, Schaer DJ, Dürrenberger F, Manolova V (2022) The oral ferroportin inhibitor vamifeport improves hemodynamics in a mouse model of sickle cell disease. Blood 140(7):769–781. https://doi.org/10.1182/blood.2021014716

Nyffenegger N, Flace A, Varol A, Altermatt P, Doucerain C, Sundstrom H, Dürrenberger F, Manolova V (2024) The oral ferroportin inhibitor vamifeport prevents liver iron overload in a mouse model of hemochromatosis. Hemasphere 8(9):e147. https://doi.org/10.1002/hem3.147

Patel AB, Masarova L, Mesa RA, Hobbs G, Pemmaraju N (2024) Polycythemia vera: past, present and future. Leuk Lymphoma:1–13. https://doi.org/10.1080/10428194.2024.2361836

Pergola PE, Devalaraja M, Fishbane S, Chonchol M, Mathur VS, Smith MT, Lo L, Herzog K, Kakkar R, Davidson MH (2021) Ziltivekimab for treatment of anemia of inflammation in patients on hemodialysis: results from a phase 1/2 multicenter, randomized, double-blind, placebo-controlled trial. J Am Soc Nephrol 32(1):211–222. https://doi.org/10.1681/ASN.2020050595

Pettinato M, Dulja A, Colucci S, Furiosi V, Fette F, Steinbicker AU, Muckenthaler MU, Nai A, Pagani A, Silvestri L (2023) FKBP12 inhibits hepcidin expression by modulating BMP receptors interaction and ligand responsiveness in hepatocytes. Am J Hematol 98(8):1223–1235. https://doi.org/10.1002/ajh.26961

Pettinato M, Aghajan M, Guo S, Bavuso Volpe L, Carleo R, Nai A, Pagani A, Altamura S, Silvestri L (2024) A functional interplay between the two BMP-SMAD pathway inhibitors TMPRSS6 and FKBP12 regulates. Am J Physiol Gastrointest Liver Physiol 326(3):G310–G317. https://doi.org/10.1152/ajpgi.00305.2023

Petzer V, Tymoszuk P, Asshoff M, Carvalho J, Papworth J, Deantonio C, Bayliss L, Wake MS, Seifert M, Brigo N, Valente de Souza L, Hilbe R, Grubwieser P, Demetz E, Dichtl S, Volani C, Berger S, Böhm F, Hoffmann A, Pfeifhofer-Obermair C, von Raffay L, Sopper S, Arndt S, Bosserhoff A, Kautz L, Perrier P, Nairz M, Wolf D, Weiss G, Germaschewski V, Theurl I (2020) A fully human anti-BMP6 antibody reduces the need for erythropoietin in rodent models of the anemia of chronic disease. Blood 136(9):1080–1090. https://doi.org/10.1182/blood.2019004653

Pilo F, Angelucci E (2024) Vamifeport: monography of the first oral ferroportin inhibitor. J Clin Med 13(18). https://doi.org/10.3390/jcm13185524

Poli M, Girelli D, Campostrini N, Maccarinelli F, Finazzi D, Luscieti S, Nai A, Arosio P (2011) Heparin: a potent inhibitor of hepcidin expression in vitro and in vivo. Blood 117(3):997–1004. https://doi.org/10.1182/blood-2010-06-289082

Poli M, Asperti M, Naggi A, Campostrini N, Girelli D, Corbella M, Benzi M, Besson-Fournier C, Coppin H, Maccarinelli F, Finazzi D, Arosio P (2014a) Glycol-split nonanticoagulant heparins are inhibitors of hepcidin expression in vitro and in vivo. Blood 123(10):1564–1573. https://doi.org/10.1182/blood-2013-07-515221

Poli M, Asperti M, Ruzzenenti P, Mandelli L, Campostrini N, Martini G, Di Somma M, Maccarinelli F, Girelli D, Naggi A, Arosio P (2014b) Oversulfated heparins with low anticoagulant activity are strong and fast inhibitors of hepcidin expression in vitro and in vivo. Biochem Pharmacol 92(3):467–475. https://doi.org/10.1016/j.bcp.2014.09.007

Poli M, Anower-E-Khuda F, Asperti M, Ruzzenenti P, Gryzik M, Denardo A, Gordts PLSM, Arosio P, Esko JD (2019) Hepatic heparan sulfate is a master regulator of hepcidin expression and iron homeostasis in human hepatocytes and mice. J Biol Chem 294(36):13292–13303. https://doi.org/10.1074/jbc.RA118.007213

Porter J, Kowdley K, Taher A, Vacirca J, Shen Z, Chen S, Chawla L, Tidmarsh G (2018) Effect of Ljpc-401 (synthetic human hepcidin) on iron parameters in healthy adults. Blood 132(Suppl 1):2336–2336. https://doi.org/10.1182/blood-2018-99-110885

Porter JB, Scrimgeour A, Martinez A, James L, Aleku M, Wilson R, Muckenthaler M, Boyce M, Wilkes D, Schaeper U, Campion GV (2023) SLN124, a GalNAc conjugated 19-mer siRNA targeting tmprss6, reduces plasma iron and increases hepcidin levels of healthy volunteers. Am J Hematol 98(9):1425–1435. https://doi.org/10.1002/ajh.27015

Preza GC, Ruchala P, Pinon R, Ramos E, Qiao B, Peralta MA, Sharma S, Waring A, Ganz T, Nemeth E (2011) Minihepcidins are rationally designed small peptides that mimic hepcidin activity in mice and may be useful for the treatment of iron overload. J Clin Invest 121(12):4880–4888. https://doi.org/10.1172/JCI57693

Ramos E, Ruchala P, Goodnough JB, Kautz L, Preza GC, Nemeth E, Ganz T (2012) Minihepcidins prevent iron overload in a hepcidin-deficient mouse model of severe hemochromatosis. Blood 120(18):3829–3836. https://doi.org/10.1182/blood-2012-07-440743

Renders L, Budde K, Rosenberger C, van Swelm R, Swinkels D, Dellanna F, Feuerer W, Wen M, Erley C, Bader B, Sommerer C, Schaier M, Meurer K, Matis L (2019) First-in-human phase I studies of PRS-080#22, a hepcidin antagonist, in healthy volunteers and patients with chronic kidney disease undergoing hemodialysis. PLoS One 14(3):e0212023. https://doi.org/10.1371/journal.pone.0212023

Richard F, van Lier JJ, Roubert B, Haboubi T, Göhring UM, Dürrenberger F (2020) Oral ferroportin inhibitor VIT-2763: first-in-human, phase 1 study in healthy volunteers. Am J Hematol 95(1):68–77. https://doi.org/10.1002/ajh.25670

Rose C, Brechignac S, Vassilief D, Pascal L, Stamatoullas A, Guerci A, Larbaa D, Dreyfus F, Beyne-Rauzy O, Chaury MP, Roy L, Cheze S, Morel P, Fenaux P, Myélodysplasies GGFd (2010) Does iron chelation therapy improve survival in regularly transfused lower risk MDS patients? A multicenter study by the GFM (Groupe Francophone des Myélodysplasies). Leuk Res 34(7):864–870. https://doi.org/10.1016/j.leukres.2009.12.004

Rottey S, Singh N, Labriola-Tompkins E, Lob HE, Mohammadi K, Hassan H, Farrelly L, Duggal S, Chin H, Hatsell S, Burczynski ME, Harari O, Perlee L, Van Lancker G (2023) Single ascending doses of REGN7999, a monoclonal antibody inhibitor of TMPRSS6, increase serum hepcidin and cause deep, sustained reductions in serum iron in healthy human volunteers. Blood 142(Suppl 1):3841–3841. https://doi.org/10.1182/blood-2023-187643

Sardo U, Perrier P, Cormier K, Sotin M, Personnaz J, Medjbeur T, Desquesnes A, Cannizzo L, Ruiz-Martinez M, Thevenin J, Billoré B, Jung G, Abboud E, Peyssonnaux C, Nemeth E, Ginzburg YZ, Ganz T, Kautz L (2024) The hepatokine FGL1 regulates hepcidin and iron metabolism during anemia in mice by antagonizing BMP signaling. Blood 143(13):1282–1292. https://doi.org/10.1182/blood.2023022724

Schett G (2018) Physiological effects of modulating the interleukin-6 axis. Rheumatology (Oxford) 57(Suppl_2):ii43–ii50. https://doi.org/10.1093/rheumatology/kex513

Schmidt PJ, Racie T, Westerman M, Fitzgerald K, Butler JS, Fleming MD (2015) Combination therapy with a Tmprss6 RNAi-therapeutic and the oral iron chelator deferiprone additively diminishes secondary iron overload in a mouse model of β-thalassemia intermedia. Am J Hematol 90(4):310–313. https://doi.org/10.1002/ajh.23934

Schmidt PJ, Liu K, Visner G, Fitzgerald K, Fishman S, Racie T, Hettinger JL, Butler JS, Fleming MD (2018) RNAi-mediated reduction of hepatic Tmprss6 diminishes anemia and secondary iron overload in a splenectomized mouse model of β-thalassemia intermedia. Am J Hematol 93(6):745–750. https://doi.org/10.1002/ajh.25079

Schwoebel F, van Eijk LT, Zboralski D, Sell S, Buchner K, Maasch C, Purschke WG, Humphrey M, Zöllner S, Eulberg D, Morich F, Pickkers P, Klussmann S (2013) The effects of the anti-hepcidin Spiegelmer NOX-H94 on inflammation-induced anemia in cynomolgus monkeys. Blood 121(12):2311–2315. https://doi.org/10.1182/blood-2012-09-456756

Scindia Y, Wlazlo E, Ghias E, Cechova S, Loi V, Leeds J, Ledesma J, Helen C, Swaminathan S (2020) Modulation of iron homeostasis with hepcidin ameliorates spontaneous murine lupus nephritis. Kidney Int 98(1):100–115. https://doi.org/10.1016/j.kint.2020.01.025

Sehgal A, Vaishnaw A, Fitzgerald K (2013) Liver as a target for oligonucleotide therapeutics. J Hepatol 59(6):1354–1359. https://doi.org/10.1016/j.jhep.2013.05.045

Sheetz M, Barrington P, Callies S, Berg PH, McColm J, Marbury T, Decker B, Dyas GL, Truhlar SME, Benschop R, Leung D, Berg J, Witcher DR (2019) Targeting the hepcidin-ferroportin pathway in anaemia of chronic kidney disease. Br J Clin Pharmacol 85(5):935–948. https://doi.org/10.1111/bcp.13877

Shen Y, Li X, Su Y, Badshah SA, Zhang B, Xue Y, Shang P (2019) Downregulation contributes to aggressive hepatocellular carcinoma via mechanism mediated by cyclin4-dependent kinase-1/STAT3 pathway. Diagnostics (Basel) 9(2). https://doi.org/10.3390/diagnostics9020048

Silvestri L, Pagani A, Nai A, De Domenico I, Kaplan J, Camaschella C (2008) The serine protease matriptase-2 (TMPRSS6) inhibits hepcidin activation by cleaving membrane hemojuvelin. Cell Metab 8(6):502–511. https://doi.org/10.1016/j.cmet.2008.09.012

Song SN, Tomosugi N, Kawabata H, Ishikawa T, Nishikawa T, Yoshizaki K (2010) Down-regulation of hepcidin resulting from long-term treatment with an anti-IL-6 receptor antibody (tocilizumab) improves anemia of inflammation in multicentric Castleman disease. Blood 116(18):3627–3634. https://doi.org/10.1182/blood-2010-03-271791

Stetka J, Usart M, Kubovcakova L, Rai S, Rao TN, Sutter J, Hao-Shen H, Dirnhofer S, Geier F, Bader MS, Passweg JR, Manolova V, Dürrenberger F, Ahmed N, Schroeder T, Ganz T, Nemeth E, Silvestri L, Nai A, Camaschella C, Skoda RC (2023) Iron is a modifier of the phenotypes of JAK2-mutant myeloproliferative neoplasms. Blood 141(17):2127–2140. https://doi.org/10.1182/blood.2022017976

Tang T, Li L, Tang J, Li Y, Lin WY, Martin F, Grant D, Solloway M, Parker L, Ye W, Forrest W, Ghilardi N, Oravecz T, Platt KA, Rice DS, Hansen GM, Abuin A, Eberhart DE, Godowski P, Holt KH, Peterson A, Zambrowicz BP, de Sauvage FJ (2010) A mouse knockout library for secreted and transmembrane proteins. Nat Biotechnol 28(7):749–755. https://doi.org/10.1038/nbt.1644

Tanzi E, Di Modica SM, Bordini J, Olivari V, Pagani A, Furiosi V, Silvestri L, Campanella A, Nai A (2024) Bone marrow Tfr2 deletion improves the therapeutic efficacy of the activin-receptor ligand trap RAP-536 in β-thalassemic mice. Am J Hematol 99(7):1313–1325. https://doi.org/10.1002/ajh.27336

Taranath R, Bourne G, Zhao L, Frederick B, King C, Liu D (2019) Regulation of iron homeostasis by PTG-300 improves disease parameters in mouse models for beta-thalassemia and hereditary hemochromatosis. Blood 134:3540. https://doi.org/10.1182/blood-2019-129309

Taranath R, Dion C, Bhandari A, Liu DY (2022) Rusfertide Analog-PN23114 as a hepcidin mimetic provides efficacy benefits in conjunction with phlebotomy in mouse model for hereditary hemochromatosis. Blood

140(Suppl 1):5345–5345. https://doi.org/10.1182/blood-2022-171139
Tefferi A, Pardanani A (2024) Momelotinib for myelofibrosis: our 14 years of experience with 100 clinical trial patients and recent FDA approval. Blood Cancer J 14(1):47. https://doi.org/10.1038/s41408-024-01029-3
Theurl I, Schroll A, Sonnweber T, Nairz M, Theurl M, Willenbacher W, Eller K, Wolf D, Seifert M, Sun CC, Babitt JL, Hong CC, Menhall T, Gearing P, Lin HY, Weiss G (2011) Pharmacologic inhibition of hepcidin expression reverses anemia of chronic inflammation in rats. Blood 118(18):4977–4984. https://doi.org/10.1182/blood-2011-03-345066
Udali S, Castagna A, Corbella M, Ruzzenente A, Moruzzi S, Mazzi F, Campagnaro T, De Santis D, Franceschi A, Pattini P, Gottardo R, Olivieri O, Perbellini L, Guglielmi A, Choi SW, Girelli D, Friso S (2018) Hepcidin and DNA promoter methylation in hepatocellular carcinoma. Eur J Clin Investig 48(2):e12870. https://doi.org/10.1111/eci.12870
Vadhan-Raj S, Abonour R, Goldman JW, Smith DA, Slapak CA, Ilaria RL, Tiu RV, Wang X, Callies S, Cox J, Tuttle JL, Lau YK, Roeland EJ (2017) A first-in-human phase 1 study of a hepcidin monoclonal antibody, LY2787106, in cancer-associated anemia. J Hematol Oncol 10(1):73. https://doi.org/10.1186/s13045-017-0427-x
Vadolas J, Ng GZ, Kysenius K, Crouch PJ, Dames S, Eisermann M, Nualkaew T, Vilcassim S, Schaeper U, Grigoriadis G (2021) SLN124, a GalNac-siRNA targeting transmembrane serine protease 6, in combination with deferiprone therapy reduces ineffective erythropoiesis and hepatic iron-overload in a mouse model of β-thalassaemia. Br J Haematol 194(1):200–210. https://doi.org/10.1111/bjh.17428
van Eijk LT, John AS, Schwoebel F, Summo L, Vauléon S, Zöllner S, Laarakkers CM, Kox M, van der Hoeven JG, Swinkels DW, Riecke K, Pickkers P (2014) Effect of the antihepcidin Spiegelmer lexaptepid on inflammation-induced decrease in serum iron in humans. Blood 124(17):2643–2646. https://doi.org/10.1182/blood-2014-03-559484
van Rhee F, Fayad L, Voorhees P, Furman R, Lonial S, Borghaei H, Sokol L, Crawford J, Cornfeld M, Qi M, Qin X, Herring J, Casper C, Kurzrock R (2010) Siltuximab, a novel anti-interleukin-6 monoclonal antibody, for Castleman's disease. J Clin Oncol 28(23):3701–3708. https://doi.org/10.1200/jco.2009.27.2377
Verga Falzacappa MV, Vujic Spasic M, Kessler R, Stolte J, Hentze MW, Muckenthaler MU (2007) STAT3 mediates hepatic hepcidin expression and its inflammatory stimulation. Blood 109(1):353–358. https://doi.org/10.1182/blood-2006-07-033969
Vinchi F, Hell S, Platzbecker U (2020) Controversies on the consequences of iron overload and chelation in MDS. Hemasphere 4(3):e357. https://doi.org/10.1097/HS9.0000000000000357
Wake M, Palin A, Belot A, Berger M, Lorgouilloux M, Bichon M, Papworth J, Bayliss L, Grimshaw B, Rynkiewicz N, Paterson J, Poindron A, Spearing E, Carter E, Hudson R, Campbell M, Petzer V, Besson-Fournier C, Latour C, Largounez A, Gourbeyre O, Fay A, Coppin H, Roth MP, Theurl I, Germaschewski V, Meynard D (2024) A human anti-matriptase-2 antibody limits iron overload, α-globin aggregates, and splenomegaly in β-thalassemic mice. Blood Adv 8(8):1898–1907. https://doi.org/10.1182/bloodadvances.2023012010
Wang CY, Xu Y, Traeger L, Dogan DY, Xiao X, Steinbicker AU, Babitt JL (2020) Erythroferrone lowers hepcidin by sequestering BMP2/6 heterodimer from binding to the BMP type I receptor ALK3. Blood 135(6):453–456. https://doi.org/10.1182/blood.2019002620
Weiss G, Ganz T, Goodnough LT (2019) Anemia of inflammation. Blood 133(1):40–50. https://doi.org/10.1182/blood-2018-06-856500
Witcher DR, Leung D, Hill KA, De Rosa DC, Xu J, Manetta J, Wroblewski VJ, Benschop RJ (2013) LY2928057, an antibody targeting ferroportin, is a potent inhibitor of hepcidin activity and increases iron mobilization in normal cynomolgus monkeys. Blood 122(21):3433–3433. https://doi.org/10.1182/blood.V122.21.3433.3433
Wu M, Wang K, MacDonald B (2022) DISC-0974, an anti-hemojuvelin (HJV) monoclonal antibody, reduced hepcidin and improved anemia in a rat model of chronic kidney disease. Blood 140(Suppl 1):8153–8154. https://doi.org/10.1182/blood-2022-169552
Xu T, Zhang X, Zhao W, Shi J, Wan S, Zhang Y, Hao Y, Sun M, He J, Jiang L, Wang H, Gao H, Luo J, Luo Y, An P (2024) Foxo1 is an iron-responsive transcriptional factor regulating systemic iron homeostasis. Blood 144(12):1314–1328. https://doi.org/10.1182/blood.2024024293
Zhang SP, Wang Z, Wang LX, Liu SJ (2011) AG490: an inhibitor of hepcidin expression in vivo. World J Gastroenterol 17(45):5032–5034. https://doi.org/10.3748/wjg.v17.i45.5032

GPSR Compliance

The European Union's (EU) General Product Safety Regulation (GPSR) is a set of rules that requires consumer products to be safe and our obligations to ensure this.

If you have any concerns about our products, you can contact us on ProductSafety@springernature.com

In case Publisher is established outside the EU, the EU authorized representative is:

Springer Nature Customer Service Center GmbH
Europaplatz 3
69115 Heidelberg, Germany

Batch number: 10762408

Printed by Printforce, the Netherlands